AF333903

The Hierarchical Genome and Differentiation Waves

SERIES IN MATHEMATICAL BIOLOGY AND MEDICINE

Series Editors: **P. M. Auger and R. V. Jean**

Published

The Hierarchical Genome and Differentiation Waves

Novel Unification of Development, Genetics and Evolution

Volume I

Richard Gordon

Department of Radiology
Adjunct Professor of Physics and Electrical & Computer Engineering
University of Manitoba, Winnipeg, Canada

Phone: (204) 789-3828, Fax: (204) 787-2080
E-mail: GordonR@cc.UManitoba.ca

 World Scientific

 Imperial College Press

Published by

Imperial College Press
57 Shelton Street
Covent Garden
London WC2H 9HE

and

World Scientific Publishing Co. Pte. Ltd.
P O Box 128, Farrer Road, Singapore 912805
USA office: Suite 1B, 1060 Main Street, River Edge, NJ 07661
UK office: 57 Shelton Street, Covent Garden, London WC2H 9HE

British Library Cataloguing-in-Publication Data
A catalogue record for this book is available from the British Library.

THE HIERARCHICAL GENOME AND DIFFERENTIATION WAVES
Novel Unification of Development, Genetics and Evolution
Volumes I and II

ISBN 981-02-2268-8 (Set)

Printed in Singapore.

Dedicated in loving memory of Jack Gordon and Pieter D. Nieuwkoop, to Natalie, and to all who turn their creativity towards the daily rekindling of life.

FOREWORD

To have been written by the late:

Pieter D. Nieuwkoop
Director Emeritus, Hubrecht Laboratory
Netherlands Institute for Developmental Biology
Utrecht, The Netherlands

We had the privilege and pleasure of having Dr. Pieter Nieuwkoop (Figures 1 and 2) stay with us on two occasions. We also corresponded for many years. Our letters and visits were filled with discussion of waves, inductions, and the history of embryology. It was during Pieter's second visit that Dick asked him to write this book's foreword. They had a lengthy discussion about how Pieter felt about the book. After some of what Pieter obviously considered a stern warning, he agreed to write the foreword. Unfortunately, he died suddenly in September of 1996, at the age of 79 (Durston, 1997; Gerhart, 1997), while *The Hierarchical Genome* was in the final stages of preparation. While he read an earlier draft, the version in your hand is one he did not live to see.

Dick and I discussed at length how best to handle the sudden loss of our friend and mentor. We also struggled with how to find a new author for the foreword on such short notice. There really was no one else but Pieter who could do it. Eventually we decided that I would write an account of the conversation they had about the book in general. In this way Pieter could still introduce the book the way I believe he would have intended to.

If anyone is qualified to write this I suppose it must be me. I have lived with *The Hierarchical Genome* since it was a skimpy hypothesis sprouting out of the slow waves we found together and Lionel Johnson's encouragement of its implications for evolution. When the rapidly growing manuscript was returned from yet another journal because it was far too long it was my idea that it should be a book. I sat in on many discussions and read much of the correspondence during its growth. I was present for many of its key events, including the point where Lionel Johnson said it had grown far too big to warrant him remaining a coauthor, through publication of a partial version (Gordon, 1992a), and finally through a sometimes laborious process of reviewing the final product for clarity and flow.

Figure 1. Pieter D. Nieuwkoop with Richard Gordon at the Fort Whyte Nature Centre, Winnipeg, in 1992.

Above all, I was present during the discussion in which Pieter warned Dick about what he would say in his foreword. I hesitate to use the words of a man with whom I can't double check that I understood him correctly or that I have remembered what he said in the way he would have said it. I also cannot know how much of what he said in a casual conversation he would have included in a formal piece of writing. Pieter was, above all else, a gentle soul, formal and careful in his writing, tactful, preferring to say nothing rather than to openly criticise. I suspect some of the things he said privately he would not have put in his written foreword. Certainly he would have said some things differently. For this discrepancy I apologise to his memory and hope he will forgive me.

The discussion began with a lengthy critique of what Dick had written about the history of embryology. It was in the midst of this discussion that Dick suddenly blurted out he wanted Pieter to write the foreword.

I remember Pieter laughing with flattered pleasure and then waving one index finger at Dick, his blue eyes sparkling. "This perhaps, you should not ask me, for I won't be kind about everything. I will be honest."

"Of course," Dick replied, "this is what I want, an honest assessment of the book so the reader will know what he's getting into."

At this Pieter sat back and nodded thoughtfully. "I should begin by saying your study of the history of embryology is quite astute. I would say that chapter alone will make the book worth buying. You have researched it well and given insights and perspectives to the problem. I especially like what you wrote about Holtfreter and Spemann. Very accurate and very much in accord with my own memories of those times and people, though I am far too young to remember some of the earlier events."

"Thank you."

"But I must say you speculate far too much. You are not an evolutionist or a geneticist but you have widely applied your ideas to include these fields. Just what you have written about in the embryology alone is revolutionary, entirely new." At this point he leaned forward and emphasised his words with a shaking of his finger once again. "If you are right of course, and in spite of what Natalie has found on the axolotl you have yet to prove your waves are the source of homoiogenetic induction. I personally would do much more research first before running off on such wild speculations. Without data to back them, they are almost irresponsible."

"I'm not an embryologist either, nor a historian, yet you like what I've done," Dick replied.

"True," Pieter said. "If you have done the evolution and genetic research as well as you have done the embryological then you have done it well indeed, at least in so far as one can learn a field and then speculate without reservation about virtually everything in it!" Pieter chuckled then. "You do not lack, what is that wonderful Jewish word?"

"Chutzpah."

"Ah yes, chutzpah, you have no lack of chutzpah, that much is certain."

"But what if I'm right?" Dick asked.

"If you are right," Pieter replied, "then you will have succeeded in unifying development, genetics, and evolution. This would be a great accomplishment. But your ideas are so far beyond the available data that you will not live long enough to know if you're right. There are generations of work here, hundreds of Ph.D. projects, with whole careers worth of research, proving you are right, or possibly that you are completely wrong. You also assume you are right, and that may be assuming much. I personally still reserve judgment on how right you are."

Now it was Dick's turn to laugh. "If my ideas are properly tested, then even if I am wrong, that's all I want because much new research will be generated and we'll accumulate a lot of new knowledge. What I want to do is provide a fresh perspective on the old problems."

"It definitely is a fresh perspective," Pieter agreed. "So fresh some people may never speak to you again!"

At this point I piped up from the corner where I was listening to all of this. "What do you like about the book?"

Pieter leaned back again, resting his hands behind his head and took a couple of thoughtful breaths. "You have succeeded in amassing an enormous number of widely divergent ideas, all referenced with astonishing completeness. In fact you reference and quote far too much in some places. It is not always easy to read because of this. Nor am I sure I appreciate the proposition format. It makes one think but it's hardly, well, an academic standard. One gets the impression one is following along with you as you think things through and in some places your thinking is a bit muddy I would have to say."

"Still, this will be a tremendous resource for those who wish a picture of the state of biological science at this time. This book, one could pick up in one hundred and fifty years and understand how the biologist of today sees the world. I can think of no major idea, concept, or theory you haven't touched upon."

Dick beamed with pleasure.

"Also, and I think this most important. You have remembered to include the work that came before the present. Many scientists seem to think the old stuff is at best quaint, and at worst mere junk. Certainly not worth reading or paying attention to anymore. In embryology, for example, I frequently despair at the bright young minds that have forgotten or never learned the classical knowledge. Today's embryologist is obsessed with gels and genes. But what is the point of knowing that a gene is expressed in a tissue without understanding even the most basic anatomy of where and how that tissue arose in the first place? Yet I see this all the time."

"I sometimes think they have forgotten the organism they study. This is where you have excelled. You never loose sight of the organism as a functioning whole, nor have you forgotten the significant older works. All the old ideas are here, right or wrong, good or bad. This book will direct the young minds who read it back to the literature and knowledge as a whole. Perhaps it is a good thing you put in so many references and quote so many people. They will find much that they otherwise would have overlooked. Yes," he finished thoughtfully, "that is what I like best about the work, its completeness."

Natalie K. Björklund
Winnipeg, 1997

Figure 2.   Pieter D. Nieuwkoop with Natalie K. Björklund at Kildonan Park, Winnipeg, in 1992, discussing early axolotl embryogenesis.

PREFACE

"Whenever a new scientific concept comes into prominence, it sends shock waves of surprise to the scholars contributing to that field" (Carlson, 1966).

It is my intention to describe a phenomenon that I seem to be the first to have stumbled across with understanding: a quiet wave that forms our very minds. It will crest as it crashes over and forever intertwines the three fields of development, genetics and evolution. The wave forming the brain is there, and many others, perhaps determining every tissue of which we are made. These waves are easily seen with simple tools in common embryos, by anyone who will take the trouble to look. This book is an interpretation of these waves. You may not agree with my interpretation, and are most welcome, and indeed encouraged, to invent other interpretations. But you will not be able to ignore these waves, which will color your view of biology from here on, as they have mine. With the discovery of differentiation waves, it is no longer true that...

"The determination or chemodifferentiation of any given part, i.e. the decision as to what any cell or group of cells shall develop into, takes place invisibly some time prior to the process itself" (Needham, 1931b).

We can now watch it happening.

This book was inspired by a lecture by Lionel (Jim) Johnson given in my Pollution Biology class in February, 1989. (The manuscript was started on June 9, 1989.) When Lionel opened my eyes to fundamentally unsolved and exciting problems in evolution and ecology, he showed how one could leap from population data on arctic charr to possible explanations of the grand sweep of evolution (Vanriel & Johnson, 1995).

I would like to thank Guenter Albrecht-Buehler, Thomas F.H. Allen, Hilary Alto, Karl J. Aufderheide, Penniman B. Austin, Lev V. Beloussov, Teresa C. Binstock, John Tyler Bonner, G. Wayne Brodland, Chris Chetland, Simon Conway Morris, James Cummins, Marc R. Del Bigio, Graham Dellaire, Daniel C. Dennett, Wim J.A.G. Dictus, Hadass Eviatar, Sergei V. Evsikov, David H.A. Fitch, Srecko Gajovic, Helen Gordon, the late Paul B. Green, Gary W. Grimes, Gerhard Grössing, Efroim Gurman, Max K. Hecht, Lewis I. Held Jr., Michael Keller, Glen R. Klassen, James A. Lake, Keith E. Latham, Michael Levin, John A. McCoshen, Andrew D. McCulloch, Ross McGowan, Leonard E. Munstermann, Vidyanand Nanjundiah, Stuart Nelson, the late Henriette C.M. Nieuwkoop-Beukers, Hugh E.H. Paterson, Andrey Polezhaev, Bruce Ratcliffe, Robert Rosen, Cynthia M. Ross, Sharon Y. Roth, Nancy D. Rumph, Tsvi Sachs, Charles B. Schom, Verner L. Seligy, Stanley K. Sessions, William Silvert, Mark Solomonovich, Roland Somogy, Arlin Stoltzfus, Stephen A. Stricker, Michael Sumner, Patrick P.L. Tam, Edward Theriot, Yuri Trusov, Leigh M. Van Valen, Michael W. Vannier, Carey Vigor-Zierk, S. Randal Voss, Richard J. Wassersug, Donald I. Williamson and Kenneth H. Wolfe for critical comments and suggested improvements of draft sections. Thanks are due to Margo Kautz, Kim Nozick, Lana M. Hunstad and Alan W. Hunstad for library assistance, and to Daniel W. Rickey for assistance with digitizations. The following people have bravely tackled reading of whole draft manuscripts: Russell W. Anderson, Natalie K. Björklund, Diana Gordon, C. Cristofre Martin, Stephen P. McGrew, the late Pieter D. Nieuwkoop, and Przemko Tylzanowski. I would not wish to suggest that any of these people endorse all the ideas put forward here. To some extent, this book has become a dialog with them.

The research behind this book was consistently supported by grants from the Spina Bifida Association of Canada (SBAC), an enlightened group of parents who see in fundamental research a possible understanding of their children's condition, which might help the next generation. Partial support was also offered by Abbott Laboratories/Winnipeg Health Sciences Centre, Children's Hospital of Winnipeg Research Foundation (especially through its former Director, Victor Chernick), Departments of Botany and Radiology and Harley Cohen and Nick Anthonisen, Deans of Science and Medicine at the University of Manitoba, Easter Seal Research Institute, Employment and Immigration Canada Challenge Program, Manitoba CareerStart, Manitoba Hydro, Manitoba Schools Science Symposium, Natural Sciences and Engineering Research Council of Canada, Spina Bifida Association of Manitoba, Spina Bifida and Hydrocephalus Association of Ontario, and the University of Manitoba Research Grants Committee of Senate. The University of Manitoba, especially my former colleagues in the Department of Botany (1984-1994, during which much of the writing of this book occurred), John Stewart and David Punter, and in Radiology, Martin H. Reed, E. (Ted) A. Lyons, I. David Greenberg, and Blake M. McClarty, provided an atmosphere of academic freedom that made this study possible. This book was written in part in the midst of a strike by professors at the University of Manitoba in defense of academic freedom (Gordon, 1996c). My predoctoral, doctoral, and postdoctoral mentors, Edward Anders, Terrell L. Hill, and Robert Rosen, respectively, taught me the importance of academic freedom by their examples.

Two of my students, Kurt B. Luchka and C. Cristofre Martin, provided much of the momentum for the underpinnings of this work. I would also like to thank the innumerable colleagues who responded to my appeals for their reprints, many of whom heard me out in seminars, conversations, and e-mail exchanges, whose ideas are woven into the fabric presented here. I would especially like to thank the staff of the Medical Library of the University of Manitoba, including Ruth Holmberg and Ju-Dai Chuang in Interlibrary Loan (over 300 items), Beverley Brown, Carol Cooke and Bill Poluha in Reference, and Jovy Beltran, Erlinda Erjavec, David Howe, Thomas Quinlan and Moira Raglan in Circulation, and their then Acting Head, Michael Tennenhouse, for their immense effort, along with Ginette Wright and Hung-Hsueh Shao in the Science Library. Thanks are also due to Oeke Kruythof, Librarian, Hubrecht Laboratory, Netherlands Institute for Developmental Biology, who supplied many of Pieter Nieuwkoop's papers, and to Sook Cheng Lim and Yolande Koh of World Scientific Publishing for their patience and editorial and production assistance in bringing this to fruition.

I have been dragged kicking and screaming by our axolotl embryos back into a now matured molecular biology which I had abandoned in 1964 as an immature graduate student. That approach did not seem to me then (nor yet, alone) to be capable of divulging 'the secret of life'. Not that it didn't make an impression on me being surrounded by the enthusiasm of Aaron Novick (Novick & Szilard, 1950), Jack Sadler (Sadler, Sasmor & Betz, 1983), Frank W. Stahl (1964), Sid A. Bernhard (1968) and the late George Streisinger (Streisinger et al., 1981) at the Institute of Molecular Biology at the University of Oregon. At that time the genetic code was being delivered by emissaries day by day, codon by codon ("Crick (1966b, 1988a) provides engaging accounts of this exciting period of biochemical sleuthing": Mange & Mange, 1990). I now understand part of what was lacking: the very physics and chemical physics I was then drawn to, with skills honed under Susan Meschel (Handler & Meschel, 1993) and Edward Anders in quantitative chemistry. As an undergraduate in his lab, I considered organic matter in meteorites and their relevance to the origin of life (Anders et al., 1964; Anders & Lewis, 1987). Under the eye of my graduate mentor, Terrell L. Hill (1956b, 1960, 1963, 1964, 1969, 1977, 1985a, 1987, 1989), I investigated equilibrium (Gordon, 1968a) and nonequilibrium (Gordon, 1968b) phase transitions and was given the freedom and encouragement to delve into embryogenesis (Gordon, 1966) while earning a Ph.D. in Chemical Physics. On my recent journey back to molecular biology, my constant guide and companion has been Natalie K. Björklund, friend, collaborator, spouse, believer in 'the wave' when it was a 'mere' prediction, and first discoverer of the ectoderm contraction wave in explants and *in vivo*.

Seven embryologists had major impacts on me. (While I nominally postdoced under Conrad H. Waddington around 1970: Waddington, 1972, he was primarily immersed in futurism at that time.) An exciting lecture on embryonic cell sorting by Aaron A. Moscona while I was an undergraduate at the University of Chicago initiated me into the fascination of embryology and the possibilities for its explanation (cf. Moscona, 1957, 1974). While I was a graduate student at the University of Oregon, I corresponded with and visited Max Braverman in Pittsburgh, who was working on hydroid colonies (Braverman, 1962, 1963; Braverman & Schrandt, 1966). We concluded at that time that it was indeed possible to generate biologically realistic patterns from 'rules' of cell behavior and interaction (cf. Gordon, 1966; Held Jr., 1992). The next problem was to figure out how embryos actually did it (a distinction that unmade leads to arguments even today: cf. Bird & Hoyle, 1994, vs. Niklas, 1994a). During my postdoc with Robert Rosen (1967, 1970, 1991), he set examples of clear and independent thinking in dissecting morphogenesis (Goel et al., 1970, 1975; Rosen, 1970, 1972a,b). I tried my hand at modelling gradients in hydra with the inspiring civil engineer turned biologist, Lewis Wolpert (Wolpert, Hicklin & Hornbruch, 1971; Wolpert, 1991a), in 1969, getting nowhere. After meeting at the 1969 Woods Hole Embryology course organized by Malcolm S. Steinberg, with whom I also later had the pleasure of collaborating (Gordon et al., 1972, 1975), Antone G. Jacobson and I then had the exhilarating and thought sharpening experience of working out how the newt neural plate changes shape (Jacobson & Gordon, 1976a; Gordon & Jacobson, 1978). More recently, I have had the honor and enjoyed the friendship, company and collaboration of that late master of classical embryology, Pieter D. Nieuwkoop (Nieuwkoop & Faber, 1956, 1975, 1994; Nieuwkoop & Sutasurya, 1979, 1981; Nieuwkoop, Johnen & Albers, 1985; Gerhart, 1987; Appendix V: Gordon, Björklund & Nieuwkoop, 1994; Nieuwkoop, Björklund & Gordon, 1996).

Three superb cell biologists, the eclectic diatom electron microscopist Ryan W. Drum (Drum, Pankratz & Stoermer, 1966; Gordon & Drum, 1970, 1994), the late Robert D. Allen (Allen & Kamiya, 1964; Allen & Allen, 1983; Allen, 1987), and the late James F. Danielli (Danielli, 1955; Stein, 1986), gave me a firm foundation in intracellular phenomena and their visualization, while early efforts at what we now call computed tomography of ribosomes with my friend Robert Bender (Bender, Bellman & Gordon,

1970; Bellman et al., 1971), started in the lab and at the suggestion of Cyrus Levinthal (1968), gave me some practical experience in biological image processing, which I parlayed into a second career in medical imaging. Lewis E. Lipkin introduced me to digital imaging in light microscopy (Lemkin et al., 1976).

While I am not an historian of science, I have a curiosity about intellectual history, which I attribute to my so inclined friend John B. Musgrave. If I am right, this work culminates a search that has been going on for well over a century, for the essence of explanation in embryology, representing a vast, worldwide intellectual effort, which I think must be acknowledged. This is why the text is woven through with quotations. I prefer to let previous and contemporary researchers, reviewers and textbook writers speak in their own words, and to retrace the history of ideas. This will exasperate or delight you, depending on your expectations of a scientific monograph. It also allows you to see if you can devise alternative interpretations to mine of what has been said by others. My interest in history does not make me skilled at it, and I have noticed subtle differences in accounts by scientists who turned late in their careers to writing about history versus professional historians of science. Thus the reader should be warned that I take accounts much at face value, only here and there noting contradictions or situations warranting further historical digging. I have not gone to primary sources (unpublished letters, etc.), the way a proper historian does.

The quotes are often extensive to add information beyond my text, and thus generally should not be skipped. Keep in mind that these quotes are selected to support (or occasionally contradict) my thesis, so they may not fully represent the authors' viewpoints. (To enhance readability, I used a colon for a full stop before a quotation, and '...' to indicate no break. If I chose to change the capitalization of a word from the original, the first letter is underlined. I have corrected obvious misspellings rather than using *sic.*) This practice of massive quotation leads to many secondary references, which I have checked when accessible. Any emphases (italics, etc.) are as in the original, unless otherwise indicated (except that genera and species names are always italicized). My connecting or explanatory words, and editorial comments, are in square brackets (although occasionally, as in $[Ca^{2+}]$, square brackets designate 'concentration of').

I use the common scientific reference style, which gives credit to each idea one builds on directly in line in the text, cumbersome as it may be. Sometimes annoying long strings of references are given, not so much to support my argument by the weight of authority, but to give the reader a broad entry into the literature. I see this text as a beginning of an explanation of embryogenesis, not the final word. There is much work yet left to do, and with a literature so vast and far flung, I prefer to pass on the benefit of my labor by pulling the literature together (nearly 7000 references). For each reader, different aspects will seem tutorial and elementary. Just skip what seems obvious.

An animated set of flip through figures of the ectoderm contraction wave has been prepared by artist and medical artist K. Jack Butler (1996).

It took six starts to decide on a style for this book. It is hard to cross so many fields and keep one's wits. I finally devised a format consisting of propositions, each followed by its justification and corollaries, by rough analogy with Martin Luther's 95 Theses posted in Wittenberg. As the jargon of the many fields is incomprehensible for the uninitiated, as I once was, I have provided a partial glossary, including repeated abbreviations. The reader should note that many common words have specialized meanings in biology, so confusion may sometimes be cleared up by checking the glossary.

I throw a wide net, one with gaping holes due to the uneven state of our intellectual successes in biology. If I seem to claim too much, it is because I am casting for a very large fish, and hope the net holds:

"Aristotle... reports that the Pythagoreans found that acoustical 'harmonies' only depend on (commensurate) ratios of lengths and on nothing else, and that they generalized from this experience in acoustics to physics and knowledge in general.... It is true that, against their own background, the Pythagoreans were recklessly imprudent when concluding so much from so little. But Isaac Newton was also 'reckless' when proposing that there is a universal gravitational law which subsumes both terrestrial falls of bodies and celestial orbitings of planets" (Bochner, 1966).

I will be accused of trying, but will not attempt, to solve everything at once in this book: there are those brave souls who have tried to see in morphogenesis (Sinnott, 1963; Edelman, 1989, 1992a) or the cytoskeleton (Penrose, 1994, 1995) a solution to the problem of self-awareness and consciousness (cf. Dennett, 1995). I have but little to offer towards the solution of this higher problem, other than perhaps a fresh look at how the brain is built, which may or may not help understand how it functions.

The ideas here are roughly hewn, for that is the imperfect nature of our science of biology and my own understanding. If fragments survive the test of time, and the scorn and mislaid praise of my peers, I will be satisfied with my efforts. If you are dusting off this quaint document a century hence, as I have admiringly done to those of my predecessors, may I wish you the best. Carry on.

Winnipeg, 1998.

FLIP ANIMATION OF THE ECTODERM CONTRACTION WAVE

The following 13 pages should be flipped through by hand to visualize the propagation of the axolotl ectoderm contraction wave. Thirteen clay models were prepared by artist Jack Butler, and each photographed in a subset of the following 6 views:

posterior
left posterior
dorsal
left dorsal
left lateral
left anterior

Each page shows one model, from a few of the above views, representing one moment in time. The time interval between pages is just under an hour (at 20°C). One can see the propagation of the ectoderm contraction wave from the beginning of gastrulation, coincident with its launching as a small dimple, through to its closing down to a small ring at the anterior end of the embryo, where it subsequently vanishes at the end of gastrulation. When it collides with the dorsal lip of the blastopore, the initially circular wave breaks into an arc. At this moment, a crimp in the surface appears to be formed, which may propagate ahead of the wave (not shown) and form a boundary for its trajectory. This wave leaves the neural plate (neuroepithelium) in its wake.

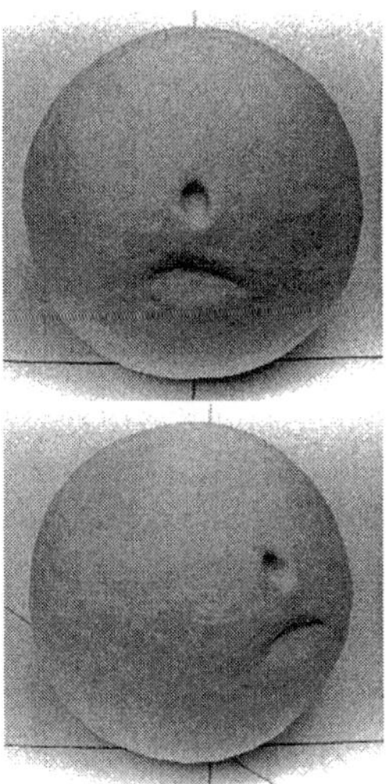

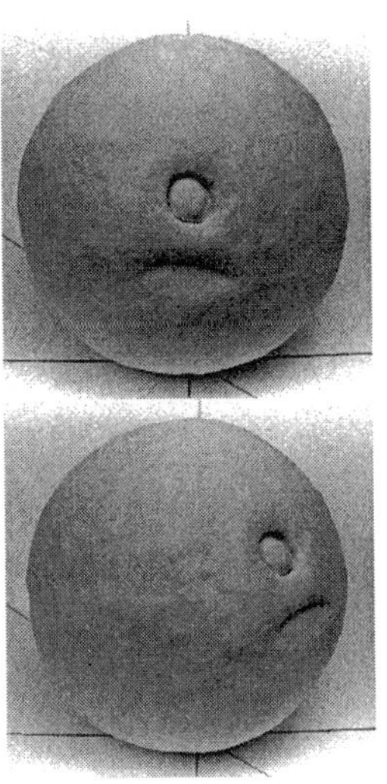

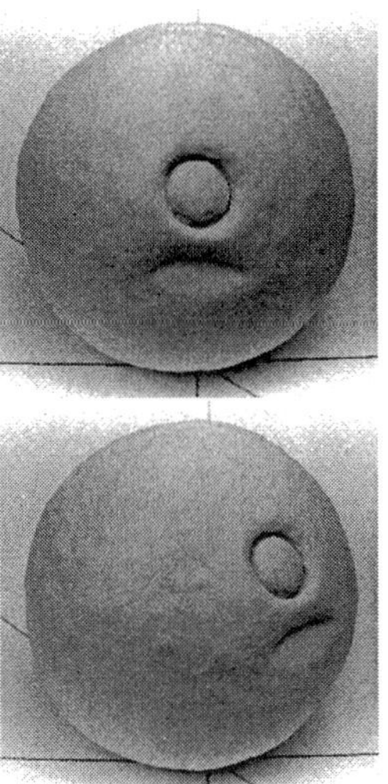

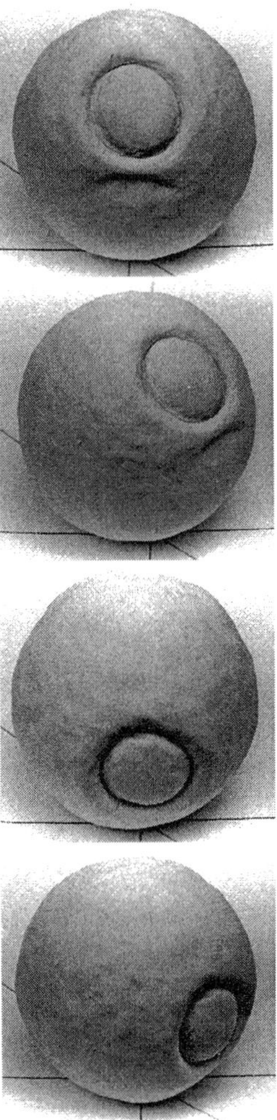

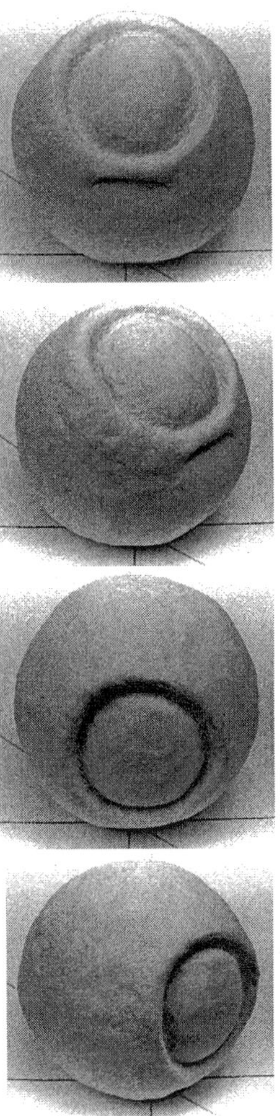

xxxi

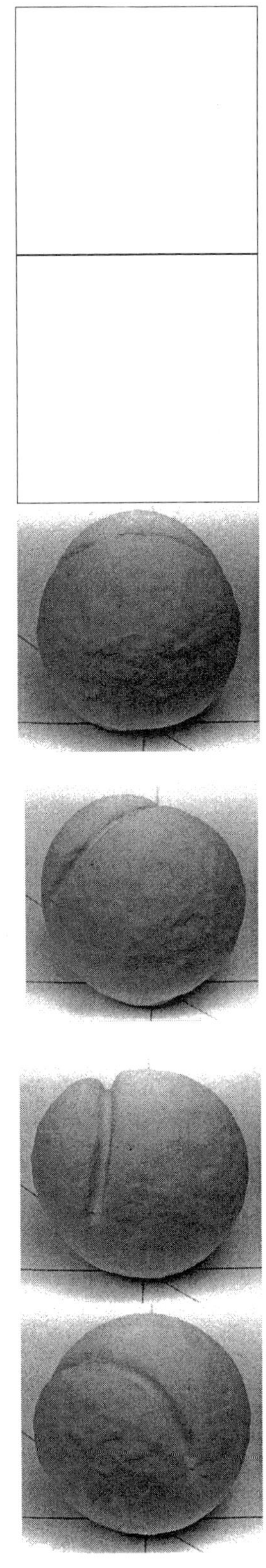

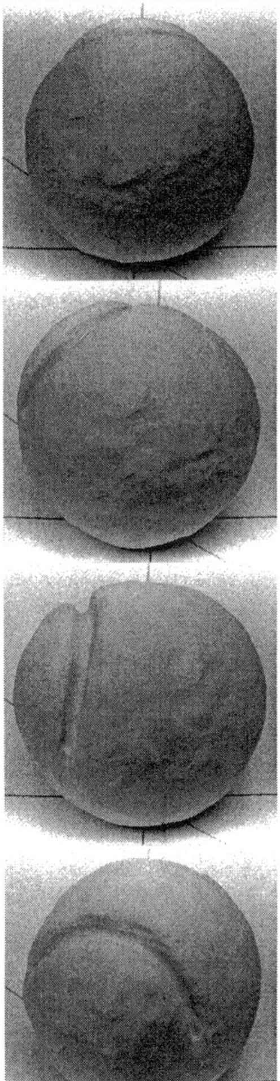

xlvii

PROPOSITION PAGE NUMBERS

1.	165	40.	263	79.	379	118.	472	157.	570	196.	659	235.	756
2.	168	41.	266	80.	380	119.	476	158.	573	197.	664	236.	757
3.	171	42.	267	81.	380	120.	478	159.	576	198.	668	237.	758
4.	172	43.	270	82.	381	121.	480	160.	581	199.	671	238.	759
5.	176	44.	271	83.	385	122.	482	161.	582	200.	675	239.	759
6.	177	45.	277	84.	388	123.	483	162.	587	201.	677	240.	761
7.	180	46.	280	85.	389	124.	493	163.	588	202.	682	241.	764
8.	181	47.	284	86.	391	125.	497	164.	591	203.	687	242.	769
9.	182	48.	286	87.	395	126.	498	165.	592	204.	691	243.	772
10.	183	49.	289	88.	401	127.	501	166.	593	205.	695	244.	776
11.	190	50.	292	89.	403	128.	506	167.	598	206.	701	245.	778
12.	194	51.	292	90.	406	129.	507	168.	599	207.	704	246.	783
13.	197	52.	294	91.	408	130.	512	169.	600	208.	705	247.	785
14.	199	53.	298	92.	409	131.	514	170.	604	209.	707	248.	788
15.	203	54.	300	93.	412	132.	517	171.	609	210.	707	249.	793
16.	207	55.	301	94.	415	133.	519	172.	610	211.	709	250.	797
17.	211	56.	304	95.	417	134.	520	173.	611	212.	714	251.	803
18.	212	57.	308	96.	419	135.	525	174.	618	213.	717	252.	809
19.	213	58.	313	97.	420	136.	527	175.	621	214.	717	253.	811
20.	215	59.	317	98.	423	137.	530	176.	622	215.	718	254.	812
21.	222	60.	320	99.	424	138.	534	177.	622	216.	719	255.	817
22.	222	61.	324	100.	425	139.	535	178.	625	217.	724	256.	822
23.	223	62.	326	101.	426	140.	538	179.	626	218.	725	257.	823
24.	232	63.	331	102.	428	141.	538	180.	628	219.	726	258.	825
25.	237	64.	336	103.	428	142.	542	181.	629	220.	727	259.	830
26.	237	65.	343	104.	430	143.	543	182.	630	221.	728	260.	831
27.	238	66.	351	105.	430	144.	544	183.	631	222.	729	261.	832
28.	238	67.	354	106.	431	145.	549	184.	632	223.	730	262.	833
29.	239	68.	356	107.	436	146.	550	185.	634	224.	732	263.	836
30.	241	69.	358	108.	437	147.	553	186.	635	225.	737	264.	840
31.	243	70.	359	109.	440	148.	554	187.	637	226.	742	265.	844
32.	244	71.	360	110.	441	149.	555	188.	640	227.	744	266.	844
33.	244	72.	365	111.	444	150.	556	189.	645	228.	744	267.	846
34.	247	73.	366	112.	446	151.	559	190.	646	229.	746	268.	847
35.	250	74.	368	113.	447	152.	561	191.	647	230.	748	269.	852
36.	256	75.	371	114.	453	153.	561	192.	647	231.	752	270.	855
37.	259	76.	373	115.	458	154.	561	193.	648	232.	752	271.	860
38.	260	77.	374	116.	464	155.	567	194.	649	233.	753	272.	863
39.	262	78.	376	117.	467	156.	569	195.	658	234.	755		

CONTENTS

Volume I

Volume II

Gordon, R. & G.W. Brodland (1987). The cytoskeletal mechanics of
brain morphogenesis: cell state splitters cause primary neural induction.
Cell Biophysics 11, 177-238.

Brodland, G. W., R. Gordon, M. J. Scott, N. K. Björklund, K. B. Luchka,
C. C. Martin, C. Matuga, M. Globus, S. Vethamany-Globus & D. Shu
(1994). Furrowing surface contraction wave coincident with primary
neural induction in amphibian embryos. *J. Morph.* 219 (2), 131-142.

Pursued by the Differentiation Wave

Björklund, N. K. & R. Gordon (1993b). Nuclear state splitting: a working
model for the mechanochemical coupling of differentiation waves to master
genes (with an Addendum). *Russian J. Dev. Biol.* 24 (2), 79-95.

1.00 INTRODUCTION

"I am a firm believer that without speculation there is no good or original observation" (Charles Darwin in Markert, 1958, from an 1857 letter to A.R. Wallace: Darwin, 1905).

"[Alfred Russell] Wallace's sales agent, back in London, heard mutterings from some naturalists that young Mr. Wallace ought to quit theorizing and stick to gathering facts. Besides expressing their condescension toward him in particular, that criticism also reflected a common attitude that fact-gathering, not theory, was the proper business of *all* naturalists" (Quammen, 1996).

"To think that heredity will build organic beings without mechanical means is a piece of unscientific mysticism" (His, 1888, reprinted in: Coleman, 1967).

"Biologists - and the rest of us too - would like to know how the brain works and how a single cell, the fertilized egg cell, develops into an entire organism" (Flanagan, 1994).

1.01 Consider a Spherical Cow

"Let us consider the egg as a physical system. Its potentialities are prodigious and one's first impulse is to expect that such vast potentialities would find expression in complexity of structure. But what do we find? The substance is clouded with particles, but these can be centrifuged away leaving it optically structureless but still capable of development.... On the surface of the egg there is a fine membrane, below it fluid of high viscosity, next fluid of relatively low viscosity, and within this the nucleus, which in the resting stage is simply a bag of fluid enclosed in a delicate membrane.... The egg's simplicity is not that of a machine or a crystal, but that of a nebula. Gathered into it are units relatively simple but capable by their combinations of forming a vast number of dynamical systems..." (W.B. Hardy in Needham, 1931b).

"The germ is a unit, in space and time. What we are interested in is... its fascinating struggle in escaping from... sphericity..." (Dalcq, 1938).

The development of a single organism takes place on a time scale of days to months. We and most of our multicellular eukaryotic compatriots (hydra, when budding, being an example of an exception) start out as a single, round cell, a fertilized egg, whose dimensions can vary from microscopic to the 1.5 kg ostrich egg (Stivens, 1974; Itô, 1980). The smaller eggs are nondescript unshelled spheres, often with no obvious spatial structure. Mammalian eggs can be manipulated in many ways, including scrambling of the cytoplasm (Evsikov, Morozova & Solomko, 1994; Humpty Dumpty notwithstanding: Baring-Gould & Baring-Gould, 1962; Shannon, 1980). They are thus essentially spherically symmetrical. Physicists, who are wont to simplify when dealing with life, are often jokingly said to start off a lecture with: 'Consider a spherical cow...' (cf. Harte, 1988). However, cows are spherical, and not just when they arc meatballs: in biology we ascribe the same species name to every part of the life cycle of an organism. Thus a fertilized, spherical cow egg is a cow. After fertilization, it divides repeatedly into cells, which somehow (our main question) become different from one another while simultaneously arranging themselves in space to form the more familiar creature we name 'cow' (or 'bull').

The fertilized human egg is one of these spheres, 0.07 mm in diameter, i.e., much smaller than the dot in '0.07' and hardly a resemblance to you or me (Figure 3). But that's how each of us started. What is it about such a small, undifferentiated cell, about the size of a drop of mist, that it can turn into one of us?

Today's accepted answer has been *genes*. Every cell in every organism contains a copy of its genetic material, an extraordinarily long strand of DNA (2 m in *Homo sapiens* : Alberts et al., 1989, 1994; Witkowski, 1994, about 3.2×10^9 base pairs of DNA, Ohno, 1984, cut and folded into the nucleus, packaged in 46 chromosomes). All of this DNA, taken together, is called a *genome*. The genome is presumed to somehow encode what the cells do to make the organism, the so called 'genetic program'. But can it be a self-contained set of instructions? The problem with such a notion is simply that, since the DNA sequence of nucleotides is considered to be the

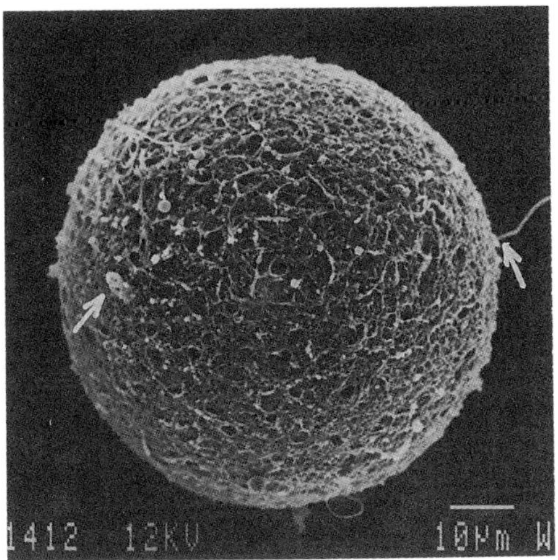

Figure 3. A scanning electron micrograph (SEM) of a fertilized human egg showing two sperm tails in the glycoprotein zona pellucida Scale bar 10 μm. From Figure 5 in Nikas et al. (1994) with permission.

same in each cell, something from 'outside' the cell must specify which subset of 'instructions' is used in each cell. But all that is outside each cell is other cells with the identical DNA and/or the outside world. Thus the existence of this DNA with its genetic code does not in itself answer the question of how the cells become different from one another. This seems self-evident, but is widely ignored.

Even at the single cell level, the self-sufficiency of DNA has been called an 'illusion':

"The simplest widespread idea around DNA sequences is that they contain all or most of the information that is necessary to build an organism. (For a critical review of such ideas see Lewontin, 1992b). Therefore, one could think that a central effort of computational biology in the near future will be the development of a methodology that could use as input 'raw' DNA sequences, and generate as output the biological 'meaning' of such sequences. The interpretation of novel DNA sequences depends on our ability to associate them with previously identified sequences whose biological function is known. This provokes the illusion that we can decipher DNA sequences directly. This illusion supports the metaphor of DNA as the molecule containing the program of regulation and development. If the process of regulation of gene activity and its physiological and developmental consequences is compared to a computer program, such a program cannot be limited to DNA but involves also other molecules and structures present in the cell" (Collado-Vides, 1995).

The 'genetic program' has been presumed to be the mechanism by which the DNA generates an organism. As we shall see, it is not simply "other molecules and structures", but the physics of the embryo, that is the essential complement to the DNA, making a genetic program possible.

The DNA also replicates and being transmitted from one organism to the next, via egg and sperm, is thus presumed to be the ultimate carrier of biological evolution. Here we have another problem regarding the presumed autonomy of the DNA, because what we regard as large and important differences between species are not reflected in obvious differences in their DNAs. While DNA can mutate in large or small chunks (point mutations through major deletions or duplications), we have not yet found any universal mapping between such changes and species differences. Morphologically similar species of reindeer, for example, have tremendous differences in their number of chromosomes (Wurster & Atkin, 1972; Short, 1976; John & Miklos, 1988; Lima-de-Faria, 1988), while our chromosomes differ only slightly from those of chimpanzees (King & Wilson, 1975; Lewin, 1987a; Diamond, 1988, 1992a) or even mice (Searle et al., 1989).

The DNA, as the basis for the genetics component of the genetic program, is what distinguishes each species in both its embryonic development and evolution. If we understood its organization as a program, we might be able to create a unified theory of the three fields of genetics, development and evolution (cf. Burian, 1993a). What I will sketch here is the outline of a *Novel Unification of Development, Genetics and Evolution,* one based on a particular, postulated underlying hierarchical structure to the DNA code. This hierarchical structure is also a basis for action: it includes a repeated, explicit interaction between the DNA in each cell and the physics of the developing organism. We shall see that such interaction is what may permit the systematic activation of different subsets of DNA in each kind of cell in our bodies.

My hypothesis, that much of the DNA is hierarchically organized for interaction with the developing embryo, explains the first part of the title of this book: *The Hierarchical Genome*. The idea of a hierarchical genome is not uniquely mine:

"Not only are these regulatory circuits more complex in multicellular organisms than in bacteria, they also fulfill different requirements.... because finding one gene among a million and not among a thousand, requires a more elaborate mechanism, such as a successive sorting of sub-sets" (Jacob, 1973).

What I add is the physics at each node in the hierarchy: each step of differentiation and its corresponding gene activation seems to start with observable, physical waves, which, sticking my neck out, I've dubbed *Differentiation Waves*. Life is short, explaining why...

"Impatient minds have sought to short-circuit the need for more detailed chemical and biological knowledge through flights of imagination, many of which stimulated fruitful discovery" (Fruton, 1972).

As a theoretical biologist, I like to look at this on the positive side:

"The basic texture of research consists of dreams into which the threads of reasoning, measurement, and calculation are woven" (Szent-Györgyi, 1960).

Thus, in this book, I will go well beyond the data, with the notion that the purpose of theory is to point a way into the wilderness of the unknown, not to provide a map of its paved streets.

Pieter D. Nieuwkoop (p.c., 1995) thought that 'differentiation' waves should be more accurately and conservatively called 'induction' waves, and I tend to agree with him. However, as we shall see, even within embryology 'induction' carries three meanings (heterogenetic, homoiogenetic and autoinduction). The very word 'induction' was probably initially chosen by making an analogy to the phenomenon of electromagnetic induction, thereby imbuing the embryological phenomenon with properties it simply does not have and de-emphasizing its distinct properties. The root meaning of 'induction' is 'to lead or tow', which, as we shall also see, may be inappropriate. The word 'differentiation', on the other hand, has a specific meaning in biology, is little used elsewhere in science (except for taking derivatives in mathematics), and includes both 'determination' and subsequent tissue specific gene expression. I could have used 'determination' wave, but 'determination' is less specific than 'differentiation'. 'Determination' carries a psychological or intention connotation (see Section 1.19), and is little used by biologists who are not embryologists, except taxonomists, who use it for another purpose. Heberlein et al. (1995) have referred to the morphogenetic furrow in the *Drosophila* eye imaginal disc as a 'differentiation wave'.

I was going to add the adjective *Ultraslow* to *Differentiation Waves,* because Lionel F. Jaffe (1995) has recognized the ectoderm contraction wave we found on axolotl embryos (Appendix II: Brodland et al., 1994) as the first member of a new class of 'ultraslow' calcium waves. As we shall see, waves on *Drosophila* (Suzuki, 1974; Ready, Hanson & Benzer, 1976), *Paramecium* (Iftode et al., 1989) and sunflowers (Hernández & Green, 1993) also fall in this category along with the many differentiation waves on axolotls (Appendix V: Gordon, Björklund & Nieuwkoop, 1994). Jaffe (1998) has recently, upon realizing that some waves are even slower, moved axolotl differentiation waves to the 'slow, multicellular developmental wave' category. I will stick with calling them 'ultraslow' for now, especially since some of the slower waves (like the *Drosophila* morphogenetic furrow) are likely to be differentiation waves. It might turn out to be better to classify calcium waves according to their rate limiting mechanism, or function, rather than speed.

1.02 The Epigenetic Problem

"The one great exception was Aristotle, whose genius foresaw what [William] Harvey [1578-1657] more explicitly declared two thousand years afterwards. Harvey quotes a sentence from Aristotle which deserves to be remembered: 'All living creatures, whether they swim, or walk, or fly, and whether they come into the world with the form of an animal or of an egg, are engendered in the same way'. And one of the most scholarly of embryologists, Prof. [C.O.] Whitman [1893, 1896a], has said 'that part of Harvey's theory which affirms that the parts of the future organism do not pre-exist as such, but make their appearance in due order of succession, and which is so often cited as the essence of epigenesis, was all clearly stated by Aristotle'" (Thomson, 1910).

The fundamental question of developmental biology is: How do cells differentiate in the right place, at the right time, into the right kinds (Driesch in Reid, 1985)? This question of epigenesis became acute in biology when Driesch (1891a,c) (cf. Driesch, 1892a, 1929; Ravin, 1977; Khaner, 1993) discovered that when he shook apart the two cells in a sea urchin embryo just after the first cell division, both cells formed whole sea urchin larvae:

"The strict atomistic, mechanistic paradigm applied to organisms led Driesch in 1891 to expect the echinoderm egg to behave like a good machine.... Driesch formed his expectations in such allegiance to the paradigm that he was certain his experiment of [separating]... the first two blastomeres resulting from the first cleavage of sea urchin eggs would result in half-embryos. The appearance of whole little animals in his dish precipitated a practical and philosophic crisis of the first rank in embryology" (Haraway, 1976).

("This work impressed me profoundly as a young biologist, and it still does": Sinnott, 1966.) Driesch's observation contradicted an earlier result by Wilhelm Roux in frogs (Roux, 1888a, translated in Willier & Oppenheimer, 1964; cf. Morgan, 1895a; Maienschein, 1991c; Papaioannou & Ebert, 1995), that one cell of a two cell embryo produced a half embryo, when the other cell was killed (Oppenheimer, 1967; cf. Spemann, 1938). Roux's conclusion was, in retrospect, premature, as Driesch (1891c) anticipated:

"Compared with Roux's results, my findings show a difference between the behavior of the sea urchin and the frog. Still. this difference may not be so fundamental. If the blastomeres of the frog were really isolated and removed from the other half - which was probably not dead in Roux's experiment - might they not behave like my echinoid cells?... I have tried in vain to isolate amphibian blastomeres; let those who are more skilled try their luck at it" (Driesch, 1891c).

"Although [Driesch]... was the initial challenger of Roux's half-embryo experiments, his opposition to Roux seems quite benign, and his regeneration experiments on more advanced stages of development supported Roux's notions of a *Mosaikarbeit* [mosaic process] and of the specificity of the germ layers, which is seldom mentioned in histories that wish to emphasize the dramatic contrast between their famous blastomere experiments.... Driesch used to stay at Roux's home when he passed through Halle" (Churchill, 1991b).

"Not only in the egg of the sea urchin and *Triton* [a newt], but also in that of the frog, a whole embryo originates from a completely isolated blastomere. That same blastomere, however, forms - at least at first - a half embryo when the other blastomere, although destroyed, remains in connection with it. This is the case not only in the frog's egg, but also in the egg of the axolotl, as we know from the work of D. Barfurth (1893).... When the isolation is absolute, a change toward the 'whole,' a regulation, takes place. When the isolation is incomplete this change does not occur" (Spemann, 1938).

"Morgan [1895a]... shocked his European colleagues by showing that the mutilation of one cell of a two-celled frog embryo does not lead inevitably to a deformed half-embryo, as they had thought. Rather, if the remaining live cell were turned upside down, a normal embryo ['*of half-size* ': Morgan, 1897a] would form!" (Carlson, 1981).

The 'shock' is a bit difficult to understand, given the first result of this kind by Chabry (1887) in ascidians (Dalcq, 1957; cf. Churchill, 1973; Fischer, 1991; Wolpert, 1991b):

"A paper of Chabry which has not become generally known is the only other investigation of this kind known to me. With a very refined apparatus constructed for the purpose, Chabry killed individual blastomeres - among others one of the first two cells.... The French investigator nowhere refers to the fundamental result of his experiment: namely, that from the unoperated blastomere there developed *not a left or right half-embryo, but always an entire embryo of half size,* from which, to be sure, certain organs of minor importance (otoliths, suckers) were missing.

"His exposition and illustrations make this certain: The result *is essentially contradictory to that of Roux.* I must note that I became aware of Chabry's work after completion of my own experiments" (Driesch, 1891c).

We also now know that Roux's form of the experiment involving incomplete separation of the two halves (when applied to sea urchins) does not yield a half embryo (Khaner, 1993).

Driesch could not see a solution to this question of how a whole could split into two wholes, and was driven towards espousal of vitalism, which had come to mean that there is a mysterious essence of life, the entelechy, which is not reducible to, but exists mingled with, physical and chemical phenomena (cf. Rádl, 1930; Schubert-Soldern, 1962; Mayr, 1976; Freyhofer, 1987; Sander, 1993a):

"Vitalism is easiest to take seriously when science is ignorant of what lies behind various biological processes. For example, before the physical basis of respiration was understood, it was possible to suggest that organisms are able to breathe only because they are animated by an immaterial life principle. Similarly, before molecular biology explained so much about the physical basis of heredity, it was possible to entertain vitalistic theories about how parents influence the characteristics of their offspring. The progress of science has made such claims about respiration and inheritance wildly implausible...

"The area of development (ontogeny) is full of unanswered questions. How can a single-celled embryo produce an organism in which there are different specialized cell types? How do these cell types organize themselves into organ systems? No adequate physicalistic explanation is available now, so why not advance a vitalistic claim about ontogenetic processes? The point to recognize is that vitalism does not become plausible just because we currently lack a physical explanation.... Although there is no reason to doubt that these phenomena *are consistent with* our current best physical theories, no one has the slightest idea how the physics might be put to work" (Sober, 1993).

Supplying the beginning of that physical theory is what this book is about. Here, in Driesch's own words, is what turned him to vitalism:

"The experiments of several years upon the power which organisms possess of regulation of form, and continual reflection on the collective results of experiments on the physiology of development, upon which I had been working since 1891, combined with a logical analysis of the concepts of 'regulation' and 'action,' brought about an entire change of my opinions and the gradual elaboration of a complete system of Vitalism....

"Analytic experimental embryology - Entwicklungsmechanik, as Roux has called it [cf. Maienschein, 1991b] - has been able to show that there are many kinds of embryonic organs or even animals which, if by an operation deprived of part of their cells, behave in the following way: of whatever material you deprive these organs or animals, the remainder, unless it is very small, will always develop in the normal manner, though, so to speak, in miniature. That is to say: there will develop out of the part of the embryonic organ or animal left by the operation, as might be expected, not a part of the organisation *but the whole*, only on a smaller scale. I have proposed the name of *harmonious-equipotential systems* for organs or animals of this type; they are 'equipotential,' because all their elements (cells) quite evidently must possess the same morphogenetic 'potency,' otherwise the experimental result would be impossible; and their elements work 'harmoniously' together in each single experimental case....

"...Take from the blastula of a sea-urchin whatever you like (but not more than three-quarters) and the rest will always develop into a very small *but complete* 'Pluteus.'

"...Experiments now show that any part of the system, however large and wherever taken, may be cut away from it without disturbing proportionate development. This proves that a 'machine' cannot be the basis of harmonious-equipotential differentiation: for a 'machine,' *i.e.* a specific arrangement of physico-chemical things and agents, *does not remain itself, if you take from it whatever you please....*

"The harmonious system, then, is not a 'machine'; it is, in fact, as it seemed from the beginning, a something that is governed by Individualising Causality. 'Entelechy,' as a non-mechanical agent of nature, is at work in the harmonious-equipotential system" (Driesch, 1914).

What we need to come to grips with is how to have our whole, harmonious equipotential system without invoking a mystical entelechy.

1.03 Wholeness and the Symmetry of the Early Embryo

One way of looking at wholeness is in terms of the symmetry properties of the early embryo. Wilson (1925) tried to dismiss Driesch's findings on the basis of the orientation of the first few cleavages to the stratification of the sea urchin egg, although this explanation must be viewed against the later observation that: "One blastomere from an eight-cell sea urchin is occasionally competent to form a miniature pluteus larva (Hörstadius & Wolsky, 1936; Hörstadius, 1973)" (McKinnell, 1978). Even if we accept Wilson's idea as a refutation, we are still left with what seem to be fully spherically symmetric eggs, such as mammals (Pincus, 1936) and the brown alga *Fucus* (Jaffe, 1968, 1969; Kropf, Berge & Quatrano, 1989; Kropf, Hopkins & Quatrano, 1989; Kropf, Kloareg & Quatrano, 1988; Novotny & Forman, 1974; Quatrano, 1973, 1990; Quatrano et al., 1991), to account for. Some embryos start out with what is probably cylindrical (axial) symmetry:

"The nine-banded Texas armadillo *Dasypus novemcinctus* produces four monozygous [cloned] offspring routinely.... In another species (*Dasypus hybrius*), however, the typical number of monozygous embryos produced ranges from seven to twelve, with a preponderance of eight....Analysis of this development strongly confirms the conclusion that the mammalian egg is completely labile until well after... blastocyst [formation] in the number of embryos that can be initiated....

"The production of two or three embryos from a single blastoderm is not infrequent in fish, reptiles, and birds, but the offspring are fated to an early death. In all of these, the two or more embryos developing from the single blastoderm inevitably share one and the same yolk sac. When the yolk is finally resorbed, the embryos become united at their abdominal surfaces as Siamese twins. In mammals the situation is different because the yolk sac is a small remnant, empty of yolk, and plays no part in later development.... It is evident that in the mammal - as in fish, reptiles, and birds - there is no preprogramming of the egg for a specific development of an embryo and no definitive determination; any part of the ectodermal blastocyst disk can establish an embryonic field capable of self-directed development into a typical embryo" (Karp & Berrill, 1981).

Classical physics starts with a system in a given, known configuration, and calculates how it will change over time. It would be nice if we could begin a physics of embryos with well defined initial conditions, especially the overall symmetry, but even defining these will take substantial work. To the best of my knowledge, no one has yet designed experiments that critically distinguish an initial spherical symmetry from an initial cylindrical symmetry, and tested any embryo with this distinction explicitly in mind, except perhaps the cytoplasm scrambling experiments of Evsikov, Morozova & Solomko (1990). This may be because of the generally loose use of the word 'symmetry' in biology, as contrasted with its rigorous definition in crystallography and group theory (MacGillavry, 1965; Shubnikov & Koptsik, 1974). Thus, there has been a general lack of interest in such questions as:

"How did the cylindrical shape arise in the evolution of each species and in the development of individuals? This question really asks two historical questions in two very different time scales and opens the door for mechanical thinking in the areas of paleontology and developmental biology" (Wainwright, 1988).

1.04 Wholeness through the Ruse of Organicism

"What as yet, unfortunately, we know practically nothing about, is the arrangement and nature of organising relations above the molecular level. Until we can form some conception of the nature of these relations little progress will be possible in our understanding of how they come into being, and embryology will never become an exact science until embryologists take organisation seriously. For embryos are not made like plum-puddings, by a haphazard congregation of ingredients" (Needham, 1931b).

Physical explanations of Driesch's mysterious entelechy were made early on although they did not seem to divert him from his vitalist course:

"The well-known biologist Thomas H. Morgan (1909a) reviewed Driesch's [1908a] published lectures [cf. Driesch, 1929].... Taking issue particularly with Driesch's antimechanist views in biology, he too attacked their poor experimental basis and additionally presented some disproving evidence from the field of crystallography [cf. Whitman, 1888]. Here recent tests on fluid crystals had shown that 'there do exist machines of which any part can reproduce the whole form.' Although his long familiarity not only with Driesch's works but also with Driesch himself [cf. Driesch & Morgan, 1895a,b; Allen, 1978] apparently softened the tone of his criticism, he eventually concluded that 'the attempt to treat the entelechy as something apart from and yet controlling the material basis will seem to most readers, we fear, to come perilously close to mysticism'" (Freyhofer, 1987).

Haraway (1976) suggests that most embryologists in the 1930s made a switch to another paradigm (Kuhn, 1970), 'nonvitalist organicism':

"Theories of tropisms, physiology, biochemistry, developmental mechanics - all illustrate both the triumphs of work conducted under the mechanistic program and the strains leading to the new paradigm. The major long-standing dualities in biology - structure $\leftrightarrow$ function, epigenesis $\leftrightarrow$ preformation, form $\leftrightarrow$ process - have all been reformulated as a result of the crisis.... The controversy that demonstrated most clearly the inadequacy of the simple machine analogy for biology centered around problems of determination and regulation in the embryo.... The sea urchin eggs of Driesch had done what a good machine should not: they had regulated themselves to form wholes from parts.... The machine paradigm had failed Driesch, but rather than abandon it, he resurrected the mechanic. His logic was impeccable, and until 1930 the embryological world occupied itself with trying to exorcise Driesch's demon. However, no spell was entirely effective until the concepts of regulation and of whole could be dealt with outside the machine paradigm. And perhaps this is the major contribution of the new organicism to embryology" (Haraway, 1976).

What is organicism? Here's the closest I found to a definition, a sort of scientific agnosticism:

"The old struggle between Vitalism and Mechanism has lost much of its acuteness. It has been recognized that it is unnecessary to decide *a priori* if the intimate processes of life are or are not resolvable in terms of our actual physical and chemical knowledge. The problems posed by early development have been faced with a complete spirit of objectivity, a philosophical climate which may be styled a pragmatic Organicism" (Dalcq, 1938).

Claiming that "the vast majority of biologists... advocate a middle position" between mechanism and vitalism, Gould (1985b) defines what seems to be organicism, without using the word:

"The middle position holds that life, as a result of its structural and functional complexity, cannot be taken apart into chemical constituents and explained in its entirety by physical and chemical laws working at the molecular level. But the middle way denies just as strenuously that this failure of reductionism records any mystical property of life, any special 'spark' that inheres in life alone. Life acquires its own principles from the hierarchical structure of nature. As levels of complexity mount along the hierarchy of atom, molecule, gene, cell, tissue, organism, and population, new properties arise as results of interactions and interconnections emerging at each new level. A higher level cannot be fully explained by taking it apart into component elements and rendering their properties in the absence of these interactions. Thus, we need new, or 'emergent,' principles to encompass life's complexity; these principles are additional to, and consistent with, the physics and chemistry of atoms and molecules" (Gould, 1985b).

Thus organicism is still very much with us, as a challenge to reductionism:

"The terms 'organismal' and 'organicism' were apparently introduced by Ritter (1919) and are now rather widely used.... Jacob's (1973) concept of the integron is a particularly well-argued endorsement of organismic thinking.

"In contrast to the earlier holistic proposals which usually were more or less vitalistic, the newer ones are strictly materialistic. They stress that the units at higher hierarchical levels are more than the sums of their parts and, hence, that a dissection into parts always leaves an unresolved residue - in other words, that explanatory reduction is unsuccessful. More importantly, they stress the autonomous problems and theories of each level and ultimately the autonomy of biology as a whole. The philosophy of science can no longer afford to ignore the organismic concept of biology as being vitalistic and hence belonging to metaphysics. A philosophy of science restricted to that which can be observed in inanimate objects is deplorably incomplete" (Mayr, 1982).

However, Hein (1971) places the organicists squarely in the vitalist camp, and Waddington (1957) seems to agree:

"...Needham (1936a) and L.J. Henderson (1917)... see organisation as one of the major categories 'which stand beside those of matter and energy'. This seems to imply a renunciation of any attempt to formulate the problems of organisation in terms of anything else but itself. It is surely premature to go so far as this until we are driven to" (Waddington, 1957).

Some critics of organicism went as far as to argue that...

"...Spemann's Nobel Prize should be revoked because organization did not seem to be a specific phenomenon describable in physical and chemical terms..." (Allen, 1975).

I do think that it is perhaps also premature to assume that "...the laws of biology, organismic laws, are primary and are not a specialization of the laws of physics" (Elsasser, 1962). We are not yet done trying to apply physics to embryos (although this will undoubtedly mean physics beyond that of nineteenth century mechanics, as pointed out by Just, 1939a).

I disagree with the assessment of Elsasser (1975) that...

"...the essential property of organisms... is their *radical inhomogeneity*.... This inhomogeneity seems to resist any description in terms corresponding to a purely mechanistic point of view.... It is utterly obscure how any set of enzymes plus rules for their appearance or cessation could be appropriate by itself to give rise to the morphological pattern of an eye....

"This brings us now to a basic question. *What is the mechanism whereby hereditary transmission of morphological properties is effected?* The answer to this question must of necessity be blunt: Nothing whatever is known about such a mechanism; we have no trace of evidence as to what it would look like or whether it even exists. Now, one might say, such a mechanism not being known at the present time will be discovered in the future. But the situation is far more difficult than such a candid avowal of ignorance would indicate. *There is extensive evidence to the effect that such a mechanism does not exist.... Biology must remain a prescience, unless we introduce a postulate that exhibits a distinction between organisms and machines in formal terms.... The organismic properties, irreducibly intermingled with the mechanistic ones, connect skeletal chemistry with morphology.... Only part of the information is supplied by the mechanistic skeleton, the remainder is maintained or regenerated by the organismic function without continuity of the information in a mechanistic sense* " (Elsasser, 1975).

In contrast to these repeated statements of our inability to explain the basic phenomena of embryogenesis, I prefer to agree with Glaser (1939):

"Back of our morphological tokens, each good for so much visible structure, lies a stretch of development complicated by all the complexities of the cell; 'organ forming substances'; the entire apparatus of genes; and beyond, a world accessible only by the methods of physics and chemistry. Either the processes going on at these levels are the remote precursors of structures and events that follow, or they have no meaning for embryology. In that event the morphologist is concerned only with miracles.... We are now free to renew the offensive initiated long ago by D'Arcy Thompson (1917) toward a general physics of growth" (Glaser, 1939).

I also disagree with Gould's (1985b) concept of 'emergent principles', and will argue that we need new disciplines analogous to statistical mechanics to derive 'emergent' phenomena from those occurring at lower hierarchical levels (Proposition 194).

Organicism, then, seems to have only substituted the word 'organization' for 'entelechy', leaving both unexplained, and thus was in effect a continuation of vitalism. In order to include mechanics, Driesch added what we might be tempted to call a supernatural element to everyday physics to explain morphogenesis, while to avoid Driesch's dilemma, subsequent organicist embryologists extirpated physics itself:

"Vitalism and organicism share basic questions and positions. From a negative point of view, both maintain that the study of the parts does not suffice to explain the behavior of the whole. The methods and conclusions of other sciences, particularly physics and chemistry, are held to be applicable to organisms but radically insufficient.... Biology is an autonomous science, not a postscript to physics" (Haraway, 1976).

1.05 The Grip of Vitalism

Perhaps a less genteel approach to vitalism is warranted , and it is indeed a...

"...dogmatic system, the chief actual representatives of which are the botanist Johannes Reinke [1901, 1911] and the metaphysician Hans Driesch. The vitalist writings of the latter, which are devoid of any grasp of historical development [but cf. Driesch, 1905], have gained a certain vogue through the extraordinary arrogance of their author and the obscurity of his mystic and contradictory speculations" (Haeckel, 1906). [Haeckel may have only been seeing his own faults in others: "Haeckel... was never shaken by doubt. No teaching altered his views" (Rádl, 1930). Nevertheless: "...Haeckel has undeniably contributed more than most; everything of value in his utterances has become permanent, while his blunders have been forgotten, as they deserve" Nordenskiöld, 1928.]

"Driesch and Roux differed greatly in philosophical viewpoints and technical approach, but both had studied with Ernst Haeckel. Driesch's interests were very broad, encompassing mathematics, physics and philosophy. Even as a doctoral candidate, Driesch had questioned the wisdom of his mentors; his work presented a direct challenge to August Weismann as well as to Haeckel and Roux. Eventually relations between Driesch and Haeckel deteriorated to the point where Haeckel advised his former student to take some time off and spend it in a mental hospital" (Magner, 1994) [presumably to recover from vitalism].

"Crick (1966a) has attacked vitalism on the grounds that life is mysterious only in proportion as one is ignorant of molecular biology: 'Provided, then, that scientific study continues on a considerable scale, we can foresee a time when vitalism will not seriously be considered by educated men... To those of you who may be vitalists I would make this prophecy: what everyone believed yesterday, and you believe today, only cranks will believe tomorrow'" (Dix, 1968).

Fruton (1972), in an excellent chapter on vitalism versus mechanism titled "The Whole and Its Parts", discussed those who...

"...believed that living organisms follow special laws, different in principle from those of chemistry and physics. This view has assumed various forms during this century, and its adherents have shifted their ground in the face of biochemical advances.... [Driesch (1908a)]... developed a theory of 'dynamic teleology', with the antimechanist argument that machines cannot do what the developing embryo (or other living things) can do; this position was opposed by Wilhelm Roux and by Jacques Loeb [1912, 1916], who called attention to the ability of physicochemical systems to do what a machine cannot do. Between the two World Wars, the vitalist position was espoused by Hans Spemann, one of the two dominant figures at that time, whereas the other (Ross Harrison) stated...:

> 'This quality of 'wholeness' in the parts of the organism, particularly the embryo, has led to much speculation and even a system of philosophy. It is the capital problem of embryology to find the physical - chemical basis for it' (Harrison, 1945).

"After 1930 the philosophical ideas of Driesch and Spemann largely ceased to have currency among experimental biologists, principally owing to the successes achieved in the chemical explanation of some biological phenomena [perhaps a reference to the critical discussion of vitalism by Needham, 1931b]" (Fruton, 1972). [But cf. Abir-Am, 1991.]

As Crick (1966a) put it: "When facts come in the door, vitalism flies out of the window" (cf. Henderson, 1913). Nevertheless, the notion of vitalism lingers on as an undercurrent in modern biology, only rarely frankly acknowledged:

"The primitive association of life with movement, which we still retain, makes it a continual source of wonder that cell movements can have a simple mechanistic explanation. Certainly for myself, even after years of research on the motile tips of growing nerve axons, I still find it hard to accept that such complex, integrated, seemingly sentient structures can arise from dumb molecules. The mystery is doubtless of my own making, a form of closet vitalism, but it was the mainspring for writing this book" (Bray, 1992).

"...Despite the oft-reported deaths of vitalism and mechanism the battle between these hoary foes rages yet.... See, for example, F.H.C. Crick (1966a): '...I believe the motivation of many of the people who have entered molecular biology from physics and chemistry has been their desire to *disprove* vitalism.'" (Ravin, 1977).

Even a life well spent in reductionism leaves what might be called a lingering hope for something philosophically special that distinguishes life from the 'inanimate' universe:

"Every biologist has at some time asked 'What is life?' and none has ever given a satisfactory answer.... Though I do not know what life is, I have no doubt as to whether my dog is alive or dead.... Life appears to be a revolt against the rules of Nature.... Life is a paradox. It is easy to understand why man has always divided his world into 'animate' and 'inanimate,' *anima* meaning a soul.... When we have broken down living systems to molecules and analyzed their behavior we may kid ourselves into believing that we know what life is, forgetting that molecules have no life at all.

"My own scientific career was a descent from higher to lower dimension, led by the desire to understand life. I went from animals to cells, from cells to bacteria, from bacteria to molecules, from molecules to electrons. The story had its irony, for molecules and electrons have no life at all. On my way life ran out between my fingers. The present book is the result of my effort to find my way back again, climbing up the same ladder I so laboriously descended" (Szent-Györgyi, 1972a).

Goldschmidt (1940) summarized the situation succinctly:

"We must confess frankly that this power of regulation [in embryos] is not yet completely understood. Otherwise it would not be the favorite haunt of vitalism and its disguised variants" (Goldschmidt, 1940).

For a considerable period, vitalism seemed to spill over into the post-Darwinian theory of evolution:

"The Cornell historian of biology William B. Provine (Provine, 1987, 1988) has pointed out that in the 1920s many, probably most, evolutionists were religious. At that time Darwinian evolution theory was in eclipse, having been temporarily replaced by the hypothesis of a purposive force which was evolving life toward more complexity. The dean of the American evolutionists, Henry Osborn [1895, 1896, 1917, 1929, 1931], head of the American Museum of Natural History, called this force 'aristogenesis'; the French philosopher Henri Bergson [1907, 1911] called it *élan vital;* the French evolutionist Pierre Teilhard de Chardin [1965] called it 'radial energy.' The terms were different but the evolutionary mechanism was the same: there was a nonphysical cosmic force guiding evolution. The existence of such a force was the consensus belief (Mayr, 1980) of evolutionists in the 1920s, and it was a small step to identify the force with God" (Tipler, 1994).

However, as we shall see with the early vitalists (Lenoir, 1982; Section 1.06), the line of thought of at least some of these people may have had considerable intellectual integrity. The following hardly smacks of "a nonphysical cosmic force":

"It is best frankly to acknowledge that the chief causes of the orderly evolution of the germ are still entirely unknown, and that our search must take an entirely fresh start.... Some kind of relation exists between the actions, reaction, and interactions of the germ, of the organism, and of the environment. Moreover, this opinion is probably capable of experimental proof or disproof" (Osborn, 1917).

The problem with the predecessors to the Modern Synthesis in evolution may have been that the Modern Synthesis purported to make irrelevant a search for the cause of directional evolution. But even here, one of the formulators of the Modern Synthesis expressed support of de Chardin's (1965) efforts (cf. Dennett, 1995):

"...He has helped us to define more adequately both our own nature, the general evolutionary process, and our place and role in it. Thus clarified, the evolution of life becomes a comprehensible phenomenon. It is an anti-entropic process, running counter to the second law of thermodynamics with its degradation of energy and its tendency to uniformity. With the aid of the sun's energy, biological evolution marches uphill, producing increased variety and higher degrees of organisation" (Huxley, 1965).

One could make fine distinctions between directional evolution and progressive evolution, and I will show in Chapters 6 and 10 how differentiation trees make at least the latter plausible in the sense expressed by Huxley (1965), and may even justify some short term directionality.

1.06 The Rise and Fall of Physics in Embryology

"Though we still know little of the 'growth energy' or the energy of life, though we are still ignorant how far it is precisely comparable (for example) with electrical energy, we can at least admit that the actual mechanical energy accompanying life obeys physical laws just as surely as its material substance obeys chemical laws" (Cook, 1914).

"...It must not be lost sight of that we still know practically nothing of the actual changes involved in differentiation. Genetics by itself will not solve this problem. To accomplish this will require all the ingenuity of the embryologist, using the most refined methods of physics, chemistry and general physiology, not only those of the present but many others still to be invented" (Harrison, 1937).

"Physics envy has long been the curse of biology" (Burian, 1988).

The relationship between physics and biology is far from settled. The conflicting points of view of vitalism and mechanism, which as we have seen persist to this day, have their roots in the early nineteenth century, when they seemed to be one and the same:

"The present study treats a period in the history of the life sciences when the imputation of purposiveness to biological organization was not regarded as an embarrassment but rather an accepted fact,... showing that a consistent, workable program of research was elaborated by a well-connected group of German biologists and that it was based squarely on the unification of teleological and mechanistic models of explanation.... It is usually assumed that the proponents of teleology are vitalists. My study attempts to show that in the early nineteenth century the central issue was not vitalism as such, but rather the more interesting problem of causality in biology.... The issue that motivated them to adopt and tenaciously defend teleological thinking in the life sciences was not religion; it was, I hope to show, a concern for good science" (Lenoir, 1982). [Perhaps a similar assessment will someday be made about today's molecular paradigm.]

Louis Pasteur was well trained as a physical chemist and made fundamental contributions to chemistry, such as the discovery of mirror image molecules (stereoisomers) (Duclaux, 1920). In Pasteur's time the word vitalism merely meant an explanation of fermentation and putrefaction in terms of the presence of living microorganisms, rather than as chemical reactions (Dubos, 1950; Asimov, 1964). Nonliving, chemical reactions were then thought to be the correct explanation for these transformations of matter, and so Pasteur's germ theory had an uphill battle:

"...Physiologists [were] wary of accepting the vitalistic theory of fermentation and putrefaction. The belief that living things were the cause of these processes was in conflict with the *Zeitgeist*, the scientific and philosophical temper of the time. Mathematics, physics and chemistry had achieved so many triumphs, explained so many natural phenomena, some of them pertaining to the mystery of life itself, that most scientists did not want to acknowledge the need of a vital force to account for these commonly occurring processes.... To appeal to a living agent as the cause of a chemical reaction appeared to be a backward step.... The priests of the new faith - Berzelius, Liebig, Wöhler, Helmholtz, Berthelot and others - were the supreme rulers of scientific thinking. The power of their doctrine and of their convictions, and the vigor of their personalities, smothered any voice that ventured to express an opinion in conflict with their own philosophy.... The new prophets had pronounced anathema on anyone who preached the doctrine of vitalism" (Dubos, 1950).

With the discovery of enzymes, the antivitalist chemists who gave Pasteur such a hard time became vindicated in a roundabout way (Dubos, 1950).

Out of this context, some momentum for a physics of embryology began to build up over the latter part of the 1800s:

"The zoologist Goette (1875)... spoke in favour of a direct explanation of embryonic development....Rauber (1880) made another attempt... to replace Haeckel's theory,... being a disciple of the new mechanistic conception of life.... He considered that [Hermann] Lotze [cf. Lenoir, 1982] was a philosopher who was underrated by the Germans, and that his work was very important for embryologists;... development is merely irregular growth, he had asserted - growth which results in a number of secondary changes of position 'which partly *appear* to be the result of displacements, protrusions, invaginations or extensions, and in part *are* really produced in one of these ways, being then the result of mechanical tensions and pressures.'

"His [1874, 1888] [cf. Maienschein, 1991c] endeavored to obtain experimental proof of his theory of foldings, but his experiments consisted merely of very rough analogies; he examined the mechanical bending of wooden plates, and drew attention to the crumpling of layers of rock [and other laminates]- phenomena so far removed from those of development that it is not surprising that his theory found few ready to accept it.... Lotze had also compared the processes of folding off of the embryo from the ovum with analogous processes in the earth's crust, as His did later" (Rádl, 1930).

However, in the later historical playing out of Roux's program for developmental mechanics, the nascent mechanics vanished into thin air:

"...The supreme guide is always Wilhelm Roux (1890). Although the biological philosophy of the founder of Entwicklungsmechanik [developmental mechanics] was thoroughly mechanistic, he nevertheless realised the difficulty of expounding the processes of development immediately in terms of physico-chemical concepts.... The aim of Entwicklungsmechanik is thus the reduction of the phenomena to the smallest number of causal processes, Wirkungsweisen [cf. Needham, 1933].... These biological generalisations would thus be as valid as those of physics and chemistry, though possessing a more complex content.... The expression 'biological autonomy'... unfortunately... is often extended to cover the purely dogmatic view that the complex components never will be translatable into physico-chemical terms, whether of to-day or tomorrow. In this latter sense it is, of course, to be rejected.... But there is sometimes now to be found a desire to replace the methods of experimental morphology by those of physiology in the hope of obtaining thereby a short cut into the arcana of biological organisation.... It is an impatience with the large amount of thinking that has still to be done on the purely biological level" (Needham, 1936a).

"...I can well remember the harsh criticism and even contempt which His's doctrine met with, not merely on the ground that it was inadequate, but because such an explanation was deemed wholly inappropriate, and was utterly disavowed.... Even the school of *Entwicklungsmechanik* showed a certain reluctance, or extreme caution, in speaking of the *physical* forces in relation to embryology or physiology" (Thompson, 1942).

We can see this process in action even in Needham (1936a) himself, despite these remarks, when he avoids physics because it is too hard to think about:

"Then there were the electrical effects, the potential differences, etc., which can be demonstrated on the amphibian egg. Their interpretation, however, is at all times so difficult that they could hardly be called upon to explain the phenomenon of the organiser" (Needham, 1936a).

This was a critical juncture for all of subsequent embryological research, because it represented an abandonment of physics approaches to embryogenesis in favor of biochemical approaches, for reasons that had nothing to do with the merits of both approaches. It did not come suddenly, for Needham, in fact, was obviously already quite ambivalent towards the role of physics in embryology (cf. Abir-Am, 1991):

"Furthermore, the Cambridge biochemist now explicitly rejected contemporary physics and chemistry as the key to biology: 'It may be very well that such questions as the mechanism of differentiation and determination in the developing egg are completely insoluble as long as we try to reduce the phenomena to terms capable of fusion with the classical physics and the classical chemistry' (Needham, 1932)" (Haraway, 1976).

Thus the predilection that Needham had towards biochemical solutions as a member of Gowlan Hopkins' new Institute of Biochemistry at Cambridge University, designed to reinforce "the identity of biochemistry as a subject worthy of study and support in its own right", is not a sufficient explanation that...

"Not surprisingly, [Joseph] Needham, Dorothy Needham, and Waddington [Needham, Waddington & Needham, 1934] decided to explore this last option,... that there was 'a single definite chemical substance, working in an almost endocrinological manner on the competent ectoderm'" (Witkowski, 1987a).

Although biochemistry was fighting for its own place under the sun and on the rise, their motivation included a turning away from physics, encouraged by an agnostic, organicist perspective. The choice made, to presume that "a single definite chemical substance" caused induction, persists to this day, for no better reasons than the original ones. In a way, it is my goal in this book to set the clock back and take the problem forward, by giving physics its due.

1.07 Can We Restore the Physics of the Youth of Embryology?

"In spite of the great progress at genetic mechanisms of embryology in recent years, we still know rather little about the physical process of morphogenesis" (Sato et al., 1992).

It is curious that this career pattern of genuflecting to physics in one's youth and then rejecting it may be traced back at least to Ernst Haeckel:

"When writing his *General Morphology* Haeckel (1866) was still under the influence of idealism: we see this in his postulate that biology admits of exact mathematical treatment [cf. Nordenskiöld, 1928]. He was looking for 'the stereometric plan of the organism; knowledge of this is as important for the organic morphologist as is a knowledge of crystallography for the student of inorganic form' [Bölsche, 1906]. Following in the footsteps of [H.G.] Bronn [cf. Mayr, 1982], he discussed the 'axes' and 'poles' of the organic body... and was convinced that all these forms could be explained as resulting from purely mechanical causes.

"Afterwards he abandoned these extremely mechanistic views of morphology, and came to disagree with His [1874] and Goette [1875] [cf. Nordenskiöld, 1928; Oppenheimer, 1967], who had tried to explain embryonic development on mechanical principles. Their explanation seemed to him 'too coarse and *mechanical* ', and contradictory to the theory of evolution.. For has not every organ a long history?... The cause is historical.... The study of evolution is the goal of all human knowledge, the principle underlying all philosophy, the key to every problem, the answer to every question" (Rádl, 1930).

Steward (1968), a plant morphogeneticist, puts his own experience in an historical context:

"The author began his own studies in the stimulating period that followed the first World War, when physical science became almost a religion, and somewhat crudely mechanistic hypotheses held their sway in biology. Organisms conveniently presented chemists and physicists with problems to be solved by the techniques with which they were then familiar; but by the imminence of a second World War, the limitations had become all too obvious. A renewed respect was born for logical organization, for the working of the dynamic machine rather than the purely equilibrium mechanism.... However, while writing... in 1940 in Richmond, Surrey, England, amid the stupidity and the exhilaration of the first major aerial war, one was then impressed with the futility that uses 20th-century physical science in ways that even threaten man's destruction.... But the sense of escape and relief is still there for those who turn from the raucous turbulence engendered by the study only of the inanimate, physical, world to the contemplation of the mystery and the quiet competence of the growth of living things. Despite all the noise and blast of rocket propulsion, of soaring satellites or even megaton bombs, man cannot yet approach or emulate the coordinated, incredibly efficient growth of a fertilized egg. All the resources of the chemical industry cannot spin a cellulose wall" (Steward, 1968).

The biases against physics range from its having given us Newton's clockwork universe to the nuclear bomb, and I sometimes get the impression that debates about physics versus biology are being fought for reasons that have nothing to do with our attempt to understand the universe through reason and science. The embryo needs a fresh look, and physics is one of our finest tools for clearing up questions of cause and effect. Physics in embryology has hardly gone past Wilhelm His' laminates. There is so much more it could do for the discipline.

The general idea that structure can arise from instabilities, now commonplace with notions of:

reaction-diffusion equations (Turing, 1952; Prigogine, 1980; Prigogine & Stengers, 1984; Murray, 1990; Schiffmann, 1991, 1994),

fractals (Mandelbrot, 1977, 1982; Feder, 1988; Prusinkiewicz & Lindenmeyer, 1990; Family & Vicsek, 1990, 1991; Birdi, 1993; Kaandorp, 1994; Tabony, 1994) and

chaos (Prigogine & Stengers, 1984; Gleick, 1987; Glass & Mackey, 1988; Seydel, 1988; Cooper, 1989; Ruelle, 1989; Stewart, 1989; Infeld & Rowlands, 1990; Mosekilde & Mosekilde, 1991; Zaslavsky et al., 1991; Lewin, 1992; Waldrop, 1992; Kellert, 1993; Cohen & Stewart, 1994; Strogatz, 1994; Malevanets & Kapral, 1996; Peterson, 1996),

should have been known to embryologists since the work of Rayleigh (1892, 1964) on mechanical instabilities. Certainly, his collected papers (Rayleigh, 1964) show that he paid some attention to matters of importance to biologists, with titles such as "Insects and the colours of flowers", "The soaring of birds", "On defective colour vision", "On the theory of optical images, with special reference to the microscope", "On the perception of the direction of sound", so here may have been a lost opportunity for reciprocation.

Perhaps physics did not persist in embryological research because of its very success in a neighboring field:

"...Animal morphology was defined in contradistinction to a new 'physicalist' program in physiology led by Carl Ludwig (1816-95), Emil Du Bois-Reymond (1818-98), Hermann von Helmholtz (1821-94), and Ernst Brücke (1819-92), who sought to redefine physiology away from what they saw as its focus on organisms as teleologically defined wholes and toward the study of those aspects or organic function measurable by physical instruments (Lenoir, 1982). The physicalist physiologists are important to the history of morphology because part of their aim was to eject from physiology the study of form, generation, and development - aspects of life that seemed at the time essentially unmeasurable" (Nyhart, 1995).

With our growing experience with computer graphics and animation, we now know how to measure form. Perhaps it is time to get on with a program of 'physicalist embryology'. Furthermore, now that we know a bit about the organization of the genome and its partial record in DNA of the history of each organism, we may be able to meet Haeckel's objection to His' attempts at a physics of embryos, by incorporating that history in our models:

"No matter how exactly His describes all the individual folds, main folds and subordinate folds, and shows these to be the substantial changes in the developing body form, nothing is in the least explained by it. For every one of these simple ontogenetic folding processes is a highly complex historical result, which is causally determined through thousands of phylogenetic changes, of transmission and adaptation processes" (Haeckel, 1875, translated in: Nyhart, 1995).

1.08 Avoiding the Spatial Component of Embryogenesis

Given a scalar cow....

Current ideas of molecular cell biology, rooted in the monumental success of biochemistry, i.e., of chemistry in biology, fail to resolve how cell differentiation is accomplished because they cannot yet explain the spatial component of pattern formation:

"While viruses and bacteria might well pour their genetic hearts out in praise of molecular biology, genes of the higher organisms are of sterner stuff, spending most of their time in a state of repressed silence. The key to development, form, and function in multicellular organisms is differential gene expression, and the most intimate knowledge of the genetic code reveals nothing about the implementation of its information in space and time" (Reid, 1985).

Burian (1986a) formulated the problem as "Lillie's Paradox" (cf. Lillie, 1927):

"The paradox is easily understood. It turns on the fact that in virtually all multi-cellular organisms, virtually all cells have an entire complement of chromosomes and thus, if you believe the chromosome theory, all cells of a higher organism have the same genes. If this is correct, the differences between nerve cells and muscle cells, between phloem and xylem, cannot be due to differences in the genes. But if that is the case, there must be something else that controls the development of nerves and muscles, bone and sinew, that determines the fate of a given cell.... Whatever that something else is, it must be inherited or largely inherited. Otherwise development would not be regular and like would not beget like. In any case, if the Mendelian/chromosomal theory of inheritance is correct, it must at best be only a partial theory, for it leaves out of account the system of inheritance that controls development" (Burian, 1986a).

The 'something else' is not simply gene regulation, but 'control' of gene regulation by something else (cf. Reinitz & Sharp, 1995). At least, that is my hypothesis: the something else is differentiation waves. I put 'control' in quotes because it is too strong a word, but hard to replace. A better word might be 'cause', in the sense of cause and effect in physics; however, in a complex sequence of events, we would then have to distinguish proximal, mediate and final causes. Ultimately, in a reductionist approach to biology, that is what we want to reach for, and if and when we get there the word 'control' may disappear from our biological vocabulary. Each differentiation wave is a wave of cell contraction or expansion, cytoskeletal changes, ionic currents, concentration changes, chemical and stereochemical reactions, mechanical effects, and changes in gene transcription. Unraveling the cause and effect and feedback links between these components of the differentiation wave represents a challenge for the future.

Burian (1986a) goes on to suggest that "the transition that tamed the paradox" was the notion of gene regulation, as put forward initially by Nanney (1958) and Ephrussi (1957), then expanded upon by Jacob & Monod (1961a,b) (cf. Monod & Jacob, 1961; Zubay, 1964), but it does not include any resolution of the spatial component of the problem:

"The 'structural' concept of gene action accounts for the multiplicity and for the phylogenetic stability of macromolecular structure. It does not account for biochemical coordination and ignores the problem of the emergence and functioning of differentiated cellular populations" (Jacob & Monod, 1961a).

There is, therefore, still some work to do, despite the early chutzpah of molecular biologists:

"It is the feeling of many that the analysis of differentiation is just a mopping-up operation following the successful beachhead established by molecular biology" (Wilt, 1972).

The presumption of success is accomplished by "substitution of the difficult, but relatively more tractable, problem of control of protein synthesis in microorganisms for the problem of differentiation in multicellular eucaryotes" (Burian, 1993c). The chutzpah continues to this day because of the (I would claim false: cf. Andersen, 1861) perception that the basic problem of embryogenesis has been solved:

"...Morphogen gradients, bistable circuits and thresholds have become facts rather than hypotheses. The central role which... models have in our field is symptomatic of the difference there must be between an explanation of regional specification and an explanation of how a particular gene is turned on.... We now really do understand what we are dealing with...

"...We shall write a *developmental program* for making a simple animal.... The main gaps in the completeness of the program lie in the acquisition of an executive function by a group of cells with a particular coding. In other words, we still do not know exactly which set of genes we need to make one cell sheet move in a particular way, or adhere to another sheet, or differentiate into a particular terminal cell type.... These are problems of morphogenesis and cell differentiation, rather than of regional specification" (Slack, 1991a).

To the contrary, I will show how the problem of cell differentiation *is* the problem of regional specification, and even with what I suggest, we are far from a complete solution. The 'developmental program' is not yet an exercise for undergraduate computer science majors. I suspect that a fair assessment is that we have yet to carry through the research path outlined by Whitman (1888):

"*The Idea of a Formative Power.* - Let us now consider whether any rational basis can be found for the idea of a formative power as a resultant from, and an expression of, physiological unity. I am fully conscious that the subject is one of profound mystery, the solution of which appears to lie as far beyond our grasp to-day as at any time in the past.... Every step in advance only brings us to a keener sense of the subtle and incomprehensible nature of the force or forces contemplated....

"The more important speculations on this subject have taken the form of theories of heredity. In most of these theories, at least those of recent date, we find a fundamental idea which must be accepted as true; namely, that the sexual cells reflect in some way in their chemico-physiological constitution all the typical structural features of the parent-organism. How all the hereditary tendencies can be contained in a single cell, and with such completeness that the developing organism repeats step for step the chief form-phases of a genealogical history stretching through countless myriads of generations, back from the present into the very dawn of life, and ultimately unfolds every detail of structure and feature of the parent-organism, is a mystery that transcends our understanding. The preformationists of last century took refuge in the celebrated inclusion (Einschachtelung, emboitement) theory, which made the real mystery unapproachable by hiding it behind an endless series of miracles. The triumph of epigenesis brought with it the reclamation of the problem, but left us with the indefinable *vis essentialis* of Wolff [1774] [cf. Driesch, 1905; Herrlinger, 1966; 'the idea of which he himself states that he borrowed from Stahl [1737]': Nordenskiöld, 1928], the *nisus formativus* of Blumenbach [1792] [who 'specifically points out that these 'forces' are merely expressions by which to denote phenomena, the cause of which we do not know, but the effects of which we can observe': Nordenskiöld, 1928]....

"The biologist does not hesitate to follow the shibboleth of 'molecular motion' to its final goal, but he must be aware of being blinded by an artifice of method to distinctions which lie beyond it.... It does not appear at all irrational to conclude that vital phenomena are the manifestation of special forces, resultants of course, and yet *quite unlike the elementary forces from which they are derived*.... Derived from them, and yet wholly unlike them, as water is something totally unlike its chemical elements or any simple mechanical addition of these elements.

"It is precisely this point which is so persistently ignored in all so-called physico-chemical theories of heredity. And yet all analysis and all observation leads to the conclusion, that molecular structure is not directly responsible for vital phenomena....

"What we do affirm is this: We cannot stop with the most complex molecules revealed or revealable by chemical or physical research; we must pass from *organic* to *living, organized* matter, not by the supervention of new laws, but by ultra-chemico-physical, or chemico-organic combinations, which are absolutely beyond the highest possibilities of chemical analysis. Inability to define these higher modes of combination is no reason for doubting the testimony of all our senses to their existence" (Whitman, 1888).

Whitman (1888) ironically also discussed "travelling waves of contraction". Present day embryology has not yet passed beyond "the most complex molecules revealed" to tackle the "formative power", just how an organism is created in space and time, so that it is more than a bottle of biochemicals.

1.09 Wholeness, the Environment, and Symmetry Breaking

"What was really dramatic about the experimental result was that the new embryo might be whole; the wholeness has never been adequately investigated..." (Oppenheimer, 1991; cf. Hollyday, 1996).

The question of the wholeness of embryos raised by Driesch's experiment, separating two sea urchin blastomeres, is usually addressed via the most widely accepted notion today that it is the 'environment' that causes cells to come to differ from one another, where this environment includes other cells (Maclean & Hall, 1987):

"...[The] activity of the genes during development [is] dependent upon the physical and chemical environment in the cell - an environment for which the genes themselves are largely responsible. A sort of cyclical feedback occurs between genome and its ever-changing chemical environment so that the changing interactions between gene and environment provide the motive power for driving embryonic cells along their diverse paths of differentiation" (Wright, 1945; cf. Wright, 1941a).

"It is a fundamental principle of developmental genetics that every organism is the outcome of a unique interaction between genes and environmental sequences modulated by the random chances of cell growth and division, and that all these together finally produce an organism" (Lewontin, 1992a).

"...Communication between a cell and its immediate environment defines, to a large extent, the phenotype of that cell; it also determines the structure and function of the tissues and organs in which the cell resides" (Stevenson, Gallin & Paul, 1992).

(Cf. Markert, 1958; Glock & Gregorius, 1984). While Ravin (1977) attributes the notion of environmental control to Troland (1917), this concept seems to go back further in the history of embryology:

"*The Doctrine of Isotropy.* - Pflüger's (1883a,b) interesting experiments with the amphibian egg to determine the influence of gravitation upon the direction of cleavage-planes, led him to conclude that the entire egg is '*isotropic* '. In other words, to quote from the author, 'the fertilized egg possesses absolutely no essential relation to the later organization of the animal, no more than a snowflake stands in any essential relation to the size and form of the avalanche which under certain conditions develops from it. That the germ always gives rise to the same form, is due to the fact that it is always brought under the same external conditions.'...

"At the beginning of any ontogenetic series, when we get the most rapid and vivid displays of nuclear energy, we see that the environment of each cell is much more potent in determining its form than the nucleus. True, certain conditions of the environment may be said to be largely the result of nuclear activity; and to this extent the nucleus may be said to determine, indirectly, the form of the cell. But this is very different from saying that the nucleus has a direct controlling power over the specific form of the cell, as claimed by Gruber [1886], Weismann [1889, 1893], and others" (Whitman, 1888).

There is no need to invoke a role for the 'extracellular environment' (Maclean & Hall, 1987), since all of the components necessary for biochemical differentiation and spatial morphogenesis are contained within the embryo. I would claim that the classical view of differentiation in time and space, as expressed here by Stern (1955), places an unnecessary sensitivity on cells to their embryonic 'environment', is probably not a robust way to build embryos, and smacks of preformationism:

"The concept which reconciles the apparent contradiction between essential genetic equality of all cells and their differentiation from one another rests on the existence of regional cytoplasmatic differences in the egg cell and its cellular derivatives. The concept implies a differential response of the genes which during cleavage come to lie in different cytoplasmic surroundings. The relatively few regional differences of the early egg are assumed to become increased when cleavage produces new topographic diversity which may lead to physiologic diversity" (Stern, 1955).

There is a grain of tautological truth here, yet something profoundly disturbing. This 'domino effect' approach to wholeness in development would seem intuitively nonrobust, in that, if we were to leave it to the dominoes to divide, reproduce and set themselves up, along with constructing (somehow) the table on which they stand, the patterns in which they would fall would not likely be very similar from one case to another. Such a concept of a developing organism effectively treats it as a "deviation amplifying mutual causal process" (Maruyama, 1963a,b; Maruyama, 1968) that has low success in generating anything approaching constant form without the inclusion of global, negative feedback mechanisms (an idea that has hardly been tested: see Gordon, 1966, in which I simulated growth of a two dimensional 'snail'; the computer generated pattern was in effect the first 'virtual ant': Peterson, 1995a). It would seem that some better controlled process is operating here, although we will be drawn back to the domino concept in Proposition 169.

While on the one hand we have the concept that the environment somehow determines which cells become what, this is offset by another concept of the not so benign environment, one that brings teratogens (Kalter, 1968; Wilson & Fraser, 1977; Okamoto, 1980; Szabo, 1989; Persaud, 1990) and more subtle phenotypic effects, such as alterations in imprinting associated with reproductive technologies (Seamark, 1994).

Wholeness through "frequency-doubling bifurcations in Turing reaction-diffusion systems" (Goodwin & Kauffman, 1989, 1990; Hunding, Kauffman & Goodwin, 1990; Kauffman & Goodwin, 1990; cf. Prigogine, 1980; Lacalli, 1990) would appear to be a variation on the domino theme, with similar stability problems, especially given their close relationship to bifurcations in chaotic systems (Bunow & Weiss, 1979; Prigogine & Stengers, 1984; Glass & Mackey, 1988; Seydel, 1988; Infeld & Rowlands, 1990; Klevecz, Pilliod & Bolen Jr., 1991; Blazsek, 1992; Kellert, 1993). Statements such as...

"...Spontaneous generation of asymmetry, epigenesis, not only conforms to experimental reality, but is the only way for robust development to occur" (Schiffmann, 1994);

"When a new cell is born... via cell division... we need additional degrees of freedom to indicate the cell's state" (Kaneko, 1995)

do not take into account the fact that there may be more than one way (or even infinitely many ways: cf. spheres in: Shubnikov & Koptsik, 1974) for a symmetry to break, and when multiple, hierarchical steps of symmetry breaking are involved, the combinatorics could lead to many alternative, morphologically untenable organisms. The combination of 'intracellular' network theory (Kauffman, 1969) with 'intercellular' Turing theory (Kaneko & Yomo, 1997a; Furusawa & Kaneko, 1998a) does not address this spatial problem. It would seem that some means of constraining the combinatorics must exist if symmetry breaking is the essence of development. In other words, we need something to bias a choice amongst the combinations, i.e., another level of symmetry breaking. But even an infinite regress will not do, since the symmetry of the set of combinations cannot be avoided. Thus symmetry breaking is far from the whole story (cf. Anderson & Stein, 1987; Goodwin, Kauffman & Murray, 1993; Hanke & Green, 1994; Tabony & Job, 1992a). While symmetry breaking may be useful to explain the origin of the universe in the big bang theory and the origin of chirality for biological molecules (Babincová· & Babinec, 1994; Babinec & Krempasky, 1994) and cell polarity (Svetina & Zeks, 1990), its usefulness for morphogenesis is circumscribed. Consequently, I do think that neither the vaguely conceived cellular 'environment' nor symmetry breaking will prove to be adequate explanations for embryogenesis, even though they may be important elements in a solution.

1.10 Wholeness through Surface Tension

"...If we move up to a large scale and consider groups of cells, the control system has to operate over a commensurately larger scale, coordinating the responses of many cells, which places it some considerable way further down the line of command from the molecule of DNA. This in turn means that the black box the geneticist has to invoke for delivering phenotype from genotype is even bigger and blacker than usual, and so has room inside it for a biochemist. To the biochemist, this black box is like a rathole to a Jack Russell terrier, even to the extent that it may be difficult to persuade him or her to extricate themselves long after it has been conclusively demonstrated that the rat is no longer at home" (Hanke & Green, 1994).

There is a fundamental idea that has been around at least since Driesch's time that points towards a solution to his dilemma and the problem of wholeness. Joseph A.F. Plateau (1873, in: Worthington, 1963) and Lord Rayleigh (1892) (reprinted in Rayleigh, 1964) showed, by a perturbation analysis (cf. Hinch, 1991; Murdock, 1991), that a cylinder of fluid will break up into a set of equally spaced drops (cf. Boys, 1959). This is a familiar phenomenon, from the spacing of dew drops on a spider web to the beading of threads of honey (or treacle pulled along a surface: Rayleigh, 1892). For mathematical analysis, ideally the fluid is regarded as suspended in space, but it can also be suspended in another fluid (Tomotika, 1935). Cells exhibit this behavior during embryonic cell sorting (Gordon et al., 1972, 1975). A form of the phenomenon, with a tapered cylinder, can be seen by running a water tap slowly: the column will break into droplets as it falls (Shaw, 1984; cf. Levich, 1962), which can be beautifully revealed if you have a tunable strobe light. In a fluorescent activated cell sorter (FACS), used to separate cells in suspension, a piezoelectric driven vibration of the falling column of water produces a stream of nearly identical droplets (Parks & Herzenberg, 1984). Amoeba pseudopodia under high pressure (Marsland, 1956) and diatom trails (Gordon & Drum, 1970) also break into droplets.

This process of a cylinder of fluid breaking into a row of drops is driven by surface tension and by the fact that such drops have a smaller total surface area than the cylinder. The cylinder is released from its mathematical point of balanced instability by thermal or mechanical fluctuations of the surface. If we consider instead of a long cylinder, a single drop of water, and forcefully split it in two, each mass of water forms into a spherical drop. Thus, except for lack of subsequent differentiation of the 'parts' after it is split, in any proportions, a water drop would seem to fit Driesch's definition of a harmonious-equipotential system. The units are the water molecules, and the 'entelechy' is the surface tension. The idea of a 'machine' made of liquid units was popularized as the robot T1000 in the movie *Terminator 2* (Cameron & Wisher, 1991), although the concept of proportionality was not explored, as the humanoid liquid robot always resumed a human size: "A very strange sort of machine indeed, which is the same in all its parts!... Therefore the 'machine-theory' of life is absurd" (Driesch, 1929).

Of course, although this is a start to solving Driesch's dilemma, it is not enough: the 'drops' that are half of an embryo don't simply round up, but go on to produce whole organisms. Alan Turing (1952) (reprinted in Mangel, 1990) put it quite succinctly, if in one step, for embryos:

"...A system which has spherical symmetry... will remain spherically symmetrical for ever.... It certainly cannot result in an organism such as a horse, which is not spherically symmetrical. There is a fallacy in this argument.... The system may reach a state of instability in which... irregularities, or certain components of them, tend to grow. If this happens a new and stable equilibrium is usually reached, with the symmetry entirely gone" (Turing, 1952).

1.11 Nonmaterial Physics as the Entelechy of Vitalism

"Something guides the disposition of matter and energy in the organism, something not yet understood and seemingly different from anything else in nature" (Sinnott, 1966).

von Uexküll (1926) took another approach to deal with Driesch's result by hypothesizing 'nonmaterial impulses', with the word 'impulse' probably to be understood by its precise Newtonian mechanics definition of "a quantity equal to the momentum produced by... forces acting for a very short time, as in impact" (Harris, 1913). Keeping in mind that 'ferments' are now 'enzymes' and other proteins, and have since been intellectually separated from the genes themselves, we can see that there is some good thinking here:

"Employing a crude but very obvious comparison, we may picture the chromosomes in the nucleus of the germ-cell as washing-lines, on which factors for the absolute and relative properties hang, side by side, like articles of clothing which the subject will put on, one by one.... In contrast to what happens with machines, the builder resides within the organism itself.... A gene or factor, then, is a ferment activated by an impulse.... It is possible that the nucleus is continually giving off to the plasma a limited number of... ferments, which lie latent within it.... The liberation of the ferments is referable to an impulse of the subject.... Driesch succeeded in showing in sea-urchin larvae... that the half-size depended on each larva having the same-sized cells but only half the number. It follows that... the shaping impulses... are to a great extent independent of the quantity of material furnished them...."

"The difficulty we experience in understanding how the form-giving genes work, lies in this, that, although they are tied to a definite place in each individual cell, yet they must act according to a system which is not present anatomically, although it embraces a whole germinal area with many hundreds of cells.

"But this difficulty disappears when we realise that it is only the material basis of the genes that is of necessity bound to a definite position in space, whereas their non-material portion, the impulse, is not bound in this way.

"The impulse always plays an active part, now stimulating a gene, and now an anti-gene. An impulse, which is not fixed to a definite position in space, may easily be connected up with other impulses into a system.

"An impulse-system can allow a whole series of cells to be simultaneously invaded by a fermentative action leading to a certain chemical change.... Simultaneous and equipotential impulses of this kind must produce in a mass of similar cells a differentiation with regard to position.... The number of cells within the mass is quite immaterial for the achievement of the final form....

"The individual impulse-system is dependent on the material only in so far as that must yield the suitable genes if the system is to become manifest. It is dependent on the adjacent systems only in so far as its fixed position is determined by its being set between them. For the rest, development within each system proceeds quite independently....

"The impulses... are fixed in space and time, but in themselves are still completely non-material. But, since they are attached to the genes, they dominate the material, for that is set in motion by the fermentative action of the genes. The genes themselves represent a union of a latent ferment with an activating impulse..." (von Uexküll, 1926).

One may read into these words an analogy between the 'impulses' and some of the properties of differentiation waves: their 'control' of the nuclei of the cells through which they pass; their activation of specific genes; their confinement to parts of tissues. After all, a wave is a physical phenomenon that passes through material, without being material itself, and so could be thought of as 'nonmaterial', or even an 'entelechy', for that matter, exonerating Driesch. We thus see that Driesch appears to have confused 'nonmaterial' with 'nonphysical'. Force, momentum, elasticity, viscosity, surface tension, and wave propagation of these phenomena were already very much part of the Newtonian physics of the time, and Driesch's failure, and that of his critics, to look to these other aspects of physics for explanations of the entelechy, shows an inadequate grounding in the physical sciences, a problem that curses biologists to this day. von Uexküll (1926) alone seems to have considered the matter without mysticism or dismissal, but unfortunately well after Driesch had irreversibly committed himself to vitalism and other biologists had more or less politely chosen to ignore him. What is ironic here is that the very concept of 'ferments' emanating from the nucleus has been attributed to Driesch:

"Long ago it had been suggested that the action of the nucleus on the cytoplasm was mediated by 'ferments' (Driesch, 1894) and by 1902 hereditary factors were associated with ferments in the minds of several early Mendelians" (Olby, 1974).

1.12 Towards a New Physics of Embryos

"If you don't even try to engage in reverse engineering, you are not in a good position to conclude that there *is no* reverse-engineering explanation to be discovered" (Dennett, 1995).

As we shall see, the battle of chemistry and physics versus vitalism may resolve itself by using a field of physics, namely mechanics, as first suggested by Wilhelm His (1888) (reprinted in Coleman, 1967; but cf. Grmek, 1972), rather than requiring the defeatism of Driesch's ultimate position on morphogenesis:

"Driesch's interesting development of this conception [of entelechy] has failed of wide acceptance partly because it is contrary to the spirit of modern scientific inquiry, and involves a practical abandonment of the problem.... The inescapable fact remains that the specific reactions of the developing egg *depend upon its organization.* Concerning the fundamental nature of this organization we are still ignorant; but we have nothing to gain by the vitalistic assumption that the guiding principle in development is not only unknown but unknowable. Existing mechanistic interpretations of vital phenomena, evidently, are inadequate; but it is equally clear, as someone has said, that they are a 'necessary fiction.'... If Mendelian heredity, at first sight so inscrutable, is effected by so simple a mechanism, we may hope to find equally simple explanations for many other puzzles of the cell that lie beyond our present ken" (Wilson, 1925).

Today mechanics is regaining a hold on thinking about the problems of differentiation and morphogenesis, an approach promulgated by His (1888) (reprinted in Coleman, 1967) in his studies of neural tube closure. It is curious that Thomas H. Morgan, perhaps influenced by His, vacillated between gradient and mechanical models for regeneration:

"Observing regeneration in animals, he is

'struck by the apparent resemblance of the change in form that they undergo to a process of expansion. The idea of expansion of a viscid body carries with it, of course, the idea of tension within the parts, and return to the former condition is brought about by a release from tension and a return to a more stable condition. If by the intercalation of new material the extended condition is fixed, a new state of equilibrium will be established' (Morgan, 1901).

"....In the [Morgan (1906a)] paper he even writes, 'The gradation is the polarity.' Nothing could be more explicit. Yet just one year later, in... [Morgan (1907)], he has abandoned all such ideas. Writing of regeneration in the earthworm he [says]...:

'...The centripetal influence is, according to my interpretation, nothing more than the tension of the outer layer of cells, and the pressure relations in general, in the rounded dome-shaped mass of new materials.'

"A similar line is taken concerning the behavior of hydra grafts with different polarities:

'Here we meet with the conception of polarity as involved in the pressure relations. The polarity from this point of view is an expression of the graded pressure relations from one end of an organism to the other, which in turn may be an expression of the gradation of the tissues, and in turn may itself, under certain conditions, be the cause of differentiation.'

"His ambivalence has come full circle. Morgan has by now abandoned gradients and regeneration and is giving all his attention to genetics. But he has laid a clear basis for a gradient theory of regeneration" (Wolpert, 1991b).

We could equally say that in Morgan's mature consideration, he preferred mechanics over gradients. That mechanics is fluid mechanics, which when combined with surface phenomena, such as surface tension and surface waves, permits a mechanics of the whole that can regulate with size. Morgan did not see all of this, but his use of concepts of mechanical stability and approach to mechanical equilibrium of fluids, and his suggestion to relate mechanical phenomena to differentiation (see also Maienschein, 1991a), place him beyond most current thinking. Even Thompson (1917, 1942), who applied mechanical concepts, and especially fluid mechanics with surface tension, to a huge array of developmental problems (Clark, 1945; Thompson, 1958; Chaplain, 1997), seems to have stopped short of considering differentiation, perhaps because "...our enquiry is not directed towards the solution of physiological problems, save only in so far as they are inseparable from the problems presented by the visible configurations of form and structure..." (Thompson, 1942). He avoided "The efforts to explain 'heredity' by help of 'genes' and chromosomes, which have grown up in the hands of Morgan..." (Thompson, 1942).

Unfortunately, early attempts to measure the physical forces of embryogenesis led to widely divergent results, which deserve reinvestigation:

"By altering the osmotic pressure within the blastula,... [Moore, 1941, 1945] found what pressure was sufficient to prevent gastrulation and in this way was able to calculate the force necessary for invagination.... This was also done by Waddington [1939a, 1942b]... for the amphibian using magnets and steel balls. Moore [for an echinoderm] obtained a force of 3.9-7.8 gms/mm^2 and Waddington [for an amphibian] obtained 0.00034 gms/mm^2" (Bonner, 1952).

(Cf. Selman, 1955, 1958, and Waddington, 1956a.) Thus we have not had a renaissance of physics in embryology.

1.13 New Tools of the Trade

"It seems to be assumed in certain quarters that we already know all the important phenomena of normal development and that mere observation is, therefore, a useless and antiquated method. If the time ever comes when every step of normal development of a single individual is known, the causes of development will not be far to seek" (Conklin, 1898).

The lack of a renaissance of physics in embryology may be because we are stuck at the fourth level of Roux's progression for embryology, i.e., stuck at the level he achieved over a century ago:

"Roux was an advocate of the position that all the sciences passed through a descriptive phase in their development toward an experimental discipline. He gave no indication of whence came this point of view and he showed no inclination to question its accuracy (Roux, 1890). According to Roux, it was altogether proper and even desirable that embryology should follow the pattern set by physics and chemistry; just as the older sciences had earlier progressed, it was now time for embryology to become experimental. There existed, in fact, a hierarchy of methods through which embryology was now passing toward a science which gave causal explanations. At the bottom of the scale stood the descriptive method.... Next up the scale was the comparative method.... In Roux's mind there existed yet another intermediate stage of investigation which he labeled the descriptive experiment.... At the top of Roux's methodological scale stood the analytical experiment.... The definition of an experiment taken from his dictionary of *Entwickelungsmechanik* concisely describes Roux's intent:

'The experiment is the artificial production of conditions of phenomena, the artificial combination of factors, in order to see what will happen because of them and in order to gain consequently a clarification of their influence' (Roux et al., 1912)'

"There is no attempt here on the part of Roux to work out the explanatory power or limitations of the experiment. He wrote, not as an analytic philosopher, but as a scientist who tried to formalize what he himself had done in the laboratory" (Churchill, 1973).

Of course, chemistry and physics have advanced well beyond the analytical experiment. Thus I think there is at least one more, fifth step, which we are just beginning to take in embryology. We formulated it as follows:

"Morphogenesis is the consequence of the action of individual cells. Studies of morphogenesis are comparable to studies in statistical mechanics: both seek explanations of the dynamics and properties of the whole (an organism or bulk matter) in terms of the behavior and interactions of its parts (cells or molecules) (Gordon, 1966). The urodele central nervous system provides an exemplary subject for such a study, because its early shaping can be analyzed in terms of the coordinated behavior of its individual cells.

"As the neural plate changes shape, the apical shrinkage of each cell pulls the whole sheet towards itself, thus contributing to the motion of the whole sheet. The result is a complex relationship between the morphogenesis of the neural plate and the behavior of its component cells. In order to unravel this relationship, we created a computer simulation which allowed us to watch, in a completely controlled manner, how the physical interactions among the cells lead to the keyhole shape of the neural plate. In constructing the logic for the computer simulation, it was necessary to formulate clear, quantitative statements of the principles governing cell behavior. Therefore, the act of creating the simulation sharpened our observations of the embryo.

"In expressing cell behavior in mathematical terms, we have derived a formal mathematical description of the flow of tissues in the developing embryo which we call *morphodynamics*. This analysis contributed to the computer simulation and also provided experimentally testable predictions not derivable by simulation. Thus our investigation has utilized a full interplay of three modes of inquiry:

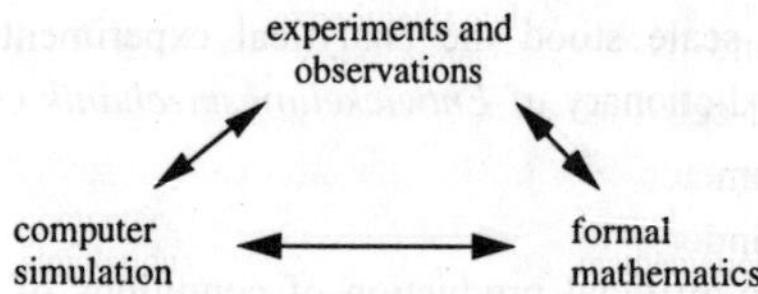

"By defining the forces that shape the early neural plate, we describe, for the first time, the cellular basis for a D'Arcy Thompson (1942) grid transformation" (Jacobson & Gordon, 1976a).

One further, sixth step is that of synthesis, the creation of an organism 'from scratch' (Fox, 1991) or from components (Jeon, Lorch & Danielli, 1970; cf. Stein, 1986). I'll comment on progress (or lack thereof) in the field of artificial life or embryonics, which aims to do this by computer simulation. A few colleagues and I have launched the Cyberworm project (cf. Kitano et al., 1997), by which we hope to accomplish computer synthesis for nematodes, at least for their early embryogenesis:

Date: Wed, 11 Dec 1996 08:03:23 -0600 (CST)
To: celegans@net.bio.net
From: gordonr@cc.UManitoba.CA (Richard Gordon)
Subject: Cyberworm

Nematodes present a unique opportunity for understanding morphogenesis in what might be called a middle-out manner: from the cells and their interactions to the organism, and from the cells down to the mechanochemical physics of the cytoskeleton and the DNA.

We would like to write a 4D database/simulation to be called Cyberworm. As a database, it would contain and accumulate coordinates, cytoskeletal arrangements, cell division times, gene expression and other data for each cell in the developing nematode. Variants, where available, either as mutants or worms other than *C. elegans,* would also be included. A rich color display mode would allow one to visualize this data in 2D, 3D, or 4D (including time), and label cells by any combination of criteria available in the database. Cross sections, time lapse sequences, etc., would be available. Access to the literature and investigators' addresses could be included. Links to the nematode and other genome databases are possible.

The Cyberworm database/simulation would also be designed to permit modeling of the mechanisms of morphogenesis and evolution. Thus a finite element/tensegrity representation of each cell, its division, and its interactions with neighbors, would be included, along with the ability to simulate cell-cell diffusion, ionic currents, genetic regulatory networks, etc., where known. Each parameter would be mutable, and simulation of a population of Cyberworms 'competing' under fitness criteria would be included. All parameters would be empirical and/or linked by current theory and compared to empirical values.

Obviously, we will need rich, accurate sources of data for such an ambitious undertaking. The very exercise of trying to simulate Cyberworm will uncover areas of ignorance, which different individuals may wish to tackle.

Your thoughts and offerings for such an undertaking are welcome. We see this as an open, modular project, where our job will be to design the skeleton and others who join the project will add their data, hypotheses and modules in a coherent fashion. The project is not funded, nor proprietary, and therefore has no constraints.

Randy Zauhar, University of Missouri, [r.zauhar@usip.edu]
Dick Gordon, University of Manitoba, GordonR@cc.Umanitoba.ca
Steve McGrew, New Light Industries, Ltd., stevem@comtch.iea.com
Przemko Tylzanowski, University of Leuven, przemko@sgi.celgen.kuleuven.ac.be
Dave Fitch, New York University, fitch@acf2.nyu.edu

Of course, it is dangerous to announce good intentions, but this statement points out one possible direction for future research on computer synthesis of embryos. Our goal is nothing less than to 'compute a worm' (cf. Wolpert, 1995a), though I suspect that full success is 10-30 years away. Synthesizing a real worm would be the next step.

Most work to date, relevant to the physics of embryos, has been concerned with the measurement and application of mechanical forces at the cellular and subcellular level, i.e., with one or more of Roux's four levels. A wide range of techniques are being explored:

acoustic microscopy (Lüers et al., 1991; Bereiter-Hahn, 1995; Bereiter-Hahn et al., 1995);
bent glass needles (Rappaport, 1960b, 1967, 1977; Harper & Harper, 1967; Hiramoto, 1975; Dennerll et al., 1988; Zheng et al., 1991; Oliver, Lee & Jacobson, 1994);
Brownian motion (Yamazaki, Maeda & Miki-Noumura, 1982; Mizushima-Sugano, Maeda & Miki-Noumura, 1983; Horio & Hotani, 1986; Hotani & Miyamoto, 1990; Isambert et al., 1995);
cell poking (Duszyk et al., 1989), with pipettes or beads (Maniotis, Chen & Ingber, 1996);
dielectrophoretic levitation (Kaler & Tai, 1988; Fuhr et al., 1995) and **optical tweezers** or **laser traps** (Berns et al., 1989; Weber & Greulich, 1992; Kreis & Vale, 1993; Simmons et al., 1993; Warrick et al., 1993; Finer, Simmons & Spudich, 1994; Finer, Mehta & Spudich, 1995);
high gravity (Epstein et al., 1988);
magnetic field gradient levitation for low gravity simulation (Valles Jr. et al., 1997);
mechanical loading (Fay, 1977; Zaner & Hartwig, 1988; Jones et al., 1991a; Adams, 1992);
microgravity (Claassen & Spooner, 1994);
micropipet aspiration (Horoyan et al., 1990);
osmotic 'restraint' (Bewley & Black, 1982);
rheology (Gordon et al., 1972, 1975; Janmey et al., 1991; Moore, Keller & Koehl, 1995; Foty et al., 1996; Srivastava & Krishnamoorthy, 1997);
scanning force microscopy (Chang et al., 1993; Ricci & Grattarola, 1994; Stemmer, 1995);
sessile drop method (Phillips & Steinberg, 1969, 1978; Phillips, Steinberg & Lipton, 1977; Hiramoto & Yoneda, 1986) and other **surface tension methods** (Stein & Gordon, 1982);
strain analysis, often using **incisions** (Downie, 1976; Jacobson & Gordon, 1976a; Gil, Murillo & Gimeno, 1992; Gilbert et al., 1994);
ultrasound (Selman & Counce, 1953; Counce & Selman, 1955; cf. Virgin, 1955; Cosgrove & Green, 1981; Niklas & Moon, 1988; Schram, 1991; Niklas, 1992; Gordon & Drum, 1994, and **'acoustic tweezers'**, Wu, 1991);
wrinkled latex (Harris, Wild & Stopak, 1980; Harris, 1984; Oliver, Lee & Jacobson, 1994; Dembo et al., 1996) and **stretchable substrates** (Lee et al., 1996; Luchinskaya, Beloussov & Shtein, 1997b) or natural supports (Kucera & Burnand, 1987; Kucera & Monnet-Tschudi, 1987);

and more magnets and metal balls (Hiramoto, 1969; Heidemann, 1993) and 'ligand-coated ferromagnetic microbeads' (Wang & Ingber, 1995).

For a table of some of these and other measurements of 'biomechanical' forces, see Fukui (1993). The 'crescograph' (Bose, 1918), with its amplification of plant movements by a factor of 10^6, may be worth reconsidering. The fields of 'cytomechanics' and 'cellular engineering' are beginning to be organized with special symposia, journal issues and whole journals (Bereiter-Hahn, Anderson & Reif, 1987; Gordon, 1994a; Opas, 1995a; Rolfe, 1995).

Various developmental problems have recently been approached on the basis of the resolution of mechanical instabilities in initially mechanically uniform tissues (Murray, Oster & Harris, 1983; Oster, Murray & Harris, 1983; Silk, 1983; Harris, Stopak & Warner, 1984; Murray & Oster, 1984a,b; Perelson et al., 1986; Nagorcka, Manoranjan & Murray, 1987; Bard, 1988; Hernández & Green, 1993; cf. Opas, 1989, 1994).

The physical sorting of embryos based on any number of optical criteria or their logical combinations may be possible with flow cytometry using a fluorescence activated cell sorter (FACS) modified for larger 'cells' (Gray et al., 1989). Respirometry methods, once popular (reviewed in Needham, 1950; cf. Witschi, 1952; Kaplan, 1980a,b; Kucera & Raddatz, 1980; Raddatz, Eyal-Giladi & Kucera, 1987; Raddatz, Kucera & de Ribaupierre, 1987; Stebbins & Cohen, 1995), indicated higher respiratory rates in the dorsal lip of the blastopore compared to other excised ectoderm. Local imaging of energy consumption in medical imaging (as accomplished in positron tomography and magnetic resonance 'functional imaging': Hibbard et al., 1987; Mora, Carman & Allman, 1989; Alavi & Hirsch, 1991; Barnes, O'Tuama & Tzika, 1993; Neil, 1993; Liotti, Gay & Fox, 1994; cf. Gordon, 1975b; Rickey, Gordon & Huda, 1992), without the need to cut or kill the embryo, might be more useful if they could be applied at microscopy scales:

"It is conceivable that when a newt embryo gastrulates, considerable work must be done in so deforming the cells against their own elasticity that the blastula tissue may pass through the narrow blastopore opening. To perform such work energy is required from the chemical energy source; it will be stored as potential energy by deformed cells as they pass through the blastopore; yet when the constriction is passed, the cells relax to their equilibrium shapes once more [or actively expand] and the [potential] energy is dissipated as heat. In such a case a knowledge of the energy-balance could tell us nothing about the work done in gastrulation.

On the other hand, if we knew about the forces required to deform the cells we should be able to estimate the work required for gastrulation..." (Selman, 1958).

We have thus begun to build the tools for a physics of embryos even though the field itself has lain dormant for over a century. Perhaps we are now ready for that much needed renaissance.

My claim is simple: Driesch used the wrong physics. Had he known about and understood Rayleigh's (1892) instability analysis of liquid cylinders, and had he considered the spatial transmission of forces (as in mere sound waves, well understood at the time) as a nonmaterial, but nevertheless physical entity, the whole subsequent history of biology might have been quite different. Perhaps Driesch did not keep up with his physics contemporaries, nor they with what was going on in biology. We need to unabashedly use whatever physics can be brought to bear on the problem of embryogenesis (and evolution, while we're at it). It is time to consider embryos as part of the real, physical world, not (only) as complex strings of symbols (the nucleotide bases of DNA).

1.14 Are We Headed for Reductionism?

"...We cannot proceed far in the study of development from any point of attack without coming face to face with the great problems of biology" (Child, 1941).

"Biology, seemingly at the point where it might possibly be 'reduced' to a phenomenon of physics and chemistry, presents an embarrassingly difficult central problem of the *organism* " (Sinnott, 1966).

Since even reductionism in physics, and especially in biology (Grantham, 1995), is a philosophical (Rosen, 1991, 1994; Sklar, 1993; Cohen & Stewart, 1994; Tipler, 1994), political (Ruse, 1988a; Rose, 1997) and moral (Sperry, 1983) quagmire, I will, for the most part, leave it to others to decide if the theory in this book is reductionist, holistic or organicist. I hope it goes beyond 'a crude reductionism' (Churchill, 1991b) or 'greedy reductionism' (Dennett, 1995), by my introduction in Proposition 194 of the concept of 'linking disciplines'. One does not have to resolve the philosophical issues of reductionism versus organicism in order to look at the embryo with a physicist's eyes:

"It is ironic that the modern reductionist, as much as the former vitalist [and the modern antireductionist: Rose, 1997], takes organization as axiomatic, rather than as the foremost focus of thought and experiment. [Joseph] Needham [1950], [Paul A.] Weiss [1968], and other organicists have been unimpressed. Schaffner [1967] dismisses the old warnings of [Erwin] Schrödinger [1945] [cf. Witkowski, 1986a] and [Max] Delbrück [cf. 'biological complementarity' in Fischer & Lipson, 1988] that life might ultimately have to be explained by autonomous principles [a misreading in the case of Schrödinger, 1945, who is, in any case, not discussed by Schaffner, 1967]. In so doing, he expresses classic hopes within a mechanistic paradigm that problems refractory to immediate solution will be solved within current frameworks of physics and chemistry. It would be difficult to find a better expression of the position that Weiss has devoted his life to refuting" (Haraway, 1976).

Thus, I am trying to set aside vitalism, organicism, and the centuries old problems that are our inheritance from René Descartes:

"Descartes'... decipherment of Nature might be crude, yet he had the courage to insist that mechanical sense could be made of the workings of Nature, throughout the realms of physics, chemistry and even physiology. By reasserting the unity and rationality of Nature, he did as much as any man to put seventeenth-century scientists back on the intellectual road first trodden by the Greeks. Where Francis Bacon had provided the manifesto for experimental science, René Descartes now did the same for scientific theory.... Before a mechanical system of natural philosophy could be made acceptable, either to the Church or to the ordinary believer, there had in fact to be a compromise; and the terms of this compromise have deeply influenced the subsequent development of science. Even today, their mark is evident in the structure of scientific ideas and institutions. For there were two topics which Descartes felt compelled to exempt from scientific enquiry: God and the human mind.... And in the years that followed most of his fellow-scientists have been prepared - though some more grudgingly than others - to accept the necessity for this division" (Toulmin & Goodfield, 1962).

While it may be true that "in the twentieth century it has at last become clear that Descartes' dualism must be set aside in favour of a unified approach"

(Toulmin & Goodfield, 1962), we are still quibbling about which side of that dualism will dominate in this unification. Besides...

"Dualism shares with its antithesis, classical materialism, a certain unassailable simplicity. What they both lack is plausibility. As a result, most scientists try to steer a course that avoids the two extremes. They face the almost impossible task of somehow demechanizing the machine, but keeping the ghosts out" (Harth, 1995).

There is, however, one aspect of reductionism that I think I adhere to consistently in this book, that is, that embryogenesis is explainable in terms of today's basic concepts of physics. I think it is more important at this stage to get on with the 'reverse engineering' (Dennett, 1995) of embryos than to seek new principles of physics (Driesch, 1914; Elsasser, 1975; Penrose, 1995), a search that excuses us from carrying out this difficult task. Along with Dennett (1995), I guess I regard these excuses as a search for vitalistic, supernatural 'skyhooks' (but cf. Orr, 1996). In this sense, I am a 'haughty old Newtonian', adhering to an uncongenial, inconciliable reductionism, albeit one that is multilevel, and thus not 'greedy' by attempting to explain everything in one step (Dennett, 1995):

"One could hardly propose a more congenial or more conciliatory form of reductionism than the argument advanced by Schrödinger (1945) as the centrepiece of *What is Life?* - for he does not advance the haughty old Newtonian claim that biological beings are 'nothing but' physical objects of high complexity, and therefore ultimately reducible to conventional concepts developed by the ' queen of sciences'. Schrödinger admits that biological objects are different and unique. They must ultimately be explainable by physical principles, but not necessarily the ones we know already [again, a misreading of Schrödinger, 1945: 'No. I do not think that.']. Therefore, biology will be as much help to physics (in providing material that will lead to discovery of these unknown laws), as physics can be to biology in finally supplying a unified explanation for all matters" (Gould, 1995a).

On the contrary, the degree to which our presently known laws of physics have been brought to bear on the understanding of embryos is minuscule. Let's get on with the real task at hand.

1.15 Chemical or Mechanochemical Instabilities?

The invocation of instability properties of reaction-diffusion equations, postulated as *the* basis for morphogenesis by Turing (1952), is now commonplace (Harrison, 1987, 1993; Foe & Odell, 1989; Murray, 1990; Holloway, Harrison & Armstrong, 1994). Most of this work ignores Turing's suggestion that "in addition to the chemical instability, there is a mechanical instability causing the breakdown of... symmetry" (Turing, 1952), i.e., that chemistry and mechanics are coupled components in an instability system (cf. Savîc, 1995). Turing (1952) used perturbation analysis analogous to that of Rayleigh (1892). (The latter has been applied to embryo cell sorting: Gordon et al., 1972, 1975). There is no indication that Turing himself was aware of Rayleigh's earlier analysis of mechanical instabilities, his inspiration for his last paper (Turing, 1952), before committing suicide, rather coming out of electrical oscillators (Hodges, 1983). ("Alan Turing [was] the British computer genius and World War II hero who broke the German code for the Allies. When Turing was discovered to be gay, he was arrested and forced to undergo injections of estrogen to control his 'deviant' sexual behavior. The injections eventually led him to commit suicide, but they did not change his sexual orientation": Burr, 1996). Rosen (1970) states:

"This idea was originally articulated by Rashevsky (1940a,b,c,d,e, 1960), and again, independently, by Turing (1952). Rashevsky bases his development on his rather complex theory of diffusion-drag forces, while Turing's is of a more straightforward dynamical nature..." (Rosen, 1970).

The pure (mechanics free) reaction-diffusion mechanism is best known for its prediction of standing waves (Turing, 1952), although propagating waves (like the nerve impulse) are also possible solutions to equations of this type (Kopell & Howard, 1973; Howard & Kopell, 1974; Karfunkel & Seelig, 1975; Hastings, 1976; Murray, 1976; Karfunkel & Kahlert, 1977; Berridge & Rapp, 1979; Field & Burger, 1985; Ross, Müller & Vidal, 1988; Nagorcka, 1989; Markus, 1990; Markus & Hess, 1990; Le Guyader & Hyver, 1991; Winfree, 1994a). Kuramoto (1984) further defines two classes of waves: 'pulse' or 'trigger' waves which leave a system in the same state before and after the wave passes, and 'kink' or 'front' waves, in which the state of the system changes, such as in a phase transition (crystallization, freezing, precipitation, etc., cf. Gordon & Drum, 1994) or chemical transition (burning, as in grass and forest fires, for example: Frisch & Hammersley, 1963; Blum, 1973; Bak, Chen & Tang, 1990; Leymarie & Levine, 1992). Differentiation waves would clearly be kink waves if we are right that the cells change type after participating in them. Like a fire, a differentiation wave is a solitary wave, but not a soliton: two of the latter can pass through one another unaffected (Scott, Chu & McLaughlin, 1973). None of the examples of test tube reaction-diffusion systems (Belousov, 1958; Zhabotinskii, 1964a, 1967; Field & Noyes, 1974; Hastings, 1976; Tyson, 1976; Zhuravlev & Trainin, 1990) are kink waves, which may explain why analogies between kink waves and developing organisms have not been made: "Although there exists a number of reaction-diffusion models which may show kinks, one has never been able to visually produce them in chemical reaction systems" (Kuramoto, 1984).

Note that a kink wave leaves a material change in the medium through which it passes, so we are not quite faced with a completely nonmaterial physics when we are dealing with wave propagation (Section 1.11). Of course, this is also true of waves in ordinary active media that have no external energy source: a wave depletes some of the chemical energy as it passes, causing a slight material change. Thus the state of the system is not exactly restored after a wave in an active medium, and the distinction between such waves and kink waves becomes fuzzy.

As an historical aside, let us note that...

"Belousov originally developed his complex reaction system [the first experimental reaction-diffusion system] to model the reactions of the Krebs' cycle, the metabolic pathway in aerobic cells by which carbon dioxide is formed and energy is captured for the synthesis of ATP. The problem was that random reactions at the microscopic level were not expected to give such sharp, periodic color transitions or waves of chemical reaction. In fact, Belousov had considerable difficulty publishing his results because the referees and the editor argued that chemical reactions could not possibly happen this way. It took Belousov two years to find a sufficiently obscure journal in which to publish his work. There it languished until its significance was recognized and championed by Zhabotinskii" (Depew & Weber, 1994).

We are thus faced with two broad, not mutually exclusive models for the spatial pattern of differentiation: reaction-diffusion mechanisms and mechanical instabilities. Clear distinctions, such as those expressed by Schiffmann (1994), may not be so facile:

"It has been suggested ([Appendix I:] Gordon & Brodland, 1987) that the initial event in differentiation is mechanical followed by the synthesis of specific proteins. Similarly, it has been suggested (Goodwin & Trainor, 1985) that it is the mechanochemical instability that is responsible for the spatially differential metabolism. By contrast, I suggested (Schiffmann, 1991) that it is the chemical (metabolic) spatial nonuniformity - as expressed by the possibility of instability of the homogeneous state in the (cAMP, ATP) system - that primarily dictates the mechanical spatial nonuniformity, and not the other way round. One example of this general thesis is the possibility of localized contraction of an apical belt of actin and myosin (Schiffmann, 1991).... My theory in which the electrical... and mechanical fields are primarily dictated by the chemical field is in the spirit of Turing's work.... I suggest that chemical instability is the primary cause of both morphogenesis and differentiation and is responsible for the correlation between them in space and time..." (Schiffmann, 1994).

One can have it both ways. After all, all mechanics at the subcellular level is mechanochemical, and every chemical reaction involving the relative motion of atoms has a mechanical component. Even when momentum is insignificant (Gordon et al., 1972, 1975), we end up with creeping motion, a respectable branch of mechanics. When macromolecular assemblies (such as the cytoskeletal components) approach the dimensions of the cell, their mechanics becomes an integral part of their chemistry. One could try to separately estimate the chemical versus mechanical energy to see under what circumstances they are similar or which predominates.

If we assume that something like a Turing mechanism based on (the minimum of two) putative morphogens such as cAMP and ATP is operating, it is likely that, at concentrations of proteins achieved in the cytoskeleton, the cytoskeletal reactions of the morphogens will themselves alter the concentrations of these morphogens. Thus, just on the basis of stoichiometry, which is cause and which is effect may be a moot point. In general, we have to solve for all concentrations, of cAMP, ATP, tubulin, actin, intermediate filament proteins, and all the polymers, connecting molecules, etc., simultaneously. (A computer program aiming towards such a goal has been written: Schaff et al., 1997, and Randy J. Zauhar and I are designing a complementary program, In Vivo, for simulating cytoskeletal dynamics.). The larger system of equations is certainly still of the Turing type. As in the computation of chemical equilibria, simplification may be possible (van Zeggeren & Storey, 1970), but it cannot be claimed *a priori*. These considerations may make it difficult to achieve the clean isolation of morphogens that biomathematicians would like to see:

"In the case of reaction diffusion models the unequivocal identification of morphogens is crucial. Until this has been done it is difficult to test reaction diffusion models experimentally" (Murray et al., 1998).

Some of the reluctance to include mechanochemistry in 'morphogen' modelling probably comes from a long history of treating molecules in cells as if they were governed by ordinary principles of elementary chemistry, which apply to well stirred solutions of tumbling molecules of random orientation whose dimensions are small compared to that of their containers: "It is well known that a [scalar] chemical reaction... cannot generate a vector force in a homogeneous system. But the myofilament system [and the cytoskeleton in general] is very asymmetric in molecular construction, so this theorem is inapplicable" (Hill, 1977).

The relative importance of mechanics versus chemistry becomes apparent, however, when cell membranes are taken into account, because not all chemical species can cross these boundaries. Thus, I tentatively take the mechanics as primary because mechanical effects are most easily transmitted from cell to cell. But even here, the mechanism of transmission is never solely mechanical. If, for example, stretch activated ion channels are involved in propagating differentiation waves, these certainly are partly chemical in nature (Appendix IV: Björklund & Gordon, 1993a) even though it is the stress that is transmitted from cell to cell. If IP$_3$ is what moves from cell to cell (Appendix IV: Björklund & Gordon, 1993a,b; cf. Sanderson et al., 1994), each cell nevertheless responds mechanically. Furthermore, since ionic calcium currents are probably involved (Appendix IV: Björklund & Gordon, 1993a; Moreau et al., 1994; Jaffe, 1995), perhaps even in cell to cell transmission, and themselves alter the concentrations of other components (if only by electrophoresis: cf. Jaffe, 1969), a differentiation wave probably has an electrical component (cf. Schlichter & Elinson, 1981; Schlichter, 1983). We could 'settle' on just calling the whole system 'electromechanochemical', which may be just as well, until we learn better how it works. All three components are actively driven, so no matter what the final analysis, differentiation waves are examples of waves in active media (Markin, Pastushenko & Chizmadzhev, 1987; Murray, 1989, 1993; Sager, 1996).

The physics that, therefore, seems to be appropriate for embryos is that of mechanical instability phenomena and nonlinear waves (Luchka & Gordon, 1998) in an active or excitable medium rather than (Just, 1939a) the 19th century concept of the machine. But we must be careful here not to close the door too soon. There are subtle effects, such as abnormal development of the axis of embryos shielded from the earth's magnetic field (Asashima et al., 1987; Asashima, Shimada & Pfeiffer, 1991), or in a weak magnetic field (Ho, 1993), for which we have as yet no explanation, certainly not one in the context of differentiation waves. But the relationship of these phenomena to differentiation waves, if any, is open for investigation.

1.16 Critique of the Theory of Self-Organizing Systems

The paradigm of living organisms as self-organizing systems (Von Foerster & Zopf Jr., 1962; Nicolis & Prigogine, 1977; Jantsch, 1980; Antonelli, 1985; Bates, 1987; Kauffman, 1987a, 1993, 1995a; Yates et al., 1987; Geyer, 1991; Bonabeau & Theraulaz, 1995; Prusinkiewicz, 1995; Resnick, 1995), while seemingly a tautology, almost never accounts for an active and changing role of the genome during differentiation and morphogenesis. Self-organizing systems are almost always conceived as consisting of units (cells or molecules) that are equivalent to one another, but whose interactions place each into a different condition (velocity, chemical state, or state of differentiation). An extended, temporally distinct history for each unit is either not included, or assumed, without proof, to follow from the very process of self-organization itself. The mere existence of multiple steady states or attractors for a self-organizing system, while permitting construction of a web or network of states accessible to one another (Kauffman, 1987a, 1993; Depew & Weber, 1994), does not give a mechanism by which single cells or units would traverse specific sequences of states, nor show how groups of cells, identical at a given moment, should, in an organized fashion, split into two spatially coherent groups and each group depart towards a different state. Something is lacking that would be needed to explain such a 'community effect' (Gurdon, 1988a; Weston, Kato & Gurdon, 1994). This is not to say that modelling of this type is impossible. It could, for example, build on the recent apparent success in network modelling of bacteriophage replication (McAdams & Shapiro, 1995; Loomis Jr. & Sternberg, 1995; Lipkin, 1995d).

Let me try a simple physical example, to explain this shortcoming of self-organizing systems as an explanatory model for embryonic development. Consider the quintessential example of the development of Bénard cells in a shallow pot of water on a stove (Chandrasekhar, 1961; Prigogine & Stengers, 1984; Rivier et al., 1984), easily made visible by throwing in some small particles, like crushed parsley:

"Convection cells are a good place to begin thinking about dissipative structures. They were first studied in a classical experiment conducted by Henri Bénard in about 1900. A contemporary version of this experiment involves heating a thin layer of oil from the bottom (source) and keeping the top of the liquid in contact with a transparent cooling plate (sink). As the temperature from below goes up, more energy flows into the system. This is reflected in a higher kinetic energy of the liquid's molecules and in a larger difference in temperature or thermal gradient across the liquid. The result is increased energy flow through the system, and increased dissipation of energy, or entropy production, to the sink. At a critical size of the gradient, or the thermodynamic force field, when the effect of viscosity is exceeded by heat transport, the intensified random motion of the molecules becomes insufficient to dissipate the energy flux. Large scale macroscopic streams of hot liquid then flow to the top, forming convection cells that are observable because of the different index of refraction between the hotter and cooler liquid. We observe convection flows when we do something as ordinary as boil a pan of water. The honeycomb pattern we see toward the bottom of the pan is the result of coupling buoyancy, thermal diffusion, and viscous forces, creating a macroscopic structure that is perceptible because it is at least one hundred billion times the scale of molecular dimensions in water itself. Self-amplifications like these are 'circularly causal' in the sense that they take the form of an autocatalytic cycle, in which the incipient formation of structure - here a convection cell - reflexively reinforces the formation of more of the structure until the system settles down to a new steady state (Swenson & Turvey, 1991). This is an example of self-amplification of a stochastic fluctuation that breaks the symmetry of the linear regime (Garfinkel, 1987). Such symmetry breaking takes the system into the realm of nonlinear dynamics (Swenson, 1989, 1991; Swenson & Turvey, 1991)" (Depew & Weber, 1994).

Now honeycomb patterns are common in nature (cf. Thompson, 1942; Lima-de-Faria, 1988; Gordon & Drum, 1994), so their explanation warrants some attention, and nonlinear dynamics probably play a role. The problem is simply this. An embryo is generally self-contained. To create a Bénard pattern from scratch, we must therefore consider the self-assembly of the pot, the stove, the heat source, the water, and the parsley, all at once, plus their proper orientation with respect to gravity. Clearly this is beyond the nonlinear dynamics forming Bénard cells. To claim that this can be done by bootstrapping (another domino theory?), with consecutive nonlinear dynamics for each item, and a grand overall nonlinear dynamics for putting the components together (however separated out from the original fertilized egg), while perhaps plausible, is undemonstrated. The new, fractal view of self-organizing systems would also not seem to remove these difficulties:

"The grand and general question is how the laws of physics - which describe processes on the microscopic scale - can lead to a world organized on all scales. The idea of 'self-organized' is that it is in the nature of nonlinear processes to organize mathematical systems into structures that have order on all length scales" (Bak, Chen & Creutz, 1989).

1.17 Protein Folding as a Deluding Paradigm

"The fact is, no one understands what 'determines' development, or quite how to attack the problem of developmental context; it is the biology of tomorrow" (Strohman, 1993).

The idea that embryos could be thought of as self-organizing or self-assembling systems probably received an intellectual boost from the presumption that protein folding is such a process. The model could be expressed by extension of the 'central dogma':

DNA $\Rightarrow$ RNA $\Rightarrow$ protein $\Rightarrow$ folded protein $\Rightarrow$ cell $\Rightarrow$ embryo.

If these first steps were so easy to explain, wouldn't the latter steps also follow as easily? (Cf., for example, the proposal for "hierarchic condensation... by a tree of local folding interactions": Rose, 1979. See also Crippen, 1978.) Well, let's just have a glance at the history of the step protein $\Rightarrow$ folded protein. Here we are going from a linear structure to a three dimensional structure, certainly a fundamental three dimensional 'form generating' process:

"Anfinsen's classical experiments on the refolding of ribonuclease *in vitro* (Anfinsen et al., 1961; Anfinsen, 1973) established that all the information required to determine the final conformation of a protein can reside in the polypeptide chain itself: the denatured enzyme can refold into its native conformation in the absence of other proteins. Similar results have since been obtained with several other small, single-domain polypeptides, and with a few larger, more complex proteins (reviewed in Baldwin, 1989; Fischer & Schmid, 1990; Kim & Baldwin, 1990; Jaenicke, 1987, [1993]; Creighton, 1990).... Acceptance of the involvement of chaperones has required some revision of long-held views about the spontaneity of the process of protein folding" (Gething & Sambrook, 1992).

"The long-held view that newly synthesized proteins in a cell fold in a largely spontaneous process has been called into question. Although the basic principle that the primary sequence of a protein is sufficient to specify its three-dimensional structure remains unchallenged, protein folding in cells, as opposed to the test tube, depends on helper proteins.... The emerging paradigm is that the binding and release of unfolded proteins is controlled through (a) the regulation of chaperones by specific protein modulators of their binding (and ATP hydrolyzing) activities and (b) through the functional cooperation of different chaperones with each other. As a result, newly made, unfolded, or partly folded proteins could be transferred in an orderly manner to the various polypeptide-handling systems of the cell" (Hendrick & Hartl, 1993).

The discovery of chaperones (but cf. Jaenicke, 1993) partly unravels a vast effort to derive the three dimensional structure of a protein from its one dimensional (primary) structure, by computer simulations of protein chemistry and physics (Skolnick & Kolinski, 1989; Nemethy & Scheraga, 1990; Somorjai, 1991; Sylvain & Somorjai, 1991; Rabow & Scheraga, 1993; Liebovitch, Arnold & Selector, 1994; Pain, 1994):

"The potential function of chaperones forces one to rethink the principles of protein self-assembly. Originally it was this theory which stated that the information carried in the protein sequence was both sufficient and necessary to specify the three dimensional structure of the native protein. Hence, the function of each polypeptide resided solely within the amino acid sequence of that polypeptide. Molecular dynamics studies show that there are many states of near equivalent energy [cf. Wurfels, Section 10.13], and that transitions between them may require fairly high activation energies. It now appears that specific chaperones may have an important and active role in the selection of the final structure. The basic theory must now be modified to state that while the information carried in the protein sequence is necessary to specify the ultimate native three dimensional structure of a protein, that same sequence is not sufficient to ensure that such a state be attained. The complexity of living systems requires multiple molecular interactions even at this most fundamental level. The central dogma of 'single gene - single protein molecule' is no longer tenable" (Tylzanowski, 1991).

The story itself is still unfolding rapidly. For example, the notion that chaperonins themselves fold spontaneously (Mascagni et al., 1991) must be juxtaposed with the hypothesis that infectious proteins (prions) are misfolded chaperones that require their own assistance in folding (Liautard, 1991, 1992, 1993). Chaperone catalysis of protein folding sometimes has "a broad target specificity" (Gething & Sambrook, 1992), suggesting that some simple physics may be operating here. Hints of what may be involved are coming from new computer simulations of protein folding that account for the presence of the chaperone without invoking specific steric interactions (Roy, Kupferschmid & Bell, 1992; cf. Ellis, 1993a, 1996a; Ellis & Hartl, 1996, who propose that spontaneous folding does occur - inside the chaperone, designated as an 'Anfinsen cage', where it is protected from interactions with the crowded cytoplasm that obfuscate folding). This may be a case of dimensionality reduction in biology (Adam & Delbrück, 1968).

Chaperonin has been shown to function by cycling between states (Jackson et al., 1993) that expose and hide hydrophobic attachment sites (Braig et al., 1994; Fenton et al., 1994) for the hydrophobic portions of incompletely or misfolded proteins, perhaps pulling hydrophobic regions of the attached protein together by a twisting action similar to that of hyaluronan (Scott, 1989a). Thus it may be that a chaperone actually carries out simulated annealing (Laarhoven & Aarts, 1987; Aarts, 1989; Azencott, 1992; Mazur, Mazur & Gordon, 1993) of a protein by adding energy to it, getting it over local minima in its energy landscape (Todd, Lorimer & Thirumalai, 1996; Corrales & Fersht, 1996). Natalie K. Björklund and I have thought of this as a transient increase in the effective temperature of the incompletely folded protein. The 'cooling schedule' of simulated annealing algorithms may correspond to the reduced probability of chaperone action as the number of hydrophobic bonds exposed on the outside of the folding protein decreases. The spacing of the ATP energy sources on the chaperone might ensure that most of the energy imparted to the folding protein goes into its large scale motions, which is possible in this nonequilibrium situation. In contrast, normal (equilibrium) thermal heating in an aqueous medium would impart most of its energy at the scale of the single amino acid residue. Chaperones have thus solved a global optimization problem that has so far evaded us in mathematics and computer science, and they may thereby point to a new class of optimization algorithms.

In some sense, chaperones provide a 'global' environment for protein folding, making the possible plausible:

"...It has been calculated that the time required for a large polypeptide to sample all possible conformations on its way to the native state would be of the order of the age of the universe" (Hendrick & Hartl, 1993).

Recent crystallographic work provides evidence that an Anfinsen cage is indeed formed, with insides that are initially hydrophobic, but which change to hydrophilic and expand in volume with a $90°$ twist of the cage (Xu, Horwich & Sigler, 1997; Rye et al., 1997):

"Presumably, a protein bound to these [internal] sites will be subject to mechanical stress during this movement, and it will unfold to some unspecified degree.... Here, the protein (which is now in a... higher-free-energy state) is free to fold (or misfold) on its own" (Lorimer, 1997).

This is consistent with a simulated annealing mechanism that puts energy into the low frequency or larger scale motions of the protein, although this is not accomplished directly by the ATPs, as we speculated. The physics, and not just the chemistry, of chaperone action may be key to understanding protein folding. The analogous lesson here is that the physics of the embryo as a whole, a physics generally ignored, may be what makes tissue differentiation plausible, not just its biochemistry or molecular biology. There is nothing autonomous or automatic about each step of the central dogma. Each one takes much disciplined explanation.

1.18 A Word on Language

"The casual assumption that biological development is *program-driven* forces use of information metaphors, not rationalized physically, to account for the wonders of biology.... The information metaphor that I claim is inimical to clear thinking in attempts to reduce biology to physics will probably never disappear; it is too convenient a shorthand. But there is little (or no) more substance in the statement that 'homeoboxes and chronogenes direct development' than there is in the statement 'balloons rise by levity'. Many explanatory constructs in conversational molecular biology remain at the low level of Kiplingesque 'Just So' stories (How the Camel Got His Hump; How the Elephant Got His Trunk...) [Kipling, 1902]" (Yates, 1987). [Cf. Gould & Lewontin, 1979.]

There is a sense, prevalent in biology, in which words substitute for understanding. Developmental biology is replete with metaphors (Woodger, 1952) and mythology. Rather than pick on individual authors, I offer the following unreferenced quotes, where the italics are mine:

"...Signals from the environment are also among the factors that *tell* a regulatory gene when to go to work."

"Regulatory genes, for example, *determine* when genes are turned on and off."

"...Genes... *lay down* tissues...."

"The greater variability of early stages may reflect an interactive, multicellular *dialog*...."

"...The molecular mechanisms used by cells... *communicate* signals which *instruct* the developmental pathways of other cells."

"...The pattern of the disc as a whole is gradually elaborated through a process of intercellular *negotiation*."

"...A single diploid nucleus... contains the genes that *designate* the properties of each and every cell type in the body."

"... The *community* effect... helps to explain the formation of blocks of tissue from sheets of cells, and could be of widespread occurrence and significance in morphogenesis resulting from embryonic induction."

"Two of the earliest *decisions* to be made by an animal embryo are which end is forward and which side is up.... Polarity in the early embryo is dependent on an intricate series of *molecular conversations* between the developing embryo and surrounding follicle cells."

"...Large numbers of cells of similar type... *collaborate* and *intercommunicate* during development...."

"...The cells of the gastrula... divide and they move, and the *objective* they *pursue* unceasingly is the construction of the primitive organs."

"Are morphogens *perceived* by groups of cells or by individual cells?"

"...Morphogens *use*... *strategy* to make stripes...."

"Clusters of homeotic genes *sculpt* the morphology of animal body plans and body parts."

"The individual cell, in a sense, *surrenders* its own wide possibilities to the production of an entire organism."

"Genes are important because they are the controllers and *executors* of developmental processes."

"HOM proteins presumably *dictate* cellular identities...."

"These homeotic selector genes are the master control genes that *coordinate* the regulators...."

"Retinoic acid *mediates* homeotic transformation of hindbrain regions."

"The importance of intercellular *conversations* comes as no surprise to vertebrate neuroembryologists...."

"Cells in their *social* context...."

"Cell *behaviour* and embryonic development...."

"...The cell makes a large-scale *assessment* of its overall relational geometry...."

And how often have you heard a lecturer saying that a molecule *'wants'* to bind to DNA? If we extract the words describing what cells and their contents are purported to do, we see that they are endowed with minds and intentions of their own:

assess
behave
collaborate
communize
communicate
control
converse
coordinate
decide
designate
determine
dialog
dictate
execute
instruct
lay
mediate
negotiate
perceive
pursue objectives
sculpt
socialize
surrender
tell
use strategy
want

In addition, they form a 'republic of cells' (Rádl, 1930; cf. Margulis, 1990a, and similar language regarding gene/cell relationships: Sapp, 1987), perhaps with analogous 'civil rights and responsibilities':

"Virchow [cf. Nordenskiöld, 1928] had treated the cell not as having an independent life but as a component part of a systematic unity. Virchow 'conceives the complex living organism as composed of multitudes of living individual cells, which remain united because they are interrelated to one another just as members of the same state' (Reichert, 1856)" (Lenoir, 1982).

"'The accomplishments of the cells do not seem to have been completed in such a way that movements of parts add up to the movements as a whole.... It is manifestly not a matter of migration of cells but of their passive subjugation to the intervention of a higher authority' (Vogt, 1923). Later Vogt (1929a)... wrote of *Staffelung,* which implies an orderly sequence of movements of more than one cell. The word is borrowed from military terminology, and Spemann (1942a), in his obituary of Vogt, commented on its military implication.... When Goerttler (1925) worked out in more detail than had Vogt the morphogenetic movements of the group of cells constituting prospective neural plate, he also compared the shifting position of cell movements to a routine military drill: his word was *Schwenkung* = wheeling in the military sense. German soldiers in the 1920's were hardly free and independent agents, but in any event, one soldier can never, in any drill in any army at any time or place, 'wheel' in the military sense all by himself, or become a row of individuals after having passed through a bottleneck" (Oppenheimer, 1970a).

The same concept is now expressed mathematically as...

"Order arising out of a complex dynamical system.... In industrial societies, the aggregate behavior of companies, consumers, and financial markets produces the modern capitalist economy.... For a growing embryo, the global consequence of the aggregate of a kaleidoscope of developmental processes is the mature individual" (Lewin, 1992).

Thus politics and economics become models for how cells interact:

"The concept of structural types [of nerve cells] was linked to that of functional differentiation, termed by Milne-Edwards (1857) the *'physiological division of labour,'* one of the dominant concepts of biology in the latter half of the 19th century (Spencer, 1898; Hertwig, 1898a)" (Jacobson, 1991b).

"Communication plays an important role in the rapid progress of modern society.... However, although we live in communities, we still treasure our individuality. By analogy, the survival of multicellular organisms depends on each cell type retaining its individuality, even though all cellular activities must be coordinated with other cells" (Kumar & Gilula, 1996).

You may protest that these are but harmless metaphors. I suggest, with Yates (1987), to the contrary, that they are obstacles to clear thinking. I don't think it helps to understand mechanisms of development by talking about them in emotion loaded social or psychological language:

"Life is a social enterprise. And one of its hallmarks is peer pressure, We become what we are to fit in with those around us. Now developmental biologists are beginning to find that even embryonic development is a social undertaking, in which constant chatter between cells in developing tissues determines their fates. To developmental geneticist Gerald Rubin [cf. Heberlein, Wolff & Rubin, 1993]... these cellular conversations are reminiscent of kids on a playground saying things like: 'I'm it, so you can't be it' or 'I'm it, so you have to be it, too.' Says Rubin, who studies eye development in the fruit fly: 'Peer pressure from neighboring cells is clearly the way most development works.'" (Schmidt, 1994).

These are widely discussed issues:

"Philosophers of science do not agree on the use of teleological explanations in biology. There are those who consider teleology to be permissible and to be entirely consistent with mechanistic causality (Nagel, 1961; Beckner, 1969; Wimsatt, 1970; Ayala, 1970; Hull, 1974). Others reject teleological explanations because of the semantic difficulties created by the use of the term 'teleology,' which has various meanings (Woodger, 1952), and because they claim that use of teleological explanations obfuscates an understanding of how goal-directed processes have evolved by natural selection (Pittendrigh, 1958; Mayr, 1965, 1974, 1982). It seems best to me to regard embryonic regulations and neuronal plasticity as *teleonomic processes,* which are defined as physiological processes that owe their goal directedness to programs that have been adjusted during evolution by the selective value of the goal or end-point (Pittendrigh, 1958; Mayr, 1982)" (Jacobson, 1991b).

However, I disagree even with the latter two approaches, because there is a presumption that the processes observed are actually goal directed. As we shall see in this book, the notion that embryogenesis is goal directed is more in our heads than in embryos.

Molecular embryology, like the field of evolution (Keller & Lloyd, 1992a; Brosius & Gould, 1992), could use attempts to clarify the meanings of words. My own approach is to introduce a few phrases where I think they represent succinct, observable phenomena, relate them to classical concepts insofar as possible, and otherwise avoid metaphorical language. (The sharp eyed reader may catch me up on this.)

1.19 The Embryology/Psychology Merry-go-round (Carrousel)

One curious aspect of metaphor in biological language is what appears to be a circular relationship between embryology and psychology. For example, stimulation of an organism with a functioning nervous system has been likened to the presumed role of the 'environment' in embryological development, which, as we have seen in Section 1.18, is discussed in psychological terms:

"...In reality the idea of a direct coding of ontogenetic patterns from the genes remains hypothetical. Its application even to early stages seems unclear, as may be gathered from statements such as the following by Weiss [1953a], an embryologist:

> 'It is evident that every new reaction must be viewed in terms of the cellular system in its actual condition at that particular stage, molded by the whole antecedent history of transformations and modifications, rather than solely in terms of the unaltered genes at the core... No cell develops independently, but... all of them have gone through a long chain of similar environmental interactions with neighboring cells and the products of distant ones.'

"....The most comprehensive and reasonable attitude for investigators of functional development therefore seems to lie along the lines of Dobzhansky's statement (Dobzhansky, 1950) to the effect that all development is both genotypic and environmental, rather than preformistic. Our premise consequently is that the hypothetical genic effects are mediated at each ontogenetic stage by systems of intervening variables characteristic of that stage in that species under prevalent developmental conditions, and that from initial stages these variables include both factors indirectly dependent on the genotype and others primarily dependent on the situation and environs of development.... It is improbable that any clear distinction can be made between what heredity contributes and environment contributes to the ontogeny of any animal.... As a guide to investigation, the view that development is a natively determined unfolding of characters and integrations is no more valid than that development is directed only by extrinsic forces or by learning.... The doctrine of a genic 'code' that is directly 'uncoded' in development is opposed by evidence concerning genotypes, not only with respect to organic growth and differentiation but also with respect to behavioral ontogeny" (Schneirla, 1964).

It should be noted that in Schneirla (1964) 'development' is also used to refer to changes in behavior after birth, hatching, etc., the common use of the word in the field of psychology. Lorenz (1981) similarly leans on embryology for models of behavioral development:

"The adapting function performed in the realization of an open program consists of receiving and exploiting information coming from external sources and correctly interpreting this information when opting for the teleonomically correct one among the finite number of possible adaptations. Experimental embryology has thrown considerable light on these important processes.... It does not matter whether the inducing influence is emanating from external factors or from adjacent organs within the embryo. What is essential is that the organism, or the organ, knows 'from within,' *how* to cope with a number of different eventualities and is informed, 'from without,' *which* of the eventualities it is that has actually arisen" (Lorenz, 1981).

In the reverse direction, Spemann leant on psychology:

"When Spemann explained in 1924 why he used the word *organizer* instead of *determiner,* he said it was because it 'is supposed to express the idea that the effect emanating from these preferential regions is not only determinative in a definite restricted direction, but that it possesses all those enigmatic peculiarities which are known to us only from living organisms' (Spemann & Mangold, 1924a). Later (Spemann, 1938) he gave as his reason that it implied psychic [psychological] analogies" (Oppenheimer, 1970a; cf. Oppenheimer, 1967).

"Embryonic regulation was generally conceived by Hans Spemann to be akin to psychic processes" (Jacobson, 1991b).

"...These processes of development, like all vital processes, are comparable, in the way they are connected, to nothing we know in such a degree as to those vital processes of which we have the most intimate knowledge, viz., the psychical ones. It was to express my opinion that, even laying aside all philosophical conclusions, merely for the interest of exact research, we ought not to miss the chance given to us by our position between two worlds" (Spemann, 1938).

The teaching of adolescent psychology to this day uncritically usurps an old embryology slogan, even with anachronistic imperialist overtones:

"In what is called a psychological theory of **recapitulation**, Hall (1904),... the father of adolescent psychology,... believed that each individual in his or her course of development retraces the history of humanity. Therefore, each person, beginning in childhood, lives through a period of animal-like primitiveness and a later period of savagery and ultimately enters a stage of maturity that reflects a more civilized nature. This view is the source of the expression '**Ontogeny** recapitulates **phylogeny**'" (Adams, Gullotta & Markstrom-Adams, 1994).

The following alternative model of instinct parallels, and may have its roots in, the concept of embryological development as a sequence of inductions:

"Results from our investigations (Schneirla, Rosenblatt & Tobach, 1963) of the maternal behavior in the domestic cat militate against assumptions of an inborn pattern, and speak in favor of progressive self-stimulative processes. Earlier development contributes certain specific organic processes..." (Schneirla, 1964).

As I will show, fundamental relationships between embryological and behavioral development may occur (cf. Grobstein, 1988), but how one thinks about both then depends heavily on what paradigm for embryological development is in vogue. The use of similar words for these disparate phenomena prejudges the issue. As Oppenheimer (1967) indicates:

"...no one since Roux has had either the courage or the conceit to try... to emulate the *Terminologie* (Roux et al., 1912)..., and modern embryology is still confronted with the old problem of using borrowed terms" (Oppenheimer, 1967).

1.20 The Cosmic Context

The evolution of biological organisms started sometime shortly after the formation of the earth about 4.6×10^9 (4.6 billion) years ago (Hoyle, 1975). It is thus a process occupying astronomical time scales, a decent portion, 35%, of the 13×10^9 (13 billion) years age of the universe (Anon., 1997b; Goodwin, Hendry & Gribbin, 1997), and cosmic distances (Davies & Koch, 1991, 1992). For instance, during this interval, our solar system has revolved around our galaxy every 250×10^6 (250 million) years (Seielstad,

1989), or about 18 times, which has probably not been an uneventful journey:

"The average length of a Warm-Cool cycle is 162 m.y. [162 x 10^6 years].... The agreement with a 150 m.y. half Galactic Year cycle is quite good,... being related to the position of the solar system with respect to the two spiral arms of the Galaxy... (Williams, 1975b).... However, this proposed cycle is undetectable in the Proterozoic (no known warm intervals between ~570 and 900 Ma) and earlier in the Precambrian" (Frakes, Francis & Syktus, 1992).

The luminosity of the sun may vary with its position with respect to the arms of the galaxy causing periods with more and less glaciation and lower and higher average earth temperatures (Williams, 1975b; cf. Hoyle & Lyttleton, 1939; McCrea, 1975; Schneider & Londer, 1984).

The appearance of life is a late phenomenon in the history of our universe since it requires solid planets, whose atoms are formed in the interior of stars. These stars exploded as supernovae towards the end of their 'life cycle', releasing clouds of interstellar gas that recondensed into new stars and planetary systems (Hawking, 1988). Thus life requires at least two rounds of star formation in a galaxy. We've had a long gestation period.

Despite periodic (Raup & Seploski, 1984; Patterson & Smith, 1987, 1989; but cf. Benton, 1995) and other catastrophic mass extinctions (Stanley, 1987; Donovan, 1989; Eldredge, 1991a; Raup, 1991; Monastersky, 1995a), once eukaryotes appeared, biological evolution gives the appearance of being an explosively accelerating process, summarized by the notion of 'progressive evolution', culminating, or perhaps terminating, with *Homo sapiens*. I will confine myself to the evolution of multicellular, eukaryotic organisms, that is, those having a nucleus in each cell, with more than one kind of cell, arranged in space in a recognizable pattern. Our eukaryotic kindred have been around for at least 590 x 10^6 years (Avers, 1989). The RNA world, the origin of life, of the cell and of the symbiotic organisms that form each eukaryotic cell, have been the subject of much fruitful analysis, speculation and controversy (Oparin, 1965; Bernal, 1967; Crick,

1981; Cairns-Smith, 1982; Hutchinson, 1983; Day, 1984; Margulis, 1984, 1993b; Margulis & Sagan, 1986a,b; Rosen, 1991; Margulis & Fester, 1991; Franklin, 1993; Gesteland & Atkins, 1993; Orgel & Crick, 1993; Breaker & Joyce, 1994; Depew & Weber, 1994; Lipkin, 1994; Mellersh, 1994; for divergent views see: Hoyle & Wickramasinghe, 1981). But this is all a backdrop for our story.

2.00 NEURAL INDUCTION AND THE ORGANIZER

2.01 A Moment of Discovery

"So confusing and chaotic did many of the facts of development appear, and so little was there in the way of any general paradigm to account for the age-old, yet burning, issue of differentiation, that the organizer concept was taken up with considerable enthusiasm and often with a less than critical eye. Methodologically and intellectually Spemann maintained his reservations. Many others did not" (Allen, 1975).

"...The experimental embryologist has gone ahead wherever a promising lead seemed open, without much thought of building up a comprehensive and internally consistent system" (Harrison, 1937).

I have always had a certain awe for those few of my colleagues who can cut and manipulate small bits of embryonic tissue, transferring grafts from one 2 mm embryo to another. This skill is almost synonymous with 'classical' embryology (Born, 1897; Harrison, 1898a; Needham, 1950; Oppenheimer, 1967). A remarkable grafting experiment, performed by Hilde Mangold in the early 1920's, captured the imagination of embryologists, and has left biology with a major, unsolved quandary. What she did, as a graduate student under the supervision of Hans Spemann, was conceptually simple, though she faced "the difficulty of raising operated embryos outside their jelly membranes, in an inadequate culture medium [lacking aseptic conditions, developed later by Holtfreter, according to Osamu Nakamura in Asashima, 1994b, which] limited the available material drastically" (Hamburger, 1988).

At the beginning of gastrulation the amphibian embryo appears quite simple on the outside: a sphere of roughly 5000 to 15000 cells looking pretty much the same, with a small dimple called the blastopore (Figure 4). Hilde

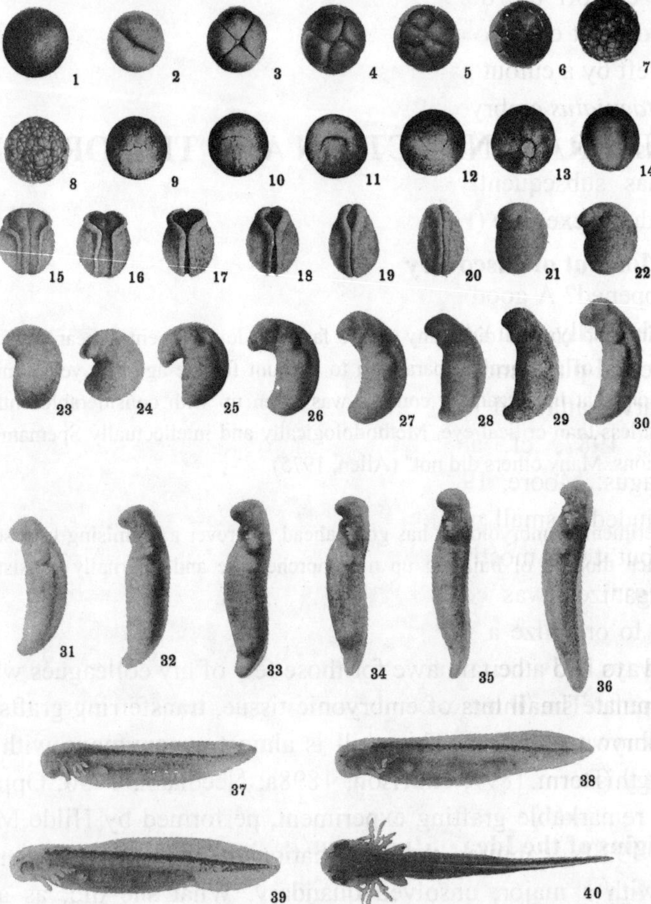

Figure 4. The stages of embryonic development of a urodele salamander, the California newt, *Taricha torosa* (previously named *Triturus torosus*), from Antone G. Jacobson, with permission. It is also printed in Twitty & Bodenstein (1948) and Lehman (1987). Note that this is a better reproduction of Figure 1 in Appendix I (Gordon & Brodland, 1987). The axolotl, *Ambystoma mexicanum,* develops in much the same way (Bordzilovskaya et al., 1989). Stage numbers are the same as those used for *Xenopus laevis* up to stage 19 (Nieuwkoop & Faber, 1994). Note that early cleavage stages are numbered 3, 4, 5, 6, 7 for 2, 4, 8, 16 and 32 cells, respectively, for both anurans and urodeles (Duellman & Trueb, 1986). A number of amphibian staging tables are referenced in Duellman & Trueb (1986).

Mangold cut off the dorsal (upper) lip of the blastopore from a lightly pigmented newt embryo of the species *Triturus cristatus* and grafted it into the hole left by a cutout patch in the ventral ectoderm of a darkly pigmented *Triturus taeniatus* embryo at the same stage. The experiment succeeded five times (Spemann & Mangold, 1924a,b; reprinted in: Fulton & Klein, 1976), and it has subsequently been repeated to the point of becoming an undergraduate exercise (Hamburger, 1960).

What happened? A good portion of the ventral ectoderm, which would have become the belly skin of the darker *Triturus taeniatus*, instead produced a second neural plate and went on to participate in the formation of a whole, second larva, attached belly to belly (Figure 4 in Appendix I: Gordon & Brodland, 1987; cf. analogous human ventrally conjoined twins or thoracopagus: Moore, 1977; Moore & Persaud, 1993). The second neural plate included a small portion of lighter tissue from the *Triturus cristatus* embryo, but it was mostly made of darker *Triturus taeniatus* tissue. Thus the name 'organizer' was coined for the dorsal lip of the blastopore, for it appeared to organize a whole embryo out of ventral ectoderm, and it was presumed to do the same with the dorsal ectoderm during normal development. The interaction between organizer and surrounding tissue became known as 'primary neural induction'. It is important to trace the history of this term.

2.02 Origins of the Idea of Induction

"More than 70 years have passed since the discovery of embryonic induction... a great challenge" (Tiedemann et al., 1996).

We can trace the idea that embryonic tissues interact back to...

"...[Pander's] conception of the germ layers....

'Actually a unique metamorphosis begins in each of these three layers and hurries toward its goal; although each is not yet independent enough to indicate what it truly is; it still needs the help of its sister travellers, and therefore, although already designated for different ends, all three influence each other collectively until each has reached an appropriate level' [translated from: Pander, 1817]....

"Whether he [Pander] intended mechanical processes, or whether in some prescient way he thought of the interplay of forces or exchanges of chemicals, we are unable to determine in his statements" (Churchill, 1991a).

Over 80 years later, experiments on a single organ, the lens, firmed up the concept that such interactions occur and gave them the name 'induction':

"...Spemann (1901a) suggested two experiments: to bring the optic vesicle in contact with foreign epidermis by transplanting it under flank epidermis or by replacing the epidermis over the optic vesicle with flank epidermis. The American embryologist Warren Lewis (1904a) had independently conceived the same idea and actually performed both experiments on the American frog, *Rana palustris.* In both instances, typical lenses differentiated from flank epidermis.... Spemann and Lewis deserve credit for the first experimental documentation of a case of embryonic induction.... The term induction... entered the vocabulary of experimental embryology much later, in the 1920's, when it was clearly understood to mean 'an embryonic activity which determined the cytological fate of the reacting cells' (Holtfreter & Hamburger, 1955)" (Hamburger, 1988).

(Cf. Spemann, 1938; Saha, Spann & Grainger, 1989; Saha, 1991; and the prediction of this result by Curt Herbst: Oppenheimer, 1991.) Allen (1978) gives Driesch some credit en route:

"Driesch (1894) had proposed the possibility of embryonic induction as a mechanism for differentiation; at the same time he proposed that such induction is the result of chemical influences on the nuclei of cells, causing those cells to produce substances that affect the cytoplasm. He even inferred that nuclear influence was mediated through 'ferments' (the older name for enzymes) [as discussed in Section 1.11].... He also concluded, in 1896, that the mesodermal tissue of sea-urchin embryos differentiated as a result of tactile stimulus from the overlying ectoderm" (Allen, 1978).

Spemann & Mangold, (1924a,b) and Spemann (1938) credit Boveri (1901a,b) with the discovery of what we now call homoiogenetic induction:

"...In the neighborhood of the blastopore, the cells are already determined at a time when the more distant parts are as yet undetermined. Hence followed the further conception of a determination which gradually advances from the blastopore forward and, as a necessary consequence, the conception of a 'center of differentiation' as the starting point of this determination.

"A similar interpretation had been considered by Boveri (1901a,b), many years earlier, for the development of the sea-urchin egg and its fragments.... It may have been working subconsciously after I had encountered it in the publication just mentioned, or even in oral communication, or it may have been part of our common stock of ideas in those invaluable years of daily intercourse with that great investigator" (Spemann, 1938).

But a qualitatively new result occurred in the Spemann & Mangold (1924a,b) experiment: a whole individual, not just a lens, was being induced from what most regarded as an 'indifferent', unremarkable tissue. Surely, if you contemplate your own belly skin with a realization that it could have created another brain, another conscious individual, you might recapture the amazement at the result which has dominated the thoughts of embryologists since that time:

"The most exciting single event of developmental biology in the twentieth century before Watson and Crick revealed the structure of DNA was Spemann's discovery of the organization center and organizer in the amphibian egg" (Haraway, 1976).

"Spemann expresses this idea picturesquely... *'We are standing and walking on parts of our body which could have been used for thinking if they had been developed in another position in the embryo* '" (Jacobson, 1991b).

It is interesting that this excitement over a 1924 publication took so much time and international funding to either be developed or turned into a new research direction outside of Spemann's realm:

"Holtfreter, brilliant and persistent in his efforts, seems to have been something of a thorn in the side of the Freiburg body of hopes for the organization center. Spemann was likely to favor slightly vitalistic explanations for the nature of his discovery; a simple chemical, least of all one of confusing specificity, was not an attractive candidate. He seems to have been less than cordial toward Holtfreter, with whom Waddington, a major collaborator with Needham and an important link with German biology, worked for a time in 1933. Waddington visited the laboratory of O. Mangold [Otto, widowed husband of Hilde Mangold] in Berlin in 1932 in order to learn techniques of amphibian operations. The British embryologist reported in a conversation on March 3, 1971, that Spemann was not interested in cooperating with his research goals at the time. Waddington had published a short paper on the chick organizer in 1930 [Waddington, 1930], after he had gone to the Strangeways

Laboratory around 1928 in order to work on such problems. He had switched his concentration from fossils to embryos in order to satisfy more readily his interest in genetics. He then focused on the organizer simply because it was the most exciting problem in the embryology of the period....

"Needham tried to construct an institute around the new paradigm commitments but was unable to obtain needed financing. Beginning in 1934 he corresponded with Dr. Tisdale of the Rockefeller Foundation, which was then interested in fostering study on the borderlines of traditional disciplines. In July Needham submitted a long memorandum outlining a plan for an Institute for Physico-chemical Morphology..." (Haraway, 1976).

The induction of the neural plate was at first called 'primary neural induction' because, at the time, it was the earliest induction known to occur, and perhaps because we regard the brain as the 'primary' organ. Nowadays, it is generally referred to simply as 'neural induction', and even in terms of research, it has taken a back seat to the earlier occurring mesoderm induction. Yet neural induction is still the only induction known to produce a whole (if attached) individual.

It has been hard to be definitive about induction:

"The processes taking place between the occurrence of an inducing stimulus and the production of its visible effect are still completely hidden from our knowledge" (Spemann, 1938).

Weiss (1935) recognized a trend in the use of the word 'induction' that continues to this day:

"A process... in which determination of one part of a developing system is brought about through decisive influences exerted by other parts, has been named 'induction.' Although inductive influences were originally thought of as specific, the term 'induction' has been stretched in more recent years by several authors to such an extent that it has no longer such distinctive meaning and has become ambiguous. It now covers all sorts of influences, from mere unspecific activation, to the very specific organizing effects laying down typical patterns in space and time" (Weiss, 1935).

2.03 Preformationism versus Epigenesis: To Be or To Become? That is the Question

"...To what extent does the egg contain a miniature and complete version of the fully developed embryo...? ...The argument rumbles on and still energises much research" (Lawrence, 1992).

"The biomedical paradigm emphasizes genetic determination of embryonic development and places great importance on flow of information from genome outward. As such it is generally criticized as a modern version of preformation, an ancient theory that inside each human egg there dwelt a tiny person and that development was merely growth. Most developmental biologists, of course, do not embrace any form of preformation but do, paradoxically, embrace a kind of DNA predeterminism" (Strohman, 1993).

Weiss (1935) wrote a short, well balanced essay critically weighing the first rush of experiments that followed publication of Spemann & Mangold (1924a,b). Many of these experiments have been forgotten as one hypothesis or result became favored over another, perhaps prematurely. First, let us note the change in thinking that was underway, probably initiated by the conflict between Driesch's and Roux's results:

"Ever since the introduction of experimental methods into embryology the strictly preformistic conception of development has been losing ground. We have come to realize that development means the bringing into existence of new characters; that the manifest increase in diversity and complexity which can be observed during development is real, not only apparent" (Weiss, 1935).

The dichotomy between mosaic and regulating embryos, a terminology apparently first applied by Driesch (Freyhofer, 1987), faded into a continuum (cf. Section 4.06). While a continuum is empirically accepted today, we have not arrived at an understanding of how we can have it both ways at the extremes, or in between cases in the middle:

"Until about 1900 it was customary to characterize embryos as 'mosaic' or 'regulative', based upon whether isolated blastomeres form only their normal portion of the anatomy (mosaic) or a larger portion (regulative) (Wolpert, 1971a; Moore, 1987; Melton, 1991). Gradually, the value of the distinction waned, as it became clear that the outcomes of such experiments depended critically upon the stage at which they were performed (Huxley & de Beer, 1934; Weiss, 1939; Davidson, 1986). However, a version of the dichotomy has persisted until only recently - namely, the notion that embryos with 'determinate cleavage' (precise sequences of cleavage planes) assign cellular fates by 'cytoplasmic determinants' (substances that are asymmetrically partitioned to daughter blastomeres), whereas those with indeterminate cleavage assign fates via cellular interactions (Wilson, 1925; Davidson, 1986, 1989; Moore, 1987). In the last few years, the conceptual wall separating these two categories has begun to crumble, thanks mainly to nematodes and sea urchins, where determinate cleavage is combined with regulative ability" (Held Jr., 1992).

The acceptance of a continuum between mosaic and regulating embryos solves nothing, unless we can find an underlying commonality between the two extremes and for every type of embryo in between. In a single issue of *Development,* we are told that "determinants for formation of the anterior-posterior" axis may be moved around with the expected effect in one cell ascidian embryos (Nishida, 1994b), while mouse embryos have "blastomeres [that are] non-polarized and equivalent" until "compaction at the 8-cell stage" (Gueth-Hallonet et al., 1994). The current paradigm handles this contradiction by using 'determinants' or 'induction' as magic wands. Determinants nicely solve the spatial problem for mosaic embryos, but leave us effectively with the homunculus, or embryo as 'little man', ready to just be inflated. Induction nicely solves the responsiveness of cells to their 'environment' in regulating embryos, but leaves out the spatial problem of how that environment happens to be just right when and where needed. Is this just another wave/particle duality? In physics the duality was resolved by discovery of a quantum mechanics whose mathematics subsumed both. Can we achieve something analogous in embryology?

This distinction, of mosaic versus regulating embryos, may be looked upon as but the old battle of preformation versus epigenesis (cf. Dinsmore, 1996; Pinto-Correia, 1997):

"The doctrine of epigenesis, i.e., that organisms developed by the successive differentiation of the fertilized ovum, was the dominant and orthodox theory of generation from the time of Aristotle. However, with the discovery of the microscope, Leeuwenhoek and Swammerdam put forward the preformationist theory, i.e., that the complete and perfect organism [the homunculus] was already preformed in the sperm, and merely grew in size.... In general, the preformationists... dominated the field in the eighteenth-century. The problem of accounting for the differentiation of parts in embryonic development seemed an insuperable obstacle to accepting the epigenetic view. It was not until the researches of C.F. Wolff [1774] were published (1759) that the doctrine of epigenesis was shown to be necessitated by the observable facts of embryonic developments; but Wolff's work was not widely known until 1812, when it was translated from the Latin by [J.F.] Meckel [cf. Nordenskiöld, 1928], another epigenesist. From that point on, the preformationist doctrine was abandoned" (Mandelbaum, 1957).

"Charles Bonnet (1720-1793) may be taken as the most thoroughgoing representative of the preformationist school, whose erroneous doctrines greatly inhibited the progress of embryological research for more than a century" (Thomson, 1910).

"Recognizing the many difficulties posed by this theory, Bonnet acknowledged that preformation theory had to be understood as a major triumph of rational thought over direct observational experience.... Those who adopt a system of preformation, [C.F.] Wolff warned, do not explain development but simply say that it does not occur" (Magner, 1994).

The battle has not been 'abandoned', but continues to this day in the context of cytological 'determinants' versus induction, often without reference to the embryological and historical realities. Some of those studying mosaic embryos have recently returned close to ideas of preformation in the form of 'cytoplasmic determinants' (Davidson, 1986), so the idea of "regional cytoplasmatic differences in the egg cell" (Stern, 1955) lingers on as if it were the only logical explanation for development:

"...The cytoplasm of zygotes is typically very heterogeneous, and daughter cells receive qualitatively different aliquots of this cytoplasm during cell division. The differences in this cytoplasm have been advanced as the essential condition for cellular diversification, and indeed they are.... It seems obvious that any acceptable explanation for cellular differentiation must involve some mechanism by which the function of the genome is, in effect, regulated by the surrounding protoplasmic environment" (Markert, 1964).

The antidote to preformation is the broad conclusion that...

"Descriptions of egg organization lose much of their force when confronted with the fact of lability in developmental potentialities" (Barth, 1964a).

We will see many examples of this lability. I would beg to differ with the...

"...emerging consensus. Biological self-organization is the gradual, cumulative epigenetic process whereby the linear information encoded in the genome takes on three-dimensional form. Small initial differences (sometimes random or nonspecific, sometimes environmental cues) are progressively amplified, generating spatial fields of one kind or another..." (Harold, 1990; cf. Harold, 1986).

This describes a 'deviation amplifying mutual causal process' (Maruyama, 1963a,b, 1968), which is not robust for form production.

In 1935, the preformation versus epigenesis dichotomy was extended to the 'parts' of an embryo:

"The perseverance in a definite course of a part of the developing organism, regardless of whether or not the typical relations to its surrounding parts persist, has been called 'self-differentiation' (Roux et al., 1912). Demonstrated by the methods of transplantation and explantation, it has, for a long time, assumed a central position in embryological experimentation and theory. This dominant position has been seriously challenged, however, by the experimental work of the last decades. Initiated mainly by the systematic work of Spemann and Harrison [1933a, 1935, 1969], continued experimentation revealed more and more examples where self-differentiation in amphibia was unmasked as due not to a primordial capacity of the particular part in question, but to an acquired capacity, derived from specific influences, directions and limitations, to which the part had been subjected during its previous ontogenetic history.... So-called self-differentiation... has ceased to play the dominant and universal rôle as *the* principle of development for which it had been proclaimed.... All that is not self-differentiation was designated as 'dependent differentiation' by Roux et al. (1912). Using his terms, development could be described grossly as a gradual transformation of dependent differentiations into self-differentiations. To indicate this general trend of development, the terms seem satisfactory; their explanatory value is, of course, nil.... This process of transformation has been called *determination* " (Weiss, 1935).

A tissue that showed 'dependent differentiation' was later referred to as 'induced', and the pendulum had now swung to the point that perhaps all tissues were thought to be induced. Even our mosaic relatives, the notochord bearing ascidians, were found to develop, at least in part, by induction (Rose, 1939), like the ancient and nearly extinct brachiopods (Freeman,

1995). The pendulum has swung again over the past few decades, for instance, in regard to nematodes, which at first were regarded as primarily mosaic or determinate (Schnabel, 1991) with "...in general, cell-autonomous mechanisms... to establish embryonic cell fates" (Kemphues, 1989; cf. Kimble, 1981; Strome, 1989), and are now touted as embryos whose tissue differences are *all* induced (Hutter & Schnabel, 1994; cf. Lambie & Kimble, 1991). The history of fruit fly (*Drosophila*) research showed the same: "The eggs of Diptera are always cited as the classical example of completely determined (mosaic) development..." (Poulson, 1940).

The same conclusion was drawn for the annelid *Nereis* by Wilson (1894), indicating an even earlier swing of the pendulum from mosaic to what we now call inductive development:

"The ontogeny is... a connected series of interactions between the various parts of the embryo, in which each step establishes new relations, through which the following step is determined" (Wilson, 1894).

A similar swing is underway for molluscs, such as the limpet *Patella vulgata* (Wilson, 1904; Kühtreiber, van Til & van Dongen, 1988; van Loon et al., 1993; Damen & Dictus, 1994). The issue is no better resolved today than it was in 1935 (cf. "spinning our wheels" in Malacinski, 1984a). Whether within an organism or between organisms, we don't yet know why some cells appear to have their fates determined 'within', while others appear to have their fates determined 'without'. It's the old, unresolved nature/nurture controversy of psychology ('genetics/environment' controversy in today's parlance) acted out in the arena of embryology. Perhaps differentiation waves, along with some clear thinking, will get us out of this impasse:

"The goal of developmental mechanics is to analyze the processes of morphogenesis of an individual organism into its physical [mechanischen = mechanical] components.... Every morphogenetic process... is a technical problem that can be analyzed into a sum of developmental and physiological problems; each process itself can be analyzed as a sum of individual effects of substances [Massen = masses] and energies [Kräfte = forces]. The forces of morphogenesis must reside in the substances, because experimental embryology has shown that effective external forces, such as gravity, light, heat, electricity, chemical influences, et cetera, can only exert environmental effects, but they are never themselves the true configurational forces. Thus it is necessary to look into the embryo itself, in order to explain the total organization and configuration in terms of material effects limited in time and space" (Vogt, 1923, translated in Oppenheimer, 1970a).

2.04 The Hunting of the Snark (The Inducer Molecule)

"'You may seek it with thimbles -- and seek it with care;
You may hunt it with forks and hope;
You may threaten its life with a railway-share;
You may charm it with smiles and soap -- '"

("That's exactly the method," the Bellman bold
In a hasty parenthesis cried,
"That's exactly the way I have always been told
That the capture of Snarks should be tried!")

(Carroll, 1876, with permission of the Electronic Text Center, University of Virginia).

"...It is probably not fantastic to picture, no doubt in the remote future, the amphibian organiser in a crystalline state" (Needham, 1931b).

It was presumed early on that primary neural induction was caused by a specific molecule emanating from the organizer. Why did the embryologists of the day (except Spemann: Section 2.02) decide that neural induction must be due to specific molecule? Why did they not consider other possibilities? For what carefully considered reason did the search for the inducer molecule, a search that occupied three generations of embryologists, begin? We would like to think it was because the science of the day pointed logically and rationally in that direction. Alas, there is no truly good reason. The reason is simply because a physical process would be harder to understand:

"...The search for the mechanism of primary neural induction was quickly narrowed to a search for a particular chemical agent, the inducer molecule. Only passing reference was made to the possibility that an alternative, nonmolecular mechanism might exist; e.g., the definition of a 'message [as]... any form of activity, chemical, physicochemical or physical, which emanates from cells, tissue or organs' (Nieuwkoop, Johnen & Albers, 1985) or the brief reference made to bioelectricity and 'mitogenetic rays' by Needham, Waddington & Needham (1934) [cf. the 'mitogenic rays', perhaps just light rays, of Gurwitsch, 1925, critiqued in Needham, 1931b, but perhaps supported by infrared 'vision' in cells: Albrecht-Buehler, 1992, 1994, 1995]. Some insight into this intellectual focussing is provided by Needham (1936a):

> 'Then there were the electrical effects, the potential differences, etc., which can be demonstrated on the amphibian egg. Their interpretation, however, is at all times so difficult that they could hardly be called upon to explain the phenomenon of the organiser.... The simplest explanation would be that induction was due to a single definite chemical substance acting in an almost endocrinological manner upon the competent ectoderm.' [Yes, this is repeated from Section 1.06.]

"The long, unsuccessful search for the inducer molecule is documented in three books (Saxén & Toivonen, 1962; Nakamura & Toivonen, 1978; Nieuwkoop, Johnen & Albers, 1985; and innumerable papers (see Witkowski, 1985a)" (Appendix I: Gordon & Brodland, 1987)

Spemann (1927a) was more careful, though he seems to have falsely assumed that nonliving materials cannot exert physical forces, and thus ended up supporting the same chemical conclusion:

"We are completely uncertain about the most elementary question: whether induction is mediated *materially* or *dynamically*, that is, whether particular [chemical] substances or particular [physical] forces awaken the latent capabilities in the parts that are to be induced. If the latter is the case, then one would expect only living cells to be capable of induction" (Spemann, 1927a) [translated in Hamburger, 1988].

Later results seemed to confirm a chemical nature for induction, although the possibility of chemical action causing a mechanical reaction in the induced cells (Appendix I: Gordon & Brodland, 1987) was not considered:

"Recently, Shen (1942) studied the effect of cultivating ectodermal explants in salt solution to which the endosuccinate derivative of 1,2,5,6-dibenzanthracene had been added. In more than 60 per cent of the explants so treated, neural structures developed in the complete absence of any solid implant so that the possibility of induction by mechanical means was completely excluded, and the adequacy, as in inductor, of chemical action alone was demonstrated" (Boell, 1942).

Holtfreter (1991) looked back at the chemical approach to induction, not realizing that the solution may have lain in the physics that he briefly considered:

"Bautzmann et al. (1932) heralded [but cf. Marx, 1931; Witkowski, 1987a] the beginning of a new era of research which may be titled: In search of the chemical constitution of the inducing agent(s). This quest expanded into a drawn-out, arduous, often tortuous campaign in which, at one time or another, investigators from all over the world became engaged. The campaign, which began in a mood of great expectations, lasted for some three decades, then petered out and ended, in the 1960s, on a note of despair and resignation. Today [1981] there is barely anybody around who is still active in this once so exciting field of research. (Ed.: Certainly, much more work is going on in this area now [1991] than when Dr. Holtfreter wrote this letter!)....

"By now, many of the old-timers have died; others, like myself, have moved off in new directions, and the younger generation of biologists has been lured away into the more fertile grounds of modern molecular biology. Unfortunately, the old and fundamental problems of ontogenesis are still waiting to be resolved. What went wrong with the campaign?....

"...Results of mine (Holtfreter, 1947a) indicated that the treatments merely operated like an unspecific trigger, setting in motion a preexisting, pent-up mechanism which, through unknown chains of events, led to neural differentiation.... Perhaps, so I thought (Holtfreter, 1948a), the situation is analogous to the one in experimental parthenogenesis of the sea urchin egg, where a variety of exogenous chemical and physical stimuli can imitate the action of the penetrating spermatozoon, namely, to activate egg development [cf. Nuccitelli, 1991]. What these activating agencies seem to have in common is that they change the structure and permeability of the plasma membrane. This, in turn, would entail the same kinds of physiological and morphogenetic chain reactions that occur in the normally fertilized egg....

"The trouble is that in all those years, when the hunt for the mysterious inductive substances went on, the real problems of induction as they are posed by the embryo were almost totally neglected.... We have no idea of the molecular machinery that is set in motion nor do we know what 'competence' means in cell-physiological or molecular terms....

"...I do not know of any solid material foundation for the assumption that gradients of any sort are present or operating in amphibian development.... To speak, in connection with induction phenomena, of vegetalizing or animalizing factors or processes (or of 'positional information') is empty rhetoric; these terms explain nothing and do not even serve as useful working hypotheses. Perhaps the most fundamental problem of embryology is: how do cytoplasmic patterns of potentialities and of actual differentiation come about and what is their relation to the genetic information?....

"During the past few decades, interest in the problems of induction has faded away. The probable reason is the failure of the workers in this field to chemically characterize the nature of the naturally occurring inductive agents. Let us hope that the new generation of embryologists will find a way out of this dilemma" (Holtfreter, 1991).

Hamburger (1988) himself mocks the physical approach, making one wonder if it was the tenor of the times, one which only Spemann himself could see past, if not transcend:

"...An amusing and rather weird idea, dating back to the early days of lens induction experiments, deserves to be rescued from oblivion. Lewis (1907a), one of the pioneers in this field, suggested that the optic vesicle, while being transformed into an optic cup, might pull the overlying ectoderm inward mechanically: the lens vesicle would originate by some kind of suction. In his definitive publication on lens induction, Spemann (1912a) took this notion quite seriously. He collected all the arguments against it and ended with the sensible suggestion that 'the stimulus is more specific and perhaps chemical in nature'" (Hamburger, 1988).

(I was told similar, cautionary tales against physical explanations by well meaning biology mentors, who were passing on the folklore of the silliness of physicists in approaching biological phenomena. Cf. Gordon, 1992d.) Thus an unproved presumption set off a search for 'the' inducer molecule, without any proof that it even existed.

2.05 A Cornucopia of Inducers

"In the 1960's, many people studying embryonic induction left the field.... At this time Dr. Yô. K. Okada and Dr. T. Fujii were attempting to identify inducing substances by the insertion method into blastocoeles. They concluded that all things, including inorganic materials such as kaolin and quinone could induce neural induction.... As a result, scientists who wanted to identify the special inducing factor for cell differentiation lost interest in the idea, and many people left classical embryology to go into modern biochemistry or cell biology.... The major people in the fields of developmental biology worked on cell differentiation and gene expression at the cellular or at molecular level. They did not return to experimental embryology which is the origin of developmental biology itself.... I have always believed that in the future many developmental biologists will return to the challenge of embryonic induction" (Osamu Nakamura in: Asashima, 1994b).

Since the discovery of the organizer, every generation of embryologists has found that its favorite substance of the day was an inducer:

"The problem is not that an inducer can't be found, but that all sorts of strange substances and conditions will cause induction:

ectoderm killed by boiling
high pH
low pH
calcium-free medium
sodium chloride
lithium chloride
basic proteins from alcohol-fixed mouse kidney
ribonucleoprotein particles from rat liver
protein from guinea pig liver
extract of 9 day chick embryos
cyclic AMP

to name a few from the latest review (Nieuwkoop, Johnen & Albers, 1985 [cf. Nieuwkoop, 1966]). (There is an equally bizarre list of inhibitors and inactive substances.)...

"We thus propose that the pantheon of artificial inducers sorts into two categories:

1) somewhat rigid substances to which ectodermal cells can adhere at their basal ends, giving them some mechanical resistance to the global tension, and biasing them towards the neuroepithelial state;

2) substances (liquids, solutions or solids) that stimulate actin deposition against the basal membranes and subsequent actin filament contraction, also biasing the affected cells towards the neuroepithelial state.

"The existence of a broad spectrum of inducers in the first category suggests that the natural inducer, namely the notochord, has only one important property in regard to its role in primary neural induction: its adhesion to the notoplate, which reduces the stress caused in the notoplate by the global tension. The mystery of why dead tissues work so well is now resolved: their proteins are coagulated, like hard boiled eggs, making them much stiffer than the live tissues. Boiling may also improve those surface properties of dead tissues that permit adhesion of the basal ends of the notoplate cells, or any ectodermal cells against which they are placed.

"The second category of artificial inducers may also have natural representatives. For example, Davis & Lemanski (1987) have shown that an RNA is involved in induction of myofibrillogenesis.

"We postulate that substances which do not induce fail to adhere and/or stimulate actin deposition. The outer surface of mature, external epithelia tend, for instance, to be hydrophobic, and may provide poor surfaces for adhesion. This could explain Brachet's (1974) incredulous remark:

'Only such very inactive tissues as banana peel or insect wings failed to elicit a reaction!'" (Brachet, 1974; Appendix I: Gordon & Brodland, 1987).

"Of much interest was the fact that Holtfreter (1934a) made a large number of implantations of plant materials, e.g. meal, potato, banana, agar, etc., always with negative results. Wax was also without effect" (Needham, 1950).

Here's another material that doesn't work: "Lens rudiments implanted into blastulae have no power of induction, either hetero- or homoio-genetic (Krüger, 1930)" (Huxley & de Beer, 1934). Hamburger (1988) further indicates that...

"Some investigators suspected that they might be dealing with simple mechanical stimuli; but this notion was ruled out by numerous experiments with implantations of agar, coagulated egg albumin, gelatine, and celloidine, almost all of which gave negative results" (Hamburger, 1988).

"Apparently first [among artificial inducers] was the discovery that the developing limb bud had some inducing power (Mangold, 1928a)" (Child, 1941). Other inducers include coagulated newt nuclei and cytoplasm

(Waddington, 1938b), electric currents (Stern & Goodwin, 1977) and "dust from the floor" (Allen, 1975, 1978). Boterenbrood (1962) adds...

"...adult tissue (Holtfreter, 1934a; Chuang, 1938, 1939; Toivonen, 1940),... alcohol- or heat-denatured presumptive notochord (Rollhäuser-ter Horst, 1953; Hori & Nieuwkoop, 1955)... [and] unspecific stimuli from synthetic culture media..., especially if they act upon the inner surface of the ectoderm (Barth, 1941a; Holtfreter, 1947a)" (Boterenbrood, 1962).

"Work [with pure chemicals]... was climaxed by Okada's (1938) finding that inorganic substances - silicon [sic: actually 'fuller's earth' and 'silica'] and calcium carbonate - could induce neural structures" (Boell, 1942). Dalcq (1946a) suggested "a kind of biochemical convergence". Holtfreter (1948a) classified many inducers as cytotoxic.

Twitty (1966) gives us a nice summary of the state of affairs:

"...Results followed thick and fast - and confusingly. In fact, many of them were so puzzling and seemingly paradoxical that Holtfreter was to remark despairingly that the analysis of organizer action was rapidly 'bringing chaos out of order.' It had been surprising enough to learn that a killed organizer could still induce [Holtfreter, 1933c], but it soon proved that other, ordinarily inert, parts of the embryo such as ectoderm became full-fledged neural inductors once they were coagulated by boiling. The roof really fell in when living or killed tissues of adult salamanders, and even from a variety of animals ranging from worms to man, were found to function in this same unexpected way (Holtfreter, 1934a). Nor did efforts to identify the active components of normal and foreign inductors by chemical fractionation serve to reduce the confusion. Separations by Needham [and colleagues: Waddington et al., 1935; Waddington & Needham, 1935; Waddington, Needham & Brachet, 1936; Waddington et al., 1936; Heatley & Lindahl, 1937; Heatley, Waddington & Needham, 1937; Waddington, 1938c, reviewed in Needham, 1950] indicated that it was the steroid-containing fraction that was active, and this seemed to be supported by positive results with chemically pure, synthetic steroids. However, similar claims, equally well supported, were soon made for totally unrelated compounds such as fatty acids and nucleoproteins. One reason why this was so discouraging was that it began to look as if almost any substance could unlock neural differentiation, and if true this would offer no clue to what is happening inside the reacting cells in response to induction" (Twitty, 1966).

(Cf. Haraway, 1976.) The litany went on and on, seemingly accepted with an uncanny equanimity:

"In amplification of the statement... that certain tissues which possess no capacity to act as organisers when alive may show this capacity when they are killed, we may refer to further recent experiments by Holtfreter (1933e). He has found that the property to induce the formation of a secondary embryo or parts of it in an amphibian gastrula are possessed by the following: *all* parts of uncleaved amphibian eggs that have been boiled to a state of hardness; *all* parts of an amphibian gastrula that have been preserved for six months in 70 per cent. alcohol, treated with xylol, embedded in paraffin and brought back to water; boiled pieces of muscle of the Annelid *Enchytraea* and of the molluscs *Planorbis* and *Limnea;* heat-coagulated cell-free extracts of the crustacean *Daphnia* and of the pupae of moths; pieces of all organs so far tested of the stickleback, fresh or boiled; living pieces of larval amphibian liver, brain and retina, and of adult liver, ovary, and heart; living pieces of liver, kidney, testis and other organs of lizards, birds, and mice; coagulated bird embryo-extract; extract of killed calf's liver, and boiled pieces of several mammalian organs; pieces of liver, brain, kidney, thyroid and tongue of a fresh human corpse" (Huxley & de Beer, 1934).

"One of the most surprising results of these experiments was the observation that some tissues which had been known to be completely ineffective while alive would give induction effects once they were killed (Holtfreter, 1933e; Spemann, Fischer & Wehmeier, 1933), for instance, any part of the ectoderm or entoderm of the blastula.... Holtfreter (1934b), Fischer & Wehmeier (1933b), Hatt (1934), Umanski (1932), [and] Woerdeman (1933a) have been successful in obtaining inductions in urodele germs with fragments taken from one or the other of the following organisms and organs:

Worms (*Enchytraeus*): body fragments.
Snails (*Planorbis, Limnaea*): foot muscles, hepatopancreas.
Daphnia: coagulated body extract.
Lepidoptera (*Deilephila*): hemolymph and ganglia of pupa.
Dragon-fly (*Libellula* larv.): fat body, ganglia.
Fishes (*Gasterosteus*): heart, liver, ovarian eggs, muscle, spleen.
Amphibia (*Triton, Salamandra, Rana*): liver, heart, ovarian eggs, muscle, cartilage, brain, retina, regeneration blastema.
Reptiles (*Lacerta*): liver, kidney, testis.
Birds: liver, kidney, testis, thyroid, fat body, brain, retina; coagulated extract of chick embryos; fragments of primitive streak.
Mammals (mouse): heart, liver, kidney, adrenals, brain, lens; calf's liver.
Man: liver, brain, kidney, thyroid, tongue, sarcoma, carcinoma.

No inductions were obtained with starch (prepared from wheat, potato, banana), agar, chick albumin, lard, wax, charcoal, gelatine, cholesterin, yeast, coagulated frog's blood" (Weiss, 1935).

Waddington & Wolsky (1936) added hydra to the 'evocator' list. It is no wonder I was inspired to quote from Shakespeare in Appendix I (Gordon & Brodland, 1987):

Eye of newt, and toe of frog,
Wool of bat, and tongue of dog,
Adder's fork, and blind-worm's sting,
Lizard's leg, and howlet's wing,
For a charm of powerful trouble,
Like a hell-broth boil and bubble.
(William Shakespeare, Macbeth, 1623)

2.06 The Snark Was a Boojum

In the midst of the word he was trying to say,
In the midst of his laughter and glee,
He had softly and suddenly vanished away -- -
For the Snark *was* a Boojum, you see.
(Carroll, 1876, with permission of the Electronic Text Center, University of Virginia).

"It has not even been demonstrated that the natural inductor is a substance" (Child, 1941).

There were some moments of expressed doubt after the initial certainty:

"In experiments with the eggs of the Anuran *Discoglossus pictus* it was found that stimuli which were probably purely mechanical could cause the formation of a sucker" (Waddington et al., 1935).

"The discovery that the production of an asymmetrical organ [the neural tube] can be entirely accounted for by a reaction between a specific substance and a homogeneous ectoderm, although it may appear to bring us immediately within the sphere of physico-chemical concepts, would actually render the process of development less, and not more, easy to understand in material terms" (Heatley, Waddington & Needham, 1937).

"It is surprising to find that, though the biologically inactive substances do not show strong evocating power, they nearly all occasionally produce an induction.... The only test we have for evocating power is to apply a substance to the ectoderm, in which the evocator is already contained" (Waddington, 1938c).

Recent claims for discovery of 'the inducer' include, for example, mRNA of the homeobox gene *goosecoid:*

"Microinjection of *goosecoid* mRNA into the ventral side of the *Xenopus* embryo, from which it is normally absent, leads to the formation of a twinned body axis. The *goosecoid*-injected cells are able to recruit uninjected neighboring host cells into this secondary axis ([Niehrs et al., 1993]). Thus, microinjection of this single mRNA is sufficient to execute the 'organizer' properties described by Spemann & Mangold (1924a,b)…. Furthermore, *goosecoid*-expressing regions of the gastrulating mouse egg cylinder have organizer-like activity when transplanted into *Xenopus* embryos" (Blum et al., 1992).

(Cf. De Robertis et al., 1992; Hahn & Jäckle, 1996.) The "Complexity of an induction event" (Lemaire & Gurdon, 1994a) has been rising with a new wave of inducer molecules, such as the TGF-β (transforming growth factor-β) family (Massagué, Attisano & Wrana, 1994), FGF (fibroblast growth factor: Gillespie, Paterno & Slack, 1989; Green, New & Smith, 1992; Kessler & Melton, 1994; but cf. Slack, 1996b), activin (Kispert et al., 1995), noggin (Sharpe, 1994), an unknown protein (Eyal-Giladi et al., 1994b), a homeobox gene *Siamois* (Lemaire, Garrett & Gurdon, 1995), plus "the activation of two pathways... induction and a second one that is cell-autonomous" and "a 'community effect', as it appears that cells that have received the inductive signal need to communicate with each other in order to progress to the next step" (Lemaire & Gurdon, 1994a).

Each generation of embryologists, trained in the biochemistry of their day, has found that the most recently discovered molecules acted as inducers. They never disproved the work of their predecessors. This cycle of events continues as I write these words, and probably will for some time after. Perhaps "The Snark was a Boojum" (Carroll, 1876).

Where do we stand today? Apart from the work presented in this book, and the papers on which it is based, the assessment of Witkowski (1987a, 1988a) is probably still appropriate:

"It was quickly apparent that what Spemann called the organizer centre cried out for biochemical analysis, and Needham (1936a) wrote that 'the nature of the organiser influence was from the first recognised to set a problem the solution of which would profoundly affect our picture of the process of development'.... The problem was then recognized as a fundamental challenge to the physico-chemical approach to the living organism. It is also a problem that continues to resist solution....

"Needham (1931a) wrote that the future of embryology lay in the closest contact between biochemistry and experimental embryology, and '...the biologist who will deserve most the gratitude of posterity will be he who finds the way to fuse these studies into one'. Eight years later, in Needham (1939), he suggested that 'it may be more like fifty years before we can expect certain knowledge concerning the chemical nature of the naturally-occurring substances involved in embryonic induction'. That fifty years is... up, but Needham's biologist has yet to put in an appearance, and the biochemical analysis of dynamic embryological events remains formidably difficult" (Witkowski, 1987a).

"...I am left with the impression that the molecular - cellular - developmental biologist will feel that the time is not yet right for an all-out assault on embryonic induction" (Witkowski, 1988a).

A curious twist to the whole story has developed over the past decade or so. The actual, natural organizer may not be the dorsal lip of the blastopore, which includes part of the nascent notochord, but rather may be the pharyngeal endoderm, as suggested by Hama et al. (1985). This tissue precedes the nascent notochord during the invagination of some of the ectoderm through the blastopore, but, despite its name, is still part of the invaginating ectoderm (Nieuwkoop & Ubbels, 1972). The position at which it touches the ectoderm from underneath approximately corresponds with the position of the launching site of the ectoderm contraction wave, about 45° dorsal to (above) the dorsal lip of the blastopore (Figure 11 in Appendix V: Gordon, Björklund & Nieuwkoop, 1994). A direct correlation between what we see in time lapse movies of the ectoderm contraction wave and histology, performed on the same embryo microtomed immediately after the wave begins, or marked with a vital dye, will be necessary to confirm or deny this correlation, or to check the following alternative interpretation:

"I must only make one correction concerning the ectoderm contraction wave, where you refer to Hama et al. (1985). Though the pharyngeal endoderm passes first underneath the presumptive neurectoderm, neural induction is exerted by the prechordal mes-endoderm which immediately follows the pharyngeal endoderm" (Pieter D. Nieuwkoop, p.c., 1993).

We are developing a microwave fixation method to make such work possible (Williot et al., 1998; Section 9.20). Since half of the initial ectoderm contraction wave actually travels in the posterior direction, if it is regarded as the cause of 'planar induction', it explains the following observation on abnormal ring embryos, which...

"...were studied in great detail by Oscar Hertwig (1892, 1906) who also gave them their name.... When during frog gastrulation epiboly is blocked either by surgical intervention or by elevated osmolarity of the medium, the yolk plug will visibly persist instead of being completely covered by the advancing blastoporal lips [leaving a ring around the exposed endoderm]. The involuted [invaginated] part of the dorsal lip [i.e. pharyngeal endoderm], then apparently induces a neural plate. This plate continues posteriorly, albeit split in two halves flanking the persisting yolk plug. Behind the yolk plug (*i.e.* in the ventral sector of the marginal zone) a more or less uniform posterior end of the neural plate will form. The causes for neural development in the parts lateral and posterior to the yolk plug have to my knowledge never been ascertained. Without independent evidence, Holtfreter (1933g) ascribed most of comparable neural differentiation in his exogastrulae to inductive influences from the lip material involuted in the respective regions, but in my view a lot of planar induction might occur as well" (Sander, 1996).

Given that so many other tissues, dead and alive ("heterologous inducers... carp swim bladder, guinea pig liver, and bone marrow... Toivonen & Saxén, 1955b; Kurihara & Sasaki, 1981": Sokol, Wong & Melton, 1990), also act as organizers, we may only conclude that there was never anything special about the dorsal lip of the blastopore. Perhaps that is not surprising, since the dorsal lip of the blastopore is hardly a distinct tissue in any classical sense. It is merely the location at which much of the ectoderm invaginates through the blastopore, and is thus a continually changing set of cells (which confounded investigators measuring respiration of the ectoderm versus that of the dorsal lip of the blastopore: Boell, 1942). There may be something temporally unique about the ectoderm cells as they change shape going over the dorsal lip of the blastopore. They temporarily become 'bottle cells' (Perry & Waddington, 1966a; Nakajima & Burke, 1996), which transiently activate specific genes (Mary Whitely, p.c.; cf. Dirksen & Jamrich, 1992; Appendix VI: Björklund & Gordon, 1994). Nevertheless, there really is no evidence that the dorsal lip of the blastopore is unique in its inducing capabilities. Thus we are led to two historically disturbing conclusions, to which I add two well known, but often ignored observations:

1. The dorsal lip of the blastopore is not the natural organizer.

2. There is no chemical inducer molecule, i.e.:

"Perhaps, like the ether [aether] through which electromagnetic waves were believed to propagate, a particular inducer substance doesn't exist" (Appendix I: Gordon & Brodland, 1987).

3. The dorsal lip of the blastopore is a wave, not a tissue, and thus does not consist of the same cells from one moment to the next. The confusion this should engender more often is well expressed by Bard (1990a):

"There is a major difficulty in trying to understand how this region controls gastrulation: its constituent cells are in a state of flux, changing as migration [invagination] proceeds with both endodermal and mesodermal cells moving through the domain. It is therefore hard to know whether the instructional properties reside within a unique group of cells or whether any cells reaching the area temporarily acquire the properties associated with this region.... This aspect of the process seems as opaque now as it was when first investigated by Spemann (1938)" (Bard, 1990a).

4. In fact, the dorsal lip of the blastopore is a succession of waves (Appendix V: Gordon, Björklund & Nieuwkoop, 1994; Appendix VI: Björklund & Gordon, 1994) and thus does not even consist, from our point of view, of cells in the same state of differentiation if one considers it over time, or even along its rim at a given time.

This state of affairs has to be brought to bear on considerations such as regional induction and the purported role of genes in 'the' organizer:

"When an early dorsal lip was transplanted, a complete axis (including a head) was formed; but at later stages the dorsal lip was able to induce trunk-tail structures only, leading to the distinction being made between a 'head' and a 'trunk' organizer (Spemann, 1931a).... Several homeobox genes that are expressed specifically in the *Xenopus* organizer have been isolated: *goosecoid* (*Gsc*), *Xlim-1* and *Xnot* (De Robertis et al., 1994)... [and] secreted signalling molecules, such as noggin (Smith & Harland, 1992), follistatin (Hemmati Brivanlou, Kelly & Melton, 1994) and chordin (Sasai et al., 1994), that are specific to the organizer.... With so many genes involved, the most straightforward way of unravelling their interactions is to clone their mouse homologues and make use of gene-targeting technology.... We can be pretty sure that [the] Shawlot & Behringer (1995) finding that *Lim1*$^{-/-}$ mice lack all head structures anterior to the hindbrain reveals the existence of a head-organizing centre [or a *Lim1* dependent differentiation wave in the anterior portion of the neural plate].... The challenge in the near future will be to cross mouse strains to obtain double mutants in organizer-specific genes to uncover how these genes interact. It seems that unravelling the molecular nature of Spemann's organizer will be a laborious but rewarding task - considering the number of players [genes] identified so far, however, it might also be prudent to continue fishing for new genes in the *Xenopus* dorsal lip" (De Robertis, 1995).

There may be so many genes involved because the formation of many tissues is occurring in the multiple waves of the dorsal lip. A redefinition of the tissues and the anatomy involved, dissecting 'the' dorsal lip into its many wave components (Appendix VI: Björklund & Gordon, 1994), i.e., getting the anatomy straight, is needed before proceeding with such fine molecular analyses.

2.07 Limb Induction: A Parallel Case?

The question of artificial inducers is rarely raised outside of primary neural induction, so it may be worthwhile to bring up the one instance that has come to my attention:

"Balinsky (1925, 1927a) found that the transplantation of an ear vesicle under the flank of an embryo in the tail bud stage can result in the formation of an extra limb in the vicinity of the graft. Practically the same effect could be obtained, however, by using as graft some dead matter, such as colloidin (Balinsky, 1927b)" (Weiss, 1935).

Supernumerary limb induction is further reviewed in Waddington (1956a) and in the sources of the following quotes:

"Depending on whether the induced limb lies nearer to the normal forelimb or hindlimb region, it may resemble a forelimb or a hindlimb in its structure (Balinsky, 1933; Perri, 1951)" Balinsky & Fabian, 1981).

"These facts, because of their fundamental significance, deserve a more thorough investigation than their fortunate discoverer [Balinsky] has so far devoted to them" (Spemann, 1938).

"A... popular method of inducing accessory limbs has been to implant various foreign substances subcutaneously into the arm, whereupon outgrowths sometimes germinate from the site of implantation. Charles Breedis (1952)..., in seeking a possible link between regeneration and cancer, discovered that if certain carcinogenic substances are put into the limbs of adult newts, tumors sometimes developed but in other cases limbs grow out from the sides of the arms.... Ruben & Stevens (1963)... found that of a wide variety of organs assayed, frog kidney worked best. Other amphibian organs such as liver, spleen, small intestine, skeletal muscle and heart, all possessed little or no capacity to promote supernumerary limbs. Even renal adenocarcinomas from frogs proved to be poor inducers. One wonders, then, what is so special about frog kidneys, especially since homografts of newt kidneys do not work.... Carlson & Morgan (1967) confirmed that homogenates of fresh frog kidney are effective in stimulating accessory limbs, as are lyophilized or frozen pieces of tissue. However, if the kidney is first boiled, its efficacy is abolished.... Implants fail to stimulate accessory outgrowths in totally denervated limbs.... Where there is the ability to regenerate, there must also be safeguards against premature regeneration. So it is that the diverse mechanisms leading to the development of a limb regenerate are all linked ultimately to injury as the first cause" (Goss, 1969).

"[K.R.] Robinson has pointed out that supernumerary limbs can be induced in amphibians simply by making an incision in the skin which will allow current to leak out (Thornton, 1954) [cf. Carlson & Morgan, 1967; Williams & Müller, 1996]. Therefore, it is possible that the outward current might act to organize or stimulate limb development" (Nuccitelli, 1988a).

"When ZPA [zone of polarizing activity] cells are grafted to the anterior margin of a budding chick wing, the result is a spectacular mirror-image duplication of the distal wing pattern. This effect can be reproduced to different extents by other signaling regions of the developing body such as Hensen's node or by molecules like retinoic acid or the chicken *shh* gene product [sonic hedgehog], which is related to the product of the *Drosophila* gene *hedgehog* (Tabin, 1991; Tickle, 1991; Krauss, Concordet & Ingham, 1993; Riddle et al., 1993).... Actually the original wing pattern... remains unmodified. Rather, an extra structure is produced" (Duboule, 1994a).

The analogies (or homologies?) of the limb story to the primary neural induction story are evident. Perhaps these are homologies, given that homeobox gene expression patterns suggest "the concept of the uniformity in the patterning mechanisms along the various axes of the body" (Dollé et al., 1991b; cf. Dollé, Price & Duboule, 1992; Dollé et al., 1993a; Duboule, 1994a).

2.08 Mesoderm and Other Inductions

"Although at the present time many findings on mesoderm-inducing factors and receptors of amphibian embryos have been reported, the general problem of the mechanism of embryonic induction remains unsolved" (Fukui & Asashima, 1994).

Comparatively speaking, I won't say much about mesoderm induction in this book for the simple reason that we have yet to define the differentiation waves involved in mesoderm induction as well as we have defined the ectoderm contraction wave for neural induction (Appendix V: Gordon, Björklund & Nieuwkoop, 1994). Furthermore, much of the mesoderm induction literature is based on explants. Since our premise is that mechanical effects are the underlying basis of induction, the act of creating explants may itself cause mechanical effects independent of the interaction between explanted or sandwiched tissues. Thus the mere act of explanting may involve the triggering of supernumerary differentiation waves (Section 9.22). Osamu Nakamura, in Asashima, (1994b), regards mesoderm induction in explants as 'an abnormal event'.

The literature on mesoderm induction is vast (Toivonen, 1953; Nieuwkoop, 1969a,b, 1970, 1992; Sudarwati & Nieuwkoop, 1971; Nieuwkoop & Ubbels, 1972; Boterenbrood & Nieuwkoop, 1973; Weyer, Nieuwkoop & Lindenmeyer, 1978; Asahi et al., 1979; Dale, Smith & Slack, 1985; Nieuwkoop, Johnen & Albers, 1985; Grunz et al., 1988; Rosa et al., 1988a; Symes, Yaqoob, & Smith, 1988; Cooke, 1989a,b; Represa & Slack, 1989; Whitman & Melton, 1989, 1992; Asashima et al., 1990a,b; Hopwood, 1990; Mitrani et al., 1990a; Roberts et al., 1990; Thomsen et al., 1990;

van den Eijnden-Van Raaij et al., 1990; Dawid, 1991; Shiurba et al., 1991; Dawid et al., 1992; Ettensohn, 1992; Gillespie et al., 1992; Godsave & Shiurba, 1992; Harvey, 1992; Hemmati Brivanlou & Melton, 1992; Kimelman, Christian & Moon, 1992, 1993; Maslanski, Leshko & Busa, 1992a,b; Mitrani, 1992; Moon & Christian, 1992; Sokol & Melton, 1992; Stern et al., 1992; Zhang & Jacobson, 1992; Christian & Moon, 1993a,b; Dawid, Rebagliati & Taira, 1993; Gurdon, Kato & Lemaire, 1993; Howard & Smith, 1993; Jones et al., 1993b; Moon et al., 1993; Slack, 1993; Smith, 1993a; Tadano et al., 1993; Cardellini, Polo & Coral, 1994; Cornell & Kimelman, 1994; Kessler & Melton, 1994; Huang et al., 1995).

It will be particularly interesting to see if the 'mesoderm inducing' gradients of activin and FGF (fibroblast growth factor; cf. Asashima, 1994a; Ariizumi & Asashima, 1995a) are caused by differentiation waves, as I would presume, rather than *vice versa:*

"The potent mesoderm-inducing factors activin and FGF are present as maternally synthesized proteins in embryos of *X. laevis*. We show that activin can act on explanted blastomeres to induce at least five different cell states ranging from posterolateral mesoderm to dorsoanterior organizer mesoderm. Each state is induced in a narrow dose range bounded by sharp thresholds. By contrast, FGF induces only posterolateral markers and does so over relatively broad dose ranges. FGF can modulate the actions of activin, potentiating them and broadening the threshold-bounded dose windows. Our results indicate that orthogonal gradients of activin and FGF would be sufficient to specify the main elements of the body plan" (Green, New & Smith, 1992).

A more somber assessment of the state of our understanding of mesoderm induction by activin, bone morphogenetic protein (BMP), fibroblast growth factor (FGF), Wnt and noggin has been given by Kessler & Melton (1994):

"Although promising candidates for endogenous mesoderm inducers have been identified, it has not yet been possible to assign them to specific inducing functions in vivo.... In addition, the existence of an activin gradient during early *Xenopus* development is as yet unsupported, although other localized factors, perhaps Vg1, could function in this manner.... Overexpression of [the activin-binding protein] follistatin in early embryos does not block mesoderm induction (Schulte-Merker, Smith & Dale, 1994; Slack, 1991b), nor does it inhibit mesoderm induction by Vg1 (Green & Smith, 1990; Slack, 1991b). These results raise questions about the role of activin in endogenous mesoderm induction.... The inhibition of notochord formation by the truncated FGF receptor is unexpected, given the inability of FGF to induce notochord efficiently in animal pole explants.... The expression pattern of *Xwnt-8* is not consistent with a dorsal modifier role during development.... The paradoxical observation of both dorsalizing and ventralizing activity is not understood.... Noggin does not induce mesoderm in animal pole explants....

"Important questions include the following: What is the order of action of endogenous inducers? How are these activities spatially regulated? How do multiple inducers result in the appropriate patterning of mesoderm into distinct tissues? An examination of downstream targets will provide some answers, and in fact, several nuclear factors have been described, including *goosecoid* and *brachyury* (*T*), that have mesoderm-inducing or -patterning activity (Blumberg et al., 1991; Cho et al., 1991a; Cunliffe & Smith, 1992; Herrmann & Kispert, 1994; Niehrs, Steinbeisser & De Robertis, 1994). This diversity of activities presents developmental biologists with the challenging task of deriving a working model of mesoderm induction from a wealth of experimental data" (Kessler & Melton, 1994).

Note that Christian & Moon (1993a) suggest we stop depending solely on "gradients of diffusible morphogens [which] provide positional information to cells within the embryo". Of course, what this book is about is our working model for mesoderm and all inductions consisting of differentiation waves, cell and nuclear state splitters, and differentiation trees. Without a working model, the search for the inducer molecule, be it for neural induction or for mesoderm induction, goes on, and on, and on:

"Almost anything you can take from your garbage can probably induce neural tissue. I am concerned that we are finding this to be the case in mesoderm induction as well" (Nieuwkoop in Appendix V: Gordon, Björklund & Nieuwkoop, 1994).

Indeed, activin itself has now been discredited as a mesoderm inducer (cf. Moriya & Asashima, 1992), with language sounding like confusion on the collapse of a paradigm, with nothing better to replace it. There is a consternation with (at least the mouse) embryo for its not doing what we wanted it to:

"Over the past five years, vertebrate embryologists have come to believe that activin, and the activin-binding protein follistatin, play a big part in the induction of mesoderm and neural tissue.... Activin and follistatin are indeed expressed during early *Xenopus* development: activin protein is present in the egg... and follistatin is expressed in Spemann's organizer, the source of natural neural-inducing signals (Hemmati Brivanlou, Kelly & Melton, 1994).

"Three papers (Matzuk, Kumar & Bradley, 1995; Matzuk et al., 1995a,b) now address the roles of activin in mouse development. They show that although mouse embryos lacking activin, follistatin or a type II activin receptor are defective in several aspects of development, mesoderm formation and neural differentiation are not among them....

"And what of follistatin? Do mice mutant for this gene lack a nervous system, as might be predicted from the work in *Xenopus*? The answer (Matzuk et al., 1995a) is 'no'. Follistatin-deficient mice are small, their muscle mass is reduced, their skin is shiny, and there are a number of other defects, but their nervous systems remain defiantly normal....

"So this work (Matzuk, Kumar & Bradley, 1995; Matzuk et al., 1995a,b) counts against the view that activin and follistatin are involved in mesoderm formation and neural induction in the mouse, although activin addicts may still argue that decidual activin A does the business, and those with a more general allegiance to the TGF-β family may point out that genes such as *nodal* are definitely required if mesoderm is to form properly (Zhou et al., 1993b; Conlon et al., 1994). But what else do the new findings tell us?

"First, they warn us that interpretation of experiments in *Xenopus* involving overexpression of gene products... is fraught with difficulty.... But we should also be reminded that for all we learn from gene knockout experiments, what is lacking in the mouse is embryological data relating to mesoderm formation. In particular, we don't even know whether mesoderm in the mouse embryo really does arise through induction, still less when induction occurs and where the signal comes from. These embryological questions must be addressed at the same time as the analysis of mice mutant for putative 'mesoderm-inducing factors'" (Smith, 1995b).

Perhaps it is dangerous to draw biochemical conclusions before the morphology and morphogenesis one is trying to explain has been observed (but cf. Tam & Beddington, 1987, 1992). On the biochemical side, our ability to reason from gene knockouts to morphogenesis has itself been criticized (Routtenberg, 1995), curiously, in the same issue of *Nature* announcing the demise of activin as a mesoderm inducer (Smith, 1995b).

I will also avoid the subject of organ induction for the simple reason that it occurs internally and we only have observational techniques developed now for observing waves in tissues on the surface of embryos (Gordon & Björklund, 1996). For a general introduction, see Nieuwkoop, Johnen & Albers (1985). For induction 'of' particular organs (which usually consist of many tissues), see:

heart (Jacobson, 1960; Jacobson & Duncan, 1968; Lemanski et al., 1985; Markwald et al., 1990; Sater & Jacobson, 1990a; Smith & Armstrong, 1990; Melnik et al., 1995; Driever & Fishman, 1996; Travis, 1996b);

kidney (Saxén, 1975; Saxén & Lehtonen, 1978; Weiss & Nir, 1979; Bard et al., 1996).

If Nieuwkoop's intuition (in Nieuwkoop, Björklund & Gordon, 1996) is correct, differentiation waves in a viewable form only occur in tissues whose cells retain at least part of the original cortex of the egg. More observational work is needed to settle this interesting question (cf. Section 9.20). The following might be taken to represent the state of the art of our understanding of organ induction, and a need to consider it from the point of view of differentiation waves:

"As a result of the extensive gene-expression data available for the kidney, we now understand nephrogenesis far less well than we thought that we did five years ago.... We... know very little about the inductive signalling mechanisms. The data to hand clearly show that each aspect of the process involves cell-surface receptors, second messenger pathways and the regulation of gene transcription, but we have little clue how to link most of these events with differentiation and morphogenesis" (Bard et al., 1996).

2.09 Regional Induction

Weiss (1935) addressed the developing concept of regional induction, which I will suggest represents the action of multiple differentiation waves:

"In order to test whether the patterns of the induced neural systems were provided by the grafts or by the affected host tissues, fragments from various regions of the 'organizer' were brought to act on host tissues of different regions.... Accordingly, both anterior and posterior fragments of the 'organizer' were confronted with either head or trunk ectoderm. The homologous combinations (head mesoderm in head region, and trunk mesoderm in trunk region) yielded corresponding fragments of secondary embryos, that is, anterior or posterior pieces, respectively.... However, the same trunk ectoderm..., when confronted with a head 'organizer,'... [gave] typical head organs... (Spemann, 1931a). Similarly, head ectoderm, when confronted with 'trunk organizer' could produce trunk characters (Hall, 1932).... Hence, the influence spreading from a grafted 'organizer' obviously consists of something more specific than mere activation.

"Additional evidence... has been brought forward... (Mangold & Spemann, 1927; Spemann & Geinitz, 1927; Bautzmann, 1928, 1929a; Mangold, 1929a, 1933).

"It is perfectly obvious that in this whole series of experiments the organizing factors contained in the grafts must have operated in a more specific fashion than by merely releasing a preformed sequence of processes dormant in the affected host tissues. Accordingly, the attempt to explain these effects as simple activation effects, as in Balinsky's [1925, 1927a,b, 1933] limb inductions, must be abandoned.

"There is one observation, however, which, recurring with a certain regularity..., seems to complicate the interpretation: under certain circumstances, one observes that the affected host tissues give rise to a formation the quality of which is not at all in accordance with the organizing tendencies of the grafted 'organizer' fragment. Thus, secondary heads were frequently obtained in the head region not only after the transplantation of head 'organizer' or of anterior parts of the notochord, but also after the transplantation of trunk 'organizer' or of posterior notochord (Spemann, 1931a). Therefore, as we see, accessory head structures may appear in the head district, no matter where the graft has come from and what its specific character may have been" (Weiss, 1935).

In the face of all the data on how so many different substances could induce the ectoderm to form the neural plate, all specificity seemed to reside in the induced tissue, not the inducing tissue. Claims that different parts of the mesoderm induced different parts of the neural plate pointed in the opposite direction. Regional induction thus has stood as an obstacle to the physical analysis of induction. Its variability, sometimes giving the 'wrong' result, is usually glossed over, but may be the key to understanding regional induction.

I do not pretend to have all the effects sorted out, but want to be sure that the reader becomes aware of their full panoply: as we see from the above quote from Weiss (1935), all the attempted tissue fragment combinations gave all possible results. In this book I suggest that regional induction actually reflects multiple, nested differentiation waves (Propositions 86-92). If this is so, the numerous experiments on regional induction or 'regionalization' should be repeated with continuous time lapse observation, to watch what these waves are doing during and after the embryo surgery involved (see Section 9.22). (For work on 'regional induction' see: Holtfreter, 1933f; Lehmann, 1938a; Dalcq, 1946a; Nieuwkoop, 1947a; Mangold & von Woellwarth, 1950; Waddington & Yao, 1950; Yamada, 1950a, 1959; von Woellwarth, 1951, 1952; Waddington, 1952c; Eyal-Giladi, 1954; Holtfreter & Hamburger, 1955; Sala, 1955, 1956; Saxén & Toivonen, 1962; Hama, 1978; Hara, 1978; Kawakami & Sasaki, 1978; Nakamura, Hayashi & Asashima, 1978; Takaya, 1978; Tiedemann, 1978; Toivonen, 1978; Nieuwkoop, Johnen & Albers, 1985; Born et al., 1989; Tam, 1989; Sharpe, 1990; Slack, 1991a; Storey et al., 1992; Krumlauf et al., 1993; von Dassow, Schmidt & Kimelman, 1993; Keynes & Krumlauf, 1994).

2.10 The Cell State Splitter

"Among the multitudes of biochemists and molecular biologists, those who study biomechanics are considered to be the lunatic fringe" (Pennycuick, 1992).

"...A new scientific discipline, perhaps best defined as molecular cell engineering, is beginning to emerge and to fill the current gap between genetic engineering and tissue engineering" (Ingber, 1993a).

"The... *Journal of Cellular Engineering*... arises out of a long history of often unplanned and sporadic linking between the worlds of the physical scientists and engineers on the one hand, and of the biologists, physicians and surgeons on the other" (Rolfe, 1995).

A remarkable subcellular structure was found in newt neural plate cells by Burnside (1971), (cf. Burnside, 1973a), consisting of a microfilament ring (Baker & Schroeder, 1967; Perry, 1975) surrounding a coplanar mat of microtubules (cf.: "the restriction of microtubules to the area within the boundaries of the microfilament rings": Zamansky, Nguyen & Chou, 1991). We have tracked its origin back to early gastrulation (Martin & Gordon, 1991, 1997a,b) and called this new 'morphogenetic apparatus' the *cell state splitter* (Appendix I: Gordon & Brodland, 1987; Gordon & Brodland, 1989). The cell state splitter's structure (Figure 5 in Appendix I: Gordon & Brodland, 1987) has been refined by more extensive electron microscopy (Martin & Gordon, 1991, 1997a,b; Figure 1 in Appendix V: Gordon, Björklund & Nieuwkoop, 1994).

The initial theory of cell state splitter functioning has been improved with our discovery of the ectoderm contraction wave (Appendix II: Brodland et al., 1994), whose existence, though not its detailed properties, we predicted (Appendix I: Gordon & Brodland, 1987):

"In order to take certain further steps in embryology we are *compelled* to invent hypotheses concerning submicroscopical cell structure which will be explanatory of the behaviour of cells during development.... It is in the highest degree likely that the required hypotheses, if and when they are discovered, will seem extremely unorthodox" (Woodger, 1948).

The history of one component of the cell state splitter goes back a while. The microfilament ring was predicted as "contractile processes near the periphery" by Moore (1941), and Burt (1943) noted that...

"Brown, Hamburger & Schmitt (1941)... and Schmitt (1941) independently have suggested that molecular interactions and desolvations in the cell surface may exert the forces necessary to cause cell elongation [during neurulation]. [Paul A.] Weiss (unpublished) has suggested further that the concentration of pigment granules which occurs in the normal folding [neural] plate indicates a contraction of the cell cortex at the free surface" (Burt, 1943).

Actually, the hypothesis of Brown, Hamburger & Schmitt (1941) is that "...evocator action resulting in elevation and folding of the neural plate may operate through an increase in the 'attractive' forces between molecules in the adjoining cell surfaces of prospective neural tissue cells so that the area of contact is actively increased", anticipating N-CAM, not microfilament rings. The cell state splitter may be Holtfreter's hypothesized "preexisting, pent-up mechanism" (Holtfreter, 1991). The history of the notion of apical contractility can be traced back even further:

"If one examines a cross-section of the blastopore lip at its earliest stage... one sees that the cells at the base of the furrow have a distinctive bottle-shaped form of their own. These were first described by Ruffini (1907), who attributed to them secretory functions. Several authors (e.g. Goodale, 1911; Vogt, 1925, 1929a; Daniel & Yarwood, 1939) have figured these piriform cells and discussed the possibility that they act in a contractile manner" (Needham, 1950; cf. Dalcq, 1938).

The observation of the microtubule mat at the apical ends of cells appears to be original with Burnside (1971). Nardi & Reynolds (1986) have also observed 'circumferential' microtubules (parallel to the apical surface) associated with apical microfilaments in explants of moth wing epithelium, which can fold either way. Folding direction may depend in part on intercellular adhesion since the "basal cell surfaces are frequently separated by spaces as large as 20-30 μm (Nardi, 1981a). As the monolayer folds, these large intercellular spaces are obliterated" (Nardi & Reynolds, 1986). Cell state splitters relax after contraction in axolotl embryos (Appendix II: Brodland et al., 1994), so there is no reason to presume that they determine the direction of tissue buckling during neural tube closure all by themselves.

In this book, I presume that cell state splitters are constructed in cells for every intermediate step of differentiation in the formation of every multicellular organism. The desire for universality is one of the driving motivations for intellectual work in science and mathematics. It is thus a perverse pleasure to have received a comeuppance from my senior colleague's insistence that we look for differentiation waves in the anuran *Xenopus laevis:*

"The surface contraction waves in the axolotl seem to be borne by cells forming part of the original surface coat of the egg and embryo (Holtfreter, 1943b). The double-layered *Xenopus laevis* embryo (Nieuwkoop & Florschütz, 1950; Keller, 1975, 1976) may, according to the first author, give us some further insight into the apparent correlation, since in *Xenopus* mesoderm as well as neural induction seems to occur exclusively in the inner, sensorial layer of the ectodermal moiety, not visibly affecting the epithelial layer before the initiation of neural plate formation. Thus in *Xenopus* we have the opportunity to observe separation of ectodermal from inductive properties. *Xenopus* is an extremal amphibian test of our model for embryogenesis via differentiation waves, because it is least like urodeles in its development (Nieuwkoop & Florschütz, 1950; Nieuwkoop, 1996). It is the most commonly used experimental amphibian and has been widely studied for biochemical analysis and gene expression. Therefore it seemed critical to attempt to generalize the wave phenomenon to *Xenopus*.

"It thus came as quite a surprise to us when we could not find homologous waves on *Xenopus* embryos" (Nieuwkoop, Björklund & Gordon, 1996).

The neat cell state splitter in axolotl ectoderm is therefore not universal unless it is beneath the surface in *Xenopus* (Section 9.20). In searching for a level at which anuran and urodele embryos share a common mechanism (cf. the axolotl/*Xenopus* limb bud grafting experiments of Sessions, Gardiner & Bryant, 1989), we are forced to look to the cortex. This exploration will take us as far as the ciliates in this book. As there is no name yet for this hypothesized common mechanism, I will continue to use 'cell state splitter', regarding the cell state splitter of the urodele ectoderm as but an example of a larger class of cortical morphogenetic mechanisms.

It has recently been recognized that the three hitherto disparate fields (Bonner, 1982a) of development, genetics, and evolution are in need of unification:

"For some years I have been concerned about the problem of the relation of genetics, evolution, and development. It has always seemed that their conventional relation is to some extent static and contrived, while in fact they must be closely integrated and part of one scheme" (Bonner, 1965a).

"Today, despite its recognized significance, the role of morphogenetic interactions in controlling phyletic transformations remains to be successfully integrated into the body of evolutionary theory.... Numerous pleas in favor of the crucial significance of development in evolutionary biology have been made (cf. Devillers, 1965; Maderson, 1975; Horder, 1981; Rachootin & Thomson, 1981...)" (Oster & Alberch, 1982).

"At the very least it is clear... that the particular embryological theory that is adopted will have a major bearing on how evolutionary events and the basis of their inheritance are conceptualized" (Horder, 1983).

Indeed, Levinton (1988) suggests that unification of development, genetics and evolution may be crucial for further progress, but warns that...

"The evolution of development is thought by many to bias evolutionary direction (Davenport, 1979; Gould, 1982a; Alberch, 1981, 1982a; Maderson et al., 1982).... While these arguments are useful and have been underemphasized in the past, it is worrisome that they will be applied uncritically to construct a new model of evolution that tends to think of change in the developmental program as a necessary or at least common progenitor to major morphological change in evolution.... This reasoning is premature and may lead us too far in the opposite direction" (Levinton, 1988).

I will indeed go as far as I can in the 'opposite direction' to show how development may be made central to our understanding of both genetics and evolution.

2.11 Meet the Axolotl

The axolotl (*Ambystoma mexicanum*) is an ugly member of the newt and salamander (urodele) order of amphibians with a face that only its cannibalistic mother could love: gills come out the side of its head (Figure 4). I (and many others: Malacinski, 1978a; Armstrong & Malacinski, 1989; Björklund, 1993; Malacinski & Duhon, 1996; Björklund & Duhon, 1998) have focussed on the axolotl as a model system for investigating development because of its year round (relative) ease of spawning in labs, and the long history of use of urodeles in experimental embryology. At the gastrula stages 10 to 12 1/2 (Figure 4; Table 1, when the 2 mm diameter axolotl embryo is a hollow ball of cells, the outer layer of cells, the ectoderm, which is nearly a full spherical shell (in the engineering sense: Timoshenko & Woinowsky-Krieger, 1959; Mushtari, 1961; Gibson, 1965; Flügge, 1973; Dym, 1974; Mollmann, 1981; Calladine, 1983; Hinron & Owen, 1984; Niordson, 1985; Brodland, 1986a, 1994; Gould, 1988a), splits into two hemispheric shells: the neuroepithelium and the epidermis. The neuroepithelium goes on to flatten into a neural plate (Gordon & Jacobson, 1978; Jacobson, 1994), which rolls up into a neural tube (Gordon, 1985c; Schoenwolf & Smith, 1990a; Van Allen et al., 1993) and then forms the brain and spinal cord. The epidermis mostly forms skin.

Table 1: Timing of early stages of the axolotl embryo, *Ambystoma mexicanum,* at 20°C (excerpted and simplified from: Bordzilovskaya et al., 1989, with permission). See Figure 4 for related figures.

Stage	Time (hrs)	Description
2	0.65	2 cells
3	2.40	4 cells
4	4.12	8 cells
5	5.22	16 cells
6	6.45	32 cells
7	8.26	64 cells
8	16.06	early blastula
9	21.28	late blastula
9 1/2	24.32	midblastula transition
10	26.00	early gastrula: dorsal lip of blastopore appears
10 1/2	32.10	early gastrula: blastopore 90° arc
10 3/4	37.00	middle gastrula: blastopore 180° arc
11	38.30	middle gastrula: blastopore 270° arc
11 1/2	41	late gastrula: blastopore 360° arc (full circle)
12	47.30	late gastrula: blastopore shrinking
12 1/2	50	late gastrula: blastopore shrunk to small oval
13-	52.15	blastopore a small slit along axis
13	55.65	early neurula: neural plate has midline groove
14	58.15	early neurula: neural folds raised
15	59.50	early neurula: broad keyhole shape
16	63.0	middle neurula: neural plate narrows and elongates
17	64.30	late neurula: 2 pairs of somites
18	66.65	late neurula: spinal neural folds nearly touching; forebrain, midbrain, hindbrain bulges visible
19	69.00	late neurula: neural folds touching throughout; optic vesicles visible; 3 pairs of somites
20	70.30	late neurula: spinal neural folds fused; 3 pairs of somites; gills start to appear
21	72.0	late neurula: neural folds fused throughout

Many people in embryology work on frogs and toads (the anuran order of amphibians). To me the major advantage of the axolotl is that the ectoderm is a monolayer sheet of cells, so it is directly observable and mathematically and computationally tractable. (However, cf. Burian, 1993b; note the claim of some syncytial axolotl neural plate cells by Baker & Graves, 1939, their Figure 1, and of multiple cell layers by Baker, 1927.) In frogs (and some other groups, such as teleost fishes, which includes the now popular zebrafish, *Danio rerio,* previously named *Brachydanio rerio*), the ectoderm is covered by an extra layer of cells, the superficial epithelium (Dettlaff, 1993). But birds and mammals, like urodeles, also have an ectoderm that is a monolayer of cells, so axolotls have a certain universal appeal for us 'higher' vertebrates. Nieuwkoop (1989, 1996) has expressed regret at the success of his publication (Nieuwkoop & Faber, 1956, 1975, 1994), staging the anuran *Xenopus laevis* (the South African clawed toad; cf. Dettlaff & Rudneva, 1991), in directing a whole generation of vertebrate and molecular embryologists towards that animal rather than the urodeles, and we are now faced with the incongruity that most classical embryology work was based on urodeles, while most vertebrate molecular embryology is based on *Xenopus:*

"First of all one has to know *when* and *where* inside the embryo certain interactions occur during its epigenetic development, before one can successfully start the biochemical and genetic characterization of the factors and genes involved.... A proper knowledge of... anatomy and histology [is] required. The latter is evidently becoming a more and more uncertain factor, as can be judged from the modern literature, which has led in several cases to incorrect and even unjustified conclusions.

"The introduction of the South African clawed toad, *Xenopus laevis,* in many medical and biological institutes during the fifties, initially for use in human pregnancy tests, led, however, to the nearly complete dominance of that species in developmental biological research....

"One of the great advantages of Urodele embryos for experimental analysis is the single-layered nature of early developmental stages, due to which the gastrulation process can easily be followed from the outside.... The Urodeles have however the *great* advantage that there is a very thorough documentation of their morphogenetic development, executed during the first half of this century, which forms a very sound and indispensable base for biochemical and genetic analysis.... However, there are some serious limitations connected with the use of *Xenopus* eggs. The main restriction relates to the double-layered nature of the totipotent animal moiety of the *Xenopus* blastula/gastrula and neurula stages (Nieuwkoop & Florschütz, 1950).... Gastrulation cannot therefore be followed properly from the outside" (Nieuwkoop, 1996).

Even the original justification for *Xenopus* had three caveats:

"The egg of *Xenopus laevis,* notwithstanding its [1] very rapid development, is quite suitable for experimental work.... The [2] rather aberrant development of this [3] systematically somewhat isolated species suggests... interesting possibilities for descriptive and comparative embryological studies" (Nieuwkoop & Faber, 1994, remarks presumably dating to Nieuwkoop & Faber, 1956).

The low fertility of triploid and especially tetraploid *Xenopus* (reviewed in Deuchar, 1975b), and lack of reports of higher ploidies, compared to up to heptaploidy in the axolotl (Fankhauser, 1955), is another advantage of the latter for studies where large cells are desirable. Axolotl adults also regenerate amputated limbs (Polezhaev, 1972; Ede, Hinchliffe & Balls, 1977; Hinchliffe & Johnson, 1980; Wallace, 1981; Sicard, 1985; Dinsmore, 1991; Hinchliffe, Hurle & Summerbell, 1991; Preachuk, 1994; Travis, 1997d) and severed spinal cords (Piatt, 1955; Butler & Ward, 1965; Wright et al., 1989; Caubit et al., 1993; Schwab, 1993; Chernoff, 1996), whereas in *Xenopus,* "the regenerative power... is more limited" (Deuchar, 1975b; cf. Hasan et al., 1991). On the other hand, the two layers of the *Xenopus* gastrula have permitted separation of inductive and morphogenetic functions of the epithelial and sensorial cell layers (Nieuwkoop & Koster, 1995; Nieuwkoop, Björklund & Gordon, 1996). It is interesting to note what happens to the outer layer (superficial epithelium) of anuran embryos:

"In the epidermis of frog embryos one distinguishes two layers. The thicker surface epithelium consists of cuboid cells that often assume a glandular character, as in the sucking glands and the frontal glands.... The surface layer is ephemerous, being completely shed at the time of metamorphosis. The deeper *basal layer* is, at the embryonic stage, a thin, squamous epithelium, usually only one cell deep.... The basal layer alone is homologous to the epidermis of amniote embryos" (Witschi, 1952).

My emphasis has been on phenomena that we can see from outside an intact embryo without intervention. In retrospect, this 'hands off' approach, contrary to the interventionist style of classical 'experimental embryology' (Roux, 1894a; Needham, 1934a, 1959; Willier & Oppenheimer, 1964; Oppenheimer, 1967), has been important in the analysis of differentiation waves. Extra differentiation waves are probably triggered by surgery in many circumstances, complicating and possibly confounding the very phenomena under investigation. This is unexplored territory warranting careful study: every major classical experiment involving surgery (Hamburger, 1960; Lehman, 1987) should be repeated under time-lapse.

The axolotl embryo is nearly opaque, which means that we have been confined to observing waves that propagate on the outside of the embryo. It is primarily for this reason that I have little to say for now about the very interesting phenomena of formation of:

blood (Knöchel, 1994; Milner et al., 1994; Walmsley et al., 1994);

bone (Hall, 1989-1992; Dixon, Sarnat & Hoyte, 1991);

heart (DeHaan, 1965; Okamoto, 1980; Butler et al., 1987; Burggren, 1988; Icardo, 1988a,b; Smith & Armstrong, 1990; Yost, 1990; Sadoshima et al., 1992a,b; Fishman & Stainier, 1994; Simpson et al., 1994a; Nascone & Mercola, 1995; Neff, Dent & Armstrong, 1996) and its regeneration (Becker & Selden, 1985), though I will cover the related problem of the origin of left-right bilateral asymmetry in Section 8.02;

limbs and their growth and regeneration (Polezhaev, 1972; Ede, Hinchliffe & Balls, 1977; Hinchliffe & Johnson, 1980; Wallace, 1981; Sicard, 1985; Dinsmore, 1991; Hinchliffe, Hurle & Summerbell, 1991; Preachuk, 1994; Akita, 1996; Kim et al., 1996; Mescher, 1996; Stocum, 1996; Zilakos, Zafiratos & Parchment, 1996; Brockes, 1997; Section 9.05);

muscle (Kedes & Stockdale, 1989);

neural crest (Hörstadius, 1950; Nieuwkoop, Oikawa & Boddingius, 1958; Le Douarin, 1982; Maderson, 1987; Hall, 1988; Ide & Akira, 1988; Le Douarin et al., 1992; Epperlein & Löfberg, 1996), which may be related to neural plate placodes (Smith, Graveson & Hall, 1994), the latter which I will consider in detail (Section 9.10);

somites (Ede, Hinchliffe & Balls, 1977; Bellairs, Ede & Lash, 1986; Primmett, Stern & Keynes, 1988; Stern, 1990a; Pourquié et al., 1993, 1995; Christ & Ordahl, 1995; Brand-Saberi et al., 1996; Richardson et al., 1998; Section 9.21);

spinal cord growth and regeneration (Piatt, 1955; Butler & Ward, 1965; Wright et al., 1989; Schwab, 1993; Chernoff, 1996).

All of these are major branches of embryological pursuit. Transparent embryos, such as zebrafish (Westerfield, 1995), may prove most useful for watching these internal events, especially if embryo rotation (King et al., 1982) were combined with computed tomography (Gordon & Herman, 1974; Gordon, Herman & Johnson, 1975; Dietzel et al., 1995; Sätzler & Eils, 1997). Alternatively, new approaches, such as:

*magnetic resonance imaging (MRI microscopy:*Aguayo et al., 1986; Cho et al., 1990a,b; Cho, Yi & Friedenberg, 1992; Johnson et al., 1993; Zhou et al., 1993a; Smith et al., 1994; Peck, Magin & Lauterbur, 1995);

x-ray microscopy (Parsons, 1980; Schmahl & Rudolph, 1984; Cheng & Jan, 1987; Schmahl & Rudolph, 1984; Sayre et al., 1988; Howells, Kirz & Sayre, 1991) with computed tomography algorithms (many developed initially for electron microscopy: Bender, Bellman & Gordon, 1970; Frank, 1992a; Shafrir, 1994);

ultrasound microscopy (Turnbull et al., 1995; cf. Kulbisky et al., 1999);

impedance microscopy (which Alvin Wexler and I hope will proceed from electrical impedance tomography: Wexler, Fry & Neuman, 1985; Wexler, 1988; Dijkstra et al., 1993; Mu, 1994);

multiphoton microscopy (Denk & Svoboda, 1997; Robinson, 1997a; Xu et al., 1996);

may allow us to watch morphogenesis inside living, visually opaque embryos. Besides, I have come to think that embryogenesis might be best approached by looking at the way the embryo does it: we need to understand the earliest stages before we can understand how the embryo builds upon them. Fortunately, at least in urodeles, most of these early events occur on the outside of the embryo.

The yolk is inside each cell of the axolotl embryo in the form of small yolk platelets. This has two important consequences. First, since the embryos contain their own food source, they only need a medium consisting of a dilute solution of salts in water (Asashima, Malacinski & Smith, 1989) or even just doubly distilled water (Barth & Barth, 1969). Second, there is no dry mass increase of the embryo until the larva has a mouth and can eat. A major convenience of the axolotl embryo is that the amount of pigment varies from cell to cell, making it easy to keep track of local coordinates (Jacobson & Gordon, 1976a; Appendix VI: Björklund & Gordon, 1994). Furthermore, the cells are large (30 µm wide and 50 µm tall at gastrula stages, Table 1, compared to 5 µm in chick embryos), so they can be seen at relatively low magnification.

The disadvantages of the axolotl are that the genome is huge, about 2.5×10^{10} base pairs (Alberts et al., 1989), an order of magnitude larger than the human genome (3×10^9 base pairs: Alberts et al., 1989), and the life cycle is long (1 year: Armstrong, Duhon & Malacinski, 1989), so that relatively few generations may be studied in the lifetime of an investigator. Thus there has been only a small amount of genetic work on axolotls (Armstrong & Muneoka, 1989; Cuny & Malacinski, 1989; Malacinski, 1989; Neff, 1989) compared to the mascots of molecular biology, the bacterium *Escherichia coli* (DNA length 4×10^6 base pairs: Singer & Berg, 1991; life cycle as short as 21 minutes: Stanier et al., 1986) or the fruit fly *Drosophila melanogaster* (about 10^8 base pairs: Alberts et al., 1989; life cycle 9 days: Doane, 1967). On the other hand...

"... the somite stages of vertebrates... may correspond to segmentation in *Drosophila* (Lawrence, 1992). *Drosophila* accomplishes the early stages as a syncytium, and thus may not provide a model for early, presomite vertebrate embryogenesis" (Appendix V: Gordon, Björklund & Nieuwkoop, 1994).

While these standard laboratory organisms have taught us much, they aren't vertebrates: it is self-knowledge that we seek.

2.12 A History of Sexism in Science Whodunit: Hilde Mangold or Hans Spemann?

Ms. Hilde Mangold...

"was an unusually gifted, vivacious, and charming young woman. Her considerable scientific talents would undoubtedly have borne fruit, had her life not been cut short by a tragic accident. She died of severe burns when a gasoline heater in her kitchen exploded. This occurred in September 1924, about the time when the organizer paper appeared in print [Spemann & Mangold, 1924a,b; cf. the posthumous Mangold & Mangold, 1929]. She was then twenty-six years old and the mother of an infant son" (Hamburger, 1988).

Spemann went on to receive the Nobel Prize in 1935 for his work with Ms. Mangold (Spemann & Mangold, 1924a,b; cf. Mangold, 1942, 1953; Hamburger, 1985; Allen, 1993), the only Nobel Prize awarded for work in embryology until 1995 (Karolinska Institute, 1995). Whether Hilde Mangold was fairly treated would make an excellent case study of women in science (Hamburger & Edsall, 1984; Hamburger, 1988; Waelsch, 1992), such as those considered by Sayre (1975), Keller (1985), Ogilvie (1986), Opfell (1986), Ainley (1990), Gornick (1990) and Bagchi (1994). (Her photograph is featured prominently in Browder, Erickson & Jeffery, 1991. More photos are in: Fässler & Sander, 1996.)

Waelsch (1992) gives a woman scientist's perspective:

"I would like to say a few words about Spemann's political attitudes without analyzing them in detail here. He was a strong German nationalist, full of mistrust towards other nationalities and sharing prejudices of his fellow nationals. I already mentioned his prejudice against women expressed also in his dealings with Hilde [Mangold nee] Proescholdt, the discoverer of the 'organizer,' who is reported not to have appreciated it when Spemann added his name to the publication of her thesis while other male students were permitted to publish their work alone. Many years later, the story of the 'dead' organizer provides another example: it was a female graduate student, Else Wehmeier, who first observed embryonic induction with a Bouin-fixed piece of upper blastopore. I was around at that time. Her name did not even appear on the first publication reporting this exciting result!" (Waelsch, 1992).

The latter story is confirmed by Spemann (1938):

"...Else Wehmeier, in my laboratory, performed the important experiment of subjecting the inductor to chemical treatment. Occupied with experiments concerning the behavior of desiccated and frozen germ parts in pockets of epidermis, she had the idea of testing the inductive faculty of fragments which had been exposed to the action of alcohol for some time (Spemann, 1932)" (Spemann, 1938).

Hamburger (1988), in the following quote, innocently shows how great consideration was given to Holtfreter's (male) friend Hermann Bautzmann by Holtfreter, O. Mangold, and Spemann, while they all ignored this apparently equal "small positive result obtained in my laboratory [by] Miss E. Wehmeier, who... had the idea" (Spemann in Bautzmann et al., 1932, translated in Hamburger, 1988). Wehmeier was never first, except when sole author (Spemann, Fischer & Wehmeier, 1933; Fischer & Wehmeier, 1933a,b; Wehmeier, 1934; Fischer et al., 1935) and later "became incapacitated by illness" (Hamburger, 1988):

"With the consent of my friend Holtfreter I quote... from his English version [of Holtfreter, 1982]:

'I had not behaved like a faithful disciple of Spemann; I had failed to tell the master what I had been doing lately, and done it without his explicit consent. Now [Otto] Mangold, who so far had been a benevolent observer of my doings, urged me to communicate these findings to Spemann. This I did with some trepidation. Spemann applauded but did not actually congratulate me. He pointed out in his letter that some three years ago, Hermann Bautzmann had been given the green light to work on heat-killed inductors. Although Bautzmann had as yet not published anything about his work, I should consider his right of priority and should not publish my findings until I had discussed them with Bautzmann. So I traveled to Kiel to see my friend Bautzmann.... His results were not very impressive....

'Then Spemann, whom I had informed about my agreement with Bautzmann, expressed his wish to join us and to report on some pertinent findings of a young student of his, Else Wehmeier.

'Thus it came to pass that the four-men communication of the inductive mechanism in embryonic development saw the light of publication (Bautzmann et al., 1932).'" (Hamburger, 1988).

"Spemann's comment on the same publication is contained in a letter to me from Freiburg, dated December 25, 1932...:

"In the December issue of *Naturwissenschaften* you will find an article by four authors, Bautzmann, [Otto] Mangold, Holtfreter, and myself. Another dream of mine has been fulfilled, and from a threatening collision [came] an amiable cooperation....'

"Apparently, Spemann's reference to a potential 'collision' in the matter of priority relates to difficulties that might have arisen if Holtfreter had published without Bautzmann's consent. This was avoided. It is clear, however, that the joint publication was orchestrated by Spemann and that his intention was to give his friends H. Bautzmann and O. Mangold and his student Wehmeier a chance to share the priority of this important discovery with Holtfreter" (Hamburger, 1988).

It was a strange way to acknowledge Ms. Wehmeier's share of the priority by excluding her from joint authorship in this "four-men communication". "[Salome] Gluecksohn-Schoenheimer recalled [Spemann] as being prejudiced against the women working in his laboratory" (Gilbert, 1991b).

Hamburger (1988) also gives a, shall we say, patronizing view of Spemann's relationship with Hilde Mangold:

"What was the nature of the partnership of the two authors of the organizer paper [Spemann & Mangold, 1924a,b]? Quite simply, the experiment was the basis for the doctoral dissertation of Hilde Mangold, whose maiden name was Proescholdt; it was one of the very few dissertations in the history of biology that was directly connected with a Nobel Prize. The conception and design of the experiment were Spemann's, based on a combination of ideas and experimental techniques that date back to his classic publication of 1918 (Spemann, 1918). Mangold's contribution was the execution of this difficult experiment, during the spring of 1921 and 1922....

"Hilde Proescholdt, who in the meantime had become Mrs. Mangold, was not happy that Spemann had added his name to her thesis publication, while Holtfreter and I and all the rest of us saw ourselves proudly in print as sole authors. Moreover, Spemann had insisted on having his name precede hers! But Spemann was perfectly right in claiming precedence, while she apparently did not fully realize the significance of her results. It was not granted to her to live to see the great impact her experiment had on the course of experimental embryology" (Hamburger, 1988).

In the case of another Spemann student, Hamburger (1988) relates:

"The reader may wonder why Spemann put his name first, both in [Spemann & Schotté (1932)] and in the publication of the organizer experiment [Spemann & Mangold, 1924a,b]. To him, the original idea deserved primacy over the execution of the experiment. This would apply to these two and to several other instances. On the other hand, in the case of O. Mangold & Spemann (1927), the authors had the idea of the experiment independently" (Hamburger, 1988).

This does not seem consistent treatment of Spemann's students, given that...

"On several occasions, Spemann turned over sectioned embryos from experiments he had done himself to graduate students who complemented the material with experiments of their own, or merely analyzed Spemann's material, for their Ph.D. theses. As in the case of Krämer (1934), Spemann did not coauthor such publications" (Hamburger, 1988).

It would seem that Hilde Mangold did much more than this, and got much less credit and independence in publication, having carried out "the execution of this difficult experiment" (Hamburger, 1988). On the other hand, here is Spemann's (1938) account of the origin of the famous experiment:

"The astounding fact that a secondary embryo may be brought about by transplantation was already observed during my first exchange experiments on the young gastrula (Spemann, 1918), but at that time it was not correctly interpreted.... The notion of a determination, the action of which advances from cell to cell, had been suggested to me by facts I had met with many years ago.... Even in this, however, as I am now aware, I had no priority. W. Roux (1888a,b)... assumes that 'differentiation proceeds by an assimilating and differentiating influence exerted by cells already differentiated on adjacent cells which are yet more indifferent.' ...it might have been possible at the time to derive from my own observations... the conclusion that those implants, completely or in the greater part, were taken from the invaginating region of the gastrula. To this fact Professor [Hans] Petersen [cf. Petersen, 1922; Witkowski, 1987a] was good enough, in a private letter, to draw my attention. This suggested the idea that possibly only the mesodermal parts of the secondary embryo were formed out of the material of the implant, while the neural tube had been formed by induction out of the ectoderm of the host. As soon as it became possible to render the implant permanently recognizable, that is, as soon as the heteroplastic exchange between the differently pigmented gastrulae of *Triton taeniatus* and *cristatus* had been effected, this question could be settled

(Spemann, 1921). This experiment was first made by Hilde Mangold (Spemann & Mangold, 1924a)" (Spemann, 1938).

The same experiment was done by Lewis (1907b), as discussed by Witkowski (1987a), but he "had transplanted the dorsal lip of the blastopore... to much older larvae..., which by that time had become too old to retain any competence for neural induction" (Needham, 1950). Fässler & Sander (1996) add:

"Lewis (1907b) had no incentive for suspecting induction of the secondary structure he observed ([possibly]... no induction was involved), because at that time everyone - Morgan and Spemann included - held the opinion that in normal development the neural plate, somites and notochord all originate by self-differentiation from the lip material and its surroundings (see Sander, 1991, 1996)....

"Ethel Browne (1909)... is said to have - on a single occasion - claimed credit for having discovered the organizer (Lenhoff, 1991)... [but] she never spoke of organization or the like" (Fässler & Sander, 1996).

Fässler & Sander (1996), after an account of how she aggressively pursued an advanced education, add yet one more excuse (without comparison to her male co-students) for Hilde Mangold receiving short shrift:

"According to his diary, in the spring of 1923 Spemann spent at least 50 h (probably much more) discussing with Hilde the interpretation of her results and giving the manuscript its final shape. That Hilde fell somewhat short of these tasks when preparing her thesis can be read between the lines of Spemann's report and in the reminiscences of Hamburger (1984); given Spemann's demands for clarity in thought and writing, this should be no surprise in someone in the field for just 2 years!" (Fässler & Sander, 1996).

An in depth analysis of the story of the organizer, taking into account what seems to be clearly prejudicial treatment of Hilde Mangold, both by commission and silent assent, has yet to be written by an historian of science. Since this was the most important experiment ever done in embryology, a critical analysis of the events surrounding the discovery would be worth having. This could have a salient effect on the continuing struggle of women scientists to achieve equal recognition for equal work.

Similarly, it is difficult to believe that Spemann's Nobel prize went unnoticed by the Nazi regime. Here is another story that appears to have been glossed over, and was undoubtedly available in the popular press of the time. After the event, we are told of "Spemann's most original and most productive student - and my lifelong friend" Johannes Holtfreter...

"He left Nazi Germany voluntarily in 1938 and found refuge in Cambridge, England, where his friends Needham and Waddington provided him with laboratory facilities. When war broke out a year later, he was declared an enemy alien, and in 1940 he was sent to Canada where he spent two years in a labor camp. In 1942 his Canadian and American friends succeeded in obtaining his release, and he worked then as a guest in the Zoological Laboratory of McGill University in Montreal: its chairman, Dr. N. Berrill, became his friend. [Cf. Berrill, 1961.]

"Holtfreter left Germany in 1939, shortly before the outbreak of the war, as a *persona non grata* of the Nazi regime" (Hamburger, 1988).

Here is Holtfreter's perspective:

"Back in München [Munich], back to my activities at the university, I encountered an entirely different world, a very ugly world in which decency, mutual trust, and Christian love were replaced by brutality, intolerance, and fear. Hitlerism was triumphant; it celebrated orgies of maniacal furor. I was full of hatred and disgust against the regime, but I felt helpless. I knew that I was spied on and that sooner or later the Gestapo would get hold of me. I saw the war coming. In 1939, shortly before the war started, I managed, 'by the skin of my teeth,' to escape from Germany" (Holtfreter, 1991).

We are told little of Spemann in regard to the Nazi regime by Hamburger (1988), a Jew (Oppenheimer, 1991), except by association with Otto Mangold and other events:

"Mangold became Spemann's successor twice (in Berlin-Dahlem and in Freiburg), and he was also his biographer (Mangold, 1953). After he had left for Dahlem in 1924, he remained in close contact with his mentor. Spemann was instrumental in the appointment of Mangold as his successor in Freiburg in 1937. At that time, the Nazi regime was firmly in power. Mangold supported the Nazi ideology and soon became active politically at the university. In 1938 he was elected by the faculty to the office of rector of the university, an honorific administrative position which, like all other institutions, had become politicized. He held it

until 1941, apparently to the satisfaction of the party and the faculty, since the term of the rectorship was traditionally only one year. Thereafter, he continued as director of the Zoological Institute and resumed his teaching and research. After the end of the war, his identification with the Nazi regime cost him his job. He was dismissed from the university by the French, who occupied South Germany in the spring of 1945. He spent the rest of his life in a small, privately financed research institute in Heiligenberg, near Lake Constance.

"Fritz... Süffert met a tragic fate. He had moved to Berlin as a member of Goldschmidt's Genetics Department at the Kaiser Wilhelm Institute for Biology, and later became an editor of Naturwissenschaften. During the last weeks of the war he and thousands of other middle-aged men were drafted and, poorly armed, lost their lives in the senseless effort to defend Berlin against the tide of the invading Russian army - one of the last crimes of the Nazi regime....

"Little progress was made in the 1940s, due in large measure to the war and its aftermath.... He [Spemann] died during the war, in September of 1941" (Hamburger, 1988).

The detailed description of the celebration of Spemann's 60th birthday in 1929 (cf. F.W. Spemann, 1943, plus the dedication of volume 117, *Wilhelm Roux' Archiv fur Entwicklungsmechanik der Organisme,* 1929), and the lack of any mention of the pomp and circumstance that must have surrounded the awarding of his Nobel Prize in 1935 (speech only published posthumously in: F.W. Spemann, 1943), offer a stark contrast in Hamburger (1988). While it is useful to have this selective personal account of events (Allen, 1993), a proper history of Hilde Mangold and Hans Spemann, in their social and historical context, deserves to be written, preferably before all mental and unpublished records have vanished (cf. Fässler & Sander, 1996).

Why did Holtfreter leave Germany? Was there political conflict between Otto Mangold and Holtfreter and/or Spemann, especially considering that there were Jews in Spemann's lab? Who financed Otto Mangold after the War? To what extent did mindsets in favor of vitalism or biochemistry influence the shift of the research from Germany to England? Did this shift lead to the hunt for 'the' organizer molecule, rather than a physical basis for induction, in the years leading to World War II? What were the roots of the bias against physical explanations? Why were all the artificial inducer experiments not met head on, with someone facing the obvious implication that there was no inducer molecule? Many intriguing questions have gone unasked, let alone answered.

If I have dwelt on the story to what may seem the point of gossip, it is because I do regard the Spemann & Mangold (1924a,b) result to be as important as the discovery of the structure of DNA, in terms of the vast amount of work it has inspired. The latter story has been expostulated (Watson, 1968; Crick, 1988a) and analyzed (Olby, 1974; Judson, 1979) with no holds barred, including considerations of sexism (Sayre, 1975). The Hans Spemann and Hilde Mangold story seems still to be covered by an anachronistic veil of politeness.

3.00 THEORY OF THE CELL STATE SPLITTER

"...<u>A</u> causal embryologist reaches much the same conclusion as do some of the best students of evolution under its other aspects: some discovery has still to be made to render really intelligible the history of life and of morphogenesis" (Dalcq, 1951a).

"The aim of science is to reduce the disorder and complexity of the natural world to coherence in the form of testable concepts. By this measure, developmental biology is in an unhappy state.... Not only do we lack the necessary unifying concepts... even worse, we are largely ignorant of the criteria for recognizing the kinds of new facts that will permit productive generalization" (Wilkins, 1986).

3.01 Overview

"The existence of phenotypic differences between cells with the same genotype merely indicates that the expressed specificities are not determined entirely by the DNA present in the cell - that other devices, epigenetic systems, regulate the expression of the genetically determined potentialities.... Both simultaneity of expression and mutual exclusion of expression imply that intercommunication and metabolic linkage are important characteristic features of epigenetic systems.... The supplementary systems were thought to occupy the cytoplasm... [but] some epigenetic control systems are located in the nucleus." (Nanney, 1958).

The history of the discovery of the cell state splitter is given in Appendix III. In this chapter I will sketch in some of the details and background, but first a quick, nonhistoric, unreferenced overview of the theory is in order.

The core of the argument is given in this and the following four sections. The rest of this chapter provides details and tutorial background that some readers may wish to skip.

The basic, new idea is that each tissue in an embryo is traversed by a wave, which splits it into two new tissues. How, where, and when that wave is launched, what kind of wave it is (expansion or contraction), and what delimits its trajectory through the tissue, are unsolved problems. Launching has been observed from points, arcs, and full circles (Appendix V: Gordon, Björklund & Nieuwkoop, 1994), which also requires explanation. Empirically, we find that the portion of tissue not traversed by one wave, is traversed by another: a contraction wave through one part, followed by an expansion wave through the other part. Thus no portion is passive: differentiation of a tissue into two new tissues involves active participation of each cell in one or the other of the two kinds of waves.

As simple as it seems, this view of tissue differentiation contrasts with standard views. The differences are spelled out in Figures 5-8. From this basic branching mechanism for differentiation, all else in this book follows.

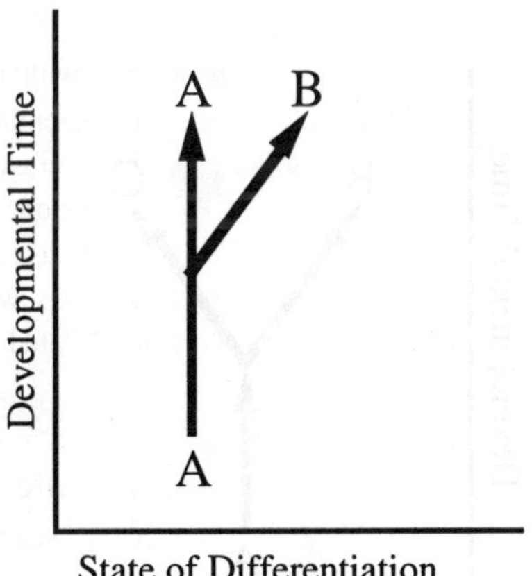

Figure 5. Classical model of differentiation, in which tissue A continues as the same tissue, except for a portion which changes to tissue B when 'induced' by an external agent.

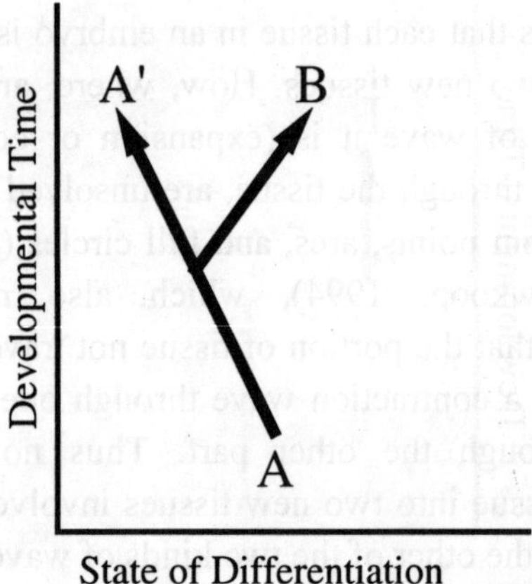

Figure 6. Alternative classical model for differentiation, in which tissue type A is continually changing towards type A', except for a portion which changes to tissue B when 'induced' by an external agent.

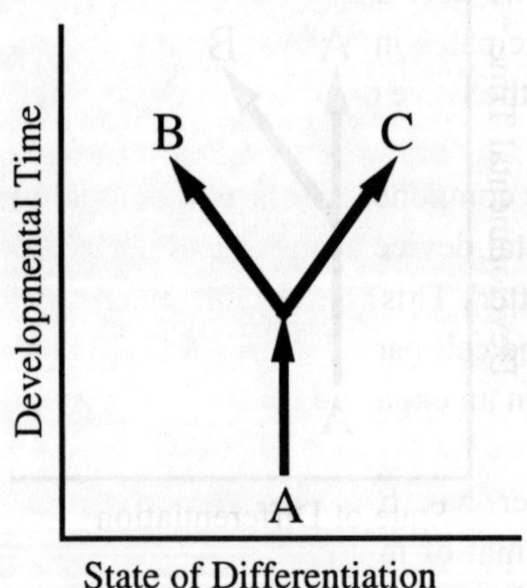

Figure 7. Our new view of differentiation, in which tissue A splits into two new tissues B and C, as contraction and expansion differentiation waves travel through complementary portions of tissue A. Tissue A ceases to exist.

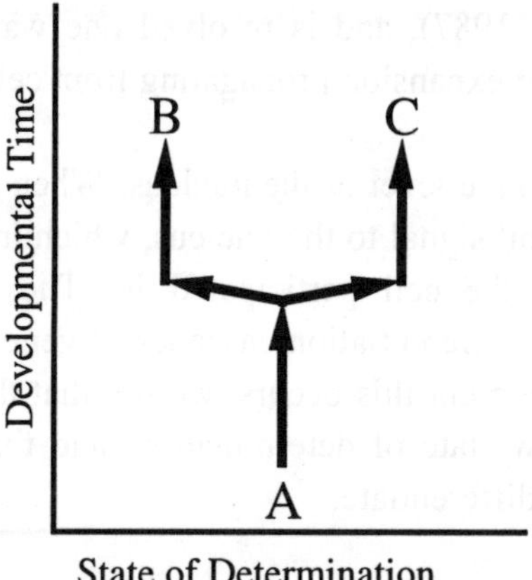

Figure 8. State of determination. If we distinguish the state of determination of a tissue from its state of differentiation, we see that the state of determination changes abruptly as a cell participates in a differentiation wave.

At the cellular level, we note that a wave passes *through* each cell, i.e., that the cell actively participates in a wave, is actually an active component of the wave, and passes the wave on to one or more neighboring cells.

We think that a major component of the cell that permits this participation in a wave is a cytoskeletal device at the apical surface of each cell, which we call the 'cell state splitter'. This is a bistable mechanochemical device, which resolves one way if the cell participates in a contraction wave, and the other way if it participates in an expansion wave.

The cell state splitter has two key components: a microfilament ring subtending an apical mat of microtubules. The microtubules are parallel to the apical surface of the cell. Like smooth muscle, the microfilament ring is contractile, while the microtubule mat can expand by polymerization. These two cytoskeletal components can thus, so to speak, engage in a radial tug-of-war. The equilibrium between them, which I take to be the physical basis of embryonic competence, is unstable (Figure 10 in Appendix I:

Gordon & Brodland, 1987), and is resolved one way or the other by the wave of contraction or expansion propagating from cell to cell.

Next, let's go down to the level of the nucleus. When a cell participates in a wave, it sends a one bit signal to the nucleus, which indicates to the nucleus what kind of wave the cell participated in. The nucleus responds by activating one of two differentiation cascades of genes, changing the cell to a new type. At the moment this occurs, we say that the nucleus and its cell have arrived at a new state of determination, and that the change in gene products causes it to differentiate.

The bistable configuration of the nucleus that makes it ready to respond to the signal with either of exactly two readied differentiation cascades is referred to as the nuclear state splitter. We do not yet know if the one bit signal transduction from the cell state splitter to the nucleus is strictly biochemical or mechanochemical, nor do we know what in the nucleus represents the states of the nuclear state splitter that receives this signal. Models for both are discussed.

Next, let's go up to the level of the whole organism. We see embryogenesis as essentially an alternation of differentiation waves and nuclear responses that set off corresponding differentiation cascades. This is the genetic program, represented as the 'differentiation tree'. All else, such as morphogenetic movement, is relegated to consequences of the differentiation cascades. This is not to minimize such effects, for they may be as crucial to the unsolved problem of the launching of differentiation waves and the shapes of their trajectories as they are important in generating form. But this view allows us to place such effects along the edges of the differentiation tree, which gives us an overall view of the embryogenesis of an organism (Figures 9, 10).

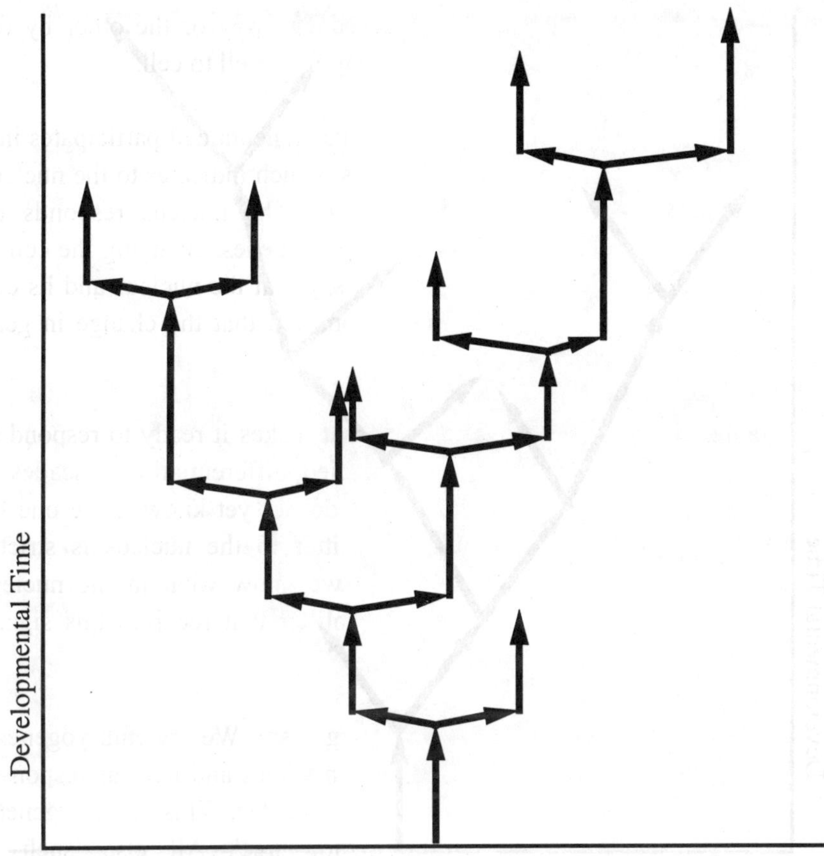

Figure 9. Determination tree. If each tissue type can, through the action of differentiation waves, split into two more cell types, then the development of an organism can be represented as a determination tree. All left arrows could represent contraction waves, and right arrows, expansion waves (or vice versa).

                                              The Hierarchical Genome

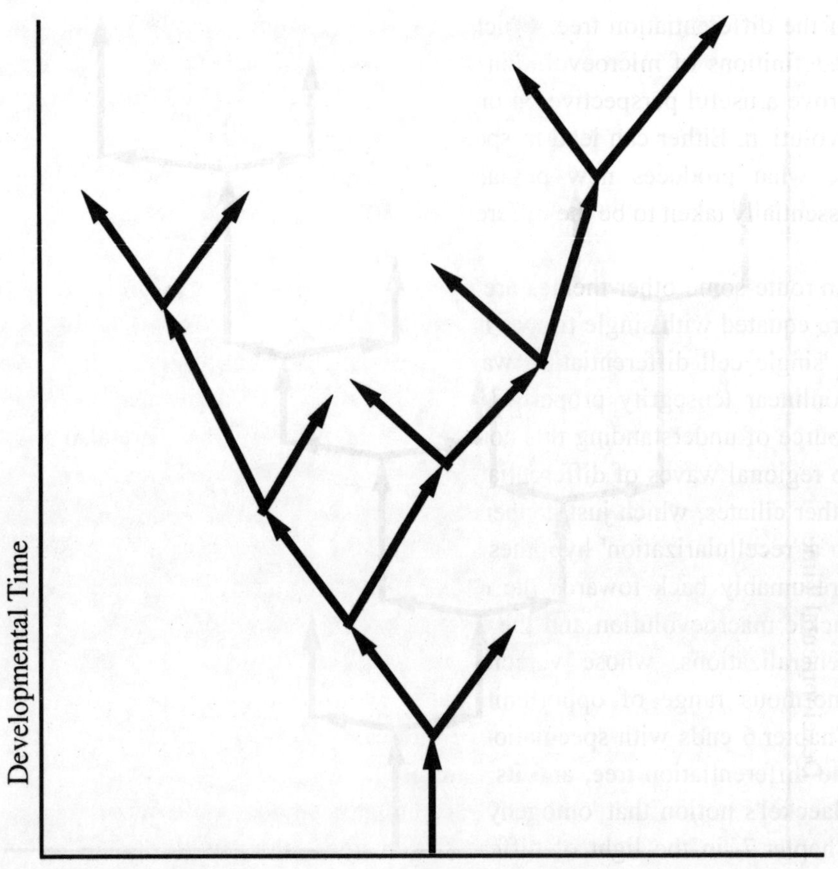

Figure 10. A differentiation tree can be derived from a determination tree, such as the previous figure, by replacing the right angles by single arrows.

I make the leap to evolution by taking the point of view that it is primarily the differentiation tree that evolves. This can happen in two major ways: 1) by changes to the edges of the differentiation tree, i.e., to the differentiation cascades, which I equate with microevolution; 2) by changes to the topology of the differentiation tree, which I will equate with macroevolution. These redefinitions of microevolution and macroevolution give what I hope will prove a useful perspective on unifying these two widely separated scales of evolution. Either can lead to speciation, but macroevolution is presumed to be what produces new phyla, i.e., changes to the Bauplan (which is essentially taken to be the differentiation tree of an organism).

En route some other themes are developed. Single cells in mosaic embryos are equated with single tissues in regulating embryos (Chapter 4). Just what a 'single cell differentiation wave' is, is not yet clear, but I focus on the nonlinear tensegrity properties of asymmetric cell division as a possible source of understanding this concept. Differentiation waves are also related to regional waves of differentiation of the basal bodies of *Paramecium* and other ciliates, which just happen to propagate at the same speed. This leads to a 'recellularization' hypothesis for the origin of multicellular organisms, presumably back towards the end of the Precambrian (Chapter 5). I next tackle macroevolution and the grand sweep of evolution, with some grand generalizations, whose veracity can at best be statistical, due to the enormous range of opportunities for variation presented by organisms. Chapter 6 ends with speculations on the evolution of the brain's portion of the differentiation tree, and its relationship to learning, instinct, and sleep. Haeckel's notion that 'ontogeny recapitulates phylogeny' is reconsidered in Chapter 7, in the light of differentiation trees, along with organisms with double life cycles and metamorphosis. In Chapter 8 I try to explain the fundamental homologies between all multicellular organisms that their molecular homologies are suggesting, and also develop some models for the mechanism of bilateral asymmetry. Chapter 9 surveys the many candidates for differentiation waves. The concluding Chapter 10 goes back to the big picture, showing that there is, through the evolution of differentiation trees, a rational approach to the question of progress in evolution. I also sketch a program of research for reading the differentiation tree from sequenced genomes, and trying to find out what on earth is going on inside a nucleus during differentiation.

3.02 How to Stop a Wave on a Sphere

"An 'infection-like' spreading of neural induction from cell to cell could not be stopped by a gradient. Indeed homoiogenetic induction is a phenomenon, which has been strangely neglected, perhaps because it does not fit into accustomed theoretical systems of thinking.... It could even be, that the gradient-concept is a dead end, distracting us from a search in other directions" (Albers, 1987).

The ectoderm contraction wave divides the ectoderm into two hemispheres: neuroepithelium (neural plate) and epidermis. We could have a quandary, because it would seem at first that this wave should proceed over the whole of the ectoderm, turning it all into neural plate. How does it stop at one hemisphere?

Our first approach to this problem, before any waves were actually observed, involved simplifying the geometry of the embryo. If we ignore the blastopore, and any nonuniformities, including the animal/vegetal cell size and yolk density gradients, we can look at the embryo as a simple spherical shell of cells, rigid enough to keep its shape due to the internal pressure. We further assumed (not knowing then that cells participating in the real ectoderm contraction wave relax after contracting), that the contraction of a cell persists. (Later contractions, evidenced by the 'height program' for the neural plate, gave us this impression: Jacobson & Gordon, 1976a.) In this idealized geometry, a wave starting from one pole of the sphere will stop when it reaches the equator (Figures 11-13). The abrupt termination of the wave would leave two hemispheres of sharply demarcated tissues, neural plate and epidermis (Jacobson & Gordon, 1976a). To distinguish this hypothetical model from the actual ectoderm contraction wave (Figure 17 in Appendix V: Gordon, Björklund & Nieuwkoop, 1994), I will call its wave the *shell model wave,* where the word 'shell' is used in the mechanical engineering sense (Flügge, 1973).

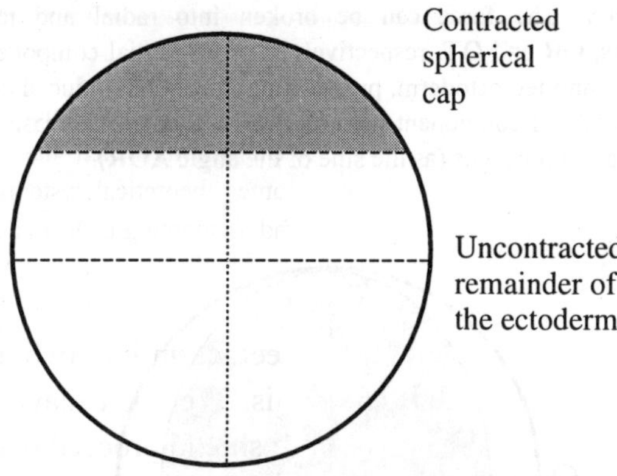

Figure 11. The contraction wave forming neuroepithelium is shown partially complete, in the simple 'shell' model. If the rate of contraction falls below a threshold, the contraction wave may stop before reaching a full hemisphere. This may explain the smaller (relative to egg size) neural plates in large, yolky eggs (Duellman & Trueb, 1986).

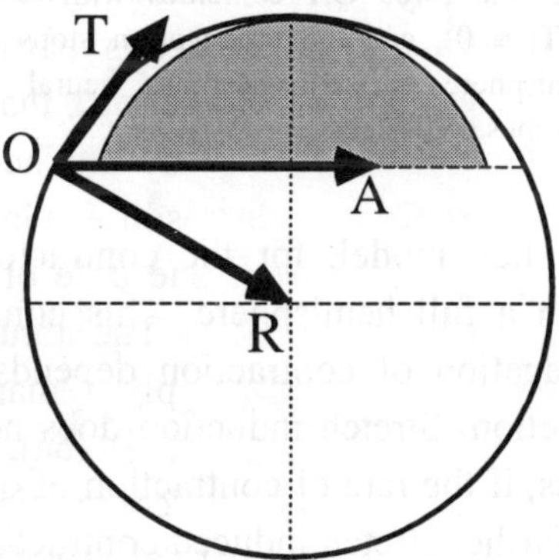

Figure 12. The spherical ectoderm of a urodele embryo is shown, in the 'shell' model. Mechanical analysis of the effect of the propagating contracted spherical cap on the remaining ectoderm can be made by doing a simple thought experiment. Suppose we cut off the contracted cap. It would shrink to some extent. We can ask what force would be necessary to rejoin the two portions of the embryo. That force may be represented by the vector **OA**, perpendicular to the (vertical) axis

of symmetry. The force can be broken into radial and tangential vector components, **OR** and **OT**, respectively. The tangential component is what pulls on the uncommitted ectoderm, propagating the stretch induced contraction. Note that the tangential component (and its rate of change) decrease in magnitude as the propagation proceeds (as the sine of the angle **AOR**).

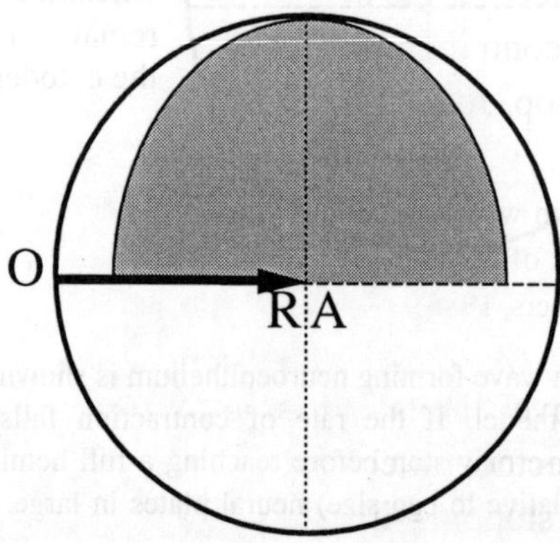

Figure 13. In the 'shell' model, if and when the contracted portion of the ectoderm grows to a full hemisphere, the force **OA** coincides with **OR**. There is no tangential component ($|\textbf{OT}| \equiv 0$), and the propagation stops (if it hasn't already). The contracted hemisphere of cells becomes neural plate and the remaining hemisphere becomes epidermis.

It is possible, in the shell model, for the contracted tissue to end up a spherical cap less than a full hemisphere. This comes about because we assume that the propagation of contraction depends on a mechanism of stretch induced contraction. Stretch induction does not occur if the rate of stretch is too low. Thus, if the rate of contraction of one cell falls below the threshold for causing further stretch induced contraction in the next cell, the contraction wave will terminate early, leaving a spherical cap less than a hemisphere (Appendix I: Gordon & Brodland, 1987).

A problem of classical embryology has been to understand the mechanism of 'homoiogenetic' induction (Nieuwkoop, Johnen & Albers, 1985; Itoh & Kubota, 1991; Servetnick & Grainger, 1991a), which is nowadays called 'planar' induction (Doniach, 1992, 1993; Doniach, Phillips & Gerhart, 1992; Keller et al., 1992a; Ruiz i Altaba, 1990, 1992; Grunz, Schuren & Richter, 1995; Duprat, 1996). Homoiogenetic induction is the propagation of 'neuralization' from one ectoderm cell to the next. We can now hypothesize that what is propagated from one cell to the next is a stretch activated contraction, and its effect on the nucleus of each cell that participates in the contraction wave. A contraction differentiation wave solves the problems identified by Nieuwkoop, Johnen & Albers (1985):

"In homoiogenetic induction we have to deal with two problems: that of the propagation of the stimulus itself, and that of how the inductive range of action becomes spatially limited" (Nieuwkoop, Johnen & Albers, 1985).

The shell model wave limits itself to one hemisphere. So does the empirically observed ectoderm contraction wave. A wave is consistent with observations such as stopping induction by a simple cut in the ectoderm (Brachet, 1927; Needham, 1931b). Its slow speed (3 μm/min) explains why homoiogenetic induction has a...

"...spread, [as] Spemann [1906a, 1912a] found, [that]... takes a measurable time, for at an early stage the ectoderm near the organiser is determined to turn into notochord, while further away it is still indifferent" (Needham, 1931b).

An aside on nomenclature is necessary. As pointed out by Hamburger (1988), Spemann unfortunately used the term 'homoiogenetic' or 'assimilative' induction in two different ways, even in one publication (Spemann, 1938):

propagation of differentiation from one cell to the next in a single, uniform tissue (Spemann & Mangold, 1924a,b; Spemann & Geinitz, 1927; cf. Lehmann, 1948a, 1953; Boterenbrood, 1962);

heterogeneous induction, i.e., the initiation of differentiation in one tissue after contact by another, when the latter happens to be of the same type (Mangold & Spemann, 1927; Spemann, 1927b).

Only the first meaning has survived in later literature, with the appellation 'homoiogenetic' or 'planar'. However, care must be taken in accepting current literature on planar induction:

"The apparently unobserved early involution [invagination] of the prechordal endo-mesoderm has led to the erroneous conclusions by Doniach, Phillips & Gerhart (1992), Doniach (1992), Ruiz i Altaba (1990, 1992, 1993, 1994) and other authors [cf. Keller et al., 1992a; Doniach, 1993; Sater, Steinhardt & Keller, 1993], published in renowned journals, like *Development* and *Science,* namely that in *Xenopus laevis* neuralization of the ectoderm should be caused purely or predominantly by a *planar* induction process. This erroneous conclusion is due to the fact that these authors have overlooked the following: The observed antero-posterior polarity of the neural structures in exogastrulae and Keller explants and sandwiches [Keller, Shih & Sater, 1992; cf. the 'sandwich' method of Nieuwkoop & Nigtevecht, 1954] is actually incompatible with a planar induction process, whose action only runs 'down-hill'. Moreover, the absence of any neural induction in Urodele exo-gastrulae pleads convincingly against planar activation of the neural anlage. They have moreover not realised that neural structures cannot develop without previous activation by means of *vertical* induction, activation being too weak in the caudal chorda-mesoderm for planar spreading into the adjacent ectoderm" (Nieuwkoop & Koster, 1995).

(Cf. Nieuwkoop, 1997.) The resolution of this situation is in recognizing that activation corresponds to launching of the ectoderm contraction wave ('vertical' induction), while planar induction corresponds to its propagation.

However, this parallel is not so facile, since we find no contraction waves *per se* on *Xenopus,* and their homology with cell intercalations can only be speculated for now (Nieuwkoop, Björklund & Gordon, 1996). Despite this observation, since much of modern embryology is based on *Xenopus,* I will continue to refer to probable 'waves' in *Xenopus* without always mentioning this caveat that *Xenopus* intercalations are what may be homologous to urodele waves. *Xenopus* ectoderm is covered by a superficial epithelium not found in urodeles. As discussed in Section 9.20, it is also possible, by analogy with the cuticle covered eye imaginal disc in *Tribolium* (Friedrich, Rambold & Melzer, 1996), that the cell state splitter is shifted to the basal end of each ectoderm cell in *Xenopus.* It's worth a look.

Homoiogenetic induction is found in *Xenopus* (Grunz, 1990b), and whatever it consists of, is probably homologous to the ectoderm contraction wave in axolotl. We have much that we need to learn about how expected homologies between *Xenopus* and urodeles work:

"Holtfreter (1944b) recognized that the isolated dorsal marginal zone of the urodele would autonomously narrow and elongate (converge and extend) when cultured under several conditions [cf. the *in vivo* description of convergence and extension by Vogt described in Dalcq, 1938]. Although the autonomous narrowing (convergence) and lengthening (extension) of the dorsal marginal zone of *Xenopus laevis,* has been described in detail at the tissue and cell level and in terms of the tissue interactions regulating these movements (see Keller, Shih & Domingo, 1992; Domingo & Keller, 1995), the process is not understood at all in urodeles... (Shi et al., 1987).... Holtfreter's description and drawings of this process suggest that the underlying cellular mechanism of *Ambystoma* convergent extension may be different from that of *Xenopus* " (Keller, 1996).

3.03 How the Ectoderm Contraction Wave Actually Stops: the Lens Model

"Epigenesis is to a very large extent a matter of morphogenesis, the creation of form. This is in the last analysis a geometrical transmutation, accomplished by physical forces" (Løvtrup, 1984a).

We thought that observation of the predicted wave would require computer vision monitoring of a short, transient event (Appendix I: Gordon & Brodland, 1987): it had been suggested that homoiogenetic induction could take as few as 5 min to cross the embryo (Nieuwkoop et al., 1955a,b; Johnen, 1956a). I was furthermore worried that the amplitude might be too small for light microscopy. The actual discovery of the wave, which turned out to be very slow (3 µm/min) and of such large amplitude (0.1 mm deep and wide) that low power microscopy was quite adequate to see it, involved a combination of some deliberate looking and serendipity (Appendix III). The initial observation was made in the summer of 1990 in a few extirpated fragments of ectoderm by Natalie K. Björklund, in the course of preparing these pieces of tissue for planned (but never successful) experiments of stretching and contraction on latex (Section 9.22), and she and I recorded the first wave in an intact embryo that December.

I am, of course, presuming that the ectoderm contraction wave is actually part of the causal chain of primary neural induction, and will try to make as good a case for that here as can now be made. Nevertheless, some humility might be in order:

"We are not unique, however, in this anticipation of imminent discovery.... If we examine the past history of our field, we learn that previous generations of biologists also have believed that the concepts most popular at their time were about to solve all the problems. Thus it is with a poignant feeling that one reads... 'Already it is known that the organising action is due to a substance which is almost certainly lipoidal and probably a sterol...' (Huxley & de Beer, 1934). Anyone acquainted with the current status of developmental biology three decades later will readily attest to the fact that the chemical nature of the organizer still is obscure. The whole character of inductive phenomena remains enigmatic despite attacks upon the problem from many different vantage points" (Barth, 1964b).

On the chance that we have actually made the critical breakthrough on the 'inductive phenomena', and at the risk of being accused of narcissism, I have included my account of the discovery of the wave written a few days after we realized we had it, as Appendix III. It is rare that one preserves moments of discovery while they are still fresh, and the adrenalin is still flowing, so I take the risk (cf. Root-Bernstein, 1989).

The wave we found has some similarity to the one we postulated, the shell model wave. It is a mechanical wave with two visible components: a 100 μm wide surface contraction similar to the surface contraction wave that accompanies the first cell division in the embryo (Yoneda et al., 1982) and a 100 μm deep furrow similar to, but deeper than, that seen in fruit fly eye differentiation (Ready, Hanson & Benzer, 1976). It travels about 3 μm/min (Appendix II: Brodland et al., 1994). The wave is launched from a focus about 45° above the blastopore. The portion of the wave that hits the region near the blastopore vanishes (perhaps due to the high tension around the blastopore?). The remaining arc travels at a nonuniform speed, with the ends travelling faster than the middle. As a consequence, the ends meet and the wave continues as a shrinking circle until it vanishes (Figures 16-17 in Appendix V: Gordon, Björklund & Nieuwkoop, 1994; see also the Flip Animation of the Ectoderm Contraction Wave). The ectoderm thus acts as a lens, bringing this solitary wave from one focus to another, while the blastopore casts its shadow. The resulting trajectory of the wave confines it to one hemisphere.

It is the strange trajectory of this wave (Figure 17 in Appendix V: Gordon, Björklund & Nieuwkoop, 1994), not anticipated in the shell model wave, that confines it spatially, contradicting the presumption that "Propagation of an all-or-none response, as in nerve impulse conduction, cannot account for homoiogenetic induction..." (Nieuwkoop, Johnen & Albers, 1985). There is something nonuniform about the ectoderm that causes this contraction wave to bend and come to a focus the way it does, and thus avoid the ventral hemisphere. The trajectory speaks against a diffusion based propagation mechanism, such as that found in colonies of *Myxococcus* and *Dictyostelium* (Sager, 1996), since diffusion in the plane of an epithelium is usually considered to be isotropic (same rate independent of direction: Jacobs, 1967; Berg, 1993). If diffusion were anisotropic (different rates in different directions: Carslaw & Jaeger, 1959; Crank, 1956; Shewmon, 1963; Stark, 1976), we would have to come up with a theory for the origin of the anisotropy. Of course, we do have an anisotropy, and I haven't got a theory for it yet. My best guess is that the anisotropy is in the mechanical strain of the ectoderm, due to nonuniform stretching of the ectoderm as it invaginates through the blastopore. There is precedence for a role for mechanical strain (Beloussov et al., 1990). Anisotropic mechanical strain is unlikely to cause anisotropy of diffusion, unless junctions are broken due to shear.

We may look at the contraction differentiation wave as a means of splitting a whole tissue (the ectoderm) into two tissues: the portion over which the wave travels becomes one tissue (neuroepithelium), while the portion it misses becomes another (epidermis). The epidermis has its own expansion wave (Figure 16 in Appendix V: Gordon, Björklund & Nieuwkoop, 1994).

Differentiation waves thus solve the problem of how a set of identical cells in a single tissue can become spatially partitioned into two adjacent, contiguous tissues.

Since differentiation (change of cell type) and morphogenesis (spatial splitting of a tissue into two tissues) occur simultaneously in the wake of differentiation waves, an answer is obtained to the enquiry of Bard (1990a):

"...Wolpert (1969) in his well-known paper on pattern formation... was able to show that the process of spatial determination across a wide range of embryos usually extended across domains some 60 cells wide and took about 10 h. It would be both helpful and interesting were morphogenesis subject to similar constraints" (Bard, 1990a).

The ectoderm contraction wave takes 9 to 12 hours (Appendix II: Brodland et al., 1994) to cross about 100 consecutive cells, measured along the curved surface of the embryo.

3.04 Internal Pressure May Synchronize Preparation of the Cell State Splitters

"How exactly an inductor selects and puts into effect its chosen competence is a question which we shall not be able to answer until we know what competence means from a physico-chemical point of view" (Needham, 1950).

We are faced with the problem of explaining how a whole spherical sheet of cells can come to set up their metastable cell state splitters simultaneously. I will presume this is the case, though one could equally imagine that they are set up in response to the oncoming wave, which affects the 2000 μm axolotl embryo's curvature 500 μm ahead (Figure 9 in Appendix II: Brodland et al., 1994), a question that could be settled by appropriately timed cytoskeletal microscopy.

At stages prior to neural tube closure, amphibian embryos have an internal blastocoele and archenteron pressure (relative to the medium in which the embryo is immersed), demonstrable by slitting experiments (Lewis, 1947;

Jacobson & Gordon, 1976a; Gordon, 1985c; cf. Honda, Yamanaka & Eguchi, 1986). However, a critical distinction between pressure in internal cavities and the "presence of a residual [growth] stress field [that] can be identified if deformation occurs when cuts are made in the [mechanically] unloaded tissue" (Rodriguez, Hoger & McCulloch, 1994; cf. Fung, 1990), has yet to be made. On the other hand, there is an 'energy-coupled water flow' that is probably responsible for generating a blastocoele and archenteron pressure (Tuft, 1961, 1965; Deuchar, 1975b; Stern & MacKenzie, 1983; Stern, Manning & Gillespie, 1985). (Baker, 1965, suggests "that, as some of the flask cells release their connection distally and retract into the interior, a temporary canal is opened... which... would allow blastocoelic fluid... into the archenteric cavity", equalizing the pressure between them.) The dynamics of this blastocoele and archenteron pressure are quite apparent in Figure 8 in Zotin (1965), if we presume a monotonic dependency of the pressure on the 'cavity' volume (archenteron + blastocoele). Such a globally increasing pressure could provide the needed property of affecting "the entire ectoderm *simultaneously* and thus may coordinate this synthesis [of cell state splitters] in all of the cells" (Appendix I: Gordon & Brodland, 1987). We could merely postulate that cell state splitters are assembled in response to a threshold value of the pressure.

This hypothesis is amenable to some obvious testing, such as relieving the pressure (as by canulation: Desmond & Jacobson, 1977), and comparing the buildup of pressure with the formation of cell state splitters, as assessed by electron microscopy (Martin & Gordon, 1997a,b) or cytoskeletal immunofluorescence. The pressure has already been reduced chemically, with a result that could be attributable to either a lack of cell state splitters or failure to launch the ectoderm contraction wave (or its cell intercalation equivalent in *Xenopus*: Nieuwkoop, Björklund & Gordon, 1996):

"An... agent which has been found to upset the fluid distribution in *Xenopus* embryos is β-mercaptoethanol. Malpoix, Quertier & Brachet (1963) found that embryos treated with this agent suffered a collapse of the archenteron cavity and an accumulation of fluid ventrally between ventral ectoderm and mesoderm. In these embryos, gastrulation was arrested. So evidently the normal movements of fluid are an important part of the gastrulation process. Interestingly, Hamilton & Tuft (1972) have found that... haploid *Xenopus*... also fluctuate in their water content during gastrulation" (Deuchar, 1975b).

The higher degree of contraction of the apical ends of bottle cells (= flask cells = Ruffini cells = piriform cells: Wilson, 1900; Rhumbler, 1902; Ruffini, 1925; Holtfreter, 1943a, 1944b; Needham, 1950; Balinsky, 1961; Baker, 1965; Oppenheimer, 1967; Nieuwkoop, 1969b; Löfberg, 1974a; Landström & Løvtrup, 1977a; Keller, 1981; Baker, 1965; Perry & Waddington, 1966a; Doucet-de Bruïne, 1973; Löfberg, 1974a; Nakatsuji, 1975a; Bluemink, 1978; Wakely & England, 1978, 1979; Kirk et al., 1982; Nieuwkoop, Johnen & Albers, 1985; Lundmark, 1986; Hardin & Keller, 1988; Bard, 1990a; Gilbert, 1991a; Bolker, 1993a; Collazo, Bolker & Keller, 1994; Nakajima & Burke, 1996) compared to cells participating in the ectoderm contraction wave (Appendix II: Brodland et al., 1994) may be due to greater effects of internal pressure on stretching the latter. The existence of mechanosensitive membrane proteins (Morris, 1990) provides a potential signalling mechanism for pressure induced macromolecular synthesis or assembly.

Microfilament ring assembly is likely to follow the two-step mechanism of Mabuchi (1990): 1) accumulation of actin, followed by 2) rearrangement into bundles (cf. White, 1990), a concept supported by analysis of a maternal mutant of *Xenopus* (Asada-Kubota & Kubota, 1991). The assembly of each cell state splitter is accompanied by an apical concentration of ribosomes, including polysomes (Martin & Gordon, 1997b; cf. Toh et al., 1980; Meshcheriakov, Zavalishina & Beloussov, 1982; Micklem, 1995), which are perhaps membrane bound (Hesketh & Pryme, 1991). It would be nice to know what messenger RNA(s) they are transcribing (perhaps actin: Kislauskis, Zhu & Singer, 1997; cf. Jeffery, Tomlinson & Brodeur, 1983). The paraxial microtubules may have a role in the apical accumulation of ribosomes (cf. Baas, Sinclair & Heidemann, 1987). It will be a challenge trying to link these observations into a coherent picture of synchronization of cell state splitter assembly.

Simultaneity of the assembly of cell state splitters cannot be due to any simple timing mechanism keyed to cell divisions, since the embryo has asynchronous cell division by the time of primary neural induction (Balinsky & Fabian, 1981; Laale, 1987). We shall see that mitotic waves correlate with differentiation waves in *Drosophila,* and thus a closer analysis of the supposed asynchrony in amphibians might prove to be a correlation with hitherto unseen differentiation waves. Alternatively, time elapsed since a cell last experienced a differentiation wave might correlate with cell state splitter assembly. In this case, we would anticipate that assembly would not be simultaneous, but would rather itself be a gene product wave behind the previous differentiation wave experienced by the same cells. If the time delay is relatively short, a cell must somehow retain its metastable state longer. But if the time delay is long, the cell might not be ready for the next differentiation wave. For this reason, I suspect that cell state splitters for the next wave are set up or reactivated shortly after the cell has participated in a wave. This idea is directly testable.

3.05 The Right Place, at the Right Time, into the Right Kinds

"The question will still need to be answered of how the appropriate reactions are triggered at the 'right' time and place for normal development" (Barth, 1964a).

Of course, the differences between neural plate cells and epidermal cells go well beyond their shape differences. We thus postulated that the genome somehow detects whether the microfilament ring or the microtubule mat won their radial tug-of-war, and responds by synthesizing the appropriate proteins (Appendix I: Gordon & Brodland, 1987). It takes only one bit to specify whether the MF ring or the MT mat won, i.e., whether the cell participated in a contraction or an expansion wave. A number of plausible mechanisms may be envisaged to explain how the nucleus gets this one bit of information from the cell state splitter (see Appendix IV: Björklund & Gordon, 1993b). I presume that it is not the carrier of this one bit of information, but rather the receiver (nucleus), that determines its specificity.

This assumption means that *the signal transduction mechanism could be the same for all steps of differentiation.*

It is important to note that there is no evidence for anticipatory synthesis of specific molecules in presumptive neural plate or epidermal cells (reviewed in Appendix I: Gordon & Brodland, 1987). Claims to this effect (Saint-Jeannet et al., 1989) do not take into account the duration of the (albeit then unknown) ectoderm contraction wave, which lasts the whole period of gastrulation. Similarly, the antigenic differentiation observed in all three germ layers of chick embryos during gastrulation and neurulation, without simultaneous morphological differentiation (Ellis et al., 1993), may correspond to early consequences of differentiation waves.

It is now apparent why I call the cell state splitter the first hypothesis that comprehensively addresses the fundamental problem of developmental biology. The role of the redefined organizer in perhaps reducing the local mechanical tension and launching the ectoderm contraction wave (Section 3.06), plus the trajectory of the wave, defines 'the right place'. The increasing embryo pressure may be what yields 'the right time' (though the minimal time, or time until the cells are 'competent', may be related to the time it takes for the previous differentiation cascade to play itself out). The instability (bistability) of the cell state splitter and the corresponding response of the nucleus provide 'the right kinds'. There is no need to invoke a role for the 'extracellular environment' (Maclean & Hall, 1987), since all of the components necessary for biochemical differentiation and spatial morphogenesis are contained within the embryo.

3.06 The Intracellular Mechanics of the Cell State Splitter Yields Ectodermal Differentiation

"The concept of unstable equilibria in early development seems to offer opportunities for advance in the mathematisation of embryology" (Needham, 1950).

I have been working under the assumption that the generation of forces and load bearing in embryonic cells is caused by macromolecules referred to collectively as the cytoskeleton. Beth Burnside (1971) discovered that the microfilament ring (Baker & Schroeder, 1967) at and parallel to the apical end of California newt (*Taricha torosa*) neural plate cells surrounds a coplanar mat of microtubules. She speculated that the function of the apical mat of microtubules was for the paraxial microtubules to push against (Burnside, 1973a), though this hypothesis could not explain the lack of a similar mat at the basal ends of these cells. I have instead taken her observations to mean that the microfilament ring and microtubule mat act together as a 'morphogenetic organelle' that I call the 'cell state splitter' (Gordon & Brodland, 1987).

At the outer (apical) end of each axolotl ectoderm cell the microfilament ring is attached to the cell membrane via rivet-like desmosomes and what we now know to be an *annular* mat of microtubules (Martin & Gordon, 1991, 1997a,b). (Figure 1 in Appendix V: Gordon, Björklund & Nieuwkoop, 1994, gives a schematic of reconstructions based on about 60 overlapping electron micrographs taken around the apical perimeter of individual cells and fit together as a montage by computer, including a few transects across a diameter, and some serial sectioning). If you try to construct such a configuration using Tensegritoy (Tensegrity Systems, Tivoli, New York), you'll see that this is a reasonably rigid structure. This exercise could also be carried out by computer using energy minimization methods.

Microfilament rings consist of actin and myosin (Mabuchi et al., 1988; cf. 'myosin rings' in Young, Pesacreta & Kiehart, 1991), and are probably contractile like smooth muscle (Stephens, 1984). It is reasonable to assume (though this needs testing), that when stretched fast enough, microfilament rings are capable of a stretch induced contraction (Appendix I: Gordon & Brodland, 1987). I do not mean to imply that the effect is direct: stretch activated ion channels and Ca^{2+} fluxes may be involved (Appendix IV: Björklund & Gordon, 1993b).

The microtubules in the apical mat should be capable of exerting forces (Appendix I: Gordon & Brodland, 1987) as they lengthen by polymerization (Dogterom & Yurke, 1997) of the free tubulin in the cytoplasm (Raff & Raff, 1978), especially if laterally supported by intermediate filaments (Brodland & Gordon, 1990, which answers affirmatively: "Is it possible that these interconnections give mechanical support to microtubules...?": Pierson & Cresti, 1992). It is likely that polymerization and depolymerization occur by the active mechanism of dynamic instability (Mitchison & Kirschner, 1984a,b, 1987; Hill & Chen, 1984, 1985; Hill, 1987; Hyman et al., 1992b; Babinec & Babincová, 1995a; Vulevic & Correia, 1997), in agreement with the observation by Burnside (1976) of their lability. Since the whole cell state splitter apparatus is within 0.5 μm of the outer, apical ends of the 50 μm tall cells, it should be possible to observe the dynamics of perhaps even individual microtubules and the microfilament ring, using dark field microscopy or fluorescence microscopy with epi-illumination. Parameters such as free tubulin concentrations have to be determined for the specific organism and cells under investigation, instead of hopping from one organism or cell type to another for analogous data, as we were forced to do (Appendix I: Gordon & Brodland, 1987). I have not considered the possibility that the apical microtubules could exert forces during depolymerization, though, for example, this could be a component of force generation in mitotic spindles (Cassimeris, Inoué & Salmon, 1988; Mitchison, 1989; Rieder & Salmon, 1994).

Burnside (1971, 1973a) made finely staged, quantitative ultrastructural observations and estimated that the volume of the microfilament ring appears to be constant, at least during neurulation stages in the urodele *Taricha torosa* (Figure 4; Table 1). Similar measurements are needed during gastrulation, and they need to be correlated with passage of the ectoderm contraction wave. If we approximate a microfilament ring as a torus of radius R at the apical periphery of a cell of radius r, then its cross sectional area is:

$$A = \pi R^2$$

and its (apparently constant) volume is:

$$V = 2\pi r A$$

In words, assuming constant volume of the mass of microfilaments in the ring, the cross sectional area of the microfilament ring is inversely proportional to the apical radius of the cell. Furthermore, the contractile force a microfilament ring can exert is proportional to its cross sectional area, so we get that the force generated by the microfilament ring is inversely proportional to the apical radius of the cell:

$$F_{MF} \propto A = V / (2\pi r)$$

The outward force exerted by the microtubule mat should be roughly independent of radius, since it is effectively dependent on the concentration of free tubulin. Tubulin concentration shouldn't change much just because the cell changes shape. Extrapolating Burnside's data back to the time of primary neural induction, we estimated that the radial forces could be equal and opposite in each cell for empirically plausible concentrations of free tubulin (Appendix I: Gordon & Brodland, 1987). This order of magnitude agreement suggests that the cell state splitter is erected in mechanical equilibrium. It deserves further testing.

Such a mechanical equilibrium would be *unstable*. This may be seen by plotting the force exerted by each component versus apical cell radius (Figure 10 in Appendix I: Gordon & Brodland, 1987; Figure 3 in Appendix V: Gordon, Björklund & Nieuwkoop, 1994). Either the microfilament ring or the microtubule mat wins this radial 'tug-of-war' (Figure 4 in Appendix V). Again, the outward force, F_{MT}, exerted by the microtubule mat at the apical end of an ectoderm cell should be approximately independent of the length of the microtubules, since it depends primarily on the concentration of tubulin monomers in the cytoplasm (Albrecht-Buehler, 1990a). The inward force, $-F_{MF}$, exerted by a microfilament ring, on the other hand, decreases as the ring is stretched out, because that force is proportional to the ring's cross sectional area, and microfilament mass appears to be constant in these cells (Burnside, 1971). Thus, at radii beyond the equilibrium radius, r_e, the microtubule force exceeds the microfilament force. If the cell's apical radius exceeds r_e by even a small amount, it will continue to increase. On the other hand, if the radius decreases slightly, the microfilament force will exceed the microtubule force, and the radius will continue to decrease. The equilibrium radius r_e is thus a point of instability at which a bifurcation occurs. Note that with cell state splitters, each cell undergoes its own internal bifurcation, rather than bifurcations occurring at the tissue level, as in Oster & Alberch (1982).

Thus a perturbation from the equilibrium apical radius will be amplified, causing the cell to change shape towards either a tall, thin column, or a flat, squat one. Since these shapes correspond to that of neuroepithelial and epidermal cells, respectively, we initially suggested that resolution of this mechanical instability one way or the other is what determines the differentiation of the cell (Figure 4 in Appendix V: Gordon, Björklund & Nieuwkoop, 1994). Thus we hypothesized that the initial event of cell differentiation is mechanical (Appendix I: Gordon & Brodland, 1987). In other words, we postulated that the first act of this differentiation (usually called 'determination': Maclean & Hall, 1987) is a mechanical event, resulting in one of two new shapes for a cell.

We also postulated that the cell state splitter has a mechanically metastable state corresponding to the classical embryological state of cellular *competence,* which means readiness to differentiate to one of two new cell types (Appendix I: Gordon & Brodland, 1987). (I will have more to say about competence in Proposition 6.) Metastability is needed to prevent each cell from resolving its instability immediately after its cell state splitter is set up. Without metastability, it would also be sensitive to the smallest random perturbations. Thus metastability is essential for the cells to 'wait' until a large enough perturbation comes along to trigger the cell state splitter. I thus presume that a differentiation wave is sufficient to overcome this instability. The notion of metastability is visualized in Figures 11-13 in Appendix I.

The mechanical metastability may be maintained either by:

1. 'steric' interactions amongst the cells, perhaps mediated by the desmosomes;

2. the intermediate filament ring that we have discovered just below the microfilament ring (Figure 1 in Appendix V: Gordon, Björklund & Nieuwkoop, 1994; Martin & Gordon, 1991, 1997a,b; cf. Blose & Chako, 1976; note that "intermediate filaments were absent in all arthropods: Bartnik & Weber, 1989, in which they are apparently replaced functionally by microtubules" Raff, 1996).

To see how the intermediate filament ring could set up a metastable state, one need only think of it as a spring with its two ends attached. Any spring has a natural resting length and behaves symmetrically with respect to expansion or compression, at least for small deformations (Friedman & Youtz, 1965). If c_0 is the resting circumference of the intermediate filament ring (approximated as a circle), then the potential energy U when it is stretched or compressed to circumference c is:

$$U = 0.5k\,(c - c_0)^2$$

Simple experiments may suffice to ascertain the force constant k. For more complex modeling incorporating viscoelasticity, which would more fully describe the constitutive properties of intermediate filaments, see Flügge (1967a), Ferry (1970), and Christensen (1982) (cf. Duszyk et al., 1989; Hart, 1989; Janmey et al., 1991).

We now know that the wave of stretch activated contraction of the microfilament rings does, indeed, exist (Appendix II: Brodland et al., 1994; Gordon & Björklund, 1996). The relaxation of the contraction in a single cell is probably not a mechanical event, but rather may be a consequence of the ectoderm contraction wave pathway (Appendix IV: Björklund & Gordon, 1993b). Contrary to our preconceptions, the contracted cell goes back to its original shape (at least to visual appearance: Appendix II: Brodland et al., 1994), perhaps due to the elasticity of the intermediate filament ring (Martin & Gordon, 1997a,b). Nevertheless, it would seem that the appropriate master gene has by then been triggered, through what we have called the *nuclear state splitter* (Björklund & Gordon, 1993a; Appendix IV: Björklund & Gordon, 1993b). The later narrowing and columnarization of the neural plate cells is probably a consequence of their specific gene cascade, perhaps due to cellular adhesion molecules, such as N-CAM (Balak et al., 1987; Becker, Becker & Roth, 1993; Edelman et al., 1983; Holst et al., 1994; Jacobson & Rutishauser, 1986; Jones et al., 1993a; Kintner & Melton, 1987; Krieg, Sakaguchi & Kintner, 1989; Levi, Crossin & Edelman, 1987; Rutishauser, 1992; Saint-Jeannet et al., 1989; Zorn & Krieg, 1992), whose expression occurs in waves (Bally-Cuif, Goridis & Santoni, 1993), and whose consequences occur *after* neural plate formation (Kintner, 1988; Livingston, De Robertis & Paulson, 1990).

To a first approximation, the cell state splitter is set up with a metastable radial force balance:

$$F_{MT} = -F_{MF}$$

A model for how a rise in blastocoele/archenteron pressure could lead to launching of a contraction wave is shown in Figure 14. We can generalize this model to see the relationship of metastable thresholds to the kind of wave that is launched. Without metastability, we get a contraction if:

$$F_{MT} < -F_{MF}$$

and an expansion if:

$$F_{MT} > -F_{MF}$$

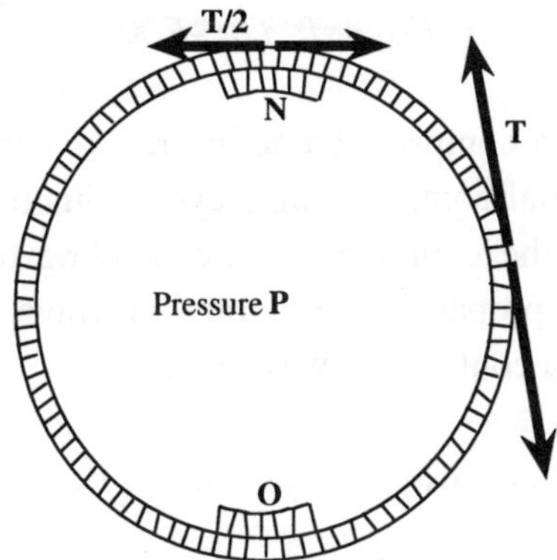

Figure 14. A pressure **P** inside an embryo leads to a global tension **T** that pulls tangentially on each cell in a monolayer epithelium, such as the ectoderm of urodele amphibians prior to primary neural induction. Because the cell state splitters are attached via desmosomes and other junctions, this tension is borne by them. This simplified diagram, treating the embryo at early gastrulation as a spherical shell consisting of a monolayer of cells, may be taken as a transverse section (Ballard, 1964) perpendicular to what will later be the axis or backbone of the adult. The pharyngeal endoderm **N** is the natural inducing tissue. Its adhesion to the ectoderm doubles the tissue thickness and thus roughly halves the tension for the overlying ectodermal cells (Appendix I: Gordon & Brodland, 1987). This leads to a force imbalance that causes the cell state splitters of these attached ectodermal cells to contract, launching a contraction wave. If another piece of tissue or an artificial substance, such as an 'organizer' **O**, is placed in the embryo ('Einsteck' method: Hamburger, 1960, 1988; Slack & Isaacs, 1994a), and it adheres to the ectoderm, it likewise launches another contraction wave. The only requirement for **O** is that it have a surface adhesive for the basal ends of ectoderm cells. In mathematical terms, the metastable mechanical equilibrium is represented by $F_{MT} + T = -F_{MF}$ for unsupported ectoderm cells, while $F_{MT} + T/2 < -F_{MF}$ for ectoderm cells attached to an 'organizer', natural or artificial.

Let T_t be the threshold tension acting radially outwards on a cell state splitter, and T the actual outward tension. This is the tension beyond which the cell state splitter resolves into the expanding state, leading to epidermal cell differentiation and a propagating expansion wave. In other words, we launch and propagate an expansion wave when:

$$F_{\mathrm{MT}} + T - T_t > -F_{\mathrm{MF}}$$

Let C_t be the threshold compression acting radially inwards on a cell state splitter, and C the actual compression. Beyond this threshold, the cell state splitter resolves into the contracting state, leading to neuroepithelial cell differentiation and a propagating contraction wave. In other words, we launch and propagate a contraction wave when:

$$F_{\mathrm{MT}} < -(F_{\mathrm{MF}} + C - C_t)$$

It would be nice to quantitatively predict the dynamics of the cell state splitter. How much tension or compression is needed to overcome the postulated metastability? Is there a symmetric relationship between expansion and contraction ($C_t = T_t$), or a mechanical bias ($C_t \neq T_t$) between them (Figures 11-13 in Appendix I: Gordon & Brodland, 1987)? Given the likelihood that the cell state splitter has some viscoelastic properties, how do rates of application of tension or compression affect the above equations? To answer such questions, we need ultrastructural details of the arrangement of the components of the cell state splitter, and their polarity:

"Handel & Roth (1971) observed microtubules structurally associated with the dense material of tangentially sectioned desmosomes in chick neural plate. They further demonstrated that microtubules in close association with structures such as desmosomes are more resistant to degradation by microtubule inhibitors such as colchicine and cold. These desmosomes may act as microtubule organizing centers (MTOCs) (Dustin, 1984), i.e., nucleation sites for tubulin polymerization [cf. Kimble & Kuriyama, 1992]. In fact, Wacker et al. (1992) have found evidence in cell culture that the microtubule-binding protein pp170 accumulates at desmosomal plaques during the establishment of epithelial cell polarity.... The tubulin would polymerize from its nucleating site at the desmosome outwards until it would reach another site for attachment, possibly the plasma membrane, and not necessarily another desmosome. This concept could be tested by ascertaining the polarity of the apical microtubules (Dustin, 1984)" (Martin & Gordon, 1997a,b).

We will also need a full mechanical analysis of the operation of the cell state splitter.

3.07 Force Generating and Load Bearing Cytoskeletal Components:
Microtubules (MT)

It is clear that for a critical test of our theory of the cell state splitter, we need to know what forces its cytoskeletal components can generate and bear. There is a small theoretical and experimental literature on direct prediction or measurement of these constitutive properties. Here I will give a brief overview of the properties of the cytoskeleton that may be important in embryogenesis and/or allow one to manipulate it. For more of the ever unfolding details, see any modern biology or molecular biology textbook covering the cytoskeleton.

Microtubules are cylinders consisting of dimers of the proteins α-tubulin and β-tubulin. These pairs are stacked helically like a spring without spaces between the turns, 13 dimers to a turn (Dustin, 1984; Warner & McIntosh, 1989; Hyams & Lloyd, 1994). Microtubules are straight, but they undulate or (rarely) kink (Cassimeris, Pryer & Salmon, 1988). They probably support compressive forces, and exert forces as they polymerize or depolymerize (Hill, 1981a; Hill & Kirschner, 1982a, 1983), sometimes in response to electric fields (Kaimanovich, Krupitski & Spirov, 1989, 1994), which could be present in cells. They also have proteins, kinesin (Vale, Reese & Sheetz, 1985; Hill, 1986a; Vale, Scholey & Sheetz, 1986; Chen & Hill, 1988; Howard, Hudspeth & Vale, 1989; Warner, Satir & Gibbons, 1989; Endow, 1991; Endow & Hatsumi, 1991; Houliston & Elinson, 1991b; Stewart & Goldstein, 1991; Romberg & Vale, 1993; Cole et al., 1994; Hunt, Gittes & Howard, 1994; Vale et al., 1994) and dynein (Mohri et al., 1976; Porter et al., 1988; Warner, Satir & Gibbons, 1989; Hays et al., 1994), that move along their length in opposite directions (cf. Thaler & Haimo, 1996). Force generation may involve rotation of the microtubule:

"Actin filaments and microtubules can slide and translocate particles along as it is well known. Moreover, bendings, corners, regions of branching and crossbridges can move in a wavelike manner along bundles of cytoskeletal elements. This has been demonstrated by microcinematography e.g. of ringlike closed F-actin bundles ('waving polygons') in cytoplasmic drops squeezed out of characean internodal cells... and by microtubule bundles of axostyles of *Pyrsonympha* (Langford & Inoué, 1979), or by the rootlet fibril (costa) of *Trichomonas* (Amos et al., 1979). Single isolated microtubules from squid giant axons that become visible by video-enhanced interference contrast microscopy can glide on glass slides and start a kind of 'fishtailing' when gliding is prevented by an obstacle (Allen et al., 1985). The described wavelike motions cannot be explained by the power-stroke or rowing-stroke model of myosin-, kinesin-, or dynein-crossbridges between filaments or microtubules - thus the problem of proper coordination and localization of the single power-strokes is unsolved. The motions can be explained and simulated in detail by macroscopic models with rotating steel helices. This indicates the existence of quickly rotating cytoskeletal elements" (Jarosch, 1992).

Cf. "...the remnant fibre stubs [kinetochore fibers = microtubules] often swing from their original orientation and the severed ends oscillate ('wobble')" (Forer, Spurck & Pickett-Heaps, 1997). Treadmilling of microtubules (polymerization at one end with simultaneous depolymerization at the other end), may also occur (Kirschner, 1980; Hill & Kirschner, 1982b; Rodionov & Borisy, 1997). Since both polymerization and depolymerization are active processes, requiring metabolic energy, these two nonequilibrium states are said to be in 'dynamic equilibrium' or 'dynamic instability' (Mitchison & Kirschner, 1984a,b; Horio & Hotani, 1986; Hameroff, Smith & Watt, 1986; Kirschner & Schulze, 1986; Kristofferson, Mitchison & Kirschner, 1986; Williams Jr., Caplow & McIntosh, 1986; Schulze & Kirschner, 1987; Walker et al., 1988; Kirschner, 1989; Hotani & Miyamoto, 1990; Bayley & Martin, 1991; Drechsel et al., 1992; Erickson & O'Brien, 1992). Rates of polymerization and depolymerization, but not rates of switching between these states, vary between cell types (Shelden & Wadsworth, 1993).

The relative stability of microtubules in some confluent epithelia (Pepperkok et al., 1990) may have a bearing on their dynamics during differentiation wave propagation. Drugs affecting microtubule polymerization include the depolymerizers:

colcemid (Lutz & Inoué, 1982; Pratt et al., 1982),

colchicine (Keppell & Dawson, 1939; Sinnott, Blakeslee & Warmke, 1939; Häggqvist, 1948; Borisy & Taylor, 1967; Löfberg & Jacobson, 1974; Zalokar, 1974; Pfeffer, Asnes & Wilson, 1976; Manes, Elinson & Barbieri, 1978; Stanisstreet & Panayi, 1980; Kubota, Yoshimoto & Hiramoto, 1993; but cf. Beebe et al., 1979),

vinblastine (Strome & Wood, 1983), *griseofulvin* (Schatten et al., 1982b; Strome & Wood, 1983; Priess & Hirsh, 1986; De Carli & Larizza, 1988; but cf. Miller & Joshi, 1996),

nocodazole (Brun & Garson, 1983; Strome & Wood, 1983; Maro & Pickering, 1984; Elinson, 1985; Fleming, Cannon & Pickering, 1986; Priess & Hirsh, 1986; Render & Elinson, 1986; Smedley & Stanisstreet, 1986; Elinson & Rowning, 1988; Anstrom, 1989; Astrow, Holton & Weisblat, 1989; Tacke & Grunz, 1990; Breed et al., 1994),

and the stabilizer *taxol* (Schatten et al., 1982a; Maro & Pickering, 1984; Collins & Vallee, 1987; Gong & Brandhorst, 1988; Breed et al., 1994; Caplow, Shanks & Ruhlen, 1994). High pressure also depolymerizes microtubules (Scharf & Gerhart, 1983). Natural inhibitors and promoters of microtubule polymerization also exist (Gard & Kirschner, 1987a,b), and the concentration of unpolymerized tubulin is apparently regulated by the concentration of tubulin mRNA (Cleveland et al., 1981).

Ultraviolet light severs microtubules (Walker, Inoué & Salmon, 1989), which may be why UV irradiation has profound effects on early embryonic development, perhaps contributing to the global decline in amphibians (Baldwin, 1915; Sergeev & Smirnov, 1939; Gurdon, 1960a; Beal & Dixon, 1975; Chung, Malacinski & Kim, 1977; Duncan, 1979; Chung & Malacinski, 1980, 1981; Cooke, 1985a; Cooke & Smith, 1987; Blaustein, Wake & Sousa, 1994; Stebbins & Cohen, 1995). The same effects occur with heavy water, D_2O (Briedis & Elinson, 1982). Natural severing occurs in some circumstances too (Shiina et al., 1994; Thaler & Haimo, 1996). Some microtubules are cold stable (Wallin & Strömberg, 1995), but most dissolve (depolymerize) at temperatures near $0°C$ (Scharf & Gerhart, 1983).

The growth of microtubule structures, such as asters, their numbers and spatial distribution, can be affected by tubulin diffusion limitations

(Dogterom, Maggs & Leibler, 1995). This is clearly a case of (reversible) diffusion limited precipitation, reviewed in Gordon & Drum (1994). Aster microtubules are nucleated by offset γ-tubulin rings (Zheng et al., 1995; Moritz et al., 1995b; Oakley, 1995; Balczon, 1996) which are perhaps constructed with the help of chaperones (Bloch & Johnson, 1995; Brown et al., 1996a,b). Tubulin itself is folded using chaperones (Mandelkow & Mandelkow, 1995). These features may provide a basis for Penrose's (1994, 1995) speculations on the role of microtubules in consciousness (but cf. Dennett, 1995). On the other hand, tubulin rings and fibers can be traced back to the bacteria (Erickson, 1997).

3.08 Force Generating and Load Bearing Cytoskeletal Components: Microfilaments (MF)

Microfilaments are made mostly of actin, with myosin present (Warn, Bullard & Maleki, 1979; Hill, Eisenberg & Chalovich, 1981; Sheetz & Spudich, 1983; Schroeder & Otto, 1988; Korn & Hammer, 1988; Otto & Schroeder, 1990) imparting contractility. The better the electron microscopy preparation, the more similar microfilaments seem in their organization to striated muscle (Stephens, 1984; Stephens et al., 1992; though "similarities between [smooth and striated] muscle are [probably] the result of convergent evolution": Goodson & Spudich, 1993; but cf. Molloy & White, 1997; Guilford et al., 1997). Modelling of microfilaments could be taken as a miniaturization of modelling of smooth muscle, to a rough approximation (Appendix I: Gordon & Brodland, 1987). One difference, however, is that the MF are usually quite labile, polymerizing and depolymerizing rapidly (Kirschner, 1980; Spudich, Giffard & Spudich, 1982; Hill & Kirschner, 1982a,b, 1983; Borejdo et al., 1983; Yonemura & Mabuchi, 1987; Ankenbauer et al., 1988; Carlier, 1989; Bearer, 1991b, 1993; Theriot & Mitchison, 1991, 1992). Their polymerization can be blocked by capping proteins (Fuchtbauer et al., 1983), and they can be depolymerized with cytochalasin (Zalokar, 1974; Manes, Elinson & Barbieri, 1978; Flanagan & Lin, 1980; Maclean-Fletcher & Pollard, 1980; Stanisstreet & Panayi, 1980), though cell-cell adhesion may also be affected by this compound

(Ghaskadbi, 1994). Microfilaments also bundle together (Otto, 1994). Microfilaments may also have a form of dynamic instability:

"In the case of microtubules and actin filaments, energy is consumed for a very unusual purpose - to make the assembled polymer unstable (Mitchison, 1992b). The individual subunits of any actin filament or microtubule will assemble and disassemble many hundreds of times [or more] in a cell's lifetime.... we will primarily discuss microtubules, where the experimental evidence for destabilization of the polymer is most complete (Kirschner & Mitchison, 1986a)" (Gerhart & Kirschner, 1997).

If both microtubules and microfilaments are subject to dynamic instability, then when one wins the tug-of-war in a cell state splitter, the other may be driven to rapid depolymerization. The 10 minutes contraction time for a cell participating in the ectoderm contraction wave (Björklund & Gordon, 1993b) is more than enough to change the lengths of the microtubules with a "...turnover... with a half time of ~5 min in interphase and 30 s in mitosis (Gelfand & Bershadsky, 1991)" (Mitchison, 1992b).

Ionophores trigger contraction of microfilaments (Schroeder & Strickland, 1974; Steinhardt et al., 1974; Steinhardt & Epel, 1974; Belanger & Scheutz, 1975; Moran, 1976; Moran & Rice, 1976; Lee, Nagele & Karasanyi, 1978; Osborn, Duncan & Smith, 1979; Robinson & Cone, 1980; Brady & Hilfer, 1982; Chernoff & Hilfer, 1982; Jeffery, 1982a; Stanisstreet, 1982; Andreuccetti et al., 1984; Mikhailov & Gorgolyuk, 1986; Battaglia & Gaddum-Rosse, 1987; Bozhkova & Voronov, 1991; Ferreira & Hilfer, 1993). (However, note that the ionophore induced cell shape changes in *Xenopus* ectoderm may not involve microfilaments: Stanisstreet & Smedley, 1986.) Microfilaments are sensitive to heat shock proteins (Miron et al., 1991), sometimes requiring stabilization (Simpson & Wieschaus, 1990). The heat shock protein "HSP27 is a component of a signal transduction pathway that can regulate microfilament dynamics" (Lavoie et al., 1993b; cf. Huot et al., 1996).

Microfilaments:

associate readily with the cell membrane ("not only at their ends, but all along their length": Mooseker & Tilney, 1975; cf. Tilney, 1976; Begg, Rodewald & Rebhun, 1978; Begg & Rebhun, 1979; Kemp, 1979; Tilney, Bonder & DeRosier, 1981; Burridge, Kelly & Connell, 1982; Weber & Glenney Jr., 1982; Cline et al., 1983; Bennett & Condeelis, 1984; Christensen, Sauterer & Merriam, 1984; Tamkun et al., 1986; Bretscher, 1993; Pavalko & Otey, 1994);

associate with the cell membrane via a family of specific proteins (Arpin, Algrain & Louvard, 1994; Marrs & Nelson, 1996);

associate with proteins which may provide the more rigid structure found in the cortex (cortexillins: Faix et al., 1996);

can be involved in "cytokine-induced reorganization of the actin superstructure" of the membrane (Heldman & Goldschmidt-Clermont, 1993);

can cause cytoplasmic streaming at the cell membrane (Hayashi, 1964; Donaldson, 1972a,b; Kamitsubo, 1972; Allen, 1974, 1980; Allen & Allen, 1978a,b; Kachar & Reese, 1988);

can interact with ion channel proteins embedded in the cell membrane (Berdiev et al., 1996; Prat et al., 1996);

change configuration on interaction with the Rho family of GTP binding proteins (Machesky & Hall, 1996) and various isoforms of tropomyosin (Lin et al., 1997).

Their genetic expression "responds to changes in cell configuration" (Farmer et al., 1983), and so would appear to have a mechanical component.

The MF ring in the cell state splitter seems stable, however, as judged from Burnside (1971, 1973a). The only MF rings that have had their physical properties measured are those in marine invertebrate eggs at first cleavage, using two glass rods stuck into the ~1 mm egg to see how much the thinner of the two rods bends (Rappaport, 1967, 1971; cf. Schroeder, 1987a, 1990). It would be useful to perform similar experiments on cell state splitter microfilament rings and other microfilament rings for comparison (Fujiwara & Pollard, 1976; Owaribe, Kodama & Eguchi, 1981; Owaribe & Masuda,

1982). The microfilament rings studied in marine embryos are involved in the first cleavage (cf. cleavage furrow rings in amphibians: Perry, John & Thomas, 1971; Mabuchi et al., 1988). However, unlike those in newt neural plate cells (Burnside, 1971, 1973a), they lose mass as they contract (Schroeder, 1972, 1975; White, 1990), presumably by depolymerization. Thus they may not be a good model for the microfilament rings in cell state splitters.

In addition to their active contractility, microfilaments can exhibit elastic (Zaner & Hartwig, 1988) and supercoiling properties (rotating polygons: Kuroda, 1964, 1990, with sharp bends: Kachar et al., 1993).

3.09 Force Generating and Load Bearing Cytoskeletal Components: Intermediate Filaments (IF)

Intermediate filaments are a family of more than 50 different, often cell type specific proteins (Osborn & Weber, 1982; Traub, 1985; Nagle, 1988, 1994; Steinert & Roop, 1988; Goldman & Steinert, 1990). Their functions have been a mystery (Geiger, 1987; Heins & Aebi, 1994), though a mechanical role for them has long been suggested (Lazarides, 1980; Goldman et al., 1985; Brodland & Gordon, 1990; reviewed in Christian, Kelly & Moon, 1991; Brown, 1993). Nevertheless, other roles must be played by intermediate filaments:

"It seems, however, unlikely that IF in the cell accomplish a merely structural role, considering the diversity of IF proteins and the complex regulation of their gene expression. In this work we primarily present electron microscopic data that points to the presence of interactions between IF and several cellular components, namely the nucleus, plasma membrane, other cytoskeletal elements, cytoplasmic organelles and ribonucleoproteins" (Carmo-Fonseca & David-Ferreira, 1990).

For example, one role of IFs is to act as a transient store of a nuclear matrix protein (Marugg, 1992). Intermediate filaments seem to be an essential component of the cortex of the outer cells in amphibian embryos (Heasman, Torpey & Wylie, 1992): when "[IF] cytokeratin filaments become depleted in the cortical cells of the [*Xenopus*] embryo... there is a loss of the 'compacted' epithelial surface of the blastula, an inability to close a wounded surface and defective gastrulation" (Torpey, Wylie & Heasman, 1992).

But the mystery of the primary functions of IFs remains: "Embryos overexpressing either wild-type or dominant mutant [IF] vimentin subunits exhibit grossly normal development" (Christian, Kelly & Moon, 1991), as do vimentin knock-out mice (Colucci-Guyon et al., 1994; Brandon, Idzerda & McKnight, 1995).

Their structure is reasonably well conserved, though apparently not always between homologous tissues of different species (Herrmann et al., 1996). Cortical wound healing in embryos (Luckenbill, 1971; Bluemink, 1972; Stanisstreet & Panayi, 1980; Stanisstreet, Wakely & England, 1980; Stanisstreet, 1982; Merriam & Christensen, 1983; Smedley & Stanisstreet, 1984a; Stanisstreet, Smedley & Snow, 1985; Hengst & Reitsma, 1976; Stanisstreet, Smedley & Veltkamp, 1986), a quite distinct phenomenon from later wound healing of mature skin (via cell migration, reviewed in Trinkaus, 1984a), may thus involve intermediate filaments. (However, gastrulation defects may not involve the same cortical intermediate filaments: Klymkowsky, Shook & Maynell, 1992.)

Intermediate filaments often bridge the space between the nuclear membrane and the cell membrane and other membranous organelles (Goldman et al., 1985; Katsuma et al., 1987; Carmo-Fonesca, Cidadão & David-Ferreira, 1987; Pavalko & Otey, 1994; Georgatos & Maison, 1996), where they bind at least to integrins (Carver & Terracio, 1993). One class, lamins, are found on the inner surface of the nuclear membrane (Aebi et al., 1986; McKeon, Kirschner & Caput, 1986; Georgatos & Blobel, 1987; Blumenberg, 1989). The lamins have a spatially nonuniform distribution (Paddy, Agard & Sedat, 1992), and their own set of IF associated proteins (IFAPs: Fuchs & Weber, 1994).

Contrary to earlier opinions, IFs polymerize, depolymerize, and copolymerize to a considerable extent (Angelides, Smith & Takeda, 1989; Lee et al., 1993; Fuchs & Weber, 1994). Their gene expression varies with cell shape (Ben-Ze'ev, 1984a; Singhvi et al., 1994) and extracellular matrix (Getzenberg et al., 1991) changes. They respond to heat shock by collapsing (Walter, Petersen & Biessmann, 1990) and to stress (Welch, Feramisco & Blose, 1985).

While intermediate filament meshes are probably viscoelastic in nature (Brodland & Gordon, 1990), over periods of up to an hour or so they may respond as nearly fully elastic materials (Martin & Gordon, 1997a,b), as plausibly shown, for example, by our experiments on the rate of rounding of axolotl gastrulae flattened from 2 mm to 1 mm for various times:

"Embryos of the Axolotl (*Ambystoma mexicanum*) were compressed at various stages from blastula to early neurula to see if gentle mechanical forces could alter the course of primary neural induction (about stages 11-11.5). The 1.8-2 mm diameter embryos, during compression to 1 mm between glass plates or Nucleopore filters for 6 hours, continued through the same stages as controls. On release, the embryos returned to a spherical shape within a few hours, and often developed a variety of morphological and movement defects. Middle stage gastrulae (stages 10.75-11.5) completed gastrulation successfully after compression. Up to 90% developed extreme spinal flexion (kyphosis, with head and tail arched upwards) at stages 10.5-11, while this defect occurred at lower frequency at earlier and later stages. Our compression induced kyphosis may thus correlate with the event of primary neural induction" (Björklund, Gordon & Martin, 1991).

It would certainly be useful to quantitatively measure the constitutive properties of the intermediate filament ring in the cell state splitter (cf. Wang et al., 1993b). The elastic component of the constitutive properties probably results in intermediate filaments being reasonably stiff (Epstein et al., 1988). This property may be important in their apparent role (presumably mediated by junctions) of "coupling somites to one another" (Cary & Klymkowsky, 1994) and perhaps other cells in tissues, such as epithelia. Keratins, which are intermediate filaments in skin (stratum corneum), are associated with various human keratinopathies (Freedberg, 1993; cf. Nagle, 1994).

A current theory for the variety and function of intermediate filaments is that they themselves are regulators of gene expression, perhaps at the level of what I will call whole differentiation cascades:

"Since cIF [cytoplasmic intermediate filament] proteins do not possess classical nuclear localization signals, its is proposed that cIFs directly penetrate the double nuclear membrane, exploiting the amphiphilic, membrane-active character of their subunit proteins. Since they can establish metastable multisite contacts with nuclear matrix structures and/or chromatin areas containing highly repetitive DNA sequence elements at the nuclear periphery, they are supposed to participate in chromosome distribution and chromatin organization in interphase nuclei of differentiated cells. Owing to their different DNA-binding specificities, the various cIF systems may in this way specify different chromatin organizations and thus the expression of distinct sets of cell- or tissue-specific proteins" (Traub, 1995). [Cf. Traub & Shoeman, 1994.]

The problem with this hypothesis is the lack of effect of single cIF gene knockouts in mice (Traub, 1995). Thus a satisfactory explanation for the existence and variety of IFs is still needed.

Microtubules, microfilaments, and intermediate filaments appear to be the basic components of the cytoskeleton. I have reviewed their properties because they have not yet been incorporated into modern molecular development, which emphasizes gene expression and gene control. It is the mechanical properties of these molecules that may bridge the gap between genome and development.

3.10 Combinations of Cytoskeletal Components

The viscoelastic properties of all three cytoskeletal proteins have been measured in isolation (Janmey et al., 1991). It is possible that these properties vary from one cell type to another, since there are cell type specific forms of each protein:

microfilaments (Davidson et al., 1982; Garcia et al., 1984; Shott et al., 1984; Shani, 1985; Gurdon et al., 1985; Mohun & Garrett, 1987; Radice & Malacinski, 1989; Mohun et al.,

1984, 1989; Rubenstein, 1990; Macias & Sastre, 1990; Brennan, 1991a; Saint-Jeannet et al., 1992);

microtubules (Raff et al., 1987a; Cleveland, 1987; Kreis, 1987; Debec, Wright & Géraud, 1994; Raff et al., 1997);

intermediate filaments (Duprey & Paulin, 1995).

However, they all interact with the cell membrane and sometimes with the nuclear membrane, and with each other, as follows:

Microtubule/intermediate filament interactions (MT/IF)

MT/IF interactions (Downing et al., 1984; Williams Jr. & Aamodt, 1985) are mediated by specific microtubule attachment proteins (MAPs : Aamodt & Williams Jr., 1984; Bloom, Luca & Vallee, 1985; Flynn, Joly & Purich, 1987; Hirokawa, Hisanaga & Shiomura, 1988; Allende et al., 1989; Hirokawa, 1991). This interaction probably allows microtubules to support up to 10,000 times the compressive force that would otherwise cause them to undergo Eulerian buckling (Brodland & Gordon, 1990; cf. Coughlin & Stamenovic, 1997). (Microtubules are already "~300 times stiffer than actin filaments": Mitchison, 1992b). This interaction with microtubules gives intermediate filaments their first clearly defined mechanical role in the cell. This hypothesis has been utilized for tensegrity theory of the cytoskeleton by Ingber et al. (1994). It alleviates some of the persistent 'mystery' or 'enigma' of their function (Fliegner & Liem, 1991; Albers & Fuchs, 1992).

Intermediate filament/microfilament interactions (IF/MF)

IF/MF interactions have been reported in:

keratinocytes (Green et al., 1987);

muscle (Lazarides, Hubbard & Granger, 1979; Gard & Lazarides, 1980; Lazarides, 1982; Wang & Ramirez-Mitchell, 1983; Christian, Kelly & Moon, 1991);

fibroblasts (Green, Talian & Goldman, 1986);

at the *cell membrane* (Fowler & Taylor, 1980; Fowler et al., 1981), including that of platelets (Bearer, 1990).

In platelets, "actin is hypothesized to polymerize from its membrane-bound end (Tilney, Bonder & DeRosier, 1981; Tilney et al., 1986; Tilney & DeRosier, 1986)" (Bearer, 1990). "As the filament grows, cross-linking proteins... bind them together and cause them to stiffen. This produces pressure on the membrane, deforming it" (Bearer, 1993). This 'stiffening' mechanism is analogous to that of microtubules linked by intermediate filaments (Brodland & Gordon, 1990) or dynamin (Shpetner & Vallee, 1989). (Note that this polymerization is perpendicular, with respect to the cell membrane, to that observed in microfilament rings.)

Microtubule/microfilament interactions (MT/MF)

MT/MF interactions (Pollard, Selden & Maupin, 1984; Mitchison, 1992a) are especially well known in the mitotic spindle (Sanger, Sanger & Gwinn, 1979; Sanger et al., 1989; Ingber et al., 1994; cf. Bowser, Travis & Rieder, 1988). They also occur at the microtubule organizing center (MTOC: Bearer, 1992). Odell & Foe (1996) have proposed a model by which such interactions allow the spindle to position a microfilament ring (cf. Chang, Drubin & Nurse, 1997). Here is what may be the first statement that the spindle is a tensegrity structure, with opposing forces:

"...It is possible to develop a theory considering two types of polymeric fibers developing in the spindle. While one contracts, the other expands" (Rashevsky, 1960).

To round out this picture, we should also consider interactions causing bundling of each cytoskeletal component with itself:

Microtubule/microtubule interactions (MT/MT)

Microtubules bundle together in many circumstances. The obvious case is that of sperm tails and cilia (Kim et al., 1987; Vasiliev, 1987; Falconer, Vielkind & Brown, 1989; Brodland & Gordon, 1990; Glomski & Tamburlin, 1990; Zheng & Chang, 1990; Tanaka & Kirschner, 1991; Wise et al., 1991; Chandra et al., 1993; Gliksman, Parsons & Salmon, 1993; Gliksman & Salmon, 1993; via dynamin: Shpetner & Vallee, 1989, 1992; Chen et al., 1991; via pp170 and desmosomes: Wacker et al., 1992). Microtubule bundles are part of the neurofibrillary tangles of Alzheimer disease (Grundke-Iqbal & Iqbal, 1989; Matsuyama & Jarvik, 1989; Pappolla et al., 1990; Steiner et al., 1990). Dynamin is a "microtubule-binding protein [that]... was found to cross-link microtubules into highly ordered bundles" (Chen et al., 1991; cf. Shpetner & Vallee, 1989, 1992; Obar, Shpetner & Vallee, 1991; van der Bliek & Meyerowitz, 1991). Interacting microtubules can form fractal dissipative structures on various size scales (Tabony, 1994).

Intermediate filament/intermediate filament interactions (IF/IF)

Intermediate filament bundling is also widely observed (Blose & Chako, 1976; Klymkowsky, Miller & Lane, 1983; Klymkowsky & Plummer, 1985; Bologna, Allen & Dulbecco, 1986; Klymkowsky, Maynell & Polson, 1987; Phaire-Washington et al., 1987; Dent et al., 1992; Martin & Gordon, 1997a,b). Bundling is sometimes temperature dependent (Herrmann et al., 1993);

Microfilament/microfilament interactions (MF/MF)

Microfilament bundles are common, especially as microfilament rings at the apical ends of epithelial cells (DeRosier & Tilney, 1981; Mabuchi et al., 1988; Sheterline, 1993; Wong & Adler, 1993; Mabuchi, 1994). MF bundling involves a diverse class of actin bundling proteins (Weber & Glenney Jr., 1982; Otto & Schroeder, 1984; Zand & Albrecht-Buehler, 1989; Otto, 1994), though it may have a simple physical chemical cause due to the polyelectrolyte nature of polymerized actin (Tang & Janmey, 1996). Microfilament bundles are capable of angular and other motions (Higashi-Fujime, 1980; Jarosch, 1992) in addition to contraction (Owaribe, Kodama & Eguchi, 1981; Nagele et al., 1989, 1991). They also have mechanical roles as stress fibers (Sanger, Sanger & Jockusch, 1983; Byers, White & Fujiwara, 1984; Wechezak et al., 1989) which can alter their configuration and apparent equilibrium with microfilament rings in response to changes in shear stress (Kim, Gotlieb & Langille, 1989; cf. Byers, White & Fujiwara, 1984).

Higher order supramolecular assemblies, such as cell adhesion molecules attached to membrane proteins, that are themselves attached to the cell membrane and the cytoskeleton, also occur and are worthy of study (Nelson, Wilson & Mays, 1992; Stevenson, Gallin & Paul, 1992; Maniotis, Chen & Ingber, 1996; Barth, Nathke & Nelson, 1997). The spindle apparatus, microtubule organizing centers (centrosomes), the bridging of cytoskeleton from nucleus to cell membrane, the cortex (Ishikawa, 1988), the nerve cell synapse (Hirokawa, 1991), and the cell state splitter, could be regarded as such supramolecular assemblies, which may be 'self-assembled' (Nédélec et al., 1997). Gerhart & Kirschner (1997) prefer to call them 'self-organized', as they are dynamic structures that exhibit what they call 'exploratory behavior'. As we consider the whole cell or tissue level, the mechanics of the cytoskeleton lead to many dynamic possibilities, such as cell motility, stress induced alignment (Sherratt & Lewis, 1993), differentiation waves, etc. A quantitative analysis at these levels may be based on tensegrity structures (Ingber et al., 1994), for which computer simulations should be possible (cf. Stamenovic et al., 1996). It has been suggested that supramolecular cytoskeletal assemblies are actually capable of 'intracellular computation' (Hameroff et al., 1992; Lahoz-Beltrá, Hameroff & Dayhoff, 1993; Dayhoff et al., 1994; Lahoz-Beltrá & Hameroff, 1995; cf. Thomas, 1991), though the most I will require for the whole cell, in this book, is just one one-bit 'computation' per differentiation step.

Much of the action that occurs at the level of the whole cell may be the consequence of discrete complexes of cytoskeletal elements. In this book I will concentrate on the cell state splitter, though two other supramolecular assemblies, the spindle and the cortex, will also be seen to play prominent roles in embryogenesis. All of these assemblies may have a common evolutionary origin.

When I consider models for bilateral asymmetry (Section 8.02), a "residual" alignment of microtubules around the whole embryo is presumed, despite the presence of cell-cell boundaries (Proposition 251). This constitutes long range order of the cytoskeleton. It has been observed that two different microtubule bundles in adjacent cells can somehow interact end to end through the pair of cell membranes separating them to form what appears to be a mechanically continuous microtubule bundle spanning the cells (Lin & Forscher, 1993). I have seen hints of this in high voltage electron micrographs of axolotl neural plate (unpublished results done with Richard Cole).

4.00 DEVELOPMENT AND GENETICS

"There are two trends of thought that have been pernicious to the development of biology. The first is that the human species is unique.... The second is that complexity dominates in the biological world..." (Lima-de-Faria, 1983).

"Classical embryologists like Lillie [1927], Child [1941], Harrison [1933a, 1937, 1969] and Dalcq [1938, 1957] did not recognize the importance of genetics for the understanding of development. Also Spemann [1938] explained the results of his experiments in epigenetic terms by assuming epigenetic interactions between groups of cells in the embryo. This is somewhat surprising, since Spemann & Schotté (1932) did one of the most spectacular experiments in experimental embryology, demonstrating the importance of genes in the control of development. Schotté succeeded in transplanting ventral ectoderm of the frog gastrula to the mouth region of a salamander gastrula and vice versa, and in raising the experimental embryos to larval stages. The resulting salamander larvae developed frog-type horny jaws and suckers, and the frog tadpoles had salamander-type balancers and dentine teeth. Clearly, the genes of the donor tissue determine the type of mouth parts that are formed. As pointed out by Hamburger (1988), the genetic implications of this experiment were largely missed by Spemann. It was T.H. Morgan (1934) who realized the importance of genetics for the understanding of development and first proposed the hypothesis that development is controlled by differential gene activity" (Gehring, 1993). [Though: "In his book Morgan (1934) considered but rejected the possibility that 'different batteries of genes come into action as development proceeds.'" (Judson, 1979).]

"In fact, little or no true headway has been made in genetical investigations which aim at showing how the gene actually brings about the transformation of a microscopically comparatively homogeneous egg into a complex heterogeneous organism" (Fothergill, 1953).

4.01 The General Cell State Splitter

"Notice that we have not concluded that selective gene expression explains cell differentiation, for we have not discussed how a cell comes to choose the appropriate subset of genes relevant to its destiny" (Maclean, 1989).

Proposition 1: every step of determination in eukaryotic cells starts with the action of a cell state splitter.

"The contrast between the conception of embryonic segregation [determination] and gradual determination in the origin of capacity for self-differentiation may be brought out by a scheme of Brandt (1929).... There is a new (epigenetic) discontinuous process at the beginning of the critical stage that makes induction of two specific segregates possible.... On the views of Vogt [1922, 1923, 1925] and Goerttler [1925, 1926, 1927], as I understand them, there is nothing corresponding to the critical phase..., but equipotency shades gradually into dynamic and material determination.... The mechanism of the next succeeding embryonic segregation exists in every cell.... It is reasonable to hold that embryonic segregation is ultimately an intracellular phenomenon" (Lillie, 1929a).

"The molecular basis of determination remains one of the major unsolved puzzles of development" (Gilbert, 1994).

If a cell state splitter actually accomplishes differentiation as a discrete, discontinuous process in each cell, then it is the first identified *morphogenetic apparatus*. The cell state splitter shares a number of properties with the spindle apparatus: similar cytoskeletal composition, mechanochemical operation as a tensegrity structure (Pickett-Heaps, Forer & Spurck, 1996, 1997), and lack of an organelle membrane. Furthermore, the cell state splitter may be a transient structure, like the spindle apparatus (Appendix I: Gordon & Brodland, 1987), sharing its monomeric components with other organelles, such as cilia (however, cf. Oka, Arai & Hamaguchi, 1990). It may involve one or more microtubule organizing centers (Appendix I), and thus perhaps, the centrioles, though this has yet to be looked for. Whether the cell state splitter can coexist with the spindle apparatus in the same cell, or they must occur at different times, has yet to be looked into (though the tight linking of mitotic waves to differentiation waves that I'll discuss in Section 9.04 would suggest coexistence is possible).

This proposition is directly testable, since I do assume that every cell state splitter is a mechanical *biodevice* (a term suggested by Glen R. Klassen, p.c.), and thus should not escape electron microscopy or immunofluorescence microscopy and mechanochemical and mechanical analysis of its functioning.

There is at least one precedent for the concept of cell state splitters: Fristrom & Rickoll (1982) discuss what appears to be a cell state splitter for the differentiation of photoreceptor and cone cells in the forming eye of the fruit fly *Drosophila* (their Figure 12). In the eye imaginal disc they found a structure resembling Figure 1 in Appendix V (Gordon, Björklund & Nieuwkoop, 1994):

"A belt of circumferentially oriented microfilaments associated with the zonula adherens...(belt desmosome or intermediate junction)... encircles the apical end of each cell.... A set of microtubules that may have a morphogenetic function occurs in the vicinity of the zonula adherens (Wehman, 1969; Fristrom & Fristrom, 1975). These tubules appear to encircle the cell just inside the adherens-associated [micro]filaments.... We can envision two possible roles for the circumferential microtubules. They may serve as antagonists to the contraction of circumferential microfilaments and thereby prevent the apical constriction of some cells, or they may, by elongating and/or sliding past each other, actively expand the cell's perimeter" (Fristrom & Rickoll, 1982).

While Fristrom & Rickoll (1982) thus recognized the mechanical opposition and possibility of a morphogenetic role for this 'cell state splitter', they were not aware of its potential for mechanical instability nor of its more general occurrence. In a recent review, Fristrom & Fristrom (1993) add the following details:

"Zonulae adherens (ZAs)... form a continuous band encircling the apical ends of cells. ZAs are typically closely associated with an apical ring of filamentous actin. In vertebrates, the actin filaments are linked, via vinculin and α-actinin, to transmembrane adhesion proteins (cadherins) located in the junctional membrane (Geiger et al., 1981). This connection between the cytoskeletons of one cell and the next provides a mechanism for transmitting forces generated by the contraction of the apical ring of actin filaments across the epithelium (Odell et al., 1981; Owaribe, Kodama & Eguchi, 1981) and is an important mechanism for changing cell and tissue shape during epithelial morphogenesis (for review, see Fristrom, 1988; Schoenwolf & Smith, 1990a). In addition to a mechanical role in morphogenesis, the

ZAs and associated cytoskeletal complex may also function in signal transduction between the extracellular matrix and the cytoskeleton.... The distribution of microtubules at different stages of disc development has been described... (Fristrom & Fristrom, 1975; Poodry, 1980; Tucker et al., 1986; Milner & Muir, 1987)" (Fristrom & Fristrom, 1993).

Crawford (1979) related 'focal contractions' caused by apical microfilaments, in cultured pigmented epithelial cells from chick retina, to their redifferentiation and cell shaping. Woods & Bryant (1992, 1993) have formulated a general model for 'cell signalling in epithelia', which nicely complements our mechanochemical model for the cell state splitter:

"The *sev*[+] [*sevenless*] product is localized to the apical-most few microns of the lateral cell membrane and to the apical microvilli of the photoreceptor precursors... [in the *Drosophila* eye imaginal disc] (Banerjee et al., 1987; Tomlinson et al., 1987). Similarly, the *boss* product is localized on the apical microvillar membrane and a narrow band of plasma membrane on the R8 photoreceptor cell just below the apical microvilli (Cagan & Ready, 1989a; Kramer, Cagan & Zipursky, 1991).... Thus, the localization of both *boss* and *sev* suggests that the signalling process occurs either at the apical lateral plasma membrane or between apical microvilli of the signaling and responding cells" (Woods & Bryant, 1992).

In other words, these 'signalling' molecules are in the cell state splitter (Figure 1 in Appendix V: Gordon, Björklund & Nieuwkoop, 1994; Martin & Gordon, 1997a,b), apparently attached to the junctions between the cells:

"Genetic analysis in *Drosophila* has led to the identification of several proteins that mediate cell-cell interactions controlling the fate and proliferation of epithelial cells. These proteins are localized or enriched in the adherens and septate junctions at the apical end of the lateral membranes between cells. The proteins localized or enriched at adherens junctions include Notch, which is important for the cell interactions controlling neuroblast and bristle patterning; Boss and sevenless, which are required for the cell interaction that establishes the R7 photoreceptor cell; and Armadillo, required for the wingless-dependent cell interactions that control segment polarity and imaginal disc patterning. Proteins localized at septate junctions include the product of the tumor suppressor gene *dlg* [cf. Bryant et al., 1993], which is required for septate junction formation, apical basal cell polarity, and the cell interactions that control proliferation" (Woods & Bryant, 1993).

Furthermore, Woods & Bryant (1993), hypothesizing a role for "apical-basal 'elevator movement' (Jinguji & Ishikawa, 1992)" of the nuclei, more commonly known as interkinetic migration, postulate a direction of information transfer from the apical end of the cell (i.e., from the cell state splitter) to the nucleus, just as we did (Appendix I: Gordon & Brodland, 1987):

"If cell communication occurs at apical junctions, the signals generated must be transmitted to the nucleus where they can be translated into effects on patterns of gene expression and/or replication. This might suggest the need for a signal transduction mechanism to relay the signal through the cytoplasm to the nucleus, and many molecules that may participate in this signal transduction have recently been identified (Pelech & Sanghera, 1992)" (Woods & Bryant, 1993).

What is lacking here is a model for the launching and propagation of the implied wave of 'cell communication' and its coordination with the interkinetic migration and cell division. The physical furrow, while acknowledged, is ignored in the causal chain (see Chapter 8).

Karlsson (1984) has considered the possibility of a mechanical explanation for imaginal disc morphogenesis and differentiation in some depth.

The implications of this proposition will increase when I discuss mosaic development.

Proposition 2: cell state splitters come in a variety of forms.

In numerous organisms other configurations of microfilaments and microtubules have been observed which are in potential mechanical opposition, and thus may qualify as cell state splitters (Figure 17 in Appendix I: Gordon & Brodland, 1987). Possible cell state splitters, or analogous structures, are found in:

the scrosal cells in the bug *Pyrrhocoris* (Enslee & Riddiford, 1981);

the diatom silicalemma (Pickett-Heaps, Tippet & Andreozzi, 1979a,b; Gordon & Drum, 1994);

blood platelets (White, 1968, 1984; Gerrard, White & Rao, 1974; White & Burris, 1984; White & Sauk, 1984; White & Krumwiede, 1985);

sea urchin ectoderm (Tilney & Gibbins, 1969; Anstrom & Raff, 1988);

nematodes (Priess & Hirsh, 1986);

chick neural tube cells (Handel & Roth, 1971);

retinal photoreceptors (Madreperla & Adler, 1989);

erythrocytes (Kim et al., 1987; Cohen, 1991).

In erythrocytes there is evidence for a protein that may bind the microfilament ring and microtubule band together (Birgbauer & Solomon, 1989): the microfilament ring appears to attach to an open ring or hoop of microtubules, which at least in molluscan erythrocytes emanate from centrioles (Nemhauser, Joseph-Silverstein & Cohen, 1983). A sequence of microfilament and microtubule cortical arrangements occur in the 'aplanosporic spherical cells' of coenocytic green algae (Mizuta, 1994). While both components start out as random mats, they end up with some parallel arrays whose orientations coincide with the change of cell shape, and give the impression of possibly being in mechanical opposition.

Using only microfilaments and microtubules, we can envisage a large number of configurations that place them in mechanical opposition (Figure 17 in Appendix I: Gordon & Brodland, 1987). Thus I do not suggest that the cell state splitter configuration observed by Burnside (1971) and Martin & Gordon (1997a,b) is the only possible one, nor that cell state splitters must consist of microtubules and microfilaments in mechanical opposition. The general idea of a mechanical balance of the cytoskeleton ('counterpoised forces': Appendix I: Gordon & Brodland, 1987; cf. 'balance of opposing forces within tissues': Harris, 1987), has been expressed in terms of 'tensegrity structures' (Ingber & Folkman, 1989a; Watson, 1991; Ingber et al., 1994; Stamenovic et al., 1996), though the possibility of mechanical instabilities and bistabilities appears not to have been considered by these authors. There are significant opportunities here for quantitative molecular dynamics computer simulations. With the crude model I built using Tensegritoy (see Section 3.06), it became quite apparent that an annular mat of microtubules bound together by elastic elements can be quite stiff and stable. Of course, what Tensegritoy cannot simulate is the contraction of the microfilament ring or lengthening of the microtubules. We need to know under what conditions cell state splitters are bistable, what it takes to make them metastable (so they don't trigger prematurely), and what forces are required to get past the metastability to resolution of the bistability one way or the other. If this can be solved for a single cell, it might then be worthwhile constructing a tensegrity model of a sheet of cells, to investigate the conditions for propagation of differentiation waves. It would also be curious to add nonequilibrium statistical mechanics, such as the dynamic instability of microtubules, to see to what extent fluctuations at this level affect the mechanics.

Other cellular components that exert mechanical forces could also function in mechanical opposition with each other and with cytoskeletal components. These could include cell-cell adhesions (Bongrand, 1988; Honda, Yamanaka & Eguchi, 1986; Appendix I: Gordon & Brodland, 1987), such as may occur in mammalian embryonic compaction (Goel, Doggenweiler & Thompson, 1986; Winkel et al., 1990) and similar adhesions involved in an 'apicobasal polarity' in starfish blastomeres (Kuraishi & Osanai, 1989). It is not necessary that both opposing components be active (ATP or GTP) force generators, in order to attain a metastable state. However, the passive mechanical components would seem to have to be away from their mechanical equilibrium to oppose other forces in such a way as to create a mechanical instability.

Proposition 3: all cell state splitters are bistable.

In general, it seems likely that all possible cell state splitters would resolve their instability in one of exactly two available ways, because they would be composed of precisely two cellular components configured in mechanical opposition. (I am assuming that intermediate filaments and other cellular components do not have biases in the directionality of forces they exert in the apical plane.) Thus the instability of a cell state splitter is, in general, a *bistability,* leading to a *bifurcation.* I will interchangeably use 'tissue' and 'cell type' to designate an embryonic or terminal (usually adult) differentiated state. The nuance is that 'tissue' refers to all cells of the same 'cell type'.

There may be a fundamental reason that cell state splitters resolve to two states and thus why differentiation is probably always binary, i.e., leading to one of two new cell states (cf. Kaletta, Schnabel & Schnabel, 1997). An initial balance between three or more force generating cellular components (a 'trifurcation') would require a much more difficult 'balancing act' than the setup of just two components in metastable opposition:

"Multiple branching is possible only for a degenerate bifurcation.... Degeneracy is always created by certain special conditions that increase symmetry. Small perturbations of the dynamics can break symmetry and lift [such] degeneracy" (Volkenstein, 1994).

Empirically, in fact, we find that...

"The contraction [C] and expansion [E] waves are often paired ([Appendix V:] Gordon, Björklund & Nieuwkoop, 1994). For example, the portion of the ectoderm (AE) that experiences a contraction wave becomes neuroepithelium (AEC). The portion that experiences an expansion wave becomes epidermis (AEE)" (Appendix VI: Björklund & Gordon, 1994). [A = animal hemisphere]

The complementary trajectories of the expansion and contraction waves in a given tissue may be a simple consequence of the refractory period, i.e., the time it takes for a cell to set up a cell state splitter and be prepared for participation in the next wave.

Proposition 4: every step of differentiation results in one of two new states for the differentiating cell.

"I have said elsewhere (Lillie, 1927) that the process of origin of segregates [determination] is dichotomous, i.e. that segregates arise in pairs by the subdivision of a preceding segregate.... The idea of dichotomy is of course implicit in Weismann's (1892, [1893]) theory of development. In discarding the nuclear and determinant theory of Weismann we seem to have thrown away also his theory of dichotomy, which, as such, is independent of any determinant theory whatever...."

"The present example is a true case of dichotomy. The indifferent ectodermal field has become divided into two fields, an epidermal and a neural; the epidermis has now the positive epidermal potencies and lost the nervous, and the reverse is true for the neural field. Each of these is both positively and negatively restricted with respect to potencies" (Lillie, 1929a).

"Ten years ago it was difficult to say anything about biochemistry and morphogenesis. Today we have to record remarkable advances in knowledge centering round the biochemical nature of the 'morphogenetic hormones' which act in normal embryonic development, forming a hierarchy of inductors" (Needham, 1939).

Classically, differentiation has been conceived to be caused by 'inductions', which are (Figure 6)...

"*...those interactions which enable a developing system to switch into a certain pathway of differentiation which differs from the pathway already partially in progress* " (Nieuwkoop, Johnen & Albers, 1985).

"In... induction, cells are made to undergo a major change in their direction of differentiation..." (Gurdon et al., 1989).

Rather, the symmetric nature of cell state splitting suggests that, in general, each cell acquires one of two new states, and departs from its old state (Figure 7), in most cases. (Stem cells may be an exception. See Proposition 66.) This is what one generally expects of bifurcations in unstable systems.

The usual justification for the common view of differentiation as a deviation from a path is Holtfreter's exogastrulation experiment (Holtfreter, 1933d; Holtfreter & Hamburger, 1955), in which ectoderm appeared to differentiate only to epidermis, presumably because it failed to contact the organizer (inducer). However, the recent results of Kintner & Melton (1987) suggest that neural plate cells *are* formed in the neck of exogastrulae: "N-CAM RNA is expressed primarily at the junction between the ectoderm and the mesoderm". In fact, exogastrulae have normal spatial and temporal patterns of homeobox gene expression (Ruiz i Altaba, 1990, 1992; Doniach, Phillips & Gerhart, 1992; but cf. Nieuwkoop & Koster, 1995) and also express neurofilament-like protein NF3 (Kintner & Melton, 1987; Dixon & Kintner, 1989). The *Xenopus* gene *Xotx1,* involved in "the early subdivision of the rostral brain", has "expression [that] can be activated in the apparent absence of vertical signals of neural induction" (Kablar et al., 1996). Holtfreter's classical experiment, based on histological criteria that would no longer be considered adequate, may have actually led him to the wrong conclusion. This was well understood by Child (1941):

"Ectoderm of these exogastrulae forms an irregular cell mass, not a definite layer.... It may become ciliated, but the ciliated beat is not definite in direction, as in the normal animal.... These exogastrulae are regarded by Holtfreter [1933d,f] as providing complete proof that there is no determination in the ectoderm before invagination of the chorda-mesoderm and, consequently, that its development is entirely the result of induction in the normal individual. This conclusion ignores, or regards as erroneous, evidence from earlier work which is not in accord with it, as Huxley & de Beer (1934) have pointed out; and there are other points to be considered. The fact that the presumptive neural-plate region of the exogastrulae undergoes more stretching than presumptive epidermis indicates a difference of some sort [presence of notoplate?].... Moreover, the Ringer solutions that bring about exogastrulation may inhibit development to some extent.... If this is the case, the exogastrulae do not provide the final proof of complete dependence of neural development on induction" (Child, 1941).

Unfortunately, Holtfreter's interpretation was accepted and created a mindset, which continues to this day, about differentiation as a process that goes in a particular direction unless diverted by an outside stimulus. The mindset prevails even when the 'default state' is reversed from Holtfreter's epidermal to neural (Green, 1994a,b; Duprat, 1996). Presumptions are often made about default states relative to some purportive induction 'signal' (Irani & Schwartz, 1994; Lans, Ho & Weisblat, 1994), and though this is sometimes supported by tissue culture (Adler & Hatlee, 1989; Meyers-Wallen, 1993; Repka & Adler, 1992), differentiation waves may be missing or highly aberrant under these conditions. Moreover, the force balance for any cell state splitters in isolated, plated cells (Repka & Adler, 1992) is likely to be highly biased towards one state, which then becomes the 'default' for artifactual reasons. Default states for sex determination exist in some taxa (Baker et al., 1989; Crews, 1993; Francis & Barlow, 1993; Horabin & Schedl, 1993a; Lovell-Badge, 1993; Meyers-Wallen, 1993), though except for gynandromorphs (Gilbert, 1991a), all cells in the organism are affected, i.e., sex determination is not a case of spatial differentiation (Proposition 53).

Due to Holtfreter's presumption, arguments now occur as to which pathway is the 'ground state', an analogy to an inapplicable concept of quantum physics:

"Contrary to the popularly held view that ventral is the ground state for all mesoderm, our results suggest that formation of ventral mesoderm requires an active signal and that, in the absence of this ventral signal, dorsal mesoderm is formed" (Graff et al., 1994).

"The induction of neural ectoderm in the amphibian embryo has typically been viewed as a standard inductive event, with signals from the dorsal mesoderm passing either tangentially or vertically to convert the upper regions of the dorsal animal hemisphere from an ectodermal to a neurectodermal fate (reviewed in Slack & Tannahill, 1992). New evidence suggests that the opposite may be true... (Godsave & Slack, 1991; Hemmati Brivanlou & Melton, 1992; Hemmati Brivanlou, Kelly & Melton, 1994)" (Cornell & Kimelman, 1994).

(Cf. Green, 1994a,b.) In fact, this 'new evidence' dates at least to Nieuwkoop (1947a):

"It is a striking fact that the boundary between prechordal and chordo-mesodermal plate coincides with the boundary between arch- and deuterencephalon in all experimental animals.

The question arises: Is the regional differentiation of the archenteron roof a primary or a secondary one, i.e. do the presumptive areas of the archenteron roof induce always the same brain parts or can the differentiation of the archenteron roof be influenced secondarily by the differentiation of the induced neural plate? [See Nieuwkoop & Weijer, 1978.].... This last possibility definitely cannot be excluded" (Nieuwkoop, 1947a).

Given the fact that the ectoderm contraction wave is twice the depth of the ectoderm cells, a mechanical influence is indeed reasonable (Appendix V: Gordon, Björklund & Nieuwkoop, 1994).

Holtfreter's experiment lead to the general notion that a cell continues inexorably on a certain course of development, unless it is 'diverted' by 'environmental' cues to another differentiation pathway. This standard paradigm is quite different from our notion that each step of differentiation involves an active process leading to one of two new cell types or tissues, after which the previous cell type ceases to exist in the embryo (Figures 6, 7). Indeed, new data supports our interpretation of differentiation being an active process whichever way it goes: "...Ventral fate requires induction rather than resulting from an absence of dorsal specification" (Suzuki et al., 1994). (For another model of differentiation, see Kupiec, 1989.)

Weiss (1953a), with typical clarity, distinguished 'double-switch action', such as we are claiming for cell state splitters, from Holtfreter's presumption of differentiation via "single-switch action...[acting on]...the autonomous course from which the cells would have to be positively diverted". The single-switch action is equivalent to that of Waddington's 'evocator genes' (Waddington, 1940, 1947; Gilbert, 1991b). The concept of active, double-switch action goes back to Noll (1892) for geotropism and Herbst (1895) for "all cases where 'an organ directly influences another in its morphological character... by contact or by means of a specific substance, or in some other material manner'" (Oppenheimer, 1991).

Proposition 5: for some steps of differentiation, all the cells in the tissue resolve towards the same new state.

It is possible that in some circumstances all the cells in a tissue will resolve the instability of the cell state splitter the same way. (In such situations, we may not be aware that a step of differentiation has occurred, if the two tissues are morphologically indistinguishable.) In 'binary differentiation' I thus include the extreme possibility that all cells in a tissue may go in one direction at a binary fork. For example, in the experiment of Beloussov et al. (1990) relieving mechanical tension over the ectoderm of *Xenopus,* much of the ectoderm turned into neural plate. Perhaps this is because a few contraction waves (or their presumed equivalent in *Xenopus:* Nieuwkoop, Björklund & Gordon, 1996) were generated whose trajectories covered the whole tissue, precluding the possibility of corresponding expansion waves going through the same cells. (It might be possible to change a trajectory by the application of electric fields, mechanical pressure or a hydrostatic pressure pulse: Steve McGrew, p.c., 1997.) With dissociated and reaggregated neural plate cells of the urodele *Triturus alpestris,* Boterenbrood (1958) observed the "striking fact that often nearly the whole neural mass of an aggregate consists of eye structures alone". Heavy water (D_2O) treatment of *Xenopus* embryos prior to first cleavage results in "enlarged dorsal and anterior structures and reduced ventral and posterior ones" (Scharf et al., 1989), probably due to a larger trajectory for the *Xenopus* equivalent of the ectoderm contraction wave. Lithium has similar effects (Kao & Elinson, 1988). The pleiotropic *Notch* gene in *Drosophila* has an analogous 'hypertrophic' effect, and is thus probably a component of the differentiation pathway for contraction waves:

"It has been shown that the lack of *Notch* activity causes cells that normally follow an epidermal pathway to be misrouted towards a neural fate (Poulson, 1937; Lehman et al., 1983). Animals homozygous for a deletion of the *Notch* locus die as embryos, showing a hypertrophy of the nervous system at the expense of epidermal structures (Poulson, 1937).... The embryological evidence gathered so far suggests that cells in the neurogenic region are equipotent and that the choice between neural and epidermal fates relies on cell interactions during the critical early differentiation phases (Hartenstein & Campos-Ortega, 1984; Technau

& Campos-Ortega, 1987; Doe & Goodman, 1985a,b).... *Notch* is one of six zygotically acting genes (the others are *Delta, Enhancer of split, mastermind, big-brain,* and *neuralized*) known to affect this process.... *Notch* is expressed in a much wider range of tissue types than those disrupted in the neurogenic mutant.... In third instar larvae, *Notch* is expressed in imaginal disks and in the central nervous system" (Johansen, Fehon & Artavanis-Tsakonas, 1989).

We need a term to deal with a step of differentiation in which all of the cells in a tissue, in response to a differentiation wave passing through the whole tissue, differentiate the same way. Since the differentiation tree becomes linear, we could call this *linear differentiation*. Here is a case of what appears to be linear differentiation (cf. Proposition 53):

"Adrenal medullary chromaffin cells, SIF [small intensely fluorescent] cells and sympathetic neurons are derived from the sympatho-adrenal sublineage of the neural crest, and represent a range of cellular phenotypes extending from endocrine to neuronal. It is suggested here that these cell types may represent different stages of developmental 'arrest' along a linear pathway whose endpoint is a cholinergic sympathetic neuron. This model explains the 'transdifferentiation' of mature cells seen in this system as simply a delayed realization of transitions that normally occur between these stages during development. Such a 'linear model' of phenotypic diversification may be applicable to other developing systems that generate closely related but distinct cell types" (Anderson, 1989a).

One must take care, however, for in some cases an apparently linear differentiation may be a branching one, where one of the cell types disappears by apoptosis (Roach, Erenpreisa & Aigner, 1995).

Proposition 6: the state of embryonic tissue competence coincides with a state of mechanical metastability of the cell state splitter.

"We believe that the understanding of the nature of competence and of its emergence and decline actually constitutes one of the key problems in developmental biology " (Nieuwkoop, Johnen & Albers, 1985).

"...The nature of tissue competence has been largely ignored.... John & Miklos (1988) concluded that the problem of ectodermal competence was primarily one of cell mechanics rather than one of simple gene regulation" (Grunwald et al., 1990).

The prolonged maintenance of the cell state splitter may require a means for producing a mechanical metastability, i.e., an energy barrier that keeps the cell from resolving its instability spontaneously. Think of a marble balanced on a hill (Figures 11-13 in Appendix I: Gordon & Brodland, 1987). If the hill had a small dent at the top, the marble would be in a metastable situation. Metastability could be provided by cytoskeletal components such as desmosomes (Appendix I). Alternatively, as we have suggested (Martin & Gordon, 1997a,b; Section 3.06), this metastability may be due to the elasticity of the intermediate filament ring in the cell state splitter.

It is of interest that early on Waddington (1936) suggested the involvement of mechanics in competence:

"...The competence for lens formation only arises in thin sheets of ectoderm, either in the epidermis of the neurula or in the thin walls of the isolated vesicles.... One can... easily suppose that the competence can be independently produced by the ectoderm provided only that the mechanical conditions are favourable, *i.e.* that the ectoderm is in the form of a thin layer" (Waddington, 1936).

Later he seems to have dropped the mechanical point of view in favor an explanation that has a more modern, but entirely chemical flavor:

"Waddington's idea of 'competence' differed from the analogous German term *Reaktionsfähigkeit,* which implies a passive ability to respond to the stimulus given it. For Waddington (1939b), competence is actively achieved by 'a complex of reactions between substances which form an unstable mixture, which may at certain times have two or more alternative modes of change'" (Gilbert, 1991b).

"Granted that competence is a state of instability in a complex system of reactants, what may we suppose these reactants to be?... A choice as fundamental as that between epidermis and neural tissue involves the whole biochemical system of the cell; the protein synthesis, the ribonucleo-protein microsomes, the respiration and the genes" (Waddington, 1956a).

There is a small discrepancy between the classically defined duration of competence of the ectoderm to form neural plate, and duration of the ectoderm contraction wave (stage 10, early gastrula, to the very end of gastrulation and beginning of neurulation: stage 13-: Bordzilovskaya et al., 1989; Bordzilovskaya & Dettlaff, 1991; Figure 4; Table 1):

"The capability of the reacting cells to respond to an inductive stimulus is restricted to a certain developmental period, the period of competence. The duration of neural competence was investigated, among others, by Chuang (1955) in *Cynops orientalis*, by Goettert (1966) in *Triturus alpestris* and by Nieuwkoop (1958a, 1960) and Gebhardt & Nieuwkoop (1964) in *Ambystoma mexicanum*. Their results for the first step of neural induction, the 'activation' or 'neuralisation' step of Nieuwkoop (1952) and Toivonen, Saxén & Vainio (1963) respectively, are that competence lasts from st. 8 (late blastula) until st. 11 1/4 (late gastrula)" (Albers, 1987).

The discrepancy regarding the end of competence for neural plate formation may simply be due to the fact that the ectoderm contraction wave has only a relatively small bit of ectoderm tissue left to travel through at stage 11 3/4 (Table 1). The discrepancy in the beginning requires more careful consideration. First of all, the ectoderm contraction wave travels through presumably competent cells, so competence most likely occurs beforehand, i.e., the launching and propagation of the ectoderm contraction wave cannot occur until at least some (and perhaps all) of the ectoderm cells have set up cell state splitters. Erection of cell state splitters can start any time after traversal of the earlier animal cap expansion wave (Appendix V: Gordon, Björklund & Nieuwkoop, 1994), i.e., beginning at stage 9. Small discrepancies may also be due to Albers' (1987) use of the Harrison (1969) staging table for *Ambystoma punctatum*. For example, Albers (1987) counts early neurula as stage 11-12, while Bordzilovskaya et al. (1989) assign this stage 13-. (Cf. also the staging table for *Ambystoma mexicanum* in: Schreckenberg & Jacobson, 1975). Thus the discrepancy between the beginning of ectoderm competence and the launching of the ectoderm contraction wave may be an artifact of differences between staging tables.

Proposition 7: competence is maintained until a cell resolves the mechanical instability of its cell state splitter, or disassembles it.

In our model, each tissue is split into two tissues by a pair of waves, which are contraction and expansion waves. Once a cell has erected a cell state splitter, it should thus remain competent until it has participated in one of these waves. For cells far from either of the wave launching domains, the wait can be many hours. There has been some work on loss of competence (Albers, 1987; Servetnick & Grainger, 1991b), which would be worth repeating while observing differentiation waves, to see whether the loss is due to passage of a wave or occurs before a wave has arrived. I would suggest the former, at least under normal developmental conditions. If launching of a wave is prevented, as perhaps in exogastrulation (Nieuwkoop, 1995) or explantation experiments (Leikola, 1963, 1965; Ohara & Hama, 1979a,b; Grainger & Gurdon, 1989), it is possible that the cell state splitters are eventually disassembled (to some terminal state?) or in some other manner lose their mechanical metastability. For example, since cell state splitters are small systems (Hill, 1963, 1964), containing, for example, perhaps as few as 65 microtubules (Appendix I: Gordon & Brodland, 1987), stochastic fluctuations in their numbers, via dynamic instability, could eventually reach a point where they disrupt the force balance between the microtubule mat and the microfilament ring. In other words, from a statistical mechanical point of view, any system in a metastable state is likely eventually to cross over the metastable barrier. While this could conceivably cause subthreshold waves or ectopic waves, in a single cell it might just cause loss of competence, without propagating to neighbors.

A combination of time lapse and electron microscopy or in vivo immunofluorescence microscopy could settle the question of the mechanism of loss of competence, including the suggestion that it "is a highly autonomous process" (Grainger & Gurdon, 1989), which would imply preset timing and thus incompatibility with differentiation waves, unless the time is 'set' to be well beyond the normal maximal waiting period:

"Competence must therefore be lost by an autonomous process, which operates in single nondividing cells as would an internal clock" (Grainger & Gurdon, 1989).

In contrast to Grainger & Gurdon (1989), I think competence may prove to be a simple mechanical phenomenon, though one dependent on the mechanical integrity of the sheet of cells forming the epithelium. Duprat (1996) notes that the rise and fall of "L-type Ca^{2+} channels expression correlates with neural competence". This may indicate that competence ends in an ectoderm cell as it participates in the ectoderm contraction wave, and that these channels then disappear as the wave passes on.

Proposition 8: the classical multistep theories of early embryogenesis may be reinterpreted in terms of multiple differentiation waves.

An embryo differentiates into many kinds of cells. This has classically been thought of in terms of 'primary' neural induction of the neural plate, followed by 'regional' inductions defining detailed segmentation of the brain and spinal cord and placode formation. This is sometimes reduced to just two steps, though it is hard to imagine how only two would suffice to generate the full complexity of the brain:

"In the classical theories of neural induction (Yamada, 1950a; Nieuwkoop, 1952, 1958b, 1963; Toivonen, Saxén & Vainio, 1963) neural induction takes place in two steps: 1) induction of the ectoderm in a general direction ('neuralisation' with Toivonen, Saxén & Vainio, 1963, 'transformation' with Nieuwkoop, 1952); in this step size and form of the neural plate are determined, and 2) regional differentiation of the neuralised ectoderm ('caudalisation' with Yamada, 1950a, 'mesodermalisation' with Toivonen, Saxén & Vainio, 1963, 'transformation' with Nieuwkoop, 1952). [Cf. 'early nonregionalization' and 'gradual regionalization': Saha & Grainger, 1992.]

"It is suggested that both steps are brought about by gradients of diffusible inductive substances. The gradient of the neuralising/activating factor determines the size of the neural plate by means of a threshold value. When the gradient sinks under this value the ectoderm no longer responds and becomes epidermis. Yamada (1950a) differs from the others in... [suggesting] for the second step of neural induction a mechanical factor, the stretching of the embryo during its growth.... Towards the end of neural competence there appears to be a short phase of 'weak' competence, during which placodal material is induced. Another possibility would be that the placodal material is not induced during a period of declining neural competence but in a subsequent period of specific placodal competence..." (Albers, 1987).

In terms of differentiation waves, Albers' (1987) step 1 consists of launching and propagation of the ectoderm contraction wave (homoiogenetic induction), the size and form of the neural plate being determined by its trajectory. There is no need to invoke a gradient. His step 2 consists of launching and propagation of the waves of the neural plate, which may be quite numerous as brain differentiation proceeds, and which may include separate differentiation waves creating the placodes. These waves are likely to be sequential and spatially nested.

Nesting of waves it not a necessary feature of their propagation. For instance, the animal cap expansion wave keeps right on going when it encounters the mesoendoderm (Figure 9 in Appendix V: Gordon, Björklund & Nieuwkoop, 1994). On the other hand:

"As the presumptive notochordal mesoderm contraction wave [itself a continuation of part of the ectoderm contraction wave] encounters the dorsal lip of the blastopore, either the bottle cells there relax for about 0.5 hr, or no new ones are formed for that duration, over a small arc at the dorsalmost portion of the dorsal lip of the blastopore" (Figure 12 in Appendix V: Gordon, Björklund & Nieuwkoop, 1994).

What permits blockage or continuation of a wave into an adjacent tissue (as in Figures 9 and 11-12 in Appendix V: Gordon, Björklund & Nieuwkoop, 1994) has yet to be determined.

Proposition 9: cell state splitters are evolutionarily conserved.

If cell state splitters are the basis for differentiation, then they are as ancient as the phenomenon itself. Because cell state splitters are made of common cytoskeletal macromolecules, whose functioning is remarkably constant through evolution (Raff et al., 1987a), they are likely to be strongly conserved. One could argue that any mutations that change their functioning are likely to be fatal at early embryonic stages (John B. Armstrong, p.c.).

4.02 Differentiation Trees

"An alternative view would be to assume that different batteries of genes come into action as development proceeds.... The idea that different sets of genes come into action at different times is exposed to serious criticism, unless some reason can be given for the time relation of their unfolding" (Morgan, 1934).

"It seems evident at first sight that a mutation affecting the expression of a 'battery' of structural genes would lead to a much more profound difference in the development of an organism, and hence of its adult morphology and character, than one resulting in a change in the function of a single protein" (Delbrück, 1986).

Proposition 10: every organism developing from a single cell has a differentiation tree whose nodes (arranged along a time axis) correspond to events of cell state splitting.

"The fact that embryonic segregation [determination] proceeds from the more general to the more special in itself implies a time sequence that is fundamental for each branch of the dichotomous system" (Lillie, 1929a).

"The day may already be said to be in sight when the laborious description of embryonic conditions in verbal terms will be superseded by elaborate yet succinct nomograms, illustrating graphically all the stages or processes through which the organism passes or may pass" (Needham, 1931b).

According to Proposition 1, every cell, at any stage of development, has passed through a sequence of cell state splittings. Each time it resolved the instability of its cell state splitter into one of two new states, it went through a *bifurcation*. Different cells may resolve their instabilities in alternate ways, so for the whole set of them we may represent the process as a bifurcating treelike graph (Gordon, 1970) in which the nodes represent the acts of cell state splitting and the lines or *edges* between them are arranged in one axis in a time sequence (Figures 10 and 15). Each cell takes one path down the differentiation tree. (When I was a non-botanist in a botany department, I was accused of not knowing which end of a tree to stick in the ground. This hiatus in my training is apparent here.)

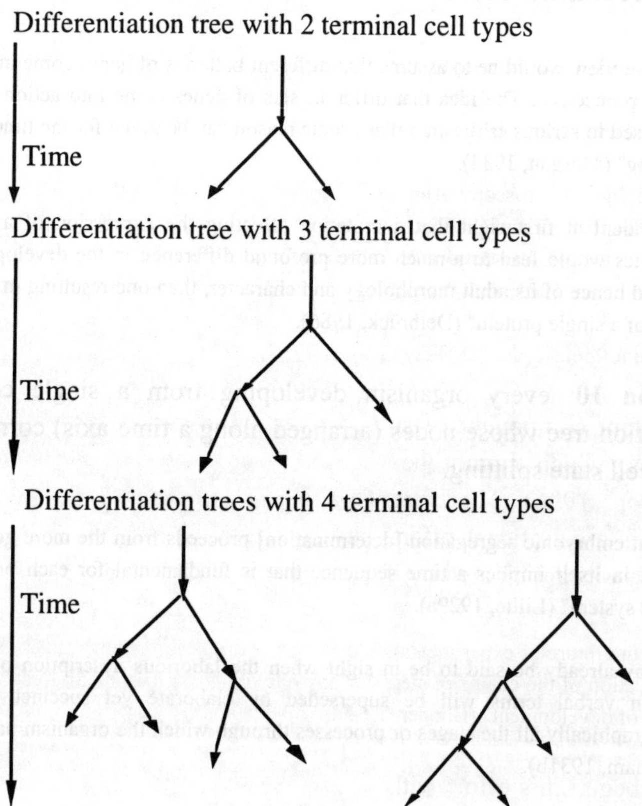

Figure 15. A few simple differentiation trees. Each node represents an act of cell state splitting. Each edge emanating from a node represents one of the two cascades of gene activity (differentiation cascades) that has occurred. Note that as developmental time goes on, the process of differentiation represented by an edge continues, until either the next act of cell state splitting, or completion of the differentiation. Any number of cell divisions might occur along an edge. (We have not represented the continued existence of terminally differentiated cells in these diagrams. Their edges could just be extended down the time axis to show this, as in Figures 1-2 in Martin & Gordon, 1995, unless cell death intervenes.)

The differentiation tree has an observable reality, because cell state splitters can be observed. But it is a difficult entity to investigate, requiring an immense effort at cell tracking, to follow each cell and find the cell state splitters within them. A full assault probably requires:

three dimensional (3D) time lapse microscopy (cf. Agard et al., 1989; Hiraoka et al., 1989, 1991; Minden et al., 1989; Kam et al., 1991; Wilson & Hewlett, 1990; Rykowski, 1991; Swedlow, Sedat & Agard, 1993; Fire, 1994; Brodland et al., 1995; Luchka & Gordon, 1998);

electron and light microscopy with serial sectioning and 3D reconstruction (Stevens & Trogadis, 1984, 1986; Frank, 1992a,b; Koster et al., 1992; Leu, Chen & Sun, 1993; Inoué & Inoué, 1994; cf. Gordon, Bender & Herman, 1970; Gordon & Herman, 1974; Gordon, Herman & Johnson, 1975; Shaw et al., 1989);

in vivo immunofluorescence (cf. Tsay et al., 1990; Hartenstein, Lee & Toga, 1995).

An important component of such research is histological identification of the tissues, which cannot be reduced to just studying gene expression (Nieuwkoop, 1996). Therefore, defining the differentiation tree for each organism is a significant effort, but experimentally possible. This program of research was anticipated by Bonner (1965a):

"Perhaps in the future of experimental embryology and development in general, the isolation and identification of the chains of steps for any one organism may be of great importance in the analysis of development" (Bonner, 1965a).

We have begun this effort with determination of the differentiation tree for axolotls though gastrulation (Figure 2 in Appendix V: Gordon, Björklund & Nieuwkoop, 1994) and its use to explain the early fate map of urodeles (Appendix VI: Björklund & Gordon, 1994).

Arthur (1984) deliberately chose morphogenesis rather than differentiation as a starting point for the role of development in evolution (which restricts the range of evolutionary change he could consider), although he later explicitly took note of a structure that is essentially a differentiation tree:

"...In one sense development is very obviously hierarchical - it is a hierarchical pattern of cell descent, or a hierarchical cell lineage, if that is not a contradiction in terms. This fact in itself is obvious and uninteresting..." (Arthur, 1987).

Eldredge (1985a) has also, in effect, dismissed the differentiation tree:

"...The somatic hierarchy, which indeed is unfolded each time an organism undergoes development, is a 'special case' hierarchy of no profound significance for evolutionary theory..." (Eldredge, 1985a).

(Cf. Davidson & Britten, 1971.) On the other hand, the 'developmental sequences' of Alberch (1985a) and Langille & Hall (1989a), the series of 'isoforms' of Caplan, Fiszman & Eppenberger (1983), "order and hierarchy between the DNA segments" (Lima-de-Faria, 1983), and the "genetic correlations within [functional] sets [of]... developmentally related traits" of Cheverud (1984) may correspond to consecutive edges of the differentiation tree, and inasmuch constitute an endorsement of the relevance of the concept. The idea that a "...detailed analysis of cell lineages should enable the different cell types to be classified in terms of a binary tree" whose nodes are 'epigenetic crises' (Stein, 1980) corresponds identically to the differentiation tree, if each 'crisis' is restricted to tissue wide assembly and resolution of cell state splitters through differentiation waves. The differentiation tree is consistent with the belief of Maynard Smith (1989a) in...

"...the most important reason why development is stepwise and hierarchical: there is a limit to the complexity of the patterns which can reliably be generated in one step" (Maynard Smith, 1989a).

McKinney & McNamara (1991) take the positive, if apologetic view that:

"...**morphogenetic 'trees'**... have a long history in developmental biology and continue to be used (Smith, 1983; Arthur, 1984).... As discussed in more detail by Alberch et al. (1979), Slatkin (1987), and Thomson (1988), traits develop from the same tissue until a transition occurs, after which they develop at least partially independently. *Cells have a shared history* (*time axis*) *until some event acts to spatially separate them* (*space axis*).... Everything from cellular activities (mitosis, migration, differentiation) to biochemical diffusion can be visualized.... the tree approach can supplement the 'closed loop' model of Oster et al. (1988)....

"As always, simplicity is bought at a price and we do not pretend that this view is analytically powerful. For more rigorous treatments of bifurcations [see]... Oster & Alberch (1982); Oster et al. (1988).... Nevertheless, the conceptual basis of developmental bifurcations is the same. Our purpose here is mainly to provide a graphical, conceptual framework. The tree directly depicts the spatiotemporal contingencies of cellular assembly. It reifies the usually vague (but often stated) idea of a cellular or developmental 'pathway.'...

"A serious drawback of the tree approach, noted by Raff (1989), is that... the idea that effects unerringly diminish monotonically is naive.... Issues as just how much development has constrained or 'channeled' evolution can never be resolved by rhetoric or diagrams with arrows" (McKinney & McNamara, 1991).

The idea hasn't quite caught on, however: the term 'morphogenetic tree' does not appear in any Medline literature title or abstract since 1983. The caveats of McKinney & McNamara (1991) notwithstanding, if we narrow the concept of morphogenetic trees down to that of differentiation trees, where the 'events' are physically observable differentiation waves, they become a powerful tool for dissecting embryogenesis and its impact on evolution.

If we add the notion of the lineage of cells to Goldschmidt's ideas on development, he may also be seen to have anticipated the differentiation tree:

"Goldschmidt (1927) accepted the concept of qualitative identity of all the cleavage nuclei and went on to... the assumption that there are specific genes for all the characteristics of each stage of development and that all other genes exist in a latent state at a given stage" (Barth, 1964a).

Waddington's (1961a) concept of 'a branching set of creodes' is effectively that of a differentiation tree, though without the physics of the differentiation waves. He realized that some sort of physics of the embryo had to be discovered and invoked:

"There seems to be no generally recognized word to indicate a path of change which is determined by the initial conditions of a system and which once entered upon cannot be abandoned. I have suggested for this idea the word 'creode' from the two Greek words $\chi\rho\eta$ necessity and $\alpha\delta o\zeta$ a path. We can say then that the hereditary materials with which an organism begins life define for it a branching set of creodes. Different parts of the egg will move along one or other of these creodes, so that they will, after a long process of progressive changes, finish up as one or other of a number of different end-results, as it might be heart, muscle, nerve, kidney and so on....

"This discussion so far has only dealt with the development of parts of the egg into cells of definite types, and we have left on one side the phenomena of individuation by which these cells become arranged into organs with definite shapes and patterns. I am afraid biologists have to confess that they still have hardly any notion of how this is done. It certainly must involve something more than purely chemical processes. Development starts from a more or less spherical egg, and from this there develops an animal which is anything but spherical; it has arms, legs, head, tail and so on, and internal organs which also have definite shapes. One cannot account for this by any theory which confines itself to chemical statements, such as that genes control the synthesis of particular proteins. Somehow or other we must find how to bring into the story the physical forces which are necessary to push the material about into the appropriate places and mould it into the correct shapes" (Waddington, 1961a).

I'll use the following definitions when referring to parts of a differentiation tree (Figure 15). In order to distinguish gene batteries corresponding to a whole step of differentiation from lower level groups of genes that are turned on/off in a concerted manner, I will call them *differentiation cascades*. *Edge* refers to one line on a differentiation tree, i.e., it represents one differentiation cascade corresponding to one step of differentiation. *Terminal branch* refers to a connected set of edges, out to the terminally differentiated cells. *Subtree* is any connected set of edges, not necessarily including terminal edges. *Subset* is any set of edges, not necessarily connected. These definitions are illustrated in Figure 16. I've avoided the term 'branch' by itself, since it has too many meanings.

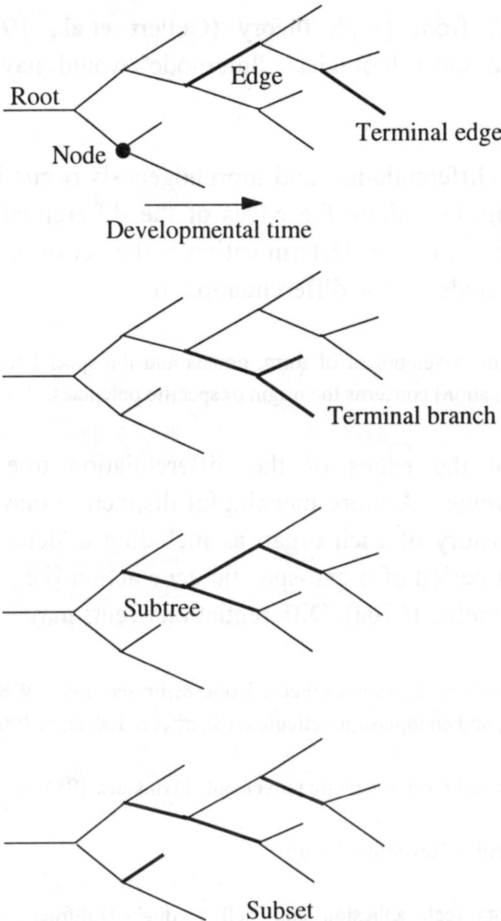

Figure 16. Terminology for parts of a differentiation tree. *Node* represents the splitting of the previous tissue into two tissues. *Root* is generally the fertilized egg cell, plus any early cleavages that do not result in cell differentiation. *Edge* refers to one line on a differentiation tree, i.e., it represents one differentiation cascade corresponding to one step of differentiation. *Terminal edge* is a differentiation cascade resulting in a terminally differentiated tissue. *Terminal branch* refers to a connected set of edges, out to the terminally differentiated cells. *Subtree* is any connected set of edges, not necessarily including terminal edges. *Subset* is any set of edges, not necessarily connected.

The term 'edge', from graph theory (Gellert et al., 1975), will seem uncomfortable to most biologists. 'Internode' would have been equally precise.

Proposition 11: differentiation and morphogenesis occur between acts of cell state splitting, i.e., along the edges of the differentiation tree, during each differentiation cascade. Determination is the act of cell state splitting, and occurs at the nodes of the differentiation tree.

"Morphogenesis is the development of form, organs and the special functions. Embryonic segregation [determination] concerns the origin of specific potencies..." (Lillie, 1929a).

This means that the edges of the differentiation tree correspond to differentiation events: "A more meaningful distinction may be to view the developmental history of each organ as including a 'determination event,' and a subsequent period of organ-specific gene action [i.e., differentiation]" (Bernfield & Wessells, 1970a). Differentiation events may include:

1: change in rate of cell proliferation (Wetts, Kook & Fraser, 1993; Wetts & Quon, 1995), mitochondrial shape, and endoplasmic reticulum (Sherbet & Lakshmi, 1967);

2: cell shape changes and morphogenetic movements (Trinkaus, 1984a);

3: tissue folding (Nardi & Reynolds, 1986);

4: cell rearrangements (cell adhesion and 'cell sorting': Holtfreter, 1948a; Townes & Holtfreter, 1955; Steinberg, 1963, 1970, 1996; Steinberg & Wiseman, 1972; Goel et al., 1975; Gordon et al., 1972, 1975; Mostow, 1975; Steinberg & Garrod, 1975; Fristrom, 1976, 1982; Harris, 1976; Jacobson & Gordon, 1976a; Phillips, Steinberg & Lipton, 1977; Keller, 1980; Poole & Steinberg, 1981; Steinberg & Poole, 1981; Kageyama, 1982; Ettensohn, 1985; Nardi & Magee-Adams, 1986; Fehon, Gauger & Schubiger, 1987; Keller & Hardin, 1987; Keller & Trinkaus, 1987; Hardin, 1989; Schoenwolf & Alvarez, 1989, 1992; Wilson, Oster & Keller, 1989; Oster & Weliky, 1990; Weliky & Oster, 1990a,b, 1992; Bard, 1991a; Alvarez & Schoenwolf, 1991; Grunwald, 1991; Wilson & Keller, 1991; Keller, Shih & Domingo, 1992; Costa, Sweeton & Wieschaus, 1993; Graner, 1993; Graner & Sawada, 1993; Duvdevani-Bar & Segel, 1994; Foty et al., 1996; Gumbiner, 1996);

5: differential adhesion brought about by new gene products such as cellular adhesion molecules (CAM's) (the events of 'topobiology': Edelman, 1988; Beddington, 1989), cadherins (Fleming, 1994), extracellular matrix proteins (Callaerts & De Loof, 1993), or integrins (DeSimone, 1994);

6: mechanical effects, such as swelling of one side of an epithelium, causing buckling (cf. Gordon, 1985c; Green, 1994c; Lane et al., 1993; Green, Rennich & Steele, 1996; Green, Steele & Rennich, 1996), probably also a result of new gene products;

7: perhaps sometimes a 'mitotic wave' of cell division (Maleyvar & Lowery, 1973, 1976, 1981; Hartenstein, 1989; Smith & Schoenwolf, 1987, 1988; Smith, Schoenwolf & Quan, 1994; reviewed in Section 9.04), etc., or asynchronous cell division (Jacobson & Gordon, 1976a; cf. possible cell proliferation defects as in: Bryant & Schmidt, 1990; Charron et al., 1992);

8: cell fusions, as in muscle formation and some other tissues (as in nematodes: Podbilewicz & White, 1994);

9: construction of basement membrane (Trelstad, 1984) and "basal cellular expansion owing to alteration of the cell cycle (Smith & Schoenwolf, 1987, 1988; reviewed in detail by Schoenwolf & Smith, 1990a,b)";

10: turning on or off ('down-regulating', as: Chong et al., 1995; Wilting et al., 1995, by 'tissue-specific extinguishers': Jones et al., 1991b) of differentiation cascades and lower order batteries of genes and gene products (Davidson, 1986), either by DNA binding proteins (Klug, 1993), by methylation and demethylation (Brandeis, Ariel & Cedar, 1993; Taylor, 1993; Martin et al., 1999; maybe via base flipping: Roberts, 1995; Travis, 1995a) or perhaps occasionally by antisense RNA (Bière, Citron & Schuster, 1992; Gehring, 1993), transcriptional enhancers (Kruse et al., 1995), RNA transcription factors (Krause, 1996) and histones (Grunstein, 1992);

11: interkinetic migration of the nuclei (Sauer, 1936, 1937; Watterson, Veneziano & Bartha, 1956; Sauer & Walker, 1959; Fujita, 1962; Watterson, 1965; Langman, Guerrant & Freeman, 1966; Messier, 1972, 1976a, 1978; Messier & Auclair, 1973, 1974, 1975, 1977; Lee, 1976; Nagele & Lee, 1979; Schoenwolf, 1982, 1986; Appendix I: Gordon & Brodland, 1987; Smith & Schoenwolf, 1988; Fischer-Vize & Mosley, 1994; Smith, Schoenwolf & Quan, 1994);

12: transfer of proteins from the nucleus to the cytoplasm (Wakamatsu et al., 1993);

13: change in nuclear rotation (De Boni, 1994; cf. Paddock & Albrecht-Buehler, 1988; Onyshko, 1989);

14: structural reorganization (Hiraoka et al., 1993) of the nucleus, such as "cell-type-specific rearrangement of functional nuclear components involved in RNA metabolism" (Lawrence, Carter & Xing, 1993), perhaps causing "transfer of the appropriate DNA from the interior of the nucleus to the region of the nuclear periphery" (Puck & Krystosek, 1992; cf. Krystosek & Puck, 1990; Park & De Boni, 1996) (though the domain model for the nuclear matrix suggests the reverse: Bodnar, 1988; Bodnar, Jones & Ellis Jr., 1989), possibly involving the penetration of cytoplasmic intermediate filaments (cIFs) into the nucleus (Traub & Shoeman, 1994; Traub, 1995);

15: transfer of proteins from the cytoplasm into the nucleus (Yamada & Kasamatsu, 1993; Li et al., 1994), including changes in the intermediate filament nuclear lamins (Nigg, 1989; Paddy et al., 1990; Burke, 1990; Paddy, Agard & Sedat, 1992; cf. Bier et al., 1988);

16: changes in inter-chromosomal domain channels through which regulatory molecules get to target genes, or don't (Cremer et al., 1993);

17: chromosome diminution (Davidson, 1986; Müller et al., 1991), so long as the germ line cells have been sequestered; turning off of telomerase synthesis in all adult cells except germ cells and constantly proliferating 'renewal' tissues (Travis, 1995c), but not embryonic cells (Mantell & Greider, 1994);

18: alternative RNA splicing or splice blocking (Rio, 1993) and other forms of posttranscriptional control (Witkowski, 1988b; Alberts et al., 1989; Ffrench-Constant & Hynes, 1989; Collett & Steele, 1993; Drevet, Swevers & Iatrou, 1995) and translational repression (Curtis, 1994).

Jacobson & Sater (1988) have noted that...

"...several aspects of embryonic induction that are critical to our understanding of induction as a developmental process have not yet been addressed using molecular approaches. The first is that embryonic induction brings about coordinated changes in cellular activities, including, though certainly not limited to, changes in gene expression. Alterations in cell motility and cell behaviour that result in morphogenetic movements are generally among the earliest responses to induction (Symes & Smith, 1987; Trinkaus, 1984a). Changes in cell cycle activity and electrophysiological properties are also observed... (Spitzer, 1979)" (Jacobson & Sater, 1988).

It is likely that the genes involved in a single differentiation cascade, the little known, so-called 'downstream' genes (Andrew & Scott, 1992; O'Hara et al., 1993), are at least partially controlled in a hierarchical fashion:

"One of the most challenging problems remaining will be to identify the... 'downstream' or 'target' genes... that we assume... bring about the differentiation of tissues and organs..." (Lewis, 1992).

However, it is important to note that these differentiation cascades, each corresponding to an edge of a differentiation tree, may not have a simple structure:

"In eukaryotes, the very large general transcription complex, which includes the RNA polymerase, can interact with several regulatory proteins simultaneously bound to distal *cis*-regulatory sites. These confer gene-specific and cell-type specific expression.... The upstream regulatory region of some genes, which can be much larger than the coding portions of the gene, resembles a battlefield, with positive and negative inputs contending for the RNA polymerase at the promoter" (Gerhart & Kirschner, 1997).

Thus the concept that differentiation cascades may be triggered by a single master gene (as we suggest in Appendix IV: Björklund & Gordon, 1993b), is undoubtedly an oversimplification. A similar story of (terminal) differentiation is unravelling for prokaryotes:

"[In the]... developmental biology of *Streptomyces* colonies... mutants... are of two general types: those lacking an aerial mycelium (*bld* mutants...) and those that produce an aerial mycelium but fail to develop the pigmentation associated with mature spores (*whi* mutants).... Remarkable... is the virtual absence of descriptions of *S. coelicolor bld* mutants that are entirely unaffected in their secondary metabolism.... Whatever the reason, clearly all sufficiently described *bld* mutants of *S. coelicolor* are pleiotropic.... The morphological phenotypes... suggested the following simple pattern of epistatic interactions: *whiG* $\Rightarrow$ *whiH* $\Rightarrow$ *whiA,B* $\Rightarrow$ *whiI* (Chater, 1975). However,... the epistatic sequence is not a simple result of a corresponding linear regulatory cascade of gene expression and moreover... the mutant phenotypes do not accurately reflect intermediate stages in normal sporulation.... The first information addressing the transcriptional regulatory interactions of sporulation genes has emerged and tends to indicate a surprisingly nonhierarchical interplay.... This observation may indicate that many physiological processes are simultaneously activated and that the biochemical interplay of these processes itself contributes significantly to the organized series of events.... Possibly, many important functions [that] are also necessary for normal growth, can be fulfilled by more than one mechanism (e.g. isoenzymes, alternative pathways, etc.)" (Chater, 1993).

In this book, I will generally assume that such networks do not cross from one differentiation cascade to another, though they may occur within each cascade (i.e., edge), as is described above by Chater (1993) (cf. Proposition 76). Considerations of induction blur this distinction, as we shall see, but also provide important clues to the evolution of differentiation.

Proposition 12: during primary neural induction, in particular, all neural plate specific genes are turned on in the wake of the ectoderm contraction wave, which leaves them arrayed as gradients in and/or behind the wave.

Gordon & Brodland (1987) (Appendix I) have reviewed a number of events that apparently come shortly 'after' primary neural induction:

"Epidermal cells develop a cytokeratin intermediate filament called the *tonofilament* (Burnside, 1973a) 'as neurulation proceeds' (Burnside, 1971).... A specific cytokeratin mRNA apparently arises a few hours after primary neural induction (Jonas, Sargent & Dawid, 1985)....

"There is also evidence for an increase in intercellular adhesion at the time of primary neural induction (Suzuki, Ueno & Matsusaka, 1986). The production of 'neural cell adhesion molecules N-CAM' seems to start immediately after induction (Jacobson & Rutishauser, 1986). Kintner & Melton (1987) have shown that the messenger RNA for N-CAM is expressed before the appearance of N-CAM, but after induction.

"It is known that the neural plate, at least in chick, is a source of a considerable ionic current (Jaffe, 1985 [cf. Jaffe & Stern, 1979; Créton et al., 1993; Metcalf, Shi & Borgens, 1994; Shi & Borgens, 1995]; Regen & Steinhardt, 1986) that is highest over the notoplate (Slack, 1984d). This observation formed the basis for a speculative mechanism of primary neural induction (Gordon, 1983b). Warner (1985a) shows that... the number of sodium pumps increases in neural plate cells after primary neural induction [cf. Shi & Borgens, 1994]....

"Slack (1985b) found a receptor for peanut lectin, epimucin (Slack, 1984b), confined to the epidermis, and absent from the neural plate. Synthesis apparently occurred within a few hours after primary neural induction. Gurdon (1987) concludes 'most molecular markers of neural induction appear rather late in development (Godsave, Anderton & Wylie, 1986; Takata, Yamamoto & Ozawa, 1981; Takata et al., 1984; Duprat et al., 1985a),' except for N-CAM and 'a *Xenopus* homeobox-containing gene X1Hbox6... [which] serves as a probe for induction as early as the late gastrula.'

"We thus see that a number of specific gene products are made shortly *after* primary neural induction, though there is no evidence for specific gene products that form just before or during the event" (Gordon & Brodland, 1987).

Bronner-Fraser, Wolf & Murray (1992) add: "In the developing nervous system, both N-CAM and N-cadherin are expressed in the neural plate shortly after neural induction (Hatta & Takeichi, 1986; Jacobson & Rutishauser, 1986; Keane et al., 1988; cf. Bally-Cuif, Goridis & Santoni, 1993)". Integrins could perhaps also be appended to this list (Bronner-Fraser et al., 1992), along with glutamine synthetase (Hatada et al., 1995). Care must be taken that "neural-specific genes" (Knecht et al., 1995) are specific to the ectoderm contraction wave, and not products of subsequent waves within the neural plate: "*cpl-1,* a marker of dorsal brain, and *etr-1,* a marker absent in much of the dorsal forebrain, are expressed in non-overlapping territories [i.e., nonoverlapping wave trajectories, probably an expansion/contraction wave pair]..." (Knecht et al., 1995).

If we include both the launching of the ectoderm contraction wave and its propagation, which takes all of gastrulation (Appendix II: Brodland et al., 1994), as part of 'primary neural induction', the we see that greater care is needed in defining what is 'after' primary neural induction. The proper distinction is what happens in/to a cell before, during, and after it participates in a differentiation wave. The timing of these events will differ from one cell to the next, depending on when the wave gets to it. Each specific gene product triggered in a cell by a wave will have a particular course of synthesis and degradation. Since the cells start these syntheses at different times, there will be a gradient in the tissue, starting at the wave.

The shape of this gradient will depend on the speed and trajectory of the wave, and on the time course of synthesis and degradation of the protein. Thus differentiation waves can create gradients. (This does not imply the reverse: some gradients may be created by other mechanisms, such as the gradient in yolk density.)

The distinction between genes involved in the act of cell state splitting from those that are turned on 'downstream' as a consequence, i.e., are part of the differentiation cascade associated with that act, can sometimes be made not simply by timing, but also genetically. For example, the *marbles* mutant in *Drosophila* exhibits "failure of nuclear migration in many cells" (Fischer-Vize & Mosley, 1994), yet proceeds with other aspects of eye differentiation "for the most part normally", including propagation of the morphogenetic furrow, a differentiation wave. (Although the nuclear migration is basal, instead of apical, this motion could probably be classified as a case of interkinetic migration.) The viability of this mutant is probably due to this being a terminal differentiation, perhaps involving a tissue specific protein, and to the protected environment of laboratory fruit flies:

"Preliminary molecular analysis suggests that *marb* is likely to encode some sort of cytoskeletal protein. One exciting possibility is that *marb* is a motor protein or a protein that interacts with one.... The role of *marb* may be to shape the cell body, thereby preventing the cell body cytoplasm, and the nucleus along with it, from flowing into the axons of the photoreceptors or into the basal regions of the pigment cells" (Fischer-Vize & Mosley, 1994).

i.e., failure of interkinetic migration. I'll discuss the *shibire Drosophila* mutant (in Proposition 248), which also permits propagation of the eye imaginal disc morphogenetic furrow, but prevents ommatidia differentiation.

Steve McGrew (p.c., 1997) has pointed out that the shape of a gradient in a tissue, left behind by a differentiation wave, can depend on the next differentiation wave, especially if the latter has a different launching domain and crosses the tissue in a different trajectory, and alters expression of the gene in question.

Proposition 13: each act of cell state splitting triggers a differentiation cascade of changes in gene activity and its consequences at transcriptional, translational, and posttranslational levels.

This triggering may be of the form suggested by Britten & Davidson (1969), perhaps modified by more modern concepts of DNA binding proteins (Klug, 1993), DNA looping, twisting, etc., and their role in gene regulation (Maniatis, Goodbourn & Fischer, 1987; Wang & Giaver, 1988; Adhya, 1989). The differentiation cascades correspond to 'functional cassettes' of gene activity (Raff et al., 1987a):

"The cell types represented in a complex metazoan like ourselves do not compose a continuous spectrum ranging from one type through fine gradations to an alternate type. Rather each cell type falls into a clearly recognizable discrete category sharply set off from other cell types. This behavior seems to imply, on a genetic level, that as one set of genes is turned on another set is, through a closely linked process, turned off" (Markert, 1964).

Hierarchically organized gene batteries, cascades, or functional cassettes are logical structures without a spatial component. The thing that is missing is why particular cells in particular locations use one and not the other of the two differentiation cascades available to them. The puzzling over this is clearly seen in Davidson (1986), who, with Morgan (1934), leans towards maternal determinants (cf. Roach & Wulff, 1987) as a 'sufficient' explanation:

"...Genetic and other experimental evidence that genomic determinants indeed control the details of the developmental process became overwhelmingly convincing. Yet there seems to have been no further conceptual advance towards the basic problem of explaining *differential* cell function until 1934, when there appeared in an essay of T.H. Morgan an explicit statement of the theory that differentiation could be caused by *variation in the activity of genes in different cell types:*

'The implication in most genetic interpretations is that all the genes are acting all the time in the same way... An alternative view would be... *that different batteries of genes come into action as development proceeds....* The idea that different sets of genes come into action at different times... [requires that] some reason be given for the time relation of their unfolding. The following suggestion may meet the objections. It is known that the protoplasm of different parts of the egg is somewhat different... and the initial differences may be supposed to affect the activity of the genes. The genes will then in turn affect the protoplasm... In this way we can picture the gradual elaboration and differentiation of the various regions of the embryo' (Morgan, 1934)....

"Genes that are not expressed until particular stages of embryogenesis, and that are activated at specific locations, are of particular interest since they take part in the processes leading to the initial appearance of differentially functioning cell types. There are... several classes of mechanism that might reasonably explain the initial appearance of regionally localized patterns of gene activity in the embryo. As proposed by Davidson & Britten (1971) particular batteries of structural genes could be specified for expression by direct interaction with maternal *trans*-regulators located in appropriate regions of egg cytoplasm, or synthesized on localized maternal mRNAs. Or they could be specified as the result of contact with an adjacent cell type that releases specific internal activators. A third alternative that is different in emphasis, because it involves an intermediate level of internal hierarchy in the genomic regulatory system, is that the initial regional specialization of the embryo is controlled by a small number of pleiotropic developmental master genes. The crucial spatial pattern of *their* expression might be determined by a few simple polarizations in the egg.... There is evidence from different biological systems that supports, to a various extent, all of these modes of zygotic gene activation" (Davidson, 1986).

Actually, the credit for the idea of differential gene expression should perhaps go instead to von Uexküll (1926), who, in the context of embryonic differentiation, albeit adhering elsewhere in his text to the idea that genes are turned off differentially, wrote:

"I must refer once more to the rough comparison I attempted to make when speaking of the way in which a subject clothes itself with properties. With regard to its functions, the subject in the germ-cell is still very simple. But in the genes it possesses a very large number of unexploited possibilities which will enable it to expand in every direction. As the possibilities are made use of, their number becomes more and more restricted.... Imagine the genes to be keys of a piano, only waiting to be struck for all manner of tunes to sound forth.... The aim of descriptive biology must be to set down, by means of a kind of musical notation, the laws

according to which the genes in various animals sound together or in succession..."
(von Uexküll, 1926).

On the other hand, the idea seems to have been around decades earlier, if we substitute 'gene' for 'ideoblast':

"It is important not to lose sight of the fact that Hertwig [cf. Hertwig, 1901-1906], no less than Roux [1883; cf. Sturtevant, 1965] and Weismann [1893], conceives the ideoplasm [cf. Mayr, 1982] (which he would locate in the cell-nucleus) as an aggregate of units ('idioblasts') which severally correspond to the hereditary qualities of the organism; and since cell-division is not qualitative, every cell must contain the sum total of the hereditary character of the species. Differentiation is conceived by Hertwig (following de Vries) as the result of physiological changes in the idioblasts, some of which remain latent, while others become active, and thus determine the specific character of the cell, according to the nature of the active idioblasts. In regeneration such of the latent idioblasts are called into action as are necessary to carry out the regenerative process" (Wilson, 1894).

Thus the basic idea of this proposition has a long history. I've just identified the trigger as the cell state splitter.

Proposition 14: the genetic program for development is the differentiation tree, which consists of alternations between genetic and mechanical events.

"...In the determination of what constitutes the ability to grow, or the morphogenetic differences between cells of different organs, the organization of the cytoplasm may - even must - play a key role. The transmission of genetic information in cells and tissues has been likened by Platt (1962) to reading out of a complex instruction manual. It is clear that the cell cytoplasm, through which the responses are mediated, also has the ability to transcribe, or to 'read,' the instructions in the DNA. Hence the cytoplasm is an integral part of the unit, or closed circuit, which responds in growth and morphogenesis" (Steward, 1968).

"The task... can be stated... in terms of a search for an algorithm that connects phenotype to genotype" (Gerhart & Kirschner, 1997).

The genetic program has been described as 'an unmoved mover' (Mayr, 1982) that contains the full 'blueprint' for an organism (cf. Delbrück, 1971; Mayr, 1976). This classical view of the DNA must be supplemented with its interactions with the physics of the cell(s) containing it, to perceive the full genetic program:

"The genome need not embody that part of the information of the real world that is contained in the very structure of matter. It is in this sense that any concept of a 'program for development' by analogy with computer programs is incomplete.... Of course, consistency between the genetic code and the physical world is still required, which may be fine-tuned by evolution. If we demand an analogy with computers, we may turn to hybrid computers, in which digital components interact with analog components (Korn & Korn, 1972)..." (Appendix I: Gordon & Brodland, 1987). [Cf. Hoffmeyer, 1975, and Pattee, 1989, as discussed in Emmeche, 1994.]

There has been some resistance to the idea of a genetic program:

"...Three of the men responsible for the molecular biological revolution have questioned the validity of the program metaphor in development (Stent, 1980; Brenner, 1981; Jacob, 1982)" (Wilkins, 1986),

But with a physical component added, the concept may be quite tenable. Thus I conclude that there is indeed a genetic program, if one includes its physical component. The orderliness of a hierarchical genetic/physical model contrasts sharply with one based on inductions:

"It appears that the realization of the phenotype is characterized by program-like, regulatory, stochastic, and historical processes that interact to form a network of relationships, rather than a linear succession of events. There is no fundamental design or program that is followed but rather the interactions taking place form the program. During the course of development, each state presupposes the foregoing state; each subsystem produces successively the prerequisites for its further development. The processes taking place are essentially reactive, depending on structural correspondences, and are therefore highly selective. Control phenomena and cascade mechanisms are not hierarchical, but self-referential. Thus, there is no hierarchy either from gene to phene or vice versa.... The genotype - phenotype relationship cannot be expected to follow strict laws, but rather to be irregular" (Wolf, 1995a).

The combination of differentiation trees with the physics of differentiation waves goes beyond 'naive physicalism' (Mayr, 1976). Differentiation waves provide the missing component needed to solve the 'mystery' of the genetic program:

"...The internal logic at work in the execution of the [genetic] program remains completely unknown. It is generally admitted that a Laplacian demon [actually 'Maxwell daemon': Dampier, 1948] able to examine the fertilized egg, its molecular structures and organization, would be able to describe the future adult. However, what kind of molecules besides DNA the demon would have to examine and what kind of algorithm it would have to use remain a complete mystery" (Jacob, 1982).

This physics then becomes the 'epigenetic system' referred to by Ebert & Wilt (1960):

"Until recently... enforced preoccupation with chemical description has widened - rather than narrowed - the gap between genetics and embryology. But, as we begin to resolve the structures and mechanisms of synthesis of nucleic acids and proteins, we see that to *make visible* the mechanisms of differentiation will require that increasing attention be paid to the interactions of genetic and epigenetic systems. Here we employ the distinction advanced by Nanney (1958): rather than the geographical dichotomy of nucleus and cytoplasm, it is proposed that the invariant determinants of the cell, wherever they may be located, be called genetic, and the variant determinants of cell fate be called epigenetic. Genetic in this sense refers to a specific nucleotide sequence, while epigenetic refers to the modifiable machinery, whatever its chemical constitution, carrying out the orders of the invariant nucleotide alphabet.... The differential physiological activity of the genetic material must reflect its interaction with epigenetic components and with their products - the epigenetic elements themselves being an expression of the heterocatalytic activity of the genetic material" (Ebert & Wilt, 1960).

Weiss (1940) wrote a particularly clear statement of this double nature of development, in which I would take 'supra-cellular powers' and 'organizing entities' to be differentiation waves, and 'fields' to be their trajectories:

"...Practically every step in development reveals the cell in a double light: partly as an active worker and partly as a passive subordinate to powers which lie entirely outside of its own competence and control, *i.e.,* supra-cellular powers. Now, it is perfectly true that some of these latter result from interactions of cell individuals and are, therefore, of cellular origin. But it is equally true - and the findings of experimental embryology are one rich store of evidence for our assertion - that many of them are supra-cellular from the beginning. They are those organizing conditions through which the fate of the individual cells - undecided... at first - is guided, controlled and progressively fixed. They are definite at a time when the individual cell fate is still indefinite. They impose order upon what otherwise would be an anarchic cell chaos. They are inherent properties of the living system, germ, as a whole, in contradistinction to the inherent properties of its constituent cells....

"One frequently refers to these organizing entities under the term of 'fields.' Their existence can be traced back to the egg. In fact, just as there is a continuity of cells from the egg to the organism through successive cell divisions, so there is continuity between the primordial organizing fields present in the undivided egg and the localized fields of the later germ. Primordial fields segregate progressively into more restricted fields, and, furthermore, induce new fields in neighboring areas of the germ [cf. wave-wave interactions in Proposition 82]. Thus, the organizing principles of a germ have an ontogenetic history of their own which is not cell history. Their possession marks the egg as an entity of the rank of the organism.... Their development is a matter of the developing system as a continuum, like tensions, currents, potentials, and the like, and they pay no heed to cell boundaries, although sooner or later the intricate interplay between them and the cells sets in....

"The existence of these primordial organizing principles in the egg has been firmly established by modern experimental embryology. No pure cell theory derived from the developed organism can embrace them, unless by a vicious circle.

"In conclusion, we may say that the cell theory is correct: The egg is a cell and it gives rise to all the successive cell generations which contribute to the organism. But the organismic theory is likewise correct: The egg is also an organism, and it passes its organization on continuously to the germ and the body into which it gradually transforms [cf. Proposition 123]. Only this dual concept seems to fit the facts, as we see them at present. To be consistent, we should supplement Virchow's well-known tenet of cell theory: 'Omnis cellula e cellula,' by its counterpart: 'Omnis organisatio ex organisatione.' If the former denies spontaneous generation of living matter, the latter denies spontaneous generation of organization" (Weiss, 1940).

The 'history' of 'the organizing principles of a germ' would correspond to the differentiation tree. Weiss (1940), by his examples, clearly places the 'organizing principles' in the realm of physics. I identify them as primarily the physics of the differentiation waves. The "intricate interplay between them and the cells" is the genetic program.

Mayr (1976) perhaps foresaw the task we have ahead of us:

"The entire concept of a program of information is so new that it has received little attention from philosophers and logicians.... The philosopher may be willing to accept the assertion of the biologist that a program directs a given teleonomic behavior, but he would also like to know how the program performs this function. Alas, all the biologist can tell him is that the study of the operation of programs is the most difficult area of biology. For instance, the translation of the genetic program into growth processes and into the differentiation of cells, tissues, and organs is at the present time the most challenging problem of developmental biology" (Mayr, 1976).

If physical factors, such as differentiation waves, are important components of development, then it is clear that no set of genes is 'necessary and sufficient' to explain embryogenesis:

"...[L.C.] Dunn [cf. Dunn, 1940] managed to draw from his students an answer to the following question: Given that mutant alleles at a certain locus result in white-eyed flies, does that mean that the non-mutated allele at that locus is responsible for red eyes? With considerable effort and patience, he instilled in his students the ideas of *necessary* and *sufficient*. The non-mutated allele at the *white* locus is necessary for the formation of red eyes, but is not sufficient to guarantee that the eyes of a fly will be red - or even, that the fly will have eyes at all.... 'Normal' alleles are only necessary, not sufficient, for normal development.... Some developmental geneticists... apparently assume that the non-mutated allele is responsible for the normal phenotype" (Wallace, 1985).

Going one step further, we see that all genes (i.e., the whole genome) are only at most necessary, not sufficient, to explain embryogenesis. Some of the 'information' necessary for development is carried outside of the DNA. A nested sequence and/or series of differentiation waves may be necessary to a full explanation of embryogenesis and the genetic program.

Proposition 15: genetic determinism cannot explain the spatial component of embryogenesis.

"There are some arguments that differential action of genes does not precede but follows pattern formation in the embryological fields of metazoa (Bennett, Boyse & Old, 1972). It is more credible that the role of the given cell in a system is related to its position in the clone, than to its initial different nuclear activity" (Kaczanowska, 1974).

"To say the least, it seems one should keep an open mind as to whether morphogenesis proceeds genetic determination, or the reverse, in any given context" (Bennett, Boyse & Old, 1972).

The concept of a "hierarchy of development-controlling genes (D-genes)" (Arthur, 1988) is only half of the story of the genetic program. That half has been magnified to the whole by much of modern molecular developmental biology. The notion that genes do everything has been labelled 'genetic determinism' by Keller (1994), who traces the history and intellectual and political power of the concept, while touching on its limitations:

"The term 'genetic determinism' refers to a belief system that locates the cause of all biological development in an organism's genes: if we only knew enough about genes (about what they are and how they 'act'), we could understand all of biology. Such beliefs - codified in what I call the 'discourse of gene action' - have been of great importance to the history of genetics and, most recently, to the launching of the Human Genome Initiative. But what does it mean to attribute... causal power to genes? ...This way of talking... has... impeded the formulation of a conceptual framework adequate to the study of developmental phenomena.... The need for such a framework has become urgent....

"Historians of biology routinely note that, for nineteenth century biologists, the term 'heredity' referred to both the 'transmission of potentialities during reproduction *and...* development of these potentialities into specific traits' (Allen, 1986). The question that compelled their interests above all was, as August Weismann put it in 1883 [cf. Weismann, 1889], 'How is a single germ cell capable of reproducing the entire body with all its details?' (in Sander, 1986). However, a crucial change occurred in the early part of this century. With the rise of the new discipline of genetics, the two aspects of heredity (transmission and development) grew apart....

"Embryologists had good grounds for concern. Not only was the status of their discipline under threat; so too was the status of their question: How *does* a germ cell develop into a multicellular organism? If the genetic content of all cells in an organism is the same, then how is one to make sense of the emergence of the manifest differences among all the cells that make up a complex organism? To them, it seemed self-evident that this problem of differentiation, so deeply at the heart of their own concerns, was simply incompatible with the notion that the gene was the exclusive locus of action.... Geneticists... interested in the relation between genes and development... changed the subject.... The discourse of gene action -... augmented with metaphors of information and instruction - exerted a critical force on the course of biological research,... on scientists, administrators, and funding agencies [cf. Sapp, 1983, 1987; Witkowski, 1990; Gordon, 1993c],... [providing] powerful rationales and incentives for the mobilization of resources, for the identification of particular research agendas, for focusing our scientific energies and attention in particular directions....

"It has... become conspicuously evident that there were all along serious problems with the discourse of gene action - besides its productive blindness to questions of development and cell differentiation.... Classical embryologists [saw]... their questions, their organisms,... and they themselves had been left behind - but a new generation of biologists had little cause to look back. The first generation of molecular biologists [and the present one] could not answer the question of how an egg turns into an organism (could say nothing, e.g., about how a gene comes to make the particular enzymes that are needed for the development of a many-celled organism, in the right amounts, at the right time and in the right place).... The simple fact is that for many years geneticists had little reason to refer to eggs and their cytoplasmic structure, and even less reason to talk about events prior to fertilization. The discourse of gene action had established a spatial map that lent the cytoplasm scientific invisibility to geneticists ('indifferent' was how Morgan [1924] described the cytoplasm) and a temporal map that defined the moment of fertilization as origin, with no meaningful time before fertilization. In this schema, there was neither time nor place in which to conceive of the egg's cytoplasm exerting *its* effects" (Keller, 1994).

A cell state splitter, in mechanically resolving its instability as it participates in a differentiation wave, determines which differentiation cascade is activated next in its cell. That gene cascade in turn sets up (or activates) the next (or reused) cell state splitter. With this view of the genetic program "in which mechanical properties and movements of cells are part of the control system that switches cell fate between alternate differentiated states (Björklund & Gordon, 1993a)" (Harris, 1994a), we can begin to talk about how genes are turned on "in the right amounts, at the right time and in the right place".

In a wide ranging review of the concept of the genetic program, Wilkins (1986) fails to discuss the spatial (in my view, physical) components of differentiation and thus of the genetic program (though he does discuss pattern formation separately). It is as if, in the triple question of where, when, and why of differentiation, 'where' has been forgotten.

The nonspatial genetic program (cf. Section 1.08) apparently grew out of bacterial molecular biology, in which the bacterium has always been regarded as a well mixed bag of macromolecules:

"The operon model was formulated to explain the regulation of genes in some of the simplest organisms, bacteria and their viruses. However, the implications for other organisms, and in particular for development, were apparent.... The idea is that an initial regulatory step can initiate a second regulated change in gene expression for one or many genes, which can, in turn, produce a third set of changes, and so on, the net effect being a highly ordered sequence of gene expression changes ensuing from a single event (Monod & Jacob, 1961; Jacob & Monod, 1963). Although the general mechanism had been proposed before, Jacob and Monod showed how it could be enacted with only the kinds of regulatory elements demonstrated in bacteria. They also made explicit a notion that had been implicit in the beginnings of molecular biology - that the characteristics of cells are determined by the proteins they synthesize. The corollary is that the source of qualitative differences between cells is found in qualitatively different (and regulated) patterns of protein synthesis. This assumption has underlain much of subsequent research on differentiation.... Nearly a quarter of a century has passed since the publication of the Jacob - Monod model, and it is pertinent to ask what its relevance is to development today.... The fundamental ideas of regulatory switches and sequential, ordered changes in gene expression remain as pertinent today as they were in 1960.... The net effect of the Jacob - Monod model on developmental genetics was to shift attention from the observable developmental effects of mutations to the underlying effects on gene expression.... A second, subtler change concerned the way that biologists began to view the 'logic' of developmental control. Development began to be seen as a 'program' encoded in the genome" (Wilkins, 1986).

The geneticists' present answer that gene cascades and maternal gradients solve all is open to question. Differentiation waves may provide the essential additional physics component of the genetic program.

As a footnote to this proposition, I must mention that Schiffmann (1994) has made me 'eat my words', which I wrote as follows:

"It is important to notice that the cell state splitter is made of ubiquitous building materials. Thus we must question any direct role of the genome in setting it up, especially if, as we postulate, the simultaneous appearance of cell state splitters in a whole tissue is a mechanochemical event. Cells may, rather, be ready to participate in such a group effort when the specific products produced *after* the last round of differentiation have, so to speak, played themselves out or fully established themselves or prepared for the next stage (cf. Warner, 1985a), with whatever morphogenetic movements, cell attachments, etc., that might entail. Thus, the cell state splitter may be set up in response to developing cytoplasmic and/or mechanochemical conditions, rather than as a consequence of a genetic program.

"This may also imply that there is no genetic program, that we have taken our analogies between 'informational molecules' and computers too far" (Appendix I: Gordon & Brodland, 1987).

Because of this, Schiffmann (1994) includes Gordon & Brodland (1987) (Appendix I) in "the trend against reduction to molecular genetics" as a 'nongenetic theory'. It is amusing to be accused of the opposite of what one intends, so a little waffling or recanting on my part is demanded. First, the 'ubiquitous' cytoskeletal components can be quite tissue specific. Second, I now presume that setting up of the cell state splitter is part of the differentiation cascade of the previous step of differentiation (Figure 17), though the coordinating mechanism to have them all ready, at least in time for the approach of the next differentiation wave, is not yet apparent, and may indeed have a physical component. Third, I compared the classical notion of the genetic program to the digital component of a hybrid (digital/analog) computer in the very same paper (Appendix I), which implied what is now explicit in this proposition: the genetic program is an alternation between the genome and the physics of differentiation waves.

Proposition 16: consideration of the spatial component of the genetic program is essential for progress in artificial life (ALife) and embryonics research.

Two switches switched two swatches. Which switch switched which swatch?

"The creation of multicelled digital organisms remains an important challenge" (Ray, 1995).

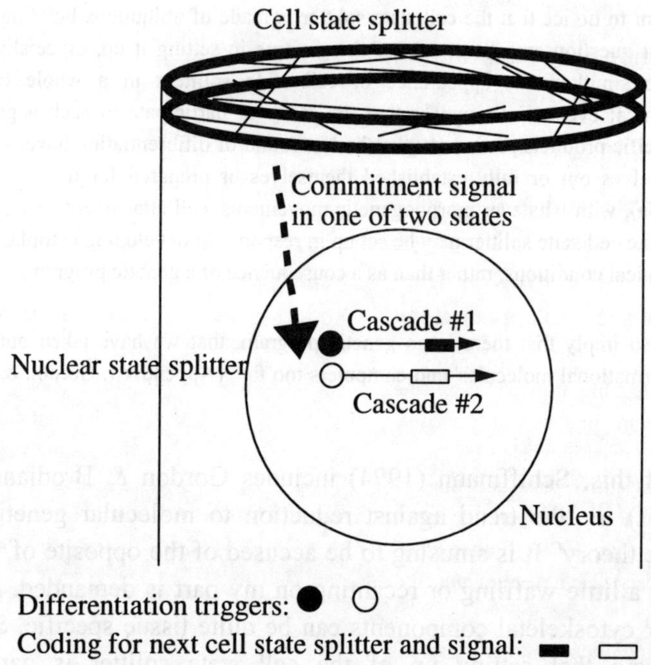

Figure 17. When the cell state splitter mechanically resolves into one of two states, a commitment signal indicating the new state is sent to the nucleus, where it causes one of two differentiation triggers to set off a differentiation cascade of gene activity. This activity includes setup of the next cell state splitter, commitment signal, and preparation of the next pair of differentiation triggers. The differentiation triggers are indicated as circles on the DNA. The mechanism in the nucleus, by which two and exactly two differentiation cascades are readied to respond, is called the nuclear state splitter.

Genetic determinism is still the prevailing belief system in biology, though some have suggested the reverse:

"Waddington [1956a]... had to believe that the cytoplasm activated the genome. This went against the grain of genetics, which held the nucleus to control the cytoplasm (Sapp, 1987; Gilbert, 1988)" (Gilbert, 1991b).

A balance is needed for a full explanation, and I think this can be found in the alternation between the physics of differentiation waves causing activation of specific genes, causing launching of further differentiation waves, etc.

In artificial life studies (cf. Loeb, 1904; Ravin, 1977; Emmeche, 1994), the physics component of the genetic program has yet to be seriously contemplated. The artificial life community has acknowledged the problem, but does not seem to appreciate how fundamental it is to development:

"...There's quite a literature about self-describing Turing machines. But that leaves out at least half the problem. In fact von Neumann [1966] mentioned this when he first talked about self-replicating machines. He said the formalization of this problem leaves out perhaps what is the most essential part, and that is the physics, the real time, space, matter, energy relationships that are involved in what I mean by construction" (Howard H. Pattee in: Rosen, Pattee & Somorjai, 1979). [Pattee also credits von Neumann with anticipating the logic of protein synthesis: "He never got any credit for this from biologists...."]

"...An intriguing direction to artificial life [is]... to begin with a single cell and, using a set of principles or rules, to grow an actual being.... [This is] the embryological path to artificial life.... [Alvy Ray Smith in 1982:] 'Consider man: How much understanding of his shape can be gleaned from treating the embryo as a living geometric computation?' [Ed Fredkin, p.c.:] 'Living things may be soft and squishy'... 'But the basis of life is clearly... a digital information process, and that life can be mimicked in its entirety by such a process. Put it another way - nothing is done by nature that can't be done by a computer. If a computer can't do it, nature can't'" (Levy, 1992). [Cf. Tipler, 1994.]

Perhaps. But computers will have to be taught or learn how to handle the physics of embryos first (cf. Jacobson & Gordon, 1976a; Gordon & Jacobson, 1978; Gordon, 1983a; Brodland, 1994; Goodwin & Brière, 1994), or come up with something logically similar to it. (This 'failure' of artificial life has been used to justify creationism: Behe, 1996.) Researchers in artificial life are now turning to 'embryonics' (Mange & Stauffer, 1994; Marchal et al., 1994, 1996). The very design of the hardware they build (Sipper et al., 1997) is based on, and thereby limited by, present developmental biology concepts of cell 'environment', 'positional information', and the structure of the genome. If we could put both the physics of differentiation waves and its specific activation of the genome into artificial life, we might attain simulations or hardware akin to real embryo development. Yaeger (1994) suggests that this represents...

"...one of the most potentially valuable directions for future ALife work. A richer, more biologically-motivated developmental process might provide... significant improvement.... [Cf. Proposition 240.] And the converse may also be true; PW [PolyWorld, his artificial life simulation] may be a very effective testbed for alternative ontogenetic theories and algorithmic models of development" (Yaeger, 1994).

Some see ALife as subsuming biology as a subset:

"Ultimately, the success of ALife will depend on the extent to which it succeeds in developing a concept of life that encompasses biology [which]... requires identification of the outstanding unsolved problems in biology and seeking their resolution.

"Our work is ultimately motivated by a premise: that there exists a logical deep structure of which carbon chemistry-based life is a manifestation. The problem is to discover what it is and what the appropriate mathematical devices are to express it.... ALife practitioners must avoid the presumption that bedeviled Artificial Intelligence - that all the needed concepts were available to be imported from the other disciplines. ALife will find its first proving ground in real biology, in its capacity for catalyzing the theoretical maturation of biology" (Fontana, Wagner & Buss, 1995).

Howard H. Pattee (in: Rosen, Pattee & Somorjai, 1979) describes the 'genotype - phenotype system' as having two components: "an internal description or self-description, [which] cannot be characterized solely... in terms of physics" and a 'self-interpreting' or 'construction' system. "...The essential requirement of life is that it is a self-describing, self-constructing system." In terms of embryogenesis, the physics of self-construction is first and foremost the physics of the launching and shaping of differentiation waves. The alternation of differentiation waves with differentiation cascades that I identify as the genetic program could be said to have been anticipated, in the abstract, by Pattee:

"My concept of hierarchy... has to do with the alternation between descriptions and constructions" (Pattee in: Rosen, Pattee & Somorjai, 1979).

Proposition 17: the nongenetic component of the generic paradigm for morphogenesis is the differentiation wave.

A new paradigm has been put forth, called 'generic' (Newman & Comper, 1990; Newman & Forgacs, 1993; Ben-Jacob et al., 1994; Goodwin, 1994a; Newman, 1994), which grows out of an understanding that many physical processes generate similar patterns (Thompson, 1917, 1942; Hertel, 1963; Stevens, 1974; Pearce, 1978; Foy, 1982; Bach & Burkhardt, 1984). The generic paradigm tends to minimize the role of genes in development, as when "Goodwin (1984a) challenged what he called 'genocentric biology'" (Sapp, 1991). Raff (1996) has provided a concise, if unsympathetic, summary of this point of view.

The generic theme has deep historical roots:

"The majority of the experimental embryologists have long been reluctant to accept an intervention of the genes in primary morphogenesis, i.e., gastrulation and formation of the main embryonic organs. According to them, genes would only control the very last stages of differentiation. For instance, while the color of the eyes is under genetic control, the formation of the eye itself (differentiation of the eye cup, induction of the lens) might not be directed by the genes. The idea was made clear by A. Brachet (1910a, 1930), when he drew a distinction between a *general* heredity, controlling early morphogenesis and involving both nucleus and cytoplasm, and a *special* heredity of the Mendelian type.

"Some of the reasons for this old skepticism are still valid.... The formation of the gray crescent itself corresponds to an extremely important morphogenetic event: dorso-ventral differentiation and symmetry plane establishment in the recently fertilized egg. Thanks to the gray crescent, we can, at this very early stage, easily recognize what will become dorsal or ventral, right or left in the adult. These all-important events are not controlled by the egg nucleus, but by its cytoplasm. As shown by Ancel & Vintemberger's (1948) remarkable experiments, the plane of bilateral symmetry can be determined at will by modifying the orientation of freshly fertilized eggs.... Another reason for doubting the intervention of genes in early morphogenesis is the demonstration that nuclei are 'equipotential' during early development.... But, on the other hand, many experiments, in which... the nucleus... was damaged, clearly show that integrity of the nucleus is also required to obtain full and normal morphogenesis" (Brachet, 1957).

My interpretation is that there is indeed a nongenetic component to development, namely the physics of differentiation waves, but that it alternates with changes in genetic expression, so that both are essential at every step of differentiation.

Proposition 18: propagation of a contraction differentiation wave from one cell to the next (homoiogenetic induction) involves stretch activated contraction of the microfilament ring in the cell state splitter.

At the time of primary neural induction, the ectoderm consists of a sheet of cells that are presumably identical in regards to their ability to become either neural plate or epidermal cells (Spemann & Mangold, 1924a,b; Spemann, 1938; Holtfreter, 1988). These cells are bound together in a monolayer epithelium (in urodeles) by desmosomes (Burnside, 1971; Appendix I: Gordon & Brodland, 1987) and intermediate filament rings (Martin & Gordon, 1991, 1997a,b). Let us suppose that one cell starts contracting at its apical end. This would stretch the microfilament rings of its neighbors. Because of the mechanical instability of the cell state splitter, one would expect that: a) the first cell would become a neural plate cell, b) its neighbors would become epidermal cells, and then c) their neighbors would become neural plate cells again, etc. But alternating rings of neural plate and epidermal cells clearly are not what forms: instead, the spherical ectoderm (ignoring the blastopore for now) becomes two sharply separated hemispheres of each tissue (Jacobson & Gordon, 1976a; Gordon & Jacobson, 1978).

To explain why a bull's-eye pattern is not generated, we made one assumption (Appendix I: Gordon & Brodland, 1987): microfilament rings in ectodermal cells are capable of stretch induced contraction, like that of smooth muscle. This is suggested empirically by the wavelike contractions or 'peristalsis' observed in isolated neural plate cells (Burnside, 1973b) and the 'puckering' of early neuroepithelium in some intact embryos (A.G.

Jacobson in Segel, 1984, based on time lapse movies; cf. the hypothesized stretch contraction waves of Odell et al., 1981, elaborated on in Beloussov et al., 1994). The "contraction wave in the chick blastoderm after muscarinic stimulation" (Drews & Mengis, 1990) may be analogous to puckering in newts.

A bull's-eye pattern is avoided if, when the first cell constricts its apical end, it stretches the microfilament rings of the adjacent cells, which then respond by contracting themselves, if they have been stretched fast enough. The contraction wave could thus propagate from one concentric ring of cells to the next. We have since suggested an alternative hypothesis in which it is the apical membrane that is stretched, setting off the ectoderm contraction wave differentiation pathway, one of whose consequences is contraction of the microfilament ring (Appendix IV: Björklund & Gordon, 1993b).

The taut appearance of an axolotl embryo during gastrulation is consistent with propagation of a contraction wave (and 'tautness' could be quantified). It would be nice if we could quantitate the whole question of bull's-eye versus smooth propagation of a differentiation wave. This will take a level of detailed measurement and modeling that has not yet been achieved. As we shall see with the *Drosophila* morphogenetic furrow, there is a third propagation pattern possible, which leads to spacing patterns: along the furrow, cells are alternately expanding and contracting. All three patterns (Figures 18-20) may someday be explicable as special cases of one nonlinear model.

Proposition 19: morphological histories of tissues or tissue lineages should be replaced by precisely observed differentiation trees.

Insofar as cells of a given tissue remain physically contiguous, the differentiation tree is also a history of the morphology of the organism. Attempts to outline such morphological histories have been made since at least von Baer's time (1828, in Patterson, 1983) up to the present day (Slack, 1983, 1991a).

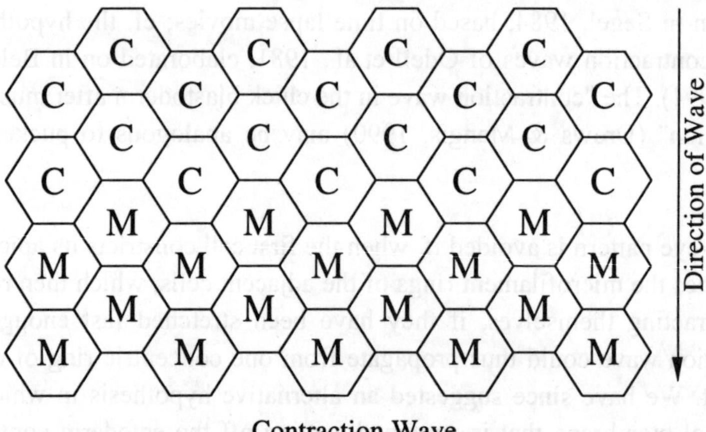

Contraction Wave

Figure 18. Smooth propagation of a contraction differentiation wave though cells with cell state splitters in the metastable state (M). Cells labelled C are undergoing, or have undergone, a contraction of the cell state splitter. This type of wave propagation probably requires stretch activated contraction of the microfilament rings in each cell's cell state splitter. Change C to E for an expansion wave.

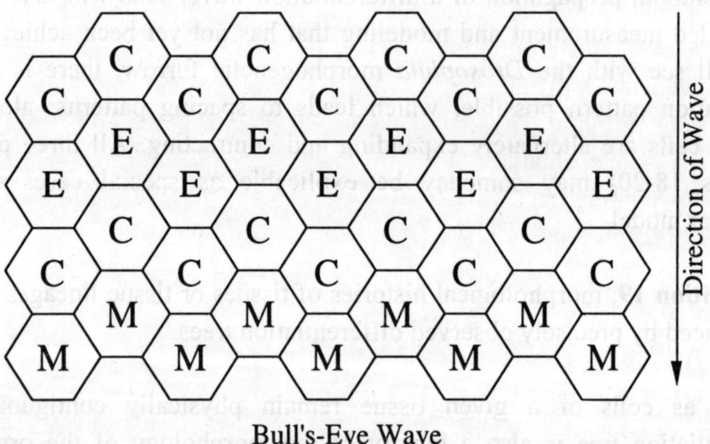

Bull's-Eye Wave

Figure 19. Propagation of a 'bull's-eye' wave, in which alternate rows of cells undergo contraction (C) or expansion (E). M = cell with a metastable cell state splitter.

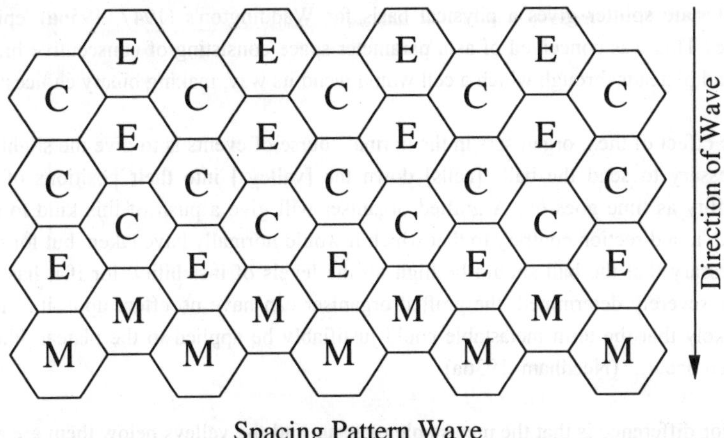

Figure 20. Propagation of a spacing pattern wave, in which along a propagating row cells alternate in expanding (E) or contracting (C). M = cell with a metastable cell state splitter. (This strictly hexagonal representation leaves something to be desired.)

It has to be noted that, since differentiation waves can cross more than one tissue, there is no one-to-one correspondence between differentiation waves and edges of the differentiation tree. We attain an artificial correspondence, because we name a differentiation wave according to the embryonic tissue it is travelling through. Thus, if it passes from one tissue into an adjacent tissue, its name changes. For instance, the presumptive notochordal mesoderm contraction wave becomes the ectoderm contraction wave (Figures 11-12 in Appendix V: Gordon, Björklund & Nieuwkoop, 1994). If we pay attention to the waves, they may be either nested or pass through adjoining tissues. If we pay attention to the names, then the waves are strictly nested.

Proposition 20: the epigenetic landscape for an organism is its differentiation tree.

As we said in Gordon & Brodland (1987) (Appendix I)...

"The cell state splitter gives a physical basis for Waddington's (1947, 1956a) 'epigenetic landscape'. This was conceived of as a parameter space consisting of consecutive branching valleys and plateaus through which a cell would wend its way, making binary choices:

> 'The effect of the... organisers in the normal course of events is to give the slight pushes necessary to send the balls [cells] down the [valleys] into their positions of greater stability as time goes on. A grafted organiser will give a push of this kind to a given ball... in a direction contrary to that which it would normally have taken, but for this it is necessary that the ball should be high up the levels of instability, for if it has already been severely determined, the grafted organiser can have no effect upon it.... It is not unlikely that the term metastable could justifiably be applied to the plateau-like states which occur...' (Needham, 1936a).

"The major difference is that the metastable plateaus and the valleys below them are actively created by the cells when they erect cell state splitters" (Gordon & Brodland, 1987).

(Cf. Gilbert, 1991b). Held Jr. (1992) gives a particularly clear picture (Figure 21) of Waddington's concept of the epigenetic landscape. One should note that: 1) the genes have no specified organization amongst themselves; 2) their relationships to the epigenetic landscape is also haphazard; 3) we can postulate that the direction the 'ball' rolls is determined by which kind of wave, contraction or expansion, the cell just participated in. If we are to meld this picture with our conception of the differentiation tree, the genes need to be brought up into the epigenetic landscape, where they are organized as differentiation cascades along each edge. In a sense, they merge with the landscape.

Let us compare the differentiation tree to the competing network model for the epigenetic landscape:

"For developing embryos, the *NK* model predicts that mutations early in development occur on a relatively uncorrelated fitness landscape. Hence the chances are low that the mutant will be more fit. Late in development, on the other hand, the fitness landscape can be presumed to be more correlated. Chances are better that a mutation will be fit rather than deleterious. Thus, the model generates von Baer's law (Kauffman, 1993; cf. Wimsatt, 1986; [Churchill, 1991a]). The von Baerian process of differentiation itself can be understood, in fact, as a response to perturbations carrying a cell into the basin of an attractor for another cell type.

Only a few of the possible attractors are in the region that can be accessed. Thus the overall process of development from a fertilized egg must follow a branching pathway. The trajectory of development is such that once a cell has begun to differentiate along a certain pathway it is no longer able to differentiate along other possible pathways. Here, Kauffman believes, are the dynamics behind Waddington's epigenetic landscapes. Indeed, Kauffman comes close in this analysis to grasping onto the synthesis of genetics, developmental biology, and evolutionary theory called for [by] Waddington and the Theoretical Biology Gathering" (Depew & Weber, 1994).

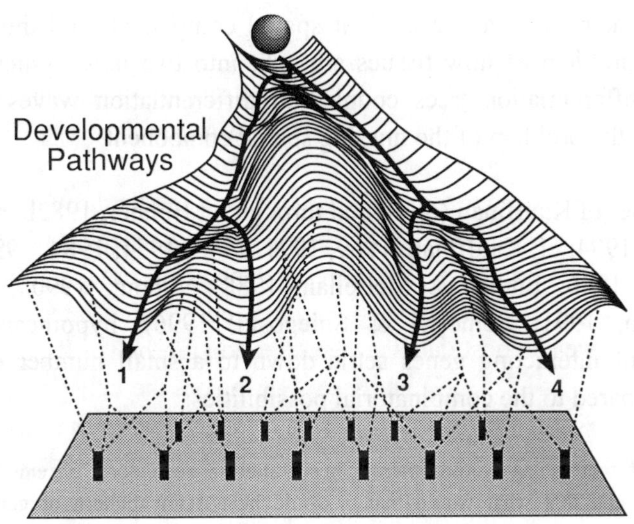

Figure 21. The epigenetic landscape, reproduced from Held Jr. (1992), who has nicely compounded two figures from Waddington (1957): "Metaphor... for the genetic control of development.... Waddington's 'Epigenetic Landscape' (Waddington, 1947, 1957). The ball rolling downhill corresponds to a cell undergoing development [differentiation]. Its fate depends upon its path. The main channel leads to a normal fate (No. 4). The contours of the canopy are maintained by an underlying network of guy wires (= interactions of gene products) anchored by pegs (= genes). Mutations will change the surface, resulting, in some cases, in an abnormal fate (No. 1, 2, or 3)" (Held Jr., 1992). Note that in Waddington (1956a), for instance, the relationship of the epigenetic landscape to the genes is not specified. With permission of the publisher, Karger.

In other words, in the network model the underlying genetic regulatory system is a nonhierarchical network, but the transitions between metastable states of this network can (but don't have to) create the appearance of a tree-like arrangement between the states. However, even for metabolism: "The biochemical and biophysical processes of the living cell do not constitute a network with random connections," but rather are hierarchical (Jensen et al., 1995). (Cf. Wijker et al., 1995.) Thus the evidence for the network model may not be there.

The network model has no explicit spatial component, and thus does not solve the problem of how tissues are split into two new tissues (Gordon, 1996e). Differentiation trees coupled to differentiation waves provide a solution to the problem of the missing spatial component.

In the model of Kauffman (discussed throughout Bonner, 1982b, and Depew & Weber, 1994; cf. Kauffman, 1969, 1971, 1973, 1974, 1984, 1993; Brooks & Wiley, 1986; Endler & McLellan, 1988; Nijhout, 1990a; Burian & Richardson, 1991; Somogyi & Sniegoski, 1996), hypothesized cyclic networks of interacting genes settle down to a small number of discrete states compared to the combinatorial possibilities:

"He realized that genes would interact in parallel in regulatory systems, rather than sequentially, and that what would matter would be a stable pattern of gene activity.... Kauffman... varied K, that is, the number of inputs per element. He found that the number of state cycles at $K=2$ is approximately equal to the square root of N, the number of elements in the system. For $N=100,000$, for example, $N^{0.5}=370$. That is approximately the number of cell types in multicellular organisms. Perhaps, then, this range of cell types had been determined not by natural selection, or even by the properties of particular materials cells are made from, but from the inherent mathematical properties of systems with a large array of elements and connections among them.... It would be remarkable if so simple a mathematical model could show properties that do resemble the dynamics of real biological systems. Kauffman's model of cell types seemed to him just such a result.... Kauffman soon realized that all the marvelous isomorphies with biological phenomena that he had found when he set $K=2$ were possible just because when $K=2$ the system is sitting, in Langton's [1986] terms, at 'the edge of chaos,' where it is subject to perturbations in accord with probabilistic schedules set by Bak's power law [Bak & Chen, 1991]" (Depew & Weber, 1994).

States of differentiation in the network model correspond to states of cycling of gene products. In the differentiation tree model, a state of differentiation consists of a set of activated genes, with activation caused by a discrete event, such as triggering of a master gene. The number of terminally differentiated cell types or tissues is set by the number of terminal edges to the differentiation tree. Differentiation trees also contain a number of intermediate, transient cell types. Network theory does not distinguish between intermediate and terminal cell types.

Note that 'the edge of chaos' would appear to correspond to a phase transition (Emmeche, 1994) or more likely to the critical point in classical statistical thermodynamics, where ordinarily small fluctuations become macroscopic. A case can be made for the states of a network model having a hierarchical arrangement (Kauffman, Shymko & Trabert, 1978; Depew & Weber, 1994). The difference between hierarchical and network models of differentiation is spelled out clearly by Wilkins (1986).

Thomas, Thieffry & Kaufman (1995) trace the history of network models for differentiation back a bit further:

"As far as we know, the first clear suggestion that epigenetic differences and, by inference, differentiation, might be ascribed to multistationarity, is found in a short comment by Delbrück (1949a).... Soon thereafter, two concrete cases of epigenetic differences were described and clearly understood. The first situation deals with bacterial populations which can be durably (150 generations or more!) blocked in either of two phenotypical states (lac operon on or off) depending on a detail of their **previous** history (Novick & Weiner, 1957; Cohn & Horibata, 1959). The second example concerns the decision for or against immunity in temperate bacteriophages..." (Thomas, Thieffry & Kaufman, 1995).

The following fills in the concept of 'multistationarity':

"Positive feedback loops (that is, direct or indirect autocatalysis) generate multiple alternative steady states; m independent positive loops can generate up to 3^m steady states, 2^m of which can be stable (i.e. m binary choices). Thus, seven autocatalytic genes are enough to account for $2^7 = 128$ stable cell types.

"The main biological application of these principles - in fact, a generalization of Delbrück (1949a) - is that differentiation is essentially the biological modality of multistationarity. One of the most challenging problems in embryology and other biological fields is: how can a transient signal result in a durable, sometimes permanent, change?" (Thomas, 1991).

Indeed, we use these concepts as part of the nuclear state splitter (Appendix IV: Björklund & Gordon, 1993b) to stabilize choices between master genes in a pair, though not to choose which pair requires stabilization next.

"What is striking about the character of... [Kauffman's] theory is that it ascribes a fundamental underlying structure to virtually all genetic systems and derives its theorem from very elementary features of that structure" (Burian, 1986b). Our theory of differentiation trees does the same, and thus presents a testable alternative with a quite different structure.

It may be appropriate to regard earlier work on oscillations and coupled oscillations in biological systems (Goodwin, 1963; Waddington, 1965; Winfree, 1967, 1980, 1987; Gilbert, 1968) as precursors to these network models for differentiation, including their spatial aspects:

"A system in periodic motion frequently has a property analogous to inertia - the oscillation tends to continue, unless some definite influence occurs to produce damping, or some other more powerful controlling system is superposed on it and suppresses it.... It is therefore one of the candidates for the 'carriers of more or less stable cellular differentiation', the identification of which is one of the major problems of present-day theoretical biology....

"...Talandically encoded information [Goodwin, 1963] might well be transmissible from cell to cell through contact.... In the study of cellular differentiation there is a whole range of phenomena - referred to in the classical experimental embryological literature as 'assimilative induction' - in which cells which are already well-set on a particular path of development succeed in imposing this on neighbouring cells, if the latter are either few in number or at an earlier and less 'determined' stage. The information which is transferred in such processes may, perhaps, not be encoded in the form of static molecular structures, but might be in the form of a particular temporal pattern of talandic oscillation....

"In any aggregate composed of entities which are oscillatory, and in which the oscillations of one entity affect those of its neighbours, for instance by diffusion, some overall pattern of oscillations will tend to build up [explaining morphogenetic fields].... This suggestion is something like that advanced by Turing (1952), but in the system he investigated the elementary processes proceeding at each point in space were not conceived of as being oscillatory, as they are here" (Waddington, 1965).

Actually, Waddington (1965) was partially wrong about this last point, since Turing (1952) did recognize that his reaction-diffusion equations had oscillatory solutions.

The differentiation tree provides an easier route to the epigenetic landscape than the network model, and has the added advantage of handling the question of the spatial component of differentiation, which the epigenetic landscape does not even depict (Figure 21).

Zelditch, Bookstein & Lundrigan (1993) "suggest that Waddington's diagrammatic presentation of the 'epigenetic landscape' may be misleading in quantitative studies of developmental regulation". It is interesting to note that their very objection hinges on their conclusion that "any adequate model of growth regulation must have a complex factor-structure, articulated at each geometric scale or region.... The diversity of ontogenetic patterns suggests that each [component] may be separately regulated". But this is precisely what we would anticipate with hierarchical development of tissues via differentiation trees, and an interpretation of the epigenetic landscape in terms of them.

4.03 Genetics and Differentiation Trees

"Driesch [1894]... formulated a model... which postulated that development was not all nuclear or cytoplasmic but was the result of an interplay between the two. His hypothesis sounds quite reasonable even today...; however, it seems to have been ignored by his contemporaries" (Raff & Kaufman, 1983).

"Great accomplishments in... chemical genetics... are undoubtedly forthcoming, and the embryologist who remains unelated by them will be a confirmed classicist indeed. It may equally well be said that it is a naive molecular biologist who thinks that his present or immediately foreseeable approaches promise a ready solution to all, or even the most central, problems of development" (Twitty, 1966).

Proposition 21: the differentiation cascade for each step of differentiation is set off by a differentiation trigger inside the cell nucleus.

Triggering may start with a Britten & Davidson (1969) 'activator gene' (cf. Davidson & Britten, 1971), which may contain the highly conserved homeobox (Gehring, 1993), found both in animals (McGinnis et al., 1984a,b; Seeger & Kaufman, 1987; McGinnis & Kuziora, 1994) and plants (Rennie, 1991a):

"...The great German poet [and naturalist] J.W. von Goethe (1790; [translated in: Arber, 1946; cf. Amrine, Zucker & Wheeler, 1987]) published his view that all plant flower organs were transformed leaves [an idea attributed by Goethe to Linnaeus: Nordenskiöld, 1928]. This transformation, a process we now call homeosis, was thought to be brought about by 'ascending juices' (plant hormones?). Some of Goethe's views might actually be true, at least in principle" (Seadler & Huijser, 1993).

Since we do not know the chemical nature of the trigger, I will simply call it a differentiation trigger. A working hypothesis for the molecular nature of differentiation triggers is given in Appendix IV (Björklund & Gordon, 1993b). In the body of this text I will keep the concept at a more abstract, general level.

Proposition 22: differentiation triggers, each of which starts off one of a pair of differentiation cascades, have two states.

Since there are two possible outcomes of an act of cell state splitting, we need two triggers to set off the corresponding two differentiation cascades. Moreover, they have to be mutually exclusive, or else both differentiation cascades may be triggered simultaneously, which would foul up the process of differentiation. A physical embodiment of these requirements is not too hard to conceive. These states may be thought of as:

pairs of molecules;

two configurations of a single, allosteric or chemically modified macromolecule;

two polarization states of the nuclear membrane;

a push/pull state of the cytoskeleton between the cell and nuclear membranes.

One might presume that differentiation triggers are most likely to be DNA binding proteins that release master genes.

Proposition 23: a commitment signal travels directly or indirectly from the cell state splitter to the nucleus and selects the differentiation trigger. The set of mechanochemical and chemical reactions transmitting this information we call the *differentiation pathway*.

"When the gene theory was in its infancy, some individuals who could not deny the evidence for genic control of organismic characteristics were willing to consider the idea that genes control the relatively unimportant details of biological specificity, but asserted that the more fundamental issues of heredity were untouched. We have reached the antithesis of this position now, and only reluctantly concede that heredity involves anything beyond the genes. Perhaps we are ready to begin moving toward a synthesis, which places a balanced emphasis on the various kinds of mechanisms" (Nanney, 1968).

How can the mechanical act of cell state splitting select a given differentiation trigger in the nucleus? Since cell state splitting is a binary event, and cell state splitters are at the cell periphery (Figure 1 in Appendix V: Gordon, Björklund & Nieuwkoop, 1994), we need a commitment signal to transmit information about what has happened at the cell surface to the nucleus. I will consider a number of models for the nature of the commitment signal. In general, we have yet to achieve a complete picture of how information is transmitted from the cell membrane into the nucleus:

"Unfortunately the field of growth factor receptor studies has not yet converged with the field of transcriptional regulation to yield a cohesive picture of the intermediate steps between the cell surface receptor and nuclear transcriptional events" (Williams, Hunter & Gallagher III, 1989).

The apical cell membrane may contain stretch sensitive ion channels (Medina & Bregestovski, 1988; Sachs, 1989; Morris, 1990; Björklund & Gordon, 1993a,b; Martin et al., 1995), which act differently when that membrane is stretched or compressed (Figure 3 in Appendix IV: Björklund & Gordon, 1993b), causing differences in ionic currents that reach the nucleus. Here nucleus associated ions may somehow provide specific changes in gene transcription:

"Evidence revealing the presence of transmembrane nuclear Ca^{2+} gradients and a variety of intranuclear Ca^{2+} binding proteins has fueled renewed interest in this key ion and its involvement in cell-cycle timing and division, gene expression, and protein activation" (Gilchrist, Czubryt & Pierce, 1994).

Fluorescently labelled fast calcium waves light up nuclei as they travel from cell to cell (Bkaily et al., 1996). In some cells the intracellular calcium wave travels inside from the apical towards the basal end (Nathanson, Burgstahler & Fallon, 1994). Ultraslow calcium waves involving the cytoskeleton, such as differentiation waves (Jaffe, 1995), might do the same, altering nuclear Ca^{2+} gradients across the nuclear membrane (Gilchrist, Czubryt & Pierce, 1994) as the calcium diffuses through the nucleus (Fox, Burgstahler & Nathanson, 1997; cf. Lin, Hajnoczky & Thomas, 1994; Koopman et al., 1997). It is possible that mechanochemical transmission through the cytoplasm carries an ultraslow wave through the cytoplasm to the nucleus, since the intermediate filament vimentin partially depolymerizes in some cells in response to a calcium wave (Inagaki et al., 1997).

As an alternative to cytoplasmic ionic currents, there may be a specific messenger molecule that travels from the cell membrane to the nucleus. (Cf. Davies, 1993, for analogous considerations in plants.) For example, one possible role of microtubules in signaling is suggested by the observation that hyaluronan receptor RHAMM (Samuel et al., 1993; Hall et al., 1995) apparently moves down microtubules right to or into the nucleus (Eva A.

Turley, p.c., 1996), perhaps because it is, or is bound to, a motor protein. The rate of movement might account for the hypothesized 10 min time delay between microfilament contraction and master gene activation (Appendix IV: Björklund & Gordon, 1993b). Ben-Ze'ev et al. (1995) put all the mechanics at the surface, at least in the following example:

"Cell-cell and cell - extracellular matrix (ECM)... interactions of the cell with its environment are mediated by numerous transmembrane receptors of the integrin [Travis, 1997c] and cadherin families of receptors (Takeichi, 1991; Hynes, 1992a). In the cytoplasmic domain, many of these adhesive interactions are linked to the microfilament system by plaque proteins which are especially prominent in adherens type junctions (AJ) (Burridge et al., 1988; Geiger & Ginsberg, 1991; Tsukita et al., 1992), thus forming a structural link between the extracellular and the intracellular domain of the plasma membrane.... In addition to proteins which are involved in establishing the structural and thus more mechanical interaction between the outside and inside of the cell, AJ contain a considerable number of regulatory molecules including tyrosine kinases, protein kinase C, and various oncogene products and proteases, which reside in the cytoplasmic side of the plaque area (Burridge et al., 1988; Geiger & Ginsberg, 1991; Luna & Hitt, 1992)... and thereby elicit long range effects on cells" (Ben-Ze'ev et al., 1995).

The intermediate filament network, which extends from the nuclear membrane (or inside it: Paddy et al., 1990; Paddy, Agard & Sedat, 1992) to the cell membrane (Alberts et al., 1989), might be capable of transmitting a mechanical push or pull (which is a binary signal), from the cell membrane to the nucleus, which is somehow transduced into selection of the differentiation trigger, perhaps by penetration right into the nucleus (Traub, 1995). Here transduction could even be based on physical forces acting on and within the DNA; it has been hypothesized that:

"...the structure of the *Escherichia coli* nucleoid is determined by DNA binding proteins and DNA supercoiling, representing a compaction force on the one hand, and by the coupled transcription/translation/translocation of plasma membrane and cell wall proteins, representing an expansion force on the other hand" (Woldringh, Jensen & Westerhoff, 1995).

Whether such compaction/expansion forces apply to eukaryotic DNA, and could be involved in differentiation, remains to be seen.

Strands or sheets of kinoplasm (streaming cytoplasm, including endoplasmic reticulum: Honda, Hongladarom & Wildman, 1964; cf. Lahoz-Beltrá, Hameroff & Dayhoff, 1993) might provide a push/pull effect. A similar idea for mechanical or mechanochemical signalling from the cell surface to the nucleus has been suggested by Ingber (1991a), Pienta & Coffey (1991, 1992), Kropf (1992), Ingber et al. (1993), Wang, Butler & Ingber (1993) and Forgacs (1995a), and is implicit in the model of Puck & Krystosek (1992):

"...The mammalian cell cytoskeleton becomes part of an information transmission system extending from the cell membrane and its specific receptor sites [for which we substitute the cell state splitter], through the cytoplasm, terminating in specific points on each chromosome, so that specific domains of exposure and sequestration result" (Puck & Krystosek, 1992).

The observation that "...congression of chromosomes (the process whereby chromosomes move to form the metaphase plate) at prometaphase occurs as a wave, starting at the top of the nucleus near the embryo surface and proceeding through the nucleus to the bottom... [in] syncytial blastoderm embryos of *Drosophila* " (Hiraoka, Agard & Sedat, 1990) is suggestive of an active pushing and/or pulling process on the nucleus during mitotic waves (which may also be differentiation waves). Note that the mitotic waves are parallel to the blastoderm surface, while the wave of congression is perpendicular to it. However, the apparent shallowness of the *Drosophila* eye imaginal disc furrow relative to cell height (Ready, Hanson & Benzer, 1976), and the dissociability of differentiation and interkinetic migration (Fischer-Vize & Mosley, 1994), would seem to make a direct mechanical link improbable. On the other hand, the observation that "Polyadenylate... RNA-rich transcript [nuclear] domains... formed a ventrally positioned horizontal array in monolayer cells" (Carter et al., 1993) suggests a fixed geometric relationship between the nucleus and the polarity of epithelial cells, hinting at mechanical links. Protein kinase C (PKC) bound to the intermediate filaments (Murti, Kaur & Goorha, 1992) or actin (Blobe et al., 1996) could, plausibly, involve these cytoskeletal components as a link to our postulated differentiation pathway (Appendix IV: Björklund & Gordon,

1993b), obviating the need for a purely mechanical link (not that anything can be 'purely mechanical' at this molecular level). Calcium fluxes are probably also involved:

"Recently, Con A has been shown to raise intracellular calcium concentration by opening L-form calcium channels in the animal cap [of *Xenopus*] (Moreau et al., 1994), and this is consistent with the idea that neural induction involves activation of protein kinase C (Otte et al., 1988)" (Hatada et al., 1995).

Mechanical effects of microtubules and microfilaments have been discounted in "Stretch caused... induction of immediate-early genes such as *c-fos, c-jun, c-myc*, JE, and Egr-1, but not Hsp70" (Sadoshima et al., 1992a; cf. Sadoshima & Izumo, 1993b), but intermediate filaments were not tested in these cardiac cells, which were stretched 20% (in arbitrary directions compared to their axes) on silicone sheets without exhibiting any active morphological response. Sadoshima & Izumo (1993a) consider three models for...

"...how mechanical stimuli are converted into intracellular signals of gene regulation (reviewed in Ingber, 1991b; Vandenburgh, 1992)....

"Our results suggest that the stretch response causes a rapid activation [peaking at 1 hr] of multiple second messenger systems, including tyrosine kinases, $p21^{ras}$, MAP [mitogen-activated protein] kinases, S6 kinase (RSK), PLC [phospholipase C], PLD [phospholipase D], PKC [protein kinase C] and arachidonic acid metabolism (particularly the P450 pathway), many of which seem to contribute to c-*fos* gene induction by stretch....

"How does cell stretch lead to the activation of multiple intracellular second messengers?...

"One possibility is that mechanical stress directly activates signal transducing molecules, such as growth factor receptors (in the absence of a ligand), G proteins, phospholipases or protein kinases. It has been postulated that mechanical forces applied to the cell surface might generate direct conformational changes in the molecules associated with the plasma membrane, which may activate downstream second messenger systems (Watson, 1991). However, at present, there is little direct evidence for this hypothesis.

"The second possibility is that mechanical stretch activates putative cellular mechanotransducers, such as mechanosensitive ion channels (Sachs, 1989), cytoskeleton or integrins (Ingber, 1991b), which in turn activate multiple second messengers. Our previous study argues against the possibility that stretch-activated cation channels (gadolinium-sensitive types), actin microfilaments, microtubules or RGD-binding integrins are involved in the stretch-induced c-*fos* induction (Sadoshima et al., 1992b)....

"The third possibility, which we favor most, is that mechanical stress causes release of some growth factor(s) and that this factor activates its receptor and subsequent second messenger cascades" (Sadoshima & Izumo, 1993a).

Cf. the effects of mechanical loading on muscle cell differentiation (Simpson et al., 1994a). Which of these three models for terminally differentiated heart cells applies to differentiating cells in early embryos is not *a priori* obvious:

"It should be noted... that stretch responsiveness may not be a general phenomenon because stretch-induced IE [immediate-early] gene expression was observed only in primary cultures but not in established cell lines such as NIH 3T3 and PC12 or in cultured pulmonary smooth muscle or aortic endothelial cells that have been passaged multiple times" (Sadoshima et al., 1992a).

"...We found several differences between the cellular responses induced by mechanical stretch (linear strain) and those induced by hypotonic cell swelling (radial strain).... Although the signaling mechanisms... are not identical, rapid activation of tyrosine kinase is one of the earliest cellular responses to both types of mechanical stress" (Sadoshima et al., 1996).

Once a signal gets to the nucleus, we have to face the questions of how it gets inside, and what it does there. There is evidence that microtubules can, in some as yet unknown way, 'reach across' the nuclear membrane during mitosis, a process that could perhaps occur during cell differentiation:

"Lauterborn [1896]... noted that in [the diatoms] *Pinnularia* and *Nitzschia sigmoidea*, chromosomes collect near the spindle even before the nuclear envelope breaks down, implying that kinetochores are responsive to microtubules even when a membrane is interposed between the two. This observation may turn out to be important..., since in certain dinoflagellate and protozoan cells, the kinetochores do apparently interact with microtubules through the nuclear envelope (*e.g.*, Kubai, 1973; Oakley & Dodge, 1976). Furthermore,

Richards (1975) has shown that movements of chromosomes at prophase can be relatable to the activity of extranuclear spindle microtubules nearby" (Pickett-Heaps, Schmid & Tippit, 1984).

I myself have seen indications of microtubules aligned from cell to cell in high voltage electron micrographs of axolotl ectoderm (unpublished; Section 3.10). This phenomenon may be involved in long range cortical organization and its potential homology to ciliates discussed in Section 5.05. A similar role for intermediate filaments has been proposed (Traub, 1995).

Finally, Maniotis, Chen & Ingber (1996) have shown that an all mechanical pulling mechanism exists that subtends the cell membrane, cytoplasm and nucleus:

"When integrins on the surface membrane were pulled by micromanipulating bound microbeads or micropipettes, cytoskeletal filaments reoriented, nuclei distorted, and nucleoli redistributed along the axis of the applied tension field.... These results indicate that cells and nuclei are literally built to respond directly to mechanical stresses applied to specific cell surface receptors, such as integrins. This property may not be limited to integrins, for example, transmembrane receptors that mediate other forms of adhesive junctions (e.g., desmosomes, adherens junctions) may exhibit similar behavior. The demonstration of direct mechanical linkages throughout living cells raises the possibility that regulatory information, in the form of mechanical stresses or vibrations, may be rapidly transferred from these cell surface integrin receptors to distinct structures in the cell and nucleus, including ion channels, nuclear pores, nucleoli, chromosomes, and perhaps even individual genes, independently of ongoing chemical signaling mechanisms.... Changes in the distribution of mechanical stress transmitted across integrins might redirect the axis of cell division" (Maniotis, Chen & Ingber, 1996).

It would be nice to see comparable studies on cells participating in differentiation waves. Perhaps the cell state splitter has a direct mechanical link to the nuclear state splitter, and the differentiation pathway is primarily a mechanical push/pull system (Section 10.13).

In the cell state splitter, when the tug-of-war is won by the microtubules, the microfilament ring may dissolve, resulting in a transient increase in free actin, which might affect one differentiation trigger (cf. Spiegelman & Ginty, 1983), while dissolution of the microtubules and increase in free tubulin might affect the other differentiation trigger. This may be thought of as ATP versus GTP mediated signal transduction from the cell state splitter, i.e., from microfilaments or microtubules, respectively, to the nucleus (Appendix IV: Björklund & Gordon, 1993b). Dissolution of microfilaments or microtubules may alternatively affect the membranes to which they are attached, which in turn may signal the nucleus. Other mechanochemical effects, such as changes in nuclear rotation (De Boni, 1988a; Fung & De Boni, 1988; De Boni, 1994) and "dramatic chromosome and nuclear reorganization" (Hiraoka et al., 1993) in response to gene activation, may also occur at the time of cell state splitting (though this may not involve bringing specific genes to the nuclear surface: Hochstrasser & Sedat, 1987; Mathog & Sedat, 1989).

Of course, it is plausible that the commitment signal and the differentiation trigger are one and the same entity. The work of Ben-Ze'ev (1985a), suggesting that mechanical shape changes imposed on cultured cells causes synthesis of alternative intermediate filaments and other proteins (cf. Ben-Ze'ev, 1986a, 1989, 1991, 1992; Farmer et al., 1983; Simpson et al., 1994b), may provide clues to unravelling the relationship between natural cell shape change and differentiation triggers (cf. Huguet et al., 1995). Ingber & Folkman (1989a) state that "cell geometry is centrally involved in the regulation of cell differentiation... (Emerman & Pitelka, 1977; Spiegelman & Ginty, 1983)..." (including apoptosis: Chen et al., 1997), and review a broad range of literature on "how physical forces provide regulatory information". The same phenomenon may be involved in manipulations of the extracellular matrix:

"Elizabeth Hay and others (Greenburg & Hay, 1986, 1988; Zuk, Matlin & Hay, 1989) have described stunning examples of the regulation of cellular phenotype by extracellular matrices. Differentiated epithelial cells (including corneal, thyroid, and kidney epithelia, notochord, and adult skin) cultured on collagen plates grow into polarized epithelia, secrete Type IV collagen, and form a basal lamina. The same cells, suspended in a collagen gel, take on the appearance of fibroblasts, become motile, and secrete Type I collagen (a fibrillar collagen)" (Morris, 1993).

Similar ideas, "that forces and/or structure can control... gene expression" are being formulated for prokaryotes (Mendelson & Thwaites, 1989). As I suggested, it is possible that the commitment signal and the differentiation trigger are one and the same entity, since they carry the same information. Alternatively, they may be at opposite ends of an involved process of information transmission. Appendix IV (Björklund & Gordon, 1993b) selects and expands some of these ideas into a full working model for the mechanochemical relationship between the cell state splitter and the nuclear state splitter (Appendix IV). Forgacs (1995a) has proposed a similar mechanochemical coupling for signal transduction in general, though he starts with membrane receptor chemistry rather than mechanics at the cell surface. Bitar et al. (1991) have shown that...

"The neuropeptides bombesin and substance P, which are present in neurons of the anorectal region [in rabbits], induce contraction of isolated smooth muscle cells from this region by activating different intracellular pathways. Substance P-induced contraction is 1,4,5-inositol triphosphate (IP_3)/calmodulin dependent, while contraction induced by bombesin is mediated by a protein kinase C (PKC)-dependent pathway" (Bitar et al., 1991).

(Cf. Bitar & Yamada, 1995.) This suggests that our proposed nuclear state pathway, combining IP_3 and PKC (Appendix IV: Björklund & Gordon, 1993b), may involve just one of these components.

Any commitment signal is a component of what we have called the differentiation pathway (Appendix VI: Björklund & Gordon, 1994). It is but one component of the chain from the cell state splitter to the nuclear state splitter. This may explain the frustration in the search for a single switch protein:

"[An]... element in several of the theories was the assumption that there exists simple unidirectional hierarchies of control of gene expression, in particular, the models of Britten & Davidson (1969) and García-Bellido (1975). The simplest conceptions of control are those in which a given regulatory element switches on the expression of a third, and so on, culminating in the synthesis of the gene products of a battery of structural genes. However, where genetic data are available, this picture of a simple linear chain of command modules often does not fit the facts.

"Certainly, hierarchical *elements* of control must exist in development. The results..., however, suggest that the *global* properties may not always be simple hierarchies. There are at least two simple alternative explanations that can be listed. The first is that many gene products might combine to give one functional unit, analogous to a multienzyme complex; inactivation of one gene product could inactivate the unit as a whole. The other possible alternative to simple hierarchies is the idea of the network, in which several genes regulate each other in an interacting circuit: if the circuit is broken because one element is missing (e.g., through mutational inactivation), the circuit ceases to operate and a mutant phenotype is produced.... Of course, networks are much more difficult to analyze, and to comprehend, than unidirectional hierarchies. However, that is a problem for the investigator, not for the organism. One possible set of network structures in eukaryotic gene control has been discussed by Kauffman (1971)" (Wilkins, 1986).

The cell state splitter, the differentiation pathway, and the nuclear state splitter, may qualify as the required hierarchical elements, within which "simple unidirectional hierarchies of control of gene expression" do indeed function.

Proposition 24: there exists a state of *nuclear competence* in which the nucleus is prepared to receive the appropriate commitment signal.

The nucleus must be ready to receive the commitment signal from the cell periphery. It must be prepared to receive and act upon that signal via one of its two differentiation triggers in the currently prepared pair of differentiation triggers. If we accept nuclear competence as an essential concept, tying the behaviour in the nucleus to that of differentiation waves, so that the physics and the genetics of the differentiation tree are coordinated, then the molecular basis for nuclear competence will be intriguing to work out:

"There's evidence that during development the structure of the chromatin, the complex of DNA and proteins that forms the chromosomes, changes so that in some cells certain genes may be primed to respond, while in other cells these same genes may be inaccessible to the transcription factors that control their activity" (Schmidt, 1994).

It is perhaps time to find out explicitly how the nuclear state splitter works. Here, then, without being able to definitively answer the question, I would like to consider some of the means by which nuclear competence may be implemented (cf. Section 10.13).

Nuclear competence may correspond to the 'primed', 'transcriptionally competent' (Stargell et al., 1993), or 'exposed' state:

"We have proposed a two-tiered scheme for the regulation of the mammalian genome which has an extra step beyond that recognized in *E. coli* (Puck, Krystosek & Chan, 1990). This scheme accounts for the mode of action of genes that are differentiation specific in the mammalian organism. The first step in the activation of such tissue-specific genes is their conversion from the sequestered to the exposed state.... The second step, then, which resembles the corresponding process in bacteria, is activation or inactivation of the exposed genes as a result of interaction with appropriate effector molecules in the surrounding medium. Thus, it is necessary, but not sufficient, for differentiation-specific genes to be exposed before they can be activated" (Puck & Krystosek, 1992).

This refines the old idea of gene batteries: a whole battery is exposed at a given step of differentiation, and then some of its genes activated, either immediately, or when they are needed. In nuclear state splitting, we would surmise that exactly two new gene batteries are susceptible to being exposed, and the signal from the cell state splitter determines which one gets exposed. The other remains sequestered.

If we allow that genes can be (sequestered or exposed) and (active or inactive), then there are four possible states for each gene:

1. (sequestered, inactive)
2. (exposed, inactive)
3. (exposed, active)
4. (sequestered, active)

State 1 may correspond to heterochromatin. Presumably the last state is not possible, but I list it for completeness, especially since none of these are absolute states. They all undoubtedly depend on binding of something to DNA, with finite binding constants, which means that there is always some chance of unbinding. In any case, we see that whatever the sequestering/exposure mechanism, it has to interact with or even compete with the active/inactive states. The latter presumably use standard operon, repressor, enhancer and activator mechanisms.

We already have some tools, such as 'recombination access mapping (RAM)' (Dellaire et al., 1997) and random gene insertion, to probe the distinction between the sequestered and exposed states:

"Repressing the activity of large blocks of genes by limiting required transcription factors seems unlikely (Singer & Berg, 1991). One means of repression is that blocks of genes are sequestered into higher order chromatin structures. This may explain why genes ordinarily not expressed by a given cell type can be expressed when introduced by transfection. Transfected globin genes that are randomly incorporated into the genome of fibroblasts are transcribed 1000 times more actively than the endogenous globin genes present in their normal repressed higher order configuration in these fibroblasts (Singer & Berg, 1991)" (Flickinger, 1994).

In this book I usually make the assumption that although "it is possible that genes accounting for cell differentiation are regulated as a group of genes" (Flickinger, 1994), this regulation mechanism is the same as that of any individual gene, i.e., that there is a master gene that regulates each differentiation cascade (Monod & Jacob, 1961; Jacob & Monod, 1963). However, now that we are considering an additional layer of mechanism, namely the states of being sequestered or exposed, it is not obvious that transitions between these two states are consequences of single genes, the same way as transitions between active and inactive. An argument could be made that single genes are involved (Natalie K. Björklund, p.c.), by assuming there is an analogy or homology with the single gene driven X-chromosome inactivation (*XIST*: Brown, 1991), or with the means by which "H1 histone silences extensive domains of chromatin DNA, not just individual genes" (Flickinger, 1994). Even then, the sequestered ⇔ exposed transition may be more of a physical than a chemical phenomenon, perhaps involving altered conformations of the chromosomes. If so, this would argue for the genes involved in a single step of differentiation being generally contiguous, as are those that are inactivated on an X-chromosome. In any case, the concept of the existence of master genes is coming under deserved close scrutiny:

"Pluripotent hematopoietic stem cells can differentiate into a number of distinct specialized cell types; however, no single lineage-specific master regulators have been identified that can activate individual patterns of gene expression. Recent evidence suggests that such lineage determination is regulated by a combinatorial matrix of regulatory proteins with overlapping tissue specificities which cooperate to define individual cell types" (Ness & Engel, 1994).

This is a vague generalization of the concept of "a multimeric protein complex (Zink et al., 1991; DeCamillis et al., 1992; Rastelli, Chan & Pirrotta, 1993; Martin & Adler, 1993)" (Sedkov et al., 1994). It contrasts with the idea that in some cases, such as eye formation, there is indeed a master gene (Halder, Callaerts & Gehring, 1995; Travis, 1997a). From the point of view of our theory, either mechanism for the nuclear state splitter is plausible, so long as it responds in a binary fashion to a one bit signal from the cell state splitter.

Inside the nucleus, the nuclear state splitter may consist of pairs of readied homeobox genes (Gehring, 1993), proto-oncogenes (Bos, 1992), nuclear receptors (Gronemeyer, 1993; Gronemeyer & Laudet, 1995), multiprotein complexes (Orlando & Paro, 1995), or other cooperatively binding molecules, in the sense of a (heterogeneous) phase transition, in which the proteins and DNA fit together more like pieces of a puzzle than like identical molecules in a crystal. Such spatial combinatorics are reminiscent of the observation that magic squares can be constructed in more than one configuration (Moran, 1981), or that polyominoes can be fit together in many ways, even with the same overall shape (Golomb, 1965).

The presumption that we have already found that homeobox genes are master genes is balanced by the concern that they themselves are controlled by other genes:

"The discovery of the homeobox marks the beginning of a new era in developmental biology in which a class of master control genes, which determine the body plan, have been identified" (Gehring, 1992).

"In early tooth germ, bone morphogenetic proteins BMP-2 and BMP-4 regulate expression of the homeobox containing genes Msx-1 and Msx-2" (Thesleff, Vaahtokari & Partanen, 1995).

"The question was then raised as to what in this environment is critical for Hox gene induction to proceed" (Grapin-Botton et al., 1995),

i.e., if Hox genes need to be 'induced', then what controls the 'controller'? Could it be a differentiation wave in every case? Rather than an infinite regress of genes controlling genes, the process has to start with something other than a gene.

A number of fascinating questions can be asked if one accepts the possibility that sequestered ⇔ exposed transitions are different in nature from active ⇔ inactive transitions:

Are all differentiation cascades sequestered at the beginning of development?

Are some differentiation cascades pre-exposed in the maternal/paternal haploid genomes, say by imprinting?

Is the midblastula transition caused by 'spread' of the sequestered/exposed mechanism to the zygotic genome?

Do previously exposed differentiation cascades become sequestered when a new differentiation cascade is exposed?

In nuclear state splitting, one of a pair of differentiation cascades will be exposed, depending on whether the cell participates in an expansion or a contraction differentiation wave. Are these pairs physically adjacent, and do they interact in a rocker switch fashion, perhaps physically, so that one and only one of the pair is exposed?

Do we need yet another level of control, which distinguishes a selected pair of differentiation cascades from all other differentiation cascades that either have already gone through an event of nuclear state splitting, or have not yet done so?

Is the ability to dedifferentiate, or the ability of a transplanted nucleus to function, related to recovery of particular sequestered/exposed states?

We thus see that the concept of nuclear competence is rich in its possibilities, but that we have much work ahead to understand what nuclear competence is and how it works.

Proposition 25: embryological competence of a cell to go through a step of differentiation requires the presence of a cell state splitter, a commitment signal, a differentiation trigger, and a corresponding state of nuclear competence.

We now see that the classical embryological concept of competence (Holtfreter & Hamburger, 1955) has these four components. This definition of embryological competence subsumes and expands that of Davidson & Britten (1979):

"Commitment would thus be due to the presence or absence of an appropriate set of sensor structures determined by the previous history of the cell lineage" (Davidson & Britten, 1979).

We do not require that this quadruplet be of identical structure from one step of differentiation to the next, though that would be more elegant. The duplication and divergence of homeotic genes from a common ancestral gene (Lewis, 1978; McGinnis et al., 1984a) suggests that this might approximately be the case, at least in metameric (segmented) organisms.

Proposition 26: a means must be available for ensuring that only the appropriate pair of differentiation triggers reacts to the commitment signal.

This restriction could take the form of a unique commitment signal for each pair of differentiation triggers. On the other hand, if the commitment signal is universal, then the nucleus must have only the appropriate differentiation trigger prepared to receive it, through some sort of masking or lack of preparation of the others. The Wurfel mechanism (Section 10.13) may provide this capability.

Proposition 27: each differentiation cascade released by a cell state splitter includes means of setting up the next cell state splitter, preparation of a pair of differentiation triggers, and the preparation of the next commitment signal.

Thus the preparation of the differentiation triggers could be part of each preceding differentiation cascade, which would remove the ambiguity about which pair of differentiation cascades was to be considered for activation next.

The above propositions in this Section flesh out the conception of Waddington (1975):

"I have argued that the elementary differentiation process of a higher organism cell (a) involves complexes or 'batteries' of genes rather than single genes [differentiation cascades]; (b) takes place in three phases: acquisition of competence [erection of cell state splitter and nuclear state splitter], in which several different [I would say exactly two] batteries of genes [differentiation cascades] become ready to enter the next phase [= nuclear state splitter]; determination, in which one of these batteries is singled out [by the one bit signal from the cell state splitter to the nuclear state splitter, signaling participation in a contraction or an expansion wave] to become dominant in the future history of the cell; activation [differentiation], in which the proteins corresponding to the structural genes in this battery actually begin to be produced" (Waddington, 1975).

Proposition 28: all gene subsets that are activated in a given cell at a given stage of differentiation are accessed via the strictly hierarchical organization of the differentiation tree.

"The embryologist... has had little success in explaining why any structure should develop just as it does.... Our knowledge of ontogeny remains incomplete... until the fields of embryology and genetics are joined by the discovery of the dynamic properties of the gene" (Driver, 1931a).

Even if an organism's genome is organized hierarchically, keeping it that way over evolutionary time requires that deviations from hierarchical organization are selected against at some level. Whether there is a feature of genome organization that prevents genes in a differentiation tree from forming cyclic network relationships remains to be determined. A mutant of *Arabidopsis* flower morphogenesis, which appears to recursively form sets of petals upon petals until it runs out of cells (Meyerowitz, 1989; Meyerowitz, Smyth & Bowman, 1989), and a mutant of the nematode *Caenorhabditis elegans* (Chalfie, Horvitz & Sulston, 1981), may have exactly such cyclical repeats. Perhaps such mutants merely have low survival value. At a higher level, I will discuss evolutionary radiation as a possible source of the feedback that keeps differentiation trees hierarchical.

Proposition 29: there is a distinction between activation of a differentiation cascade and activity of genes within that differentiation cascade.

It is this distinction between differentiation cascade activation and gene activity which is lacking in the network theory of differentiation initiated by Kauffman (1969). Kauffman (1993) states his opinion against hierarchical control, as via a differentiation tree, in the following way:

"A familiar conceptual model of genetic regulatory architectures has been a military hierarchy, with a few generals commanding overlapping battalions of officers and troops below them (Britten & Davidson, 1969; Davidson & Britten, 1979). In these models, a few master genes control overlapping downstream cascades, or batteries, of genes. Control of alternative developmental programs lies in unleashing the different possible cascades of activities from the master control gene at the top of each hierarchy. The overlapping activations spreading through the overlapping downstream cascades would create different cell types, each with a unique battery of gene activities. In one extreme form, such regulatory networks are entirely free of feedback loops. The terminal structural gene 'twigs' of the hierarchical cascades, once activated, might maintain a stable pattern of activity by virtue of a stable unmethylated state.

"The hypothesis that genetic regulatory systems are hierarchical and free of feedback loops is an important one. If true, propagation of control is particularly simple. Yet even with only modest insight into the statistical properties of scrambled wiring diagrams, it is clear that such a hierarchical system is highly improbable. Genome regulatory networks are very likely to be rich in feedback loops....

"...To attain and maintain an improbable hierarchical wiring diagram in the face of mutations tending to randomize connections and creating a richly webbed architecture, selection would have to struggle against powerful forces.... Selection is very likely to fail in such an endeavor....

"...We shall find conditions... which permit evolution of hierarchical command structures even in genomic systems rich in feedback loops....

"Canalyzing Boolean networks crystallize orderly dynamics - in other words, they spontaneously lie in the ordered regime - because a specific kind of subnetwork, called an *extended forcing structure,* literally crystallizes out of the dynamics of the network (Kauffman, 1971, 1974; Fogelman-Soulie, 1984). The genes making up the forcing structure fall to fixed active or inactive states; the forcing structure percolates through the genomic network and typically leaves behind one or more unfrozen islands of genes free to turn on and off in complex patterns....

"Consider a small isolated island of genes headed by a single $K = 1$ live feedback loop which has descendent tails of regulatory and structural genes. Such a loop is logically identical to the master gene in charge of the descendant battery of genes.... If some form of hierarchical regulation is often useful, which seems plausible, but must occur in genomic systems rich in feedback loops, then the existence of a large frozen component is almost certainly an essential requirement" (Kauffman, 1993).

To discuss Kauffman's (1993) doubts of a hierarchical genome, I have to anticipate elements of my argument. As an organism develops into many tissue types, the presumed 'feedback loops' would have to cross cell boundaries. This is, of course, the essence of the view that inductive interactions are primary to embryological development. However, I will discuss the secondary nature of inductive interactions between tissues (in Section 5.02). To the extent that induction is secondary, the differentiation of cells must follow their lineage, which is itself a hierarchical, tree structure. In other words, there is a fundamental tree structure to development, imposed by the fact that most cells arise via cell division (syncytial cells excepted). (This is not a characteristic of the military hierarchy, which does not follow a family tree, making the metaphor of dubious value.)

Overlaps of downstream genes, when they occur, are attributed in my model to genes in modules off the differentiation tree, the so-called housekeeping genes (Proposition 144). The differentiation tree is but an elaboration of the lineage tree. Within a single cell, feedback loops can indeed occur. These can be between genes in all of its so far activated differentiation cascades or batteries. At this intracellular level, but not for the whole organism, webbed gene networks may prevail. The same comment applies if the postulated webbed networks are organized into island hierarchies.

Kauffman (1993) does not further address distinctions between genes that could be classified as activated or sequestered. His island model for gene hierarchies does not explain the spatial component of differentiation.

4.04 A New Definition of 'Tissue'

"...Gene expression represents an *effect*, rather than a *cause*, of cell, tissue, and embryonic morphogenesis" (Malacinski, 1990d).

Proposition 30: a tissue consists of the equivalence class of all cells that have followed the same path through the differentiation tree.

All cells that have followed the same path along the differentiation tree are fundamentally the same, and it is on this basis that I define them as being in the same tissue. The concept of a tissue is thus not based on microscopic morphology or histology, though I would generally anticipate that all cells in a tissue, as redefined, would be similar in these regards.

Note that the definition of a tissue as a differentiation tree equivalence class permits histologically indistinguishable tissues to be nonequivalent in the sense of Lewis & Wolpert (1976) (cf. Wolpert, 1982; Maclean & Hall, 1987; Nagorcka, 1989; and the concept of the 'second anatomy' of Slack,

1982a, 1983, 1986). This means that histological criteria may not be sufficient to define differentiation trees. Lewis & Wolpert (1976) have shown how certain bilaterally symmetric diseases require this concept of nonequivalence, and how standard reaction-diffusion models of morphogenesis (à la Turing, 1952; cf. Meinhardt, 1982a; Murray, 1990; Yates & Pate, 1989) ignore it. Consider, for example, that the new curly ear mutant of domestic cats, affecting ear cartilage only, may be an example of a differentiation tree mutation affecting only that terminal branch leading to ear cartilage (Robinson, 1989a; Weiss, 1990a), as distinct from other cartilage tissues. Here we have an example of histologically indistinguishable tissues that are nevertheless not equivalent. Of course, 'histologically indistinguishable' is a matter of the state of the art in histology. We would anticipate that some discernible differences could be found, if we were to look hard enough.

Mosaic organisms often have pairs of symmetrically dividing cells, which have been considered as being in an equivalence group (nematodes: Kimble, 1981; leeches: Weisblat et al., 1987). These pairs would form examples of tissues with two equivalent cells in embryos of mostly one cell tissues Proposition 55).

Cells in an equivalence group do not necessarily arrive into that group simultaneously, if only because differentiation waves are slow, and reach cells in the source tissue at different times. They may also be spatially separated into noncontiguous portions, as in left-right symmetry, though in this case there is the possibility of an early node in the differentiation tree that distinguishes left from right (Levin et al., 1995). The combined concepts of differentiation trees and equivalence groups might resolve the problem "...that the expression domains of many of the key genes that lay down tissues do not... respect existing anatomical boundaries.... [This creates difficulties in] the construction of [Internet] databases that link domains of gene expression to developmental anatomy" (Bard & Davies, 1995).

Proposition 31: tissue synchronizing mechanisms initiate the next act of cell state splitting in a tissue consisting of contiguous cells.

An example of a possible tissue synchronizing mechanism is the buildup of embryo pressure hypothesized in Section 3.04 for primary neural induction. This is perhaps the function of purported cell-cell 'communication' via gap junctions (Christ, Brink & Ramanan, 1994) during embryogenesis (Warner, Guthrie & Gilula, 1984; cf. Sneyd, Charles & Sanderson, 1994; Cox, Kirkpatrick & Peifer, 1996). Clearly, much work is needed to find out what kinds of physical and/or chemical tissue synchronizing mechanisms could be operating. The following observations can be made:

"In order for a second wave of transitions to pass through the set of systems [cells], the systems must all return to a metastable state before the wave reaches them. This defines the synchronization issue: some mechanism must make sure that all the systems are potentiated before the next wave is launched.

"There is no issue if the systems all re-potentiate at the same speed, and if it takes a lot longer for a new wave to be triggered than it takes for the systems to re-potentiate. If there *is* a need for synchronization, it can be accomplished by any sort of signal that influences the rate at which each system re-potentiates.... To serve as a synchronization signal, this signal would have to propagate much faster than the differentiation wave. It would affect potentiation without triggering a state transition" (Steve McGrew, p.c., 1997).

Pressure has the property of much faster propagation than a differentiation wave, but so would other synchronization mechanisms one could imagine. It is also possible for a differentiation wave itself to potentiate the cells ahead of it, though this would imply that embryological competence is not reached until just before homoiogenetic induction occurs, and thus is a short lived state. As this would offhand not seem to be compatible with classical results on competence, I consider it unlikely. However, until we can define competence at the cellular level (as in erected, metastable cell state splitters) and observe its spatiotemporal distribution, it is best not to prejudge the possibilities.

Proposition 32: if the cells in a tissue lack a synchronizing mechanism, then the precise positioning of some of the nodes of the differentiation tree along the time axis would have to be replaced by a statistical ensemble of paths of its cells along the tree.

Consider a dispersed tissue such as blood:

"During the second and third days of incubation, in the case of the chick,... groups of densely packed mesodermal cells appear in the area opaca all around the sides and the posterior edge of the area pellucida. These groups of cells are known as **blood islands**. Next, the cells of the blood islands become differentiated into two kinds: the cells on the periphery join to form a thin epithelial layer, the endothelium of the future blood vessels; the central cells, on the contrary, become separated from one another and are differentiated as blood corpuscles. Thus from the beginning, the blood corpuscles lie inside the blood vessels" (Balinsky & Fabian, 1981).

If we may presume that each blood island is not contiguous with the next, then there may be some temporal independence of each island in terms of when further differentiation occurs. Looking at the whole embryo, we would see a dispersal of the corresponding node of the differentiation tree along the time axis. In this sense, then, the differentiation tree has to be replaced by an ensemble of differentiation trees, with this node, and its terminal branch, scattered to some extent over the time axis.

Proposition 33: the community effect is caused by differentiation waves.

Gurdon et al. (1989) found what they considered to be...

"...a novel mechanism, referred to as a 'community effect', in which differentiation of muscle cells, and perhaps of other solid [sic] embryonic tissues such as notochord or nerve depends on like cells responding to induction in the same way at the same time" (Gurdon et al., 1989).

"Most vertebrate structures are formed from large numbers of cells of similar type, which collaborate and intercommunicate during development, as suggested by the 'community effect' concept (Takeichi, 1991; Gurdon, Lemaire & Kato, 1993). What this means functionally... remains mysterious, though the problem lies close to the bone of what metazoan tissues are" (Davidson, 1994).

The community effect (Gurdon, 1988a; Stüttem & Campos-Ortega, 1991; Gurdon, Kato & Lemaire, 1993; Gurdon, Lemaire & Kato, 1993; Gurdon et al., 1993; Kato & Gurdon, 1994) is probably nothing more than the action of a differentiation wave:

"Like muscle cells, notochord cells show a temporary requirement for proximity to their normal neighbours, and become independent of this type of cell interaction well before expression of genes which reflect their differentiated state. We also see that independence of the community effect, as indicated by the ability of singly implanted cells to activate their cell-type specific genes, takes place earlier for the notochord (mid-gastrula) than for muscle (late gastrula). This correlates with the earlier morphological differentiation of the notochord, which is visibly demarcated from the rest of the mesoderm by stage 13 (Nieuwkoop & Faber, 1956) [cf. Table 1]. Somites, which in *Xenopus* consist very largely of muscle, first become demarcated 5 h later at stage 16 (Nieuwkoop & Faber, 1956). From these two examples, we suggest that the community effect is required at a stage when cells become determined for their particular type of gene expression, and that it may represent an important stage in the process of cell differentiation.

"We do not know how the community effect is mediated in muscle or in notochord. Several *Xenopus* genes are strongly expressed in the notochord of *Xenopus* during the early gastrula stage. These include *X. goosecoid* (Cho et al., 1991a), *Xlim-1* (Taira et al., 1992) and *noggin* (Smith & Harland, 1992). Of these, *noggin* encodes a small peptide, which may be involved in the notochord community effect" (Weston, Kato & Gurdon, 1993).

Certainly, cells must stay in proximity for a differentiation wave to pass through them, but once that has happened, under gentle conditions, or when they are placed amongst other cells, we could anticipate that the differentiation cascade initiated by that wave would play itself out. This corresponds to the common observation (or definition) that determined cells differentiate independently of their (new experimental) locations. It would be curious to know, and I would predict, that implanted cells, if they respond at all to the next wave through the tissue into which they are implanted, take the next step appropriate to the kind of wave it is (i.e., expansion or contraction). In other words, do they differentiate according to the edge of the tree on which they were before they were explanted and embedded? Such experiments, performed between embryos of the same species, between embryos defective for wave propagation, and between embryos of different species (cf. Oppenheimer, 1939), could open a whole new line of research into exactly how cells transmit and respond to differentiation waves (cf. Heberlein, Wolff & Rubin, 1993). It could even lead to a rapid method of constructing differentiation trees (Natalie K. Björklund, p.c.), since cells from any embryonic tissue at any stage might be transplantable, say, to ectoderm, where they would either experience a contraction or an expansion wave.

Unfortunately, this is not the way the 'community effect' experiments were done. Instead of implanting the cells into intact embryos, more convenient 'sandwiches' were used (Weston, Kato & Gurdon, 1994), as originally proposed by Holtfreter (1933c). In such explant sandwiches, ectopic differentiation waves could have been generated, or normal waves never launched. The variable results obtained in ectoderm/endoderm sandwiches, ranging from single tissues, to two tissues with a sharp boundary, or alternatively random mixtures (Savage & Phillips, 1989), may be attributable to variable mechanical boundary effects that either permit or prevent differentiation wave launching and/or propagation (Section 9.22).

Sandwich experiments are examples of *in vivo* transplantation, which has a long history. In these experiments of classical embryology a phenomenon called 'neighborwise' development would appear to be equivalent to the 'community effect':

"Born (1897) was the first to accomplish the fusion of parts of amphibian embryos belonging to different species and genera" (Holtfreter & Hamburger, 1955).

"By his brilliant extensions of Born's transplantation method, Spemann (1918) showed in the case of the newt that up to a certain stage of gastrulation the fates of most of the embryonic regions are not irrevocably determined.... Exchange transplantations (*Austauschversuche*) performed on the amphibian embryo by Spemann and his school (Spemann, 1918, 1919, 1921; O. Mangold, 1922, 1923, 1925, 1928a,b) showed that before gastrulation has begun all parts of the embryo are interchangeable with other parts.... [Consider] an exchange between presumptive epidermis [ventral ectoderm] and presumptive neural plate [dorsal ectoderm].

Material which would have formed skin [epidermis] formed brain [neural plate], and vice versa.... After the completion of gastrulation, on the other hand, the behaviour of transplanted pieces is no longer neighbourwise (*Ortsgemäss*) but selfwise (*Herkunftsgemäss*)" (Needham, 1950).

The ectoderm transplantation exchanges were repeated by Jacobson & Gordon (1976a), with the same results.

Further experiments are needed to see if "single embryonic cells in a tissue can complete their differentiation without interacting with their normal neighbors" (Kato & Gurdon, 1993). With today's ability to transplant single cells (Wylie et al., 1986; Jacobson & Xu, 1989; Becker & Technau, 1990; Kato & Gurdon, 1993), we can ask how a cell, implanted in front of an oncoming differentiation wave, will respond. Will it traverse the next step of differentiation when the differentiation wave comes by? Will this step correspond to one of the two edges at the position along the differentiation tree that it had attained prior to transplantation? Will there be any influence of its neighbors, other than their transmission of the expansion or contraction wave to it?

Proposition 34: at each event of cell state splitting, the so-called positional information for a cell is increased by precisely one bit, namely, which side of the future boundary between the two new tissues the cell happens to be on.

"It is still unclear how neural cells sense small differences in the concentration of inductive factors and respond with the generation of distinct cell types" (Tanabe & Jessell, 1996).

Since each tissue splits into two tissues, each of which contains cells all on the same edge of the differentiation tree, we can see that in terms of spatial position, the only thing that is important is which side of the boundary between the two new tissues a given cell ends up on.

Since only one bit of positional information is involved when a tissue splits into two new tissues, there is no need for gradients of morphogens or morphogenetic fields to explain tissue formation (cf. Slack, 1982a). The commitment signal need only transmit one bit of information to the nucleus, namely whether the microfilament ring or the apical microtubule mat has won the tug-of-war, i.e., whether the cell has participated in a contraction or an expansion wave. The distinction "that a positional signal is graded and can produce multiple outcomes..., whereas an inductive signal is all-or-none..." (Smith, 1989d; cf. Wolpert, 1989a) thus vanishes. In fact, since n steps involving consecutive cell state splitters could lead to 2^n cell types that are spatially separated from one another, we see that each cell in effect then acquires n bits of spatial information in the process (Appendix I: Gordon & Brodland, 1987). The small discrete differences between neighboring tissues in this regard could readily be mistaken for a gradient. Thus, without close observation taking into account the ever decreasing sizes of the trajectories of differentiation waves (assuming waves that don't cross tissue boundaries), it is possible to imagine gradients where there are only small, stepwise differences across an embryo.

The concept of positional information, as originally expounded by Wolpert (1969), assumed that:

1. A cell can determine where it is.
2. It can do anything at all, in response to this 'knowledge' of its coordinates.

The concept is somewhat akin to that of a lookup table in computer programming:

1. The program reads in a data value.
2. It then looks up what it is to do when this value is encountered.

The manner in which the lookup table is generated, i.e., its values and its length, are not considered, as long as they remain within the constraints of the word size and computer memory. Thus there are no *a priori* constraints: any pattern whatsoever can be explained, nor are there any limits to the precision with which 'positional' values can be discriminated:

"The two strengths of the positional information concept are that (1) it provides a framework for understanding homologous replacements... and (2) it liberates the observed pattern from the shape of hypothetical underlying morphogen gradients.... The possibility that there are few, or perhaps only a single, universal form of positional information is appealing.... The principal problem is the obverse of its strength: it is so flexible that it can explain *any* pattern or pattern change as an alteration in the interpretative response of the cells. In effect, the idea is not subject to disproof. A second weakness is that it may serve to conflate certain position-associated properties of cell state with position per se and thereby obscure the basis of differential cell responses" (Wilkins, 1986).

Concepts of positional information in one form or another are not new. The model for differentiation of Rose (1952a) is a simple linear hierarchy (not a branching tree) with particular gene products peaking at various intervals along a pre-existing gradient, in effect, an early positional information (Wolpert, 1969) model. (See Arbib, 1972a, for a model of relative, rather than absolute, positional information.) Barlow & Carr (1984b) suggest...

"The idea that position within the whole guides the development of cells was explicitly stated by the zoologist Hans Driesch and the botanist Hermann Vöchting in the last century....

'The relative position of a blastomere in the whole may well determine in general what develops from it; if it lay elsewhere, it would give rise to something different;... the prospective value is a function of position' (Driesch, 1893a)....

'From this it follows that no vegetative cell in the plant possesses a specific and unchanging determination... the developmental function of the cell is determined primarily by its position in the living unit' (Vöchting, 1877)....

'To put it bluntly a cell knows where it is...' (Wolpert, 1970a)" (Barlow & Carr, 1984b).

Positional information, this idea that "a cell knows where it is" (Wolpert, 1970a), has become a widely accepted concept (Wolpert, 1971a, 1989a; Wolpert, Hicklin & Hornbruch, 1971; Cooke, 1972c; Lawrence, Crick & Munro, 1972; Wolpert, Hornbruch & Clarke, 1974; McMahon & West, 1976; Adler, 1978; Smith, Tickle & Wolpert, 1978; Duranceau, Glenn & Schneiderman, 1980; Barlow & Carr, 1984a; Wolpert & Stein, 1984;

Walthall & Murphey, 1986; Tomlinson et al., 1987; Wolpert & Hornbruch, 1987; Meinhardt, 1988; Rubin, 1988b; Kay & Smith, 1989a; Frohman, Boyle & Martin, 1990; Allaerts, 1991; Schiffmann, 1991; Summerbell, Smith & Maden, 1991; Mlodzik et al., 1992). Distinctions are made between Cartesian (perpendicular) and polar coordinate systems (French, Bryant & Bryant, 1976; Bryant, French & Bryant, 1981; Lewis, 1981a; Lubinsky, 1991; Bryant & Gardiner, 1992; Kondo, 1992; Bryant, 1993a; Cohen, 1993; Couso, Bate & Martinez Arias, 1993; Held Jr., 1992, 1993; Girton & Jeon, 1994). Positional information inspired a generation of embryologists, and led to a tremendous amount of interesting work. It is often taken for granted uncritically as a true and complete explanation of embryogenesis, and has become so ingrained that it is invoked even without reference. Positional information has been a major paradigm since its launching in Wolpert (1969), with doubts only rarely raised (Ragsdale Jr. & Brockes, 1991).

I assume, contrary to the concept of positional information, that a cell never knows where it is in an embryo: all it can 'know' is its local environment (the general assumption for cellular automata: Eden, 1958, 1960; Gordon, 1966; Codd, 1968; Gardner, 1971; Ransom, 1981; Wolfram, 1983, 1984a; Langton, 1986; Toffoli & Margolus, 1987; Markus, 1990; Gutowitz, 1991; Markus & Kusch, 1995; Peterson, 1995a), or even less: just whether it last participated in a contraction or expansion differentiation wave. It is not a sentient being, and it certainly doesn't have an overview of the embryo and its place in it. Its responses are also quite limited: if we take the cell state splitter model seriously, at any given moment a cell has at most two 'choices' before it, increasing its developmental information by one bit.

Proposition 35: positional information does not exist.

"...The real question is whether chemical specificity comes first in any given embryonic segregation [determination], or whether the various chemical specificities that evidence themselves by variety of chemical products are consequences of some preceding biological segregation" (Lillie, 1929a).

To see how this proposition can be justified, considered the matured views (Wolpert, 1989a) of the originator (Wolpert, 1969) of the idea in an article titled "Positional information revisited":

"The basic idea of positional information is that there is a cell parameter, positional value, which is related to a cell's position in the developing system. It is as if there is a coordinate system with respect to which the cells have their position specified. The cells then interpret their positional value by differentiating in a particular way.... Perhaps the least attractive feature of positional information, if the basic idea is correct, is that it places a great burden on the process of interpretation..." (Wolpert, 1989a).

With differentiation waves, the process of interpretation is simple, since the 'signal', whether a cell participated in an expansion or a contraction wave, is just one bit. Here Wolpert (1989a) makes various distinctions between positional information and induction, which I have summarized in parallel with differentiation waves in Table 2:

"A positional signal should be distinguished from induction in a number of ways [Table 2]. Induction is defined as an interaction between two different tissues, one inducing, the other responding (Gurdon, 1987). This at once makes it different from positional signalling, for which two different tissues need not be involved. There are other differences. In general, induction specifies just one cell state such as muscle or neural tissue or cartilage, whereas a positional signal, which need not involve only one substance, specifies, by definition, multiple cell states. Related to this a positional signal is graded whereas induction is basically all-or-none. Along the same lines, induction is short range whereas a positional signal is usually specifying cell states over a greater distance. Again, a positional signal has 'polarity' whereas induction is not related to the polarity of either tissue. A positional signal is provided by, or linked to, a boundary region and as such is localised; inductive signals need not be. Finally, a positional signal is always instructive whereas an inductive signal may be either permissive or instructive. The distinctions made here are somewhat exaggerated to emphasise the differences" (Wolpert, 1989a).

Wolpert (1989a) is left with no firm examples of positional information:

"It is thus far from clear to what extent the patterning of muscle in early amphibian development involves primarily induction or whether positional signalling is involved (Smith, 1989d)....

Positional signal	Induction	Differentiation wave
Multiple states	Single state	One of two states
Graded	All-or-none	One of two states Multiple states via multiple waves
Polarity	No polarity	Polarity by wave asymmetry
Long range	Short range	Launching sometimes by contact, sometimes by global buckling and/or ionic currents Propagation range up to whole tissue
Boundary region	Diffuse	Limits defined by trajectory
Same tissue	Different tissue	Launching may be initiated by a different tissue Propagation within same tissue Wave can cross into other tissues
Instructive	Instructive or permissive	Don't know yet what determines the kind of wave (expansion or contraction)

Table 2: Here I copy (with permission) and extend Table 2 of Wolpert (1989a), which compares positional information with induction, to a third column making a comparison with differentiation waves. For the column on induction, he implicitly ignores homoiogenetic induction.

"Molecular genetics has enabled the activity of a wide variety of genes to be mapped during early insect development (Ingham, 1988). The question is whether any of these can be regarded as providing the cells with a positional value. In general the answer seems to be in the negative....

"Homeobox genes... (Holland & Hogan, 1988a)... [have] patterns of expression [that] are consistent with, but of course do not in any way prove, a role for homeobox genes in establishing domains of positional value in the mammalian embryo..." (Wolpert, 1989a).

An unnamed prepattern mechanism by which cells are 'specified as a group' creeps into the argument:

"...Even with a positional field in which each cell is uniquely specified, it is necessary to interpret positional values so as to generate a pattern. A formal solution is that each cell contains a complete specification of the behaviour of every cell in every position. There must be a complete list of all the cells that will differentiate as type A, and another list of those that form type B and so on. For systems like the vertebrate skeleton it seems unlikely that there is a positional specification for every cell that develops into, for example, cartilage. This could partly be resolved if contiguous cartilage cells were specified as a group, and a prepattern mechanism can provide just such a mechanism..." (Wolpert, 1989a).

(See my critique of prepatterns in Section 9.24 or Gordon & Drum, 1994.) There is also a hint here that the DNA may not be able to store enough information to tell each individual cell what to do, a dilemma that has long been recognized (cf. the 5×10^{25} bits needed to 'beam up', Carey, 1988, or specify an adult human at molecular resolution: Dancoff & Quastler, 1953, with the mere 6×10^{9} bits specifiable by our DNA: Raven, 1961, the latter computations reviewed and criticized by Apter, 1966):

"...The very concept of differentiation *via* interpretation - is that states of cell differentiation are, as it were, specified as a single event. While it is not an absolute requirement, the implication is of a single switch and is in strong contrast to any combinatorial mechanism of specification. It is far from clear whether or not the specification of cell differentiation is combinatorial..." (Wolpert, 1989a).

Positional information neatly 'solved' the regionalization problem in one step, which goes against the thrust of this work and that of many others on the hierarchical nature of development. Wolpert's own doubts would appear rather comprehensive:

"In one case, at least, there is good evidence that positional information alone is not sufficient to specify the pattern, and that is the chick limb (Wolpert & Stein, 1984)....

"In looking for universal mechanisms and signals, it is likely that the systems in which they are least likely to be found are eggs and very early development" (Wolpert, 1989a)

Thus key systems, in which one would have thought positional information should be crucial, are found to be anomalies for the theory.

What is strange is, given all these reasonable doubts about positional information by Wolpert (1989a), the organizers and participants of the symposium in which this appears ("The Molecular Basis of Positional Signalling": Kay & Smith, 1989a) seem to take positional information for granted. What we have here appears to be a paradigm in crisis. Consider...

"...what scientists never do when confronted by even severe and prolonged anomalies. Though they may begin to lose faith and then to consider alternatives, they do not renounce the paradigm that has led them into crisis. They do not, that is, treat anomalies as counterinstances, though in the vocabulary of philosophy of science that is what they are.... Once it has achieved the status of paradigm, a scientific theory is declared invalid only if an alternate candidate is available to take its place. No process yet disclosed by the historical study of scientific development at all resembles the methodological stereotype of falsification by direct comparison with nature.... The decision to reject one paradigm is always simultaneously the decision to accept another, and the judgment leading to that decision involves the comparison of both paradigms with nature *and* with each other" (Kuhn, 1970).

The three differentiation mechanisms alluded to by Wolpert (1989a), namely positional signalling, induction, and prepatterns, and the unnamed fourth, can probably all be subsumed within the new paradigm of differentiation waves (Table 2). A change to the paradigm of differentiation trees may therefore be in order.

In the context of plant development, Lintilhac (1984) has used some strong language to dissuade us from using the concept of positional information:

"My concern is that the original concepts put forward by Wolpert [1971a], having served so well in directing thinking about development in animal science, will be applied uncritically to plant systems, which I consider to be organized in a fundamentally different way.

"The question is not whether these models can be applied to plant development, because they can. I doubt that they could be proved wrong.... I regard the theory of positional information as a metaphor which has been stretched to cover many developmental problems, but which will not in the end contribute to our ability to predict and manipulate the real variables of plant growth and development.... The application of this model to plant systems may serve only to obscure their extraordinary sensitivity to structural and mechanical stimuli....

"Holder (1979)... considers that 'The central problem of the study of morphogenesis is how the genetic potential of undifferentiated cells is expressed in a controlled manner to produce the specific spatial array of cell types characteristic for the species'.... Placing the entire responsibility for development with the genome undercuts a whole level of events.... Many aspects of spatial regularity, and even some instances of structural differentiation, may arise as a direct and highly sophisticated architectural response to specific mechanical stimuli....

"The question now becomes how the growing tissue as a whole maintains its own stress-mechanical environment. The cells [in a tissue] are all the same and have no need to know their locations on the apex [of an apical meristem], or their positions relative to each other, in order to determine what to do, because the stress-mechanical environment varies in predictable ways from location to location within the apex....

"The investigation of these problems has been severely hindered by the lack of appropriate experimental tools. In particular, the lack of a nondestructive method for mapping the stresses in living meristematic tissue is a constraint which leaves us with no direct way of correlating specific orientational and developmental events with their stress-mechanical causes" (Lintilhac, 1984).

I agree with this assessment of positional information, though I would add that it applies equally to animals, and that gene expression is, in the first instance, governed by 'stress mechanics'. Furthermore, the transmission of the 'stress mechanics' from cell to cell may occur, at least when differentiation is involved, via mechanical differentiation waves.

Hooper & Scott (1992) see a lack of appropriate positional information mutants:

"...How are the high-resolution patterns within [insect] larval segments and imaginal discs generated so reproducibly? Some theories have posited that positional information is a continuously graded quantity. This mechanism requires graded distributions of something, either a morphogen such as a diffusible small molecule or a graded distribution of a physiological property such as the level of phosphorylation of some protein. In either case, all pattern evolves from a single system, so genes should exist whose product is everywhere, whose mutants abolish all pattern, and whose mutant effects are nonautonomous in mosaics. None of the known genes fit these criteria" (Hooper & Scott, 1992).

The simplest way out of these problems is to accept the possibility that positional information, while a clever solution to the problem of morphogenesis, does not exist in real organisms. This was the ultimate approach to the question of the existence of the aether, whose conceptual beginning with Isaac Newton in 1675 had curious biological undertones, and which was let go with reluctance in the 20th century (Gillispie, 1960).

Proposition 36: the basis of a cell's 'memory' of the path it took along it's genome's differentiation tree may be a subset of the accumulation of activated genes or gene products generated at each stage of differentiation ('memoron set').

In order to obtain 2^n cell types or tissues from n consecutive acts of cell state splitting, each cell must have some sort of 'memory' of its previous states (Appendix I: Gordon & Brodland, 1987; cf. the 'memory circuits' of Kauffman, 1973, the "record... [of] the sequence of binary commitments" of Kauffman, Shymko & Trabert, 1978 and the "internal records of the effects of past external influences" of Lewis & Wolpert, 1976). A 'memoron set' is in effect a record of the path the cell took along the differentiation tree:

"It is clear that in the absence of any epigenetic crisis the current epigenetic message for a cell line must be faithfully transmitted to the daughter cells, generation after generation" (Stein, 1980).

"It is remarkable that eucaryotic cells and their progeny will usually persist in their differently specialized states even after the influences that originally directed their differentiation have disappeared - in other words, these cells have a memory. Consequently, their final character is... determined... by the entire sequence of influences to which they have been exposed in the course of development. Thus as the body grows and matures, progressively finer and finer details of the adult body pattern become specified, creating an organism of gradually increasing complexity whose ultimate form is the expression of a long developmental history that is remembered by the cells.... Cell memory, as manifested in the phenomenon of determination, presents one of the most challenging problems in molecular biology: what makes certain patterns of gene expression self-sustaining?" (Alberts et al., 1989).

(See Meins Jr. & Wenzler, 1986, for a discussion of memory of the differentiated state in plant cells.) What can a memoron set consist of? How does a cell 'remember' what it has been through in the course of embryonic development? The solution of this problem ("cracking of the epigenetic code": Stein, 1980) might be found in the following observation:

"...Three [sex determining] genes... exhibit a clear epistatic hierarchy of the form *dsx > tra > ix*, which is consistent with their sequential involvement in a single developmental pathway regulating sexual differentiation.... [Such] homeotic loci in *Drosophila melanogaster* share three features in common:

(a) they function in a cell autonomous manner;

(b) while they begin functioning early in development, their wild type products are also needed late in development for normal differentiation; and

(c) each controls a binary decision in a given developmental pathway" (John & Miklos, 1988).

In other words, the 'need' for early products later on in development could be precisely as a form of memory of the early differentiation events. If so, we can conceive of a memoron set as a set of particular gene products, at least one for each step of differentiation, that represent together the full history of differentiation steps a cell went through. I will presume that a memoron set is a set of macromolecules, and call the latter 'memorons'.

I presume the memorons are replicated, or at least persist in sufficient numbers, so that at each cell division both daughter cells end up with the same memoron set. Unlike 'determinants', memorons are presumably partitioned to both cells, even in an asymmetric cell division. Each step of differentiation adds at least one memoron molecule. The one added would be different depending on whether the cell participated in a contraction or an expansion differentiation wave. However, considering, say, two consecutive contraction waves, there need be no necessary chemical relationship between the memorons they add. While a uniform mechanism from one step of differentiation to the next might be more elegant, it is not dictated by the logic of the situation.

In *Drosophila,* the homeobox genes or gene products may correspond to the postulated memorons:

"Each of the 17 known homeo box genes shows a unique pattern of expression during early embryonic development. Thus, virtually every embryonic cell [tissue?] contains a unique combination [memoron set] of active and inactive homeo box genes" (Levine & Harding, 1987).

This idea has also been suggested for vertebrates:

"...If the differentiated state of a particular cell is specified not by a single protein but by a combination of factors [memoron set], significantly fewer regulatory proteins are required to constitute a unique genetic program for a single cell type" (Dressler, 1989).

Methylation of DNA has been postulated as the mechanism of 'cell memory' (Riggs, 1989, 1990; cf. Bird, 1993; Riggs & Pfeifer, 1992) as have "stable changes in the higher-order constitution of chromatin" (Paro, 1993; cf. Zuckerkandl, 1974; Spradling, 1993; Patterton & Wolffe, 1996) and the nuclear membrane (Danchin, 1987). One could possibly consider the alternate states of nuclear tensegrity, hypothesized in Section 10.13, as the memorons. (Since some states may be accessible only via others, each state could also represent the memoron set.) The solution to cellular memory may lie in the master genes that may be arranged as a biochemical nuclear state splitter that is resolved as a 'rocker switch' coinciding to the resolving states of the cell state splitter (Appendix IV: Björklund & Gordon, 1993b). A memoron set could be a stable circuit of reactions such as postulated by Kauffman (1969).

'Forgetting', in the sense of earlier differentiation cascades in the history of a cell no longer being active, could also occur. However, as each terminal branch of the differentiation tree is likely to be unique in its set of differentiation cascades, so long as at least one differentiation cascade is active, cell types will be distinguishable despite some 'forgetting'. Moreover, most embryonic cell types probably do not persist, since they give rise to terminal cell types. Thus the adult cells could 'forget' about these earlier cell types, in the sense of not retaining the earliest memorons of their memoron sets.

The accumulation of memorons, different for cells from one tissue to the next, but the same for all cells in a given tissue, is what guarantees that each pathway along the differentiation tree leads to distinct kinds of cells and tissues. Note that in Proposition 270 I postulate a set of 'higher order, key-like memoron molecules' or 'homeokeys' involved in the colinearity of homeobox genes and dedifferentiation.

We are thus left with an unresolved array of possibilities for memorons and memoron sets. It seems clear that they must exist, yet it is hard to decide, amongst all the macromolecules, configurational states of the chromosome and/or nucleus, etc., exactly what is going on here. Some good thinking and critical experiments are needed.

Proposition 37: the memorons in a cell are triggered in response to a sequence of contraction and expansion waves, which therefore forms the *differentiation code*.

"...The ability of myoD... to initiate muscle gene expression was found to depend on the prior history of the cell in question, indicating that myoD was operating interactively within a particular cellular context (Saitoh et al., 1992; Schäfer et al., 1990)" (Strohman, 1993).

We now have evidence that each step of differentiation is in response to either a contraction or an expansion wave going through one or the other portion of a differentiating tissue (Gordon and Björklund in Appendix V: Gordon, Björklund & Nieuwkoop, 1994). If we designate these events by E = expansion and C = contraction, we can formulate a 'differentiation code' for each cell at any stage of differentiation.

An example of a differentiation code is 'ACECC' for presumptive notochord, where 'A' refers to the 'animal' or dorsal hemisphere of the embryo, a distinction we found necessary to maintain (Figure 3 in Appendix VI: Björklund & Gordon, 1994). The differentiation code for a given tissue begins with 'V' if it derives from the vegetal or ventral hemisphere. If we assigned $C = 0$, $E = 1$, then the above code becomes A0100 (or 0010A), which may be thought of as a binary number.

The idea of an 'activation code' has been put forward by Jack & McGinnis (1990) in regard to the hierarchy of segmentation genes in *Drosophila*, and would be comparable to our differentiation code, though the differentiation code has a physical basis.

Of course, the differentiation code only formally represents the memorons, whose nature has yet to be worked out.

Proposition 38: the initial distinction between animal and vegetal hemispheres in amphibian fertilized eggs is due to the hydrodynamics of the yolk.

The animal/vegetal distinction seems to have disappeared with the evolution of mammals (Evsikov, Morozova & Solomko, 1994). Thus I need to discuss the nature and source of this distinction in amphibians. It could be that 'A' and 'V' ultimately refer to some kind of maternal 'determinants' (cf. Baldwin, 1915). But since embryos with the opposite of the normal A = black, V = white pigmentation are attainable, this indicates at least that such 'determinants' do not move with the cortically bound pigmentation:

"In one of my experiments, in which the uninjured blastomere had been *reversed* after the operation, it developed into a half-embryo, and not into a whole embryo of half-size. Moreover, in this embryo the medullary folds appeared on the white surface of the egg, showing that a rotation of the contents of the blastomere must have taken place" (Morgan, 1897a).

For further evidence that whole, intact, but inverted eggs often (but not always: Wakahara, Neff & Malacinski, 1984a, 1985) develop normally, albeit with this dorsoventrally reversed pigmentation, see: Chung & Malacinski (1982, 1983); Neff & Malacinski (1982); Neff et al. (1984b).

There are differences in the membrane viscosities of the two hemispheres in amphibian oocytes, which become accentuated at fertilization (Dictus et al., 1984; see Jacobson, Sheets & Simson, 1995; Saxton & Jacobson, 1997, for a possible molecular/cortical basis). Whether these membrane viscosities are reversed for inverted eggs is not known. The differences between the hemispheres found by Dictus et al. (1984) do not prevent both the animal and vegetal hemispheres from supporting later differentiation waves (the ectoderm and contraction wave and the vegetal yolk mass contraction wave, respectively, for instance: Appendix V: Gordon, Björklund & Nieuwkoop, 1994). Thus the difference in membrane viscosities may not be critical for development.

The explanation for these phenomena may lie with the hydrodynamics of the yolk in the fertilized egg prior prior to first cleavage:

"A new blastoporal lip appears in close contact with - or surrounding - any place where some important mass of yolk has fallen, and has been condensed (Penners, 1929; Penners & Schleip, 1928a,b)" (Dalcq, 1938).

Some of these motions have been simulated hydrodynamically using immiscible fluids in glass spheres (Flint et al., 1989). If the yolk comes to occupy the lower hemisphere, that hemisphere becomes ventral ('V'), independently of pigmentation, cortical properties, or 'determinants'. The nearly yolk free hemisphere becomes the animal hemisphere ('A'). The yolk's juxtaposition to the cortex may, then, alter cortical properties (as more vegetal microtubules: Elinson & Rowning, 1988; Elinson & Palecek, 1993), membrane viscosity and the location of 'determinants', while leaving the pigmentation where it was. No one has looked into these possibilities yet.

Proposition 39: the position of a cell is important, because that determines which kind of differentiation wave it will participate in next.

A lesson that has been learned repeatedly (Minden et al., 1989) is that "...fate depends upon blastomere position rather than lineage" (Moody & Kline, 1990). Blastomere cell fate maps have been made for embryos selected to have 'stereotypical' cleavage patterns, "but even within our selected population displaying stereotypic cleavage patterns, furrow positions never were in the *exact* same place, and named blastomeres did not circumscribe the *exact* same cytoplasmic domain" (Moody & Kline, 1990; cf. Moody, 1987a,b). Gravitational effects altering the plane of first cleavage that have...

"...a decreased... animal/vegetal cleavage ratio (AVCR), defined as the ratio of the height of the animal blastomere to the height of the *Xenopus* embryo at the 8 cell stage... exhibited reduced survival in early developmental stages, indicating serious difficulties in cleavage, blastulation and/or gastrulation" (Yokota, Neff & Malacinski, 1992).

Here is a detailed study of the variability of early cell lineages:

"To trace the cellular origins of the *Xenopus* Organizer, we labelled dorsal blastomeres of three of the four tiers (A, B and C) of the 32-cell embryo with green, red and blue fluorescent lineage tracers.... The typical early gastrula Organizer is composed of approximately 10% A1 progeny in its animalmost region, 70% B1 progeny in the central region, and 20% C1 progeny in vegetal and deep regions. Variability in the composition of the early gastrula Organizer results from variability in the position of early cleavage planes and in pregastrulation movements.... Experiments to perturb later developmental events by molecular or embryonic manipulations at an early stage must take this variability into account.... At a given stage, a cell's position in the embryo rather than its lineage may be more important in determining which genes it will express" (Vodicka & Gerhart, 1995).

The proper tissue lineage of a cell is not from mother to daughter cells, which can be separated from one another by shear and morphogenetic movements (Jacobson & Gordon, 1976a), but the sequence of differentiation waves that a cell has participated in, i.e., its differentiation code.

4.05 The Relationship between Cells and Tissues in Regulating Embryos

"...The nature of 'correlative differentiation' [now called induction] remains almost wholly unknown. It has been vaguely ascribed to 'interaction' of the cells (O. Hertwig), to the 'development of the organism as a whole' and the like; but such phrases are only restatements of a fact which in its essential nature still remains unexplained" (Wilson, 1925).

Proposition 40: the spatial component of differentiation, i.e., the splitting of a tissue into two new tissues, is accomplished by robust mechanical means that impart some independence from cell and embryo size. This is the essence of embryonic regulation, attributable to the properties of differentiation waves.

We have seen one example of this in our shell model for the propagation of the wave in primary neural induction (Section 3.02): it is robust in the sense that the geometric result, division of a spherical tissue into two hemispherical tissues, is independent of the detailed parameters for the wave propagation and internal pressure, over a wide range of those parameters (Appendix I: Gordon & Brodland, 1987). Such robustness could give the embryo the ability to regulate, i.e., to form well proportioned tissues despite assorted experimental interventions.

This may also be the source of the ability of an embryo to accommodate its development to the material and cells at hand. The tissue available has been experimentally decreased (Spemann, 1903a; Ruud & Spemann, 1922; Schmidt, 1933) or increased (Bruns, 1931; Waddington, 1938a) with little effect. A wide range of independence from the number of cells has been found (Cooke, 1981a; Harris & Hartenstein, 1991). Mostly normal development occurs when the size of the cells is varied by changing the degree of polyploidy (Fankhauser, 1955). The natural variation in egg size, as in the toad *Bufo vulgaris,* may be quite useful in this regard:

"The volume ratio between extreme size classes amounts to 1:5, which seems to be the widest variation on record.... Since in the large eggs the cell number at the time of gastrulation is smaller than in the small eggs, the average cell size is very much greater in the former. However, the numbers of cells in the ectodermal and mesodermal areas of the two classes almost coincide..." (Dan, 1960).

Perhaps the actual, focussing mechanism for the real ectoderm contraction wave (Figure 17 in Appendix V: Gordon, Björklund & Nieuwkoop, 1994; Section 3.03), when we achieve a physical analysis of its propagation, will also prove to be similarly robust. Clearly other waves, with simpler trajectories, such as the vegetal yolk mass contraction wave (Figures 7-11 in Appendix V), will proceed until they annihilate or encounter absorbing or transmitting boundaries. In the sense that they keep going until they run out of territory, they are also robust, and 'regulate' to the amount of tissue and number of cells available.

An interesting exception 'proves the rule'. In tetraploid mice...

"The embryos are essentially normal [as are triploid humans, which can come to term: Niebuhr, 1974; Mange & Mange, 1990] save that the forebrain and its associated tissues fail to develop properly.... Close examination of sectioned material, in contrast, showed that tetraploid morphology and morphogenesis were indistinguishable from those of controls, except in forebrain-associated material. This conclusion gives some insight into an important developmental question, how fine can the developmental map be for normal cellular differentiation to proceed? As tetraploids have only about half the expected number of cells, the ability of these embryos to develop normally in all regions except the forebrain and its derivatives argues that pattern formation mechanisms can cope with the abnormally small number of cells in all regions except the forebrain. The results as a whole argue for size regulation in mammalian embryos being achieved by assaying absolute size rather than counting cell numbers" (Henery, Bard & Kaufman, 1992).

What may be happening here is that the trajectories of consecutive, nested waves in the forebrain get so small that they cannot be supported by the smaller number of large tetraploid cells available. (We see a similar phenomenon in the termination of recursive mutants: Chalfie, Horvitz & Sulston, 1981; Meyerowitz, 1989; Meyerowitz, Smyth & Bowman, 1989). Differentiation waves may thus explain "the grain of the developmental map" (Henery, Bard & Kaufman, 1992). This concept may be understood by analogy with digitized images. The latter cannot represent detail any finer than the size of a pixel. If we use too large a pixel when digitizing an image, detail is lost. If an organism is made of a relatively few large cells, its development lacks 'detail' compared to that of an organism of the same size (with a similar genome). Since the tissues of embryos tend to be defined by 'broad strokes' in a hierarchical development scheme (such as differentiation trees represent), the finest details, i.e., some of the terminal cell types, are precisely what we would expect to be lacking in large celled organisms. A direct look with time-lapse at development and differentiation waves in organisms of different cell sizes could be well worthwhile.

The trajectories of differentiation waves would seem, then, to be reasonably independent of the number of cells and overall size of an embryo. This independence is what is meant by 'regulation'. Differentiation waves therefore represent a possible solution to the problem of regulating embryos that drove Driesch to a neovitalistic position:

"[W. Bateson in B. Bateson (1928)] concluded, 'The accommodatory mechanism [regulation] is the thing to go for. I don't believe it is generally recognised as existing, though when stated it seems obvious.' This accommodatory mechanism, (which allowed a viable equilibrium to be attained in the face of disequilibrating changes) was to intrigue Driesch and the later systems theoreticians: it remains a mystery, and there is still no general recognition of its existence" (Reid, 1985).

"This justly famous discovery by Driesch began as a piece of empirical research, but the more he reflected upon it, the more it seemed a colossal refutation of all mechanistic philosophies in biology" (Fleming, 1964).

Much hope has been placed on Turing reaction-diffusion mechanisms for morphogenesis. However, these require enzyme rates that are dependent on organism size to achieve size independence of the pattern (Hunding & Sørensen, 1988; cf. Othmer & Pate, 1980), and they are also sensitive to surface curvature (Spirov & Boinovich, 1985), so that the undulations of a tissue could alter its pattern. Bunow et al. (1980) state flatly that this class of models is not robust:

"Kauffman, Shymko & Trabert (1978) have presented a model for sequential compartment formation in *Drosophila* wing disks based upon reaction-diffusion instabilities on an elliptical model for the wing disk. Using... a more accurate representation of the shape of the wing disk [we find that the]... nodal lines... are very sensitive to the shape of the domain.... The patterns generated by these instability models are generally quite sensitive to changes in the shape of the domain and physical-chemical parameter values. This sensitivity makes it difficult to imagine how pattern form and sequence could be maintained in the face of normal biological variation. We conclude that the use of instability models in this aspect of morphogenesis is an interesting speculation, but one whose validity will be difficult to demonstrate convincingly" (Bunow et al., 1980).

Others have expressed less precise doubts:

"Beautiful as Turing's ideas are, they probably have at best a very attenuated application to real biological systems. John Maynard Smith (p.c.) recalls being entranced by Turing's (1952) paper (which his supervisor, J.B.S. Haldane, had shown him), and for years he was convinced that 'my fingers must be Turing waves; my vertebrae must be Turing waves' - but he eventually came to realize, reluctantly, that it could not be that simple and beautiful" (Dennett, 1995).

Some patterns do vary quite a bit, such as "the positions of macrochaetae and ocelli" on the head of *Drosophila* (Sondhi, 1963). A quantitative comparison of the robustness of two dimensional Turing patterns versus differentiation waves is in order. This could be done by computer simulation, or perhaps, by perturbation analysis.

Proposition 41: cell state splitters may form the basis for fractal morphogenetic processes.

Consider a fractal morphogenetic process, such as development of the lung:

"It seems reasonable to suspect that the morphogenesis of the lung is itself a dynamic fractal process, a type of critical phenomenon evolving over time (West, Bhargava & Goldberger, 1986)... In this regard, the new fractal scaling mechanism has interesting implications for the development of complex but stable structures using a minimal code based on self-similarity. One of the many challenges for future research will be to unravel the molecular mechanisms whereby self-similar scaling information is encoded and processed. The evidence for these scaling principles that we have seen in the bronchial data from several species suggests the working of some potentially universal embryological principle" (West & Goldberger, 1987).

Cf. the review of Iannaccone (1990) and work on the fractal nature of brains (Hofman, 1991) and of arterial patterns in kidneys of Gil-García, Gimeno-Domínguez & Murillo-Ferroll (1992). Shlesinger & West (1991) have shown that fractal morphogenetic processes can produce more uniform structures than deterministic branching processes. This independence of physical scale, typical of fractal patterns, may, for instance, partially explain the ability of siphonous green algae to exhibit the same degree of morphogenesis for varying degrees of division of the organism into cells (Kaplan & Hagemann, 1991). The robustness of a differentiation wave model for the division of a tissue into two tissues, such as a sphere into two hemispheres, persists over a wide physical size range. (See Avnir, Biham & Malcai, 1998, for a discussion of acceptable empirical size ranges versus the mathematical concept of a fractal.) Thus in the broad view, it is a self-similar, fractal process. Perhaps in fractal morphogenesis, the termini of the differentiation trees are actually recursive, as in morphogenesis of a mutant *Arabidopsis* flower (Meyerowitz, 1989; Meyerowitz, Smyth & Bowman, 1989) and analogous mutant nematodes (Chalfie, Horvitz & Sulston, 1981).

Proposition 42: all morphogenetic movements are secondary to differentiation waves.

There are, of course, numerous mechanisms of morphogenesis (Trinkaus, 1984a). I place emphasis on tissue splitting by differentiation waves as the primary morphogenetic event, because that is the one morphogenetic event that ties the formation of tissues to the differentiation of cells. All other morphogenetic events, such as cell rearrangements, I take to be consequences of the differentiation cascades associated with differentiation. Thus I place an even greater burden on the edges of the differentiation trees, since they also stand as symbols for all secondary morphogenetic events. Wolpert (1989b) agrees with the secondary nature of morphogenetic movements:

"I would argue that in all changes in form, whether it is the folding of the neural tube or gastrulation, the mechanical changes are preceded by a patterning mechanism which specifies the cells that generate the forces. This can be, for example, the specification of adhesive molecules or the concentration of microfilaments" (Wolpert, 1989b).

This reinforces our notion that genetic and morphogenetic events alternate. I would take the 'patterning mechanism' to be the differentiation waves. Perhaps the full cycle should be:

differentiation wave (expansion or contraction)
⇒ corresponding master gene activation (determination)
⇒ differentiation (specific gene products)
⇒ secondary morphogenetic movements
⇒ launching of next pair of differentiation waves

Consider, for example, the concept of a cell surface 'morphogenetic code' (or 'adhesive code': Redies & Takeichi, 1996), which I would regard as being of secondary importance:

"Because of the large number of adhesive systems used by an individual cell, nearly every cell type will share at least one cell-cell adhesion system with every other and therefore bind to it with some affinity.... The mixture of specific types of cell-cell adhesion molecules and matrix receptors present on any two cells, as well as their concentration and distribution on the cell surface, will therefore determine the total affinity with which two cells bind to each other and to the matrix. It is this mixture that presumably constitutes a 'morphogenetic code' to help determine how cells are organized into tissues" (Alberts et al., 1989).

The secondary nature of cell adhesion requires some discussion, especially because it is central to some theories of embryogenesis, such as topobiology (Edelman, 1988). What I mean to imply by this is that the model in which embryogenesis is caused by specific cell-cell adhesions is fundamentally in error. Rather, tissues are formed by differentiation waves, and then their cells can move relative to one another, change shape, develop mechanical tensions, etc., as a result, in part, of specific cell adhesion molecules that are but one of the many tissue specific gene products triggered by differentiation waves. 'Secondary' only means in regard to placement on the differentiation tree. The production of a cell adhesion molecule is placed on an edge of the differentiation tree, as part of a particular differentiation cascade. 'Secondary' does not mean 'unimportant'. Cell-cell adhesion is certainly essential to construction of many organisms, as seen in a peculiar mouse mutant:

"At early gastrulation stage, the embryonic ectodermal cell layer is subjected to increased cell proliferation and morphogenetic movements, such as induction of mesoderm, primitive streak formation, and the establishment of the anterior-posterior axis. This all requires continuous integration of rapidly dividing cells into the epithelial cell layer. On the cellular level, a constantly occurring transition between polarised and dividing, non-polarised epithelial cells has to be assured with a precise control of cell adhesion, cytoskeletal organisation, and cell polarity.

"A possible explanation of the phenotype in β-$cat^{-/-}$ embryos is that the lack of β-catenin affects the adhesive properties of E-cadherin in embryonic ectodermal cells, leading to an improper integration of dividing cells into the ectodermal cell layer.... Cells detached from the ectodermal cell layer and were dispersed into the proamniotic cavity (Haegel et al., 1995).

On the other hand, the mild effects of gene knockout of 'cell adhesion genes' (Hynes, 1996) suggest that they are indeed of secondary importance in early embryogenesis.

There has also been an unresolved theoretical problem with cell adhesion as a morphogenetic mechanism. One approach, dating back to the work of Aron A. Moscona, starts from the assumption (well justified by now) that the surface of a cell is studded with specific molecules, and that these molecules bind in specific ways to those sticking out of adjacent cells (Moscona, 1957, 1960). The other basic approach, epitomized by the work of Malcolm S. Steinberg (Steinberg, 1963, 1964, 1970, 1996; Steinberg & Garrod, 1975; Steinberg & Poole, 1981, 1982; Foty et al., 1996), assumes that the surface of a cell binds nonspecifically to the surface of a neighboring cell. This approach successfully predicts the so-called transitive relationships: if, in 3D culture, embryonic tissue A engulfs embryonic tissue B, and B engulfs embryonic tissue C, then A will engulf C. Tissues are treated as if they were immiscible fluids (cf. Gordon et al., 1972, 1975).

Thus both approaches make accurate testable predictions, but no rapprochement has yet been achieved (cf. Grunwald, 1991).

Proposition 43: secondary morphogenetic events are robust.

Secondary morphogenetic events can have a precision and robustness of their own: despite the fact that the cells are small systems (Hill, 1963, 1964) they can respond and interact in ways that lead to stable shapes (Gordon, 1966) and patterns:

"Random fluctuations in cellular performance can be viewed as contributing to the random inputs that drive IP [inherently precise] morphogenetic machinery.... This idea differs from the Turing model in the fact that random disturbances are incorporated into the driving of morphogenesis, not just in the triggering of it (Turing, 1952).... Embryonic development is impelled by the sequential injection of new cellular specificity as a result of the reading of the genetic information.... New proteins mean new cells, with new types of interactions, and hence new pattern-forming situations" (Elsdale, 1969).

Waddington (1957) also saw the operation, in what he called 'developmental noise', of a mechanism that is robust in the face of fluctuations in a developing system:

"It can hardly be expected that any epigenetic mechanism can operate with complete precision.... For instance, when the melanoblasts from the neural crest distribute themselves over the flanks of an amphibian tadpole, they do so in a pattern which is more or less definite but not absolutely the same from individual to individual. Again in the primitive streak of the chick, the invaginating mesoderm becomes dissociated into cells which are loosely arranged and can move with some freedom relative to one another... (Spratt Jr., 1955)....

"This kind of looseness or 'play' in the epigenetic machine might be referred to as developmental 'noise', using that word in the sense given to it in information theory. There is some degree of imprecision even in developmental processes which involve large numbers of cells. It is certainly noteworthy that embryonic organs are often somewhat variable at the time of their first appearance, and only gradually become regulated towards the standard. The early somites of a chick embryo, for example, are often variable both in size and in outline.... It seems that any mass of neural tissue, which at the time of its formation has a rough resemblance to some part of the brain region of the neural tube, will eventually develop into a relatively well-individuated segment of that organ (cf. Waddington, 1952c; Nieuwkoop, 1955a)" (Waddington, 1957).

This robustness may be essential not just to form organs, but to set up launching of the next differentiation wave at a reasonably precise location and time. Here is a whole field waiting for investigation. Surface tension or sintering phenomena might be a good place to start, as these robustly produce a sphere from a wide range of initial blobs of isotropic fluids (Gordon et al., 1972, 1975; cf. Section 1.10).

Proposition 44: the shape of a tissue is determined by: a) the location and shape of the launching domain of the differentiation wave that creates it, b) the trajectory of its differentiation wave, including possible refraction effects; and c) the manner of termination of that wave: self-annihilation, physical barriers (absorption), annihilation upon meeting another wave, and/or propagation into another tissue (transmission).

Here lies the work for the next generation of embryo mechanics, for I suspect that the spatially and temporally varying constitutive properties of the embryo determine the 'weak' spots, the places that launch differentiation waves. (Cf. the formation of the blastopore from problastopores: Appendix V: Gordon, Björklund & Nieuwkoop, 1994.) We have no firm answers. As we have seen in the axolotl embryo, launching domains can be circles (the endoderm contraction wave, Figures 7-11 in Appendix V), 'points' (the animal cap expansion wave, Figures 8-9 in Appendix V, and the ectoderm contraction wave, Figure 17 in Appendix V), or circular arcs (waves over the dorsal lip of the blastopore, Figures 13-15 in Appendix V). Although only single, solitary waves travelling about 3 µm/min are observed, their launching may, for example, be similar to the 'pulsatile foci of Ca^{2+} release' from which multiple 1300 µm/min waves emanate in *Xenopus* oocytes (Jouaville et al., 1995), or to the launching of the activation wave by sperm (Créton & Jaffe, 1995).

The animal cap expansion wave appears to travel across a few tissues, though, as a result, we change its name as it travels, so that we call it the mesoendoderm expansion wave as it enters the mesoendoderm (Figure 9 in Appendix V: Gordon, Björklund & Nieuwkoop, 1994). Expansion waves have been classically referred to as epiboly motions (Kageyama, 1980, 1982; Keller, 1980; Baumann & Sander, 1984; Trinkaus, 1984a,b; Betchaku & Trinkaus, 1986; Keller & Trinkaus, 1987; Sguigna, Fluck & Barber, 1988; Boterenbrood & Narraway, 1990; Warga & Kimmel, 1990; Weliky & Oster, 1990a; Gevers & Timmermans, 1991; Strähle & Jesuthasan, 1993; Wilson, Helde & Grunwald, 1993), including that of the animal cap (Bauer, Huang & Moody, 1994). Recent evidence in chick (Mareel et al., 1984) and zebrafish (Solnica-Krezel & Driever, 1994) points to a 'pushing force' exerted by microtubules, similar to what we proposed for expansion waves (Appendix V). Whether all epiboly involves differentiation has yet to be investigated.

The trajectory of the ectoderm contraction wave is peculiar, to say the least (Figure 17 in Appendix V: Gordon, Björklund & Nieuwkoop, 1994), and may be determined by nonuniform tensions in the ectoderm and/or other factors:

"A wave trajectory will generally be determined by the delay each cell imposes between receipt of the wave and passing the wave on to neighbors. That delay may in turn be caused by mechanical properties of the cell and its environs: size, rigidity, shape, diffusion rates, etc.; or it may be caused by chemical or genetic properties or state of the cell, or it may be caused by the topology of the array of cells.... It would be interesting to model development starting with the assumption that the primary function of the IFs [intermediate filaments] is to modulate the tempo of contraction & recovery, and to set the position and stability of the metastable state. Changes in the IFs would then have a profound effect on differentiation wave trajectories" (Steve McGrew, p.c., 1997).

The ectoderm contraction wave self-annihilates near the anterior end of its trajectory (Figure 17 in Appendix V: Gordon, Björklund & Nieuwkoop, 1994). There is a possibility that en route it may be bounded by a physical barrier, a mechanical 'crimp', lasting about half an hour, that precedes it

(Figure 17 in Appendix V). This 'crimp' may be the external manifestation of the 'gastrular cleavage', described here for the frog *Rana temporaria:*

"The first external indication of gastrulation is the appearance of a slight irregular groove, approximately horizontal, lying across the sagittal plane on the posterior side of the egg, just at the lower margin of the germ ring, *i.e.,* just below the equator. Thus located, the groove lies just between the animal cells and the yolk cells, and therefore comes to be lined by both kinds of cells on its opposite faces. From subsequent development we know that the formation of this groove is the beginning of invagination, the groove itself the beginning of the archenteron, its upper margin the rim of the blastopore, and the cells lining it above and below, ectoderm and endoderm respectively. Hemisection of this very early gastrula shows that the elevation of the floor of the blastocoel is very rapid at this time, and one of the first results of the arching up of the yolk cells (endoderm) is the formation of a narrow groove all around the margin of the blastocoel, between the base of its wall and its rising floor. As the yolk continues to rise, this groove soon becomes quite marked and compressed into a narrow slit which, though originating in the manner just described, seems to be continued ventrally all around the central cells as a definite splitting or delamination. In effect this narrow space separates definitely the ectoderm and endoderm in the region within (above) the restricted invaginating region, which of course also gives rise to an ectodermal and an endodermal layer. This original groove is called the *gastrular groove,* and the delamination which extends it is the *gastrular cleavage;* the formation of these is not limited to the region of the dorsal lip of the blastopore, but extends entirely around the gastrula, even to the side opposite that of invagination" (Kellicott, 1913).

Cf. Michniewska-Predygier & Pigón (1957). The ectoderm contraction wave annihilates upon meeting the dorsal lip of the blastopore. This very encounter may generate another wave at that location (Figure 12 in Appendix V: Gordon, Björklund & Nieuwkoop, 1994). K. Jack Butler (1996), in preparing clay models for the Flip Animation of the Ectoderm Contraction Wave, found it necessary to place a horizontal crease tangentially to the dorsalmost aspect of the blastopore, at the very moment when the ectoderm contraction wave meets the dorsal lip of the blastopore. His predicted crease may initiate the observed crimp. The crimp also corresponds to the intersection of the plane of the floor of the blastocoele (Figure 13 in Appendix V) and the ectoderm, so some sort of buckling, creating a physical barrier for the ectoderm contraction wave, may be occurring.

The question of a barrier to 'spread' of neural induction has been raised before:

"Deuchar (1970) has demonstrated that only a few mesoderm cells are necessary for neural induction, and ectoderm cells, once signalled, are able to transmit this signal to other ectoderm cells. In the light of these experiments one can ask what limits the spread of the inducing signal to the area of the neural plate? One possibility is that the information in transferred only slowly from one cell to the next; Kelley (1969) has suggested that 3 hr is necessary for mesoderm to induce the neuro-ectoderm.... There could be a real barrier to the spread of the inducing signal beyond the edge of the neural plate, which would be responsible for the change in cell morphology at the boundary between presumptive neural tissue and surrounding ectoderm (see Schroeder, 1970) and allow divergence between the electrical properties of the two groups of cells. The structural and functional nature of such a barrier remains to be identified, but the problem is similar to that in the insect epidermis, where each segment is capable of independent behaviour developmentally, yet low electrical resistance connexions are present across the intersegmental border (Warner & Lawrence, 1973)" (Warner, 1973).

Whether the barrier is physical and/or electrical, and whether it precedes the ectoderm contraction wave, or the wave is constrained by other means and itself creates the barrier, have yet to be ascertained. "Current is driven out the lateral margins of the neural folds and out of the blastopore, returning through the general body surface" (Shi & Borgens, 1995). However, this current does not appear until gastrulation is complete, and vanishes at neural tube closure, so unless a weaker or more localized current has yet to be detected during gastrulation, in the region of the crimp, it may not be involved in barrier formation. In any case, in some sense the crimp or barrier is secondary: while it can limit the spread of the ectoderm contraction wave, it cannot alter its trajectory as it propagates. That strange trajectory (Figure 17 in Appendix V: Gordon, Björklund & Nieuwkoop, 1994) may be quite sufficient to explain the confinement of the ectoderm contraction wave to approximately one hemisphere.

The ectoderm expansion wave (Figure 16 in Appendix V: Gordon, Björklund & Nieuwkoop, 1994) crosses a short distance into the trajectory of the earlier ectoderm contraction wave. We suspect that the cells that have experienced both the ectoderm contraction wave and the ectoderm expansion wave become sensory placode cells (Appendix VI: Björklund & Gordon, 1994). High resolution time-lapse will be needed to resolve this. While the ectoderm expansion wave is on its way, the neural plate is flattening, and by the time the wave arrives, it encounters a $90°$ sharp edge. Perhaps such an edge is a physical barrier preventing further propagation; it is, in effect, an absorbing barrier. This edge corresponds to a line of shear and loss of junctions due to the (developing) sudden jump in apical cell diameters (Figures 28 and 29 in Jacobson & Gordon, 1976a), perhaps involved in formation of migratory neural crest cells:

"Although the perimeter of the [neural] plate remains about constant in total length, the cells that compose the perimeter are those that reduce their [apical surface] diameter the most. At the boundary between the epidermis and the plate, while the epidermal cells are enlarging in diameter [due to the ectoderm expansion wave], the plate cells are shrinking [in response to the ectoderm contraction wave]. In the normal embryo the boundary is very abrupt. The probable consequence is that cells change neighbors at the boundary. Thus the boundary may be a line of shear:...

"Shear may... sever tight junctions or other intimate cell contacts at the boundary between epidermis and neural plate. This could, in turn, lead to subsequent changes in cell behavior (Loewenstein, 1968). The eventual appearance of a special population of cells, the neural crest, at this boundary could well be due to shear" (Jacobson & Gordon, 1976a).

We have speculated that shear is involved in the formation of germ cells at an earlier stage:

"...Ventral mesoderm precursor (ACEE) would be equivalent to lateral mesoderm precursor (ACEE) at the end of gastrulation except for those cells that are torn free and undergo an independent [single cell] contraction [wave?]. These cells could be the same cells that are known to arise in this precise region at this time and eventually form migrating germ cells (Nieuwkoop & Sutasurya, 1979)" (Appendix VI: Björklund & Gordon, 1994).

It may be of some importance, also, that apically stretched neural plate cells lie nearby in the neural plate, suggesting highly anisotropic stress and strain (Figure 12 in Jacobson & Gordon, 1976a). A model of 'cortical tractoring' has been proposed that permits easy changing of neighbors in an epithelium (Jacobson, Odell & Oster, 1985; Jacobson et al., 1986; cf. Weliky & Oster, 1990a).

One property of differentiation waves, as examples of waves in an excitable medium, is that they take the shortest path from one point to another (as demonstrated in the use of waves in excitable media to solve mazes: Steinbock, Toth & Showalter, 1995; Kaiser, 1995). This is, in effect, an application of Fermat's Principle:

"The laws of rectilinear propagation, of reflection and of refraction can be subsumed under a single law, first stated by [Pierre] Fermat (1608-65) as the Principle of Least Time. Indeed, the whole theory of geometrical optics can be deduced from this single principle. Fermat maintained that nature is economical and that light travels from one point to another by the path of least time....

"[Christiaan Huygens (1629-95)]... most fundamental and original contribution was a geometrical method of representing wave propagation, known as *Huygens' principle*.... Huygens assumed, without any satisfactory explanation, that... secondary wavelets destroy each other except where they touch their envelope..., and that they are not propagated in a backward direction.... This was the decisive weakness that led [Isaac] Newton to reject the theory.... [Huygens]... found support in Fermat's principle, which he had rejected earlier as a 'pitiable axiom', until he saw how it confirmed his own theory" (McKenzie, 1962).

Note that: "Fresnel, making use of the principles of interference, was able to resolve the difficulties and provide a satisfactory explanation" (Morgan, 1953) to Newton's objection. Fermat's Principle will, for instance, permit us to define a refractive index for a differentiation wave. Such a concept could be helpful in understanding the physics of nontrivial trajectories, such as that of the ectoderm contraction wave (Figure 17 in Appendix V: Gordon, Björklund & Nieuwkoop, 1994). The refractive index may be related to the physical strain in the tissue, which probably varies over the ectoderm due to nonuniform invagination through the blastopore. In any case, the refractive index to a differentiation wave is a measurable property of a tissue, deserving of explanation. It is curious that Huygens' principle, in its original formulation, at least, applies more accurately to waves in an excitable medium than to electromagnetic waves.

Proposition 45: morphogenesis via tissue splitting explains compartmentalization of a tissue into two sharply demarcated new tissues.

Compartmentalization is widely studied in insects (García-Bellido, Ripoll & Morata, 1973, 1976; Crick & Lawrence, 1975; García-Bellido, Morata & Ripoll, 1976; García-Bellido, Lawrence & Morata, 1979; Martinez Arias & Lawrence, 1985; Allaerts, 1991; Nijhout, 1994d). It is analogous to regionalization in vertebrates. During compartmentalization a barrier is erected to cell migration (Matela, Ransom & Bowles, 1983; Blair et al., 1994; cf. Birgbauer & Fraser, 1994). This prevents clones from interpenetrating, which makes the compartment boundary visible (Kuhn et al., 1983; Fishell, Rossant & van der Kooy, 1990). Compartments are also coincidentally barriers to electrical contact (Blackshaw & Warner, 1976; Trinkaus, 1984a; cf. Warner & Gurdon, 1987) or diffusion (Serras & van den Biggelaar, 1987; Kalimi & Lo, 1988; Bagnall, Sanders & Berdan, 1992; Laird et al., 1992). This happens despite the fact that two tissues involved otherwise mechanically form a continuous epithelium. Compartmentalization appears to occur hierarchically, dividing each tissue into smaller tissues (Kauffman, 1977; Kauffman, Shymko & Trabert, 1978; Lawrence & Morata, 1979; Cooke, 1980b; Brower, 1985).

Differentiation waves offer a possible mechanical explanation of how compartmentalization comes about in tissues. Such waves would, of course, travel over any complex surface (such as the irregularly shaped *Drosophila* imaginal discs: Bryant, 1979), though much has to be learned about what directs their trajectories. The combination of cell state splitters within cells and tissue synchronizing mechanisms allows us to resolve the following controversy over the level at which 'decision making' occurs, without ascribing higher intelligence to cells or tissues:

"Are morphogens perceived by groups of cells or by individual cells? The accepted view is that the uniformity in a sheet of cells in the early embryo is converted into discontinuities, termed compartments or polyclones. The decision-making unit is a group of cells.... However, an alternative viewpoint is that individual cells themselves are the basic decision-making units (Gergen, Coulter & Wieschaus, 1986)" (John & Miklos, 1988).

Whether other effects that accompany compartmentalization, such as duplication or intercalation (Bryant, 1979), can be explained mechanically, by differentiation waves, remains to be seen. If tissue splitting is a mechanical process, then *ad hoc* assumptions of cell sorting "for maintaining the sharp and precisely placed boundaries at compartment borders" (Lewis, Slack & Wolpert, 1977; cf. Lawrence, 1975) may not be necessary. "In *Oncopeltus* the segmental border is a straight line in a simple epithelial monolayer and is marked by a change in pigmentation that is visible in the electron microscope" (Lawrence & Green, 1975). Both of these features may be consequences of different differentiation waves on both sides of the compartment border, the first causing "an abrupt change of cell adhesiveness", as the authors suggest. Fehon, Gauger & Schubiger (1987), in a critical review of cell sorting in *Drosophila* imaginal discs, conclude that that phenomenon may not be the cause of compartmental boundaries, but rather that 'lineage restriction', i.e., differentiation, precedes boundary formation:

"If anterior and posterior cells do not differ in their cell-surface properties, then the hypothesis that all compartment boundaries are maintained by sorting out because of compartment-specific affinities (Crick & Lawrence, 1975; Morata & Lawrence, 1975; Lawrence & Morata, 1976a) must be reevaluated. In fact, sorting out does not seem to be adequate to explain compartment boundaries, given that all cells within the disc are firmly bound by junctional complexes and therefore do not normally move more than a cell diameter or so relative to one another except by cell division (Oberlander, 1985). Taking these constraints into consideration, it is difficult to imagine how sorting out could maintain a compartment boundary. On the other hand, it is easy to imagine that a line of nondividing cells could serve this function (Brower, 1985). O'Brochta & Bryant (1985) have shown that the dorsal-ventral compartment boundary coincides with a zone of nonproliferating cells (ZNC) that appears at approximately the same time as the lineage restriction" (Fehon, Gauger & Schubiger, 1987).

While Bryant (1987) indicates: "We have argued (O'Brochta & Bryant, 1985) that the lineage restriction may be simply a result of the presence of a zone of nonproliferating cells in this position", the reverse is possible. Perhaps we should look for potential reductions in cell proliferation at all boundaries of differentiation wave trajectories. Certainly, if differentiation waves are accompanied by much cell division, as a remnant of their possible evolutionary derivation from cleavage waves (see Chapter 9), then those few cells that miss a wave, by being at the edge of a trajectory or just off it, may have reduced proliferation rates.

Lawrence (1990) has summarized the current concept of compartmentalization, which now appears to also be applicable to rhombomere formation in late neural tube development in vertebrates (Lewis, 1989; Fraser, Keynes & Lumsden, 1990; Guthrie, 1995; Dubbeldam et al., 1995):

"The compartments have precisely the same borders in each individual, but within those borders the founder cells generate clones of diverse sizes and shapes that differ from one individual to another (García-Bellido, Lawrence & Morata, 1979). The founder cells are selected by the positions they occupy, not by ancestry.... After the founder cells are allocated, clones of cells remain confined to one side of a boundary or the other. The link with genetics... was proposed and developed by (García-Bellido, Ripoll & Morata, 1973, 1976; García-Bellido, 1975), who suggested that the developing compartment would be coextensive with the realm of action of a specific class of regulatory or 'selector' genes (García-Bellido, 1975; Lewis, 1963, 1978). The selector genes would determine the pattern made and the developmental pathway (Lewis, 1963, 1978) followed by each compartment. Thus the compartment theory envisages the animal as being made in modules, each growing and shaping itself independently of the others and each being subject to somewhat independent genetic control" (Lawrence, 1990).

The hypothesized coincidence of compartmental boundaries with gene expression has been found in a mollusc, the limpet *Patella vulgata,* at least to the extent that "dye compartments coincide with different presumptive regions" (Serras et al., 1985). We thus see that the concept of the differentiation tree and the mechanical nature of tissue splitting appear to account for the features of compartmentalization. Moreover, we may suggest that primary neural induction itself may illustrate compartmentalization: we would expect similar cell lineage restrictions after the 'compartmental boundary' between epidermis and neural plate is formed:

"We have found that the abrupt boundaries between cell domains in the stage 15 neural plate [Figure 4; Table 1] correlate with lines of shear in the notochordal region and at the boundary between the neural plate and epidermis. These lines of shear could break junctions between cells and serve to isolate domains of cells from one another, allowing them to pursue different developmental pathways" (Jacobson & Gordon, 1976a).

(Cf. Gordon & Jacobson, 1978). This boundary between epidermis and neural plate also represents a boundary at which electrotonic coupling is decreased and then broken in axolotls (Warner, 1973). I would now attribute the shear boundaries as a consequence, rather than a cause, of compartmentalization by the ectoderm contraction and expansion waves. Similar boundaries may form between rhombomeres (Fjöse et al., 1992; Heyman, Kent & Lumsden, 1993). As cell specializations, such as neural crest differentiation and rhombomere boundary cells (Heyman, Faissner & Lumsden, 1995), occur at such interfaces, it will be interesting to learn whether differentiation waves are involved here also.

Proposition 46: whether a wave is launched as an expansion wave or a contraction wave depends on two thresholds of mechanical tension.

The idea that stretching of the ectoderm prior to launching of the ectoderm expansion wave causes that wave to be an expansion wave occurred to Natalie K. Björklund (p.c.). Dennerll et al. (1989) found what may be exactly the required dynamics while exerting forces upon axons:

"We conclude that neurite/axon length is regulated by axial tension in both elongation and retraction, as previously suggested by Campenot (1985). More specifically, our data suggest a tension-sensitive, three-way control of neurite length via a reversible growth process... [analogous to] a double-pole, double-throw electric switch. Our data for PC12 cells suggests that at neurite tensions above some threshold, towed growth (Weiss, 1941; Bray, 1984) occurs.... At neurite tensions below a different, lower threshold, the neurite is stimulated to generate tension, shortening the neurite.... At neurite tensions between the two thresholds the neurite responds with passive viscoelastic behavior" (Dennerll et al., 1989).

(Cf. "Biphasic responses, with both thresholds and upper limits" for modelling differentiation: Robertson & Grutsch Jr., 1987). This work is reviewed by Heidemann & Buxbaum (1994), who emphasize "...the view that mechanical tension plays a role in axonal development roughly analogous to that of a second messenger" and attempt to relate their observations to tensegrity structures (Fuller, 1961; Ingber & Folkman, 1989a; Heidemann, 1993; Wang, Butler & Ingber, 1993; Ingber et al., 1994; Pickett-Heaps, Forer & Spurck, 1997). Note that...

"Existing models for chemical second messengers would appear to be inapplicable and the notion of force as a 'second messenger' is almost unexplored" (Heidemann & Buxbaum, 1994).

The evidence we have for an analogous phenomenon during early embryogenesis is only correlative: both the animal cap expansion wave (Figures 8-9 in Appendix V: Gordon, Björklund & Nieuwkoop, 1994) and the ectoderm expansion wave (launched as the presumptive notochordal mesoderm contraction wave, Figure 11 in Appendix V) are launched at a time that the embryo is under tension, while contraction waves may be launched at times of lesser tension. See: Lewis (1947); Jacobson & Gordon (1976a); Gordon & Jacobson (1978); Gordon (1985c); Beloussov et al. (1990), for slitting experiments demonstrating such tension.

One could imagine doing appropriate experiments, even on the intact axolotl embryo, for instance. It might be possible to apply local 'towing' by blowing medium through a small pipette tangentially onto the outer surface of the ectoderm (after removing the vitelline membrane: Hamburger, 1960; Asashima, Malacinski & Smith, 1989). Sucking could possibly generate a local compression, or at least counter the global tension. I would predict the launching of an expansion or contraction wave if the hypothesized thresholds were, respectively, crossed, by sufficiently rapid currents. Another technique for applying tangential forces could be developed using antigen coated beads (Maniotis, Chen & Ingber, 1996), which perhaps could be magnetic (Steve McGrew, p.c., 1997).

Viscoelastic media respond with characteristic time constants, apparently of the order of 10 min for neurites, which has been related to the characteristic time for microtubule elongation or depolymerization (Dennerll et al., 1989; Zheng, Buxbaum & Heidemann, 1993). This is about the same time scale as the "...diffusion of [actin] rods from adjoining domains interdigitating with each other within 5 min, thus 'knitting together' the domains into a gel state" (Kerst et al., 1990). This time also corresponds to the time an ectoderm cell remains contracted during passage of the ectoderm contraction wave (Appendix V: Gordon, Björklund & Nieuwkoop, 1994), which is also roughly the time that we suggest is needed for the nuclear state splitter to be activated, leading to a change in gene expression (Appendix IV: Björklund & Gordon, 1993b). There is need here for a detailed mechanical analysis, which would probably reveal that not only tensions, but their durations, are involved in setting thresholds, as we should expect for a nonlinearly viscoelastic medium. In any case, there is a possible analogy, or even homology, between the three states of axonal response to tension (elongation, retraction, and passive) and the three states of the cell state splitter (expansion, contraction, metastable), i.e., in particular, the 'passive' state of axons corresponds to the 'metastable' state we have attributed to the cell state splitter while it is in the embryonic state of competence (Appendix I: Gordon & Brodland, 1987; Proposition 6).

The mechanism by which the launching domain turns mechanics into a wave will include mechanochemical components. Stretch activated ion channels are present in amphibian membranes from the oocyte stage (Yang & Sachs, 1989, 1990), and mechanosensitive channels are found in embryonic rat neuroepithelium (Mienville, Barker & Lange, 1996). Although their physiological importance has been questioned (Morris, 1992; Appendix IV: Björklund & Gordon, 1993b), they continue to appear in apparently nonartifactual situations (Levina, Lew & Heath, 1994). It is possible that their role is to provide a nonlinear feedback mechanism that gets a cell or group of cells started as the launching domain of a differentiation wave (Natalie K. Björklund, p.c.), and one could speculate that stretch-inactivated ion channels and stretch-activated ion channels (Morris & Sigurdson, 1989) may correspond to the two kinds of wave launchings. Whether there is a fundamental distinction between the mechanism needed to launch a wave, compared to that needed to propagate it from cell to cell, remains to be seen. For either phenomenon, the direct regulatory link of actin to ion channels (Berdiev et al., 1996; Prat et al., 1996) could be important.

While we do not yet know how to manipulate which kind of wave gets launched, a preparation has been found in which the type seems random:

"Various attempts to culture isolated parts of amphibian blastulae and gastrulae in organic media proved unsuccessful [cf. Frost, Neff & Malacinski, 1989], but it was discovered that pieces introduced into the coelem of an older larva would live and develop. Development in these coelemic cultures differs in certain respects from that in salt solutions, as regards presumptive epidermis and neural plate (Holtfreter, 1929, 1931a). Presumptive neural plate may develop either into pure nervous tissue or into epidermis with practically equal frequency; and presumptive epidermis, into epidermis with differentiation of gland, ciliated, and pigment cells or into nervous tissue" (Child, 1941).

This result confirms, by the way, the equivalence of the whole ectoderm. With some chemical analysis and effort, we might duplicate coelemic conditions, and try to learn why waves get launched as contraction or expansion waves.

There would be a temptation to relate the launching of an expansion versus a contraction wave to the positive versus negative polarity (or vice versa) of an ionic current directed through the cells that launch a wave:

"It is not commonly appreciated that steady, polarized DC currents [direct current] are driven through developing animals and plants for hours, even days. These endogenous ionic currents (and their associated voltage gradients) may be controls of development and not accidental to it (Jaffe & Nuccitelli, 1977; Borgens, Vanable Jr. & Jaffe, 1979a; Jaffe, 1981a, 1982).

Endogenous currents often foreshadow the locus [launching domain?] of cell growth, orientation, or migration" (Metcalf, Shi & Borgens, 1994).

However, electrical polarity is not associated with all steps of differentiation (Shi & Borgens, 1995) and is an inconsistent indicator of morphological polarity:

"What [morphogenetic] fields might there be? The most familiar candidates are endogenous electric fields, generated by transcellular electric currents, which could serve as a localizing mechanism by the electrophoresis of organelles or membrane proteins. This proposition, originally put forward by Lund (1947) and articulately championed by Jaffe and his associates (Jaffe, Robinson & Nuccitelli, 1974; Jaffe & Nuccitelli, 1977; Jaffe, [1969], 1981a) has lost luster recently because there appears to be no constant and obligatory relationship between the polarity of the transcellular electric current and that of growth (Harold & Caldwell, 1990; Gow, 1990). A more appealing version of the idea, that ionic currents localize morphogenetic events, identifies the spatial signal with a localized influx of calcium ions" (Harold, 1991).

Clearly we need experimental investigations correlating localized stretching/compression, direction of ionic currents, state of junctions and ionic channels, physical buckling, and extracellular matrix components, etc., to determine exactly which are involved in whether a differentiation wave is launched as a contraction or expansion wave, and what determines when launching occurs.

Proposition 47: natural 'inducers' are involved in the location and type of differentiation waves launched.

While we can be amused by the long history of so-called inducing molecules (Sections 2.04-2.06), amongst them may be many candidates for molecules involved in allowing two tissues to make contact in a defined region, from which a differentiation wave is launched. Here I just indicate a few possibilities from the modern repertoire. If the hypothesis that inducer molecules are involved in wave launching is to be taken seriously, then we will need experiments correlating their expression patterns with type and location of the wave launched. Ectopic launchings might also be investigated from this viewpoint. A relationship of wave launching to both vertical inducer molecules and ionic currents probably exists. 'Vertical' refers to contact between two layers of tissue at a specific domain, as in the contact between pharyngeal endoderm and presumptive notochord (Figure 11B in Appendix V: Gordon, Björklund & Nieuwkoop, 1994). (Of course, it is possible that some wave launchings just require ionic currents, without any vertical induction, i.e., without any underlying tissue contact. This may, for instance, be what 'autoneuralization' involves: Mattsson, Nakatsuji & Løvtrup, 1987.)

Unfortunately, vertical and horizontal induction have not been separated yet in the gene expression literature. Therefore, I have just suggested that some molecules that are involved in cell-cell interactions at the time of inductions are candidate wave launchers:

"**Dick**: Embryonic induction is the process of setting off a wave of activation of the cell state splitters. We don't yet know how these waves get started, but from the point of view of cell state splitters, all we need is a force imbalance (Figure 3 in Appendix V: Gordon, Björklund & Nieuwkoop, 1994). It is possible that the very act of adhesion of two tissues over an area of a few cells is sufficient to generate the tension needed for the force imbalance we are postulating (Pesciotta Peters & Mosher, 1989, [on fibronectin]). Regarding the question of specificity of induction, I don't see why we can't 'have our cake and eat it too'. The mechanism starting off a wave may be merely physical. However, because cell-cell adhesion is sometimes required to get the physical effect, and numerous specific molecules (fibronectin [Duband & Thiery, 1982; Boucat et al., 1984, 1985; Lee, Hynes & Kirschner, 1984; Darribère et al., 1985, 1988; Tuckett & Morriss-Kay, 1986; Smith et al., 1990b; DeSimone, Norton & Hynes, 1992; DeSimone et al., 1992; George et al., 1993; Collazo, 1994], collagens [Wessel & McClay, 1987; Knibiehler, Mirre & Le Parco, 1990; Otte et al., 1990b], N-CAMs [Saint-Jeannet et al., 1989], etc. [L-CAM: Thiery et al., 1984; PDGF: Ataliotis et al., 1995]) could be involved in setting up a very specifically localized cell-cell adhesion, specificity is possible" (Gordon, Björklund & Nieuwkoop, 1994).

Some of these adhesion molecules may be involved in epithelial continuity, morphogenetic movements, or propagation of differentiation waves (*hedgehog:* Ma et al., 1993), so that it will be necessary to see exactly what is being expressed and utilized right at the vertical tissue-tissue interfaces where waves are launched. Hints of vertical induction mechanism(s) may be seen, for instance, in the role of integrin as a "transmembrane linkage between fibronectin and actin" (Tamkun et al., 1986) and its relationship to changes in cell shape (Ingber, 1990).

Proposition 48: an embryonic field corresponds to the trajectory of an expansion or contraction wave.

There is a long history to the usual view that mechanisms operating over 'embryonic fields' (Waddington, 1956a; Goodwin, 1982a, 1984a, 1985a, 1987; Davidson, 1993; Gilbert, Opitz & Raff, 1996; Raff, 1996) cause pattern formation, positional information (Wolpert, 1969), birth defects (Opitz & Gilbert, 1982; Lubinsky, 1987), etc. The concept stems from physics and goes back at least to Von Baer:

"Driesch (1891a,b) had drawn an analogy between the polar structure of the egg and a magnet: On this basis, egg fragments should behave similarly, particularly if they are part of a 'harmonious equipotential field'.... About... 1921..., the 'field' concept was introduced into the theory of development (Huxley & de Beer, 1934). Spemann [1938] was probably the first to use it, in relation to his work on the organizer; borrowing the field concept from physics, he remarked that the organizer established a 'field of organization.' The idea was also introduced by Gurwitsch [1922, 1927] and by Weiss [1939], the latter applying it to the behavior of the blastema of the regenerating limb" (Wolpert, 1991b).

"Von Baer's laws, according to which an embryo develops from an undifferentiated, homogeneous state to a highly articulated and differentiated one, which have already played an important role in the story we have been telling, were conceived within this [teleomechanist] research program. Von Baer [1828] saw development as driven by a distinct vital force (*Lebenskraft*), physically emergent from a physico-chemical base and then reacting on it. Operating from the center to the periphery of the embryo, this force guides development toward maturity. The 'life-force' is supposed to be a force like other Newtonian forces, spreading out from a center after the fashion of an inverse square law, maintaining equilibrium as it goes, but irreducible to gravity, electricity, magnetism or other canonical Newtonian forces" (Depew & Weber, 1994).

Needham (1950) traces it back even further, though without spelling out the connection between light and magnetic fields:

"The conviction that the magnetic field has a part to play in guiding our ideas about morphological patterns is now of some seniority. As a matter of fact it goes back to light-mysticism of Persian and Jewish origin. The Kabala [Matt, 1983, 1995], which was studied by all seventeenth-century biologists, led in one case to the extremely interesting work of Marcus Marci, of Kronland, a Czech. Marci (1635) was a mixture of purely scientific contributions to optics, and speculative theories about embryology [cf. appendix of Pagel, 1967]. Thus he explained the production of manifold complexity from the seed in generation by an analogy with lenses, which will produce complicated beams from simple light sources. The formative field radiates from the geometric centre of the embryonic body, creating complexity but losing none of its power. Monsters originate from accidental doubling of the radiating centre, or from abnormal reflections or refractions at the periphery (cf. mirror reduplications, accessory organizers, etc.). Marcus Marci, in his forgotten work, thus links together the following trends of thought: (a) the old Aristotelian theory of seed and blood, (b) the new Cartesian rationalistic mathematical attitude to generation, (c) the new experimental approach, in his work on optics, (d) the cabbalistic mysticism of light as the fountain and origin of all things; finally, by his brilliant guess of centres of radiant energy, (e) he anticipates the field theory of modern embryology. Cf. Needham, 1934a" (Needham, 1950).

The assessment of the biggest promulgator of the concept of morphogenetic fields may be the most *a propos:*

"In fact at present we lack any generally applicable theory of the physico-chemical nature of the so-called 'morphogenetic field[s]', which, in the absence of such a theory, remain no more than a convenient descriptive device with no explanatory value" (Waddington, 1965).

Cf. the same remark made 30 years earlier:

"The emphasis, that there is such a thing as 'field' activity, deserving attention and experimental attack, as well as the description of whatever of its properties can be deduced from its manifestations, is about all there is to be gained by the introduction of the term. It has proved to be of descriptive, but not yet of ultimate explanatory value. It presents a problem rather than solves one.... The activities playing in the whole [field], or rather the conditions releasing those activities, e.g., potential differences, may they be of mechanical (tension) or electrical or chemical or whatever nature, are graded and distributed in some typical way, and this pattern of gradation and distribution tends to persist as such" (Weiss, 1935).

Hamburger (1988) also indicated that...

"The field concept was introduced [but cf. above] by A. Gurwitsch (1922) in a theoretical study and by P. Weiss (1925) as a means to interpret his experiment on limb regeneration in adult salamanders. The field concept figured prominently in the 1930s; this is reflected in the textbooks of Weiss (1939) and Huxley & de Beer (1934).... But neither Spemann (1938) nor Harrison (1969) made much use of the term.... I suppose that the field concept had no appeal to them because it is an abstraction taken over metaphorically from physics, with no explanatory value" (Hamburger, 1988). [Cf. the review of gradients and Child's, 1941, remarks discussed by Witkowski, 1992a. A counterclaim is made that "It is difficult to realize how powerful the concept of the morphogenetic field used to be.... Harrison's friend, Hans Spemann, 1921, reinvented this concept as an *Organisationsfeld* and said that the dorsal blastopore lip established such a 'field of organization'": Gilbert, Opitz & Raff, 1996.]

Unfortunately, morphogenetic fields have been raised by some to an extracorporeal status transcending even the time and space of physics:

"...What Sheldrake (1981) [cf. Sheldrake, 1995] offers is an approach to the problem of development that combines the old idea of a morphogenetic field with the radically new notion that such fields arise from the forms of previous organisms and, moreover, survive the death of the organism to influence in some fashion the forms of its progeny.... Morphic resonance... takes place on the basis of similarity, and involves a nonenergetic transfer of information. Thus, it is a kind of action at a distance in time and space.... It is a kind of field currently unknown to physics - a field of information rather than one of matter or energy.... It would appear that Sheldrake's theory is at least a contender in the developmental biology sweepstakes" (Casti, 1990).

Real ionic currents do form measured electric fields over the surface of neural plate stage amphibian embryos (Shi & Borgens, 1995), and when they are countered by Na^+ channel blockers (Borgens & Shi, 1995) or DC (direct current) electric fields (Metcalf & Borgens, 1994), development of the central nervous system is severely disrupted, even to the extent of massive cell disaggregation. However, no such effect occurs during gastrulation, and so electric currents and electric fields are not likely to represent a universal mechanism for morphogenetic fields.

A morphogenetic field has been regarded as a larger region of an embryo from which a given structure arises (Waddington, 1956a). Spatially, if differentiation waves define tissues, then a field must correspond to a whole tissue. As such, the field can be redefined by the trajectory of the wave that set it off as a distinct tissue. The definitive structure it produces (as the limb bud from the limb field: Gurwitsch, 1922; Swett, 1923) either represents morphogenetic movements within the field, or one or more subsequent steps of differentiation, i.e., the trajectories of further nested waves. This suggests that the portion of the field that does not become the definitive structure is nevertheless different by at least one step of differentiation from the surrounding tissue. This should have observable consequences, for instance via differential gene expression.

Proposition 49: each change in a component of the differentiation pathway, and each gene product specific to the state of cell differentiation being triggered, will occur in the wake of the differentiation wave that initiates it, and track that wave over the embryo.

This proposition makes a simple, robust, testable prediction. Every measurable gene product should track the physical differentiation wave that initiated its synthesis. For instance, a fluorescent antibody to N-CAM should show up as a propagating wave behind the ectoderm contraction wave:

"Differential adhesion, if and when present, could be a consequence rather than a cause of differentiation, as shown for primary neural induction by Jacobson & Rutishauser (1986) (cf. Duprat et al., 1985a; Balak et al., 1987). N-CAM (neural cell adhesion molecule) could be a specific gene product resulting from the differentiation. This and other specific proteins... all appear *after* primary neural induction" (Gordon & Brodland, 1987).

Expression of such gene products should follow the same trajectory as the wave, with a delay that depends on the sensitivity of our imaging equipment and the time from determination to synthesis of, for example, N-CAM. If we alternated between bright field and fluorescent imaging, we should see the N-CAM fluorescence follow the ectoderm contraction wave at a small (or zero) distance behind it. With waves having a peculiar or distinctive trajectory, such as the ectoderm contraction wave (Figure 17 in Appendix V: Gordon, Björklund & Nieuwkoop, 1994), this is a robust prediction. If the spatiotemporal order of tissue specific gene expression failed to follow the wave that corresponded to formation of that tissue, the basic premise of this book would be undermined. To me this observation thus represents a critical experiment.

Of course, if N-CAM persists or continues to be synthesized in later tissues deriving from the neural plate (which forms in the wake of the ectoderm contraction wave), its expression pattern may not correlate with any later waves going through the neural plate.

This view of CAMs in particular is opposite the conclusion reached by Edelman & Gallin (1987) that...

"...a primary regulatory factor in vertebrate development is the CAM-dependent coupling of cells into cooperative populations that will produce and respond to inductive signals" (Edelman & Gallin, 1987).

I'd suggest that differentiation waves (which I think are the initial acts of induction) set off the territories for gene expression, including CAMs, via their trajectories. The trajectories could, however, be influenced by the distribution of CAMs in the tissue through which they pass, placed there during the previous step of differentiation: CAMs would be expected to alter morphogenetic movements and constitutive relationships between cells.

The huge literature on the anatomy of specific gene expression shows mostly the whole of each tissue within which a specific protein is expressed, after the wave triggering it has come and gone (cf. hints of *Xenopus* neural plate waves in Papalopulu & Kintner, 1994). No studies of which I am aware would have caught a wave mid-trajectory and demonstrated a gradient of a protein in the wake of the wave, except for those of the *Drosophila* eye imaginal disc morphogenetic furrow (Saint et al., 1988; Kidd et al., 1989; Moses & Rubin, 1991; Okano et al., 1992; Poeck, Hofbauer & Pflugfelder, 1993).

The deepest part of the furrow of a contraction wave is not the leading edge of the wave. It is merely the location at which the microfilament rings reach their maximum contraction. Events correlated with a differentiation wave can thus seem to run 'ahead' of the wave, if the deepest part of the furrow is taken as the origin of a coordinate system that moves with the wave. Note, for instance, in the poster of the eye imaginal disc of *Drosophila* by Wolff (1993), that some ommatidia formed ahead of the furrow. This may be due to a local mechanical irregularity. Since the ectoderm contraction wave of axolotl proceeds with a somewhat irregular front (Appendix II: Brodland et al., 1994), such an effect is not surprising. Time lapse in such a case might reveal the local mechanical irregularities. On a larger scale, note again the flattening of the 2000 μm diameter axolotl embryo 500 μm ahead of the furrow (Figure 9 in Appendix II; cf. the review of Whitman, 1888, which discusses analogous flattening phenomena at earlier stages), and the observation of *hedgehog* expression 30% of the diameter (20 μm) ahead of the *Drosophila* eye morphogenetic furrow (Ma et al., 1993; cf. Heberlein, Wolff & Rubin, 1993; Brown et al., 1995). Indeed, Lebovitz & Ready (1986), found that the eye imaginal disc can be dissected into an 'undifferentiated' region, followed by a determined region representing 35-40% of the eye disc length, followed by the morphogenetic furrow, followed by the region in which the ommatidia are differentiating. This can explain why...

"...rudiments appear on schedule in unpatterned epithelium even when it is dissected away from the oncoming pattern wave (Sengel, 1975; Lebovitz & Ready, 1986)" (Wolff & Ready, 1991a).

There is a long stretch of determined tissue that is 'unpatterned', as its cells have not yet undergone or completed the contraction that creates the visible furrow. By analogy with the spread of cortical rigidity from the sperm entry point (Kubota, 1967), or during cleavage (Wilson, 1951), we might anticipate changes in cortex rigidity in and/or around the advancing furrow of a contraction wave. In terms of the mechanics of the cell state splitter, the long range flattening of the tissue (at least in axolotls) and the relatively sharp profile of the furrow suggest highly nonlinear dynamics. Entrance of a cell into a contraction furrow is an abrupt event. Determination precedes the sudden and obvious furrow by some distance.

Proposition 50: the differentiation pathway, and the sequence of syntheses of the specific gene products, can be partially worked out from the delays between the physically observable differentiation wave and the fluorescence waves for those products.

This is an old trick, going back to Melvin Calvin when he used radioactive labelling to work out the temporal sequence of appearance of the intermediates in photosynthesis (Bassham, 1962; Keeton, Gould & Gould, 1986). The time delay between the physical and the fluorescence wave depends, of course, not only on the actual delay in synthesis, but also on the sensitivity of the observational technique. The shape of the wave of fluorescence will depend on the time course, within a single cell, of production and degradation of the gene product being followed. Thus quantitative explanations of the various concentration profiles of gene expression 'ahead', within, and trailing behind a differentiation wave (Ma et al., 1993) ought to be forthcoming.

Proposition 51: some morphogenetic movements, and their aberrations, may be directly caused by differentiation waves.

This follows from our observation that the depth (amplitude) of the ectoderm contraction wave, 100 µm, is twice the thickness of the cells it passes through (Figures 9-10 in Appendix II: Brodland et al., 1994). One would expect such a large amplitude wave to have mechanical effects over at least a few wave widths, and perhaps, due to nonlinear buckling or perturbation phenomena, over a whole tissue or even adjacent tissues. The following account of furrows in chick and mouse neural plate closure could be read this way if these furrows are differentiation waves homologous to those we observe in axolotl neurulation:

"...[In] chick embryos... neurulation occurs in four distinct but temporally overlapping stages (see reviews by Karfunkel, 1974; Schoenwolf, 1982; Gordon, 1985c; Schoenwolf & Smith, 1990a; Schoenwolf & Alvarez, 1992). The first stage of neurulation is formation of the neural plate.... The second stage, shaping of the neural plate, commences shortly after.... The third stage is bending of the neural plate. This stage, which begins while shaping is still in progress, involves two important morphogenetic events: furrowing and folding. Furrowing (the formation of one midline and two dorsolateral longitudinal furrows) occurs as the median and paired dorsolateral hinge points form. These are localized areas of neural plate that become anchored to adjacent tissues, the prechordal plate mesoderm or the notochord for the median hinge point (MHP) and the surface epithelium of the neural folds for the dorsolateral hinge points (DLHPs). After the hinge points form and the neural plate within the hinge points furrows, the remaining neural plate (i.e., that part of the neural plate not participating in hinge point formation) undergoes folding around each hinge point.... The fourth and final stage of neurulation involves closure of the neural groove and cranial and caudal neuropores.... It is probable that the mouse MHP, like that of the chick, provides a locus for subsequent bending of the neural plate and that MHP cell wedging provides the force necessary to generate the longitudinal midline furrow around which subsequent folding of the neural plate (i.e., neural fold elevation) occurs" (Smith, Schoenwolf & Quan, 1994).

See also Moury & Schoenwolf (1995). If differentiation waves were directly implicated in the mechanics of neural tube closure, then any aberrations in their amplitudes or trajectories (Appendix V: Gordon, Björklund & Nieuwkoop, 1994) might lead to failure of neural tube closure and thus to the birth defects (Moore, 1977) of spina bifida (failure of closure along the spinal cord) or anencephaly (failure of closure in the brain, but cf.: Andres & Bixler, 1981). For instance, retinoic acid leads to microcephaly in *Xenopus* (Durston et al., 1989), which may correlate with a reduced total ectoderm contraction wave trajectory (or its *Xenopus* equivalent: Nieuwkoop, Björklund & Gordon, 1996). Similarly, reduced invagination due to ultraviolet (UV) irradiation is likely to drastically alter the trajectory of this wave, which is how I reinterpret the following idea:

"We speculate that the size of the organizer [diminished by UV] determines the amount of convergent extension, which in turn determines how far tissue can escape from the caudalizing signal at the posterior end" (Stewart & Gerhart, 1990).

If invagination is driven by differentiation waves at the dorsal lip of the blastopore, and these waves are damaged by UV, then we would have here an example of an aberrant wave-wave interaction (cf. Proposition 82). The large amplitude of differentiation waves means they can have a direct, mechanical effect on nearby tissue, and probably on each other, even at some distance.

Proposition 52: in amphibians, the first cell state splitter is the apparatus for cortical rotation and gray crescent formation.

Prior to first cleavage, the whole cortex of an amphibian embryo undergoes a remarkable rotation of about $30°$ about a horizontal axis (Gerhart et al., 1989; Ubbels, 1997a) (Figure 6 in Appendix V: Gordon, Björklund & Nieuwkoop, 1994). This motion is correlated with the appearance of a dense, parallel array of microtubules near the cell membrane, 90% of which have a common orientation (Houliston & Elinson, 1991b; cf. Elinson & Rowning, 1988; Elinson & Palecek, 1993; Larabell, 1993), and which appear to be distinct in their tubulins from other microtubules in the fertilized egg (Chu & Klymkowsky, 1989). Two main hypotheses, polymerization of the microtubules and sliding between them or other molecules, are presently under investigation, though perhaps rotation of microtubules around their long axes (Jarosch, 1992) should also be considered:

"The occurrence of directed microtubule polymerization... raises the possibility that some of the force of the cortical rotation could be provided by the polymerization itself.... Such a phenomenon has been observed in sea urchin eggs. A heat-induced rotation of cytoplasm relative to the cortex is accompanied by polymerization first of a large microtubule aster and then of a spirally arranged cortical array (Harris, Osborn & Weber, 1980a; Schroeder & Battaglia, 1985).... Another proposal for force generation is that the microtubules of the array may act as tracks along which translocation occurs mediated by a microtubule-associated motor molecule (Vincent, Scharf & Gerhart, 1987; Elinson & Rowning, 1988)" (Houliston & Elinson, 1992).

There is an initial 'choice' of $360°$ for the direction of rotation of the cortex (Ancel & Vintemberger, 1948, 1949), which perfectly correlates with the later direction of the axis (at least when only one axis forms) (Gerhart et al., 1989). The relationship of the vegetal parallel array of microtubules to the direction of rotation is probably also precise. However, in anurans, which ordinarily permit entry of only a single sperm, there is about an $8°$ 'wobble' in the correlation between sperm entry and the subsequent embryonic axis (Gerhart et al., 1989). The entering sperm generates an aster, whose microtubules are 'independent' from those in the vegetal cortex, at least in the sense that isolated fragments of vegetal cortex form parallel arrays of microtubules whether or not a sperm aster is present (Houliston & Elinson, 1991a; Elinson & Palecek, 1993). This probably explains why urodeles, which permit multiple sperm entry, also successfully undergo cortical rotation (Chung & Malacinski, 1981). One would predict that the successful sperm, amongst those that enter the egg, is the one that grows the aster and that the position of this aster correlates, with some wobble, with the direction of the axis. On the other hand, in normal cortical rotation in the intact fertilized egg, the "Vegetal cortical array of parallel microtubules [incorporates]... aster microtubules, cytoplasmic microtubules and vegetal cortical microtubules" (Palecek & Ubbels, 1997).

The parallelization of the cortical microtubules could be by either or both of two mechanisms: polymerization or flow. The degree to which polymerization could cause alignment depends on the mechanical stiffness of the microtubules (reviewed in Brodland & Gordon, 1990). This has yet to be investigated. Flow induced alignment of microtubules is a likely candidate (Gerhart et al., 1989), since reorientation of the egg relative to gravity, prior to cleavage, reorients the axis:

"A frog's egg, in the two-cell stage, was compelled to maintain an inverted position, after having been previously compressed vertically to the egg's axis (O. Schultze, 1894a) or after pricking one blastomere (Morgan, 1895a). Each of the two blastomeres in the first case, or the surviving one in the second case, developed into a whole embryo.... Schultze's experiment was repeated by Wetzel (1895) and, after him, a number of times, most thoroughly by Schleip & Penners (1925, 1926). These experiments suggested that the stratified egg substances, being of different weight [density], began after inversion to flow in different ways, due to minor accidental influences; and because of their rearrangement double and multiple monsters resulted.... Upon the complete removal of one of the blastomeres, an internal rearrangement of structure takes place in the remaining one.... There follows a regulation of the 'half' to the 'whole,' so that a bilateral twin embryo results" (Spemann, 1938). [Cf. Chung & Malacinski, 1983; Neff et al., 1983, 1984a,b, 1993; Neff, Malacinski & Chung, 1985; Neff, Wakahara & Malacinski, 1990; Flint et al., 1989.]

We have demonstrated the density stratification of axolotl embryos prior to first cleavage by sectioning the hard boiled, fertilized egg perpendicular to its animal/vegetal axis. The sections retain their order in a sucrose density gradient (Russell W. Flint & Richard Gordon, unpublished results, 1989).

Abnormalities and anomalies of development caused by microgravity typically involve failure at or during gastrulation (Claassen & Spooner, 1994). It will be interesting to see if the same holds for the abnormalities of mouse embryos whose development arrests in space (Anon., 1997a; Schenker & Forkheim, 1998).

Of course polymerization, or even sliding, for that matter, once the cortex starts rotating, causes flow, which can feed back on either mechanism to cause further alignment. Flow alignment has a remarkable property: it provides a mechanism for interaction between polymeric molecules over a long distance compared to chemical forces. For this reason, we have coined the term 'supercooperative' phase transition to describe the impact of flow induced orientation on the generation of parallel arrays of microtubules (Gordon & Brodland, 1988). Supercooperativity permits a precision of 'crystallization' in a short time that is unattainable in ordinary crystallization, in which multiple domains normally occur. Large crystals, whether artificial or natural, are only formed by very slow crystal growth (Laudise, 1970). Since supercooperativity is unique to cortical rotation, flow alignment is probably an essential component to the alignment mechanism. (The active, 'dissipative' 1 mm domains of microtubules obtained by Tabony & Job, 1990, 1992a,b; Tabony, 1994, may be a related phenomenon.) This alignment of cortical microtubules probably also has an electrical component:

"In every case where correlations were possible, the primary axis of the organism came to lie in the plane of the greatest voltage drop from the vertex... of the animal pole. In other words, it was possible to predict from the voltage pattern where the head of the organism would develop. It is hard to escape the conclusion, therefore, that the electrical pattern is primary and in some measure at least determines the morphological pattern... the same pattern was to be seen in the fertilized but undivided egg. Moreover, it can be followed through the early stages of segmentation, the formation of the morula, the gastrula, and even shows the same characteristics in the early medullary plate stage. It seems clear from the above that, at least in the frog's egg (and in the egg of *Amblystoma,* for some preliminary studies in the latter confirm these findings), there exists an electrical field which appears prior to the appearance of the formed structures and also predicts the longitudinal axis of the future embryo" (Burr, 1941).

The liquid crystal nature (cf. Brown & Wolken, 1979; Khabibullaev, Gevorkyan & Lagunov, 1994) and polar orientation of the cortical microtubules, if not their global and dynamic organization, could be said to have been anticipated by Harrison (1945) (reviewed in: Witkowski, 1980a). It would be interesting to know if the polar alignment of microtubules at cortical rotation is somehow preserved by the cortical cytoskeleton, i.e., apical microtubules of later cell state splitters. This could solve the problem of exactly what it is that persists and preserves bilateral symmetry from first cleavage to the appearance of the blastopore (cf. Section 8.02). An alternative could be an ionic current through the grey crescent.

There are hints that cortical rotation or some similar gravitation dependent phenomenon operates at least in birds (cf. Gerhart & Kirschner, 1997):

"In birds, the blastodisc is initially radially equipotential. The anterior-posterior axis of the blastoderm is oriented with respect to gravity late during cleavage as cells accumulate at the lowest end of the blastocoele, which becomes the head. This occurs during passage of the egg down the oviduct so that experimental rotation of the egg in the oviduct results in duplication of the embryonic axis and the production of twins (Clavert, 1960a,b; Eyal-Giladi, 1969, 1984; [cf. Kochav & Eyal-Giladi, 1971; Eyal-Giladi & Fabian, 1980; Bellairs, 1989; Eyal-Giladi et al., 1994a]). In mammals, embryonic polarity is established as a result of asymmetry produced by implantation of the embryo in the uterine endometrium (Smith, 1985c)" (Jacobson, 1991b; cf. Clavert, 1962).

In summary, then, cortical rotation splits the state of an embryo from axisymmetric to bilaterally symmetric. The apparatus is cortical and involves at least a set of microtubules parallel to the surface of the fertilized egg, which are in the same location as later apical microtubules involved in the cell state splitters of ectoderm cells. It seems reasonable to regard this as the first cell state splitter in the embryo. Whether any nuclear state splitting occurs at first cleavage is unclear.

Proposition 53: the sex of an individual is determined by triggering of an expansion or a contraction differentiation wave, corresponding to female and male (or vice versa).

In mammals, the reproductive system starts out as a sexually 'neutral' tissue, the fetal gonad, a tissue competent to begin sexual differentiation into either a female or male. I would like to entertain the hypothesis that male and female correspond to an expansion or contraction wave traversing the whole fetal gonad (linear differentiation, Proposition 5). While a male 'Y chromosomal gonadogenesis gene' (Platt, 1990), perhaps the 'Y-linked gene Sry' (Lovell-Badge, 1993) may exist (cf. Eicher, 1988; Mittwoch, 1989), our species occasionally produces an XY female (Kingsbury, Frost & Cookson, 1987; Hipkin, 1993) or an XX male (Zakharia & Krauss, 1990). It is possible that these will be attributable to Y-chromosomal DNA translocations and deletions (Muller, 1987; Levine & Suzuki, 1993), but...

"It is concluded that the notion of a single testis-determining gene being responsible for male sex differentiation lacks biological validity.... The genes involved in testis development may function as growth regulators in the tissues in which they are active" (Mittwoch, 1992).

Thus, let us suppose that some 'accident' of development, such as a viral infection or fever of the mother, occurs after the tissue is competent, but before the 'appropriate' wave is launched somehow by zygotic gene products. Sexual development would then proceed apace independent of the so-called sex determining gene. Cases of 'mixed gonadal dysgenesis' (Cantrell et al., 1989) may be due to both expansion and contraction waves proceeding through complementary portions of the fetal gonad, rather than one 'sex-determining' wave traversing the whole of the tissue. The demonstration of embryonic induction by grafts (Rashedi et al., 1990; cf. Lesimple et al., 1989) and experimental chimeras (Burgoyne, Buehr & McLaren, 1988), and a genitourinary, pleiotropic gene of 'mesenchymal-epithelial transitions' (Pritchard-Jones ct al., 1990; cf. Marcantonio et al., 1994), may implicate a differentiation wave.

It is curious that in crocodiles (Smith & Joss, 1994a), geckos (Gutzke & Crews, 1988) and turtles (Wibbels, Bull & Crews, 1991) temperature, rather than the genome, 'determines' whether an egg develops as a male or female, with an abruptness akin to a phase transition:

"As in many other turtles, the sexual differentiation of gonads in embryos of *Emys orbicularis* is temperature-sensitive, 100% phenotypic males being obtained below 27.5 degrees C and 100% phenotypic females above 29.5 degrees C.... Both sexes differentiate at 28.5 degrees C, suggesting that at this intermediate (threshold) temperature, sexual differentiation of gonads conforms with sexual genotype" (Zaborski, Dorizzi & Pieau, 1988).

Within a critical temperature range, genetic determination of sex occurs, while above or below, males or females may be produced, depending on species (Dournon, Houillon & Pieau, 1990), and hormone dependent sex reversal can only be attained in a critical period (Desvages, Girondot & Pieau, 1993) (with the exception of placental mammals: Wolf, 1988). It appears that this temperature sensitive process can be biased by exogenously applied sex hormones (Gutzke & Chymiy, 1988; Gahr, Wibbels & Crews, 1992). These observations may give a clue to the unknown mechanism by which it is determined whether my hypothesized sex determining differentiation wave launches as an expansion or as a contraction wave. If so, it will be interesting to know if the type of wave launched is in strict correlation with the resulting sex, independent of species and temperature. However, it will take some effort to distinguish this initial step of sex determination from the multiple subsequent steps of sex differentiation. Many of these are hormone dependent (Voutilainen, 1992), are perhaps involved in cases of sexual ambiguity (New & Josso, 1988; Crews, Wibbels & Gutzke, 1989; de la Chapelle et al., 1990; Wilker et al., 1994) and may be the "alternative pathways for the development of the same final sex" in fish (Shapiro, 1992a).

Proposition 54: differentiation in tissue and organ culture proceeds via differentiation waves.

Tissue culture has long been taken as a model for the *in vivo* situation for cells, so much so that it is sometimes hard to remember that cells that are cultured for many generations frequently have gross chromosome abnormalities, which, at least in plants, correlate with loss of normal morphogenetic capabilities (Murashige & Nakano, 1967; Torrey, 1967a). Thus, along with my general predilection to avoid explant work (Section 9.22), goes a general discounting of the relevance of tissue culture studies to embryogenesis, and especially to differentiation.

Nevertheless, differentiation of some sort does occur in tissue culture (Earle & Demarly, 1982; Taub, 1985; Vasil & Thorpe, 1994), and some success has occurred in organ culture (Lipkin, 1995c), such as that of heart valves (Lipkin, 1995a) and skin (Arons, Wainwright & Jordon, 1992), which may involve differentiation. It would not be surprising if these phenomena, which involve cell interactions such as contact inhibition of movement (Gilbert, 1991a) or cell overlapping (Steinberg, 1973; Martz, Phillips & Steinberg, 1974), also involve physical differentiation waves. Changes in microtubule configurations are certainly involved (Nagasaki, Chapin & Gundersen, 1992; Nagasaki, Liao & Gundersen, 1994). Of course, the constraint of adhesion to an inelastic substrate, such as the bottom of a Petri dish, may severely damp the mechanical amplitude of tissue culture waves (cf. "...whether the spindle shortens... or not... [on UV-microbeam irradiation] depends both on the tension across the spindle and the extent to which the poles are anchored to the cytoplasm and/or substrate": Forer, Spurck & Pickett-Heaps, 1997). This could be alleviated by use of an elastic substrate (Harris, Wild & Stopak, 1980), and indeed, methods for measuring force versus time developed by a tissue culture are being worked on (Eastwood, McGrouther & Brown, 1994). Consider, for example, "that growth on ECM [extracellular matrix] relieves the need for MTs [microtubules] to serve as compressive supports for neurite tension" (Lamoureux et al., 1990). I would therefore predict the existence of differentiation waves in tissue culture, and appropriate time lapse studies should prove me right or wrong. A good place to start might be with the ontogeny of calcium waves in differentiated tissues, such as liver (Robb-Gaspers & Thomas, 1995). As differentiation waves, at least in our experience so far, proceed through two dimensional tissues, it will be interesting to see how three dimensional tissue cultures fare in this way (Hoffman, 1991, 1993; Li et al., 1993; cf. Armstrong & Armstrong, 1973).

4.06 The Relationship between Cells and Tissues in Mosaic Embryos

"Almost a century has passed since embryology started on its search for a solution of the problem of animal development through a jungle of bold theoretical adventures and daring technical enterprises. But embryologists of the present day are still lost in a valley between the peaks of regulation and mosaicism.... But however primitive the concept of mosaicism may be, it cannot be denied that the advantage of mosaicism lies in its concreteness" (Dan, 1960).

"Regulative and mosaic behavior are not mutually exclusive, and both phenomena are observed in most embryos, usually depending upon cell type or stage of development. Nevertheless, regulation is often possible even if maternal localization processes are disrupted or if cells have to follow an altered cell lineage pathway. This phenomenon has been difficult to explain mechanistically and is still poorly understood" (Raff, 1996).

Proposition 55: the fundamental difference between mosaic embryos and regulating embryos is that tissues in strictly mosaic embryos consist of one cell while tissues in regulating embryos consist of many cells.

"In determinate cleavage there is a point for point correspondence between cell-division and embryonic segregation [determination] in the early part of the life history..." (Lillie, 1929a).

Mosaicism is one way for organisms to avoid the need for tissue synchronizing mechanisms: each cell, at each step of differentiation, could be unique. In other words, in mosaic embryos, each cell is its own tissue. In regulating embryos, each contiguous tissue requires a means of coordinately moving a connected group of cells through the next step of differentiation (i.e., a differentiation wave). In mosaic organisms such coordination is not required, but I will nevertheless find it useful to introduce the concept of the 'single cell differentiation wave'. With this, the only fundamental difference between mosaic and regulating embryos is the number of cells per tissue (one versus many, respectively).

The correspondence or homology between tissues in regulating organisms and single cell tissues in mosaic organisms is implicit in the...

"... *concept* of a *stepwise determination*... proposed for vertebrate development (Hadorn, 1965). For example, the blastula ectoderm in amphibians still has the capacity of forming both epidermal and neural tissue. Following primary induction these capacities are segregated and the cells of the neural plate are determined for neural structures. In a second step the different types of neural and epidermal structures are determined. However, in vertebrates the evidence is largely based upon experiments involving tissues rather than cells, which makes their interpretation more difficult. In *Drosophila* where determination can be analyzed at the cellular level, the evidence is strongly in favour of stepwise determination" (Gehring, 1976a).

This proposition has recently been nearly arrived at by alternative reasoning:

"The modes of development of animals can be distinguished first on a descriptional level. Animals with so called 'determinate' development show an 'embryonic lineage', i.e. there is a clear correlation between descent and fate on the single cell level. By contrast in 'indeterminate' development no such correlation exists. It is usually implied that 'determinate' development reflects cell autonomous or 'mosaic' development while indeterminate development is often used synonymously with 'regulative' or non-autonomous development. *Drosophila* and amphibians are well studied paradigms for indeterminate development. Nematodes were considered to be a major paradigm for a determinate development. That determinate development only uses cell autonomous specification was challenged by the demonstration of inductions during early development in *C. elegans* (Priess, Schnabel & Schnabel, 1987; Priess & Thomson, 1987; Schierenberg, 1987; Schnabel, 1991; Wood, 1991; Bowerman et al., 1992; Goldstein, 1992; [Sternberg, Hill & Chamberlin, 1993]) which showed that these two modes of development are not mutually exclusive. The results presented here show that the establishment of a large part of the body plan of *C. elegans* depends on a hierarchy of inductions. Thus it appears that the major difference between the determinate and indeterminate modes of development is the number of cells involved in embryogenesis" (Hutter & Schnabel, 1994).

This conclusion may be an exaggeration, even if it supports my argument: while some inductions occur, other portions of the nematode lineage tree appear 'innate', i.e., mosaic, because they unfold in ectopic positions (Moskowitz, Gendreau & Rothman, 1994). One component of the reproducibility of the spatial locations of cells in nematodes may be their limited cell migration (Sternberg, 1991). On the other hand, a comparative study of 'free-living soil nematodes' shows some cell migrations (Skiba & Schierenberg, 1992).

There is an interesting parallel here between nematode cells and plant cells, that might have wider ramifications. For instance, Schmalhausen (1949) has suggested an evolutionary reason for the distinction between mosaic and regulating organisms, that could use justification in developmental mechanisms:

"Environmental variations may be periodic or seasonal so that seasonal polymorphism is produced in animals and plants. Under certain conditions, the existing lability or individual adaptability of the organization has a decisive, vital importance for the organism. *Stable* species, including animals with an autonomous morphogenesis, are characterized by a much greater individual stability. Particularly stable are individuals with mosaic development (nematodes, annelids, arthropods, and the higher mollusks)" (Schmalhausen, 1949).

Nevertheless, it is clear that many plants are capable of regulative development, with many cells per tissue, which they either start with or build up to, at least in tissue culture:

"...The tendency of a tissue placed in culture to form an organ or an embryo is apparently species dependent.... Surrounding cells are often induced to divide and become incorporated in the new structure, as has been shown, for example, in *Torenia* (Chlyah, 1974) and radiata pine (Smith & Thorpe, 1975), and through the use of plant chimeras (Norris, Smith & Vaughn, 1983). In somatic embryogenesis, it appears that single versus multiple-cell proliferation represent varying expressions of the same underlying morphogenetic phenomenon (Williams & Maheswaran, 1986).... Nomura & Komamine (1985) were able to establish a high frequency and synchronous somatic embryogenic system in carrot [cf. Steward, 1968; Osuga & Komamine, 1994] by selecting competent single cells. These cells were small, round and cytoplasm rich, and went on to produce embryogenic cell clusters..." (Thorpe, 1994).

As an example of plant regulation *in situ,* here is the description of what is, in effect, the differentiation tree of the apical meristem in bryophytes, by Bopp (1984):

"...At each of the first four division steps, a decision is made between two courses of differentiation, maybe by turning on or off the specific sets of genes responsible for the further differentiation of a cell or of all its descendants, thus determining the pattern along the shoot axis. Regarding this scheme, one has to bear in mind that each step is accompanied by several divisions transverse to the longitudinal axis, which enlarge the surface of all organs but do not change the future destiny of the cells" (Bopp, 1984).

It is these cell divisions between the nodes of the tree that distinguish a differentiation tree from a lineage tree. This is fundamentally what characterizes a regulating organism. For a strictly mosaic organism, the lineage tree is its differentiation tree, and each tissue consists of one cell. This is essentially the conclusion reached by Kaletta, Schnabel & Schnabel (1997), who seem stuck on the differences rather than similarities between nematodes and other organisms:

"We propose that in *C. elegans,* most of the lineage is built by a stepwise binary diversification of blastomere identities. This general strategy for cell specification in early embryogenesis appears to be very different from that in *Drosophila,* where regional identities are established using a multitude of gradients (St Johnston & Nüsslein-Volhard, 1992; Rivera-Pomar & Jäckle, 1996)" (Kaletta, Schnabel & Schnabel, 1997).

On the contrary, if we view the single blastomeres as tissues unto themselves, nematode and fruit fly development can be seen from the same perspective.

Proposition 56: a cell state splitter is set up at every step of cell division in a strictly mosaic embryo, so that its cell lineage is identical to its differentiation tree.

The parallel rings of microfilaments and microtubules in cells of developing nematodes (Priess & Hirsh, 1986), which may be in mechanical opposition (Figure 17D in Appendix I: Gordon & Brodland, 1987), may be evidence for this proposal. If so, then the lineage trees of cells that have been worked out for developing nematodes (Sulston et al., 1983; Wood, 1988a) represent the first fully known differentiation trees (see the comparison between nematode 'equivalence groups' and *Drosophila* compartments and polyclones made by Kimble, 1981). Comparisons of cell lineages between species, mutants, and experimentally manipulated individuals (Ambros & Fixsen, 1987) may, therefore, be direct studies of the evolutionary history of differentiation trees. The role of microtubules in positioning of the nucleus before an unequal cell division (Lutz & Inoué, 1982) may reflect cell state splitters in mosaic organisms. The ability of ectopic lineages to duplicate the spatial relationships of cells in nematodes (Moskowitz, Gendreau & Rothman, 1994) speaks to a geometric autonomy within at least some subtrees of the nematode that is hard to imagine being other than cytoskeletally based. We have yet to address this spatial component.

Of course, if mosaic embryos undergo differentiation of their cells with each cell division, then we must explain the coexistence of a spindle apparatus and a cell state splitter in the same cell. This may not be as difficult to achieve as the coexistence of cilia and a spindle (Rieder, Jensen & Jensen, 1979; Masuda & Sato, 1984b; Buss, 1987; though flagellates have solved this problem: Patterson & Hedley, 1992), since unlike the cilia, the cell state splitter may not require a microtubule organizing centre (MTOC). On the other hand, if cell division and the erection of cell state splitters are in strict temporal alternation in the cells of mosaic embryos, then they may both use the same (or replicated) MTOCs. Unicellular eukaryotes (protists) which change cell shape suddenly, followed by new protein synthesis as they enter a new phase of the life cycle, may use structures analogous to cell state splitters (Glen Klassen, p.c.; cf. Barr, 1981; Barr & Allan, 1985), and may represent evolutionary precursors of cell state splitters. Heterovalvy in diatoms, the ability to switch phenotype in one cell division (Stoermer, 1967; Stoermer, Kingston & Sicko-Goad, 1979; Håkansson & Stoermer, 1984), may have such a basis. The same may be speculated about the apparatus that causes cortical rotation and gray crescent formation prior to first cleavage in amphibian embryos (Houliston & Elinson, 1991a,b). This possible evolutionary continuity between the cytoskeleton of single celled organisms and cell state splitters may underlie the ability of some cleavage arrested, multinucleate cells, such as *Chaetopterus,* to exhibit a so-called single cell 'multiple differentiation' (Wilson, 1925; Crowther & Whittaker, 1986; Jeffery, 1985a; Wolpert, 1988). Alternatively, we may consider this capacity to be a contradiction of the idea that cell state splitters are the basis for differentiation. It may be critical to undertake a detailed ultrastructural and mechanical analysis of this phenomenon and other cases of so-called 'organismal morphogenesis' (Kaplan & Hagemann, 1991).

The facts:

that animals with mosaic embryos have regulating adult tissues (Grant, 1978; Maclean & Hall, 1987), or even embryonic tissues (Minden et al., 1989; Damen & Dictus, 1994),

that "legs of insect-larvae are easily restored, but after pupation no further growth or regeneration takes place (Krízenecky, 1914, 1917)" (Thompson, 1942),

that animals with mostly regulating embryos have some mosaic stages (Kumé & Dan, 1968a), that the ascidian ancestors of highly regulating vertebrates are mosaic (Jeffery, 1985a; John & Miklos, 1988), though not completely so (Rose, 1939; Reverberi, Ortolani & Farinella-Ferruzza, 1960; Nishida & Satoh, 1989; Jeffery, 1992a),

that 'primitive' insects can fully regulate, while 'advanced' ones cannot (Akam & Dawes, 1992), and

that mosaic embryos are not always strictly so (Grant, 1978; Stent & Weisblat, 1985; Maine & Kimble, 1990), or have some tissues with apparently more than one cell and adjust that cell number to embryo size (Gerhart et al., 1982),

indicate that we are dealing with more of a continuum: "...The distinction between regulative and mosaic embryos has lost its significance (Melton, 1991)" (Evsikov, Morozova & Solomko, 1994). Yet this persistent and exasperating problem will not go away so long as we lack a theory, such as that of differentiation waves, that sees regulating and mosaic embryos as variants of one and the same 'strategy' for embryonic development, not two:

"It remains unclear why today some organisms like nematodes rely quite heavily on highly determinate patterns of cell cleavage which are often coupled to asymmetric cell divisions, while others, like vertebrates and insects, rely almost exclusively on cell interactions. Why should these two different strategies have evolved? ...the puzzle is compounded by the fact that organisms that are specified by determinate cell lineages... can also reproduce asexually and have considerable powers of regeneration. Such animals include Cnidaria, Platyhelminthes, and polychaete Annelida (Barnes, Calow & Olive, 1988; [Barnes et al., 1993]).... It seems that cell interactions represent the primitive condition and that the requirement for asymmetric cell divisions for specification was secondary. Both processes would set up a body plan in which interactions were dominant, and it is very difficult to see how cell interactions could evolve from asymmetric cell divisions" (Wolpert, 1994b).

As we shall see (Proposition 68), the view of which came first comes down on the side of mosaicism, when the problem is seen and perhaps resolved from the point of view of differentiation waves, and the notion that in mosaic organisms, each cell is a separate tissue.

The strictly mosaic organism may be a mere mental construct:

"Even if 'mosaic' is restricted to animals with 'invariant' cell lineages, I still think that the real situation is much more complex than this rather simplistic characterization, even in *C. elegans*. First, there are cells that stochastically adopt cell fates (e.g., which cell becomes the anchor cell or the ventral uterine precursor cell... is random, and depends on lateral signaling via *lin-12*, the *C. elegans Notch* homolog). This variability also occurs for how the left and right 'Pn' cells will interdigitate along the ventral midline: sometimes the left cell will be anterior, sometimes the right. Secondly, the 'invariance' may also result from the observed population being isogenic. Certainly non-isogenic strains of other [nematode] species show a surprising amount of cell lineage variability (surprising to *C. elegans* geneticists, that is) (e.g., see Sommer & Sternberg, 1995; [cf. Fitch & Emmons, 1995]).... I think the one-cell-one-tissue notion does not entirely apply. There are many examples where cells from very different parts of the cell lineage contribute to the same tissue [but are they identical in terms of activated differentiation cascades?]. For example, the 'hypodermal' cells that form the 'hyp7' hypodermis are derived from very different lineages [Podbilewicz & White, 1994]. In fact, these cells fuse together to form a syncytium that covers nearly the entire animal! Also, you refer to bilateral pairs of cells as a single tissue, which I would agree with" (David H.A. Fitch, p.c., 1995).

Nevertheless, despite the numerous exceptions, a nematode can be said to consist mostly of one-cell tissues: 807 asymmetric out of the 949 non-gonadal cell divisions (Horvitz & Herskowitz, 1992). Evolutionarily, this may have been the starting point for tissues that have multitudes of cells.

Proposition 57: asymmetric cell divisions in mosaic organisms may correspond to a pair of contraction and expansion differentiation 'waves' that propagate across complementary portions of the dividing cell.

"The origin of the putative asymmetry signal is completely unknown" (Rhyu & Knoblich, 1995).

This prediction follows directly from the previous proposition. Asymmetric divisions tend to be defined morphologically, as in stem cells (Wolpert, 1988) and *Drosophila* neuroblasts (Goodman & Doe, 1993). Testing of the proposed homology, presumably of large cell to expansion wave, and small cell to contraction wave, will have to await the time when we can test homologies of single cell tissue differentiation pathways to those in multicellular organisms. In light of our Proposition 46 that which kind of differentiation wave is launched depends on mechanical tension, it is worth considering the possibility of mechanical effects in the reversal of which cell becomes which in *lin-44* mutants of the nematode *Caenorhabditis elegans,* hinted at in this description:

"Of the 949 non-gonadal cell divisions that occur during the development of the *C. elegans* hermaphrodite, 807 are asymmetric, generating sister cells that differ in their fates (Horvitz & Herskowitz, 1992). We have discovered a mutation that does not disrupt the generation of asymmetry in this cell lineage but causes the orientation of some of these cell divisions to be reversed with respect to the body axis of the animal.... Reversals in cell polarity caused by *lin-44* mutations probably occur at the stage of the mother cell rather than after the two sisters have been generated.... The suggestion that polarity reversals might function in the evolution of nematode cell lineage was made previously... (Sternberg & Horvitz, 1981, 1982).... How does *lin-44* affect the B, F, U, T, and P12 cell lineages? These cells are all located in the tail and far posterior body region. One possibility is that the defects in these cell lineages are secondary consequences of defects in another cell or cells. For example, the polarity reversals that occurred in the F and U lineages of *lin-44* animals were identical to those previously seen by Chisholm & Hodgkin (1989) in wild-type animals in which the B cell was killed with a laser microbeam.... Alternatively, *lin-44* might act more directly to specify cell polarities. Polarity may be determined by a polarizing signal from a neighboring cell. *lin-44* mutations might disrupt this signal.... A third possibility is that *lin-44* encodes a protein that acts within polar cells to specify their orientations" (Herman & Horvitz, 1994). [Cf. Herman et al. 1995.]

While "*lin-44* encodes the *C. elegans* homolog of *wingless,* and is thus most likely a cell-cell signal" (David H.A Fitch, p.c., 1995), its 'target' could be the cytoskeleton/membrane complex.

Indeed, mere micromanipulation at the 6 cell stage can completely, and harmlessly, reverse the left-right asymmetry of nematodes (Wood, 1991):

"The conclusion that Abal-pl and ABar-pr cell pairs are equivalent relates to an old controversy (see Ludwig, 1932, for review) between, on the one hand zur Strassen [1896] and his students, for example Dunschen (1929), and, on the other, Schleip (1929) and his students, for example Bonfig (1925). The zur Strassen school maintained that *Ascaris* embryonic development was strictly mosaic in character and that asymmetry and *l-r* [left-right] differences, like all other differences in cell fates, must be dictated by determinants present in the zygote. Schleip (1929) defended the alternative possibility that the early AB cells are equivalent, becoming different only later as a result of cellular interactions. Bonfig (1925) had reported, for example, that the entire quadrant of AB cells at the 6-cell stage could occasionally spontaneously rotate 90° (reversing handedness) or even 180° (not reversing handedness) without impairing viability of the embryo. The results presented here add to previous evidence from Priess & Thomson (1987) in clearly supporting the view of Schleip and Bonfig... [which was] not widely accepted at the time" (Wood & Kershaw, 1991).

Perhaps the following is a single cell differentiation wave homolog of the ectoderm contraction wave in axolotls:

"Each AB granddaughter divides to produce one cell that has the potential to make abundant epidermis and one that instead produces primarily nervous system" (Gendreau et al., 1994).

The mechanics of asymmetric cell division, which we have seen both in stem cells and in mosaic organisms, is an opportunity for fascinating exploration. Current models for the cytoskeleton as a tensegrity system (Ingber et al., 1994) do not consider the possibility of nonlinearities that cause 'snapping' between alternate stable states, such as we have postulated for cell state splitters. It is easy to construct small structures with such a property, using the construction toy Tensegritoy™ (Tensegrity Systems, Tivoli, New York), by analogy with small molecules that snap between different conformations, such as ammonium ion and sucrose (Geoffrey Hunter, p.c., 1995), so one would anticipate that much larger snapping conformations should be constructible (cf. Brodland & Cohen, 1987). Asymmetric interactions with the cell membrane may be a key component of such processes (Oliver & Berlin, 1982; Simons et al., 1992). Configuration snapping may also use cytoskeleton binding molecules that shift the overall organization from one mechanical state to another (Ridley & Hall, 1992), perhaps stabilizing alternative tensegrity configurations. The asymmetric cell divisions in nematodes involve both cortical and cytoplasmic flows (Hird & White, 1993), calcium ion redistribution (Centonze & White, 1995), and the action of the cytoskeleton (Strome & Wood, 1983; Strome & Hill, 1988; Strome, 1989):

"Our experiments were designed to explore three different ways that microfilaments may be involved in establishing asymmetries: (1) Proper microfilament structure may be required before any of the events themselves take place to establish the proper cellular 'context' for the events. (2) Proper microfilament structure may be required only during the execution of events. (3) Proper microfilament structure may be required after the events have occurred in order to maintain the asymmetries that have been generated. We have tested these three possibilities by disrupting microfilament structure before, during, and after zygotic events and analyzing subsequent development of the one-cell [nematode] embryo. The results of our experiments show that proper microfilament structure is required during pseudocleavage, pronuclear migration, and germ-granule segregation in order for these events to occur normally ((2) above). Disruption of microfilaments before or after the actual occurrence of these events has no effect on them. Proper microfilament structure is required prior to the generation of an asymmetric spindle ((1) above). Microfilaments do not appear to be necessary to maintain asymmetries that have already been established" (Hill & Strome, 1988).

In plant cells (cf. volvox: Kirk et al., 1993), asymmetric actions of the cytoskeleton have been implicated in cell differentiation/divisions and cell polarity, without explanation of how this could come about:

"Formation of the stomatal guard cells occurs late in leaf development.... The division preceding the formation of the stomatal complex is asymmetrical. These formative divisions are said to be unequal because the mitotic spindle is not centered in the cell....

"When highly polarized, mitotically quiescent parenchyma cells, such as stem or root cortical tissue, are wounded, some of the cells in the vicinity of the wound divide, enlarge, and differentiate. Many of the cells stimulated by wounding develop a new axis of polarity at right angles to their original polar axis. The first indication of this altered polar axis is a reorientation of the cortical microtubules and microfilaments which begins shortly after wounding, before the cells begin to divide....

"Cells that have developed axial polarity also may have an asymmetrical distribution of cytoplasmic organelles and/or other cellular components.... One of the functions of the cytoskeleton is to localize or distribute cytoplasmic structures. This has been demonstrated most clearly in the gravity-perceiving cells in the root cap known as **statocytes**....

"The position of the nucleus also is determined by the cytoskeleton.... Nuclear and organelle positioning is particularly important during unequal cell divisions... [before which] the sorting and separating of cell components... such as mRNAs... are brought about by the cytoskeleton" (Fosket, 1994).

These asymmetric cytoskeletal actions might be worth considering as analogous to miniature 'differentiation waves' due to tensegrity instabilities. It is interesting to note that asymmetric cell divisions also occur during pattern formation in prokaryotic blue-green algae or 'cyanobacteria' (Mitchison & Wilcox, 1972). Although they have generally been presumed not to have a cytoskeleton, which might seem an essential component of differentiation waves, it was suspected that they do (Jensen & Ayala, 1976; Jensen, 1994), and prokaryotes have now been shown to divide both symmetrically and asymmetrically using a tubulin homolog polymer (Erickson, 1997).

Hart & Trainor (1990) have shown that an isotropic continuum model with microfilaments and microtubules exhibits instabilities. Cf. Trainor & Goodwin (1986). This could present a starting point for a tensegrity model whose instabilities result in an asymmetric cell division. Propagation of the instabilities across the mother ⇒ large daughter + small daughter cells could represent the differentiation waves.

In summary, the snapping of the cytoskeleton in a single cell is postulated to be the basis for asymmetric division. The cell may be thought of as having one cell state splitter which produces a contraction wave over one portion of the cell and an expansion wave over the other. Those two parts become the larger and smaller daughter cells, respectively. The result is two new kinds of cells, just as differentiation waves in regulating embryos give rise to two new kinds of tissues (Figures 22-23). In some sense, the two asymmetric daughter cells may be considered to carry over their surfaces, as they separate, two 'single cell differentiation waves'. One may be a contraction wave due to a travelling microfilament ring. The other could be an expansion wave corresponding, for example, to sequential impingement of microtubules in an aster on the cell membrane (Steve McGrew, p.c., 1997). These are testable predictions. If you'd like to play with an analogy, take a partially inflated balloon and squeeze it in the middle: you get two equal 'cells'. If you squeeze it off center, however, one part inflates (expands) and the other deflates (contracts). Bubbles will do the same (Isenberg, 1978). Another analogy is Ostwald ripening, in which large crystals grow at the expense of small ones (Sugimoto, 1978; Voorhees, 1992; Chen & Voorhees, 1993; Gordon & Drum, 1994). Not all these phenomena are interpretable in terms of waves, but they do indicate that there is a wide range of instability phenomena that resolve either into two equal portions or asymmetrically into two unequal portions. There is even a sense in which symmetric cell division seems to be the evolutionarily derived case. If we liken asymmetric division to a greased pig, it will squeeze out from one's hands one way or the other, but not likely both ways at once.

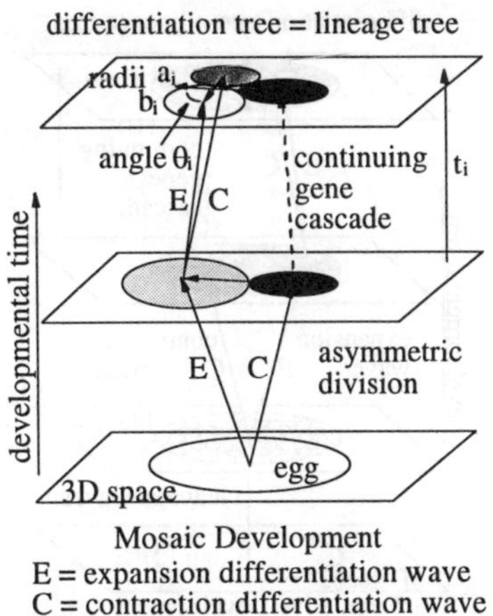

Mosaic Development
E = expansion differentiation wave
C = contraction differentiation wave

Figure 22. Four dimensional geometry of the development of a mosaic organism, such as a nematode. The waves are single cell differentiation waves.

Proposition 58: so-called determinants are positioned by the cytoskeleton during asymmetric cell divisions by single cell differentiation waves.

The fact that microfilaments have only a transient role in an asymmetric cell division in nematodes (Hill & Strome, 1988, 1990) and in the establishment of polarity in fucoid algae eggs (Quatrano, 1973; Brawley & Quatrano, 1979; Kropf, 1992) is consistent with the idea that some kind of cell state splitter is operating here (cf. proposed corresponding asymmetric distribution of calcium ions, Jaffe, 1986, and calcium ion stimulation of actin aided transport of Golgi vesicles bearing wall softening enzymes in fucoid eggs: Brawley & Robinson, 1985; Nuccitelli, 1988a). The notion of the cytoskeleton acting as a cell state splitter may be better than presuming that an 'asymmetry determinant' is shuffled around when microfilaments are disrupted (Hill & Strome, 1990). Since the 'determinant' must be membrane bound (Laufer & von Ehrenstein, 1981), it would move with the membrane, which is exactly where the microfilaments are bound. In fact, a role for cytoskeletal proteins in determining asymmetry has been suggested:

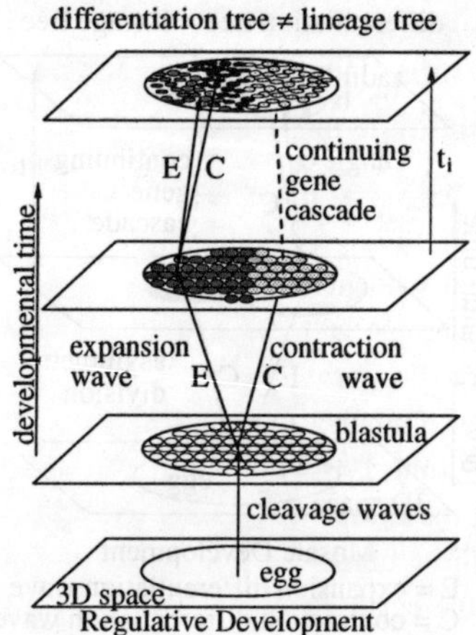

Figure 23. Four dimensional geometry of the development of a regulating organism, such as a vertebrate. The waves are multicellular differentiation waves.

"She1p/Myo4p [a myosin or 'minimyosin'], with the help of other She proteins, might generate Ash1p asymmetry [in yeast] by moving substances from mothers into their buds" (Bobola et al., 1996; cf. Sil & Herskowitz, 1996). "What was 'rather surprising,' [Ralf-Peter] Jansen says, was that... none [of the SHE = *Swi5-dependent HO Expression* genes] encoded transcription factors or other proteins that might affect Swi5p directly. Instead, one of the genes, *SHE1*, encoded a cytoplasmic 'myosin' protein called Myo4" (Roush, 1996a).

Without hinting where 'cytoskeletal asymmetry' itself might come from, Amon (1996) concludes:

"Although the origin of this [yeast] asymmetry is not understood in detail, actin-dependent transport may be responsible. In the *Drosophila* oocyte, polarized microtubules and the actin cytoskeleton are thought to provide a structural basis for localizing cell fate determinants (reviewed by Lehmann, 1995; Erdélyi et al., 1995; [Lin & Spradling, 1995]). Perhaps the polarized actin cytoskeleton plays a similar role in facilitating localization of Ash1p. Utilizing cytoskeletal asymmetry to generate developmental asymmetry may prove to be a common mechanism by which a cell generates two daughter cells of different fates" (Amon, 1996).

Visible granules and other so-called cytoplasmic determinants, including calcium (Centonze & White, 1995), are probably thoroughly enmeshed in and moved by a microfilament network attached to the cell membrane (cf. Longo, 1988):

"...The weight of experimental evidence at present available favors the concept that the random motion of the echinochrome granules of the unfertilized *Arbacia* [sea urchin] egg may be associated with contractile fibrils in the cytoplasm. The data strongly suggest that each echinochrome granule has a number of such contractile fibrils attached at many points on its spherical surface.... Within 1 to 2 min of fertilization... echinochrome granules... quickly travel from deeper in the cytoplasm directly to... rest in the outermost cytoplasm of the egg.... The other ends of these fibrils may be attached... to the extensive villar structures of the plasma membrane of the egg... [which] seems justified from the rapid manner with which an echinochrome granule 20-30 μ deep in the cytoplasm is 'pulled' to the subsurface layer [the cortex],... the upper 2 μ of the cytoplasm" (Parpart, 1964).

Analogously, the cytoskeleton has been invoked as a carrier of neurotransmitters purported to "regulate morphogenesis of early embryos":

"Serotonin and norepinephrine are synthesized and released by yolk granules following fertilization of sea urchins and chicks, and are present in blastulae of these animals as well as rodents (Buznikov, [1990], 1991; Emanuelsson, Carlberg & Löwkvist, 1988; Burden & Lawrence Jr., 1973; Ignarro & Shideman, 1968).... It has been proposed that intracellular receptors mediate [regulation of cleavage]... and may be associated with cytoskeletal elements located in the cleavage furrow (Buznikov, 1991; Renaud et al., 1983). This is supported by the finding that 5-HT binds to microfilaments and microtubules in the neural tube of the chick embryo (Emanuelsson, Carlberg & Löwkvist, 1988).... It has been suggested that 5-HT, located in the region of the blastopore [of sea urchins], initiates gastrulation, whereas acetylcholine released from ectoderm controls invagination of the archenteron (Gustafson & Toneby, 1970).... Norepinephrine and 5-HT are synthesized by the notochord in the chick and frog (Lauder, 1988; Wallace, 1982; Strudel, Recasens & Mandel, 1977; Godin, 1986). During neurulation, these transmitters are transiently taken up into the floorplate of the neural tube and the neural folds where they are accumulated in non-overlapping regions, which cover most of the neuraxis (Lauder, 1988; Wallace, 1982)" (Lauder, 1993).

The role of microfilament rings and microtubules in shaping some early mosaic eggs is becoming apparent (leech: Astrow, Holton & Weisblat,

1989). It may be well worthwhile to see if these are in mechanical oppositions indicative of cell state splitters at each step of differentiation.

The localization of so-called determinants could be analogous to "...the translocation of signal-transducing molecules... to the actin-based cytoskeleton during platelet activation" (Ozawa et al., 1995). The origin of cytoplasmic localization during asymmetric cell divisions may go back in evolutionary time to the bacteria (Milhausen & Agabian, 1983). Raff (1996) has summarized work on ascidians, which also implicates the cytoskeleton:

"When ascidian eggs are fertilized, they exhibit an amazing behavior. Several colored cytoplasmic domains sort themselves out and move to defined locations in the egg.... These colored particles fortuitously mark the localization of determinants of autonomous cell differentiation, and map the sites of origin of particular cell types....

"The region from which primary muscle cells arise is marked by the cytoplasmic domain of the myoplasm. Experiments by Whittaker (1980, 1982) and by Deno & Satoh (1984) that involved moving myoplasm into neighboring cells showed that the maternally loaded myoplasm determines muscle cell fate. The myoplasm was shown by Jeffery & Meier (1984) to consist in part of a microfilament-like cytoskeleton. Swalla, Badgett & Jeffery (1991) have found that a particular microfilament protein is present in the eggs of both tailed and tailless species, but localized to the myoplasm only in tailed species" (Raff, 1996).

The localization of 'determinants' and other molecules to portions of the cytoskeleton may be part of the more general...

"...concept of *microcompartmentation* (Bereiter-Hahn, 1988; Clegg, 1984a, 1988; Friedrich, 1985, 1991; Jones & Aw, 1988) which means that the mobility of enzymes and metabolites is restricted, thus giving rise to locally different metabolic activities without a membrane limiting the reaction space.... Therefore the patterns of cytoskeletal elements changing during morphogenesis or by external influences may manifest themselves in changes of mechanical properties and of cellular metabolism as well" (Bereiter-Hahn, 1994).

Unfortunately, microcompartmentation specifically to cortex is not discussed in Jones (1988). Nor has it been investigated as a potential source of genetic control.

Microtubules, the other component of a cell state splitter that we anticipate, may not be involved in the segregation of cytoplasmic determinants (Strome & Wood, 1983; Kropf, 1992), though in axons, the microtubule associated protein MAP2 is 'compartmentalized' to dendrites and the cell body (Hirokawa, 1991). Thus I note the observation of Wilson (1894), suggesting a primary role for microtubules or their attachment proteins in asymmetric cell division:

"In *Nereis*,... a mosaic-like character appears from the beginning, because of inequality of the first cleavage, which conditions the entire subsequent development through the peculiar relations established by it.... A study of the first cleavage-spindle shows that the inequality is unmistakably foreshadowed before the least outward sign of division appears; for the asters at the spindle-poles are conspicuously unequal in size, the larger aster [a microtubule based expansion wave?] corresponding with the future larger cell" (Wilson, 1894).

Thus we can conclude that at least actin and some of its associated proteins have a role in determining how an asymmetric cell division takes place. Since the cytoskeleton is then asymmetric, I would like to tentatively suggest that this phenomenon is equivalent to the cells having a single cell differentiation wave:

"In our model microtubules are captured by a specialized structure at the cortex, and stabilized against rapid depolymerization.... If the sites of attachment to the cortex are located asymmetrically, microtubule depolymerization will induce a torque and cause the rotation toward the attachment site.... At present we know little about regional variation in the cortex that could generate a local site of microtubule association. The results with cytochalasin D suggest a possible role for actin" (Hyman & White, 1987).

Proposition 59: the spatial separation of so-called 'cytoplasmic determinants' of mosaic and regulating eggs may be caused by cell state splitters.

As we have seen in Proposition 58, some so-called cytoplasmic determinants tag along with the cytoskeleton. If mosaic embryos are to be subsumed under the umbrella of organisms using cell state splitters to accomplish differentiation, then we must come to terms with this notion of 'cytoplasmic determinants'. This idea is an old one, here well expressed in the language of the beginnings of biochemistry:

"The really fundamental problems to be determined... are, first, whether the fertilized ovum is to be regarded as a composite structure made up of various system- or organ-anlagen, or the chemical progenitors or 'ferments' of such, distributed in definite and, perhaps, constant positions throughout the cytoplasm. Or, second, is the ovum to be considered in the sense of a unicellular organism, differing in no great respect from the physiological structure of unicellular organisms in general but possessing that specific potential to elaborate 'ferments' and pro-anlagen at successive genetic stages, and, ultimately the anlagen of the later developmental stages? In the former instance it is presumed that the embryonic parts are pre-localized in the cytoplasm of the ovum and make their appearance, in the words of Lankester [1877], as 'a sequel of a differentiation already established and not visible.' In the second assumption the embryonic parts are unrepresented in the ovum, the regions of the cytoplasm being then, so far as the future embryo is concerned, of equipotential value" (Baldwin, 1915).

I have already critiqued cytoplasmic determinants in terms of predetermination and the homunculus (Section 2.03). But let's look at the claims for them in some detail. In general, the localization of identifiable molecules in the cortex of mosaic eggs (and others: Jeffery, 1985a, 1989a) is to areas of the egg that are too large to correspond to single cells at later stages (Davidson, 1986). A proper homunculus should be more detailed.

A search has been ongoing for 'determinants' in regulating embryos, in the form of localized messenger RNA or 'transcripts':

"A number of laboratories have attempted to detect localized RNAs and proteins in oocytes and eggs of the amphibian *Xenopus laevis* with the goal of identifying molecules that are important in cell type determination or in regulating specific cellular processes (Moen & Namenwirth, 1977; King & Barklis, 1985; Rebagliati et al., 1985; Dreyer, Scholz & Hausen, 1982; Miller et al., 1989; Kloc et al., 1989). Transcripts [were] localized to either animal or vegetal hemispheres... (Weeks & Melton, 1987a,b; King & Barklis, 1985; Kloc et al., 1989, 1991; Reddy, Kloc & Etkin, 1991a; Miller et al., 1991)" (Reddy, Kloc & Etkin, 1992).

Again, the localization is to rather large areas, i.e., here to whole hemispheres. Some of the localized transcripts have no obvious importance to development:

"...It is clear that the mRNA component for xlgv7 in stage 6 oocytes is redundant and is not required for further development. It will be interesting to examine the situation for other maternally stored gene products" (Kloc et al., 1989).

Despite this clear expression of doubt of a role for any of these maternal 'determinants', the same authors later claim that...

"...it is likely that such localized maternal gene products play important roles in the specification of cell type or in a cell-specific function. Therefore analysis of these genes will aid in our understanding of the strategy used by the embryo to regulate early developmental events" (Reddy, Kloc & Etkin, 1992; cf. similarly Kloc et al., 1991).

Therefore we must be careful when a claim is made that something is a 'determinant'. It need not be so, just because we believe it has to be.

Evsikov, Morozova & Solomko (1994), using "A new micromanipulation technique [that] permitted the scrambling of the zygote cytoplasm", suggest that they have laid to rest the notion of cytoplasmic determinants in mammals. At early stages, cells can be rearranged at will, and even their contents can be seemingly thoroughly mixed. Embryogenesis nevertheless continues normally:

"Observations of the zygote cytoplasm uniformity are supported by apparently normal development of single 2-8-cell stage blastomeres (Moore, Adams & Rowson, 1968; review by Fehilly & Willadsen, 1986), chimeric embryos (reviews by Gardner & Rossant, 1976; McLaren, 1976a) and by embryos that had undergone centrifugation (used for pronuclear visualization when producing transgenic farm animals, also: Mulnard, 1970; Tellez et al., 1988), or cytoplasm removal or its addition (Petzoldt & Muggleton-Harris, 1987; Evsikov, Morozova & Solomko, 1990). Contradicting the data obtained for other animal groups, these results demonstrate that blastomere recombinations or cytoplasmic perturbations have no effect on embryo development...."

"A needle with a piezocrystal pressed to its side was used for cytoplasm scrambling.... The needle was inserted into the zygote, the voltage on the piezoelement was applied, and the egg cytoplasm was completely scrambled in 10-20sec. Cytoskeletal inhibitors effectively prevented puncturing of the plasma membrane or the nuclear membranes of pronuclei. The scrambling was assumed to be complete when cytoplasmic granules and pronuclei had rotated a few times around the cytoplasm and had changed their position according to the egg centre...."

"Normal preimplantation development of such zygotes proves that the first steps of cell differentiation do not involve any factors of autonomous cell specification...."

"The necessity for preformed regulatory factors, which permit the high speed of initial stages of development (Davidson, 1989, 1990) disappears..." (Evsikov, Morozova & Solomko, 1994).

Thus we are left with the prospects that cytoplasmic determinants do not exist, or that they can be resegregated after such thorough mixing. Of course, (re-) attachment to specific cytoskeletal components of an asymmetrically acting cell state splitter allows one to have it both ways, though then it is the cell state splitter that is determining the spatial arrangement, not the 'determinants'. Alternatively, we may think of the cytoskeleton as the 'determinant'. If "Microtubules are responsible for chromosome segregation and the movement and reorganization of membranous organelles" (Vallee & Sheetz, 1996), why shouldn't the resolution of the instability of a cell state splitter containing microtubules and actin be able to selectively bind and segregate the so-called determinants to the expanding/contracting cytoskeletal regions?

Proposition 60: cytoplasmic determinants may not exist.

Before the discovery of the cytoskeleton, Lillie (1929a) had already reached this conclusion. In the following one need only substitute 'cortical cytoskeleton' for 'matrix' or 'area':

"It must be emphasized that the formed substances in question are merely constituents of areas of protoplasm, and that there is nothing either in the observations or experiments to prove that the different prospective functions of the areas are due to such constituent parts rather than to the remainder of the area. It would be equally consistent to regard the formed substances as merely 'raw materials' and to ascribe the specific potencies of the areas to the 'matrix'....

"In the experiments of Lillie (1906, 1909) on *Chaetopterus,* Morgan & Spooner (1909) on *Arbacia,* Morgan (1909b) on *Cummingia,* Conklin (1910) on *Limnea* [Meshcheriakov, 1990b] and Conklin (1917) on *Crepidula,* it was shown that all the formed substances of these eggs might be displaced from the areas in which they normally occur without detriment to the prospective potencies of the areas in question, and conversely without modification of prospective potency of the areas into which they were driven. This result, if fully established, would completely invalidate the observational and experimental basis on which the theory of organ-forming substances rests. It should therefore be examined critically....

"The theory of organ-forming substances... is thus deprived of all factual basis... and the mechanism of origin of the segregate remains unknown.... Morgan (1909b) is 'inclined to adopt the view that a differential cleavage is one in which a physical condition is reached [a cortical cytoplasm instability?] that marks a further step in development..." (Lillie, 1929a).

It is possible that the scrambling experiment of Evsikov, Morozova & Solomko (1994) did not scramble the cortex and any determinants it might contain. The experiments described by Lillie (1929a) take care of this caveat. So do experiments on cortical rotation before first cleavage in amphibians:

"In contrast to Curtis' cortex transplantation experiments (Curtis, 1960, 1962a)... egg rotation experiments [Ubbels et al., 1983] suggested to us that neither the dorsal egg cortex nor the DYFC [dorsal yolk free cytoplasm] contain particulate dorsal determinants essential for axis formation (Kirschner et al., 1980; Gerhart et al., 1981)" (Ubbels, 1997a).

Perhaps the current set of purported 'cytoplasmic determinants' should be subjected to a similar gentle separation from the cortex. At present we only have demonstrations that they move with the cytoskeleton in normal (Jeffery & Meier, 1983, 1984; Jeffery, 1984a; Gilbert, 1994) and mutant (Kirby, 1992; Gilbert, 1994) development. In the words of Morgan (1909b), the 'determinants', originally identified with 'visible inclusions', are but tagging along with something he could not see, which we can now identify as the cytoskeleton:

"...The positive results show, I think, with all clearness that the visible inclusions in the cytoplasm... are not organ-forming substances; their role in development is of secondary importance. Behind them lies an organization that is the chief director of the series of events that characterize development. The visible materials of the egg follow and do not lead in the development....

"Perhaps I should add, in order to prevent misunderstanding, that, although I dispute the view that the visible substances in the sea urchin's egg acted upon by the centrifuge are organ-forming, I hold that development is the outcome of physical changes in the egg" (Morgan, 1909b).

At stronger centrifugation, the 'areas' move, suggesting to Lillie (1929a) "that they possess different specific gravities and that they are coherent, which implies other physical differentiating properties such as viscosity". We may presume (or test) that stronger centrifugation began to not merely dislodge the visible inclusions from the cytoskeleton, but collapse the cytoskeleton ('areas') itself. These classical experiments may be worth repeating with modern observational techniques.

Muller (1947) proposed a physical mechanism for differentiation based on the *Chaetopterus* observations:

"HOW GENES ALIKE IN ALL PARTS CAN CAUSE THOSE PARTS TO DIFFERENTIATE.... This embryological mystery, or rather, series of mysteries, though for the most part unsolved in specific chemical terms, is nevertheless not so surprising in principle as it seems to be at first sight. As oil separates from water under the influence of gravity [and surface tension], and as water itself in the form of snow crystals settles into very elaborate patterns, so in the very complicated protoplasmic mixture whose composition was ultimately determined by the genes, some of the constituents [perhaps of the cytoskeleton?] may be expected to separate from one another in definite arrangements and to take on very specific forms....

"Now each such laying down of special material or the putting of the material into special forms in a given region, differentiating that region from the rest, may serve as the starting point for still further differentiation as development proceeds.... Thus the pattern of the developing organism tends to become more and more complex, the differentiation to become ever finer and more diversified, and the degree of specialization of each part to become greater, even though all parts have been provided with (and probably still contain) identical groups of genes....

"...The same group of genes may give rise to a very different group of final products, according to the type of medium and other conditions that surround it. Whether under given circumstances certain genes tend to be more aroused to activity than others, so as to give rise to relatively more of their primary products, whatever these may be, we do not know..." (Muller, 1947).

We may take this as a first attempt at a physical explanation of the spatial component of differentiation. Muller (1947) does not specify how his intracellular mechanism of separation of mutually immiscible components would apply to whole cells in tissues, although later work on cell sorting (Townes & Holtfreter, 1955; Steinberg, 1963) shows how this could be possible. He also leans a little on the 'environment', perhaps unnecessarily:

"Which group of potentialities is brought out in any given region after a given period is decided by the conditions already provided within that region by all the differentiation which preceded that period, plus influences presently and thereafter impinging on that region from other regions and from the outside environment" (Muller, 1947).

Nevertheless, here we have a model in which a coupling occurs between a reiterated physical process and reaction of genes to this process. Thus this is a model in the same spirit as that of differentiation waves, one in which genes are not the sole controls of development, and one which is reasonable for both mosaic and regulating organisms.

To summarize, it is possible that a good portion (or all) of the cytoskeleton in an asymmetrically dividing cell is acting as a cell state splitter. A mechanical instability is set up, whose resolution is asymmetric. Of course, there is an overall symmetry, in the sense that cell A could equally give rise to cells BC or cells CB (as an example), if all else were constant. However, all else is generally not constant: the influence of previous cell divisions, leaving behind 'orientation scars' (Proposition 64), the presence of adjacent cells ('induction'), etc., can provide a bias. This bias is not absolute, given the various mutants and interventions that can reverse it. Once the asymmetric instability resolves itself, different proportions of cytoskeletal components end up in the two different daughter cells, and these drag the so-called 'determinants' along with them, presumably by specific binding to particular components of the cytoskeleton. In this manner, 'determinants' become nonuniformly partitioned between the unequal daughter cells. These, then (or just the cytoskeleton itself, *a la* Lillie, 1929a) could produce differing gene expression in each cell. Presumably different differentiation cascades are triggered in each cell.

There is one factor that I have not taken into account, namely doubling of the cytoskeleton during cell division:

"What do you think of the possibility that the wave observed on a single cell is the result of a changing pattern of cytoskeletal components making contact with the cell membrane? When a cell is getting ready to divide, it's got to assemble a second cytoskeleton. There's got to be some sort of surface effect as that second cytoskeleton is being assembled, or perhaps as it's being separated from the first; and a very likely effect would be a wavelike disturbance at the surface. The two most likely patterns would be an oval or circular wavefront emerging from a point and stopping somewhere short of halfway to the opposite side, and a furrow that grows from a point until it encircles the cell" (Steve McGrew, p.c., 1997).

Whether growth of the cytoskeleton is primary or secondary to single cell differentiation waves will be interesting to determine. Quantitative analysis of the amount of cytoskeleton versus stages of the cell cycle, and localization of the mass increase within the cell, has just begun (microtubules: Elinson, 1985; microfilaments: Rao et al., 1990; Heil-Chapdelaine & Otto, 1996; Wong, Allen & Begg, 1997).

Proposition 61: compaction is a phenomenon occurring in both regulating and mosaic embryos, at early, few cell stages, that uses both a contraction and an expansion wave to separate blastomeres into external and internal cells.

Let's compare two classically extreme embryos, mammals and molluscs. Here is a description of mouse compaction:

"Following the third cleavage in the mouse, a process termed *compaction* occurs in which the cells become pressed against one another along their lateral surfaces causing the embryo to appear almost as a single, highly irregular-shaped cell.... The microvilli... become very unevenly distributed.... They may function in bringing together the surfaces of adjoining cells. The apposition of adjoining surfaces leads to the formation... of desmosomes, gap junctions [Kidder, Rains & McKeon, 1987; Becker, Leclerc-David & Warner, 1992; Kumar & Gilula, 1996], and tight junctions [Pratt et al., 1982] between the cells.... As cleavage progresses, a ball of cells termed the *morula* is formed.... Among the cells of the morula, a very few occupy completely internal positions (3 to 5 at the 32-cell stage), the remainder forming a layer of cells at the outer surface of the embryo.... Fluid is secreted into the spaces between the cells, and the embryo grows rapidly in volume forming a blastocyst" (Karp & Berrill, 1981).

I would guess that mammalian compaction is a contraction differentiation wave and that the internalized cells (later contributing to the embryo proper) have participated in an expansion wave (cf. Levy et al., 1986). Note that microtubules are involved (Maro & Pickering, 1984; Houliston, Pickering & Maro, 1989), suggesting use of a cell state splitter. Note also: "Indirect immunofluorescence with a monoclonal antibody to E-cadherin indicates that PKC activation causes a rapid shift in the localization of this cell adhesion molecule, which coincides with the observed compaction" (Winkel et al., 1990). Thus cell adhesion might suffice, so that microfilaments are not necessarily needed for the contraction event of compaction. This is consistent with my broader definition of a cell state splitter.

A similar 'compaction' process occurs in the highly mosaic limpet, a mollusc, suggesting a similar interpretation in terms of differentiation waves:

"The first and decisive step in the formation of the dorsoventral axis in equally-cleaving molluscs appears to be the centralisation of one of the equipotential macromeres (van den Biggelaar, 1977; van den Biggelaar & Guerrier, 1979, 1983; Verdonk & Cather, 1983; Boring, 1989)..., the presumptive 3D-macromere.... Centralisation of a macromere is not a passive [contrast Goel, Doggenweiler & Thompson, 1986; Lewis III, Goel & Thompson, 1988] but an active process.... Kühtreiber et al. (1986) observed an extracellular matrix in the blastocoelic cavity of *Patella,* which... is required for the centralisation.... Possibly, the extracellular matrix is involved in generating differential cellular adhesiveness....

"...The presumptive 3D-macromere... has, as a result of its original eccentric position in the future D-quadrant, different contacts with animal micromeres from different quadrants (cf. Arnolds, 1982; Damen & Dictus, 1993).... Since gap junctions are present between the centralising macromere and animal macromeres, these channels [i.e., junctions] may play a role in the intricate mechanism that is responsible for the simultaneous induction of the dorsoventral axis in both the animal and the vegetal hemisphere by mediating the exchange of differentiating-inducing signals... in agreement with the idea that [the] 3D [macromere] acts as an organiser in molluscan development (Clement, 1962; Cather, 1971; Verdonk & Cather, 1983)" (Damen, 1994).

While these two examples of compaction are not quite at the single cell level, they suggest that some phylogenetically widespread contraction/expansion wave phenomena occur at the earliest developmental stages.

Proposition 62: in mosaic organisms a differentiation wave travels only over a single cell, along its cortex.

"The existence of segregates in unsegmented eggs, and the facts of cell-lineage, both observational and experimental, prove that the mechanism may operate within the confines of a single cell, i.e. that it may be an intracellular phenomenon.... In fact the origin of segregates in general in vertebrates and in animals with indeterminate cleavage is undoubtedly multicellular, i.e. between groups of neighboring cells. Yet the phenomenon is precisely the same in all of its biological implications. A mechanism of one type seems to be implied whether early intracellular, or relatively late intercellular, embryonic segregation [determination] is concerned" (Lillie, 1929a).

There is no direct evidence for this proposition, so I will have to build the case from a few observations. To begin, let's make the connection with the cortical cytoskeleton, by considering the work of van Loon et al. (1993) on the actin genes in early development of limpets, which is "...initiated by...

the cell line-specific localization of filamentous (F) actin in two cell lineages...: the trochoblasts and the dorsal macromere 3D (Serras & Speksnijder, 1990, 1991)":

"...In [the limpet] *Patella,* actin is present as a multigene family.... In sea urchin, *Styela* and *Drosophila,* actin genes encoding cytoplasmic actins have been isolated and shown to be expressed at a specific developmental stage (Beach & Jeffery, 1990; Burn, Vigoreaux & Tobin, 1989; Cox et al., 1986). However, no specific function could be attributed to these cytoplasmic actins. We think that the actin genes in *P. vulgata* would enable us to demonstrate the direct correlation between the differential expression of actin genes and their function" (van Loon et al., 1993).

These actins may be components of the cell state splitters responsible for each early step of differentiation in these embryos. If so, the variety and cell specificity of the actins suggests that each cell state splitter may be somewhat different in its cytoplasmic components, and that at least the actin genes for each cell state splitter are part of each differentiation cascade, rather than being off the differentiation tree as gene modules (Proposition 144).

To make the connection of these actins to the cortex, I turn to recent work on development of another mollusc, the freshwater snail, *Lymnaea* (Browder, Erickson & Jeffery, 1991), where we see an actual single cell contraction wave that propagates over part of the cell:

"The spatial organization of the egg is probably established and maintained by the cytoskeleton. In particular the cortical domain, which is very rich in cytoskeletal elements, is thought to play an important role in providing the framework for the forces that structure the egg (van den Biggelaar & Guerrier, 1983; Dohmen, 1983).... The eggs of *Lymnaea* show a polar [ionic] current during meiotic divisions (Zivkovic & Dohmen, 1989).... We provide evidence that the polar currents during the second meiotic division consist of two components: an inward current associated with the site of cytokinetic constriction and an outward current associated with the vegetal cortex of the egg.... Centrifugation results in the displacement of the meiotic apparatus, which subsequently leads to the displacement of the contractile ring.... The inward current is associated with membrane constrictions rather than with karyokinesis.

"Interactions between intrinsic plasma membrane proteins and the cytoskeleton have been observed in a large number of cases, for example, in the rat brain, where Srinivasan et al. (1988) demonstrated that sodium channels are directly bound to the cytoskeleton [cf. Berdiev et al., 1996; Prat et al., 1996].... On the basis of centrifugation, con A and cytochalasin experiments, we suggest that constriction of the *Lymnaea* egg membrane locally activates an inward current. This activation of ionic currents may originate from stretch-sensitive ion channels, which have been described previously in *Lymnaea* neurons (Morris & Sigurdson, 1989). The inward current could, in its turn, act back upon the constriction. The constriction of the cleavage furrow has been shown to be critically dependent on the Ca^{2+} concentration in the cytoplasm....

"The spatial correlation of inward current with membrane constriction, as found in this study, and the evidence implicating that Ca^{2+} at least partially carries the polarized current during meiosis (Zivkovic & Dohmen, 1989), suggest that the contractile ring might propagate and maintain itself via inward ionic currents through the plasma membrane" (Créton et al., 1992).

Hird & White (1993) have obtained detailed three dimensional time lapse images of nuclear and cortical motions during the first few cell divisions of nematode embryos (cf. Thomas et al., 1996). Before the first full cleavage, they observed a midcell 'pseudocleavage', in which "cytoplasm flows towards the posterior end of the egg and a transient cleavage furrow (the pseudocleavage furrow) forms and then regresses (Nigon, Guerrier & Monin, 1960)". The event they describe could be interpreted as a single cell tissue splitting itself into two new single cell tissues using the pseudocleavage furrow as a differentiation wave. This wave has a speed comparable to the 3 μm/min speed of the axolotl ectoderm contraction wave and produces movements of the cortex. We are thus treated to a detailed glimpse of what may well be differentiation waves traversing single cells, including electrical effects and attachment of 'determinants':

"We have examined the cortex of *Caenorhabditis elegans* eggs during pseudocleavage (PC), a period of the first cell cycle which is important for the generation of asymmetry at first cleavage (Strome, 1989). We have found that directed, actin dependent, cytoplasmic, and cortical flow occurs during this period coincident with a rearrangement of the cortical actin cytoskeleton (Strome, 1986a). The flow velocity (4-7 microns/min) is similar to previously determined particle movements driven by cortical actin flows in motile cells. We show that directed flows occur in one of the daughters of the first division that itself divides asymmetrically....

"During PC, the anterior cortex exhibits ruffling activity and the PC furrow forms. This furrow is unusual in that it does not bisect a mitotic spindle. The oocyte pronucleus begins to migrate from the anterior end of the egg towards the sperm pronucleus at the posterior end (Albertson, 1984). Simultaneously, there is a flow of cytoplasmic material from anterior to posterior (Nigon, Guerrier & Monin, 1960) that stops by the time the maternal pronucleus migrates through the PC furrow....

"Granules adjacent to the plasma membrane in the posterior half of the [50 x 25 μm] embryo streamed from posterior to anterior towards the site where the PC furrow was starting to form. This granule movement appeared as a flow of the entire field and was restricted to a layer of 2.5 μm thickness subjacent to the surface (corresponding to the cortex) in the posterior of the embryo. Cortical granules moved at speeds of 5.6 μm/min ± 1.3.... Those in the anterior cortex... drifted slowly....

"At the same time... there was an oppositely oriented flow of granules in the cytoplasm... at ~4.4±0.8 μm/min.... Granules in the cytoplasm anterior to the PC furrow did not exhibit this dramatic streaming, but did drift slowly posteriorly.... Granules could be followed as they flowed through the cytoplasm towards to the posterior end, and then moved back towards the furrow in the opposite direction along the cortex.... [Cf. 'fountain streaming': Rappaport, 1970, and 'cortical tractoring': Jacobson, Odell & Oster, 1985.]

"*Cortical Granule Flow during Pseudocleavage Does Not Depend on Microtubules, but Does Require Microfilaments....*

"*The Flow Patterns of PC... Are Determined by the Position of the Centrosomes....* There is evidence that the cytoplasmic flow carries determinants and other factors (such as P granules) to the posterior end of the egg so that they are inherited by the posterior daughter, P1, at first cleavage (for review see Strome, 1989).... How is the polarity of the flows specified?... It is... possible that the sperm asters induce and orient the cortical and cytoplasmic flows and the transient furrow....

"*The Flow Patterns May be Due to Movement of the Actin Cytoskeleton....* The actin component of the contractile ring is probably formed from preexisting cortical actin microfilaments, which flow towards the equatorial region along the cortex (Cao & Wang, 1990a,b).... The force responsible for the actin movement is probably actomyosin based, but may also be driven by localized actin polymerization (for review see Bray & White, 1988; Mitchison & Kirschner, 1988; Smith, 1988c)....

"...Some component of the asters induces the movement of the cortical actin network away from the area in which they are located,... either... the centrosomes or the astral microtubules [with] very short microtubule 'stubs' [cf. Moritz et al., 1995a]....

"*Interaction between the Centrosomes and Cortex during Cytokinesis....* Just before furrowing, the cortex of a [general] cell is in a state of uniform tension, but as cytokinesis is initiated, tension falls off at the poles and increases at the equator (Hiramoto, 1987). This gradient of cortical tension is initiated by the asters (Rappaport, 1986a), and would result in the movement of the actin network (and associated cortical material) away from the lower tension polar regions towards the higher tension equatorial region where filaments align circumferentially to form the contractile ring (White & Borisy, 1983; [cf. Harris & Gewalt, 1989]). The asters could create this tension gradient by one or a combination of two mechanisms: (*a*) stimulation of cortical tension at the equator (Rappaport, 1986a; Devore, Conrad & Rappaport, 1989); and (*b*) relaxation of cortical tension away from the equator (Wolpert, 1960; Schroeder, 1981a; White & Borisy, 1983). Both result in the same final tension differential.... The flow patterns that are seen when the MA [mitotic apparatus] is displaced with nocodazole [a microtubule disrupter] are more consistent with the relaxation model" (Hird & White, 1993).

A full mechanical or mechanochemical analysis would seem to be in the offing. In the meantime, genetic analysis of nematode unequal cleavages proceeds apace. For example, Rose et al. (1995) showed that the "...maternal effect mutation, *nop-1* (*it142*),... abolishes the anterior cortical contractions and the pseudocleavage furrow" yet usually leaves the nematode embryo with normal development. The regular cleavage furrow is still present, and we should note that: 1) other asymmetric cleavages are not necessarily accompanied by a pseudocleavage wave; 2) the first asymmetric cleavage is not preceded by a cleavage, and may need something special to get the process started.

A differentiation wave is not just a travelling furrow or expansion wave. It includes many events inside a cell, some of which undoubtedly precede appearance of the contraction or expansion itself. When we look at the possibility of each asymmetric cell division itself being an act of cell state splitting, including a differentiation 'wave' (or pair) that only traverses the single cell, or even only part of it, it is easy to lose sight of cause and effect.

Thus this proposition is, in a way, a statement that there is a homology between single cell differentiation waves in mosaic organisms, and many cell differentiation waves in regulating organisms. Exactly how this homology works will take careful investigation and comparison of both systems. We don't yet have the data to ask what components are homologous. For example, Fagotto & Gumbiner (1996) define 'juxtracrine signaling' as induction occurring only between single cells in contact, and suggest that the "best documented example is found in neural development" of *Drosophila*. Yet, if we are to believe that neural development in insects is homologous with that in regulating vertebrates, it cannot be juxtracrine signaling that is the fundamental homology.

Here is one way to think about single cell differentiation waves:

"I guess you're proposing that at each cell division, a cell state splitter is constructed and then triggered, so each daughter cell has only two possible states that it can end up in. That idea is plausible and attractive because it assumes a single, conserved, underlying differentiation mechanism for both mosaic and regulatory embryos, perhaps triggered in different ways. However, the connection between a 'differentiation wave' on the surface of a single cell and the creation, then asymmetric triggering of, two cell state splitters is pretty muddy at this point. The single-cell differentiation wave seems unnecessary. It's plausible that before the two daughters are separated, physics forces the two cytoskeletal state splitters to flip to opposite states, which would then result automatically in asymmetric division. In proliferation without differentiation, perhaps the conditions are different so that probabilities are biased in favor of the two splitters to start out in the same state. In that case, maybe the observed 'differentiation wave' on the surface of a single cell is 'caused by' the underlying construction of a new cytoskeleton, and echoes the underlying asymmetry or symmetry, rather than 'causing' asymmetrical division." (Steve McGrew, p.c., 1997).

Yes, the idea is muddy at this point, until we start to observe metastable and bistable cytoskeletal biodevices, measure their mechanical properties, and confirm what we see by computer simulation.

Proposition 63: the cell state splitter in asymmetric divisions also causes the intracellular morphogenesis that segregates components to the two distinct daughter cells.

"*Asymmetric Mitosis*.... This type of cell reproduction is extraordinarily significant for the differentiation of multicellular systems, yet we know little or nothing about the fundamental origin of the asymmetry" (Mazia, 1961).

Many questions are left unresolved, but with the background of this Section, are perhaps now approachable:

1. *How is the initial asymmetry generated?*

Is there a bias by the sperm, such as is found in *Xenopus* (Gerhart et al., 1989)? This seems to be the case for nematodes:

"When a sperm enters an oocyte in the abnormal end, development proceeds normally, but with the a-p [antero-posterior] axis completely reversed. The result suggests that the site of sperm entry specifies the a-p axis.... The results imply that unlike most other invertebrates, the unfertilised oocyte in *C. elegans* has no axis pre-specified in the form of asymmetrically localised cytoplasmic determinants" (Goldstein & Hird, 1995).

Since the "Enoplida [nematodes] demonstrate apparently equal cleavages to the 8-cell stage (Malakhov, 1994)" (Fitch & Thomas, 1997), one questions whether the lack of a 'pre-specified' axis might, indeed, extend beyond nematodes to other invertebrates:

"In the cleavage of... free-living marine nematodes of the order Enoplida... there is no early establishment of bilateral symmetry nor rigid determination of the fate of blastomeres, such as takes place in the development of most nematodes. Cleavage among marine members of Enoplida is variable. At the four-cell stage, we observe various geometric configurations: tetrahedral, rhombic, and T-shaped. All of these configurations are encountered in the development of nematodes of other groups, but in Enoplida they exist in development in the same species within the context of individual variability" (Malakhov, 1994).

2. *Why does cortical streaming only occur to one side of the pseudocleavage?*

Is this the result of some sort of symmetry breaking instability process involving the cortex? Is gravity involved in breaking this symmetry (Souza

& Black, 1985; Flint et al., 1989; Ubbels et al., 1989; Tabony & Job, 1992a; cf. Ubbels, 1997b; cf. review of gravity and the cytoskeleton in: Jones, Leivseth & Tenbosch, 1995), or perhaps even the earth's magnetic field (Asashima, Shimada & Pfeiffer, 1991)? What is the role of the physics of asters (Bjerknes, 1986)? Dan (1960) reviews the work of Motomura (1935) suggesting a "permanently superficial position taken by the cortex" involving "an active cytoplasmic movement within the blastomeres themselves, which recurs with the division activity at each cleavage". Is there some supercooperative (Gordon & Brodland, 1988) cortical and/or aster interaction (accompanied by an asymmetric voltage gradient: Burr, 1941) that is inherently unstable, and whose resolution must be an asymmetric cortical flow? Perhaps there is an actual homology here with the breakdown of symmetry driven by cortical microtubules in amphibians prior to first cleavage, resulting in grey crescent formation (Elinson & Rowning, 1988; Elinson & Palecek, 1993), even though microfilaments are apparently the active component in nematode pseudocleavage. What is the role of genes whose mutations prevent this symmetry breaking (Levitan et al., 1994)?

3. *Does the membrane move with the streaming cortex?*

Is the microfilament ring stationary only in the laboratory coordinate system (like the amphibian blastopore)? For that matter, does the cortex move at all, or just granules attached to it? Marking the membrane with India ink particles (Drum & Hopkins, 1966) or by fluorescence (Edidin, 1974) might resolve this.

4. *Which components of this interplay of nucleus, aster, cortex, and granules actually cause differential gene expression?*

(Cf. Hird, Paulsen & Strome, 1995, 1996.) Differential gene expression in nematodes can be traced back to the first full cleavage after the pseudocleavage (Bowerman et al., 1993; Evans et al., 1994). What is the relationship between the distribution of maternal mRNA and subsequent expression of the zygotic genome (Edgar, Wolf & Wood, 1994)? What is the cause/effect relationship to the observed asymmetry of asters in unequal divisions?

"Applying the technique for isolating the mitotic apparatus (Mazia & Dan, 1952a), it was found that there is a striking difference between the two asters in the unequally cleaving eggs; however, the difference is not in their diameters but in their shape (Dan, Ito & Mazia, 1952), the aster forming the larger blastomere being the usual radiate sphere, while that of the smaller blastomere has a flattened radiate shape like a daisy" (Dan, 1960).

"This relation between size of aster and size of daughter cell in unequal cytokinesis is exactly as predicted by Dan... in his theory of mechanisms of cell division" (Dan, Ito & Mazia, 1952).

"...There are... eggs, like those of the annelids and molluscs, which undergo unequal cleavage from their first division. If the metaphase mitotic apparatus of a cell which is in the process of cleaving unequally is isolated, it is found that while the aster which will belong to the larger blastomere is spherical, that of the smaller blastomere is flat on its outer side as though it had been sliced off (Dan & Nakajima, 1954a). This condition is not restricted to the first cleavage of annelids and molluscs, but occurs in exactly the same way at the fourth cleavage of the Echinoidea, when the mesoderm-forming micromeres are formed [in 'the sea urchin *Hemicentrotus pulcherrimus*' (Dan, 1960)]. In the eggs of annelids and molluscs, also, the size difference among the blastomeres eventually shows itself to be connected with a difference in developmental fate; this strongly suggests that the largeness and smallness of the blastomeres, far from being chance characteristics, are of basic significance" (Kumé & Dan, 1968b).

"The [aster]... which is to go into the bigger cell is of usual radiate shape, while the other which is to be included in the smaller cell is flat; the astral rays being arranged in a single plane like a daisy.... The vegetal interphase nuclei move *in toto*, headed by the centrosome, to the inner side of the vegetal pole. As a consequence, when the nuclear membranes are disrupted, the astral rays which radiate out from the attached centrosomes can distribute themselves only on one plane; since the centrosome is squeezed between the cell surface and the spindle pole. Probably, here lies the cause of the daisy-shaped aster" (Dan, 1978).

For instance, does the daisy or cup shaped aster launch an asymmetric single cell contraction wave, when its microtubules hit the cell membrane? Does the other, spherical aster them launch a single cell expansion wave?

Many of the predicted components of the nuclear state splitter (Appendix IV: Björklund & Gordon, 1993b) are here, and the opportunity for working out the physics of the interactions between the microfilament ring, the cortical flow, and the microtubules (asters) is obvious. The pseudocleavage furrow is probably a differentiation wave, which, moreover, travels about the same speed as the ectoderm contraction wave in axolotls (3 µm/min). Its function appears to be to segregate at least some mRNAs to one side prior to the first cleavage. In this sense, then, mechanics precedes subsequent gene expression, which is the fundamental assumption of cell state splitter theory. To see how this changes one's point of view, consider...

"Studies of early *C. elegans* embryogenesis... [that] imply the existence of cytoskeletal architecture dedicated to the process of moving or preventing movement of molecules involved in the process of determining cell fate" (Levitan et al., 1994).

If we assume there is a discrete determination event, then it is that motion itself (or what initiates it), that causes partitioning of these molecules. This movement is the first event, and thus represents determination. The so-called 'determinants' are then secondary, and are probably involved in the subsequent process of differentiation, rather than determination. Once we find an asymmetric movement of the cytoskeleton, we can break the infinite regress (Amon, 1996) of gene controlling gene controlling gene controlling gene.... In other words, once we trace a signalling pathway back to mechanochemical proteins, we should pause and examine the possibility that those proteins are part of a mechanochemical device that itself 'determines' the course of differentiation. Ben-Ze'ev (1992) gives an explicit model for this situation:

"A... mechanism whereby [a] cytoarchitectural component can... regulate signal transduction is by transiently anchoring regulatory molecules and by controlling their ordered transport to different locations in the cell (Ben-Ze'ev, 1991). For example, molecules regulating nuclear activity can be kept in the cytoplasm in association with the cytoskeleton and thus be unavailable in the nucleus. In response to an extracellular signal initiated by growth factors, or by changes in cell contacts [or by differentiation wave propagation?], the state of assembly of cytoskeletal elements is often altered. This process may allow the release of cytoplasm-sequestered nuclear factors and their subsequent translocation into the nucleus [especially after an asymmetric division], as demonstrated for several protein kinase subunits (Nigg, 1990), or certain transcriptional factors (Gilmore, 1990). Compartmentalization of various cellular components in association with the cytoskeleton was demonstrated for several mRNAs in the cell, including Vg1, actin, and certain development-regulating genes in *Xenopus* and *Drosophila* (reviewed in: Ben-Ze'ev, 1991; Gottlieb, 1990; Singer, 1992). There is ample evidence suggesting an association between mRNA and the actin-cytoskeleton (reviewed in: Ben-Ze'ev, 1991; Singer, 1992; Hesketh & Pryme, 1991).... Cytoarchitecture may thus influence the spatial regulation of gene expression through targeting and positioning mRNAs in the cytoplasm (Hesketh & Pryme, 1991)" (Ben-Ze'ev, 1992). [Cf. Ben-Ze'ev, 1994.]

With these ideas and the four questions in mind, the methods of Hird & White (1993) could, if applied step by step to subsequent cell divisions (i.e., acts of cell state splitting), lead to an understanding of how the nematode embryo constructs itself. Observation of and intervention with cellular and cytoskeletal events are a necessary complement to work on single cell gene expression (such as Mango et al., 1994; Mello, Draper & Priess, 1994) if development consists of interactions between differentiation waves and the portion of the genome organized as the differentiation tree. Full understanding may require a systematic approach, in which we build up our understanding the way the organism is built, cell by cell, at least until we are satisfied with any general principles that might emerge. Differentiation trees at least provide an approximate theory with which to begin this ascent.

There are some interesting timing problems here, and a possible explanation of why cell division and differentiation are so tightly coupled in mosaic organisms: they do not have the ability to propagate differentiation waves over more than one cell. This is an easily testable prediction.

Proposition 64: asymmetries in the cortical cytoskeleton in cells undergoing asymmetric division are passed on in some manner to the daughter cells, through 'orientation scars', which may be the launching sites for the next set of single cell differentiation waves.

If every asymmetric division in a mosaic organism occurred at a random orientation, the resulting organism would differ greatly from one individual to the next. Most would resemble a disoriented mass of cells. The spatial organization of asymmetric division, at least in such organisms, is much more highly ordered. Launching of single cell differentiation waves presents similar problems (and, I would suggest, homologous problems) to the launching of differentiation waves in regulating embryos. The latter wave launchings also do not occur at random. The launching and propagation of waves in active media over nonspherical surfaces, whose shape can alter in response to wave propagation, such as in the single cell *Acetabularia* (Goodwin, 1994a; Goodwin & Brière, 1994), give the flavor of the kind of work that may need to be done to understand mosaic development at the single cell level. For nematodes in particular, we could speculate that wave launching is coupled to 'persistent remnants of previous cell divisions' which I call orientation scars:

"In those cells that divide on the same axis, there is an additional directed rotation of pairs of centrosomes together with the nucleus through well-defined angles. Intact microtubules are required for rotation..." (Hyman & White, 1987).

"During *Caenorhabditis elegans* embryogenesis, specific cells in the P_1 lineage rotate their duplicated centrosome pair onto the anterior-posterior axis; this rotation is correlated with and necessary for a differential inheritance of cytoplasmic determinants in the daughter cells. Centrosome pair rotation is sensitive to inhibitors of actin and microtubule polymerization and may require microtubule attachment to a specific cortical site ['orientation scar']. We show that actin and the barbed-end binding protein, capping protein, transiently accumulate at this cortical site, possibly by assembly onto persistent remnants of previous cell divisions ['orientation scars']. Based on these observations, we propose a model for the molecular basis of centrosome rotation that is consistent with the dependence of rotation on actin filaments and microtubules" (Waddle, Cooper & Waterston, 1994). [Cf. White & Strome, 1996.]

There are many phenomena that could be associated with scar formation and give us hints as to the composition of an orientation scar and how it gets positioned at a particular site on the cell membrane. The observation of spiral patterns of 'surface folds' in scanning electron micrographs (SEM) of the beginning of polar body formation in molluscs (Dohmen, 1983; Domen in: van den Biggelaar, 1991) suggests that a localized, *asymmetric* membrane specialization can exist (see also the spatially localized 'subcortical microtubule network' of Bilinski, Klag & Kubrakiewicz, 1995, and the rearrangement of cortical microtubules in conifer tracheids during differentiation: Abe et al., 1995). Polarity mutations of yeast affecting both actin and microtubules (Verde, Mata & Nurse, 1995) may yield some understanding of orientation scars (cf. Chant, 1996). The nucleus itself may be polarized and pass on transcripts to only a localized portion of the intracellular microtubules (Francis-Lang, Davis & Ish-Horowicz, 1996). Hints of such a mechanism are seen in the fission yeast *Schizosaccharomyces pombe* (Chang, Drubin & Nurse, 1997), which forms scars at symmetric divisions (symmetric by the measure of equal volumes: Johnson et al., 1982; cf. Calleja et al., 1980). Ion currents associated with mitosis (Coombs, Villaz & Moody, 1992) could be related to orientation scar formation. Orientation scars may be products of transient 'connecting cords' or 'telophase bridges' containing spindle microtubules, which are remnants of cell divisions (Bellairs & Bancroft, 1975; Nagele & Lee, 1979; Everaert et al., 1988).

Some asymmetric membrane phenomena may also offer clues to how orientation scars form. For example, formation of the 'dimple' initiating the activation wave in frogs might offer an electromechanochemical model for orientation scars:

"In *Discoglossus pictus* egg sperms fuse with the egg only in the animal dimple, an indentation located at the center of the animal hemisphere.... At this time electrophysiological studies have detected inward currents at the dimple center, outward current at the rest of the egg surface and an eight-fold increase in $[Ca^{2+}]_i$ which has been found to propagate from the site of activation to the whole egg" (Campanella, Kline & Nuccitelli, 1988).

Further modelling for orientation scars could be based on membrane capping of some membrane protein in the fluid plasmalemma membrane of a cell (Pimenta et al., 1994). For instance, the capping of ion channels in *Fucus* eggs has been proposed as the mechanism of their polarity generation (Jaffe, 1968, 1969). Such capping (Fromherz & Zeiler, 1990) is probably the result of self-focusing of the ion channels through their charge and electrophoresis, brought about by the very ions they pump. Self-focusing especially ought to occur at points where cells meet (Fromherz, 1995). Predicted self-organization patterns of membrane capping (Fromherz & Kaiser, 1991; Fromherz & Zimmermann, 1995) might be related to the spatial angle of an orientation scar. Criteria for proving that ionic currents, such as Ca^{2+}, are causal of cell polarity are discussed and tested by Harold & Caldwell (1990) and Hyde & Heath (1995).

If orientation scars have a surface asymmetry, then they need to be characterized not only by location on the surface of the cell (3 coordinates), but also by a vector normal to the surface and a direction of twist (4 more coordinates). Restoration of these coordinates may be what is behind the following observation:

"We have investigated the hypothesis that the abnormal [leech] cleavage pattern of isolated CD blastomeres is due to removal of mechanical constraints normally imposed by cell AB. We find that when cell CD is constrained in vitro to mimic its in vivo shape, it cleaves more normally" (Symes & Weisblat, 1992).

An example of a membrane structure that could serve as an orientation scar with twist is the...

"...asymmetric unit membrane (AUM) [which] forms numerous plaques covering the apical surface of mammalian urinary bladder epithelium. These plaques contain four major integral membrane proteins called uroplakins... which form particles arranged in a well-ordered hexagonal lattice with p6 symmetry and a lattice constant of 16.5 nm.... Examination of the luminal face of freeze-dried/unidirectionally metal-shadowed AUM plaques established a left-handed vorticity of the 16 nm protein particles, whereas the cytoplasmic face exhibited no significant surface corrugations" (Walz et al., 1995).

Uroplakins are highly conserved and 'differentiation-related' (Wu et al., 1994).

An orientation scar could perhaps be associated with an underlying centrosome with "important functions in cellular morphogenesis" (Hoops &

Witman, 1985) such as determination of the plane of the cleavage furrow (Johnson & Porter, 1968), i.e., with triggering of the next single cell differentiation wave. Any form of attachment of an orientation scar to the underlying cytoskeleton could provide or reflect a polarity of the cell.

This concept of nematode development via orientation scars is a bit different from the current paradigm, expressed, for example, by Wood & Edgar (1994):

"Recent studies reveal preliminary insights into the mechanisms of embryonic patterning in *Caenorhabditis elegans*. It appears that both embryonic axes and early blastomere fates are determined by a combination of segregating determinants and cell interactions, under the control of maternally expressed genes" (Wood & Edgar, 1994).

I can use the same language, but turn it around in the following way to offer an approximate guess at what is going on. It appears that both embryonic axes and early blastomere fates in nematodes are determined by nonlinear tensegrity instability phenomena involving the cytoskeleton (microfilament furrows, ionic currents, cortex, aster, and their interactions, i.e., single cell differentiation waves and cell state splitting). This causes the segregation of so-called determinants and maternal mRNA, and differential gene expression between daughter cells of the (often) asymmetric divisions that result. The 'determinants' may segregate because they bind to just one of the cytoskeletal proteins, i.e., to a cytoskeletal component that differs between the larger (expansion wave) and smaller (contraction wave) daughter cells. The waves should perhaps be called the actual determinants. Which genes are expressed in each daughter cell depends on the differentiation tree and its transcriptional regulators. The launching site of each differentiation wave is determined by 'remnants' of previous cell divisions (orientation scars) and/or interactions with adjacent cells. This is the basis for cell to cell induction in nematodes. As with launching domains in regulating embryos, we do not yet know what determines the exact location and timing of launching of single cell differentiation waves. It would be worth learning if orientation scars always correspond to such wave launching, or if cell-cell interactions play a role. This is similar to the dichotomy of vertical induction versus autoinduction in regulating embryos.

Consider the spiral development in the snail *Ilyanassa*. We could regard the microfilaments as the 'morphogenic determinants', instead of merely harboring them, as implied by Gilbert (1991a):

"The morphogenic determinants sequestered within the polar lobe are probably located in the cytoskeleton or cortex and not in the diffusible cytoplasm of the embryo... (Clement, 1968; Verdonk & Cather, 1983)..." (Gilbert, 1991a).

The variety of actins certainly makes it plausible that they at least correlate with steps of differentiation, and thus may be causal of those steps. Perhaps tubulins will also be found to be involved in steps of mosaic development, giving us a handle on single cell expansion waves. Certainly, cell flattening is a common phenomenon, and a place to look.

Of course, actin does not act alone. For instance, in the asymmetric process of budding in yeast, specific proteins are involved in nucleating actin assembly (Li, Zheng & Drubin, 1995). How these proteins are localized, and whether localization depends on the actin and/or previous 'orientation scars', etc., is starting to be worked out:

"The pathway for development of cell polarity during bud formation starts with an intrinsic spatial cue, set up during the previous cell cycle by cortical actin or septin cytoskeletal proteins. A signaling complex containing at least two RAS-related GTP-binding proteins assembles at the site marked by the cue, then a polarized actin and septin cytoskeleton assembles, and the secretory apparatus becomes oriented toward the spatial cue" (Drubin & Nelson, 1996).

Perhaps the orientation scars or 'cues' correspond precisely to the "Attachment of one spindle pole to the cortex in unequal cleavage" (Dan & Tanaka, 1990). The rotation of the spindle prior to asymmetric cleavage (Cheng, Kirby & Kemphues, 1995) may involve a specific voltage-sensitive ion channel whose "mechanism of... segregation... is presently unknown"

(Machaca, DeFelice & L'Hernault, 1996). Laser microbeam "experiments demonstrate the central importance of a correct partitioning of cytoplasmic components during early embryogenesis and suggest a stepwise, binary segregation mechanism associated with the unequal cleavages" (Schierenberg, 1989), concordant with the notion that each cell (or equivalence pair) in a nematode is a tissue. Schnabel (1991) even goes so far as to "propose that the body plan of the *C. elegans* embryo may be established by two primary signals followed by secondary interactions. The suggested mechanisms are reminiscent of those involved in amphibian development." (Cf. Hutter & Schnabel, 1994, quoted in Proposition 55.) Those 'induction signals' may be differentiation waves. All of these phenomena may give us an observational basis for unraveling the cause/effect relationships between asymmetric cleavage, differentiation, differentiation waves, ionic currents, membrane capping of membrane proteins, and the geometry of morphogenesis of mosaic organisms. Certainly, we need something to break the current impasse:

"Asymmetric cell divisions, in which two daughter cells differ in their fates, [are] presumably due to the asymmetric distribution of some factors at the cell division.... The general problem of how a cell divides to generate daughter cells with distinct fates remains unsolved.... In *C. elegans* progress has been made recently by the identification of genes necessary for asymmetric cell divisions" (Sternberg, 1991).

Grimes & Aufderheide (1991) state quite well what is needed:

"...How do localized determinants become localized?... The localization... may indeed involve processes at least phenomenologically similar to global or local patterning in protozoa. Herein lies the crux of our argument. The localization of determinants in the developing egg must depend upon genic expression (the determinants must first be synthesized). Nevertheless, it also depends upon the specific architecture of cytoskeletal components already present in the cortex" (Grimes & Aufderheide, 1991).

We have yet to build spatial models of asymmetric divisions, to start to understand the relationship between the mechanics of unequal cleavage and the differential gene expression between the two daughter cells.

Treinin & Feitelson (1993) offer a simple model of unequal cell division in which a master gene is expressed much more in one of two unequal daughter cells, merely because of the size difference: "Thus the daughter cells may follow distinct differentiation pathways even if there were no localized determinants in the mother cell". Of course, this means that an earlier, mechanical event would be needed to make the cells different in size: "It should be noted that the question of what causes unequal cell divisions is left open". We still need models for that initial event.

Returning to regulating organisms, we would predict an analogous set of mutations, in which expansion waves were substituted for contraction waves, and *vice versa.* These would likely be embryonic lethals, unless they occurred near the tips of the differentiation tree.

If embryonic development is primarily a cortical phenomenon, somewhat independent of the underlying cells, then induction, i.e., 'fate determining' interactions between cells, is a phenomenon of secondary importance, a conclusion also reached in Proposition 74 by evolutionary reasoning.

Proposition 65: orientation scars are directly related to the cortical attachment site of the asymmetric spindle in unequal cleavage.

Dan & Tanaka (1990) not only summarize the evidence for attachment of the 'daisy' spindle in unequal cleavage to the cortex, but suggest that this attachment is fundamentally involved in the asymmetric differentiation that ensues. Furthermore, the asymmetric location of the cleavage furrow, which I would take as a single cell differentiation wave, is also implicated in these events:

"We have shown that when a cell divides unequally, the mitotic apparatus moves to one side of the cell, bringing one pole of the spindle very close to or in contact with the cortex (Tanaka, 1976; Dan, 1979; Dan & Ito, 1984; Schroeder, 1987b). A daughter cell on the side to which the spindle approaches becomes a smaller blastomere and the other a larger blastomere....

"...The significance of mechanical anchorage of a spindle pole to the cortex is to hold the off-median furrow in a constant position.

"As a matter of fact, as far as we are aware, there are only two cases in which the anchorage is unequivocally established. One is Conklin's (1917) old report on centrifuging *Crepidula* maturation spindles. In ordinary circumstances, the spindle, which is heavy, tends to be thrown down by centrifugal force. But if it is centrifuged after the maturation spindle is attached to the animal pole by one end, even though it is stretched across the cell, the attachment point does not come loose; the result is the formation of a giant polar body.

"The other example is ganglion cell formation in the grasshopper, *Chortophaga viridifasciata.* Kawamura (1977) stretched the spindle with a microneedle after it attached to the ganglion cell pole. The spindle was stretched to 2.5 times the original length, yet it did not detach.... [Cf. Inoué, 1990.]

"...It is clear that the interaction between the cortex and the mitotic apparatus (MA) must be made via astral rays [microtubules] (Rappaport, 1986b).... The role of the spindle in unequal cleavage is to fix the position of the cleavage furrow. Once the position of the spindle is established, a furrow is formed whether the furrow is median, off-median, or even close to one end of the cell....

"...What we are contending in this study is that such an attachment is a *sine qua non* for the realization of unequal cleavage....

"A further fascinating point is that if embryos were suppressed from forming micromeres..., none of the treated embryos could form larval spicules. This spectacular result (Tanaka, 1976) teaches us the real meaning of unequal cleavage.... Unequal division is not a simple instrumental device for differentiation. Its purpose is to insure future differentiation long before differentiation ever begins" (Dan & Tanaka, 1990).

"It can be inferred... that equal divisions which are repeated according to the rule of $90°$ rotation are rather means for increasing the number of cells of a similar nature; hence, are not directly connected to differentiation. It is unequal cleavage, violating the $90°$ rule, which gives rise to cells of dissimilar natures; hence, this is the first step to differentiation" (Dan, 1978).

In other words, the mechanism of unequal division is the mechanism of determination. At the level of the single cell, we are thus beginning to get a dissection of the events of determination. These include:

migration of the nucleus to one side of the cell;
distortion of one aster to a daisy (or other asymmetric) shape;
rotation of the spindle (cf. Rose & Kemphues, 1998);
attachment of the daisy shaped end of the spindle to the cortex;
construction of a septin scaffold for the cleavage furrow (Chant, 1996);
launching of the cleavage furrow (differentiation wave?);
segregation of so-called determinants;
unequal cleavage.

I have not numbered these, because the exact time sequence has yet to be worked out, and the earliest events, such as possible ionic currents, have as yet even to be looked for. Each of these features may have a homolog in regulating organisms or tissues, where launching of a differentiation wave may be readily distinguished from cell differentiation itself. Despite intracellular spatiotemporal dissection and genetic separation of some of the events and gene expressions, we still have a way to go, as exemplified by this comment on one localized gene:

"Both the actin dependence of Inscuteable localization and the presence of a putative SH3 binding domain suggest a tight connection between Inscuteable localization and cytoskeletal asymmetry. We wondered where the information for asymmetric localization of Inscuteable comes from" (Kraut et al., 1996).

The problem here, again, is to stop the infinite regress at the cytoskeleton, whose nonlinear tensegrity instability may be where "the information for asymmetric localization... comes from" (Kraut et al., 1996). To aid in the unravelling of cause and effect in these events, I list here a few ancillary phenomena:

1. There is a relationship between the asters, and the cleavage furrow, which needs investigation and may differ for equal and unequal cleavages. The 'self-propagating' nature of this mechanism is reminiscent of the resolution of the bistability of the cell state splitter:

"The cleavage contraction ring is often not detectable until the cell starts to furrow, or, in any event, until the cell starts to elongate along the spindle axis at the beginning of cleavage....

Perhaps this fact, as well as the sharp delineation of the width of the cleavage furrow, can be better understood in the light of Dan's (1988) 'astral mechanism.' The compression exerted on the furrow cortex by criss-crossing astral rays, radiating from the opposite poles of the spindle extending in anaphase-B, could impose the anisotropic deformation of the cortex localized to the furrow region. That would tend to align the actin filaments, which were previously oriented randomly, along the presumptive contractile ring. Once such an anisotropic alignment is initiated, the contractile ring could become self-propagating" (Inoué, 1990). [But cf. White, 1990.]

2. There is a relationship between the animal pole, a common launching site for differentiation waves (Figure 8 in Appendix V: Gordon, Björklund & Nieuwkoop, 1994), cleavage waves (Hara, 1977), and other waves (Speksnijder, Sardet & Jaffe, 1990a), and spindle attachment:

"The experiments... demonstrate that the cell surface (membrane? cortex?) at the animal pole of the oocyte acts as a unique attachment site.... This poses an interesting chicken-and-egg question - which came first: animal pole differentiation or spindle attachment to the animal pole?" (Inoué, 1990).

3. This chicken-and-egg problem might have a resolution in favor of the cortex, since there is an apparently independent role of the cortical actin in *Tubifex* (Shimizu, 1990). Kawamura & Yamashiki (1990) observe the spindle moved to an eccentric position by a contraction wave, and Jansen et al. (1996) find specific actin motors in asymmetric division, and put this phenomenon right in the line of causing cell differentiation:

"Grasshopper neuroblasts repeat unequal cytokinesis along the dorsoventral axis of the embryo to produce a large daughter neuroblast to the ventral side and a small daughter ganglion cell (GC) to the dorsal side.

"The eccentric position of the spindle at the time of furrowing was caused by a contraction wave (CW) that occurred in the cell cortex during mitosis.... By its squeezing action, the CW pushes the cell contents, including the spindle body, towards the GC-side during the first half of its propagation. After the pole of the spindle anchors itself to the GC-pole cortex, the cell contents on the GC-side are squeezed towards the [cap cell] CC-side, leaving the spindle body on the GC-side" (Kawamura & Yamashiki, 1990).

"It is becoming increasingly clear that the unequal segregation of very specific cell components has an essential role in many cases of cell differentiation. How some cellular components are asymmetrically segregated while most are segregated evenly is a question of great interest. It was not known at the outset of this study whether nuclear or cytoskeletal processes were responsible for mother cell-specific *HO* [endonuclease] expression. Our results clearly point to the latter.... The identification of a specific myosin subtype (Myo4p) in the asymmetric accumulation of Ash1p is direct evidence that actin-based motors have a crucial role in such processes" (Jansen et al., 1996).

4. Something, perhaps involving both microtubules and microfilaments (Allen & Kropf, 1992), rotates the spindle, in a process that smacks of being 'spontaneous symmetry breaking' (Steve McGrew, p.c., 1997) or 'exploratory' (Gerhart & Kirschner, 1997):

"The orientation of the mitotic spindle is critical for an unequal cell division. Cells that undergo highly unequal cell divisions have spindles that are oriented orthogonal to the plasma membrane, with the peripheral spindle pole abutting the cell cortex (Conklin, 1917; Carlson, 1977; Dan, Endo & Uemura, 1983; Dan & Ito, 1984; Schroeder, 1987b; Lutz, Hamaguchi & Inoué, 1988)....

"During each successive cell cycle the axis of migration of the centrosomes shifts by 90° relative to the previous axis of migration. This predictable migration of the centrosomes leads to a default pattern of cell division in which each division occurs at right angles to the preceding division.... To deviate from this pattern an additional mechanism is required to reposition the centrosomes. For example, the centrosomes and nucleus of each P blastomere of *C. elegans* embryos undergo an additional 90° rotation, which allows these cells to divide successively along the same axis (Hyman & White, 1987).

"During the early stages of leech development the cell divisions follow the default mode in which the axis of the spindle is shifted 90° relative to that of the previous spindle. Around the start of stage 5 [Weisblat et al., 1980a], however, the spindles of some of the D'-derived blastomeres undergo 45° shifts in orientation rather than 90° shifts (Sandig & Dohle, 1988)....
"...D'-derived cells of the α-amanitin-treated embryos... [have] spindles [that] are not oriented orthogonal to the cell membrane, suggesting that the centrosomes are unable to rotate to new positions. Thus, we hypothesize that zygotically transcribed gene product(s) may be required to rotate the centrosomes..." (Bissen & Smith, 1996).

"[In]... the current working hypothesis to explain rotational alignment [for the first cleavage of nematodes]... a localized region in the cortex or on the membrane acts to capture and shorten microtubules.... Any slight force imbalance will produce a turning moment on the MA [mitotic apparatus], causing it to rotate. As this occurs, more microtubules from the centrosome nearest the attachment point are likely to be captured, thus increasing the torque and causing the MA to rotate further. This process will continue until the centrosome from which the captured microtubules emanate is adjacent to the cortical capture region. Interestingly, very similar models have been proposed for the process of MA alignment in the alga *Pelvetia* (Allen & Kropf, 1992) and the budding yeast *Saccharomyces cerevisiae* (Palmer et al., 1992)" (White & Strome, 1996).

5. Some details of attachment of the smaller aster and its reshaping into a daisy or 'umbrella' form have been noted:

"The smaller aster becomes deformed into an umbrella shape, as though being pushed against the cell cortex (or being pulled by a shortening astral ray linking the centrosome to the nearest point on the cortex). Concurrently, the outer spindle pole starts to oscillate up and down, with the flattened aster appearing to glide beneath the cortex.... In a few minutes, the rocking of the spindle becomes less vigorous and eventually ceases. It is as though the spindle had hunted (with overshoot) for an attachment site located about 30 degrees above the equator. (The smaller, outer asters on spindles isolated at this stage are capped with tightly adhering cortical cytoplasm: Dan & Ito, 1984.)" (Inoué, 1990).

"In *C. elegans* both of the centrosomes in the P cells initiate rotation towards a defined region of the anterior cortex until microtubule-mediated contact is established by one (Hyman, 1989). A spindle pole that is closely apposed to the cortex displays a flattened or truncated aster (Czihak, 1973; Kawamura, 1977; Dan, 1979; Dan & Inoué, 1987; Schroeder, 1987b; Lutz, Hamaguchi & Inoué, 1988) [cf. Czihak & Kojima, 1992], as well as a smaller and elongated centrosome (Holy & Schatten, 1991). Spindles isolated after attachment to the cortex have a detergent-resistant membrane component associated with the truncated aster (Dan & Ito, 1984)" (Bissen & Smith, 1996).

6. One could ask whether intranuclear mechanical and electrical forces speculated to be present in the intact nucleus have a bearing on spindle movements:

"It has been argued that directed movements of chromosome territories and chromosomal domains occurring in three dimensions throughout the karyoplasm depend on specific intra-nuclear forces, possibly based on an actin/myosin machinery (for review see De Boni, 1994). In addition (or alternatively) extra-nuclear vectorial forces could exert effects on the structure and positioning of chromosome territories (and nucleoli) via interconnected filament networks that extend through the nuclear envelope (Ingber, 1993b; Maniotis, Chen & Ingber, 1996). Beside mechanical forces, long range electric fields could also play a role in directed movements of chromosome territories of chromosomal domains (Cremer et al., 1993)" (Cremer et al., 1995).

7. There is likely to be a relationship between cell polarity and asymmetric division. For example, the "columnar epithelial progenitor cells in the ventricular zone [of mammalian brain]... undergo... symmetric divisions (producing two new progenitor cells)" for cleavage perpendicular to the epithelial surface, or "asymmetric divisions (producing one progenitor and one neuron) (Chenn & McConnell, 1995)" (Rhyu & Knoblich, 1995). The mere geometric location of the nucleus can create asymmetric spiral calcium waves in cells (Dupont, Pontes & Goldbeter, 1996), and any other asymmetries in the cell or its cortex might do the same. Both cell polarity and asymmetric division may be based on interactions between ionic currents and microtubules:

"...Polarized, 8-cell stage mouse blastomeres drive an ionic current through themselves, into their apical ends and out of their basal ends (Nuccitelli & Wiley, 1985). Such an apical-basal current is found in all polarized epithelial cells but this was the first measurement of such a current in a cell in the process of polarizing [in this case during compaction, which may be a contraction differentiation wave: Proposition 61].... This current... enters apically and leaves basolaterally, leaking to the outside of the embryo by passing between the cells. This leakage current is responsible for generating the voltage measured across the epithelial layer, and may act back on the outer blastomeres to further polarize them.... After 2 h in a field of only 4 mV/cell diameter, there is a microtubule-dependent elongation of the blastomere along the field direction and segregation of organelles with many of the mitochondria shifting toward the positive pole.... Thus the mitochondria are moving in the same direction [basally] that they do *in vivo*.... This is the first time that the polarity of a higher animal cell has been found to be influenced by an imposed electric field, and this result strongly supports our hypothesis that the endogenous electric field plays a role in the development of polarity in the mouse system" (Nuccitelli, 1988a).

8. There may be polar traffic of molecules along specific microtubules and/or intermediate filaments extending from the nucleus to the cell membrane. This is suggested by combining the study of Maniotis, Chen &

Ingber (1996) on directed physical linkage from the cell membrane to the nucleus with the reverse demonstration of localized, directional export of transcripts out of the nucleus, probably to localized microtubules (Francis-Lang, Davis & Ish-Horowicz, 1996), along which they could be transported to localized portions of the cell membrane. Perhaps this is how orientation scars are formed.

9. It has been argued that the end of the line, so to speak, for bacterial cell division (for all bacteria and chloroplasts, but not mitochondria) is the cytoskeleton in the form of the tubulin homolog FtsZ, which forms the 'Z-ring' in both symmetric and asymmetric cleavage (Levin & Losick, 1996):

"FtsZ can assemble into the Z-ring without the assistance of any protein so far identified.... The primordial cell-division machine may have consisted of FtsZ alone.... We are now obtaining evidence that FtsZ-GTP and FtsZ-GDP favour the straight and curved conformations [Erickson & Stoffler, 1996], respectively.... The 15000 molecules of FtsZ per cell (Dai & Lutkenhaus, 1992;...) could form a protofilament that encircles the cell 30 times. In order to achieve septation, the protofilaments of the Z-ring must be attached to the membrane... perhaps [by] direct attachment of FtsZ to the lipid bilayer.... FtsZ may be not just the scaffold but also the motor. I suggest... a mechanism for constriction based on the conformational change from a straight protofilament to the sharply curved miniring (Erickson et al., 1996).... This could produce the force to power constriction and achieve septation" (Erickson, 1997)

Modelling of the spindle apparatus, so far only attempted for symmetric cleavages (Bjerknes, 1986; Czihak et al., 1991; Czihak & Linhart, 1992; cf. Valdés-Pérez & Minden, 1995; Valdés-Pérez, 1996), may be of some help in unravelling the course of unequal cleavage. But modelling asymmetric sporulation (Levin & Losick, 1996) or swarm cell differentiation (Quardokus, Din & Brun, 1996) in bacteria might get answers first (Erickson, 1997). In the tubulin homolog protein FtsZ we may have the fundamental origin of symmetric cleavage, asymmetric cleavage, and the cell state splitter. However, caution is advisable, since the postulated contractile role of FtsZ filaments (Erickson, 1997) is opposite the expansion role we hypothesize for microtubules in the cell state splitter.

In summary, so far, there is evidence for roles for both microfilaments and microtubules in asymmetric divisions. Within the cell, we could conceive of part of it undergoing a contraction wave, while the other part undergoes an expansion wave. At least the contraction wave, if perhaps not the expansion wave, is cortical. These cellular regions would presumably produce the smaller and larger daughter cells, respectively. Strictly speaking, then, the single cell differentiation wave would be a pair of subcellular waves, one contraction wave and one expansion wave. It is possible that this pair is homologous to what we will see in Proposition 250 as a pair of waves in the morphogenetic furrow of the *Drosophila* eye imaginal disc.

Proposition 66: stem cells in regulating organisms are similar to asymmetrically dividing cells in mosaic organisms. They are also members of stem tissues which have some means of postponing setup of the next cell state splitter, pair of differentiation triggers, and/or commitment signals.

If stem cells differentiate singly, then they are reminiscent of cells in mosaic organisms, which are also capable of undergoing either symmetric ('proliferative') or asymmetric cell divisions. In the case of blood, observation to date...

"...does not provide sufficient evidence to resolve the dilemma whether the development of haemopoietic lineages proceeds according to schema of 'progressive and stochastic restriction' or more limiting 'developmental tree' with some random control process involved.... It is postulated that commitment is an intrinsic process, which in the early stages of hemopoiesis may not be under the influence of external factors (Ogawa, Pharr & Suda, 1985).... While we still do not fully understand the processes of commitment, proliferation and differentiation, we now know they are subject to a complex regulatory mechanism" (Novak & Stewart, 1994).

That 'complex regulatory mechanism' may just be single cell differentiation waves. Perhaps stem cells belie our descent from mosaic ancestors.

Embryonic tissues and terminal tissues are distinguished by whether or not their cells are at a terminal edge of the differentiation tree. This definition permits us to regard stem cells (Wolpert, 1988; Potten & Loeffler, 1990) as fundamentally embryonic in nature, even though they occur in adult organisms. (Note that I am using the word 'terminal' to indicate cells that are at the terminal ends of the differentiation tree. They may or may not have also gone through a 'terminal differentiation' in the sense of an inability to undergo further cell divisions. Cf. Reiner, 1983.) This view of stem cells thus gives us particular mechanisms to seek that maintain cells as stem cells. It is not necessary in the definition of a tissue that the cells be physically contiguous, though the latter would seem necessary for the transmission of physical forces from cell to cell within a tissue. One might suspect that stem cells that appear in noncontiguous tissues have a physical instability mechanism that resolves into two cell states for each of the daughter cells, one the same as before, and the other different. Blood stem cells, and their progeny, for example, seem to follow most of the properties we ascribe here to differentiation trees:

"The differentiation of the different kinds of blood cells can be thought of in terms of the branching pathway model... the branching pattern is also a cell lineage which starts with the multipotential stem cell. The key feature in the programme of differentiation of the stem cell is one of asymmetry [cf. Novak & Stewart, 1994].... Is this asymmetric behaviour an intrinsic property of the stem cell...? ...As the cells proceed from the stem cell to the various mature blood cell types they divide many times. While some of these divisions represent a branch point, many of the divisions are only proliferative, and serve to amplify the number of cells at that particular stage in the lineage" (Wolpert, 1991a).

In human adults blood cell differentiation occurs in a 'microenvironment' in the bone marrow (Quesenberry, 1995; Abboud & Lichtman, 1995; Stern, 1996). Whether differentiation occurs in groups of stem cells or singly appears to be an open question, though the involvement of gap junctions (Rosendaal et al., 1994) would appear to favor groups, perhaps permitting some sort of localized, multicellular differentiation waves.

One fascinating recent observation is that central nervous system stem cells occur in particular, contiguous tissues (Reynolds & Weiss, 1992; Vescovi et al., 1993). This suggests the possibility that they are a contiguous tissue that results from a particular sequence of differentiation waves. Neuroectodermal tumors in children (Peringa et al., 1995) act as if they share this and other properties of stem cells (Trojanowski, Tohyama & Lee, 1992; cf. 'embryonic neoplasms': Witschi, 1952). Thus they may be the result of an 'aberrant sequence' of differentiation waves.

Some nearly terminal embryonic tissues and stem tissues (as in blood cell differentiation), for which coordination with other tissues and tissue contiguity are lacking or may not be so important, might lack global synchronization mechanisms. This effect may also be involved in the formation of some organs containing mixed cell types (Fristrom & Rickoll, 1982).

The key to stem cells, whether they operate individually or as locally contiguous tissues, is perhaps in some as yet unnamed and unidentified mechanism for suspending the process of travelling along the differentiation tree, so that at least some of them do not go on to terminal differentiation. Stem cells thus have a 'pause' mechanism at a point along their differentiation code. This may be as simple as cell type $A \Rightarrow A + B$, instead of $A \Rightarrow B + C$, as discussed further in Section 9.26.

5.00 DEVELOPMENT AND EVOLUTION

"Gene expression is not only significant for the homeostasis of the mature organism, but at the developmental level is the basis of both normal development and evolutionary change. With the notable exceptions of Waddington's *Strategy of the Genes* (Waddington, 1957), S. Løvtrup's *Epigenetics* (Løvtrup, 1974) and R. Riedl's *Order in Living Organisms* (Riedl, 1978), remarkably little attention has been paid to the theoretical aspects of this central and fundamental problem" (Reid, 1985).

"No Darwinian will ever question the importance of development in evolution, but evolutionary interpretation is constrained by the extent to which the proximate causations of development have been elucidated by the embryologists.... If someone were to ask me which frontiers of evolutionary biology are likely to see the greatest advances in the next ten or twenty years, I would say elucidation of the structure of the genotype and of the role of development" (Mayr, 1991)

"At least one speaker believes that a central problem impeding a successful synthesis and integration of developmental and evolutionary biology is the absence of a comprehensive theory of development" (Wake et al., 1991).

"The last several years have witnessed a coming together of developmental and evolutionary biology.... It is far from a total intellectual merger" (Hanken, 1993).

5.01 Evolution of Cell State and Tissue Splitting

"Since we do not understand how the plan on which all these structures are built changes over evolutionary time, we cannot claim to have explained the evolutionary mechanism" (Taylor, 1983).

Proposition 67: *synchronization mechanisms* for coordinating the appearance of cell state splitters in all of the cells of a tissue developed in the course of evolution in response to selective pressures for larger tissue sizes.

It is generally presumed that larger organisms come to occupy the top ranks of the food chain or web, and that niche space is always open for larger organisms (Bonner, 1988; but cf. Hanken & Wake, 1993; Krebs, 1994). This results in an evolutionary push, pull or drift (depending on one's theory) of at least some phylogenetic lines towards larger sizes. There are basically three ways to make larger organisms: 1) increase the number of cell types; 2) make larger cells; 3) increase the number of cells of each kind. Nematodes, with around 900 distinct kinds of cells, and about the same number of cells (Horvitz & Herskowitz, 1992), may represent the limit of the first strategy. Some plants and salamanders make larger cells and organisms through polyploidy or by increasing their junk DNA (Martin & Gordon, 1995).

The third solution, to make more cells of each kind, requires some evolutionarily novel tricks. A distinction has to occur between cell proliferation and cell differentiation. A cell has to 'remember' what state of differentiation it is in, while proliferating, so that the daughter cells remain of the same type. If the cells are to form a contiguous, many celled tissue, then differentiation from a previous cell type has to be more or less synchronized: larger tissue sizes present the problem of how to coordinate the differentiation of all cells within a tissue. Once a multicellular tissue is formed, the cells may need mechanisms to maintain their differentiated state. Here I will concentrate on the problem of synchronizing differentiation, a role possibly played by cell state splitters and differentiation waves (cf. Section 3.04).

It is possible that synchronization mechanisms involve junctions between cells:

"...Low-resistance junctions make it difficult for an individual cell to set up and maintain a higher concentration of small molecules than is found in its immediate neighbors.... Only metazoan cells have so far been found to have low-resistance junctions. It is therefore tempting to suggest that the development of junctional communication, with its potential effect of stabilizing cell populations, has been a major factor in the evolution of metazoa from protists" (Sheridan, 1974).

We can thus imagine a sequence of events in which multicellular organisms arose via cellularization of single cell organisms (see Section 5.05). This cellularization became rather complete, in the sense that open connections, making the cellularized organism essentially a syncytium, nearly vanished (or transport between cells was severely curtailed). Differentiation waves that previously traversed maybe the whole (single-celled) organism were now confined to the individual cells. New mechanisms then evolved that permitted limited junctions between cells, but did permit differentiation waves to proceed from one cell to the next, at least within a tissue.

Proposition 68: the need for cell state splitter synchronizing mechanisms explains why regulating embryos are a later evolutionary development than mosaic embryos.

It is always difficult to argue what came first in evolution, without having a clear fossil record (Raff, 1996). Let us suppose that cells in mosaic organisms are indeed single cell tissues, and that induction between those cells is an evolutionarily secondary phenomenon. We can see then that synchronization between cells in a tissue is not needed by mosaic organisms. If it is 'added on', then it likely occurred later. This proposition is contrary to...

"the principle that developing organisms are self-organizing wholes which we will call fields. The occurrence of mosaic properties is then due to the presence of forces resisting the field property..." (Goodwin, 1982a).

Thus Goodwin (1982a) would place regulating organisms first, and derive mosaic organisms from them. There is evidence in coelomates that induction is the more primitive condition (Freeman & Lundelius, 1982). On the other hand, the suggestion that mosaic organisms such as nematodes evolved well before the Cambrian explosion, which probably included all the known regulating organisms, supports this proposition:

"According to Wray, Levinton & Shapiro (1996), there is new data suggesting that animal lineages are far older than the Cambrian explosion [Gould, 1989a]- perhaps another 500 Myr older! Nematodes probably diverged from other animals before protostomes diverged from deuterostomes (Sidow & Thomas, 1994; cf. Conway Morris, 1993; Winnepenninckx et al., 1995)" (Fitch, 1996).

This is consistent with the concern of Raff (1996):

"It is tempting to assume that the Cambrian radiation and divergence of regulatory genes go hand in hand. It seems more likely that most of the evolution of regulatory genes and their patterns of expression predate the Cambrian radiation and even most of the radiation of body plans" (Raff, 1996).

The support for this proposition is basically one of reasonableness: it would seem easier to spread differentiation waves from single cells to multiple cell tissues, over evolutionary time (via the evolution of cell-cell junctions?), than the reverse. Raff (1996) clearly states the general problem of concluding which way evolution went. All of my propositions to this effect must thus be rather tentative until proper phylogenetic work is done:

"The point is that without an unequivocal phylogeny or evolutionary history, we cannot determine the direction of evolutionary change in development. For most situations in which evolutionary conclusions are drawn from comparative data, the direction of change must be inferred. In many of the most interesting cases, such as consideration of the evolution of developmental processes among animal phyla, the problem is compounded by the fact that the phylogenies rest heavily on comparative embryology. Circularity fails as a method" (Raff, 1996).

There is an element missing from this argument. Sometimes we are confronted with a step in evolution, say $A \Rightarrow B$, that would seem to be irreversible. In such a case, the reverse, $B \Rightarrow A$, is unlikely, and we can say that A must have preceded B. I would like to suggest that this proposition is of this type, though collectively we have a long way to go to learn when such argumentation is reasonable and when it is not.

Present day organisms with mosaic embryos may be divided into two classes: those with mosaic adults and those with regulating adults; the latter being generally larger are probably evolutionarily more advanced. This is suggested by the observation that some mosaic embryos switch to regulation once they become adults (Grant, 1978; Maclean & Hall, 1987).

Proposition 69: a major step in the evolution of multicellular organisms was the ability to generate the next cell state splitter and accompanying differentiation triggers and commitment signal, i.e., the invention of *continuing differentiation* with postponed terminal differentiation.

Buss (1987) has observed that...

"Of the some 23 monophyletic protist [sic] groups, fully 17 have multicellular representatives.... Of the 17 multicellular taxa, only 3 groups - the plants, the fungi, and the animals - have developed cellular differentiation in more than a handful of species.... We know virtually nothing as to why this transition was closed to all but a few taxa. Minor taxa, which possess only a small number of species with cellular differentiation, provide valuable clues to the roadblocks in the path to the differentiated state..." (Buss, 1987).

In our model for the mechanochemical basis of differentiation of ectoderm using cell state splitters (Appendix IV: Björklund & Gordon, 1993b), a link is made between the contraction of the microfilament ring in each ectoderm cell and the selection of a state for a master gene. The regulation of genes by genes is a ubiquitous phenomenon in prokaryotes (Stanier et al., 1986; Singer & Berg, 1991), but actin is found only in eukaryotes (Stanier et al., 1986; however, cf. Minkoff & Damadian, 1977; Nanninga, 1998).

Thus continuing differentiation, which I define as the ability to indefinitely produce additional cell types, may have been confined to eukaryotes precisely because both master genes and actin (and/or microtubules and intermediate filaments) had to be present in the same cell first, before mutation could lead to their linking. Perhaps we could trace the origin of the cell state splitter (and the spindle apparatus: Pickett-Heaps, Forer & Spurck, 1997) to a particular symbiosis: at least tubulin may have had its origin in spirochetes (Hinkle, 1991; Margulis, 1993b; Erickson, 1995) or just about any other bacteria (Erickson, 1997). Furthermore, such a link may thus be the basis for all continuing differentiation. One component, the master genes, may have an ancient history:

"Although the plant homeoboxes differ considerably from those found in animals and fungi, the similarities still suggest that all homeoboxes descended from a gene in the organisms' common ancestor, which lived a billion years ago. The role of a homeoboxlike gene in that unicellular ancestor is open to speculation. One possibility, McGinnis [cf. Shephard et al., 1984] notes, is that the gene regulated the transformation of those one-celled creatures into a variety of forms. After multicellularity evolved, that function could have been co-opted to produce different cell types in different body regions" (Rennie, 1991a).

A hint of the molecular basis of continuing differentiation is in the offing:

"The wide representation of tyrosine kinases in the *Drosophila* [and vertebrate] genome is in sharp contrast to their absence in unicellular organisms, suggesting that they are a metazoan 'invention' created for purposes of communication between cells in multicellular organisms" (Schejter & Shilo, 1989).

As we ascertain the full spectrum of molecules that are in multicellular but not in single cell organisms, we may also assemble a list of the components of the earliest cell state splitter, the differentiation pathway, and the nuclear state splitter.

Proposition 70: the evolution of continuing differentiation coincided with the ability to duplicate terminal branches of the differentiation tree.

The transition from simple terminal differentiation to multistage, continuing differentiation has yet to be explored and understood. It is possible that this step of evolution coincided with the evolution of mechanisms for the duplication of multigene families to produce gene superfamilies (Arnheim, 1983; Raff et al., 1987a). Gene superfamilies may themselves reflect the ability to duplicate terminal branches of the differentiation tree. We occasionally see what may be duplications of differentiation tree terminal branches (cf. Proposition 134), as in the ectopic appearance of whole sublineages in nematode mutants:

"Our lineage studies revealed striking reiterations of wild type lineage patterns in ectopic positions. These lineage patterns accurately reproduced not only the proper numbers of cell divisions and terminal cell morphology, but frequently the relative timings of terminal cell divisions as well. Thus, the lineages of the AB^8 cells appear to represent sublineage modules or cassettes that can be executed as complete units" (Moskowitz, Gendreau & Rothman, 1994).

Raff et al. (1987a) have recognized that regulatory duplication precedes protein diversification, as would be expected for duplication of terminal branches of the differentiation tree:

"...It appears likely that the primary selective pressure toward genetic multiplicity in eukaryote evolution is in the establishment of separate control mechanisms for batteries of genes that act in differentiation of specific cell or tissue types. It seems likely that diversification of regulatory repertoires preceded selection for modulated protein function....

"That regulatory requirements are the primary constraint in evolution of a multiple gene family is also implied if one considers the possible pathways of joint selection for pattern of expression and for specificity of protein function. The more logical, and, one assumes, more likely, pathway is that only after a duplicate copy of a given gene is brought under differential control will selection operate to fix changes in the two protein gene products that may be advantageous in the specific context in which each gene is expressed" (Raff et al., 1987a).

Duplication of a differentiation tree terminal branch guarantees that the 'differential control' is in place when the 'given gene' is duplicated, though, as shown by Figures 24-26, this itself may take some evolutionary time, so there could be overlap between the regulatory and protein intervals of evolutionary change. The latter could continue until the homologies between 'daughter' proteins are no longer apparent (Kafatos et al., 1987).

Proposition 71: the evolution of continuing differentiation coincided with the origin of imprinting.

Another fundamental, and perhaps equally widespread mechanism, imprinting, in which a gene is expressed differently depending on whether it is supplied maternally or paternally (Howlett, 1991; Martin, 1996), may have evolved at or about the same time as continuing differentiation (cf. Chandra & Nanjundiah, 1990):

Terminal branch of a differentiation tree

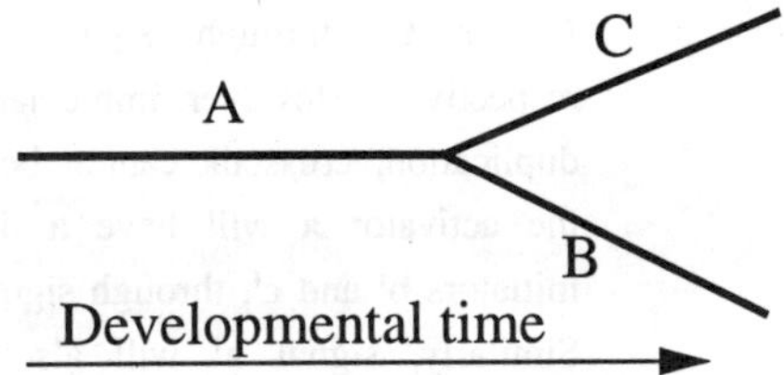

Its possible representation in DNA

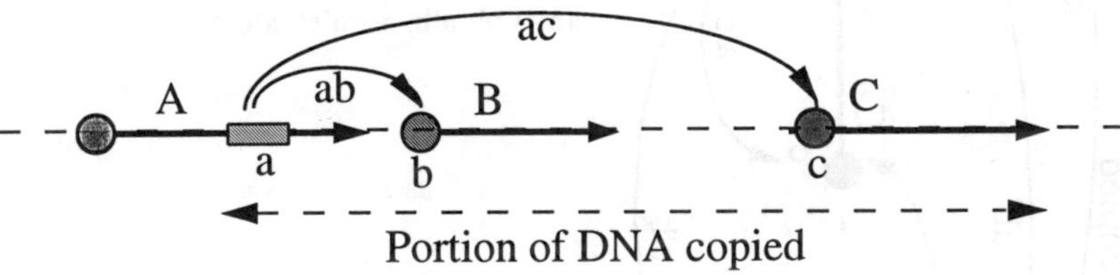

Figure 24. DNA basis for a differentiation tree branch duplication. Tissue type **A** erects cell state splitters in all of its cells, that resolve their mechanical instabilities as either cell type **B** or **C**, which are considered terminally differentiated cell types in this example. The activator for gene cascade **A**, **a**, sends out one of two signals in each cell: either **ab** or **ac**, depending on whether the cell has participated in an expansion or a contraction wave. These signals activate either gene cascade **B** or **C**, respectively.

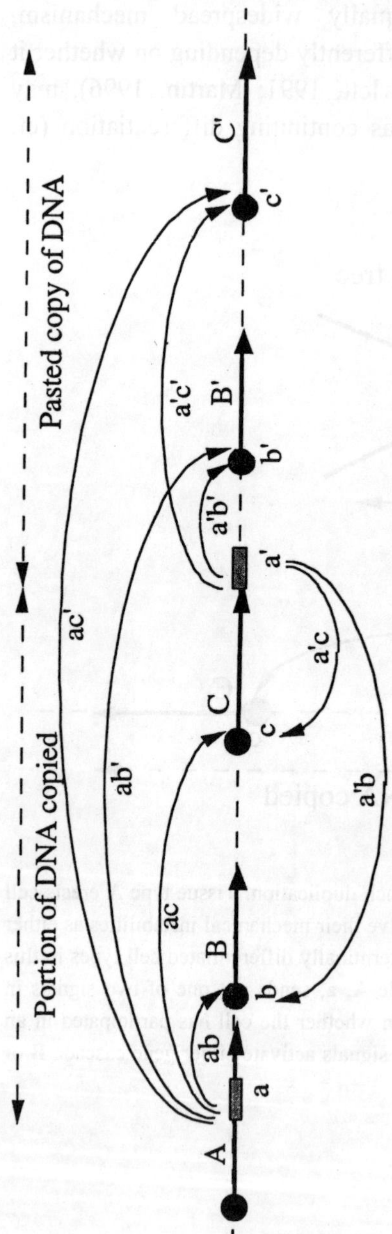

Figure 25. State of the DNA after a duplication. Here I assume the simplest case, namely that the copied terminal branch is inserted at the end of the formerly terminal **C** cascade (a "reduplication": Darlington, 1958), extending the terminal branch from which it was copied, in such a way that activation of cascade **C** also leads to the copy of the activator **a**, namely **a'**, also being activated. This in turn activates either **B'** or **C'** through signals **a'b'** or **a'c'**, respectively. However, immediately after such a duplication, crosstalk cannot be avoided. Thus the activator **a** will have a direct link with initiators **b'** and **c'**, through signals **ab'** and **ac'**. Similarly, signal **a'** will also back react to cascades **B** and **C** through the signals **a'b** and **a'c**, respectively. Of course, in terms of molecular structure, **a**=**a'**, **b**=**b'**, **c**=**c'**, and **ab**=**a'b'**=**a'b**, **ac**=**a'c'**=**a'c**.

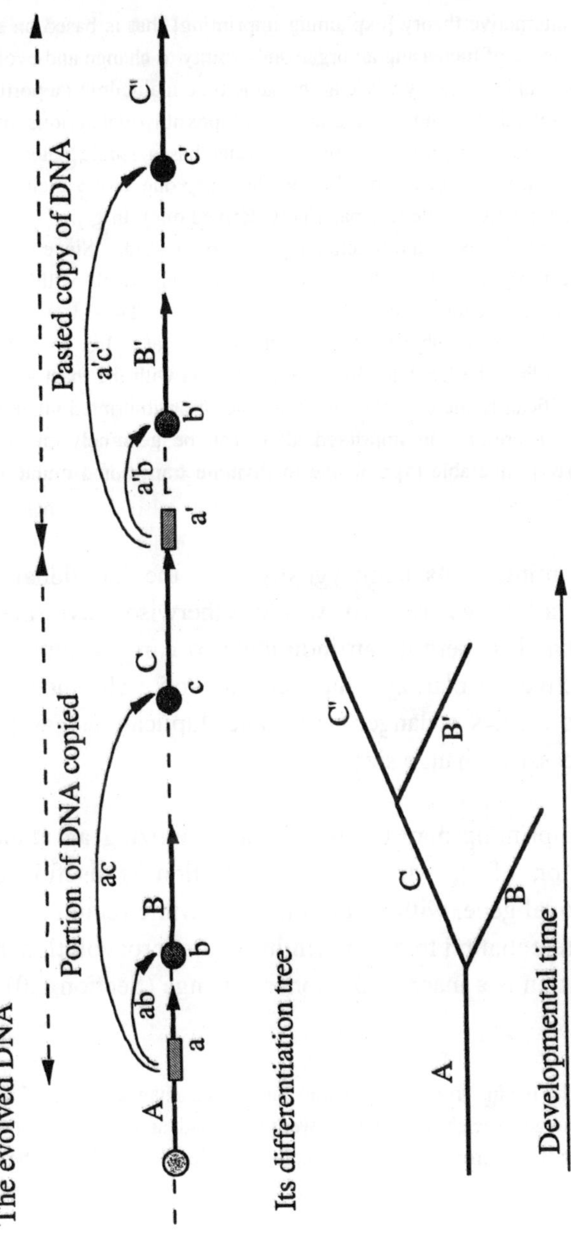

Figure 26. Coevolution of DNA after a duplication. Activator **a'** is assumed to coevolve with initiators **b'** and **c'**, so that a new, cleanly separated terminal branch of the differentiation tree results. In this diagram, the old cascades are shown unchanged, though they, too, could continue evolving. Thus we can end up with **a'**≠**a**, **b'**≠**c**, **c'**≠**c**, **a'b'**≠**ab**, **a'c'**≠**ac**, **B'**≠**B**, and **C'**≠**C**.

"I wish to suggest a possible alternative theory [explaining imprinting] that is based on the idea that imprinting may be a means of increasing an organism's ability to change and evolve its genetic loci relatively quickly and efficiently while at the same time maintaining a portion of the population safe from selection against the altered and possibly deleterious new alleles.... Within a sexually reproducing group of organisms that has a single paternally imprinted locus,... every individual within that population would carry one active allele (the maternally derived one) and one inactive allele (the paternally derived one). In general terms imprinting causes a locus to essentially exist in a functionally hemizygous state. Since in this basic model there is only one type of allele in the population, all individuals within the population would have the same phenotype elicited from that allele. They would all, however, have a second silent allele (not subject to natural selection) that is free to mutate differently in each individual. In fact, the DNA methylation associated with the inactivation of imprinted alleles would significantly increase the mutation rate of an imprinted allele by 5-methylcytosine deamination. Therefore, an imprinted allele can be genuinely called a 'mutational hotspot', particularly [vulnerable to] cytosine to thymine transitional mutations (Wiebauer et al., 1993)" (Martin, 1994).

A disadvantage of imprinting is its hemizygosity, i.e., the individual is dependent on a single allele where two would otherwise have been available. On the other hand, it permits an individual to carry a mutation without the reproductive isolating mechanism of chromosome incompatibility, which is always a danger with gene duplications (as the chromosome must, of necessity, change size).

It is worth noting that imprinting may be useful for optimizing an allele, replacing an older version of it, while gene duplication is useful for introducing an entirely novel gene, with a new function, while retaining the old allele. In terms of differentiation trees, imprinting is a microevolutionary change, and gene duplication is a macroevolutionary change (Section 6.01). Imprinting is an example of...

"*The storage hypothesis* [which] begins from the fact that mutational changes, even if not beneficial, could be stored in the genome so long as they were not disadvantageous. As soon as all the changes necessary for the improvement of a complex functional system have been stored they will all act together and produce the improvement by collaboration" (Riedl, 1978).

Imprinting is based on DNA methylation, which can be highly localized along the DNA (Taylor, 1984; Riggs, 1989, 1990; Bestor, 1990; Wilkins, 1990; Brandeis, Ariel & Cedar, 1993; Bestor & Verdine, 1994; Razin & Kafri, 1994), though the mechanism for its 'starts' and 'stops' is not yet understood. One could imagine the intervals of methylation covering subtrees of the differentiation tree, consisting of a number of differentiation cascades, not just isolated alleles. In this case, an individual could be carrying two versions of a subtree of the differentiation tree: the original and a mutated copy. The mutation would have some chance of spreading in the population, and since imprinting can act on either DNA strand, the mutant would be 'tested' for viability and increased (or decreased) fitness. A subsequent duplication event might then incorporate the mutant subtree as a new subtree, supplementing the original subtree.

It will be interesting to learn if duplication of multigene families and/or imprinting are the capabilities lacking in simple organisms that may have only one step of (terminal) differentiation available to them:

"Just as the slime mold *Dictyostelium* displays spontaneous reversion to the multicellular state from one of cellular differentiation, so does *Volvox*. Mutants of *Volvox* are known in which cellular differentiation is lost; all cells are capable of generating a new colony (Kochert, 1975). In a single step, cellular differentiation has been lost and *Volvox* has returned to its initially multicellular state" (Buss, 1987). [Cf. Desnitski, 1993, 1995.]

Are there any organisms with continuing differentiation that can make such a complete switch to lack of cellular differentiation? If so, they may contain the key to understanding continuing differentiation.

Proposition 72: continuing differentiation required the evolutionary invention of the genetic program.

Here I am using my redefinition of the genetic program as an alternation of physical and gene cascade events (Proposition 14), in which a particular cell lineage wends its way through the differentiation tree. Harold (1990) has reviewed the genetics of morphogenesis of unicellular organisms and concluded that they lack "transcription of [particular] genes at particular stages" and that they thus lack "a genetic program in [the]... rigorous sense [of Stubblefield (1986)]". But this appears to be an essential component of differentiation in multicellular organisms: as discussed in Proposition 88, with rare, disputable exceptions, genes specific to a given type of cell are not activated until that cell type differentiates from a previous cell type. It will be interesting to try to find organisms of intermediate capability: perhaps some of the colonial microorganisms will be found capable of continuing differentiation. Such organisms could give hints as to whether the physical or genetic component of the differentiation tree came first in the early evolution of multicellular organisms.

To test if continuing differentiation evolved more than once, we'll need to do comparative work on the differentiation trees of organisms that have only a few cell types (Bonner, 1988).

Proposition 73: the diversity of potential differentiation trees exceeds the number of individual organisms that have ever existed.

"...Complexity is in the eye of the beholder" (Robert Rosen in: Rosen, Pattee & Somorjai, 1979).

The division of development into edges and nodes of a differentiation tree has consequences for complexity. Each edge, in representing the playing out of the characteristics of a given differentiated cell type through its differentiation cascade, effectively operates at the cellular, histological level. The number of edges (generally terminal edges in the adult, except for stem cells), and their topological arrangement, correlates with the morphology of the organisms, and thus determines what we usually take as a measure of complexity, i.e., its visual and anatomical form. This point has also been made for 'morphogenetic trees' (Arthur, 1988; Rieppel, 1990).

The degree of complexity for a given size of differentiation tree is incredible. Butler (1990) derived the number of rooted binary trees with n distinguishable terminations as...

$$T_n = (2n - 3)!! \approx [1 + O(n^{-1})]\, n!\, 2^{n-1}/(n\,[\pi n]^{1/2})$$

or

$$T_n \approx 2^{n-1/2}\, n^{n-1}\, e^{-n}$$

For example, with, say, $n = 200$ cell types we get $T_n \approx 10^{431}$ (cf. Bjerknes, 1993). (Bonner, 1988, guesstimates $n = 120$ to 500 for vertebrates.) Thus potential diversity, for our size of differentiation tree, greatly exceeds the number of individual organisms that ever existed. In fact, it exceeds the number of atoms in the universe (around 10^{81}: Tyson, 1994) many times over. The number of earthly organisms may be overestimated by dividing the total biomass on earth (1.8×10^{18} g total dry organic matter: Ramade, 1984) by the dry mass of a bacterium (2×10^{-13} g: Stanier et al., 1986), assuming a short generation time for a bacterium (0.14 hours: Stanier et al., 1986), and assuming that unique individuals have been produced in every generation since the origin of the earth 4.6×10^9 years ago. The upper bound is about 2×10^{45} organisms since the origin of life. Thus the potential number of differentiation trees far exceeds the number of individual organisms that have ever lived on earth. Certainly, many more differentiation trees are possible than the estimates of 10 to 100 million (10^7 to 10^8) species that have ever existed on earth (Wilson, 1992):

"In spite of the enormous diversity of life, with many millions of species through the years, it represents only a minute fraction of the possible forms of life" (Simpson, 1964a).

As an example of the diversity available via differentiation trees, let's consider nematodes, for which the lineage tree is nearly identical to the differentiation tree. Even though the size of nematodes appears to be limited, perhaps due to propagating errors related to orientation scars (Propositions 64 and 170), they have managed to carve out a significant number of alternative, successful differentiation trees:

"Nematodes are one of the very largest groups in the animal kingdom, with a vast range of exploited environments. The number of nematode species according to estimates by various authors is between 100,000 and 1 million. With regard to numbers of individuals, nematodes undoubtedly exceed all other multicellular animals" (Malakhov, 1994).

Differentiation trees therefore place no significant restraint on the diversity of organisms.

5.02 The Secondary Importance of Embryonic Induction

"Science isn't just a matter of putting a puzzle together,... it's more a matter of determining what pieces belong in any particular puzzle" (Root-Bernstein, 1989).

Proposition 74: embryonic induction is an evolutionarily opportunistic and thus secondary interaction between two tissues. Its function may be to help coordinate the timing of their development.

"The time-order of embryonic segregation [determination] is autonomous; the localization is dependent on induction" (Lillie, 1929a).

Some authors have proposed that induction is the primary cause of embryogenesis in regulating embryos and the primary developmental fact to be considered in relating development to evolution (Horder, 1983; Arthur, 1988):

"...Rather than uncovering a gene-encoded program, the goal of developmental neurobiology must be the discovery of the functional relations, or algorithms, that govern the nonprogrammatic, contextually determined intra- and intercellular interactions underlying the historical phenomenon of metazoan ontogeny" (Stent, 1985a).

Figures showing complex networks of tissue inductions may be seen, for example, in Seidel (1953) and Coulombre (1965) (cf. Riedl, 1978). It is supposed that the embryo builds itself through these inductions of one tissue by another.

Our work has, however, driven us to different conclusions. We found that the most parsimonious description of the role of an inducing tissue, such as the 'organizer' that is believed to function in primary neural induction (Spemann & Mangold, 1924a,b; Hamburger, 1988), is as a passive mechanical component, at best a component in whatever determines the launching domain of a differentiation wave. It is the differentiation wave that triggers a gene cascade. Induction of a tissue 'by' another, when it exists, serves merely to specify the launching domain for that wave.

It is widely observed that events that require induction in one organism proceed spontaneously in related organisms, and this has led to suggestions of 'self-induction' or 'self-differentiation' when no inducer can be found:

"Eakin (1973) has shown that pineal lens induction is not dependent on contact with specific surrounding tissues: the inductive stimuli responsible for lens differentiation are likely to be the result of interactions between cells in the vesicle itself..." (Horder, 1983).

Induction with or without a neighboring inducing tissue has been raised to a principle of 'double assurance' (Filatow, 1925; Spemann, 1927b; Needham, 1931b), rather than being seen as a problem for the concept of the organizer. Analogous results are found when tissue cultures of ectoderm are made: some amphibian species exhibit 'autoneuralization' while others do not (Mattsson, Nakatsuji & Løvtrup, 1987; cf. 'autodifferentiation' in Grunz, 1990a). (This variation has been attributed to a threshold phenomenon: Pieter D. Nieuwkoop, p.c., 1995; Duprat, 1996). If we compare phyla we find that...

"In insects no inductive interaction with mesoderm is required to form the neuroectoderm.... However, inductive interactions among the neuroectodermal cells are required for their segregation into neuronal and epidermal lineages" (Denburg & Norbeck, 1989).

"In ascidian embryos, the presumptive area of the endoderm resides at VP [the vegetal pole], but many organs can differentiate autonomously without VP area (reviewed in Satoh, 1994)" (Kuraishi & Osanai, 1994).

"In two... species [of nematodes], the vulva is not induced by the gonad, but instead relies on intrinsic properties of precursor cells. Thus, evolution of organ position involves changes in induction and competence" (Sommer & Sternberg, 1994).

Hall (1983a) shows that Meckel's cartilage is induced by entirely different tissues in different classes of vertebrates (cf. Wagner, 1989b,c). Jacobson & Sater (1988) review and give a "List of [37] species tested to determine whether the lens will (+) [25 species] or will not (-) [12 species] form in the absence of the retina" contravening the classical result that initiated the concept of induction:

"Herbst (1901), reporting that the number and position of lenses corresponds to the number and position of abnormal optic cups in cases of cyclopia and near cyclopia, was among the first to propose this causal relationship. Spemann (1901a)... concluded... that 'lens formation is only possible under the influence of the optic cup' (Spemann, 1938). Further experiments failed to support this conclusion" (Jacobson & Sater, 1988).

Successful neurulation has been observed in notochordless embryos (Lehmann, 1934; Youn & Malacinski, 1981; cf. Goerttler, 1926; Hoadley, 1938; Witschi, 1952; Hamburger, 1988; Sato & Sargent, 1989; Placzek et al., 1990c; Clarke et al., 1991; Yamada et al., 1991; Darnell, Schoenwolf & Ordahl, 1992; Artinger & Bronner-Fraser, 1993; Goulding, Lumsden & Gruss, 1993; Ang & Rossant, 1994; Weinstein et al., 1994; Ruiz i Altaba et al., 1995; Pennisi, 1996; Psychoyos & Stern, 1996; Swalla & Jeffery, 1996; cf. Greenspoon et al., 1995, for a laser ablation technique and contrast van Straaten & Hekking, 1991). This suggests the secondary nature of even primary neural induction. The facts gathered here contrast with the usual view that...

"Embryonic induction is probably the most widespread, and perhaps most important, mechanism of cell differentiation in vertebrate development. Almost every major organ or tissue is known to require one or more inductions in its formation" (Gurdon et al., 1989).

The weight of the evidence thus leads us to suggest this proposition. The concepts of 'induction' and 'self-induction' may be similar in status to that of epicycles within epicycles as an explanation for the movement of the planets (Sarton, 1959). All heavenly motion was supposed to be perfect, i.e., circular. When this didn't work, it was conveniently found that just about any motion could be approximated as circles moving on circles moving on circles, ad infinitum. But small discrepancies were found to persist by Kepler. When he replaced epicycles by simple ellipses, we had a breakthrough (Kuhn, 1957). (Cf. the criticism of Goldschmidt (1954), of the 'statistical philosophy' in genetics with "selection involving more and more modifier systems, without inquiring whether this explanation is necessary".)

McKinney & McNamara (1991) summarize the relationship between induction and heterochrony as follows:

"...For novelties we must consider... differentiation. Here rate or timing changes result in spatial arrangements that can lead to new cellular juxtapositions (configurations) and sometimes even alteration of induction" (McKinney & McNamara, 1991).

In other words, induction between two tissues can only occur during embryogenesis if those tissue coexist, and heterochronic changes over evolution can cause tissues to coexist or not, and change their temporal (let alone spatial) relationship. For this reason I suspect that induction is opportunistic: if there is some advantage to the use of an available tissue interaction, it may be selected and becomes an 'induction', i.e., a means of launching a particular differentiation wave.

The extensive study of Richardson (1995), showing heterochronic relationships between most somite-stage vertebrate embryos, does not simply attest to lack of a 'phylotypic' period, but also suggests a lack of universality in inductive relationships between tissues.

We see then that induction is secondary because: 1) the primary event, in terms of gene expression, is triggering by a differentiation wave; 2) comparative embryology and considerations of heterochrony show that induction need not exist for a given tissue differentiation to occur.

Proposition 75: high rates of speciation via heterochronic changes may not only disrupt inductive relationships between tissues, but also tend to keep the differentiation tree from establishing crosstalk or webbing between edges. Reduced webbing permits a higher rate of species radiation.

"...Hierarchical organization is an outcome of hierarchical construction, which in turn is an evolutionary strategy that must itself have been shaped by natural selection" (Layzer, 1990).

'Crosstalk' means that the genes in two edges of the differentiation tree become regulatorily intertwined. The heterochronic 'sliding' of edges relative to one another over evolutionary time, breaking and making inductive interactions between tissues, reduces the chances that this will happen. For example, in discussing direct development and adultation in larvae, Freeman (1982) speculates that...

"While it is difficult to imagine how one of these transitions took place, it is probable that the change in egg size created physical stresses that led to a change in the pattern of morphogenetic movements during early embryogenesis. In some cases this may have facilitated new modes of germ layer formation and provided the basis for new kinds of inductive interactions during early embryogenesis" (Freeman, 1982).

Heterochronic sliding helps retain the tree structure of the differentiation tree, preventing it from becoming a *differentiation web,* "an intertwined set of genetic instructions" (Smith & Morowitz, 1982), or equivalently, a "buildup of contingencies in developmental programs... called 'hardening'" (McKinney & McNamara, 1991). This idea could be tested, for instance, by working out differentiation trees for the placental mammals, which have had a recent major radiation. A comparison of their differentiation trees might reveal numerous cases of altered relationships between many branches, including changes in inductions.

In terms of graph theory, there is a sharp distinction between trees and the general graph (Berge, 1985). Unfortunately, there seems to be no general term in graph theory for a non-tree graph, so I will refer to such a structure as a 'web'. In phylogenetics, Mindell (1992) and Margulis (1993a) use the term 'net'. I prefer 'web', because of its connotation of entanglement. A web reflects an inductive relationship that cannot disappear, no matter how the organism evolves. We will have to do a fair amount of comparative embryology before we can identify which parts of given differentiation trees appear to be inexorably tied up as webs, and which parts can slide relative to one another, permitting heterochrony and a change in inductive relationships.

There is a general, apparently unasked question about biological systems lurking here. While we freely acknowledge the hierarchical nature of life, are there actually forces (such as success via radiation) that keep the hierarchical levels that we perceive relatively cleanly separated from one another, or is the hierarchical structure of nature more a mental construct? As this question merges with epistemology, and is intimately involved with the idea of emergent phenomena, I will not venture further in this pursuit, at least for now.

Proposition 76: the regulatory levels of the differentiation tree and the differentiation cascades forming the edges of the tree might be distinguishable on the basis of the amount of web formation.

While evolutionary radiation may act on differentiation trees, to keep them (for the most part) from becoming differentiation webs, no source of such evolutionary constraint is apparent at the next lower hierarchical level, i.e., that of the individual differentiation cascade (corresponding to one edge of the differentiation tree). Thus I would anticipate that within each differentiation cascade there would be numerous entanglements, making the gene networks within a differentiation cascade more web- than tree-like (cf. Figures 27-28). While each master gene initiating a differentiation cascade might have many downstream genes (including the next pair of master genes), there need be no simple hierarchical arrangement of genes within a differentiation cascade. The differentiation tree would then be a tree whose edges correspond to webs. If this is so, then we may be able to distinguish master genes for differentiation from lower level genetic 'controls'. The latter will usually be involved in webs of genetic interactions. Master genes for differentiation will not. To distinguish the two cases of web formation, which occur at different hierarchical levels, I will use the terms:

differentiation tree webbing (Figure 29);
differentiation cascade webbing (Figure 28).

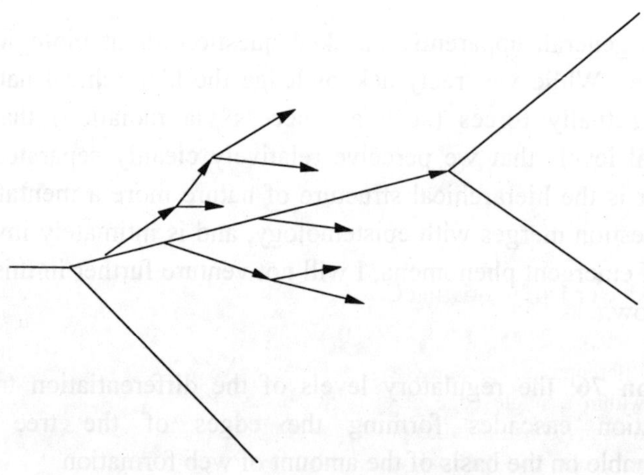

Edge or Differentiation Cascade with Tree Structure

Figure 27. Hierarchical differentiation cascade. Each edge or differentiation cascade of a differentiation tree represents 50 to 300 individual gene activations. These may be triggered in a neat hierarchical pattern, as depicted here.

Proposition 77: developmentally earlier edges of the differentiation tree represent differentiation cascades that tend to be simpler.

It is a general observation that the cell cycle time increases in the course of development. I assume that a given step of differentiation must be completed before determination of the next step of differentiation can occur. Thus the short cell cycles in early embryogenesis place constraints on what can be accomplished in early steps of differentiation. Two constraints that have been identified are a requirement for shorter genes, and a requirement for few sequential activations of genes:

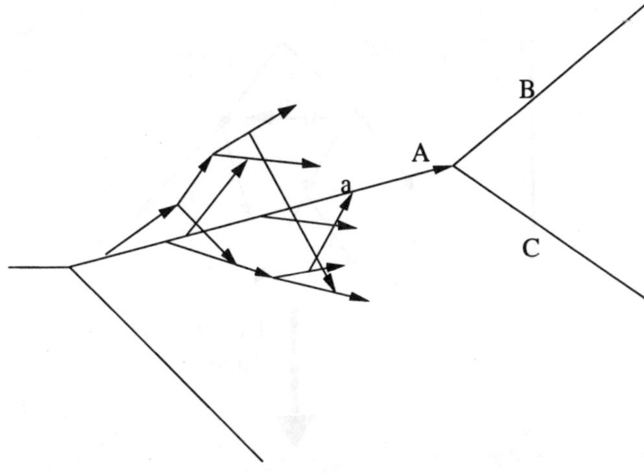

Edge or Differentiation Cascade with Web Structure

Figure 28. Differentiation cascade as a web. The more likely arrangement for genes activated within a single edge or differentiation cascade is a web, since there are no obvious constraints keeping these particular interactions hierarchical. Note in particular that internal webbing (at **a**) in the differentiation cascade may be involved in setting up the next cell state splitter and readying the next pair of master genes, in preparation for the next bifurcation of the differentiation tree, $A \Rightarrow B + C$. Cf. Proposition 11.

"The length of a gene may have a more important impact on gene function than previously thought... (O'Farrell, 1992). During the early development of *Drosophila*, mitotic divisions take place very rapidly, every 8 min for the first 10 divisions. As the time necessary to transcribe a long gene may be longer than the period of the cell cycle, the transcription of this long gene will stop and the nascent transcript will be eliminated when the cell [*sic:* these nuclear divisions in *Drosophila* are syncytial] enters mitosis" (Gronemeyer & Laudet, 1995). [Cf. Rothe et al., 1992; Bolker, 1995.]

"It must, however, be kept in mind that significant time is needed for a [long] eukaryotic gene to be transcribed and its product to reach a high enough concentration to activate another gene. The rate of elongation during transcription is about 15 nucleotides per second in most eukaryote organisms investigated (Davidson, 1986). A long transcript of 100 kb would therefore take nearly 2 h to synthesize.... It is, therefore, unlikely that in eukaryotic cells there can be a very rapid succession of entirely dependent sequential gene activations" (Gurdon, 1988b,c).

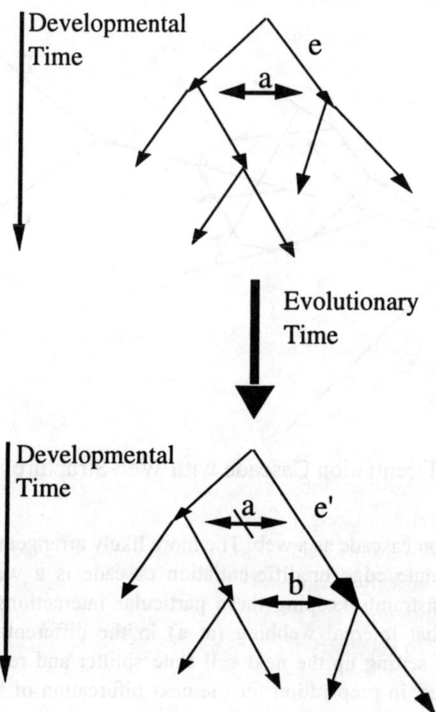

Figure 29. Induction is secondary. Edge **e** of a differentiation tree is subject to an inductive interaction **a**, partly turning the tree into a differentiation web. In the course of evolution, heterochrony leads to a modification **e'**, so that induction **a** can no longer occur, but a new inductive relationship **b** with another tissue may form.

Maternal RNA can also shorten the time needed to complete a step of differentiation, though since it may be diluted amongst many cells as development proceeds, this may only be important in early embryogenesis. The minimum length of time needed for a step of differentiation also places constraints on heterochronic alterations of the differentiation tree.

Proposition 78: on an evolutionary time scale, it may be that heterochrony forces the breaking and making of new inductive relationships.

The breaking and making of new inductive relationships is illustrated schematically in Figure 29. It is a necessary consequence of Cope's view of heterochrony:

"Haeckel thought in terms of the whole organism: the condensation of ontogeny proceeds equally for all characters and brings the total configuration of the ancestral adult into earlier and earlier stages ['exact parallelism']. Cope [1887] applied his concepts to *individual organs* and recognized that they may be accelerated (or retarded) at different rates ['inexact parallelism']" (Gould, 1977a)

(cf. Løvtrup, 1978; Gould, 1992a; Raff, 1996). Bonner (1965a) notes, for example...

"Therefore in heterochrony we can assume that the initiating stimulus for a chain of steps must have shifted in time, or there must have been a shift of the ability to respond.... Despite this criss-crossing of internal effects, many of the steps are in units or chains, and a whole chain can be moved forward or backward in time in the life cycle, or even eliminated completely, without necessarily having devastating effects upon the other steps of the cycle. This bunching of steps into units is a tremendous help in producing viable variations in large multicellular organisms, so necessary for evolutionary change" (Bonner, 1965a).

The 'bunching[s] of steps' may be interpreted as edges of the differentiation tree that are sufficiently labile over evolutionary time that they can break any existing inductive relationships and move with respect to one another along the whole differentiation tree (Figure 29). We could follow Bonner (1965a) further, and bunch edges of the differentiation tree together with 'unbreakable' inductions, with the bunches able to slide with respect to one another because of 'breakable' inductions (Figure 30). Stearns (1986) notes...

"Another sort of developmental constraint arises when a change in the timing of events is favored. For changes in timing to evolve, processes that were previously integrated must become uncoupled in such a way that function is preserved (Raff & Kaufman, 1983). This is probably not possible for all processes, but we do not yet know in general what separates developmental processes that can be uncoupled from those that cannot" (Stearns, 1986).

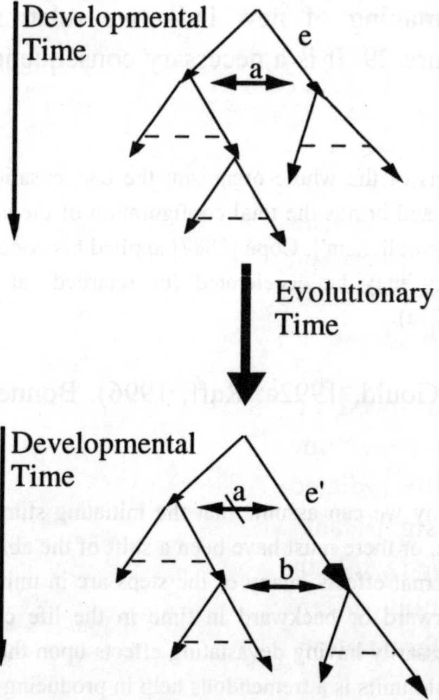

Figure 30. Unbreakable inductions. Two groups of edges or differentiation cascades are bunched together by 'unbreakable' inductive interactions (dashed horizontal lines). While these groups may slide past one another over evolutionary time, there is no sliding within them. **a** and **b** are as in Figure 29.

Differentiation trees provide a separation between "processes that can be uncoupled from those that cannot" or equivalently 'dissociation' into 'developmental modules' (Raff, 1996). Cf. also Gould (1977a), Wagner (1986) and Lord & Hill (1987). The generally 'unbreakable' processes are the individual differentiation cascades, and possibly their bunching by 'unbreakable' inductions (Figure 30).

Raff (1992a) provides an example of heterochronic changes in inductive relationships and their secondary importance:

"Gastrulation is the most fundamental movement in animal development, and establishes the primary germ layers and their topological relationships for subsequent inductive interactions.

Gastrulation varies between classes, but would be expected to be conserved between closely related species, because these would share many inductive interactions and downstream morphogenetic processes that would constrain evolutionary changes in gastrulation. It is evident from experimental manipulations of indirect-developing sea urchins that abundant inductive interactions occur prior to and during gastrulation (Ettensohn & McClay, 1988; Henry et al., 1989; Hardin & McClay, 1990). However, the timing and topological differences in morphogenesis in [direct developing] *Heliocidaris erythrogramma* versus [indirect developing] *H. tuberculata* indicate that cell-cell communication has been substantially modified despite the clear role for such inductive interactions" (Raff, 1992a). [Cf. Henry & Raff, 1990.]

To test these ideas, we need a careful comparison of differentiation trees in related organisms and their inductions (cf. Raff, 1996). If we consider a group that has recently radiated, I'd anticipate that different species within the group will have significant differences in inductive relationships. Is the tree basically preserved while the inductive relationships change? Does edge bundling occur, creating differentiation tree webs that stay together while other parts of the differentiation tree change via heterochrony?

Proposition 79: organisms whose differentiation trees evolve into differentiation webs cease to radiate.

This proposition is the converse side of the previous one. Thus, contrary to the widespread belief that induction is the basis of morphogenesis and its evolution, I suggest that induction, by creating crosslinks between parts of the differentiation tree, actually inhibits evolution. In those cases where these crosslinks become essential to survival, the differentiation tree becomes at least in part a differentiation web, and the options for further evolution are severely reduced. Perhaps it was a differentiation web in dinosaur genomes, rather than 'ecological stress', that caused the final decline and extinction of the dinosaurs:

"The dinosaurs of the final Maastrichtian... generated no speciation, no creation of new species.... I do not wish to rule out *a priori* the possibility of a meteoritic impact at the end of the Cretaceous. But there can be no doubt that the dinosaurs, like probably most other groups that became extinct, were suffering at the end of the Maastrichtian from 'ecological stress,' to use the neat phrase coined by the American paleontologists William Clemens and David Archibald" (Tassy, 1993).

Proposition 80: organisms whose differentiation trees have largely converted to differentiation webs either have persisted with little apparent change over long geological times (exhibiting evolutionary 'stasis') or become extinct.

Of course, it will be difficult to determine differentiation trees of extinct species, except by analogy with living relatives. On the other hand, there are many taxa with long evolutionary histories that are still extant, with accessible differentiation trees. Stasis is discussed in greater detail in Proposition 163, where an alternative mechanism is suggested. Mayr's (1982) 'cohesion of the genotype' might have its origin in differentiation webs:

"It is my own conviction that extinction is somehow correlated with the cohesion of the genotype. Surely, the rate of mutation ought to be approximately the same in different species of organisms. However, some of them have a genotype that is so well integrated, and thus has become so inflexible, that it can no longer produce the departures from the traditional norm that might permit a major switch in resource utilization or an answer to the challenge of a competitor or pathogen. These, of course, are only words until we have learned more about the structure of the eukaryote genotype and its regulatory system" (Mayr, 1982).

Proposition 81: homeothermic (warm-blooded) organisms may rely less on induction mechanisms to synchronize embryonic development. This may allow heterochronic mutations to be successful more often, and thus provides a basis for faster evolutionary radiation.

The selective advantage of induction may be that it leads to more frequent success of embryogenesis independent of fluctuations in temperature. For example, the wide temperature range for normal development in amphibians (Duellman & Trueb, 1986) and insects (Lamb, Gerber & Atkinson, 1984) may be possible because of inductive relationships between key tissues that ensure their near developmental synchrony, so that some unfolding edges of the differentiation tree do not get ahead of other edges. On the other hand, the constant body temperature of mammals and egg incubation by birds may free them in part from the need for such inductive relationships for developmental timing purposes, allowing more often successful heterochrony. Thus homeothermy and the consequent reduction in the developmental constraint of induction may be one key to why certain phyla radiate rapidly. A comparative study of radiation rates of homeothermic phyla versus poikilothermic phyla (cold-blooded organisms) might produce interesting results. Whether these considerations apply to the possibly warm blooded dinosaurs (Desmond, 1976; Thomas & Olson, 1980; cf. Ruben et al., 1997) we may never know.

Proposition 82: the launching of one differentiation wave may have a mechanical dependence on the completion of the previous differentiation wave. This provides a new lockstep wave-wave inductive mechanism, and may explain the broad temperature independence of development in poikilothermic organisms.

Beyond ordinary induction of one tissue by another, we have hints in our data on axolotl differentiation waves that there is a lockstep interaction between some consecutive waves. This speculation is based on an impression:

"On a broader scale, Natalie [K. Björklund] has observed that the termination of one wave and the initiation [launching] of another sometimes seem to be correlated [Figure 2 in Appendix V]. Perhaps wave-wave interactions are responsible for the coordinated development of the embryo, and thus for the near independence of the course of development over a wide temperature range (Duellman & Trueb, 1986)" (Appendix V: Gordon, Björklund & Nieuwkoop, 1994).

Careful observations are needed of the differentiation waves of a cold-blooded (poikilothermic) organism, and their relative timing, versus temperature, to see if this correlation holds over the whole range of normal development. If so, then one wave may set up the physical conditions requisite for the launching of the next.

Wave-wave inductions would provide a lockstep mechanism for embryonic development. Since physical momentum is unimportant in the mechanics of embryos (Gordon et al., 1972, 1975), all mechanical effects depend on creeping motion (low Reynolds number: Happel & Brenner, 1973). Insofar as viscoelastic effects can be ignored, the *sequence* of mechanical changes (stress, strain, elasticity, etc.) may be independent of temperature. (Note I said sequence, not rates.) Thus, while we know that enzyme catalysis, and metabolism in general, are very temperature dependent, the mystery of the temperature independence of normal development may be solved via embryo mechanics, i.e., while the rate of change of the mechanical state of the embryo may depend directly on the metabolic rate, the sequence of mechanical changes may not. In a sense, development may be analogous to a race in which a baton is passed on from one runner to the next. Each runner must wait for the previous one, no matter what the pace of the race.

One way to test these ideas is by use of thermal gradients across embryos (Huxley, 1926, 1927; Spemann, 1927b, 1938; Vogt, 1927, 1928a,b, 1932; Dean, Shaw & Tazelaar, 1928; Gilchrist, 1928, 1929a,b, 1933; Child, 1941; Glade, Burrill & Falk, 1967), which was originally done in an attempt to test the idea that gradients cause morphogenesis (reviewed with a conclusion against "the gradient theory in the sense of Child [1928], Huxley & de Beer [1934]" by Spemann, 1938; cf. history of arguments pro and con in Witkowski, 1980a). One result is a curious challenge to our understanding of differentiation waves:

"A new method introduced (Gilchrist, 1933) consists in applying a lateral temperature-gradient for a certain (not too great) length of time, and then to reverse the sign of the gradient for an equal length of time [to reduce the effect of dealing with 'chimeras of age': Spemann, 1938]. Eggs of the Urodele *Triturus* thus treated, and heated first on the right during the late blastula and then on the left during the early gastrula stages, show abnormally large neural folds on the right side" (Huxley & de Beer, 1934).

Is this a second order effect, due to an imbalance of mechanical tensions between portions of a differentiation wave (either of the blastopore lip and/or the ectoderm contraction wave) travelling at different speeds? Or does it bring into question the idea that development is independent of temperature? A mechanical effect would appear consistent with the observation that an animal/vegetal thermal, and therefore axially and bilaterally symmetric gradient has little effect (Huxley, 1927; Spemann, 1938). Furthermore, Spemann (1938) emphasized...

"...that a *duplicitas anterior* [two headed embryo] results (Gilchrist, 1928).... It is not impossible, in my opinion, that the disturbance of gastrulation has led, in a purely mechanical way, to a fission in the material of the archenteric roof with the further consequence of a duplication of the medullary plate, like that in a *duplicitas anterior* produced by mechanical constriction" (Spemann, 1938).

(Rare spontaneous caudal duplications also occur: Dominguez et al., 1993.) Clearly, a closer look at the result of Gilchrist (1933), under time lapse microscopy, is warranted.

The breakdowns in normal development that occur at the temperature extremes for a poikilothermic species (Lamb, Gerber & Atkinson, 1984; Duellman & Trueb, 1986) may be quite useful in pointing to critical stages for wave-wave interactions. Consider...

"...the effect of temperature variation (18-32°C) on the development of meristic characters in the snake, *Natrix fasciata* (Osgood, 1978). The frequency of gross abnormalities in vertebrae was increased at both high and low temperatures.... Extreme temperatures have also been associated with floral abnormalities in field populations of *Linanthus* (Huether, 1969)" (Hoffmann & Parsons, 1991).

The effect of temperature on "...fluctuating asymmetry [which purportedly] provides a measure of developmental stability (Mather, 1953; Beardmore, 1960; Reeve, 1960; Parsons, 1961, 1962; Van Valen, 1962; Soulé, 1967)" (Hoffmann & Parsons, 1991) may be related to wave-wave interactions. (Cf. Soulé, 1979; Palmer & Strobeck, 1986; Parsons, 1990; Møller & Pomiankowski, 1993.) These tissue fluctuations are perhaps attributable to variable left-right timing of differentiation waves and their interactions. Thus left-right variations in the launching, trajectories, or interactions of differentiation waves, somehow due to 'environmental stress', may be the underlying cause of fluctuating asymmetry. The latter has been correlated with stress and the current worldwide decline of amphibians (Alford, Bradfield & Richards, 1997).

Chimeras (Le Douarin & McLaren, 1984) may be one way to force developmental asymmetries (as in the pigment patterns of chimeric zebrafish: Lin et al., 1992) without environmental or genetic stress, and test our proposed lockstep wave-wave inductive mechanism under otherwise normal conditions. Such an experiment would be particularly interesting since "no aberrant development is known to be caused by chimerism per se" (Mullen, 1984). (However, note that Binstock, 1995, has suggested that mammalian females with X-chromosome mutations causally linked to brain function can show variable penetrance of 'X-linked neuropsychological phenomena' due to variable, i.e., chimeric 'in-brain X-inactivation', Tan et al., 1995, analogous to the pigment variations found between calico cats: Robinson, 1971.) Appropriate chimeras could be made, for instance, from salamanders of differing ploidy (Fankhauser, 1945a; cf. Sinnott, 1939a; Roth, Blanke & Wake, 1994).

One might anticipate some relationships between these considerations and developmental effects on longevity and senescence (Finch, 1990).

There is a wide variance in individual insect development times at constant temperature (Lamb & Loschiavo, 1981), even for genetically identical organisms, such as aphids (Lamb, MacKay & Gerber, 1987). This suggests a stochastic component in insect development that could lie in variable differentiation wave speeds, but more likely, is caused by stochastic variations in the time it takes to build up whatever mechanical strain or other factor launches a differentiation wave. Direct time lapse observations of individual organisms that develop slowly or quickly under identical conditions might provide a test for our idea of wave-wave interactions, or show what else might be going on. It may be worthwhile looking for and following similar individual variations in the development of cloned amphibians (Di Berardino, Hoffner-Orr & McKinnell, 1986), zebrafish (Streisinger et al., 1981), or mammals (Willadsen, 1989; Prather & First, 1990a; First & Prather, 1991). The following general comment of Schmalhausen (1949) may be worth exploring in this context:

"Mutations frequently manifest considerable individual differences in their expression even when the environmental factors are the same (e.g., vestigial, bar, abnormal abdomen, beaded, aristopedia, tetraptera, etc. [in *Drosophila*])" (Schmalhausen, 1949).

If there is a stochastic component to the launching of a wave, and there may well be if there is a positive feedback (Milsum, 1968; Resnick, 1995), avalanche-like (Cheng et al., 1989) mechanism operating here, then variable development of even genetically identical individuals would be expected. Thus wave-wave interactions may be both lock-step and stochastic.

Of course, wave-wave inductions add one more layer of inductions to the differentiation tree. At least for those tissues involved in such inductions in poikilothermic organisms, we would then expect little in the way of heterochronic evolution.

Proposition 83: some major birth defects, partial inductions by so-called regional inducers, and incomplete or disorganized embryos (embryoids), could be due to a failure of the proper sequence of wave-wave inductions.

This proposition is based on the observation that the breakdown of normal development at the high and low ends of the temperature range of a species (Duellman & Trueb, 1986; Dettlaff, 1991a) occurs via the common major birth defects, such as the neural tube defects of spina bifida and anencephaly (Lillie & Knowlton, 1898; Atlas, 1935; Moore, 1939, 1942b; DuShane & Hutchison, 1941; Løvtrup, 1953; Brown, 1976; Nelson, Mild & Løvtrup, 1982), although other effects, such as polyploidy (Fankhauser, 1942), notochordlessness (Hoadley, 1938), meristic effects (Kwain, 1975; Osgood, 1978), malformed adults (Lamb & Gerber, 1985), embryo death (Gerber, 1985), etc. (Howe, 1967; Anderson, 1972; Bursell, 1973), may also occur.

Again, direct observation of the differentiation tree for failing embryos, at the temperature extremes, would test this idea. Failure may be related to the exponential dependence of speed on temperature for calcium waves (Jaffe, 1993; yet to be measured for ultraslow waves). Expression of specific genes, such as homeobox genes, may be expected to be lacking in defective embryos, especially if and when birth defects correspond to altered or pruned differentiation trees. Teratogens may also cause embryos to have aberrant differentiation trees.

Consider an experiment such as the following:

"Injection of activin mRNA induces dorsal mesoderm in animal pole explants. Unlike expression of processed Vg1, activin mRNA injection organizes only a partial dorsal axis, lacking notochord and head structures (Thomsen et al., 1990)" (Kessler & Melton, 1994).

What could the difference be between two artificial inducers of differentiation waves? Since once a differentiation wave is launched, it propagates essentially from cell to cell through a tissue, three possibilities present themselves:

1. The kind of wave launched, expansion or contraction, could differ between artificial inducers.

2. The launching domain, which may involve an electric current feedback mechanism, could vary between artificial inducers.

3. The trajectory of the wave, even if launched from the same domain, could vary between artificial inducers.

Obviously, all three possibilities are directly observable and distinguishable by time lapse microscopy. Once one of these conditions occurs, the whole sequence of wave-wave interactions could be disrupted, leading to a partial playing out of the differentiation tree, analogous to incomplete falling down of a domino setup. For example, "...the muscle-specific determination genes myf5, myogenin, myoD, and myf6 are expressed in... embryoid bodies in a characteristic temporal pattern which precisely reflects the sequence observed during mouse development *in vivo* " (Rohwedel et al., 1994), as if only a subtree of the differentiation tree were accessible in embryoids. Here may lie the fundamental explanation of 'regional inducers', so-called 'head inducers' and 'trunk inducers' (Nieuwkoop, Johnen & Albers, 1985). Phenomena such as "'dorsalization' of mesoderm, resulting in formation of ectopic dorsal axial structures (Graff et al., 1994; Suzuki et al., 1994)" (Kessler & Melton, 1994) could be nothing more than extra waves gone awry. Claims such as the following are less certain if we see the possibility that the relevant differentiation waves perhaps just don't get launched in some explants (cf. Section 9.22):

"...The failure of... [Keller & Danilchik (1988)] explants to undergo differentiation or morphogenesis indicates that planar signals are not sufficient for all aspects of neural induction (Dixon & Kintner, 1989; Ruiz i Altaba, 1992), and that underlying chordamesoderm is required for neural tube formation (van Straaten et al., 1985a,b, 1989; Yamada et al., 1991)" (Kessler & Melton, 1994).

How could the third case occur, a difference in trajectories? Look at the physical conditions the tissue is subjected to:

"Treatment of animal pole explants with soluble activin protein from a variety of sources induces dorsal mesoderm tissues (Asashima et al., 1990a,b; Smith et al., 1990a; Sokol, Wong & Melton, 1990; Thomsen et al., 1990; van den Eijnden-Van Raaij et al., 1990). At high doses, 'embryoids' are formed that display a rudimentary axial pattern and head structures (Sokol, Wong & Melton, 1990; Thomsen et al., 1990)" (Kessler & Melton, 1994).

Why high doses? A physical effect, perhaps due to membrane binding properties of the activin (Mathews & Vale, 1993a,b), could be implicated. Since the tissue is "incubated in protein [at]... high dose" (Kessler & Melton, 1994) we can imagine that the physical strain in a tissue supporting a differentiation wave might be different from normal development, perhaps resulting in an altered trajectory. (Cf. the multiple, disorganized neural plates when the strain was released by Beloussov et al., 1990.) We thus get histological differentiation, but the tissue is not the right shape, and the physical conditions needed to set up the launching domain for the next wave may not be met.

This proposition gives an explicit model for the concept "that one of the first reactions of an embryo when approaching the limits of optimal temperatures is desynchronization of individual developmental processes" (Dettlaff, 1996a).

Proposition 84: induction helps break spatial symmetries in a tissue in a way that is consistent with previously broken symmetries, again assuring greater success of embryonic development.

Embryos of many animals tend to be bilaterally symmetric. If embryology consists of repeated breaking of symmetries, there must be some constraint that keeps these consecutive symmetries from going every which way. That such disorder is possible can be seen in the famous experiment of Spemann & Mangold (1924a,b) (cf. Waelsch, 1992), which showed that a transplanted dorsal lip of the blastopore (nascent notochord, or 'organizer') could induce a secondary embryo wherever it was placed, in any orientation (Holtfreter & Hamburger, 1955). This shows the essentially spherical symmetry of the ectoderm. However, the position of the natural inducer, along the midline of the invaginating tissue (Hama et al., 1985), may help guarantee that the neural plate follows the same bilateral symmetry as the gastrula and the earlier gray crescent.

While this interaction would specify the position of the neural plate, we do not yet understand how the orientation of the neural plate is also forced to be along the bilateral axis of symmetry. This may involve asymmetries surviving, perhaps as oriented microtubule organizing centres (MTOCs), from the time of grey crescent formation (Flint et al., 1989). Alternatively, it may be a mechanical result of continued invagination movements, as suggested by Holtfreter & Hamburger (1955):

"The cephalocaudal axes of the secondary embryo and of the host may be oriented at any angle to each other, but several workers observed a striking preference of the graft for developing its main axis parallel to that of the host, even if its original orientation was at a wide angle to it (Geinitz, 1925a; Spemann, 1931a; Lehmann, 1932). Probably the invaginating host tissues tend to deviate the gastrulation movements of the graft, carrying the latter along with their own cell streams" (Holtfreter & Hamburger, 1955).

This is consistent with in vitro experiments demonstrating a positive feedback, if one assumes that the stretch provided by the main axis 'artificially' stretches the graft:

"Data have... been provided (Beloussov, Lakirev & Naumidi, 1988) on the reorientation of cell intercalation as a result of the corresponding reorientation of artificially applied tensions: the cell intercalation is always directed perpendicularly to the tension.... Artificial stretching of the explants generates their active deformation in the same direction" (Luchinskaya, Beloussov & Shtein, 1997b).

The role of cell interactions between individual cells in embryos hitherto thought to be completely mosaic (Nishida & Satoh, 1983; Maclean & Hall, 1987) may also be to split symmetries, perhaps again initially just in a mechanical sense. For example, the orientation of cell division in the nematode EMS cell can be manipulated at will during a critical period, by placing a P_2 cell anywhere next to it (Goldstein, 1995b). In the intact embryo, the P_2 cell would be in a particular location before the interaction. Disruption of inductive relationships may be the basis for the development of morphological asymmetries in extreme phenotypes or 'phenodeviants' (Williamson, 1987).

The relationship between induction and differentiation waves, considered from the point of view of symmetry, is that inductive interactions launch differentiation waves at specific launching domains that break or maintain existing symmetries.

Proposition 85: while a process of induction may require mere physical adhesion between two tissues, establishment of that adhesion may involve specific interactions at the interface between their cells.

The specificity of inductions is usually believed to lie in specific 'inducer molecules', but if there is any universality to the process of primary neural induction, its amazing *nonspecificity* (Twitty, 1966; Appendix I: Gordon & Brodland, 1987; Section 2.05) may be paramount. We can tentatively resolve this apparent conflict by considering the possibility that specificity, where needed, may reside in the establishment of tissue-tissue adhesions. This model would permit specific roles of membrane proteins, extracellular matrices, symbionts (Montgomery & McFall-Ngai, 1994), etc., in the process of induction. Their physical effect would then tip the balance of mechanical forces on the cell state splitters in the 'induced' cells so that the resolution of the instability went a particular way for those cells involved in the adhesion (Figure 14). Homoiogenetic propagation of this change from the site of adhesion would then cause contiguous cells to resolve to the same state.

The need for specificity in induction is probably greater as embryogenesis proceeds, since there are more pairs of tissues that may come into adventitious contact. If they all adhered and had inductive relations, the embryo's development might mechanically grind to a halt. The lack of need for specificity at primary neural induction, because only a few tissues coexist at that stage, may explain why there is no protection against just about any artificially introduced tissue working as an 'inducer'.

Raff (1996) has suggested that embryos can vary substantially from one species to another before and after a 'phylotypic' stage, which in vertebrates occurs after neural tube closure. At the phylotypic stage, all vertebrate embryos supposedly look pretty much the same (but cf. Richardson, 1995; Richardson et al., 1997). His explanation of this empirical observation is basically one of specific inductions being somehow required at this stage:

"The phylotypic stage really represents a midpoint in development, at which the early embryo has blocked out the primary germ layers and the modules characteristic of late development are just beginning to appear. It is at this developmental midpoint that ultimately widely separated and distinct modules interact with one another in complex and pervasive ways.

What could be more starkly illustrative of this than the observation of Jacobson (1966) that heart mesoderm helps to induce the vertebrate eye?" (Raff, 1996).

To the extent that this idea is correct, then the common existence and specificity of inductions should peak at the phylotypic stage, for all embryos in a taxon. If inductions are universally the cause of launchings of differentiation waves, then all phylotypic embryos should exhibit the same tissue juxtapositions when differentiation waves are launched. However, the evidence for a large degree of heterochrony between related embryos at the so-called 'phylotypic' stage (Richardson, 1995; Richardson et al., 1997) implies that launching of differentiation waves may not always depend on coexisting tissues. As the phylotypic stage has been taken as roughly coincident with somite formation, and divergence in development increases even more later, this implies that specific adhesions used in tissue-tissue contacts that launch differentiation waves may vary considerably, both between induction events and between species.

Proposition 86: 'regional' inductions are actually multiple steps of differentiation.

"Personally I believe that embryonic segregation [determination] is an event of a single phase, and not of two or more successive phases. The assumption of two or more phases confuses the issue, for it leaves no simple criterion.... It must be emphasized that in... the young neurula, there are no other potencies in the ectoderm. The ectoderm at this time has only two *specific* potencies, though a rather considerable number of *prospective* potencies" (Lillie, 1929a).

Our cell state splitter model suggests that a tissue divides into precisely two new tissues, i.e., that a bifurcation occurs at each step of differentiation. How then are we to account for the nature of 'regional inducers' (Nieuwkoop, Johnen & Albers, 1985)? Regional inducers are inducing tissues for which the response of the induced tissue seems to vary with the nature of the tissue that induces it. For example,...

"...with regard to homoiogenetic induction by the neural tube, it is found that anterior portions induce the formation of anterior cephalic structures (e.g. eye), middle portions induce posterior cephalic structures (e.g. ear), while posterior portions induce structures characteristic of the trunk and tail (Mangold, 1929b, 1932)" (Huxley & de Beer, 1934).

Let us note first that there is no one-to-one correlation here:

"...Trunk-organiser grafted at head level in the host can also produce... cephalic structures" (Huxley & de Beer, 1934).

Spemann (1927b) put it nicely:

"The organizer may induce more than it ought to do; it has the tendency to build up a whole, behaving in this like a harmonious equipotential system of Driesch" (Spemann, 1927b).

To add to the confusion...

"...the properties of the organiser seem to be attached to a certain region of the embryo, regardless of the identity of the cells which occupy it" (Huxley & de Beer, 1934).

Finally...

"If three extra organisers are grafted into the close vicinity of an organiser in an intact embryo so that their polarities all converge to a point in the centre of the host-organiser, there is no inductive effect of any kind.... This annihilation of the inductive effect is difficult to understand" (Huxley & de Beer, 1934).

How an organizer is supposed to emit an 'active substance' and yet not be the same cells from one moment to the next (due to invagination) is not questioned.

Waddington (1956a) ended his book with a review of the unresolved question of regional induction. He emphasized a few salient points:

1. Regional induction is at best a tendency, not a well defined spatial pattern over either the inducing or induced tissues (ter Horst, 1947). At one extreme...

"Waddington & Yao (1950) showed that if presumptive anterior or posterior portions of the organiser are exchanged in young newt gastrulae, completely normal individuals may be produced" (Waddington, 1956a).

"The evidence presented... makes it clear that a fully normal embryo [of the urodele *Triton alpestris*] may be formed after the reversal of the organizer, or the exchange of its anterior and posterior parts, when the operations are made in early gastrula. Such a result involves the complete regulation of the topographically altered organization centre.... It was pointed out... that this differs sharply from the result of similar experiments in the Anuran *Discoglossus* (Waddington, 1941b)..." (Waddington & Yao, 1950).

2. In experimental manipulations "the brain is always fully formed up to a certain level, anterior to which it is altogether absent" (Waddington, 1956a), as shown by experiments of Nieuwkoop (1947a) and Lehmann (1948a).

3. Most experimental work ignores the role of the...

"...energetic morphogenetic movements [which] are going on.... Nothing very definite is known about the significance of these movements for the processes of regionalisation, but it seems almost impossible to believe that they are in fact without importance" (Waddington, 1956a).

4. The duration of induction has no apparent effect (Waddington & Deuchar, 1952).

A proper study of the relationship of these observations to differentiation waves requires repetition of the experiments under the watchful eye of time lapse. Lacking that, I can at best speculate. First of all, there is an obvious relationship between regional induction and segmentation of the brain. Thus if I am correct that segmentation is accomplished via hierarchical differentiations using differentiation waves, all of these experiments should fall in place in such a scheme. This notion fits with the observation that regional differentiation in experimental conditions is discrete: either a wave is launched or not, and so a portion is present or not. The presence of posterior portions first, then anterior, is consistent with the general observation on the order of appearance of segments, and with the correlation between this order and the order of corresponding homeobox genes in the genome (which I discuss in Proposition 270). Until the waves are looked for and possibly observed, this is as far as one can go with this multiple correlation.

The role of morphogenetic movements, namely invagination through the blastopore, can go beyond that of induction from the pharyngeal endoderm, and other potential mesoderm inductions. Invagination generates a nonuniform strain over the ectoderm, which may influence the trajectory of the ectoderm contraction wave and subsequent waves within the neural plate. With experimental manipulations, we should assume that these strains will be altered, and thus that wave trajectories will be altered. A physical understanding of strain generation could make prediction possible.

If induction corresponds to launching of waves, then the duration of contact is irrelevant, so long as it exceeds the minimum time for the launching to occur (as little as 0.5 h 'minimal induction time', as reviewed by Nieuwkoop, Johnen & Albers, 1985). With this, all of Waddington's (1956a) concerns are addressed, and we at least can see our way to a program of study to unravel regional differentiation.

The net zero interaction of multiple organizers (Huxley & de Beer, 1934) might be due to creation of a condition under which the ectoderm contraction wave is not launched (a testable idea). These phenomena are more in the realm of physics than chemistry. For instance, Nieuwkoop (1947a) has suggested that interactions between two dorsal lips of blastopores on one embryo (*duplicitas*) are mechanical. Similarly Waddington (1952c) argues that heat killed inducers have an effect that is primarily mechanical:

"The chaotic arrangement of the induced neural tissue in its early stages, and the great variability of its appearance later, suggests that in these inductions by dead parts of the organizer we are not in fact dealing with any sort of specific regional influences at all. It seems more likely that the implant merely evocates masses of neural tissue, but that if some region of the mass happens to have a shape roughly like that of the neural system, forces arise within it which cause it gradually to approach that part in its development" (Waddington, 1952c).

The idea that regional induction simply reflects multiple rounds of differentiation is consistent with the conclusions of Dias & Schoenwolf (1990) "...that the role of neural induction is to produce neuroepithelium of unspecified regional character, and that the formation of regional character depends on subsequent morphogenetic events." Examples of regional induction include the forebrain/hindbrain split of the neural plate (Sharpe & Gurdon, 1990), the formation of the eyes (optic sulci or grooves: Moore, 1977), perhaps the ears and nose (neural tube portions of the otic and olfactory placodes: Balinsky & Fabian, 1981; Hilfer, Esteves & Sanzo, 1989), rhombomere formation (Lewis, 1989; Fraser, Keynes & Lumsden, 1990; Lawrence, 1990; Heyman, Kent & Lumsden, 1993; Heyman, Faissner & Lumsden, 1995), neuromere formation in general (Alvarado-Mallart, 1993), so-called 'cordones' (Steindler, Faissner & Schachner, 1989), spinal motoneuron differentiation in the neural tube (Hamburger, 1977; Bloch-Gallego et al., 1993), and possibly cell wedging of the notoplate (called the *median hinge point* in chick: Smith & Schoenwolf, 1987, 1988, 1989, 1991; Schoenwolf, Bortier & Vakaet, 1989; Schoenwolf & Smith, 1990b; Schoenwolf, 1994: cf. Brodland & Clausi, 1995; and *noninvoluting* [noninvaginating] *marginal zone* in *Xenopus*: Jacobson, 1991b). If each one involves another round of cell state splitting, then perhaps a wave like the ectoderm contraction wave, with specific genes expressed, is associated with each one:

"There is little question that the brain is actually many organs in one, so mapping the distribution of mRNAs and proteins should add to our anatomical and physiological understanding of brain function" (Hahn et al., 1983).

Regional induction clearly needs a fresh look, with physical phenomena such as differentiation waves in mind.

Proposition 87: regional induction of the eye placodes may be caused by a single differentiation wave in the neural plate that splits in two waves.

Björklund and Kurt B. Luchka (in Gordon, Björklund & Nieuwkoop, 1994) noticed a circular wave at the anterior end of the neural plate that breaks into two circles as it shrinks:

"During normal development, the prechordal mesoderm, in the midsagittal plane, induces a monocular eye field in the midventral floor of the neural [plate].... This eye field then divides into two rudiments which move apart and give rise to two bilateral eyes" (Laale, 1987).

"**Natalie**: The eye anlage contraction wave begins as a circle on the edges of the rising neural ridges of the anterior neural plate at stage 14 [in axolotls: Figure 4; Table 1]. Upon reaching the central area of the anterior neural plate, the circle breaks into an arc and the two ends close into two separate circles corresponding with the two eye rudiments [eye placodes]. The circle vanishes and the region disappears laterally under the rising neural ridges.

"**Pieter**: Although the primary extension of this wave may correspond with the area with eye differentiation tendencies, the wave probably reflects true morphogenetic movements in the anterior neural plate" (Appendix V: Gordon, Björklund & Nieuwkoop, 1994).

(Cf. *Drosophila* segmentation of optic lobes in: Schmidt-Ott, González-Gaitan & Technau, 1995). Here is a possible marker gene for this 'eye anlage contraction wave':

"In late gastrula [zebrafish] embryos, *hlx-1* transcripts are detected within a circular area in the region of the presumptive rostral brain" (Fjöse et al., 1994).

and another:

"α_6 integrin in avian embryo... is first observed on neuroepithelial cells of the cranial neural plate" (Bronner-Fraser et al., 1992).

If this is the first contraction wave occurring in the neural plate, it should be named the 'neuroepithelium contraction wave', to conform to our convention of naming a wave after the tissue through which it travels. While I will speak of this wave in definitive terms, Pieter D. Nieuwkoop's caveat above must be taken seriously. The cells of the axolotl neural plate are smaller than those of early gastrulation, and any waves amongst them have much shorter distances to travel. Thus time lapse microscopy at higher magnifications than our initial studies, with automatic focussing, will be necessary to definitively study them.

Kurt B. Luchka and Natalie K. Björklund (p.c., 1992) suggested that when one eye forms (cyclopia) instead of two, the wave does not split into two waves. This is perhaps due to lack of some sort of interaction of that wave with the floor plate (Mangold, 1931a; Hatta et al., 1991b; Strähle et al., 1993) towards the end of its trajectory:

"According to von Woellwarth (1952) the [normal] duplication of the eye anlagen occurs at the time of closure of the neural tube. Adelmann (1932, 1936, 1937) concluded from the results of partial and complete excision of the prechordal substrate that the bilaterality of the prosencephalic structures is linked with the bilaterality of the subjacent mesoderm, mainly as a result of eye-inhibition in the mid-line, due to an influence of the underlying prechordal plate. Its influence, however, does not act until later stages of development" (Boterenbrood, 1962).

Cyclopia is a readily created abnormality (Adelmann, 1934, 1936; Eguchi, 1957; Sato & Eguchi, 1960; Ingalls & Murakami, 1962; Moore, 1977; Magrassi & Graziadei, 1987; Gilbert, 1991a; Hatta et al., 1991b; Thisse et al., 1994):

"If the prechordal plate is excised, or damaged,... the optic rudiment remains single and a midventral cyclopian, or synophthalmic, condition develops" (Laale, 1987).

Cyclopia may be the primitive condition, as in *Amphioxus* (Lacalli, Holland & West, 1994; Raff, 1996). Failure to launch or propagate the eye wave may correspond to the eyeless condition:

"In some mutants, the neural ectoderm of the optic rudiment is incapable of responding to prechordal plate induction, and no eyes will form. This... is referred to as an eyeless condition, and is attributable to a recessive mutant gene in the ectoderm. Normal eyes, however, will develop if the mutant presumptive eye ectoderm is replaced by wildtype ectoderm. Wildtypes, similarly, may be rendered eyeless by substituting mutant ectoderm for normal ectoderm" (Laale, 1987).

"A variety of anterior segment congenital eye abnormalities result from heterozygosity for mutations in the mouse, rat or human *Pax-6/PAX6* gene. These include *Small eye* [*Sey*] (Hill et al., 1991, 1992; Matsuo et al., 1993; [and aniridia: Glaser, Walton & Maas, 1992]).... *Small eye* is a semidominant mutation. Homozygous mutant mice have no eyes and no nasal cavities. Since, like the lens, nasal cavities form from invagination of an ectodermal placode, these phenotypes have been suggested by Hogan et al. (1986) to result from a failure of early placode differentiation. *Sey/Sey* embryos also have no olfactory bulbs (Hogan et al., 1986) and have defects in neuronal differentiation and migration..." (Grindley, Davidson & Hill, 1995).

The recent results by Halder, Callaerts & Gehring (1995) on 'ectopic' formation of eyes by targeted expression of the *eyeless* gene in *Drosophila,* "homologous to the mouse *Small eye* (*Pax-6*) gene", are interesting to try to interpret in terms of nuclear state splitters (cf. ectopic wings: Kim et al., 1996). Suppose that a differentiation wave crossing a wing, leg or antenna imaginal disc primordium encounters a cell or clone of cells with an ectopically 'readied' master gene, such as *eyeless*. Both the normal and the ectopic genes would set off their differentiation cascades in the *same* cell(s):

"Data for several homeotic genes indicate that there is competition between the ectopically expressed gene and the genes normally expressed in a given segment (Kuziora & McGinnis, 1988a; Mann & Hogness, 1990; Gibson et al., 1990; Gonzáles-Reyes & Morata, 1990; Gonzáles-Reyes et al., 1990; Lamka, Boulet & Sakonju, 1992; Castelli-Gair et al., 1994)" (Halder, Callaerts & Gehring, 1995).

"How could genes compete for the opportunity to be promoted by a cell state splitter? Would *both* genes be coupled to the same state of a cell state splitter? Would the cell state splitter couple to one gene or the other, randomly, in different cells? mRNA studies might provide the answer" (Steve McGrew, p.c., 1997).

Each ectopic eye...

"...probably involves the formation of an ectopic morphogenetic furrow.... The induction of ectopic eyes in *Drosophila* is reminiscent of the classical experiments of Spemann (1938) in which he induced ectopic eyes by transplanting the primordia of the optic cup to ectopic sites in amphibian embryos. Our experiments extend these observations and identify the gene that is necessary and sufficient to induce ectopic eyes at least in imaginal discs" (Halder, Callaerts & Gehring, 1995).

The restriction of these ectopic eyes to imaginal discs suggests that a differentiation wave is also necessary for eye morphogenesis. The development of eyes 'instead of' other tissues may, for example, have something to do with the ectopic *eyeless* gene product binding more strongly to the commitment signal than other master genes. A gradation in binding constants occurs in homeobox genes (Pellerin et al., 1994).

The floor plate (Coghill, 1924a) is itself an example of regional induction in the neural plate (Griffith & Sanders, 1991a) and specific gene expression (Moos Jr., Wang & Krinks, 1995) and the site of further regional induction (Hirano, Fuse & Sohal, 1991; Hynes et al., 1995; Jessell et al., 1989; Jessell & Dodd, 1993). Artinger & Bronner-Fraser (1993) find that in chick this region is reasonably independent of induction from the mesoderm (notochord), a result corroborated in zebrafish (Halpern et al., 1993). Such independence is consistent with a nested set of differentiation waves all *within* the neural plate. These waves I postulate are probably then responsible for "homeogenetic floor plate induction (Hatta et al., 1991; Yamada et al., 1991; [Placzek, Jessell & Dodd, 1993])" (Klar, Jessell & Ruiz i Altaba, 1992).

As only a little information on the floor plate is available in some standard texts (Bailey & Miller, 1911; Witschi, 1956; Romanes, 1964; Arey, 1974), and it is sometimes confused with the 'notoplate' (Gordon, 1983a), here is a description of the incompletely defined relationship between the two (cf. Jacobson, Odell & Oster, 1985; Jacobson et al., 1986):

"The notoplate becomes the initial floor plate. In the newt *Taricha torosa*, the convergence-extension movements of this tissue continue after neural tube formation, until about stage 23 [Figure 4]. At stage 23, the floor plate is a single file of cells down the ventral midline of the neural tube with each cell stretched perpendicular to the long axis and connecting the two basal plates. In later stages, the floor plate is several cells wide. The question is, are all these cells of the now-wider floor plate formed from the original notoplate cells seen at stage 23, or has the notochord (or the floor-plate cells themselves) induced additional floor plate cells from the edges of the basal plates?" (Antone G. Jacobson, p.c., 1995).

It may be important, in the subsequent migration of neurons from the floor plate, to note:

"The massive amount of shear in the notochordal region [of the neural plate, i.e., in the notoplate]... may have a role in dividing the neural field into two bilaterally symmetrical equivalents. The region of the neural plate that overlies the notochord approximately maps into the future floor plate of the spinal cord and hind brain" (Jacobson & Gordon, 1976a).

As with germ cells and neural crest cells, shear may be involved in the separation of individual, migrating cells from this epithelial sheet.

Each example of regional induction of the neural plate, starting with the floor plate and the eyes, deserves careful scrutiny in terms of differentiation waves. As we shall see, as much or more differentiation may occur in the formation of the central nervous system as in the rest of the body. All of this development has been lumped into the term 'regional induction' in classical embryology, without much attention to the need for hierarchical models of such a rich process, and usually with disregard for the fact that this tissue goes on developing all the way to the adult brain. Thus regional induction cannot be treated as a single step added to primary neural induction.

As the neural plate closes to a neural tube in the salamander *Taricha torosa*, it does not close smoothly, but rather in steps (Jacobson, 1978a,b; cf. Selman, 1955, 1958; Stern & Goodwin, 1977; Rapp, 1979). These pulsatile movements may be caused by the comings and goings of multiple differentiation waves on the neural plate.

The recent discovery that there are four "separate initiation sites for neural tube (NT) fusion" in human and other mammalian embryos, causing four separable classes of neural tube defects, possibly "controlled by separate genes" (Van Allen et al., 1993), might have its explanation in the formation of, lack of formation of, or aberrant timing of the multiple differentiation waves causing regional differentiation of the neural plate. A snapshot of these differentiation waves ('concave and convex bending sites') and their apically located cell state splitters is possibly seen in this description:

"The neuroepithelium of the mouse exhibits prominent regional variations in size and shape along the embryo axis. The complex shape of most of the cephalic neural tube (e.g., forebrain and midbrain) is due to the coexistence of concave and convex bending sites whereas more caudal regions (e.g., hindbrain and spinal cord) generally lack sites of convex bending and have a relatively simple shape. The apical morphology of neuroepithelial cells was found to be correlated more closely with the local status of bending of the neuroepithelium than with the specific region of the neural tube in which they are located. In areas of enhanced apical constriction, microfilament bundles were particularly prominent" (Bush et al., 1990).

"*In situ* precipitation of calcium (Ca^{2+}) with fluoride and antinomate shows that Ca^{2+}-specific precipitate is localized almost exclusively within lipid droplets of neuroepithelial cells during neural tube formation in chick and mouse embryos. The density of Ca^{2+} precipitate within lipid droplets is generally greater in the apical ends of cells [at the cell state splitters] situated in regions of the neuroepithelium that are actively engaged in bending" (Bush, Lee & Nagele, 1992).

We thus have hints of multiple differentiation waves in regional inductions, which deserve a direct view. Correlation of possible underlying differentiation waves with finite element approaches to brain shaping (Fujita, 1986) may be a useful way to work out the morphogenesis of the CNS (central nervous system) and its neural tube defects.

Proposition 88: regional 'predetermination' does not occur.

Sharpe et al. (1987) have suggested that the main idea of primary neural induction, that the cells of the ectoderm are all equivalent in regards to their ability to become neural plate or epidermal cells, is fallacious, that the region of the ectoderm that becomes the neural plate is different ('predetermined') before primary neural induction. However, their work on a regional 'predetermination in neural induction' is actually only concerned with 'predisposition', the word 'predetermined' not being used in their text at all, only in their title. They discuss an *XlHbox6* homeobox containing marker but it does not appear before primary neural induction has occurred. While they have shown "that the part of the ectoderm destined to form nerve in normal development already possesses a substantial predisposition to respond to mesodermal induction", this predisposition may be related to minor physical effects, such as:

the variation in cortical properties (membrane viscosity: Dictus et al., 1984; intermediate filament organization: Klymkowsky, Maynell & Polson, 1987; Klymkowsky & Maynell, 1989; Dent et al., 1992; microtubule distribution: Elinson & Rowning, 1988; Elinson & Palecek, 1993);

cell size (Gillette, 1944);

or yolk content and cytoplasmic mobility (Neff, Ritzenthaler & Rosenbaum, 1989).

These could alter the force balances or rates of change (cf. Figures 11-13). Similar predispositions towards neural versus epidermal differentiation have been discussed by Holtfreter & Hamburger (1955) for amphibians (cf. the mechanical explanation in Appendix I: Gordon & Brodland, 1987), fruit flies (Campos-Ortega, 1990), and leeches (Stent & Weisblat, 1985). For further discussion of regional predispositions see Savage & Phillips (1989), Otte & Moon (1992a,b), Kinoshita, Bessho & Asashima (1993), and the review by Good, Richter & Dawid (1990a). The regional confinement of markers to parts of the neural plate may merely reflect another, later round of differentiation. Kessler & Melton (1994) summarize the situation as follows:

"The presumptive ectoderm is prepatterned for neural induction (Sharpe et al., 1987). This prepattern can be visualized during cleavage stages with the epidermal marker, Epi1, which is expressed in presumptive non-neural ectoderm but not in future neural plate (London, Akers & Phillips, 1988; Savage & Phillips, 1989). This bias is reflected in a differential response of dorsal and ventral ectoderm to neural induction. Recombinants of chordamesoderm and dorsal ectoderm result in strong expression of neural markers, whereas ventral ectoderm expresses low levels in response to the same signals (Sharpe et al., 1987). However, ventral ectoderm can produce the full spectrum of neural tissues, demonstrated most dramatically in the organizer transplant [Spemann & Mangold, 1924a,b] and also observed in planar recombinants of dorsal mesoderm and ventral ectoderm (Bolce et al., 1992; Doniach, 1993)" (Kessler & Melton, 1994).

Perhaps due to the physical differences noted above, the various explant combinations have different probabilities of launching differentiation waves. The main point is that all cells of the ectoderm can go either way, to neural plate or epidermis, and none of this recent work contradicts this basic observation of Spemann & Mangold (1924a,b).

Proposition 89: the relationships between the ectoderm and the underlying mesoderm may be mutual, and affect the trajectories of differentiation waves in these tissues.

Consider the following observation:

"Ruiz i Altaba & Melton (1989b; cf. Ruiz i Altaba & Melton, 1989a,c) looked for genes that might be responsible for the anterior-posterior polarity of the embryo.... They found that the *Xhox3* mRNA was seen shortly after midblastula transition. It peaked during the late gastrula - early neurula stages and disappeared shortly thereafter. Moreover, a series of dissection experiments showed that *Xhox3* mRNA was found only in the mesoderm and that it formed a gradient whose posterior end had from 5-10 times more *Xhox3* message than the anterior end. Since the regional specificity of the neural ectoderm is thought to reflect underlying differences in the mesoderm, the *Xhox3* gene product looked like an excellent candidate for the molecule that would specify positional information" (Gilbert, 1991a).

There is an alternative interpretation of these results. The ectoderm contraction wave takes the whole period of gastrulation to get from the posterior to the anterior end. (This may be related to the fact that induction has come to be seen as a long, drawn out process, requiring continued interaction between tissues: Jacobson, 1966.) The wave is 100 μm deep in a 50 μm thick ectoderm, and thus must mechanically alter the shape of the underlying mesoderm (if it is indeed present while the wave is propagating). Let us suppose that synthesis of the *Xhox3* mRNA starts when the wave goes by (N.K. Björklund, p.c.), and assume for now no mRNA degradation. This mRNA and its gene product will thus have a higher concentration at the posterior end than at the anterior end, i.e., the wave leaves a gradient in its wake. If this scenario proves correct, then we have an interesting reversal of classical dogma: changes in the mesoderm are induced by a phenomenon occurring in the ectoderm. (Cf. "Transplantation and ablation experiments have led to the generalization that in insects the mesoderm is naive, and that pattern is imposed upon it by the ectoderm" in Greig & Akam, 1993). More complex gradients would be produced if degradation of the mRNA occurs.

Nieuwkoop, Johnen & Albers (1985) have also discussed "The reciprocal nature of the neural induction process". In particular, I would like to note an effect that raises the possibility of presumably two way electrotonic coupling:

"Ito & Ikematsu (1980) found that in the embryo of *Cynops pyrrhogaster* neuroectoderm cells are electrically coupled to adjacent chordamesoderm cells during gastrulation. The coupling generally decreases at somite stages" (Nieuwkoop, Johnen & Albers, 1985).

Perhaps this occurs only within the furrow of the ectoderm contraction wave and later differentiation waves:

"Neural competence within the presumptive epidermal ectoderm of the newt embryo decreases abruptly between stage 12c and stage 13b (Suzuki & Ikeda, 1979) [cf. Figure 4; Table 1]. Our data show that all of the P.N.E. [presumptive neuroectoderm] cells are coupled to underlying mesodermal cells at stage 12c. Uncoupling begins at stages 15-16 and at stages 22-23 the P.N.E. is totally insulated from Ch.M. [chordamesoderm]" (Ito & Ikematsu, 1980).

Despite earlier indications (Eakin & Lehman, 1957; Yamada, 1960), there appear to be no direct connections between ectoderm and mesoderm cells, at least in *Xenopus* (Kelley, 1969). It may be worth reopening this question in regard to the furrow of the ectoderm contraction wave in urodeles.

It is known that the mesoderm in urodele amphibians undergoes similar, though not identical, movements while the overlying neural plate is reshaping from a hemisphere to a keyhole shape (Jacobson & Löfberg, 1969; Jacobson & Gordon, 1976a; Keller, Shih & Sater, 1992). In *Drosophila:* "Homeotic genes are expressed in the mesoderm, and are regulated in a segment-specific pattern analogous to, but different from, that seen in the ectoderm" (Greig & Akam, 1993). Perhaps the ectoderm contraction wave reflects a similar wave in the underlying mesoderm during gastrulation. This may prove difficult to unravel: Jacobson & Gordon (1976a) found that the underlying notoplate and notochord in newt embryos changed shape as one and were difficult to mechanically separate (see Sausedo & Schoenwolf, 1993, for a comparison of the behavior of 'median hinge-point cells' in chick embryos, presumably also notoplate, with notochord cells).

The question of the separability of mesoderm from ectoderm movements is perhaps clarified by observations of similar movements in uncleaved eggs. Here there is no underlying mesoderm, and thus autonomy of the movements in the cortex:

"The stretching process leading to gastrulation involves the surface material of the egg and embryo as a whole, and not merely the region around the invagination center. Since the tissue movements are clearly of a supracellular nature, the fundamental question is: What is the nature of the directing force or substance? Even in aging, unfertilized frog eggs, according to Holtfreter (1943a), a simulation of the stretching and convergent movements of the cortical material takes place without any cell division taking place. Diffuse streaks of darker pigment radiate from the animal zone across the dorsal grey crescent toward the imaginary blastopore; and other cortical patterns recall the cellular subdivision and convergent stretching of the dorsal mesoderm of a developing embryo. The cortical expansion in the aging, unfertilized egg consists of a centrifugal movement of particles along lines of flow toward the blastoporal region. Thus 'the segmentation of the egg substance seems to be a superimposed process that interferes with the tendency of expansion which is of a submicroscopic nature' (Holtfreter, 1943a)" (Berrill, 1961).

These strange convergence/extension movements of the cortex 'in aging, unfertilized frog eggs' are important because it is the animal hemisphere of the cortex that becomes the ectoderm, at least in normal urodele embryos. (The unresolved relationship of animal cortex in urodeles versus anurans, Nieuwkoop, Björklund & Gordon, 1996, unfortunately confounds the meaning of these observations.) Analogous observations have been made in *Drosophila:*

"Analyses of mutations that block cellularization indicate that the mechanisms driving the dorsal movement of the posterior midgut plate can operate in the absence of cells (Rice & Garen, 1975; Rickoll & Counce, 1981; Swanson & Poodry, 1981; Merrill, Sweeton & Wieschaus, 1988).... In most acellular mutants, the anterior movement of the posterior midgut primordium is limited to about 25% egg length, suggesting that further movement along the dorsal side is driven by mechanisms requiring a subdivision of the embryo into individual cells" (Costa, Sweeton & Wieschaus, 1993).

Thus the movements of the ectoderm may be autonomous. It will perhaps be necessary to carry out a precise mechanical analysis, to see if the obviously active contraction of the ectoderm cells is accompanied by passive or active deformations of the underlying mesoderm cells. Confocal microscopy, perhaps with albino axolotl embryos (Frost, 1989), may allow *in vivo* observations of both layers of cells at once, so that the surgical approach of Jacobson & Löfberg (1969) could be avoided.

Proposition 90: induction of the neural plate may not require the mesoderm at all.

We have already seen cases of 'autoinduction' of neural plate or its homolog, which have no classical explanation. I would like to review a bit of history that may be relevant to this possibility in urodeles. The work of Goerttler (1926, 1927) hints at the possible independence of neural induction from the underlying mesoderm:

"...The time at which the material of the later medullary plate proved to have acquired capacity of self-differentiation coincided with the time when the contact with the underlying 'organizer' material was established (Marx, 1925). One could safely assume, then, that the invaginated layer of 'organizer' material contained factors capable of inducing the overlying ectoderm to undergo neural differentiation. It had been claimed by Goerttler (1926, 1927) that the sector of ectoderm of the blastula destined for the later formation of the medullary plate could accomplish this formation also in the absence of underlying mesoderm, providing the necessary movements of the ectoderm were not disturbed. This contention, however, lost its convincingness when Holtfreter (1933a,d,f) showed that the experimental basis of Goerttler's thesis was inadequate. Thereupon, there was no more serious objection to the view that the first formation of the neural system was really due to influences of the 'organizer.'" (Weiss, 1935).

Given subsequent 'inductions' of neural plate obtained without the organizer (Barth, 1939a, 1941a) or notochord (Youn & Malacinski, 1981), the dismissal of Goerttler's work may have been premature:

"...Forceful arguments in favor of neural predetermination were advanced by Goerttler, a student and colleague of the German anatomist Vogt [1925, 1928b, 1929a]. The design of the experiments was strongly influenced by Vogt's theoretical inclination toward a preformistic view of development.... How did Spemann react to these provocative assertions of a much younger and rather brash colleague? He gave them serious consideration and devoted to them several pages in Spemann & Geinitz (1927).... The work of Goerttler has not withstood the test of time. Years later, J. Holtfreter repeated two of his experiments and refuted all of Goerttler's claims.... These brilliant experiments ended once and for all the claims of predetermination of the neural plate, and they also buried Spemann's favorite idea of the spreading of a neuralizing agent within the ectoderm" (Hamburger, 1988).

Thus planar or homoiogenetic induction got thrown out in the course of the controversy. Perhaps Spemann or Goerttler would have pursued deeper investigation of planar induction, i.e., what we now know to be differentiation waves, had Holtfreter, Spemann's 'ingenious student' (Hamburger, 1988), not defended Spemann so successfully. The more sympathetic view of Child (1941) may indicate what happened:

"From his experiments Goerttler (1925, 1926, 1927) concluded that orientation of an inductor implant may be favorable or unfavorable, according to its relation to the regional cell migrations; consequently, he regarded those migrations as factors in determination of the neural plate and the presence of the underlying chorda-mesoderm as not absolutely necessary. Holtfreter (1933a,b) found orientation without effect and concluded that induction, not the migrations, is the essential factor in development of the neural plate. However, Goerttler has described and figured cases in which more or less development of neural plate and neural tube occurred when the whole dorsal lip region had been removed before its invagination" (Child, 1941).

A possible interpretation of the latter result is that the ectoderm contraction wave had already been launched, or that it was launched by the very act of cutting out the 'organizer'. Another possibility is that launching can occur by global means, such as ionic currents, without a necessary presence of an underlying tissue. Repetition of Goerttler's work under time-lapse might be well worth pursuing, to resolve questions of whether or not the mesoderm is essential to primary neural induction, and whether the orientation of implants is important.

Proposition 91: regional inductions are often caused by launching of differentiation waves by mechanisms not requiring an underlying, adjacent tissue.

The neural plate undergoes a number of nested differentiations that appear not to involve the underlying mesoderm:

"It has been proposed that the A-P [anteroposterior] pattern in the nervous system is derived from a parallel pattern of inducers in the dorsal mesoderm which is 'imprinted' vertically onto the overlying ectoderm. Since it is now known that planar signals can also induce A-P neural pattern, this kind of model must be reassessed.... Planar neural-inducing signals from the mesoderm must move through the presumptive neurectoderm from posterior to anterior.... Recent results from Saha & Grainger (1992) indicate that there is a self-organizing ability in the ectoderm after induction; the pattern information that the ectoderm receives first appears to be very crude, and the ectoderm can go on to refine this pattern independently of further contact with the mesoderm" (Doniach, 1993).

See the following for further evidence that indicates that so-called regional induction doesn't require any heterogeneous induction at all:

in explants (Servetnick & Grainger, 1991a; Doniach, 1992; Doniach, Phillips & Gerhart, 1992);

in exogastrulae (Ruiz i Altaba, 1990, 1992);

in transplant experiments (Servetnick & Grainger, 1991a);

in cell reaggregation experiments (Boterenbrood, 1962).

The spatial and temporal expression of homeobox genes is preserved compared to the *in situ* (normal) situation, "even in ventral ectoderm" explants (Doniach, Phillips & Gerhart, 1992). In summary, these assorted manipulations show that regional inductions can proceed without underlying tissues. This does not preclude a role for those tissues in normal development, such as in determining exactly where wave launching occurs (i.e., launching domains), but does show that once regional differentiation waves are launched, they propagate autonomously.

Proposition 92: hierarchically organized homeobox genes correspond to regional differentiations by being activated by corresponding differentiation waves.

There is an interesting inconsistency in the language applied to homeobox genes in regard to regional induction. On the one hand...

"The emergence of regional pattern along the anteroposterior axis of the nervous system appears to depend on the expression of position-specifying genes. Perhaps the single most important discovery in recent years has been that a set of genes that specify positional values along the axis of the fly embryo is conserved in vertebrates. These homeotic selector genes are the master control genes that coordinate the regulators of all processes involved in the development of structures appropriate to axial position" (Lumsden, 1993).

This is strong language regarding the Hox genes: 'depend', 'position-specifying', 'master control', 'coordinate'. On the other hand, how is it then that the same author asks...

"If Hox genes encode positional value, what positional signalling mechanism is responsible for activating their specific regional control elements?... The next major challenge is to understand how individual cell phenotypes are encoded and assigned.... So, it is hoped, molecular mechanisms will entirely supplant our familiar concepts, although the price, measured in terms of complexity, may be high" (Lumsden, 1993).

Something is master to the 'master' Hox genes, and that something is probably simply a set of sequential, nested differentiation waves, i.e., the physical component of the genetic program, rather than a complex, infinite regress of 'molecular mechanisms'.

Here is the same idea of a nested process expressed in gradients, in a form we now call 'positional information' (Wolpert, 1969). I would merely substitute 'differentiation waves' for 'gradients', 'signals' and 'interactions':

"The signal that passes in... [neural] induction is in its formal pattern rather simple.... All that is required to specify the different parts of the two-dimensional sheet of ectoderm is a pair of gradients in the intensity of two different stimuli emanating from the chordamesoderm, so arranged as to provide distinct coordinates for every part of the ectoderm that needs to be initially specified....

"There has been a gratifying convergence of views among the major research groups studying neural induction on just such a picture of the process.... One may mention the groups associated with Toivonen [Toivonen & Saxén, 1955a], with Yamada [1962], with Nieuwkoop [Nieuwkoop et al., 1952a,b; Nieuwkoop, 1952], and with Tiedemann [1966], as active in the experimental work that has led to this conclusion....

"It seems clear that the gradient system of the primary inductors does not specify at one stroke within the responding ectoderm all the different kinds of neural cell.... Probably a few major subdivisions of the induced ectoderm initially become distinguishable by the emergence of differences in the competence of their cells to respond to signals, and the differences in the signals they generate. Further interactions then subdivide the major areas in a similar way..." (Abercrombie, 1965).

As this example shows, and we see often in this book, sometimes the clearest insights to a problem are in its initial investigation. Afterwards, this early work becomes subsumed into the currently fashionable paradigm, and contact is lost with the actual observations.

The development of limbs is apparently tied to a nested cascade of homeobox genes (Dollé et al., 1989; Gardiner & Bryant, 1996), and to a hierarchical scheme of development that is evident from comparative, cladistic studies, which suggest an 'underlying... structural hierarchy' (Shubin, Wake & Crawford, 1995).

Nieuwkoop (1947a) produced a number of *duplicitas* larvae of the newt *Triturus taeniatus,* following the original method of Spemann & Mangold

(1924a,b). He reconstructed these larvae in three dimensions from serial sections, and noted which neural tissues were present and absent. In his description of his work, we can see the stepwise (cf. Saint-Jeannet et al., 1993), hierarchical nature of differentiation of the brain, the sequential division of the neural plate, and the correspondence between differentiation and gross anatomy of the brain:

"...Incomplete nervous systems give a... point of attack for the analysis of the correlations between the sense organs and the brain and between different parts of the brain.... After specific staining these incomplete nervous systems may form a very important source of data for the causal analysis of the development of the tracts and nuclei in the nervous system, especially in conjunction with the exhaustive investigation of the normal development by Herrick (1937, 1938a,b,c, 1939a,b)....

"All the experimental animals show the important fact that the induced parts of the brain have normal relative proportions.... This means that *the nervous system is completely formed up to a certain level.* This level may be situated either exactly between two successive parts of the brain or within one of the parts.... The most anterior part of the brain can be formed in any degree of reduced size without any influence upon the following parts....

"This characteristic development of the nervous systems of the duplicitas allows the following conclusions: (1) *The central nervous system is determined in a large number of successive zones,* at least seven in the brain alone, viz. tel-, di-, mes-, met- and three zones in the myelencephalon. (2) *All these brain parts are qualitatively 'equivalent';* each of them can form the most anterior part of the brain induced. This implies that the formation of any part is independent of the presence of another part anterior to it.... (3) *Each of these parts of the brain must be determined by a* (qualitatively or quantitatively) *different value of the organizing 'agent'....*

"Finally, I will point out the interesting parallelism between the position of the boundaries of the successive zones of determination and the anatomical, regional structure of the brain. The brain as well as the spinal cord is supposed to consist of a number of neuromeres. Tel- and metencephalon correspond to one neuromere, di- and mesencephalon consist of two and finally the myelencephalon is thought to be formed from a number of neuromeres. This supposition agrees in the main with the configuration of the zones of determination in the brain....

"In my opinion, the character of the induction process changes gradually during gastrulation, viz. from a 'regulative' into a 'mosaic' induction. This implies that the successive stages of the determination process must be analysed separately - a very important fact which, however, has received little consideration hitherto" (Nieuwkoop, 1947a).

Similar results and conclusions were reached by Dalcq (1946a) for double embryos formed by a $180°$ rotation of the animal hemisphere at early gastrulation. It would be nice to see this kind work repeated with modern gene expression techniques for identifying tissues, 'master' genes, and downstream genes, and with three dimensional computer rendering (cf. Salisbury & Whimster, 1993; Salisbury, 1994) of the normal and duplicitas embryos. Of course, I would also predict that time lapse studies would show a one to one correspondence between differentiation waves and steps of differentiation. When brain development is incomplete, I would predict this is because of failure of launching of a particular differentiation wave. With that failure, subsequent steps of differentiation dependent on that wave do not occur, and certain tissues just don't get formed.

What happens in the long run to tissues 'suspended' at a nonterminal state of differentiation deserves study in itself, especially since so many evaluations of results are done at stages much after a given differentiation should have occurred. Competence probably does not last forever, but what cell 'state' it is replaced by under these circumstances is not obvious. We will see that stabilization of differentiation may only apply to terminally differentiated cells, so that cells that have not reached a terminally differentiated state may be unstable and unpredictable (except for stem cells). Perhaps here is a basis for specific types of childhood cancer distinct from bases for adult cancers.

5.03 Dedifferentiation and Redifferentiation

Proposition 93: dedifferentiation and redifferentiation may occur by cells traversing down the differentiation tree and then back up via another route.

For this to happen, some sort of 'back pointers' may be needed. I adopt some of the language of computer science (Jensen & Wirth, 1978; Schneider & Bruell, 1981): each nonterminal differentiation cascade is a 'pointer' to two other differentiation cascades, so that the differentiation tree is a 'linked structure'. At a lower level, Dawkins (1986) noted that: "It is as if each exon ended with a pointer...." Note that a pointer, being a molecule binding to a specific DNA sequence, points to that sequence, and not to a particular position along a particular chromosome. If that specific sequence is moved, as by a chromosome inversion or transposition, it will still be pointed to. If it is duplicated, both copies will be pointed to simultaneously (unless one is sequestered), as I implicitly assume that pointers can diffuse to any physical (exposed) position within the nucleus. (For a discussion whether of diffusion is likely to be along the DNA, versus through the nuclear medium, see Berg, 1993.)

When a differentiation cascade has a 'back pointer' we say that it is 'doubly linked'. Some evidence, going back to 1879 (Nyhart, 1995), that regeneration 'recapitulates' normal development (Anderson & Waxman, 1985; Carrino et al., 1988; Hoffman & Cleveland, 1988; McDowell et al., 1990; Bonner-Weir et al., 1993) suggests the existence of back pointers. While the processes of regeneration and normal development are not identical (Maden, Gribben & Summerbell, 1983; Ferretti & Brockes, 1991; Stewart & Atwood, 1992; Bates & Meyer, 1993), the differences may be indicators of the back pointer mechanism.

The ability of cells to take on a limited range of new commitments upon redifferentiation (Maclean & Hall, 1987) would be explainable if they have arrived at a node part way down the differentiation tree, and can only go forward then in a few directions (Figure 31). The observation that "...the developmental program for cytoskeletal gene expression is recapitulated during axonal regeneration" (Hoffman & Cleveland, 1988; cf. Price & Porter, 1972) is consistent with dedifferentiating and redifferentiating cells traversing the differentiation tree.

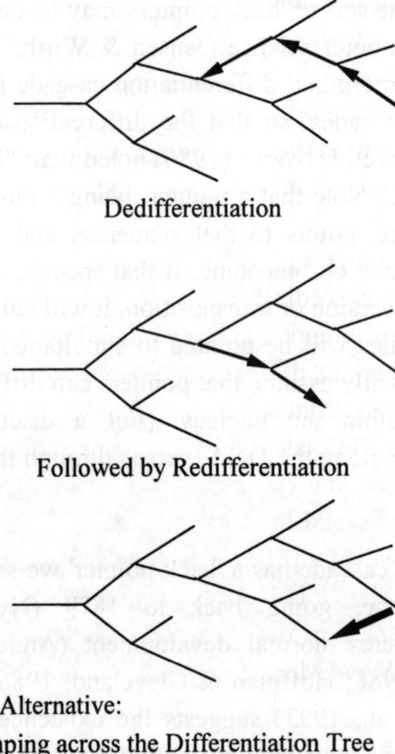

Figure 31. Two models for transdifferentiation. In the first, a cell dedifferentiates, changing its state of differentiation in steps, and stopping at some particular state. It then goes forward (redifferentiates), in the case shown, to a (possibly) new terminal cell state. The alternative is that each cell somehow 'jumps' across the differentiation tree to a new terminal type.

The alternate model is that...

"...a differentiation pathway can jump across from one branch to another.... We should maybe be thinking in terms of networks as an alternative to branches" (J. Paul in Paul et al., 1982).

We have no experiments at this time that distinguish these models, though cell tracking might solve this problem. The similarities between regeneration and normal development speak against branch jumping in the differentiation tree.

Proposition 94: differentiated cells capable of dedifferentiation may have sequential back pointers to earlier cell state splitter differentiation cascades, making the differentiation tree at least partly a doubly linked structure.

By a 'sequential back pointer' I mean one that links a differentiation cascade to the one immediately preceding it. The need for back pointers places a limit on how much dedifferentiation is possible, and may be related to the limited abilities of nuclei in differentiated cells to exhibit totipotency in nuclear transplantation experiments (Grant, 1978; McKinnell, 1978). It is not clear whether or not the latest memorons (Propositions 36, 37) would have to be removed to allow dedifferentiation. If so, the mechanisms by which they are removed could logically operate as the back pointers. "One important point is that when a path is traversed from the trunk of a tree out to a terminal branch, a choice is made at each intermediate branch; whereas when the path is traversed in the opposite direction there are no choices available to be made" (Steve McGrew, p.c., 1997), except when to stop.

Transplantation to eggs is the extreme case in which the state of nuclear competence is completely reset to the initial edge or differentiation cascade (root) of the differentiation tree:

"Dedifferentiation on the Epigenetic Address Model:... After union of the two nuclei, the egg must begin its epigenetic development by what one might call an 'initializing' of the epigenetic program (a setting of all the regulatory markers back to zero on the time scale of epigenetic crises). In the same way, and perhaps by the same process, the transferred nuclei in Gurdon's experiments (Gurdon, 1974) might become 'initialized'" (Stein, 1980).

This 'initialization' could be quite a different process from dedifferentiation using back pointers.

The literature on nuclear transplantation appears to deal solely with transplants to eggs (Gurdon, 1977a; Grant, 1978; McKinnell, 1978). It would seem fruitful, to test the notions suggested here regarding differentiation trees, dedifferentiation, and redifferentiation, to transplant nuclei between cells at different stages of differentiation. Experiments of "nuclear transplantation from one somatic cell to another" were suggested by Ephrussi (1953) (quoted in Burian, Gayon & Zallen, 1991), and by Barth (1964b):

"...Why not... simply [exchange]... nuclei between fully differentiated adult cells in tissue culture? In this case, their cytoplasms being different, if their nuclei were identical, unchanged from the nucleus of the fertilized egg from which they arose, there should be no incompatibility.... If the nuclei are different, however, such an exchange... could not distinguish between nuclear as opposed to cytoplasmic differentiation....

"...The 'ideal' test for a nucleus from a cell at an advanced stage of differentiation would be to examine its ability to support development of an enucleated egg only after this test nucleus had been serially transplanted into successively younger cytoplasms of earlier stages of development. In other words: Run the 'film' backward, instead of throwing a 'differentiating' nucleus back immediately into the relatively undifferentiated cytoplasm of the activated ovum. Such an experiment appears as 'fantastical' in terms of feasibility as was Spemann's (1938) of nuclear transplantation" (Barth, 1964b).

It is still difficult, and has not been attempted (Marie A. Di Berardino, p.c., 1996). We must ask, then, are more subtle shifts in nuclear competence possible than whole genome 'initialization', and what are their consequences? Someone will have to attempt experiments in which nuclei are transferred between cells at different points along the differentiation tree. There are three general cases:

1. transfer into a cell at a predecessor position along the differentiation tree;

2. transfer into a cell at a successor position along the differentiation tree;

3. transfer into a cell whose differentiation state is on a different terminal branch of the differentiation tree.

Such experiments may allow a refined dissection of the process of dedifferentiation and its relationship to normal development. They would also set bounds on the molecular mechanisms for dedifferentiation and redifferentiation, and the nature of back pointers. If their results turn out to have a simple relationship to the topology of the differentiation tree in a few species, then nuclear transfer may provide an alternative to cell lineage analysis for working out differentiation trees in other species.

Proposition 95: the success of nuclear transplantation between two differentiated cells is inversely dependent on the differences between their differentiation codes.

The success of dedifferentiation during nuclear transplantation may depend on the number of steps of backtracking that are necessary. This may be why, in general, nuclear transplants from later differentiated cells to an egg are less successful: "the percentage of normal development decreases dramatically when the nuclei are derived from progressively older donor embryos" (Browder, Erickson & Jeffery, 1991).

In somatic embryogenesis in plants (adventive embryogenesis: Raghavan, 1976), plant embryos are formed in tissue culture by presumed dedifferentiation of adult somatic cells, with a success that appears to depend on distance along the differentiation tree between the embryonic state and the current differentiated state of the cell:

"Somatic cells which are themselves embryonic or not far removed from the embryonic state are generally more easily induced to undergo somatic embryogenesis than differentiated vegetative cells.... The block to regeneration may be epigenetic, involving stable, but potentially reversible constraints in the functioning of genes required for growth and organized development (Halperin, 1986).... An explanation based on stable transcription complexes that repress and activate genes has been proposed for this phenomenon (Brown, 1984).... Halperin (1986) points out... the basic assumption under which we work, namely, that explanted tissues are capable of regeneration, and that it is only a matter of finding the right combination of chemical and physical stimuli for achieving the goal" (Thorpe, 1994).

There is at least one plant case of "...apomictic reproductive structures on the leaf of the bog orchid, *Malaxis paludosa* (Taylor, 1967)... [in which] the potentiality of vegetative cells to regenerate embryoids is expressed within the organization of an intact plant" (Raghavan, 1976), though of course the analogous phenomenon occurs in the budding of hydra (Graf & Gierer, 1980; Sinha & Mookerjee, 1992; Meinhardt, 1993; Pérez & Berking, 1993; Shenk et al., 1993).

This line of research, whole organ or organism regeneration from explants, has not been attempted for animals, beyond the cloning of early mammalian embryos by splitting (Seidel Jr., 1983; Willadsen, 1989; Prather & First, 1990a; Rumph, 1995) or of chopped up hydra (Chalkley, 1945; Gierer, 1974; Bode & Bode, 1984; Shimizu, Sawada & Sugiyama, 1993; cf. Streisinger et al., 1981). It attests to a remarkable ability to do without or restore whatever cortical factors are necessary for embryogenesis, and apparently proceeds without cytoplasmic 'determinants'. Indeed, for plants, Halperin (1986) refers to the problem as one of restoring embryonic competence, which we attribute to cortical cell state splitters. This may be another case of work on plants being ahead of that on animals, as in the epigenetic control of tumor cells (teratomas: Braun, 1959, 1978; Kahl & Schell, 1982; cf. Mintz & Illmensee, 1975).

Cloning animals from somatic tissues, without nuclear transplantation, may be well worth attempting, especially since progress has been made in getting past what was regarded as a block to cloning at compaction (reviewed in: Rumph, 1995, who notes that cattle embryos compact at 32-64 cells, compared to 8-16 cells for mouse: Van Soom & de Kruif, 1992; Van Soom et al., 1992, 1993), and dedifferentiation after compaction has been observed (Gurth-Halionet & Maro, 1992). Rat embryos sometimes compact at the 4-cell stage instead of 8-cells (Piscopo, 1995), so there is quite a range for the relative timing of compaction. Pluripotent to totipotent embryonic stem cells and embryoid bodies (Conlon, Barth & Robertson, 1991; Graves & Moreadith, 1993; Sukoyan et al., 1993; Surani et al., 1993; Iannaccone et al., 1994; Seamark, 1994; Sims & First, 1994; Wheeler, 1994; Dushnik-Levinson & Benvenisty, 1995; Rosenberger, 1995) parallel the capabilities of many plant cells, and may represent excellent material for the study of dedifferentiation and redifferentiation.

We could count the number of dedifferentiation steps needed by using the differentiation tree or the differentiation code. Transplants from frog larvae to eggs, for example, generally succeed (albeit only 2% of the time) only for nuclei from larval intestinal epithelium (reviewed in Karp & Berrill, 1981). While this is attributed to the relatively large size of these cells, Natalie K. Björklund suggests that it is due to the small number of steps of differentiation that these particular cells have gone through. They are derivatives of the wall of the archenteron (Laale, 1987), whose differentiation code at neurulation is just VC (yolk mass endoderm in Figure 3 in Appendix VI: Björklund & Gordon, 1994). The major problems with this proposition are the small numbers of partially successful nuclear transplantations from larval and adult (Lust et al., 1991) tissues to eggs or oocytes and the uncertainty in older work that differentiated cells, rather than stem cells, were the source of the donor nuclei (Di Berardino, Hoffner-Orr & McKinnell, 1986). 'Partial success' means development of larvae that did not feed, except in the latter study:

"...Many... experiments clearly reveal that while nuclei from juvenile and adult amphibians are pluripotent, 'to date, no nucleus of a documented specialized cell nor of an adult cell has yet been shown to be totipotent' (Di Berardino, 1987)" (McKinnell et al., 1993).

Since in this book we are dealing mostly with early development, the partial success of nuclear transplantation would appear to be nearly complete success of at least early development. Besides, full success has now apparently been achieved in mammals with the cloning of sheep (Wilmut et al., 1997). Thus, given standardization of the nuclear transplant procedure, and many transplants, statistically significant comparisons of the relative success of transplants between cells with different differentiation codes might be reasonable to expect.

Proposition 96: the success of natural dedifferentiation and redifferentiation (i.e., regeneration) depends inversely on the number of backtracking steps needed to get to the node of the differentiation tree from which redifferentiation has to start.

The simplest assumption is that each step of dedifferentiation, whatever the mechanism, has, in general, a probability $p < 1$ of succeeding. If n dedifferentiation steps are required to get from a terminal to an earlier node, then the success rate would be p^n, i.e., a fall off by a geometric progression. This is a rough way of quantifying a general impression:

"I don't think any of us envisage the possibility of a neurone becoming an erythrocyte or an erythrocyte becoming a neurone. That is quite out of the question, but the question we are asking is about a population of flexible cells and how far [back] up does it go in the tree.... [Do they] still have the flexibility to go back and enter into another pathway and is this a fairly universal phenomenon?" (J. Paul in Paul et al., 1982).

If we compare differentiation trees of two taxa, say amphibians and mammals, we may find in general that the number of dedifferentiation steps required for mammals is greater than for amphibians, for homologous tissues, perhaps due to topological differences in their differentiation trees (Natalie K. Björklund, p.c.). The differentiation tree may thus be a guide to the obstacles we need to overcome to achieve reliable regeneration of our limbs, spinal cord (Seil, 1994; Sternberg, 1996), etc., feats readily accomplished by amphibians (Piatt, 1955; Polezhaev, 1972; Becker & Selden, 1985; Clarke, Alexander & Holder, 1988; Muneoka, Bryant & Gardiner, 1989; Dinsmore, 1991; O'Hara, Egar & Chernoff, 1992; Zammit et al., 1993; Chernoff, 1996; cf. Travis, 1996a; Lazarov-Spiegler et al., 1996, for progress in mammalian nerve regeneration).

Proposition 97: transdetermination occurs via dedifferentiation followed by redifferentiation.

There is a large literature on 'transdetermination', 'transdifferentiation', or 'metaplasia', the process by which cells make 'unexpected' changes in their state of differentiation (Beresford, 1990; Bode et al., 1986; Bosco, 1988a; Boukamp, 1995; Clayton, 1982; Eguchi & Kodama, 1993; Lopashov,

1983a, 1991; Grindley, Davidson & Hill, 1995; Ide & Akira, 1988; Kawamura & Fujiwara, 1995; Kodama & Eguchi, 1994; McDevitt, 1989; Okada, 1991; Opas & Dziak, 1994; Polezhaev, 1972; Rao et al., 1988; Schmid et al., 1993; Zhou & Opas, 1994; Patapoutian, Wold & Wagner, 1995; see the discussion about trying to distinguish the former two terms at the end of Eguchi, 1976). If we look at transdifferentiation from the viewpoint of differentiation trees, we are faced with two general possibilities:

1. transdifferentiation occurs by cells backing down the differentiation tree and then coming forward again;

2. transdifferentiation occurs by jumps across the differentiation tree (cf. Proposition 93).

There is evidence for jumping (Figure 31), and it deserves further investigation as a possible challenge to our theory of differentiation waves:

"...That the determination of presumptive epidermis or neural plate has not become highly specific and fixed, even in the gastrula, is indicated by... results of transplantation. Presumptive ectoderm at the beginning of gastrulation, implanted in presumptive mesodermal or entodermal regions, becomes mesoderm or entoderm (Mangold, 1923)" (Child, 1941).

On the other hand, the production of 'mesoderm or entoderm' may have only been production of tailbud tissues from ectoderm (Appendix VI: Björklund & Gordon, 1994), and thus not a dedifferentiation process at all. Similarly, one "model explains the 'transdifferentiation' of mature cells... as simply a delayed realization of transitions that normally occur" (Anderson, 1989a). Reh, Jones & Pittack (1991) found in shaker culture versus plating that: "The ability of chick pigment epithelium to undergo transdifferentiation... appears to be dependent on the physical configuration of the cells".

We need to know the differentiation tree for the organism in question and the shortest route along that tree between two cell types related by a transdetermination. Then we could examine whether the cells traverse the tree or jump across it. Transdetermination occurs along quite specific 'pathways', so the question is the degree of correspondence of these pathways with the terminal branches of the differentiation tree:

"Hadorn (1966a) maintained *Drosophila* imaginal disc tissues in adult abdomens as vigorously growing undifferentiated cells for many transfer generations.... Transdetermination was not entirely random; certain patterns emerged:

1. A shift to a new state was abrupt; there were no intermediate stages containing mixtures of cells of two phenotypes [indicating discrete differentiation waves?].

2. Transdetermination involved a shift to broader phenotypic expression [dedifferentiation waves?]....

3. Some transdeterminations followed a predictable sequence of phenotypes [redifferentiation?] before a terminal state was stabilized....

"Transdetermination correlates with rapid proliferation [but cf. Filoni et al., 1995].... Transdetermination... occurs in a single cell developing as a clone (Hadorn, 1966b)" (Grant, 1978).

Some authors avowedly acknowledge a process of dedifferentiation followed by redifferentiation (Sparks et al., 1986; Raymond, 1991a; Hass, 1994; Tsonis, Washabaugh & Del Rio-Tsonis, 1995), consistent with this proposal:

"Retinal pigmented epithelial cells (PECs) of chicken embryos extensively and almost synchronously transdifferentiate into lens cells in medium containing phenylthiourea and testicular hyaluronidase, passing through the bipotent dedifferentiated state" (Agata et al., 1993).

Of course, the very act of cutting a piece off an imaginal disc might set off a differentiation wave. Perhaps such a wave could occasionally lead to transdetermination. The possibility must be kept in mind, due to analogous phenomena occurring with small pieces of amphibian tissue:

"In the early gastrula, the prospective endoderm is least responsive to inductive stimuli. Its ventral yolk cells may be induced to form a second intestinal tube with an extra liver and other secondary differentiations; but only under very special conditions has it been possible to obtain mesodermal or ectodermal structures from prospective endoderm. On the other hand, the great importance which inductive factors can assume is best demonstrated by small pieces of indetermined prospective ectoderm. Dependent upon induction from outside, it is competent to respond to a wide range of inductors. Any piece may produce epidermis, neural tissue, sensory organs, or even mesodermal tissues, according to the character of the inductive influence to which it becomes subjected" (Witschi, 1956).

Some of these changes may simply represent differentiation stopping at different points along the subtree of the differentiation tree that the excised tissue has already reached, or dedifferentiation followed by redifferentiation (Lo, Allen & Brockes, 1993), rather than edge hopping or true transdetermination. Further investigation, with the differentiation tree in mind, is warranted.

Proposition 98: transdetermination in imaginal discs could be a process of revertible, somatic mutation.

Transdetermination in insect imaginal discs happens in experiments in which prolonged proliferation occurs. Perhaps, in the course of multiple transfers of the imaginal discs, and the ensuing cell proliferation, an occasional point mutation of a differentiation trigger occurs:

"When [imaginal] discs are transplanted into *adult* flies, ... no differentiation takes place. Rather, the imaginal disc cells continue to proliferate. These proliferating cells can be continually cultured by transplanting them from adult fly to adult fly. At the same time, they can also be tested for their state of differentiation by removing pieces of the growing discs and placing them back into metamorphosing larvae.... Transdetermination happens more frequently after several passages through adult flies and occurs preferentially in certain directions. A genital disc, for instance, can generate leg structures, but leg discs have not been seen to differentiate into parts of the genitalia. Although the cause of this directionality remains unsolved, it is clear that determined cells can give rise to cell types other than those dictated by their normal fate" (Gilbert, 1991a).

The cycle of transdifferentiations (Figure 12.18 in Kauffman, 1993) may be little more than the result a few irreversible somatic mutations and a bulk of reversible somatic mutations, each with its own rate, just as in bacteria:

"What happens when... a [mutated] population grows indefinitely? At first thought, one might expect the proportion of mutant cells to increase until it reaches 100 percent. This is prevented, however, by the phenomenon of *reverse mutation*. Many mutations are capable of mutating back to the original state, and this reverse mutation will have its own characteristic rate" (Stanier et al., 1986).

Imperfect revertants, where two nearby point mutations cancel each other enough to produce a viable protein, are also possible (Watson, 1965). The idea of transition probabilities between differentiated states (Kauffman, 1973) is consistent with point mutations and their revertants.

Proposition 99: transdifferentiation or transdetermination involves differentiation tree edge jumping at the level of the nuclear state splitter.

The morphogenesis of a leg where an antenna should be is one of the startling feats of insects whose adult tissues grow out of imaginal discs. These discs are both terminal tissues for the larva and the initial, or root tissues for the adult structures (cf. Proposition 224). From the point of view of differentiation trees, transdetermination (García-Bellido, 1966a; Tobler, 1966; Hadorn, 1966a,b, 1968, 1980; Schubiger, 1971; Kauffman, 1973; Lee & Gerhart, 1973) seems to be a jump across the differentiation tree (Figure 31), though we have shown in Proposition 93 that some transdifferentiation can be regarded as dedifferentiation followed by redifferentiation. At first sight, it would seem that this should not happen. However, the homeotic genes are all quite similar to one another, so that errors in the process are plausible (Natalie K. Björklund, p.c.).

In terms of the nuclear state splitter, the pair of differentiation cascades that is 'readied' could be prepared in one of at least two ways:

1. All DNA not to be used could be masked in some way, except for the pair of master genes in waiting.

2. The signal could be specific to the pair of master genes in waiting.

Either way, errors are certainly conceivable. In the latter case the differentiation triggers are probably a family of proteins, whose specificity may in some cases partially overlap.

Proposition 100: dedifferentiation and transdifferentiation occur via two dimensional dedifferentiation waves and transdifferentiation waves.

The observation of a "highly differentiated state of... hepatocytes after... prolonged [3D = three dimensional] culturing..., in contrast to the rapid dedifferentiation known to occur under conventional monolayer culturing conditions" (Li et al., 1993) makes one wonder if dedifferentiation requires two dimensionality, and if, on the other hand, retention of a terminally differentiated state requires three dimensionality. Since dedifferentiation is a change in which genes are turned on and off in a whole tissue, it is plausible that it is accomplished by what we may call a dedifferentiation wave:

"Cells whose transdifferentiation is known or suspected can be viewed in isolation in single-cell culture. Isolation makes visible any division and may lessen any 'community effects' (Gurdon, 1988a) [presumably waves], thereby bringing notice a role of neighbouring cells or stroma in aiding or hindering (Moscona, 1986) the transdifferentiation" (Beresford, 1990).

A little more may be involved here. When 3D culture is achieved by forming spheroids, they "showed an organized structure consisting of squamated cells on the outermost layer and cuboidal cells in the interior" (Li, Colburn & Beck, 1992). There are thus two cell phenotypes, and perhaps differentiated types, in some 3D cultures. Which, if not both, is the cell type that grows and then dedifferentiates in monolayer culture is not obvious. In any case, it may be worth looking for waves of dedifferentiation in monolayer cultures, as a possible model for dedifferentiation dynamics *in situ*. The relationships to memorons, the differentiation tree, whether the 'dedifferentiation pathway' is the same or similar to the differentiation pathway, etc., are all open and approachable questions, especially in tissue culture.

Transdifferentiation has been modelled as what may be interpreted as a 'transdifferentiation wave':

"The final product of transdifferentiation is usually observed as a population of cells with similar phenotypes, not as a single isolated cell. This may result from the clonal growth of progenitor cells. A different interpretation was proposed by Pritchard (1981, 1986) and was based on close observations of transdifferentiation of melanocytes from NR [neural retina] glia cells. In this model, transdifferentiation events may be pioneered [launched] by 'leader' cells whose response to a set of triggering conditions occurs at a low threshold. The leaders may then influence their less responsive neighbours to follow by discharging small effective molecules [a homoiogenetic wave?]. This model seems to fit well in the change of NR to RPE [retinal pigmented epithelium].... Is it too speculative to assume that a spatial distribution of transdifferentiation cells may be controlled by some homoeotic genes?" (Okada, 1991).

A dedifferentiation wave might be the answer to the question: "Which signals trigger an individual cell to decide on whether undergoing retrodifferentiation or terminal commitment?" (Hass, 1994). Consistent with this idea is the observation that "the signals inducing RPE transdifferentiation are very local" (Zhao, Rizzolo & Barnstable, 1997). While there is no firm evidence for dedifferentiation and redifferentiation waves, they may be well worthwhile seeking.

Proposition 101: all cells involved in a given homeotic transformation or transdetermination have experienced the same sequence of expansion and contraction differentiation waves.

This idea, perhaps the main theme of this book, would clear up the confusion about how cells that are in proximity can transform together, even though they may not have the same lineage:

"...Mutations in several *Drosophila* genes can cause a partial transformation of the antenna into a second leg. Strangely, a clonal analysis of leg-tissue islands in the antennae of *Antennapedia* flies showed that the transformations occur concomitantly in groups of neighboring cells, i.e. via *proximity*, not via *pedigree* (Postlethwaite & Schneiderman, 1969). Similarly, another type of homeotic transformation (termed 'transdetermination': Hadorn,

1965, [1968, 1980]), which is routinely encountered during long-term culture of nonmutant *Drosophila* tissues, also affects nonclonal groups of cells (Gehring, 1987).... The perplexing implication is that organ states such as 'legness' or 'wingness' may not be properties of single cells, but rather of cell *clusters* (cf. Chandebois, 1977). *Xenopus* muscle differentiation likewise appears to involve a 'community effect': a cell will only shift its fate (in response to an inducing signal) if a sufficient number of its neighbors also does so (Gurdon, 1988a)" (Held Jr., 1992).

The same recruitment of cells, I would claim by differentiation waves, can be visualized in the answer to:

"Does lineage determine the dopamine phenotype in the tadpole hypothalamus?...

"The ancestry of dopaminergic (DA) neurons in the *Xenopus laevis* hypothalamus was investigated by combining intracellular lineage dye injections of 16- and 32-cell blastomeres with the immunofluorescent detection of tyrosine hydroxylase at tadpole stages. At these stages, DA neurons in the hypothalamus comprise a discrete nucleus that contains from 22 to 45 cells on each side.... The DA nucleus descends from only four of the 16-cell blastomeres.... The participation of some of the progenitors in the DA lineage is only probabilistic. The number of DA neurons generated by the same blastomere varied greatly in different animals.... Bilateral deletion of the major 32-cell progenitor (D1.1.1) resulted in a nearly complete restitution of the DA nucleus in 74% of the embryos that successfully completed gastrulation and neurulation. In the rest, the hypothalamus was smaller than normal or missing, and the DA nucleus was significantly reduced in size or absent. These results show that the DA nucleus can be restored after its normal lineage is deleted, but complete regulation is not always accomplished. Several blastomere progenitors dramatically altered their contribution to the DA nucleus after D1.1.1 ablation, including two blastomeres that normally do not contribute to the DA lineage. Thus, the fate to produce DA neurons is not determined at cleavage stages" (Huang & Moody, 1992).

This hearkens back to our basic assumption: cells do not know where they are or perhaps even who their neighbors are. In terms of differentiation, they primarily respond to the sequence of differentiation waves that they participate in.

Proposition 102: some cancers can be controlled via dedifferentiation.

"Cohenheim (1877) postulated that cancer cells could be caused to mature into normal nonmalignant cells if their local environment could be made into a duplicate of the embryonic state" (Becker, 1992b).

Are cancer cells in some sense a kind of differentiated cell? If so, then their 'control' by neighboring cells may be seen as an example of dedifferentiation followed by redifferentiation. This sequence of events is seen in the 'control' of teratoma cells by normal mouse embryo blastula cells during development of teratoma/normal allophenic (chimeric) mice (Mintz & Illmensee, 1975; Balinsky & Fabian, 1981) and in plants (Braun, 1959). However, in this case the dedifferentiation is all the way back near the root of the differentiation tree. A more instructive example is the phenotypic recovery of amphibian cancer cells implanted into a regenerating limb blastema (Rose & Wallingford, 1948; Mizell, 1965; reviewed by Becker, 1992b; cf. Wolsky, 1978).

Proposition 103: cancer may involve differentiation and/or dedifferentiation waves.

We can loosely reason in the following way. If the change of cells from normal to cancerous is a change in state of differentiation, then if all steps of differentiation involve differentiation waves, so should cancer. Preventing cancer could then involve preventing initiation and/or propagation of the purportive waves. It is probably at least worth looking for differentiation and/or dedifferentiation waves during experimental transformation of cells. One model for cancer is by "failure of mature cell-cell junctional communication" (Phipps et al., 1990). Such failure could alter some aspect of differentiation wave propagation.

One important application of this idea may be to provide a mechanism for multicentric breast cancer (McDivitt, 1984; Sarnelli & Squartini, 1986), perhaps by spread of a differentiation wave along the milk ducts, though it is difficult to know whether the additional foci are metastases of one primary tumor or if there were multiple primaries. (Current evidence suggests that breast cancer is not 'systemic': Tabár et al., 1992; Hellman, 1994; cf. Sivaramakrishna & Gordon, 1997a, implicating a single primary tumor).

The relationships between embryonic development and cancer are widely postulated (Einhorn, 1983, 1991; Deschamps & Meijlink, 1992), and, as we have seen, speculations go back at least to Cohenheim (1877). Here is another example of the hints that there may be a relationship between cancer and differentiation trees:

"In several malignant disorders, imprinted genes are, again, unfolded. Characteristically, expression follows the same tissue presentation as during embryogenesis.... We hypothesize that imprinted genes not only accompany cancer but may play a causative role as well" (Biran et al., 1994).

Sugahara & Watanabe (1994) propose an epigenetic model for cancer that could be consistent with differentiation waves:

"The present paradigm of radiation-induced carcinogenesis has been based on somatic mutation theory initiating a mutation in somatic cells derived from DNA damage by ionizing radiation. A non-threshold linear model has been assumed from high doses to low doses. As an alternative to this, we are proposing a new model that epigenetic changes induced by low dose irradiation may be a first step of carcinogenesis" (Sugahara & Watanabe, 1994).

The BRCA1 gene, implicated in human breast cancer, when knocked out in mice, produces neural tube defects (Gowen et al., 1996; Seachrist, 1996). As the breast is a gland of epidermal, and thus of ectodermal origin, a relationship of BRCA1 to differentiation waves may be worth pursuing. In general, since differentiation waves have so far been observed only in embryonic epithelia (Appendix V: Gordon, Björklund & Nieuwkoop, 1994), the following observation could help tie cancer, differentiation, and differentiation waves together:

"Abnormal proliferation of epithelial cells, leading to skin, lung and mammary carcinomas, causes the vast majority of cancer deaths (Berg & Hutter, 1995; Travis, Travis & Devesa, 1995)" (von Kalm, Fristrom & Fristrom, 1995).

As an aside, let me note that the complexity of the immune system may be proportional to the number of adult cell types and the number of ways cancer can arise, and thus to the complexity of the differentiation tree. There may be a tie-in with the suggested more efficient DNA repair systems in long-lived organisms (Britten, 1986) and their lower mutation rates (Gibbons, 1995).

Proposition 104: the ability to deactivate single differentiation cascades is an evolutionary advancement comparable to that of continuing differentiation, and made dedifferentiation and regeneration of complex morphology possible.

If this proposition is correct, it raises the possibility that organisms could be found with a spectrum of intermediate dedifferentiation and regeneration capabilities (cf. Polezhaev, 1972). To investigate this, we would need to determine organisms' differentiation trees, and then see what happens to terminal differentiation cascades during regeneration. It will be most interesting to compare the kind of dedifferentiation sometimes seen in regeneration with the kind seen in metamorphosis (see Section 7.02).

Proposition 105: activation and deactivation of differentiation cascades during differentiation and dedifferentiation involves not only master genes, but also a physical mechanism.

If we were only dealing with sequential differentiation, a purely logical mechanism would seem sufficient, in which master genes activated whole cascades of genes by ordinary repressor/activator mechanisms. The only distinction between master genes and other regulatory genes would be their triggering by the cell state splitter. But with dedifferentiation we have the problem of turning off a whole, already functioning differentiation cascade. In other words, if we assume that dedifferentiation actually means return of a cell to a previous state of differentiation, then it follows that a whole differentiation cascade must be deactivated or sequestered for each step back down the differentiation tree. This is a nontrivial matter. One could imagine that a process analogous to X-chromosome inactivation (Lyon, 1988; Chela-Flores, 1987; Romagnano et al., 1987; Willard et al., 1993; Eils et al., 1996) is occurring. However, differentiation cascades are probably distributed across a few chromosomes, making such a mechanism problematic. We will develop a simple 'Wurfel' model for the nucleus (see Section 10.13) that may account for resequestering during dedifferentiation and redifferentiation.

If there are dedifferentiation and/or redifferentiation waves, then cell state splitters should be involved. Here is a hint of involvement of cytoskeletal and/or membrane components and thus perhaps of the cell state splitter in permitting cells to dedifferentiate:

"The significance of cell-cell contacts in maintaining the stability of a differentiated cell is implied by the requirement for their disruption before transdifferentiation can occur" (Clayton & Truman, 1986; [cf. Schmid, 1992]).

There may be, in general, obstacles to the evolution of back pointers. Even in metamorphosis, little or no dedifferentiation may be occurring (Alberch, 1987). As dedifferentiation and redifferentiation involve much cell proliferation and a kind of return or at least approach to an embryonic state, we may anticipate that telomerase synthesis resumes, as in germ cells and renewal cells (Travis, 1995c).

Proposition 106: the contradiction between atavistic development and the decay to background of unused genes can be resolved via the (nongenetic) persistence of the physics of differentiation waves.

The existence of atavisms or throwbacks contradicts our modern understanding that, in the face of frequent mutation, genes must be used to be retained, by constant selection. It is possible that atavisms involve genes that are ordinarily, in a given species, used for other purposes (cf. Sibatani, 1990) and are, indeed, highly conserved (cf. the involvement of homeobox genes in teeth formation: MacKenzie, Ferguson & Sharpe, 1992, and the idea that reconstituted 'mosaic proteins' are involved: Lima-de-Faria, 1994,

1995). To rescue the concept of atavistic genes, it has thus been suggested that they *must* have other functions:

"The embryonic epidermis of a [chick], when grafted onto mouse dermis, develops scales in a pattern typical of mouse hairs (Dhouailly, 1973); the 'signal' from the dermis is conservative enough to evoke a response across class lines. In toothed vertebrates, teeth form from epidermis in response to induction by jaw mesoderm. Kollar & Fisher (1980) reported that jaw epithelium from chick embryos developed teeth when placed on the jaw mesoderm of mice - a striking example of the conservation of a response to induction that has not been expressed for more than 80 million years.... The genes governing these mechanisms must have other developmental functions, for if they were nonfunctional they would degenerate by mutation and genetic drift" (Futuyma, 1986).

Indeed, the ordinary expectation is that 'useless' genes should vanish:

"Moles and mole rats... have a long evolutionary history in subterranean habitats and, presumably, a long evolutionary history of eye reduction, yet they may still retain genes for three and one lens-specific crystallin proteins, respectively (Quax-Jeuken et al., 1985). With evolutionary times for eye degeneration to occur of 25 million years for the mole rat and 45 million years for the mole, the crystallin genes should, in the absence of selective constraints, be expected to have accumulated enough mutations to have been rendered silent (Quax-Jeuken et al., 1985). Their continued expression is an enigma" (Fong, Kane & Culver, 1995).

Kauffman (1993) asks:

"The existence of such atavisms as hens' teeth (Kollar & Fisher, 1980) and whales' legs (Andrews, 1921) and the evolution of differentiation raise an obvious question: Can there occur mutants which alter developmental pathways but not cell types?" (Kauffman, 1993).

The one case of legs in a whale is not an atavism in the same sense as hens' teeth, because, as Andrews (1921) discusses, hind limb rudiments are ordinarily found in embryos of a few species (Kükenthal, 1914), and in this single adult, resorption (apoptosis?) may be what failed. (See also the reviews of Hall, 1984a, and Van Valen, 1982a, and cf. Philip D. Gingerich on whale legs in Asimov & Asimov, 1993.)

In the context of his network model for differentiation (in which "differentiation is induced by exogenous stimuli or by asymmetric distribution at cell division", i.e., by the environment or maternal determinants), Kauffman (1993) finds that "mutations can tune one or a few transitions between cell types without altering all pathways at once". Emmeche (1994) interprets this to mean that...

"During the course of evolution, mutations may induce the choice of new pathways of cell differentiation; the old pathways remain as silent, potential alternatives" (Emmeche, 1994).

Whether all the genes involved would be kept active, and thus prevented from decaying to background, in this scenario, is problematic (cf. Lande, 1988; Raff, 1996).

There is another possibility when not just a gene product, but a whole atavistic morphogenetic event is involved, coming from the need to make a careful distinction between genetic and physical components of development:

"The genome need not embody that part of the information of the real world that is contained in the very structure of matter. It is in this sense that any concept of a 'program for development' by analogy with computer programs is incomplete. As is common practice in artificial intelligence and robotics, data and/or algorithms describing the real world must be included in computer programs. The genome needs no such programming, because it operates in the real physical world" (Appendix I: Gordon & Brodland, 1987).

"If D'Arcy Thompson's [1917, 1942] perspective is true, as I think it is, then growth in size and the attending expression of organic form depend on two types of information, one contained in physical and chemical 'laws' and another obtained from genetic 'algorithms'" (Niklas, 1994b).

The apparent persistence of unused 'genetic' information over geological time may simply be due to the likelihood that the laws of physics haven't changed in the duration: "The 'general laws' of geometry, physics and chemistry... need not take message space" (Cohen & Massey, 1983). This viewpoint is reasonable in the chick/mouse graft (Dhouailly, 1973; Kollar &

Fisher, 1980), given the plausibility of a physical ('cell traction') explanation for the spacing of hairs and feathers (Oster, Murray & Harris, 1983) and scales (Nagorcka, Manoranjan & Murray, 1987). It is less reasonable, though worth exploring, in the case of atavistic chicken teeth (for which "Cummings et al. (1981), Lemus et al. (1986) and Fuenzalida et al. (1990) have reported conflicting results": Raff, 1996). Our suggestion that induction is mostly a physical phenomenon (Appendix I: Gordon & Brodland, 1987), so that the 'signal' is a physical wave, also fits in with the idea that atavisms have a significant basis in the stability of physical laws. Even the genetic code itself, hitherto regarded as probably an historical accident (Crick, 1968, 1981; which in a sense created history: Smith & Morowitz, 1982), may have a physical basis (Taylor & Coates, 1989; but cf. the genetic code variations of mitochondria: Alberts et al., 1989, and ciliates: Fox, 1985). This implies the possibility of more than one (physics based) origin of life:

"I have formulated the tree of life hypothesis as a very strong (i.e., logically ambitious) claim. It says that *all* present-day organisms on earth are genealogically related to each other. But why accept the tree of life hypothesis in this strong form?.... One standard line of evidence used to answer this question is the (near) universality of the genetic code.... Biologists believe that the code is *arbitrary*.... If the code is not arbitrary [and]... happened to be the only (or the most functional) physical possibility, we might expect all living things to use it *even if they originated separately* " (Sober, 1993).

In more general terms, "...the essential elements of organic design are inherent in the material properties of the universe" (Thomas & Reif, 1993; cf. Niklas, 1994c). The complementarity between the biochemistry and the physics of embryos is but a specific example of what Henderson (1913) meant by *The Fitness of the Environment*.

The obvious contradiction between apparent persistence of 'genes' and their expected decay due to nonuse (Futuyma, 1986) does not make its way into some textbooks and monographs:

"Remarkably, although chicken cells have not formed teeth for millions of years, they still retain the genetic information to do so" (Levine & Miller, 1994).

"...As cases of atavism remind us, many pathways of development have persisted within the genomic repertoire, despite the fact that their end products are not functional. There is, for example, a case on record of a humpback whale, *Megaptera,* which externally carried two tapered cylinders, approximately four feet in length, which included a femur, tibia and vestiges of a tarsus and metatarsus" (John & Miklos, 1988).

"The ability to induce atavisms tells us that they need not arise as the result of gene mutations but that epigenetic events such as timing of development, tissue interactions and/or growth and morphogenesis can activate previously quiescent portions of the genome" (Hall, 1984a).

To the contrary, we would anticipate that "quiescent portions of the genome" will decay to background DNA. Given the rich fossil record for whales, it should be possible to estimate how rapidly the genes involved in limb development should have disappeared. (Rates of disappearance of unused genes might also be ascertainable from blind cave animals; cf. visual pigment genes in blind cave fish: Yokoyama & Yokoyama, 1990.)

Genes don't make tissues. Tissues are constructed via the interaction of genes with the physics of the embryo, via a genetic program expanded in concept to include differentiation waves organized via the differentiation tree. For any given atavism, we need to understand which parts of it are due to the genetics and which due to the physics. Insofar as the laws of physics don't change over evolutionary time, the latter component is 'immutable'. This does not solve the problem entirely, however:

"Yes, okay-- but I don't think the laws of physics know about 'teeth'! I'm sure you don't mean to imply that they do. Each tissue develops in ways subject to the same immutable laws of physics-- but some tissues make brains and others make teeth. It even seems unlikely that the laws of physics *restrict* tissue development to those kinds of tissues that currently exist. I guess my problem is that there does not seem to have been a discernible line drawn yet to separate physics from genetics in this context" (Steve McGrew, p.c., 1997).

The key here may be that the genetics is in the responding tissue, and the physics is in the 'stimulus' (i.e., induction). Until the question is investigated from this point of view, we won't know where to draw the line, and how much morphogenesis occurs on each side.

Proposition 107: germ cells are set aside early in development so that they avoid participation in most differentiation waves, which keeps their DNA accessible (exposed?) for the next generation.

This notion could be tested by comparing the totipotency of freshly set aside germ cell nuclei with nuclei of other cells that have experienced the same number or more differentiation waves. For example, in axolotls...

"...at the end of gastrulation... cells... are torn free and undergo an independent [single cell] contraction. These cells could be the same cells that are known to arise in this precise region [ventral mesoderm precursor] at this time and eventually form migrating germ cells (Nieuwkoop & Sutasurya, 1979)" (Appendix VI: Björklund & Gordon, 1994).

These ventral mesoderm precursor cells in axolotl have experienced three differentiation waves (Figure 3 in Appendix VI: Björklund & Gordon, 1994), which would seem about the right maximum number for easy 'dedifferentiation' back to totipotency.

This process by which axolotl germ cells may be set aside, tearing in a confluent sheet (Appendix VI: Björklund & Gordon, 1994) apparently due to high mechanical strain (cf. Jacobson & Gordon, 1976a), may have some similitude to the later separation of neural crest cells. There might even be a homology here, through duplication of the terminal branch of the differentiation tree leading to germ cells. Perhaps the effect could be duplicated and studied in culture. We also need careful studies tracking the origin of germ cells and the differentiation waves they participate in.

5.04 The Selfish Differentiation Tree

"...Some kinds of evolution can actually become easier with time. This is because each step in the process draws on the accumulated wisdom of the genes.... Organisms in the natural world... have a history of growing expertise" (Wills, 1989).

Proposition 108: the differentiation tree, seen as a hierarchically organized set of differentiation cascades, is the fundamental germ-line replicator on which adaptation and thus evolution proceed.

"One gets the impression that the cohesion of the genotype becomes stronger and stronger with time.... Just exactly what controls this cohesion is still puzzling" (Mayr, 1984).

The differentiation tree fits Dawkins' definition of an active germ-line replicator:

"The optimon (or selecton) is the 'something' to which we refer when we speak of an adaptation as being 'for the good of' something. The question is, what is that something; what is the optimon? ...I define a *replicator* as anything in the universe of which copies are made.... An *active replicator* is any replicator whose nature has some influence over its probability of being copied.... A *germ-line replicator* (which may be active or passive) is a replicator that is potentially the ancestor of an indefinitely long line of descendant replicators.... The active germ-line replicator... is, I suggest, the 'optimon', the unit for whose benefit adaptations exist" (Dawkins, 1982).

This concept is part of the more general 'units of selection' debate (Lewontin, 1970; Brandon & Burian, 1984a; Lloyd, 1988; Depew & Weber, 1994):

"Ever since Darwin (1859) published the *Origin of Species,* a nagging and persistent set of problems has plagued evolutionary theory. What is it that evolves, what is it that natural selection acts on, what are the fundamental units of reproduction? ...For Darwin selection acts primarily at the level of individual organisms.... On the other hand there are those... who hold that selection acts on groups, populations, and/or species as well as individuals.... The other relevant departure from Darwin is [that]... it is genes, not organisms, which reproduce differentially....

"Biologists have long recognized that nature is arranged in a hierarchy. Genes are located on chromosomes and chromosomes are located in cells. Cells form organs and whole organisms. Organisms form families, groups, demes, and finally species. Species or local populations form communities that, in turn, enter into ecosystems. The controversies in which the proponents of various forms of group selection, individual selection, and genic selection have become embroiled concern the location in this hierarchy of the target or targets of natural selection" (Brandon & Burian, 1984b).

I would like to argue that the differentiation tree is the fundamental level of selection. The genes constituting the differentiation tree obviously all work together (albeit in different cells at different times) in a hierarchical scheme to help construct an organism. In this sense they are 'cooperating genes':

"...The sense in which genes may be said to 'caucus' and form 'alliances' is the following. Selection favours those genes which succeed *in the presence of other genes, which in turn succeed in the presence of them.* Therefore mutually compatible sets of genes arise in gene-pools" (Dawkins, 1982).

Ohta (1988b) similarly points out: "Most... gene families constitute complicated interacting systems of genes... [that] cannot be independent of one another..." She "...treats the gene family rather than individual loci belonging to the family as a unit of selection." Lewontin (1974) refers to...

"The genome as unit of selection.... Genes in populations do not exist in random combinations with other genes. The alleles at a locus are segregating in a context that includes a great deal of correlation with the segregation of other genes at nearby loci. The fitness at a single locus ripped from its interactive context is... [irrelevant] to real problems of evolutionary genetics..." (Lewontin, 1974).

Dawkins (1976, 1989a) spoke of 'the selfish gene':

"Williams' (1966) brilliant critique of group adaptation was not entirely negative.... He argued that it is not the group nor even the organism that is the unit of selection. For Williams, the unit of selection is the 'meiotically dissociated gene.' This is the idea that Dawkins (1976) subsequently popularized in his book *The Selfish Gene* " (Sober, 1993).

This idea of 'gene as organism' has been attributed to Troland (1917) by Ravin (1977), who indicated that "Sewall Wright (1941a) also came to adopt the notion..." (though this concept is not discussed in the latter paper). But if the genes work together, then the whole interacting group of them is what is selected for. Thus we could even better speak of 'the selfish differentiation tree'.

One curious consequence of viewing the differentiation tree as the fundamental germ-line replicator is that, insofar as there is a one-to-one correspondence between differentiation trees and individual organisms, it restores the individual as the object of natural selection ("...organisms always evolve as wholes": Roth & Wake, 1985a). This is the full circle that Dawkins (1986) seems to be working up towards from his earlier concept of the selfish gene:

"Embryos are put together by all the working genes in the developing organism, in collaboration with one another.... We have a picture of teams of genes all evolving towards cooperative solutions to problems" (Dawkins, 1986).

Indeed, Dawkins (1982) devotes a chapter to "Rediscovering the Organism", suggesting that "the very existence of [individual] organisms should be treated as a phenomenon deserving of explanation in its own right", on the basis of 'cooperating' genes. (Cf. the epistatic artificial life simulations of W.D. Hillis in Levy, 1992 and the current revival of the concept of 'group selection': Bower, 1995.) Maynard Smith (1989a), in a review of Lewontin (1974), goes back to the old gene centered viewpoint:

"...Genes in natural populations tend to be in linkage equilibrium.... It is usually not the case that groups of cooperating genes are held together, so that they can collectively spread through a species, replacing an earlier group of genes" (Maynard Smith, 1989a).

Empirically, he may be right, and 'the unity of the genotype' (Mayr, 1975), if it exists, may have to consist of something higher than genes and their correlations. Thus I suggest that it is not particular groups of genes, but rather the topology of the differentiation tree, i.e., their cooperative relationships, that is the inheritable, transmissible 'collective' of genes. Individual alleles can come and go without affecting the topology of the differentiation tree.

There are assorted hints that this idea may be right. Correlations at the level of the differentiation tree may be the cause of 'multiplicative epistasis' observed in experimental evolutionary 'bottlenecks' in houseflies (Bryant,

Combs & McCommas, 1986). The independent evolution of "structural genes and regulatory elements for the insect chorion" (Kafatos et al., 1987) should extend to some independence of differentiation cascade genes from the regulatory structure of the differentiation tree.

Proposition 109: the differentiation tree is a logical, rather than physical, organization of the DNA, which is, for the most part, undisturbed by chromosomal rearrangements.

"There is more and more data from molecular biology about the extraordinary conservation of genes and molecules important in development. If so much is conserved at this level, the proximate causes of phenotypic diversity may lie at higher levels of organization" (Bolker, 1994).

A major advantage of the differentiation tree as a replicator is that it is a logical organization of the DNA, rather than a physical segment of it (Dawkins, 1984) per se. Thus the differentiation tree can survive many kinds of chromosomal rearrangements. The muntjac deer show how much rearrangement is possible:

"An extreme case of chromosomal evolution is presented by the two muntjac species *Muntiacus muntjac* (Indian muntjac, 2n = 6 [females], 7 [males]) and *M. reevesi* (Chinese muntjac, 2n = 46). Despite disparate karyotypes, these phenotypically similar species produce viable hybrid offspring, indicating a high degree of DNA-level conservation and genetic relatedness" (Levy, Schultz & Cohen, 1992). [Cf. Johnston, Church & Lin, 1982.]

"...Most genes seem to function quite happily almost anywhere in the euchromatic portion of the genome. We may reasonably anticipate, therefore, that their actual location within a given genome is simply a reflection of their past history of movement, either as a result of structural rearrangement of the chromosomes themselves or else by other means.... The muntjac situation [Lima-de-Faria, 1988]... shows us that massive rearrangements of the genome, this time in a form of structurally rearranged chromosomes, leave metabolism, development and phenotype... undisturbed in this multicellular eukaryote..." (John & Miklos, 1988).

I picture such rearrangements as essentially preserving the logical structure of the differentiation tree. This ability of chromosomes for major rearrangement and change in numbers without much impact on the differentiation tree may account for the lack of any universal correlation between chromosome properties with the progress of evolution (White, 1975; Cavalier-Smith, 1985a). An example of this lack of correlation is given by the relationship of one species of muntjac deer to us:

"The Indian muntjac, an Asiatic deer, has the lowest diploid chromosome number among mammals (female 2N = 6; male 2N = 7). Using flow cytometric quantification of propidium iodide-stained cells, we determined the DNA content of muntjac cells to be 94% that of human" (Levy et al., 1993).

The differentiation tree may be a key to the...

"...quest to explain life in any of its possible manifestations, without restriction to the particular examples that have evolved on earth. [Artificial life]... includes biological and chemical experiments, computer simulations, and purely theoretical endeavors. Processes occurring on molecular, social, and evolutionary scales are subject to investigation. The ultimate goal is to extract the logical form of living systems" (Christopher G. Langton in Levy, 1992).

Proposition 110: chromosomal repatterning is subject to the constraint that a viable differentiation tree is preserved.

When a chromosome rearrangement, i.e., a transposition, inversion, or duplication, occurs, it could disrupt the differentiation tree of an organism. Chromosomal rearrangements, or 'chromosomal repatterning' (White, 1975), are subject to a number of identifiable restraints, the net result of which is that relatively few are viable:

"The major chromosomal rearrangements such as inversions, fusions and dissociations are subject to a number of restrictions.... They must be capable of surviving through mitosis without leading to chromosome breakage. And in sexually reproducing species they must also be able to survive meiosis without leading to the production of aneuploid gametes. They must not be of types which produce deleterious position effects. The latter restriction may be more important in many organisms than has been thought in the past. Thus the parthenogenetic grasshopper *Moraba virgo* has a genetic system in which the 'meiotic barrier' should not operate against newly arisen rearrangements, provided that they can pass unharmed through somatic mitosis. Nevertheless, although this is a widespread species that clearly has been in existence for tens of thousands of years at least, structural heterozygosity in the natural populations is restricted to about half a dozen major rearrangements. The fact that we do not find hundreds of rearrrangements in the heterozygous state in this species is thus evidence that natural selection against newly arisen rearrangements is extremely severe, and in a species with this type of genetic system that can only be because the overwhelming majority of these rearrangements has associated 'position effects' of a deleterious kind" (White, 1975).

Violation of these constraints is, for example, deliberately made in "the use of translocations in insect control (e.g., Curtis & Hill, 1971; Foster et al., 1972; Hoy & McKelvey, 1979)" (Hedrick, 1981). Some of these constraints operate during embryogenesis: about 5% of human early spontaneous abortions (miscarriages) show "de novo and inherited structural [chromosome] rearrangements" (Kalousek et al., 1993) that may violate one or more of the constraints, including that of producing a viable differentiation tree:

"...Three points are especially important. (1) Something like one in 500 individuals, in populations as diverse as lilies, grasshoppers, and man, carries a newly-arisen chromosomal rearrangement. (2) Most rearrangements are deleterious, so that they are rapidly eliminated by natural selection. (3) For practical purposes, all chromosomal rearrangements must be regarded as unique, unlike gene mutations, which have a predictable rate of recurrence....

"The third conclusion arises from the following considerations. If we imagine the *Drosophila* karyotype as having 5,000 bands (in the polytene chromosomes) and 5,000 interbands.... The frequency of recurrence of the same [two break] rearrangement would be one in $5,000^2$ or 25,000,000.... Each chromosome rearrangement only gets a single chance to establish itself.... [This is overly pessimistic, since the cumulative population of *Drosophila* certainly exceeds $500 \times 25,000,000 = 12.5 \times 10^9$. Thus each chromosome rearrangement, at band resolution, can occur repeatedly in *Drosophila*.]

"The opposition of some evolutionary geneticists to the idea that chromosomal rearrangements play a primary role in speciation has been based on the tacit assumption that these obey the ordinary algebraic rules of population genetics, and that they would consequently be eliminated from the population by natural selection whenever they diminished the fitness of the heterozygote. But since the chromosomal rearrangements that establish themselves in speciation are only a minute fraction of the total number that occur, it would not be surprising if some or even many of them exhibited anomalous properties.

"...When we speak of a chromosomal rearrangement giving rise to a genetic isolating mechanism, we are not necessarily implying that the fecundity of the heterozygote is reduced to zero, or even to 50 percent of that of the homozygotes. In certain cases a reduction in fecundity by only 5 or 10 percent might be quite enough to initiate divergence. Failure to appreciate this point has led more than one evolutionary biologist to deny the importance of chromosomal rearrangements in evolution.... [It must just]... be able to pass through the 'heterozygosity barrier'" (White, 1978).

Cancer is one possible outcome of chromosome rearrangement (Neumann et al., 1993; Nowell, 1993) that obviously reduces fitness. In some cases large numbers of chromosome rearrangements are consistent with viability: "Considering the incredible array of cytological anomalies during honeybee meiosis and fertilization known to be compatible with subsequent development" (Sander, 1993c). For the muntjac deer species, in which the chromosomes are relatively fragmented, the constraint seems to be that recombination occurs primarily at noncoding DNA sites:

"For two deer species to be so closely related and morphologically similar, one would expect little or no changes to occur within the euchromatic chromosome arms of the ancestral, Chinese muntjac-like species during the formation of the [tandemly fused] Indian muntjac chromosomes. This would be possible if chromosomal breakages occurred only within the repetitive DNA sequences of the centromeric or telomeric regions of the ancestral chromosomes. The observed interstitial centromeric heterochromatin and telomeric DNA sites substantiate this notion" (Lee, Sasi & Lin, 1993).

Hedrick (1981) has shown a number of conditions under which, even in the presence of lowered heterokaryotypic fertility (White, 1978), a new chromosomal variant could spread through a whole population:

"Closely related species frequently differ by some chromosomal change. Because chromosomal hybrids generally have reduced fertility, some mechanism is necessary to enable a new chromosomal type to become fixed in a population.... There appear to be four situations which from a theoretical point of view appear to be potentially important mechanisms. These situations are meiotic drive alone, meiotic drive in combination with genetic drift, inbreeding in combination with a selective advantage of the new homokaryotype, and inbreeding in combination with genetic drift" (Hedrick, 1981).

All of these considerations for chromosomal rearrangements apply to differentiation trees. The constraints that allow passage of a new differentiation tree through the heterozygosity barrier may include retention of isochoric structure of chromosomes, at least in warm blooded vertebrates (Bernardi, 1989; Bickmore & Sumner, 1989; Li & Graur, 1991). Inbreeding would aid in getting past a heterozygosity barrier, if a chromosome rearrangement is passed on from parent to more than one offspring, and would be more likely in small, isolated populations, a factor that makes geographic isolation and chromosomal speciation complementary, synergistic mechanisms. Nonviable differentiation trees can be considered as lethal 'position effects', though generally not very simple ones.

This proposition is consistent with the suggestion of Dobzhansky (1975) that...

"The interrelations of gene and chromosomal changes in evolution became less, rather than more, plain as relevant information accumulated.... The... problem may well be related to that of the integration of structural and regulatory genes in the chromosomes" (Dobzhansky, 1975).

The 'integration of... genes' is precisely their organization in a differentiation tree.

Proposition 111: the high correlation of chromosomal repatterning with speciation suggests that reorganization of the differentiation tree is a common basis for speciation.

"In populations which became temporarily isolated and in which a dominant member had a major adaptive chromosome change, this rearrangement could be rapidly fixed in the homozygous condition. Rearrangement such as fusions, fissions, whole-arm translocations, or pericentric inversions, provided a rapid means of reorganizing regulatory mechanisms and thus had an enormous impact on developmental changes. They permitted a homozygous population to penetrate and exploit unsuitable habitat at the periphery of a species' range.

Bush (1975) equated these rearrangements with Goldschmidt's (1940) 'macromutations'" (King, 1993b).

As White (1975) indicates...

"...in most groups of animals even the most closely related species differ in karyotype.... Such a situation certainly suggests that in many instances the origin of karyotypic differences and the origin of new species may have been related events. But the exact causal relationship remains in most instances obscure" (White, 1975).

Chromosomal repatterning is a phenomenon that occurs at a physical scale up to many orders of magnitude larger than that of the single gene. Changes in the differentiation tree can possibly also occur up to this size scale. This match in physical size scales may account for this correlation, i.e., successful chromosomal repatterning (that survives mitosis, meiosis, and natural selection) may often produce differentiation trees representing nascent species. As long as the physical repatterning of the chromosomes is invariant with respect to viability of the differentiation tree and its other restraints, there is nothing to prevent this from happening. Thus the bizarre array of karyotypes in organisms is not under much selective restraint, explaining why the causal relationships are 'obscure'.

Consider an analogy with a current technology, the 'hard disk' for computers. As I wrote this manuscript, it appeared to me on my computer monitor as one logical whole, and it was even written with a hierarchical word processor, so that it had a tree structure. Yet it was probably fragmented into many linked pieces on the disk, and the size and locations of these pieces had no necessary relationship to the edges and nodes of the tree. As the manuscript evolved, with deletions, insertions, transpositions, and even duplications of text that then diverged, the physical pieces probably went through many changes in appearance. Again, these changes had no necessary relationship to the manuscript tree. As a writer, I find the physical distribution of my manuscript across the disk irrelevant to the logic of my presentation, so long as no physical events (hard disk crashes) occur that disrupt that logic. Similarly, the karyotype of an organism is irrelevant, so long as it is physically intact and retains the logic of the differentiation tree. If a deleterious mutant (a computer disk crash) occurs, I resume with my backups on other disks. Nature goes on with other individuals having viable differentiation trees, losing the 'crashed' individuals to embryonic inviability or natural selection.

An encyclopedia can be bound as a large number of small volumes or a small number of large volumes. The logic of its hierarchical (by alphabet) and cross reference structure would not suffer, though 'reproducing' the whole thing would present difficulties if one had a mixed set, especially if you are not permitted to read and understand (= transcribe) it during xeroxing. This may be exactly the problem with similar species containing different numbers of chromosomes:

"In attempting to explain the morphological differences between species, biologists have turned their attention to... related species [that] often differ in either the number or the structure of their chromosomes. In recent years, such differences have led to the assumption that they are in some way fundamental to both altered morphology (Bush, 1981) and to speciation (White, 1978; [cf. King, 1993b; Sites Jr., 1995])....A simple, but very convincing, example, involving a comparison of two species of deer, provides data on both these points. One species [the Chinese muntjac, *Muntiacus reevesi*] has 46 chromosomes in the female, the other only six [the Indian muntjac, *M. muntjak*]. Despite this massive disparity in the organization of the genome, there is very little morphological difference between them.... The two hybridize in captivity and can be reciprocally crossed.... This implies that the developmental circuits are, in all essentials, identical within these two species, and that it is irrelevant whether the genes which underlie these circuits are in six or 46 chromosomes.... Although somatic development is completely normal... (Liming, Yingying & Xingsheng, 1980),... meiosis is arrested in prophase I (Liming & Pathak, 1981)" (John & Miklos, 1988).

Some of these ideas can be found in the analogy between books and genomes by Platt (1962), though he assumes that cell-cell interactions are what cause different 'chapters' to be 'read' by different cell types.

Proposition 112: chromosome segments involved in repatterning are probably split from one another at nodes of the differentiation tree or at lower level pointers within differentiation cascades.

The analogy between the hierarchical computer manuscript and the differentiation tree breaks down on one fine detail. On the computer disk, there have to be two kinds of pointers: those involved with the hierarchical structure of my manuscript, and those that link physically separated pieces of the manuscript. One might presume that the latter type of pointer is lacking in genomes, because there is presumably no mechanism that can 'recognize' that two parts that have been separated must be linked, and that generates the linking, DNA binding proteins. Thus the logical components and the physical pieces of DNA that represent them must have some correspondence, to preserve differentiation trees. This may account for the evidence behind the...

"...hypothesis that similar rearrangements may have similar effects on the phenotype and hence be adaptive to the same environment.... Internal causes seem much more plausible than external ones" (White, 1975).

5.05 The Ciliate Origin of Multicellular Organisms

"Taken together, the facts encourage us to see in cortical polarity a historical invention that ought to be analyzable in terms of structure and process. They do not encourage us to think that the task of molecular biology is finished, even at the cellular level. In Tartar's (1961) words, *'Our greatest lack and most fruitful opportunity in biology lies in conceiving and testing the nature and capabilities of persistent* [cortical] *supramolecular patterns* '" (A.D. Hershey in Ephrussi, 1972).

Proposition 113: ciliates pattern their surfaces via cortical waves.

"The cortex... is believed to be the seat of future differentiation (Just, 1939a)..." (Dan, 1960).

There is a remarkable set of waves that traverse the cortex of the single celled ciliate *Paramecium* (Iftode et al., 1989; Le Guyader & Hyver, 1991; Bouck & Ngô, 1996), which leave in their wake a few alternate patterns of basal bodies (approximately = attraction sphere = attractophore = blepharoplast = central body = central corpuscle = centriole = centrosome = centrosphere = centrum = cortical unit = division center = idiozome = kinetid = kinetoplast = kinetosome = microtubule organizing center

(MTOC) = nucleus associated body = parabasal body = spindle pole body, etc.: Robertson, 1927; Mazia, 1961; Buss, 1987; Lynn, 1988) and associated structures, from which cilia grow. The basal bodies are arranged as...

"...different areas of... cortex... known to display different physical properties... as seen in a variety of cell-breakage experiments.... We have followed the evolution of five major cortical structures [regions of distinct basal bodies] throughout [cell] division.... The most striking result is the visualization of the progressive invasion of the cell surface by superimposed morphogenetic waves, which successively trigger the duplication, assembly or reorganization of each structure...." (Iftode et al., 1989).

I will attempt to use the nomenclature of Guenter Albrecht-Buehler (p.c., 1998). He suggests that the centrosome, an ill-defined organelle apparently without a membrane, is the primary reproducing entity. In most animal cells, but not usually in higher plant cells, the centrosome contains centrioles, which should be referred to as basal bodies only when they are at the base of cilia. Because of the mixed usage in the literature, I will use 'centriole' whenever possible, but will often revert to 'basal body'. I will try to use 'centrosome' whenever the larger, more encompassing entity is intended, though many authors do not make the distinction.

From Figure 13 of Iftode et al. (1989) we can estimate 'first' and 'second' wave speeds in *Paramecium* at 1.5 and 3 μm/min, respectively, putting them in the 'ultraslow, mechanical' calcium wave speed category (Jaffe, 1995). Calcium stores are certainly present (Stelly et al., 1991, 1995) These waves travel about the same speed as axolotl differentiation waves (3 μm/min), and "can affect physically unrelated structures over an approx. 0.5 μm thickness of the 'cortex'" (Iftode et al., 1989), which coincidentally(?) corresponds to the thickness of the axolotl cell state splitter (Figure 1 in Appendix V: Gordon, Björklund & Nieuwkoop, 1994). The trajectories of these *Paramecium* waves are confined to specific regions of the cortex, and the cortical surface is left morphologically differentiated in their wakes. The...

"...first wave of proliferation in fact proceeds in two 'rounds'. First, in both 1-bb [single basal body] and 2-bb [double basal body] units, one new basal body is generated. In a second round, another new basal body is added yielding the observed groups of three or four. Then, as the cell elongates, the packed basal bodies move away from each other and become equally spaced along the longitudinal axis.... As the dividing cell elongates, an increased spacing of duplicated basal bodies spreads anteriorly and posteriorly" (Iftode et al., 1989).

A similar crowding and then expansion phenomenon (expansion wave?) has been beautifully visualized in time lapse movies of mitotic waves of duplication of the subcortical nuclei in *Drosophila* during the syncytial phase of embryogenesis (Theurkauf, 1994a; cf. Baker, Theurkauf & Schubiger, 1993; Sullivan & Theurkauf, 1995; Theurkauf, 1994b; Matthies et al., 1996). Crowding and subsequent expansion are also apparent in the *Drosophila* eye imaginal disc morphogenetic furrow, which is accompanied by a wave of cell division (Wolff, 1993). Whether morphogenesis via 'growth pressure' due to cell proliferation (Hendrix, Madra & Johnson, 1993) might correlate with an expansion wave is worth considering. Whether there is a surface furrow, either in *Paramecium* or the *Drosophila* syncytial blastoderm, has apparently not been examined.

But let us go on with the waves in *Paramecium:*

"In certain regions, such as the anterior end of the posterior division product, basal bodies... duplicate again and this proceeds as a second wave. The resulting pairs of basal bodies will not separate and will be included in 2-bb units. Third, in a restricted area of the left anterior ventral field, pre-existing 2-bb units elongate and the two previously paired basal bodies separate. Later on, each of these basal bodies will duplicate to reconstitute 2-bb units. This zone (zone 2), intercalated between an invariant field (zone 3) and a 'normally' duplicating one (zone 5), can therefore be singled out by its specific mode of duplication.... Finally, in well-defined zones [3 and 4] of both future daughter cells, no basal body duplication at all will take place" (Iftode et al., 1989).

An actual differentiation tree is shown by Iftode et al. (1989) in their "Figure 7D [which] recapitulates the different types and steps in development of units according to their localization on the cell and initial 1-bb or 2-bb status". But the analogy goes further, in that *Paramecium* exhibits regulation. The steps of differentiation through which a given cortical unit goes depend on the sequence of waves it participates in, i.e., the morphogenesis is dependent on the trajectories of the waves, just like we propose in embryos:

"The fate map shows that the presumptive territories differ quantitatively and qualitatively in their contribution to the surface of the daughter cell. In this differential expansion [of the cortex], the fate of units is independent of their lineage.... The global pattern does not rely on heritable properties of particular units or files of units. The 'progeny' of cortical units and their role in the formation of the pattern of the daughter cells then mostly depend on their localization on the cell.... The same conclusion has been drawn from previous observations on *Tetrahymena* (Nanney, 1967, [which critically reviews earlier literature]) and *Paramecium* (Beisson & Sonneborn, 1965) showing that, due to an unexplained 'slippage' of rows around a cell, a particular row can contribute to any part of the cellular pattern.... Altogether, it appears that all cortical units are potentially equivalent [cf. Nanney, 1972] and that their different fates are controlled by their position on the cell" (Iftode et al., 1989).

Indeed, just like cells in regulating embryos, we could almost describe basal bodies as readied and competent for differentiation, without their course being predetermined:

"Lwoff (1950), after examining in detail the morphological changes occurring in ciliates with complex life cycles, concluded that cortical elements are indeed 'polyvalent,' and that their morphogenetic behavior is primarily determined by their positions in time and space" (Nanney, 1967).

Regulation to overall size occurs, when size is varied by starvation and refeeding of the ciliate *Tetrahymena* (Lynn & Tucker, 1976; cf. Lynn, 1977), analogous to size regulation in regulating embryos (Proposition 40). The corresponding ciliate surface waves probably cover proportional areas.

It is curious and perhaps not coincidental that the mitotic waves in the blowfly *Calliphora,* and perhaps in *Drosophila,* occur from specific launching domains, and spread in an asymmetric manner over the syncytial blastoderm, analogously to the trajectories of the waves in *Paramecium:*

"I was delighted when Dr. Anthony Durston pointed out that in the higher dipteran *Calliphora* (Agrell, [1962], 1964) the 9th to 12th syncytial cleavages occur as four mitotic waves.... The 9th and 10th show synchronous syncytial metaphase, and anterior or posterior anaphase waves. The 11th and 12th show metaphase waves from the ends to the middle.... Our preliminary results in *Drosophila* indicate mitotic synchrony before the 9th cleavage, and mitotic waves like those in *Calliphora* in the last four cleavage divisions, 9th-12th. We are hopeful, but not yet persuaded, that these waves are a stable aspect of early *Drosophila* development [cf. Sonnenblick, 1950, as discussed in Agrell, 1964" (Kauffman, 1975; cf. Kauffman, 1973).

Another striking analogy between metazoans and *Paramecium* is apparent: growth of a transient parallel array of microtubules. This creates the 'cytospindle' in *Paramecium* "at a very early stage" (Iftode et al., 1989). The microtubules are the 'tracks' along which basal bodies reproduce (Fleury et al., 1993), and a similar structure causes cortical rotation in amphibians, before first cleavage (Elinson & Palecek, 1993). Further homologies to kineties (rows of basal bodies) could perhaps be sought in the linear strings of cells differentiating in leech (Kenyon, 1994) and zebrafish (Kimmel, Warga & Kane, 1994) embryos (Proposition 244).

If the cortex of *Paramecium* in any sense contains cell state splitters, then we need to identify its actin and intermediate filaments. Indeed, in ciliates there is a "vast family of non-actin calcium-binding contractile proteins" (Fleury et al., 1992), intermediate filament-like proteins (cortical 'epiplasmins': Nahon et al., 1993) and microtubules (Iftode et al., 1989; Fleury et al., 1993; Jeanmaire-Wolf et al., 1993) that could be available to set up mechanical oppositions and carry ultraslow waves:

"Centrins are supposedly 20 to 25 kD acidic proteins that bind calcium and are involved in contractile systems in lower eukaryotes (Bhattacharyya, Steinkötter & Melkonian, 1993). They are also known as caltractins, and are supposedly homologous in sequence to the yeast cdc31 locus.... Certainly paramecia have [actin], but it seems to be used for motility in the cytoplasm instead of structure" (Karl J. Aufderheide, p.c., 1995).

Thus, although *Paramecium* waves have been modeled as calcium/kinase reaction-diffusion waves (Le Guyader & Hyver, 1991; Hyver & Le Guyader, 1995; Laurent & Fleury, 1995), a differentiation wave model based on cortical contractions and/or expansions might also be reasonable.

Wave propagation by means other than diffusion has been recognized as possibly necessary for modelling ciliate morphogenesis:

"If diffusion of a 'morphogen' is the basis for maintenance of [purported] gradients in ciliates, then the conveyance of such a substance must somehow be restricted to the surface layer [cortex]. An alternative, suggested by Kaczanowska (1974), is that the basis for morphogenetic gradients lies in the propagation of events such as configurational transitions over the cell surface, possibly through the membrane itself. This suggestion is at least plausible, since pronounced local differences in membrane properties are known to underlie the differential receptivity of different regions of the ciliate surface to mechanical stimuli (Naitoh & Eckert, 1969a)" (Frankel, 1974).

The "centers [that] specify the sites of differentiation of primordia of the contractile vacuole pores" (Kaczanowska, 1974) could be the launching domains of ciliate cortical waves. Much faster waves of contraction of microfilaments, probably accompanied by 'long-ranged Ca^{2+} waves' (Allbritton & Meyer, 1993), appear to coordinate cilia beating (González Santander & Martínez Cuadrado, 1980; González Santander et al., 1984; see metachronic waves in Proposition 118), though it is not clear that these microfilaments have any direct relationship to those that may be involved in the ultraslow 'morphogenetic waves' (Laurent & Fleury, 1995).

The launching domain for the *Paramecium* morphogenetic waves is the fission furrow, which is probably underlain by a contraction ring similar or homologous to a microfilament ring:

"When cells enter division, the fission furrow develops as an equatorial break in the continuity of longitudinal rows of cortical units" (Iftode et al., 1989).

"In *Paramecium*... the presence of feeding organelles is essential for regeneration (Tartar, 1954, 1964)" (Suhama, 1975).

Even without my present consideration of differentiation waves, the analogies between ciliates and multicellular organisms have been strong enough to warrant the following remarks:

"In contrast to *Stentor* which can reconstruct its whole cellular pattern from a small piece of cortex (Tartar, 1960), amputated paramecia can only recover their shape slowly through several successive divisions (Tartar, 1954; Chen-Shan, 1969, 1970; Suhama, 1975). The ciliates serving as major experimental organisms (reviewed in Frankel, 1984, [1989]) have therefore been used either to illustrate the importance of pre-existing local patterning ('cytotaxis' in *Paramecium*, see Sonneborn, 1963, 1970b) or, in contrast, to stress the possibility of long-range global regulation (which is more readily observed in *Stentor* and Hypotrichs [Grimes et al., 1980]). A distinction akin to that made in embryology between 'mosaic' and 'regulative' eggs could therefore be used to characterize ciliate morphogenetic strategies" (Iftode et al., 1989).

In summary, with regional differentiation, restricted trajectories (Iftode et al., 1989), cytoskeletal origin, regulation, and appropriate speeds, the waves on *Paramecium* resemble differentiation waves on the axolotl. It is plausible that an early ciliate was the common ancestor of all multicellular organisms (cf. Müller, 1995, 1997, suggesting a common ancestor for all animals; see Ruvinsky, 1997, for the contrary hypothesis of multiple origins for multicellular organisms). But is there a genetic component to ciliate differentiation waves? What would that mean in a single celled organism?

Proposition 114: the cortical waves of ciliates may involve changes in protein expression, which may also be changes in cortical gene expression.

"Heredity is not the exclusive property of nucleic acid molecules, but may be a property of biological systems at any level of the hierarchical array" (Nanney, 1972).

Let's examine the possibility that the waves of cortical differentiation in ciliates are actually homologous with differentiation waves in multicellular organisms, by being accompanied by changes in protein and maybe even gene expression. There are many possibilities at each node of the *Paramecium* differentiation tree:

1. $wave_i \Rightarrow protein_i$ (activation);

2. $wave_i \Rightarrow RNA_i \Rightarrow protein_i$ (translational control);

3. $wave_i \Rightarrow DNA_i \Rightarrow RNA_i \Rightarrow protein_i$ (transcriptional control);

4. $wave_i \Rightarrow RNA_i \Rightarrow DNA_i \Rightarrow RNA_i \Rightarrow protein_i$ (reverse transcriptase);

5. wave$_i$ $\Rightarrow$ protein$_i$ (activation) $\Rightarrow$ DNA$_i$ (promotional control);

6. none of the above.

Here the subscript represents either an expansion or contraction wave, triggering expression of a particular protein. Went (1977) suggested the fourth possibility, and Steve McGrew (p.c., 1997) suggested the fifth. Wheatley (1982) points out that our ability to entertain all these hypotheses "points out the need to identify the sites in which information molecules for centriole morphogenesis lie, and to elucidate the pathways of molecular events leading to the construction of these organelles...".

Let me take the last, sixth hypothesis, that nothing changes but the cortex, first. This is expressed as follows:

"...Information concerning supra-molecular patterns may in some instances be maintained and transmitted through grosser structural assemblies and by mechanisms distinct from those operating in nucleic transmission" (Nanney, 1966a).

We can think of this as a form of crystallization (Sonneborn, 1970a, calls it 'microgeography'), and the cortical waves as kink waves (see Section 1.15), i.e., waves of change of state. This might be so, but I will not pursue it further, since it is perhaps subsumable under hypothesis 1.

The first possibility, then, of protein 'activation' (allosteric changes, phosphorylation, etc.) allows all the genetics to occur in the micronucleus and/or macronucleus, with gene product uniformly available to all basal bodies. It is not implausible. The surface pattern of basal bodies replicates in a nearly crystalline array by an unknown mechanism (see Albrecht-Buehler, 1990b, for a model of the process), which is often referred to as 'crystalline'. Given that there is "a certain autonomy of the basal bodies and, more generally, of the proteins of the cortical units" (Hyver & Le Guyader, 1995; cf. Palazzo et al., 1992), it is plausible that ciliate waves are triggering specific protein changes via, for example, chaperones (Rattner, 1991; Perret et al., 1995; Brown et al., 1996a,b) in each cortical unit. Chaperones are present and involved in nucleating microtubule polymerization at the γ-tubulin rings (Zheng et al., 1995; Moritz et al., 1995b; Oakley, 1995). At least one chaperone is probably transported with attached proteins from the base "down the length of the flagellum [cilium] to the flagellar tip" (Bloch & Johnson, 1995). Assorted waves of phosphorylation (cf. Kaczanowska et al., 1995) accompany duplication and disassembly of cortical components (Sperling et al., 1991). Thus a direct relationship between differentiation waves and specific changes in some protein(s) must be considered a possibility. Frankel (1974) proposed a model for ciliate morphogenesis in which "positional information is interpreted primarily on the basis of differential protein-protein interactions rather than on differential activation of genes...".

The second possibility, translational control of RNA produced by the micronucleus and/or macronucleus, or (unlikely) self-replicating RNA (Hager, Pollard & Szostak, 1996), is supported by the presence of RNA in the cortex or centrosome (Stich, 1954; Mazia, 1961; Argetsinger, 1965; Hoffman, 1965; Stubblefield & Brinkley, 1967; Dippell, 1968; Brinkley, & Stubblefield, 1970; Hartman, Puma & Gurney Jr., 1974; Hartman, 1975; Dippell, 1976; McGill et al., 1976; Berns et al., 1977; Heidemann, Sander & Kirschner, 1977; Peterson & Berns, 1978; Dodson, 1979; Berns et al., 1981; Wheatley, 1982; Tiedtke, 1985; Frankel, 1991). To date "There is no evidence that this [centrosome] RNA is messenger, ribosomal, or transfer [RNA], but it may play a structural role in the ontogeny of the basal body (Chatton & Brachon, 1935; Hartman, Puma & Gurney Jr., 1974)" (Aufderheide, Frankel & Williams, 1980; cf. Peterson & Berns, 1978). In the latter case, the RNA could be ribozymal or even self-replicating, functions which themselves could be triggered by differentiation waves. Differentiation of the single cell amoeba *Naegleria gruberi* to a ciliated form involves microfilament mediated transport of mRNA for tubulin and other proteins from the nucleus to the basal body (Han et al., 1997; cf. Yang et al., 1997). Phillips & Rattner (1976) found "that the formation of procentrioles was dependent upon protein synthesis immediately before their appearance". Hartman (1975) suggests a simple nucleating role for RNA:

"The centriole or basal body is formed like a ribosome in which RNA forms the nucleating site for the organization of proteins into a complex structure (Nomura, 1970)" (Hartman, 1975). [Cf. the resonance hypothesis of McLardy, 1981.]

Can we show a chain of DNA $\Rightarrow$ RNA $\Rightarrow$ protein, i.e., that the third possibility, full gene expression, occurs in centrosomes? A link between the centrosomes, the nucleus, and what may be differentiation waves or their precursors, is seen in protozoans with a single cilium, such as *Astasia,* a colorless euglenid:

"Division in most flagellates begins with duplication of the flagellar structures, followed by mitosis, and then by progression of the division furrow from the front to the back of the cell" (Patterson & Hedley, 1992).

This is an old observation:

"Martin & Robertson (1909)... found that axostyles arise after division quite independently of the nucleus or of centrodesmose, and regarded them as independent... self-perpetuating organoids which may be the first to divide in the processes of reproduction (*Giardia*) or the last to divide (*Trichomonas*)" (Calkins, 1933).

Evidence is slowly building that there is DNA in kinetosomes at the base of cilia (Sukhanova & Nilova, 1965; Smith-Sonneborn & Plaut, 1967, 1969; DeFoor & Stubblefield, 1974; Berns et al., 1977; Hall, Ramanis & Luck, 1989; Hyams, 1989; Hall & Luck, 1995; Walther & Hall, 1995; Dolan & Margulis, 1996; Margulis & Dolan, 1997), and thus in the cortex of ciliates. Claims for DNA in blepharoplasts, bodies in some plants that produce centrioles, go back to Lee (1955), and in the basal bodies of the ciliate *Tetrahymena,* to Seaman (1960), Argetsinger (1965) (who suspected nuclear DNA contamination) and Randall & Disbrey (1965):

"Randall & Fitton Jackson (1958) [using Feulgen staining] have... observed DNA in the kinetosome of *Stentor*. In this respect kinetosomes resemble the kinetoplast of flagellates which also contains Feulgen positive material (Bresslau & Scremin, 1924; Robertson, 1927; Lwoff & Lwoff, 1931; Barrow Jr., 1954), and which rapidly incorporates thymidine (Steinert, Firket & Steinert, 1958)" (Seaman, 1960).

About the negative results of Yuasa (1935), Mizukami & Gall (1966) note that: "A negative Feulgen reaction merely sets an upper limit to the concentration of DNA possibly present". There are many other negative results and opinions regarding DNA in centrosomes (Rampton, 1962; Grimstone, 1966; Sorokin & Adelstein, 1967; Pyne, 1968; Hufnagel, 1969; Sonneborn, 1970a; Flavell & Jones, 1971; Friedländer & Salet, 1971; Fulton, 1971; Hartman, Moss & Gurney Jr., 1972; Margulis, 1973; Hartman, Puma & Gurney Jr., 1974; Hartman, 1975; Younger et al., 1972; Rattner & Phillips, 1973; Frankel, 1974; Heidemann, Sander & Kirschner, 1977; Peterson & Berns, 1978; Dodson, 1979; Berns et al., 1981; Wheatley, 1982; Johnson & Dutcher, 1991; Holmes, Johnson & Dutcher, 1993) which suggests either 'controversy' in the field, a true variation in DNA content between centrosomes of different species, contamination, or the need for definitive studies. Perhaps it is time to apply highly sensitive PCR (polymerase chain reaction) techniques to the question. Certainly, if the "DNA at each basal body site corresponds to a length of double helix of about 67 μm" (Randall & Disbrey, 1965), then modern techniques could readily be brought to bear on confirming or denying its existence. It will also be interesting to learn whether the cortical pattern mutants of ciliates (reviewed in: Jerka-Dziadosz, Garreau de Loubresse & Beisson, 1992) are mutants of the purported DNA or RNA in the basal bodies.

The search for centrosome DNA continues to this day:

"Whereas *Chlamydomonas* cells have only a pair of kinetosome-centrioles per cell, we know from Kirby's and subsequently Brugerolle's [1976] work that calonymphids may have hundreds of kinetosome sets, each organized with (karyo-) and without (akaryo-) its nucleus. Thus, calonymphids make ideal material for probing any putative kinetosome-centriole DNA" (Kirby Jr. & Margulis, 1994). [Italics reversed.]

Indeed, such DNA is now reported (Dolan & Margulis, 1996; Margulis & Dolan, 1997). If DNA in centrosomes is incontrovertibly confirmed, cortical inheritance may come to have the same symbiotic status as inheritance via DNA in mitochondria and chloroplasts. Guenter Albrecht-Buehler (p.c., 1998) adds two criteria before such a conclusion is justified: "No, the presence of DNA is not sufficient. At least the DNA should be replicated outside the nucleus and in synchrony with the organelle".

Since the cortex bears differentiation waves, cortical inheritance and these waves may be closely linked, and share a common evolutionary origin. However, given the apparent equivalence of the basal bodies (Lwoff, 1950; Nanney, 1967; Iftode et al., 1989), any cortical inheritance of pattern must revert to those factors that determine the launching domains and trajectories of the ciliate differentiation waves. As with eukaryotic embryos, this is an open question.

Since basal bodies are different in different regions of the *Paramecium* cortex, and these differences are correlated with the waves they participate in, changes in the proteins in basal bodies must be occurring at some level. We cannot yet eliminate any of the mechanisms listed in this Proposition, and others might be possible. This situation is unfortunate, because we may be seeing in *Paramecium* the basic mechanisms underlying all differentiation.

Proposition 115: centrosomes may be symbiotic organelles.

"The concept of reproductive continuity in centrioles is an old one, dating back at least to Boveri (1900)" (Went, 1977; cf. Wilson, 1925).

As we have seen in Proposition 114, DNA or RNA may be present in the centrosomes of all cilia bearing multicellular organisms, which suggests that centrosomes are symbiotic organelles, just like mitochondria and chloroplasts:

"Indeed, proponents of the symbiotic theory [for the origin of undulipodia = cilia: Margulis, 1981, 1993a,b; Barth, Stricker & Margulis, 1991; Munson et al., 1993; Bermudes, Hinkle & Margulis, 1994] have long predicted the presence of extranuclear nucleic acids as a means of explaining such non-Mendelian phenomena as cortical inheritance in ciliates (Grell, 1973) and motility mutants in *Chlamydomonas* (Dutcher, 1986). Although the semiautonomy of plastids and mitochondria had long been recognized, the regular existence of DNA in mitochondria and plastids proved to be the convincing evidence supporting their symbiotic origin. The discovery of kinetosomal DNA and the genetic autonomy of kinetosomes argue for symbiosis as the source of this evolutionary novelty" (Hinkle, 1991).

The origin of this genetic material may be spirochete bacteria (though with the discovery of FtsZ 'microtubules' in *E. coli* and other bacteria, we may have to open the field a bit wider: Erickson, 1995, 1997; Yu & Margolin, 1997; Lowe & Amos, 1998; Mukherjee & Lutkenhaus, 1998; Nanninga, 1998):

"Members of the phylum Ciliophora, a relatively well-known group of protoctists, are primarily single cells.... Ciliates are defined as eukaryotic cells with dimorphic nuclei (macro- and micronuclei) and a cortex, the fairly rigid surface layer about one micrometer thick. The cortex is composed of kinetids [centrosomes]. Kinetids are best known in ciliates, but they are characteristic of all undulipodiated [ciliated] cells.... Kinetids, in all their patterned detail, reproduce, but not by direct division. They develop offspring kinetids often in proximity to parent ones [Sonneborn, 1970b].... Because ciliates, unlike most other eukaryotes, have so many kinetids, they display extremely regular and observable patterns on their cortices... arranged in rows called *kineties*.... It seems highly likely to us that the cortical phenomena of differentiation so well known for ciliates also occur in other protoctists, animals, and plants....

"...Let us assume that the kinetids directly originated from spirochetes and their attachment sites. Some elegant experiments on the determination of cortical patterns in ciliates have been done (Grimes, 1976, 1982b; Grimes et al., 1981; de Terra, 1974, [1985a,b]).... Grimes summarizes them in terms of three phenomena: global pattern determination, short-range pattern determination, and developmental assessment. It is our contention that the basis of their understanding lies in the recognition that these processes are fundamentally processes of microbial ecology and evolution.... New kinetids are oriented with respect to old ones because they are essentially offspring, products of remnant bacterial cell reproduction, of the old ones. Developmental assessment is a process in which kinetids regulate their number, position, and interactions.... We interpret this behavior to be the interaction of spirochete populations growing to a maximum number determined by the limits of their environment (their host cells)....

"By this reckoning the cortices of ciliates should be considered extremely obvious examples of the relation between microbial population phenomena and development that exist throughout the eukaryotic world.... Differentiation is fundamentally a manifestation of microbial population biology..." (Margulis & Sagan, 1986a). [Cf. the "narrow range of basal body numbers in *Tetrahymena* species": Nanney, Chen & Meyer, 1978.]

Margulis (1993b) has further elaborated the spirochete theory for the origin of kinetosomes (attributed by her to Kozo-Polyasnki, 1924, in Margulis & McMenamin, 1990; cf. Sagan, 1967; Margulis, 1971a, 1991, 1993c). Here is a selection from her Table 9-2 of "Evolutionary Question/Hypothesis" regarding "Symbiotic origin of microtubule organizing centers (MCs) and their products":

"Are these symbioses obligate?
  Yes, in all mitotic organisms, because of the utilization of 'motility proteins' from the free-living spirochete for mitosis and other intracellular motility processes.

What genetic legacies of free-living prokaryotes do MCs and undulipodia retain?
  Evidence for DNA (Hall, Ramanis & Luck, 1989) and kinetosomal RNA, most likely in the lumen of centrioles and kinetosomes (Stubblefield & Brinkley, 1967; Dippell, 1976); new RNA synthesis accompanies new kinetosome production (Younger et al., 1972); and kinetosomal RNA may function in MC formation (Heidemann, Sander & Kirschner, 1977). [Cf. Frankel, 1989; Dirksen, 1991.]

What did the protoundulipodia lose after becoming symbionts?
  Peptidoglycan walls and some membranes; most biosynthetic functions, including synthesis of DNA and ribosomal proteins, probably were integrated to form the nuclear gene-controlled nucleocytoplasmic system.

How are offspring MCs produced?
  Usually they form from parent MCs; for example, new kinetosomes form at right angles to the parent in a complicated sequence with many variations (Mizukami & Gall, 1966; Dirksen & Crocker, 1965; Dirksen, [1982], 1991). Centrioles and associated organelles that duplicate in mitotic organisms before becoming visible must have undergone nucleic acid replication, by hypothesis.

What do the genomes of the original spirochete symbionts code for now?
  Not for tubulin [cf. Erickson, 1995, 1997], but otherwise unknown; they may code for a tubulin-based herbicide resistance (James et al., 1988) or RNAs or proteins of the 80S ribosomes. Membrane-associated RNA or DNA synthesis may be related to kinetosome reproduction (Dippell, 1976).

What is the evidence for genetic autonomy of kinetosomes?

Inheritance of stentor undulipodia independent of the macronucleus (Tartar, 1961, 1968); inheritance of cortical structures in *Paramecia* independent of the macronucleus, nucleus, mitochondria, and soluble cytoplasm (Beisson & Sonneborn, 1965). Growth and development of undulipodiated bands and related microtubular structures of hypermastigote gametes in absence of the nucleus (Cleveland, 1956). See Jinks (1964), Bermudes, Margulis & Tzertzinis (1987), To (1987) for reviews. [The latter provides alternatives to a spirochete origin of kinetosomes.]

Why do the numbers of centriole-kinetosomes and undulipodia per cell vary, yet the unit sizes of microtubules, kinetosomes, and axonemes remain constant?

At first, they varied as a consequence of the fact that the symbiont/host ratio is seldom 1/1; as control evolved, they varied because of differentiation for motility and the evolution of mitosis. Size (0.25 x 1-100 μm) is determined by the original spirochete genome; like the diameter of prokaryotic cells in general, individuals vary very little" (Margulis, 1993b).

To this I must add an important physical consideration: "The repeating unit of cortical structure in *Paramecium* has a volume of the order of one cubic micron" (Sonneborn, 1975a), which is the same as that of a many small bacteria (Margulis & Schwartz, 1987; cf. Margulis, Schwartz & Gould, 1998). Much can go on in such a volume: "Preparations of basal bodies contain more than 200 polypeptides" (Dutcher & Lux III, 1989). Thus the centrosome is possibly seen to be a symbiotic bacterium, much like the chloroplast or mitochondrion, but lacking its own membrane. Research on it is also decades behind these other symbionts (cf. Thorsness & Weber, 1996). One property of centrioles, their apparent ability to detect infrared light (Albrecht-Buehler, 1994), and perhaps use it in a form of cell-cell communication (Albrecht-Buehler, 1992), would be worth looking for in spirochete bacteria, to see if it bolsters this purported line of descent. It has been suggested by Pickett-Heaps, Forer & Spurck (1997) that the spindle apparatus for separating chromosomes had its origin in prokaryotic cell division (cf. Nanninga, 1998), in which case it too might be traceable back to spirochetes:

"We suggest that the principles of 'cellular tensegrity' (Ingber, 1993b) derived from the behaviour and organization of the interphase cell apply to the spindle. In an evolutionary context, this argument further suggests that the spindle might originally have evolved as the mechanism by which a single tensegral unit (cytoplast: [Porter & McNiven, 1982]) is divided into two cytoplasts; use of the spindle for segregating chromosomes might represent a secondary, more recent development of this primary function. If valid, this concept has implications for the way the spindle functions and for the spindle's relationship to cytokinesis" (Pickett-Heaps, Forer & Spurck, 1997).

Insofar as the spindle apparatus and the cell state splitter are analogous organelles, they might then share this common origin (cf. Erickson, 1995, 1997).

Caveats to the symbiotic theory for centrosomes (cf. Gray, 1992) are summarized by Frankel (1991):

"Normally, new centrioles or basal bodies form near old ones, but there are some well documented exceptions... (Fulton & Dingle, 1971; Grimes, 1973a; [cf. Mazia, 1961, for an earlier review]).... So you don't need an old basal body to make a new one. There could be something left that's asymmetrical that you simply cannot distinguish using the electron microscope.

"There were claims around 1970 that there was DNA in basal bodies. These were pretty well refuted (Fulton, 1971; Hartman, Puma & Gurney Jr., 1974) [but see Proposition 114]. It was recently claimed that the *uni* linkage group [in *Chlamydomonas*], which contains a number of flagellar markers, was located in the basal body... (Hall, Ramanis & Luck, 1989). This has also been disputed (Johnson & Rosenbaum, 1990).

"There is fairly reasonable evidence that there is RNA in basal bodies. In ciliates, Ruth Dippell (1976) showed that the dense material in the lumen of the basal bodies is dissipated by RNAse but not by DNAse or proteases. There were other demonstrations in this and other systems (e.g. Hartman, Puma & Gurney Jr., 1974). But there are no ribosomes in the basal body and nobody knows what that RNA does. Even when new basal bodies form near old ones, they don't divide: a new one forms at a right angle to the old one, several hundred Ångstroms away (Dippell, 1968). So I do not think that one can call basal bodies self-replicating" (Frankel, 1991).

Thus for now we are in the uncomfortable position of not even being sure whether centrosomes and/or their basal bodies or centrioles are self-reproducing. There are cases where centrioles do not seem to arise from previously existing centrioles (Schuster, 1963; Dirksen & Crocker, 1965; Mizukami & Gall, 1966; Dippell, 1968; Fawcett & Phillips, 1969; Fulton & Dingle, 1971; Grimes, 1973a; Frankel, 1974; Calarco-Gilliam et al., 1983; however, cf. Batisse, 1968; Kalnins & Porter, 1969, to the contrary, and the possibility of "Centers for the production of the formed kinetosomes... in the shape of complex aggregates of endoplasmic reticulum": Schuster, 1963). Of course, the centriole precursors may just not have been detected (Fulton & Dingle, 1971): they may be part of the amorphous centriolar rim (Fais, Nadezhdina & Chentsov, 1986), essentially invisible when microtubules are not present. But to this discomfort I can add that the presumption that *preexisting* cortical structures are necessary for the development of patterns is challenged by the existence of the spherical auxospore (zygote) stage or "gametes [that] are amoeboid and morphologically isogamous" (Bold & Wynne, 1978) in the highly ornate diatoms (Gordon & Drum, 1994), sometimes showing mirror symmetry (Pickett-Heaps, 1983), whose nuclei undergo specific spatial arrangements that are clearly related to overall cell shape and symmetry and structure of the silica shell (Pickett-Heaps, Schmid & Tippit, 1984; Pickett-Heaps, Schmid & Edgar, 1990; Pickett-Heaps, 1991a):

"In [diatom] species producing 'free,' rounded gametes and spherical, apolar zygotes, the causes of the origin of polarity are quite unknown.... The external shape of the auxospore often deviates from that of the definitive vegetative cell, and since the valves of the primary cells arise inside the auxospore as a copy [in overall form: see Gordon & Drum, 1994], these and their direct offspring are abnormally shaped. Mostly the middle of the auxospore is abnormally inflated, and this bulge disappears only after many cell divisions.... In *Meridion* the most bizarre and often irregular and repeatedly inflated shapes of the auxospores are copied exactly and without harm by the primary cells and their progeny" (Geitler, 1969).

Diatoms go through many vegetative cell divisions before entering the sexual phase (Jewson, 1992a). Ciliates, in similarly going through a number of cell divisions (even thousands: Grimes & Aufderheide, 1991), would seem to show that cortical inheritance lasts all this while. But we do not yet have proof that cortical inheritance survives through sexual reproduction in ciliates. Perhaps the nucleus does, ultimately, reign supreme, and at some stage may synthesize the very 'seed' that carries out phenomena of cortical inheritance (in the sense of a seed crystal that initiates or nucleates growth of a crystal; cf. 'reversion' of L-forms in bacteria to walled forms: Wyrick & Gooder, 1977). Of course, on the other hand, some cortical units that reproduce could persist in diatom auxospores, and perhaps impose their effect on cell shape through subsequent generations, as their numbers increase. Starting from red algae, which have MTOCs (microtubule organizing centers) but lack centrioles, Pickett-Heaps (1975) suggested the evolutionary sequence nuclei ⇒ intranuclear MTOCs ⇒ cytoplasmic MTOCs ⇒ centrosomes. As summarized by Mizukami & Gall (1966):

"...Many schemes, ranging from complete nuclear dependence to complete organelle autonomy are, in fact, compatible with the morphological data.... One could imagine that the centriole proteins are formed in a typical fashion, under ultimate nuclear control, and migrate to the cortex in which they are organized next to the ciliary basal bodies (centrioles). The distribution of the new centrioles on the surface of the organisms would be dictated by the already existing distribution, and in this way the pattern could be 'self-replicating' [like crystal growth] without the individual units themselves being so. Hypotheses of this type have been discussed by Sonneborn (1964); they differ fundamentally from the strict self-replication of ciliary basal bodies postulated by Lwoff (1950)" (Mizukami & Gall, 1966).

The notion that basal bodies are symbiotic remnants of spirochete bacteria is, then, plausible, but far from proven with anywhere near the certainty we have about the symbiotic nature of mitochondria and chloroplasts. Whether centrosomes can reproduce, whether or not they must contain centrioles or some minimal nucleating structure for centrioles, and are the primary reproducing unit rather than centrioles, apparently has not yet been demonstrated.

Proposition 116: maternal determinants may be related to centrosomes.

I would like to suggest the possibility that the homolog of centrosome RNA in metazoans is maternal RNA, the so-called 'maternal determinants':

"Maternal messages have been of great interest to biochemical embryologists in recent years (see Davidson, 1976, [1986]) and are known to exist in masked form as ribonucleoprotein particles (Spirin, 1966; Gross et al., 1973) most of which become active after fertilisation.... There is often some cytoplasmic localisation of substances which determine some elements of the body plan and there is also often little genomic activity, most protein synthesis being directed by maternal mRNA.... Not only [is]... protein synthesis in the egg... directed by maternal mRNA but... protein species which appear at later stages may be formed from maternal mRNA as well" (Slack, 1983).

Thus the unmasking of some maternal mRNA probably coincides with the fertilization wave (Grainger & Winkler, 1987; Kelso-Winemiller & Winkler, 1991), and perhaps the remainder is unmasked by subsequent cleavage waves, or even differentiation waves (cf. unmasking at gastrulation: Born et al., 1989; and specificity of unmasking: Tafuri & Wolffe, 1993). (For recent progress on masked RNA see: Marello, LaRovere & Sommerville, 1992; Ranjan, Tafuri & Wolffe, 1993; Standart & Dale, 1993; Bouvet & Wolffe, 1994; Simon & Richter, 1994; Spirin, 1994.) Of course, this mRNA is associated with polysome formation (Davidson, 1986; cf. Martin & Gordon, 1997a,b), which is apparently lacking in the ciliate cortex, so the possible homology between RNA in cilia basal bodies and maternal mRNA must be taken as no more than a clue to be followed up, perhaps starting with a search in the centrosomes for the...

"Y-box proteins [which] are the most evolutionarily conserved nucleic acid binding proteins yet defined in bacteria, plants and animals.... The structure of this domain consists of an antiparallel five-stranded beta-barrel that recognizes both DNA and RNA. The diverse biological roles of these Y-box proteins range from the control of the *E. coli* cold-shock stress response to the translational masking of messenger RNA in vertebrate gametes" (Wolffe, 1994).

Until more evidence is in, we can thus speculate that the ultraslow waves in ciliates are differentiation waves that unmask the RNA in the centrosomes.

There is one peculiar symbiont that has an apparently nonspatial relationship to differentiation. These are the gamma particles in *Blastocladiella* (Cantino & Mills, 1983). Margulis & Sagan (1986a) suggest that their differentiated cells differ in their proportions of symbionts, with meiosis somehow guaranteeing a minimal complement of each:

"Cell differentiation in animals and plants originated from the population dynamics and interaction of interdependent members of former bacterial communities.... This protoctist, *Blastocladiella,* has provided us with examples, at least in principle, of the phenomenon of differentiation as a manifestation of differing ratios of genetic components of cells.... The elegant research work on them has been for the most part ignored by mainstream students of differentiation phenomena.... Gamma particles contain DNA and messenger RNA. They are cytoplasmic organelles about one micron across that display a life cycle of their own. They are the sites, in the cell, of chitin-synthetase.... Those zoospores that contain only one or two gamma particles produce the tiny chytrid bodies; those that contain some eight to ten form ordinary colorless thalli. The zoospores that contain sixteen to thirty-two gamma particles form the thick-walled resistant sporangia.... [In] the microbial communities that we recognize as embryos... quantities of DNA vary in different types of tissue and at different ages of the cell or organism. The organelles themselves are present in variable quantities per cell.... Biochemical feedback systems must ensure that at least one copy of each gene, organized into sets of chromosomes and organellar genomes, is present at the beginning of each new generation of individuals. If this hypothesis is correct meiotic sexual cycles may be required as an error-correcting process, a sort of bio-inventory checking and adjusting the mechanisms necessary for differentiation and development" (Margulis & Sagan, 1986a).

Perhaps the evanescent 'determinants' for rotation of the whole cortex in fertilized amphibian eggs prior to first cleavage are related to these spirochete origins of the microtubules (Margulis, 1993b):

"Originally the UV effect [on cortical rotation prior to first cleavage] was thought to be due to the destruction of a dorsal or axial determinant, but it was later found by Scharf & Gerhart (1980) that embryos which had been subjected to UV irradiation just after fertilization could still form a normal pattern if they were tipped [relative to gravity] before the normal time of subcortical rotation. So the present majority view is that the UV effect involves damage to a component required for the dorsoventral polarization, perhaps the subcortical microtubule array, rather than destruction of a cytoplasmic determinant for the dorsal region. However, Elinson & Pasceri (1989) showed that eggs which had been UV irradiated as *oocytes* underwent the normal subcortical rotation and could not be rescued by tipping. The pattern defects from irradiation of oocytes are similar to those from irradiation of eggs and this result has given the dorsal determinant theory a new lease of life" (Slack, 1991a). [Elinson & Pasceri (1989) add: "In addition, UV activates unfertilized eggs and eliminates primordial germ cells".]

Hartman (1975) concluded that the "RNA molecules involved in the MTOCs will be called morphic RNA as they determine the interior structure

of the eukaryotic cell... [and] egg", so this proposition is not a new hypothesis.

Proposition 117: bacterial colonies capable of pattern formation are homologous to the cortex of multicellular organisms.

If there is an evolutionary continuity between spirochete or other bacteria and centrosomes of eukaryotic organisms, then we should look further back in evolution to ask if bacteria themselves are capable of patterning and functioning at least equivalent to that found in ciliates. The answer is 'not quite', but there are many intriguing hints, and we have just begun to look at bacteria this way.

There is a wide repertoire of spatial patterns that colonies of separate and attached prokaryote cells are capable of achieving. Bacterial colony patterns are often fractal, diffusion limited aggregation patterns of unattached cells, and some of these patterns can be described via reaction-diffusion equations (Eden, 1958, 1960; Budrene, 1985; Budrene & Berg, 1991, 1995; Shapiro, 1992b; Ben-Jacob et al., 1993, 1994, 1995; Lipkin, 1995b; Woodward et al., 1995; Mendelson & Salhi, 1996; Murray et al., 1998; reviewed in Gordon & Drum, 1994; see Berg, 1996a, for the most intricate patterns so far found by Elena O. Budrene). Patterns formed by colonies of prokaryotes, in which the cells are attached to one another, include the one-dimensional spacing patterns of heterocysts (nitrogen fixing cells) and akinetes (spores) in blue-green algae (= cyanobacteria: Wolk, 1967; Mitchison & Wilcox, 1972; Wilcox, Mitchison & Smith, 1973a,b; Wolk & Quine, 1975; Hirosawa & Wolk, 1979; Elhai & Wolk, 1990; Fay, 1992; Liang, Scappino & Haselkorn, 1992; Leganes, Fernandez-Pinas & Wolk, 1994; Singh et al., 1994; Carrasco, Buettner & Golden, 1995). Some blue-green algae also form filamentous branching patterns (Fritsch, 1935). Some colonial bacteria can alternate between attached and detached forms (James et al., 1995). Even *Escherichia coli* may simultaneously differentiate and form spatial patterns:

"Differentiation is often associated with morphological change, so in the past the morphological uniformity of many bacterial cell populations... was taken as tacit evidence of their inability to differentiate. More recently, this misconception was corrected.... Stationary-phase *Escherichia coli* cells differ from rapidly growing ones in many positive attributes, and there may be more than one kind of stationary-phase *E. coli* cell (Siegele & Kolter, 1992). *E. coli* colonies even exhibit multicellular differentiation, inasmuch as they display heritable pattern formation (Shapiro, [1984], 1988, [1995a])" (Chater, 1993).

The most complex prokaryote structures known include the attached mycelium (strawberry runner structure) of the Actinobacterium *Streptomyces* and the stalked reproductive body of the Myxobacterium *Stigmatella aurantiaca,* which is formed after cell aggregation (Margulis & Schwartz, 1987). Prokaryotic branching patterns and fruiting bodies have two or more differentiated cell types present (Fritsch, 1935; Bai, 1972; Dworkin, 1985; Margulis & Schwartz, 1987; Shimkets, 1990), though these are examples of terminal, not continuing, differentiation. The myxobacteria originated 900 million to 1 billion years ago (Shimkets, 1990), suggesting that fruiting body capabilities took some time to come about in the evolution of bacteria, and was also in place as a possible precursor to differentiation of eukaryotes:

"Many aspects of myxobacterial biology have no other known parallels in the bacterial world and provide an evolutionarily ancient solution to many global developmental and behavioral problems.... The manner in which the genetic code specifies such remarkable three-dimensional structures as the *Stigmatella aurantiaca* fruiting bodies or such synchronized cell movement as the ripples [see Proposition 118] is a fundamental problem that is relevant to development of all higher organisms" (Shimkets, 1990).

Most research on colonial bacteria has been done on artificial substrates under unnatural conditions, in which other organisms are not present. An even greater variety of patterns should be anticipated from mixed prokaryote colonies, especially when symbiotic relationships form between different prokaryotic species. (Mixed myxobacterial colonies sort out, like embryonic tissues: Shimkets & Dworkin, 1997.) The recent discovery of tubular connections between plastids (Kohler et al., 1997) may indicate that some procaryotic colonies have a syncytial character, which could be important in pattern formation. These situations are yet little explored, and may be relevant, for instance, to the patterns of different kinds of perhaps symbiotic kinetochores in the ciliate cortex (Iftode et al., 1989). Especially germane is the observation that the protozoan *Mixotricha paradoxa* has both a spirochete and another bacterium attached in one-to-one proportions at each 'bracket' on its surface (Cleveland & Grimstone, 1964). Examples can be found in the "325 species of [human oral] bacteria" which aggregate in specific manners (Kolenbrander & London, 1992), including a 'cobweb' pattern of spherical bacteria along a single filamentous bacterium (Jones, 1972; Listgarten, Mayo & Amsterdam, 1973). Eukaryotic diatom cells are also found studded with rod-shaped, nonmotile bacteria (Rosowski & Langenberg, 1994). We have just scratched the surface in investigating the repertoires of colonial bacteria.

It is curious that we may actually have available a 'living fossil' in the form of spirochete bacteria that are attached to a protozoan and yet act almost identically to cilia, a possible evolutionary intermediate between independent bacterial colonies and the bacteria derived cilia:

"Many... spirochetes are attached by one of their cell ends to the body surface of protozoa indigenous to the termite gut (Breznak, 1973; Kirby Jr., 1941). Adherent spirochetes are either localized on certain areas of the protozoan [cell] body or are distributed over most of the body surface (Cleveland & Grimstone, 1964; Kirby Jr., 1941). Usually the attached spirochetes perform a continuous, undulating type of movement readily apparent in living specimens observed by light microscopy. Although this movement differs from the strokelike motion of cilia (Cleveland & Grimstone, 1964; Kirby Jr., 1941), some investigators have mistaken the attached spirochetes for cilia.... Each spirochete is attached by one of its cell ends to a small bracket protruding from the surface of the protozoon.... The spirochetes attached to actively swimming cells of *Mixotricha* undulate vigorously in a well-coordinated manner, whereas in moribund protozoa, which move slowly or are not motile, the coordination is lacking or the undulating motion is altogether absent (Cleveland & Grimstone, 1964).... Other consortia involving termite gut protozoa and spirochetes have been studied by various investigators (Bloodgood & Fitzharris, 1976; Bloodgood et al., 1974; Smith & Arnott, 1974; Smith, Buhse Jr. & Stamler, 1975; Smith, Stamler & Buhse Jr., 1975)" (Canale-Parola, 1978). [Note in contradiction: "...The fact that the flagella are held securely to the body throughout most of their length could indicate that the flagella contribute little to the forward motion": Smith & Arnott, 1974.]

"Attachment of spirochetes to protozoa, and the demonstration that adherent spirochetes serve a propulsive function in *Mixotricha paradoxa,* have prompted Margulis (1970a) to suggest that spirochetes are likely forerunners of contemporary eukaryotic flagella" (Breznak, 1973).

Hinkle (1991) (his Figure 2) shows the similarity in the ultrastructure of attachment of a clam gill cilium to attachment of a spirochete to a protist from the hindgut of a subterranean termite. One wonders if the specific cortical attachment sites ('small brackets': Canale-Parola, 1978) for spirochetes on some cells are comparable to the *nuclear* attachment sites of the microtubule containing axonemes of the heliozoan *Sticholonche zancles,* which also exhibit metachronic waves (Cachon et al., 1977). (More will be said on metachronic waves in Proposition 118.)

The patterns we have found so far are not as rich as even the cortical behavior of *Paramecium,* so whether bacterial colonies themselves are capable of the cortical morphogenetic mechanisms in ciliates and multicellular organisms is an open question. Perhaps their incorporation into a cell was essential for an emergent phenomenon of greater potential, namely the differentiation wave, to arise. The essential ingredient might have been the meshwork of microfilaments and intermediate filaments (Arima, Shibata & Yamamoto, 1985), which ordinarily one would not anticipate in colonies of bacteria. But the example in Proposition 118 shows that bacterial colonies may indeed have appropriate waves. If we take as a hint the ability of protoctistan slime nets (Labyrinthulomycota) to secrete external 'actinlike proteins' (Margulis & Schwartz, 1987), then we ought also to keep our eyes open to the possibility of secretory proteins of morphogenetic effect in bacterial colonies.

We also know so little about differentiation and differentiation waves of the ciliate cortex, that there is much work yet to do. Perhaps the ciliatologists have the best model organism, or perhaps the patterned behaviors of bacterial colonies (Shapiro, 1995a) afford the best place to start studying the evolution of morphogenesis. Given the possibility of some kind of cytoskeleton, stress or stretch activated proteins, and spatially distinct membrane domains in *E. coli* (Nanninga, 1998), it is plausible that the basic patterning mechanisms are present even in single bacteria. The problem with deciding such matters is that we are not sure exactly what was added at each stage of this course of evolution, and what was already present in a previous stage:

"It would be surprising indeed if multicellular organisms did not exploit new opportunities for patterning provided by multicellularity, such as cell interactions and the possibility of different active genes in different cells. However, it would also be most surprising if they totally abandoned mechanisms operating in Ciliates and other unicellular organisms" (Sonneborn, 1975a).

Pleiotropic mutations affecting cortical morphogenesis in ciliates via the orientation of basal bodies and the trajectories of "morphogenetic [differentiation?] waves" (Jerka-Dziadosz, Garreau de Loubresse & Beisson, 1992) may be an excellent place to start, though the attempts to 'explain' this phenomenon in terms of positional information (Iftode et al., 1989; Frankel, 1974, 1989, 1990; Jerka-Dziadosz, Garreau de Loubresse & Beisson, 1992) would best be dropped in favor of observation and direct physical modelling. For example, the model of Le Guyader & Hyver (1990, 1991, 1994) is of the pure (nonmechanical) reaction-diffusion type, "...an autocatalytic system [yielding]... periodic solutions the peaks [of concentration of hypothetical morphogens] of which are assumed to represent the location of cortical units" (Le Guyader & Hyver, 1994), despite the suggestion that: "The generation of Ca^{2+} waves might rely on the Ca^{2+}-sensitive mechanoreceptors whose polarized distribution in the plasma membrane has been demonstrated (Ogura & Machemer, 1980)" (Jerka-Dziadosz, Garreau de Loubresse & Beisson, 1992). Sonneborn (1975a) critically assessed the applicability of the positional information model to ciliates and concluded that many observations are...

"...difficult to reconcile with Wolpert's (1969, 1971a) hypothesis and are at least as reasonably interpreted as based on [propagating, i.e., wave-like] nearest neighbor interactions.... There is nothing to support the view that the gradients operate by conveying abstract positional information which is then translated into developmental events" (Sonneborn, 1975a).

A mechanical component seems to be a necessity for ultraslow waves (Jaffe, 1995), such as ciliates exhibit (Iftode et al., 1989). Continuum cytoskeletal models, such as that of Hart & Trainor (1990), might be applicable to ciliate ultraslow waves. In any case, all such models leave out the nature of the link to the basal bodies and that of their reproductive mechanism and reproduction of the centrosome.

All in all, it is plausible that the eukaryote cell surface (i.e., the cortex) contains symbiotic bacterial colonies whose not so limited morphogenetic capabilities were carried into that symbiosis.

Proposition 118: coordinated beating of cilia (metachronic waves) in ciliates and in multicellular organisms are homologous, perhaps evolving from surface colonies of primitive spirochetes.

One further, behavioral observation suggests homology between ciliates and eukaryotes, the coordinated beating of cilia, which both possess:

"The embryo is perfectly integrated before it has a nervous system. Not only are the processes of growth within its tissues co-ordinated in such a way that development of all parts progresses normally, but there is, in *Amblystoma* and many other Amphibia, a co-ordination of ectodermal cilia, the rhythmic motion of which keeps the embryo in constant movement within the egg membrane. Between this earlier non-nervous and the later nervous condition of the embryo there appears superficially to be a hiatus in development, a point where one condition ends abruptly and something entirely different takes its place" (Coghill, 1929).

Such coordinated beating of cilia is found in both plants (volvox: O'Kelly & Floyd, 1983) and animals, including parts of us (oviduct: Blake, Vann & Winet, 1983; trachea: Arima, Shibata & Yamamoto, 1985; Dirksen & Sanderson, 1990). We don't yet fully understand how the coordination develops (Burke, 1983; Boisvieux-Ulrich, Laine & Sandoz, 1985; Boisvieux-Ulrich & Sandoz, 1991; Chevillard et al., 1991; König & Hausen, 1993) or functions (González Santander & Martínez Cuadrado, 1980; Dalen, 1983; Eshel & Priel, 1987; Gheber & Priel, 1989, 1990; González Santander et al., 1984; Holley, 1984; Preston, 1990; Priel & Tsoi, 1990; Rosen & Rosen, 1990; Sanderson & Sleigh, 1984; Tamm, 1984; Wong, Miller & Yeates, 1993; Wulffraat, Veerman & Stamhuis, 1985):

"The coordination of waves of ciliary action is now relatively clear (Párducz, 1967; Machemer, 1969a,b), and mechanisms underlying this coordination are being elucidated (Kinosita & Murakami, 1967; Naitoh & Eckert, 1969b; Kuznicki, 1970; Kung & Eckert, 1972; Eckert & Naitoh, 1972). However, at present extensive comparative information is not at hand" (Hanson, 1977).

"An analysis in hydromechanical terms by Machin (1963) shows that structures such as flagella, undulating in proximity, will tend automatically to synchronize their movements" (Grimstone, 1966). [Cf. Tamm, 1984; Priel & Tsoi, 1990, for further work on hydrodynamic entrainment of cilia.]

In multicellular organisms, this cortical phenomenon is one that clearly cuts across cell boundaries. In *Volvox* itself: "The fact that the two cell lineages [somatic cells and gonidia] can establish and maintain different transcript populations despite the extensive cytoplasmic continuity that exists between them is somewhat surprising..." (Tam, Stamer & Kirk, 1991) and suggests an effectively syncytial arrangement. On the other hand, we must be careful here, in that each nucleus, perhaps through its attached intermediate filament meshwork (Goldman et al., 1985; Katsuma et al., 1987), may not need a discrete cell membrane to have its own local influence:

"Evidently, cellular individuality can survive protoplasmic confluence.... Since each nucleus keeps protoplasm within a certain radius under its control, protoplasmic territories have the value of cells.... Any change in the colloidal consistency of the protoplasm attended by biochemical and bioelectrical differences will necessarily produce... some degree of physiological isolation" (Weiss, 1940).

We could speculate that some early spirochete colonies attached to surfaces and were capable of coordinated beating. This could have been as useful to them as coordinated beating is to sessile and colonial protozoans (Patterson & Hedley, 1992; cf. Harwood & Canale-Parola, 1984). It is clear that the hundreds of internal microtubule-like flagella of a single modern spirochete cell are somehow coordinated, and there is some suggestion that separate, adjacent bacteria can exhibit coordinated swimming:

"Observing swarm colonies under a microscope, one sees thousands of flagella on dozens of swarmer cells moving in synchrony, creating oscillating waves as the cells spread outward from the periphery. Such exquisite coordination prompted Alexander Fleming, the discoverer of penicillin, to wonder in print if *Proteus* [*mirabilis*] could possibly have a nervous system!... [Cf. Williams & Schwarzhoff, 1978; Rauprich et al., 1996.] Swarming is strictly a multicellular activity; an individual cell that gets separated from the rest is unable to advance over the agar unless it is engulfed by another swarmer group, at which time it starts to move again" (Shapiro, 1988).

Some kind of wave seems to also be going on in swarms of bacteria (cf. Rauprich et al., 1996):

"When the swarmer cells finish one phase of spreading..., a series of visible waves composed of more densely aggregated cells moves from inside the swarm zone toward the perimeter" (Shapiro, 1988).

Certain gliding bacteria demonstrate swarming motility waves that travel at the same speed as differentiation waves, and about the same speed as that of their gliding motility, whose mechanism is still unsolved (Shimkets, & Kaiser, 1982), though pili have been implicated (Rosenbluh & Eisenbach, 1992):

"...Myxobacteria... show an intriguing periodic behavior called 'rippling' which was investigated in detail (Reichenbach, 1965).... Ripples are a series of evenly spaced ridges, parallel to one another, moving in a pulsating fashion leading to wave propagation from the center to the edge of the colony. These periodicities occur in 5-hr cycles interspersed with a quiescent state. The ripples seem to have a wavelength of 45 to 70 µm and repeat about every 20 min. They travel at 2.2 to 3.7 µm/min (Shimkets & Kaiser, 1982; Shimkets, 1990). The cells are arranged at an angle of 40° to the direction of ripple propagation. Movement of cells in adjacent tracks is in opposite directions.... The mechanism and control of rippling is not well understood" (Wimpenny, 1992).

"Fluorescently labeled cells were tracked with confocal laser microscopy and were found to move in the direction of wave propagation, reversing direction after about one wavelength (approximately 10 cell lengths) (Sager & Kaiser, 1994)" (Shimkets & Dworkin, 1997).

The transition from a colony to a syncytium, or *vice versa,* is readily imaginable, given the existence of bacterial cells that reach a ploidy of 100 or more during the stationary phase of growth (Maldonado, Jimenez & Casadesus, 1994). Thus the basic features of two dimensional arrays of coordinately beating cilia, in ciliates and multicellular organisms, could conceivably have come from spirochete or other bacterial colonies. A search for such behavior in present day bacterial colonies might be rewarding. We saw in Proposition 117 that colonies of spirochetes beat in synchrony on certain protozoans (Canale-Parola, 1978). Perhaps these or other modern spirochetes, under appropriate culture conditions, could do this without a host cell. Even the common *E. coli* attaches to induced 6 µm 'pedestals' on intestinal cells, and then glides at up to 4.2 µm/min, somehow propelled by the actin beneath the host cell's membrane, as the pedestals "bend and undulate, alternately growing longer and shorter while remaining tethered in place on the cell surface" (Sanger et al., 1996).

Critical observations and analysis of spirochete swarming behavior are needed, to see if the apparent synchronization is real: all periodic systems with comparable frequencies will seem in phase for short periods at least, whether or not they are actually coupled. The apparently obligate nature of cell-cell adjacency for swarming does not prove whether or not the synchrony is real.

It would also be nice to know whether the "Microtubule-like structures oriented toward microtubule-organizing center-like structure in the cytoplasm of *Diplocalyx cryptotermitidis* (Ashen, 1992)" (Munson et al., 1993) or any other spirochete have any sequence or at least secondary, tertiary or quaternary structural homology (cf. Grimstone, 1966; Fracek Jr., 1984; Obar, 1985; Szathmáry, 1987; Tzertzinis, 1989; Barth, Stricker & Margulis, 1991; Hinkle, 1991; Charon et al., 1992; Bermudes, Hinkle & Margulis, 1994) or functional homology (Berg, 1976; Fosnaugh & Greenberg, 1988) whatsoever to microtubules. Morphological similarity alone is not satisfying. The rotation model of Jarosch (1992) for microtubule motion and force generation could be related to the mechanics of motility in spirochete bacteria (reviewed in Berg, 1991).

All told, coordinated beating of spirochete colonies, ciliate cilia, and epidermal cells in some plants and most animal embryos, may be one and the same phenomenon, with a common evolutionary origin.

Proposition 119: phenomena such as twinning and regeneration appear to have a universal cortical basis.

If we switch our thinking to multicellular organisms and consider cell divisions occurring during embryogenesis as being equivalent to ciliate 'directed assembly' (Grimes & Aufderheide, 1991), their effects on cortical inheritance need only carry on long enough to complete an embryo. In this sense, directed assembly, as found in ciliates, may be of substantial importance in embryogenesis of multicellular organisms. The 'persistence time', which may depend on the 'energy' barriers between alternate global cortical configurations (Brandts, 1993a), may vary from one species to another. For example, twinning, which occurs in ciliates (Grimes & Aufderheide, 1991), is perhaps an alternate global organization to the cortex of the early mammalian embryo, i.e., an alternative to formation of a single embryo. Armadillos, which produce up to 12 or so genetically identical twins, may form twins at a different time from human embryos (Gilbert, 1997), via a cortical mechanism:

"I am inclined to believe that duplicate human twins become physiologically isolated at a considerably earlier period than do armadillo quadruplets, and my reason for this belief is founded on the fact that there is so little mirror-imaging in the former and so much in the latter. It appears to be a good general rule that the earlier the separation the more complete is the reorganization of symmetry relations in the separate individuals and the less residuum of the original common symmetry" (Newman, 1916, as quoted in Newman, 1917).

"*Mirror-imaging between individuals limited to integumentary structures.* - When examples of symmetry reversal were first discovered to occur so frequently in the armor characters of the armadillo, it seemed probable that similar reversals would be found in the visceral arrangements. One would expect to find occasionally the heart apex turned to the right instead of to the left [*situs inversus*] and the greater curve of the stomach turned to the right. An extensive examination, however, shows that no visceral reversals occur. Bateson (1913) calls attention to the same situation in human duplicate twins and says: 'If anyone could show how it is that neither of a pair of twins has transposition of the viscera the whole mystery of division would, I expect, be greatly illuminated.'

"From what we know of the process of polyembryonic fission in the armadillo, it would appear that the most likely solution of this problem lies in the fact that twinning is initiated and carried out in the ectoderm, and the endoderm becomes involved only passively and considerably later. What more natural, then, than to look for evidences of twinning - mirror-image reversals - only in ectodermal and closely associated structures of the integument? In human duplicate twins it is true also that reversals are confined to the friction ridges [dermatoglyphics], which are quite homologous in origin with the armor characters of the armadillo" (Newman, 1917). [Cf. Newman, 1921.]

Simple mechanical intervention in mosaic animals, such as the polychaete *Chaetopterus,* has the same result:

"...If the cleavage pattern is altered by compression so that some polar lobe material enters both AB and CD blastomeres then mirror-symmetrical double embryos can be produced (Titlebaum, 1928; Tyler, 1930; Guerrier, 1970b). These bear an obvious resemblance to the double embryos produced in insects and amphibians by disturbance of the early cytoarchitecture.... This type of mirror duplication has been encountered several times: as an outcome of organiser grafts in amphibians, following equal cleavage in the annelid *Chaetopterus,* induced by ultraviolet irradiation in the midge *Smittia* [Kalthoff & Sander, 1968; Kalthoff, 1971a,b, 1973] and as an outcome of the maternal-effect mutant *bicaudal* in *Drosophila* " (Slack, 1983).

Mirror image phenomena are also observed just after cell division in the nuclear location of corresponding, active genes (Montijn et al., 1994; cf. Rabl, 1885) and in the locomotion of daughter cells (Albrecht-Buehler, 1977a,b, 1978).

Twinning also occurs in parts of organisms, including insect abdomens and vertebrate limbs (Hinchliffe & Johnson, 1980; Duboule, 1994a). It may thus be an indicator of a fundamental, common, cortical patterning mechanism. In terms of regeneration, the ability of both hydra (Chalkley, 1945), and the ciliate *Stentor* (Tartar, 1960), to reconstitute themselves when minced, also suggests a common cortical basis. We need to observe differentiation waves during twinning to see if they form the basis for twinning, and to begin to understand how the symmetries between twins develop.

Proposition 120: there is an evolutionary continuity in the cortex as the seat of differentiation waves.

"*The Biology of the Cell Surface*, the publication which capped the career of expatriate black American biologist E.E. Just (1939a), promoted the view that the cell surface was a dynamic structure with a key role in developmental and evolutionary processes.... To Just, the role of the cell surface was paramount, superseding even that of the genes:

> 'The ectoplasm stands not simply as a barrier of the cell against the outside world; it is also the medium of exchange between cytoplasm and environment. As such, it is the first cell-region to receive impressions from the outside world; through its delicacy of adjustment and fineness of reaction, it constitutes the first link in the chain of cytoplasmic reactions and sets the path for the orderly succession of events comprising the course in the differentiation of development' (Just, 1939a).

"However, for several reasons discussed by Manning (1983) and by Gilbert (1988), Just's work was largely and unfortunately ignored [cf. Gould, 1985b]. The initial analysis of vertebrate embryonic cell adhesion in its modern form has been traditionally attributed to Johannes Holtfreter (1939a). However, although certainly aware of the... studies by H.V. Wilson [1907], [F.R.] Lillie, and Just, Holtfreter makes no reference to these studies..." (Grunwald, 1991).

It is generally assumed that continuity of descent in evolution is through the nucleus. It may be time once again to challenge the completeness of this notion. In dealing with embryos we usually take ambient conditions dictated by our presence at the surface of the earth as both given and unimportant. But we now need to examine evidence for a role of the cortex in development, and therefore perhaps in evolution, that comes from *Drosophila* and snails in two peculiar results involving magnetic fields:

"In our laboratory, we have shown that brief exposures of synchronously-developing early *Drosophila* embryos to weak static magnetic fields - during the period when cryptic pattern determination processes are taking place - resulted in a high proportion of characteristic body pattern abnormalities in the larvae hatching 24 hours later (Ho et al., 1992b). The energies of the magnetic fields are well below thermal threshold, so we conclude that there can be no significant effect unless there is a high degree of cooperativity or coherence among the molecules involved in the pattern determination processes reacting to the external fields....

"The normal pattern consists of separate consecutive segments, whereas a continuous helical pattern tends to appear in larvae exposed to the magnetic fields. It must be stressed that this transformed morphology, which I call 'twisted', is highly unusual. I have worked with *Drosophila* for 15 years and never have I, or anybody else come across this kind of body pattern under a variety of perturbations such as heat or cold shocks or exposure to chemicals or solvents (Ho et al., 1987). Similarly, tens of thousands of genetic mutants of body pattern have been isolated by *Drosophila* geneticists over the years and so far, not a single one of them has the 'twisted' morphology. This indicates that the static magnetic field exerts its effect in a most specific way, perhaps via orientation of membrane lipids on a global scale....

"It is also highly significant that in all live organisms examined so far, which includes the *Drosophila* larva, *Daphnia,* rotifers, nematodes, larva of the brine shrimp, hatchling of the zebra fish and the crested newt, the anterior-posterior body axis invariably corresponds to the major [optical] polarizing axis of all the tissues in the entire organism. This is a further indication that some global orienting field is indeed responsible for determining the major body axis.... We do not yet know how to interpret the [polarizing microscope] colours [Ho et al., 1992c] precisely in terms of molecular structures, except that as with ordinary crystalline and liquid crystalline material, they do tell us about long-range order in arrays of molecules.... The earliest stages associated with [*Drosophila*] pattern determination are some of the most colourful ones" (Ho, 1993).

"Interestingly, the GMF [geomagnetic field] seems to have a relationship with LR [left-right] chirality (Anderson, 1988). The geological fossil record shows a clear correlation between flipping of the GMF polarity and reversals of the chirality of several types of molluscs such as *Globorotalia menardi* (Harrison & Funnel, 1964; Dubrov, 1978)" (Levin, 1997).

Perhaps the magnetically induced helical pattern in snails will be shown to be caused by a helical 'mitotic wave'. If so, it is possible that the pattern also involves the cortical cytoskeleton, which, for example, is oriented in amphibian eggs prior to first cleavage (Elinson & Rowning, 1988). Effects of low magnetic fields on tissue culture cells have been recognized and duplicated (Raloff, 1997), if not yet understood. Certainly, gene expression would be worthwhile investigating in 'twisted' blastoderms. But let's take stock:

The cortex is the location of the cell state splitter in axolotl ectoderm.

The cortex is the location of the apparatus causing cortical rotation prior to first cleavage in amphibians, for which the long range orientation of microtubules is important.

Ciliates contain an elaborate, oriented cortex.

The cortex is the probable location of the magnetic and optical polarizing effects.

Turbellarian worms embody characteristics of both ciliates and metazoans, suggesting that it is their cortex that is key to their morphogenesis, not their cellularization (Proposition 123).

There apparently are differentiation waves in the cortex of the ciliate *Paramecium*.

Launching domains for these waves may be related to "discontinuities anywhere within the cortical cytoskeleton [that] can trigger basal body assembly and oral development [in the ciliate *Stentor*]" (de Terra, 1985b).

Taken together, these observations suggest a continuity by descent of cortical differentiation waves. This fits with the general notion that cells, and by implication nuclei, are secondary in development, which I will examine in Propositions 121-124:

"One must therefore conclude that the essential continuity of the organism is independent of the cells" (Wigglesworth, 1948).

Proposition 121: during eukaryotic evolution, cortical differentiation via differentiation waves preceded cellular differentiation.

"In stentors or in other unicellular forms it is obvious that differences between different regions of the cortex or cytoplasm develop without corresponding nuclear differentiation simply because all parts of the cell draw upon the same nucleus.... A large part of the 'information' or basis for epigenesis resides in the semisolid ectoplasm, with the nucleus merely supplying essential substances" (Tartar, 1962).

Some forms of differentiation can occur in single celled organisms, such as sex determination in yeast via homeobox genes. Moreover, 'combinatorial circuits' of molecular interactions (Alberts et al., 1994; cf. Shephard et al., 1984), which are involved in many steps of differentiation in multicellular organisms, also occur in yeast (Bürglin, 1988; Johnson, 1992a). Beyond ciliates, examples of local differentiation of the cortex of single cells include the case of *Chaetopterus* (Lillie, 1902; discussed in Proposition 126), the complex morphologies of coenocytic algae, which represent some of the largest organisms on earth (Niklas, 1992), and, of course, *Drosophila,* in which many regional differences occur in the blastoderm before cellularization (Lawrence, 1992). The latter may be a 'reversion', to speed embryogenesis. These observations of spatial differentiation in single celled organisms suggest the possibility that cortical differentiation is the 'primitive' mechanism of spatial differentiation. This is the theme of Frankel (1989):

"As pointed out by Grimstone (1961), 'Protozoa are comparable, in different contexts, with both metazoan cells and whole metazoan organisms'.... The ciliates... allow us to ask to what degree the *intracellular* mechanisms of protozoan development resemble the mixture of intracellular and intercellular mechanisms observed in the development of multicellular organisms.... [The] principal ideas are: first, that there exists in ciliates an intracellular hierarchy of qualitatively different systems of spatial control nested within one another; second, that these systems of spatial control can be inherited cytoplasmically; and third, that the most global level in this intracellular hierarchy is analogous and possibly homologous to positional information in developing embryos" (Frankel, 1989).

Perhaps the 'as yet undreamt of' universal mechanism is the differentiation wave:

"The reader will probably be correct in viewing this as a speculative house of cards. In reality, the molecular mechanism for large-scale patterning in ciliates - and other organisms - is unknown, and, as Pringle et al. (1986) have pointed out for the equally unsolved problems of localization of the budding site in yeast, '...mechanisms [may be] involved in morphogenesis that are as yet undreamt of in the philosophies of biochemists and cell biologists'. In a condition of uncertainty with regard to mechanism, research should proceed along as broad a front as possible and should include a search for new theories (Holliday,

1988).... My own view is that the true mechanisms are not yet understood, not even in principle, and that answers will be found by scientists who are willing to think and work unconventionally" (Frankel, 1989).

As the mechanism in common between ciliates and multicellular organisms, perhaps the differentiation wave is the original one, shared by their common ancestor (Figure 1 in: Prescott, 1994). Thus cortical differentiation via differentiation waves in early ciliates may have been the evolutionary origin of cellular differentiation.

Proposition 122: cortical inheritance and differentiation waves preceded nuclear inheritance in the origin of eukaryotes.

"The fact that physiological unity is not broken by cell-boundaries is confirmed in so many ways that it must be accepted as one of the fundamental truths of biology" (Whitman, 1888).

If the analogy between possible differentiation waves of ciliates (Iftode et al., 1989; Le Guyader & Hyver, 1991) to differentiation waves in multicellular organisms actually turns out to be a homology, then it is plausible that cortical inheritance preceded nuclear inheritance at the beginning of eukaryotic evolution, i.e., that those genes that are involved in pattern formation entered eukaryotic cells through their centrosome symbionts (presumably some early spirochete or other bacterium, as discussed in Proposition 115). This potential homology is enhanced by the remarkable case of the turbellarian flatworm *Stenostomum incaudatum* discussed in Proposition 123, whose development uncannily resembles that of a ciliate, even though it is a multicellular organism (Grimes & Aufderheide, 1991; cf. Hanson, 1977). Thus not only am I suggesting that the nucleus is, in the first instance, 'controlled' by a cortical phenomenon, namely the differentiation waves, during pattern formation, but also that cortical morphogenesis, at least at a level of complexity suggested by the ciliates, preceded this control of the nucleus in the course of evolution. For example, cortical asymmetry in ciliates is the primitive state (Iftode & Adoutte, 1991; Baroin-Tourancheau et al., 1992), and may have also carried through evolution and be the root cause of the phenomenon of left-right heart and brain asymmetry, discussed in Section 8.02.

Proposition 123: multicellular organisms are descended from ciliates via recellularization of their cortical bacterial symbionts.

"Darwin was able to write the *Origin of Species*... without any reference to cellular structure. This I find one of the most remarkable facts of the history of science" (Jacobson, 1991b).

I speak of the process of deriving multicellular organisms from ciliates as 'recellularization', because the ancestors of the ciliates presumably contained bacterial colonies with their own cell membranes and walls for each bacterial cell, which have been lost, at least in modern ciliates. Furthermore, as far as is known, there are no membranes separating present day centrosomes from the rest of the cytoplasm in any eukaryote. This suggests that loss of the symbiont's cellular membrane was an early event. Since cellularization of syncytial organisms, such as *Drosophila* at the blastoderm stage, involves curtains of cortical membrane coming down between the nuclei, after they are arrayed near the cortex (Fullilove & Jacobson, 1971, 1978; Bresgen, Czihak & Linhart, 1994), we can imagine that recellularization involved some similar process.

The hypothesis of recellularization has four major evolutionary components:

1. Ciliates arose via symbiosis of an early pre-eukaryotic cell (perhaps a 'protoeukaryote' with the nucleus derived from an eocyte and the cytoplasm left over from a 'gram-negative eubacterium': Lake, 1989; Rivera & Lake, 1992; Lake & Rivera, 1994), with spirochete or other bacteria colonizing the surface.

2. These bacteria lost their cell membranes at some point ('cryptic endosymbiosis': Henze et al., 1995), and were reduced to self-replicating centrosomes (undoubtedly with transfer of many or all genes to the cell nucleus).

3. The eukaryotic cell became multinucleate.

4. New cell membranes separated the multiple nuclei.

It is possible that all of the genes that came in with the early spirochetes were transferred to the nucleus (in addition to "...the well-known pattern wherein an intracellular symbiont loses functions redundant with those possessed by the host": Fontana, Wagner & Buss, 1995). Protein transport across an organelle membrane is no obstacle if there is no membrane, unlike the situation in mitochondria and plastids (cf. von Heijne, 1986), so that 'reductive evolution' of organelles, whatever drives it (Kurland, 1992), could have gone all the way for centrosomes, and quite rapidly on an evolutionary time scale. While it may seem paradoxical to speak of a 'self-reproducing' organelle that has no nucleic acids of its own, mitochondria and plastids are no more self-reproducing, since they rely on the nucleus for many or most of their genes. ("Since the discovery of prions a 'self-reproducing' organelle that has no nucleic acids of its own is no longer paradoxical": Guenter Albrecht-Buehler, p.c., 1998.) Besides, mitochondria have been totally lost from some cells, leaving only traces in the nuclear genome and a ('self-reproducing'?) organelle containing no DNA, the hydrogenosome (Clark & Roger, 1995; Embley et al., 1995; Bui, Bradley & Johnson, 1996; Germot, Philippe & Le Guyader, 1996, 1997; Horner et al., 1996; Roger, Clark & Doolittle, 1996; Hirt et al., 1997; van der Giezen et al., 1997; Roger et al., 1998). Centrosomes would still fit the definition of a 'replicator' (Dawkins, 1982), and the 'conflicting views' might be reconcilable:

"Conflicting views hold that there is no compelling evidence to regard centrioles as genetically independent organelles. Pickett-Heaps (1971) argued that their observed distribution simply reflects an efficient mechanism whereby the cell can partition centrioles at mitosis. In this model no genetic autonomy is necessary to explain centriolar replication" (Johnson & Dutcher, 1991).

"...Nucleic acids... are not necessary in all conceivable models of self-duplication" (Hoffman, 1965).

Reductive evolution of centrosomes may have progressed to a different extent in different evolutionary lineages (cf. Brennicke et al., 1993), so that reports of DNA or RNA in centrosomes, at least of different species, could be both true and not universal. In this case, some portion of the nucleic acid inheritance in a eukaryotic cell may still be borne by the cortex, in some species. Some of these questions might be resolved if we attempted to culture centrosomes *in vitro* (Steve McGrew, p.c., 1997).

It is difficult to speculate on whether the multicellular organisms originating from early ciliates were mosaic or regulating. We have seen that ciliate basal bodies are capable of regulation (and thus the statement that *Paramecium* represents "an extreme case of mosaicism" is incorrect: Iftode et al., 1989). What was the relationship between the trajectories of differentiation waves on these early ciliates and the trajectories of differentiation waves on early multicellular organisms derived from them? At one extreme, we could imagine that one trajectory corresponded to one cell, which is our model for a mosaic organism. At the other extreme, each trajectory crossed many of the recellularized cells, corresponding to our model of a regulating organism. If both scenarios occurred, then mosaic and regulating organisms could have a common ciliate ancestor, but not be derived one from the other. These are interesting questions which, once raised, might yield to answers via molecular evolution. In either case, the hypothesis of recellularization does not require an increasing physical range of 'genetic control' (i.e., trajectories of differentiation waves) to control ever more cells, over evolutionary time, as hypothesized by Ray (1995), since in a crude sense the cortex of a ciliate maps to the whole (including the involuted) surface of a eukaryote:

"The transition from single- to multicelled existence involves the extension of the control of gene regulation by the mother cell to successively more generations of daughter cells" (Ray, 1995).

With this background, let me examine the old question of 'cortical inheritance' (Nanney, 1966a), which at least makes us pay attention to the cortex as the possible vehicle of morphogenesis:

"All the protozoological work pointed to the same conclusion: the guiding mechanism for the elaboration of formed parts was to be sought in the most solid portion of the cell, the ectoplasm. When Tartar (1961) posed the question 'What are stentors good for?' he readily found his answer in the cytoarchitecture of the cell and the relevance of its study for the 'great unsolved problem or organic form'.... Thus protozoologists divided the cell into two parts: genes and gene products on the one side and, on the other, a cell cortex or skeleton that seemed to manifest supramolecular properties.... From a genetic point of view, the important point was whether the evolutionary changes in the cell cortex were due to genomic changes, independent cortical changes, or parallel series of independent but selectively correlated changes in the genome and cortex... (Sonneborn, 1963)

"The phenomenon of cortical inheritance in ciliates was placed beyond doubt.... [But] there were few cases of experimental evidence supporting cortical inheritance in Metazoa.... Embryological literature was replete with arguments for the importance of the cell cortex in cell-cell interactions. Cortical localization was also important in the pattern of some mosaic eggs. The many parallels, Sonneborn (1974a)... argued, suggested that the roles of the cortex and the principles of their operation may be fundamentally alike in unicellular and multicellular animals" (Sapp, 1991).

The most remarkable case of multicellular cortical inheritance I know of is summarized here by Grimes & Aufderheide (1991):

"The turbellarian flatworm *Stenostomum incaudatum* has an anatomy typical of members of its phylum, with a head containing sensory organs and cerebral ganglia, and a ventrally located pharynx leading to a blind digestive system (Sonneborn, 1930a). The worm propagates asexually by longitudinal growth followed by transverse fission [cf. Marcus & Macnae, 1954; Hanson, 1977]. Prior to fission, a new pharynx and new cerebral structures develop in the posterior fission product; these new structures are derived from undifferentiated stem cells (Sonneborn, 1930a,b). The pattern of growth is thus quite similar to that of ciliated protozoa. Sonneborn (1930b) treated populations of worms with lead salts, causing various morphogenetic distortions. When released from the lead, the deformed worms showed three responses: some regulated back to a typical form, some died and a few developed into a homopolar doublet form. Homopolar doublet worms possessed duplicate sets of structures, including two heads and two pharynxes. The doublet worms propagated as doublets upon continued vegetative growth (Sonneborn, 1930b). Sonneborn argued that the doublet condition could not be genically conferred, because the treatment leading to formation of the doublet worms was not mutagenic, nor did it involve meiosis or fertilization. Furthermore, without continued selection, some doublet lines regulated to singlets. Such singlets did not produce doublets again, even though one would have expected them to do so if the doublet condition were determined by a genic mutation. Because new structures arose from undifferentiated stem cells, the doublet condition did not depend upon simple division of existing differentiated structures, but rather upon some mechanism whereby an existing set of structures specified the formation, several tens of micrometers away, of a new set (Sonneborn, 1970a). This phenomenon is certainly analogous to the stable propagation of various ciliate phenotypes (including doublets).... The propagation of doublet *Stenostomum* can be considered to result from directed assembly events, but in a multicellular organism" (Grimes & Aufderheide, 1991).

In the spirit of "treasure your exceptions!" (Bateson, 1908; cf. Luft et al., 1995), we have to take this case of a multicellular organism developing like a ciliate very seriously, for it represents a possible evolutionary stepping stone for all multicellular eukaryotic organisms. Here is a summary of the idea that metazoans are derived from multinucleate ciliates by the formation of cell boundaries between the nuclei:

"The ciliates are a magnificent product of evolution.... They have reached a level that, the greatest student of these forms says, raises them '...to the rank of pseudometazoans' (Fauré-Fremiet, 1952). Such a comment now poses the inevitable question: How did the real multicellular animals, the eumetazoans, actually arise?...

"Turbellarian worms [flatworms] vary in size from adult individuals no more that a few hundred micrometers long to those 50 cm. or more on length (the land planarians).... Cilia commonly cover the entire surface of these worms.... Three body layers [are]... found.... These layers are the outermost epidermis..., the innermost gastrodermis lining the gastrovascular cavity or intestine - only in the acoels is this digestive epithelium lacking - and an intermediate layer of solid mesenchyme or parenchyma. The careful work of electron microscopists (Pederson, 1961a,b, 1964, 1966; Skaer, 1961; Dorey, 1965) has shown that all tissues of the flatworms studied thus far are cellular.... The epidermis is an epithelium one cell-layer deep. Its external border is usually ciliated, and it often rests internally on a basement membrane.... Dorey's (1965) plates show regularity of ciliary distribution from one epidermal cell to the next.... There are many cilia associated with each cell.... One expects continuity of the rootlet system regardless of cell boundaries, though it then becomes difficult to understand how the separate cells cooperate in developing shared rootlet structures [but cf. the end-to-end parallel microtubule bundles in adjacent cells, observed by Lin & Forscher, 1993; see Section 3.10].

"There remains now the proposal deriving the acoels from ciliatelike ancestors.... The history of this idea is much older than its recent attribution to Hadzi (..., 1953, 1963) and Steinböck (1954) [cf. Willmer, 1990]. This idea has been variously formulated by von Jhering (1877a), Kent (1880-1882), and Sedgwick (1895). These historical facts are summarized elsewhere

(Hanson, 1958).... The analysis of ciliate - acoel homologies... was carried out by Hanson (1963). That analysis will be... revised in the light of two new points. First, there are the findings from electron microscopy that acoels are completely cellular and not syncytial as was previously thought to be the case. Second, the study of homologies is made more critical by the use of plesiomorphs, so that now we are explicitly making comparisons between *Haploposthia rubra* and the ciliate form most like it, which is thought to be *Stephanopogon mesnili,* the ciliate plesiomorph.

"For the ciliates, the basic organizational pattern was multinuclearity made up of a highly polyploid nucleus along with one or more diploid ones and enclosed in a highly differentiated cortex.... Cellularization may have initially arisen simply as a result of developing permanent boundaries of endoplasmic reticulum around nuclei lying in the central part of the organism or located eccentrically at the inner edge of the outer endoplasm.... This placing of nuclei in as different parts of the body as possible increases the variety of local environments around the different nuclei, which is reinforced by presence of cell membranes, and provides, therefore, the opportunities of different gene action in different cells.

"This would be the point of emergence of the eumetazoan eotype. Here there would be a bilaterally symmetrical, multicellular animal capable of that cellular heterogeneity that is the forerunner of tissue differentiation....

"The transitional organisms gave up the rapid unicellular reproduction of their ciliate forebears for the genetic variability of sexual reproduction, which was useful in pursuing the epigenetic potentials of embryogenesis. All of these - genetic variability, cellular construction, developmental plasticity - together endowed the flatworms with an extraordinary evolutionary potential, which in hindsight we see has been realized in the diversity of eumetazoan forms" (Hanson, 1977).

This old idea of the origin of multicellular organisms via what I would now call 'recellularization' of ciliates is not popular today (cf. Willmer, 1990; Strickberger, 1996, for the theory of 'cellularization of a multinucleate protozoan' and alternatives). Its history is tied to a battle between proponents of the 'cell theory' and the idea that there is an entity called 'the organism as a whole', starting from the...

"...conclusion of von Siebold & Stannius (1846-1848a,b) [cf. 'Siebold on Infusoria' in: Nordenskiöld, 1928] that in the Protista or lowest forms of life the whole body consists of but a single cell; for this suggested the view that the multicellular body of higher forms is equivalent to an assemblage or colony of one-celled individuals; and from this grew the further conception that the multicellular organism may be regarded as a 'cell-state' the one-celled members of which have undergone a physiological division of labor. (A considerable group of modern authorities have sought the origin of Metazoa in syncytial or multinucleate rather than actually colonial forms:... Sedgwick, [1894, 1895]; Delage [1903].) Elaborated especially by [Henri] Milne-Edwards, [Rudolf L.C.] Virchow and [Ernst] Haeckel, this conclusion offered a simple and natural point of attack for the problems of cytology, embryology, and physiology, and revolutionized the problems of organic individuality. Its value as a means of biological analysis needs no other demonstration than the immense advances it made possible. Inevitably in practice we treat cells as distinct, though closely coördinated, elementary organisms or organic units; and although some writers have questioned the validity of this procedure it nevertheless remains an indispensable means of analysis....

"We shall therefore proceed upon the assumption, if only as a practical method, that the multicellular organism in general is comparable, to an assemblage of Protista which have undergone a high degree of integration and differentiation so as to constitute essentially a cell-state.

"This view has been vigorously assailed by many writers, especially by those who have emphasized the conception of the 'organism as a whole.' See, for instance, Whitman (1888, 1893), Sedgwick (1894), Dobell (1911), Child (1915a) and especially Ritter (1919)" (Wilson, 1925).

Sedgwick (1895) is particularly wry, in the course of distinguishing the fact of cellularization of many organisms from the evolutionary 'cell theory' of how they got that way:

"A theory to be of any value must explain the whole body of facts with which it deals. If it falls short of this, it must be held to be insufficient or inadequate; and when at the same time it is so masterful as to compel [people]... to look at nature through its eyes, and to twist stubborn and inconformable facts into accord with its dogmas, then it becomes an instrument of mischief, and deserves condemnation, if only of the mild kind implied by the term inadequate" (Sedgwick, 1895).

The multicellular organism was thus seen by most biologists to be a collection of Protista cells, rather than derived from one of them. This paradigm was supported by the weight of a major political/economic theme of the times, division of labor in the 'cell-state', a focus that we have inherited, usually without questioning how or why (cf. Section 1.18). Here is the opposite view:

"There is from Rohde's [1923] point of view no fundamental difference between 'unicellular' organisms and multicellular. The Protozoon, for example, is not to be regarded as homologous with the tissue cell of a Metazoan, but as the equivalent of the whole body of the latter.... The cortical layer of the Protozoon is from this point of view strictly comparable with the ectoderm of Metazoa....

"The contrast between Rohde's view of the cell and the organism and the classical cell-theory is sufficiently striking. Instead of the composite, colonial cell-state, made up of semi-independent, collaborating units - essentially an analytical view - we have the organism conceived as a unitary and unified whole...

"...Rauber (1883)... maintained that... the Protozoon was to be homologized with the whole body of the Metazoon. The essential difference between them was that the Protozoon had only one surface, the external surface of the body, while the Metazoon possessed, in addition to an external surface, many internal cell-surfaces....

"Calkins [1933] [and] the distinguished Russian protozoologist Awerinzew (1910)... called attention to the close analogy between the intracellular differentiations of the higher Protozoa and the specialized tissues and organs of the Metazoa. We may conveniently conclude... by quoting his considered opinion that between the Protozoa and the Metazoa there exists no qualitative, but only a quantitative difference" (Russell, 1930).

The idea that cellularization is incidental in some embryos was suggested for plants since the early 1800's (also in reaction to the cell theory: Kaplan & Hagemann, 1991; Niklas, 1994b). In his work on plant development, Sinnott (1939b) concludes...

"Evidence from... various sources therefore supports the view that the development of an organ proceeds with little relation to the manner in which it is cut up into organized cellular units.... If the organization of an organ or body is *not* the result of cellular interactions, however, and if organization at one level does not spring from that at a lower one, we are forced to look elsewhere for an explanation of the control of development. It must be admitted that the search has not as yet been very fruitful.... We are perhaps mistaken in seeking explanations of organized development which are purely biochemical rather than exploring those which are more strictly physical in character" (Sinnott, 1939b).

No disproof of the concept of an origin of multicellular organisms by recellularization has been put forth. The idea merely contradicts the prevailing paradigm, and is generally ignored. Perhaps it goes against the grain in appearing holistic, while the cell theory would seem to automatically be reductionist. Thus an origin of metazoans via the recellularization of ciliates contrasts almost philosophically with the alternate cell aggregation hypotheses, which are now automatically presumed to be true without considering the alternative of recellularization:

"...Metazoans (animals) have been divided into three major taxonomic units, namely: animals with three embryonic layers (triploblasts), animals with two embryonic layers (diploblasts) and animals with extremely loose tissue differentiation (sponges).... The reconstruction of the evolutionary pathways followed from unicellular organisms to these diverse body plans, remains highly controversial. In particular, analyses of the origin of metazoans using methods of 'molecular phylogeny' and ribosomal RNA (rRNA) have recently revived the controversy as to whether all metazoans can be viewed as successive offshoots within a single monophyletic unit, or if different processes of cell aggregation led to parallel radiations and different body plan organizations (i.e. sponges, diploblasts and triploblasts).... Molecular data seem to imply that an embryonic organization in three layers arose early in the history of metazoans.... Considering the relatively short distance that separates the protozoan world from the triploblast - diploblast divergence, [this]... implies a rapid evolution from unicellular organisms to complex structures" (Christen et al., 1991).

One probe for phylogeny in the recellularization hypothesis for the origin of multicellular organisms might be the observation that some ciliates have a somewhat distinct genetic code (Fox, 1985, 1987; cf. Prescott, 1994; Jimenez-Sanchez, 1995; Tourancheau et al., 1995; Schultz & Yarus, 1996). On the other hand, genetic code changes, such as those in mitochondria (Kurland, 1992), can be of relatively recent origin. Using small subunit ribosomal RNA sequences, Wainright et al. (1993) conclude that the Animalia are monophyletic and have a choanoflagellate as the common ancestor, along with the fungi.

The association between nuclei in metazoans and centrosomes suggests that there may have once been a one-to-one correspondence between ciliate centrosomes and the first cellularization of ciliates to form the 'eumetazoans', though present day flatworms have many more basal bodies than nuclei (Dorey, 1965; Hanson, 1977):

"...In vertebrate cells centrioles may be converted into basal bodies and form flagella either normally or after drug treatment (reviewed by Fulton, 1971; Weber & Osborn, 1981a). Similarly, in flagellates the flagellar basal bodies tend to become associated with nuclear division, in some cases becoming physically detached from the flagella (Johnson & Porter, 1968; Hoops & Witman, 1985) and migrating toward the cell interior to act as centrioles (Coss, 1974). In ciliates, however, micronuclear mitosis is *always* noncentriolar (summarized by Heath, 1980, 1986), and no cases of close association of basal bodies with dividing macronuclei have been reported. Indeed, instead of centrioles migrating to the interior, in certain ciliates nuclei move to the surface during division, sometimes forming loose associations with the fibrillar system, although not with specific basal bodies (Jaeckel-Williams, 1978; Tucker et al., 1980)" (Frankel, 1989).

There seems to be no evidence suggesting that the peculiar macronucleus of ciliates, derived from the micronucleus (Prescott, 1994), is homologous with the nucleus of other eukaryotes, except for "Some foraminiferan protozoa [that] also show nuclear dimorphism... (Raikov, 1985)" (Herrick, 1994). See also Orias (1991). The eukaryotic nucleus is more closely related to archaebacteria than eubacteria (Gray, 1992), while spirochetes are eubacteria (Margulis & Schwartz, 1987). This problem may be resolved by the suggestion of Golding & Gupta (1995) that the eukaryotic nucleus may be an amalgamation of both an archaebacterium and a eubacterium (cf. Margulis, 1996). If the recellularization hypothesis proves correct, then metazoan cell nuclei may be homologous with ciliate centrosomes, so that our nuclei are direct descendants of spirochete bacteria. Alternatively, there may have been an association of centrosomes with nuclei of another origin, as part of the spindle apparatus (cf. Roos, 1984).

Of course, the amusing aspect of all of this is that the major organizational problems of developmental biology are now placed on the symbiotic colonies of spirochete or other bacteria and their descendants, i.e., we are back to 'independent' cells and their interactions and 'division of labor', and we can again face the temptation of a reductionist analysis of all the 'organismal' properties of ciliates (discussed in: Russell, 1930). The arguments against the cell theory, in favor of the wholeness of the organism, may have been wrong, if we now have to turn to a cellular level, that of prokaryotes, for an explanation of pattern formation. Furthermore, confident statements such as "Life on earth is hierarchically organized" (Grantham, 1995) become questionable if the catechism of cells $\Rightarrow$ organisms $\Rightarrow$ populations, etc., gets unstuck at the first step, and we are all but overgrown bacterial colonies. Now we have prokaryotic cells $\Rightarrow$ colonies $\Rightarrow$ single cell ciliates $\Rightarrow$ cellular organisms $\Rightarrow$ populations, etc., which is not a neatly nested hierarchy.

Given the apparent presence of differentiation waves travelling through the cortex of ciliates, it may be time to dust off the old hypothesis that multicellular organisms are descended from ciliates via recellularization. Recellularization has the advantage over aggregation in that the regional differentiation apparent in ciliates was immediately available for cell differentiation, so that regional differentiation did not have to be reinvented. It is not likely that it was reinvented:

"Over the last thirty years, in a number of comprehensive reviews (Aufderheide, Frankel & Williams, 1980; Frankel, 1989; Grimes & Aufderheide, 1991; Sonneborn, 1963, 1975a; Tartar, 1961), prominent ciliatologists have discussed the problems of ciliate morphogenesis and its general significance with respect to cell biology and development" (Beisson, 1994).

The 'general significance' may indeed be that ciliates demonstrate, by homology, the major mechanisms of development.

Proposition 124: control of the nuclear genome by differentiation waves came relatively late in evolution.

"Driesch (1908a) long ago recognized that the basic pattern of reconstitution of the protozoan *Stentor* following injury is comparable to that observed in operated embryos or coelenterates, and concluded that 'we have therefore here an instance where the so-called 'elements' of our harmonious-morphogenetic system are not cells, but something inside cells'.... The possibility that in all systems cellularity is not fundamental to the establishment or even the interpretation of positional information merits serious consideration" (Frankel, 1974).

Let us consider syncytial organisms, such as coenocytic algae (Ray et al., 1983; Niklas, 1992), fungi (Raven, Evert & Eichhorn, 1986), and even higher plants ('symplasm': Burgess, 1985). They too may have centrosomes in some form, as Margulis (1993b) argues...

"The failure to discover [9(3)+0] centrioles in plant cells (except prior to sperm formation) implies that the MC [microtubule organizing center] homologue of plants generally lacks this phenotypic expression. However, the presence of (1) standard mitosis, (2) colchicine-sensitive microtubules during the formation of the phragmoplast or cell plate (Hepler & Jackson, 1968), and (3) in some groups, gametes with undulipodia suggests that an MC remnant of the former spirochete genome is present in plants and eventually will be identified" (Margulis, 1993b).

In a syncytial organism of reasonably complex morphology, the numerous nuclei move long distances by fast protoplasmic streaming (Raven, Evert & Eichhorn, 1986), orders of magnitude faster than the speed of any changes in morphology. Thus, if they are present, we cannot expect that differentiation waves would have particular effects on particular nuclei, as we expect in cellular organisms. (In fact, coenocytic algae nuclei are often equivalent, in the sense that all give rise to gametes: Bonner, 1965a.) Thus there would be no morphogenetic link from the cortex to the travelling nuclei, i.e., we have no reason to expect regional differences between the nuclei in terms of which genes they have turned on and off, or sequestered versus exposed. (Cell fusions may return some nuclei in multicellular organisms to this state: Podbilewicz & White, 1994.) Thus there is not likely to be any differentiation pathway in these organisms. Since coenocytic organisms often belong to taxa containing cellular organisms (Bold & Wynne, 1978), the cellular taxa may also lack the differentiation pathway, or show only the beginnings of its formation. (This also raises the possibility that differentiation pathways have arisen multiple times in evolution.) We thus arrive at the conclusion that nuclear involvement in differentiation waves is an evolutionarily late phenomenon. It is not necessary for pattern formation on ciliates. Nor is it necessary for multicellular organisms. Since cytoplasmic streaming of nuclei also occurs to some extent in the syncytial blastoderm stage of *Drosophila* (von Dassow & Schubiger, 1994), it is important to know whether its nuclei are at all determined at this stage. The observations of Ho (1993), discussed in Proposition 120, suggest that cortical mechanisms are quite important in *Drosophila* blastoderms, so maybe they're not.

This proposition unfortunately contributes to the ongoing battle between the proponents of nuclear versus cytoplasmic inheritance, with its politics of 'the power relations in the cell' (Sapp, 1987; cf. Witkowski, 1990). However, the proposition is reinforced by the notion that the nucleus is 'just another' symbiotic organism in the eukaryotic cell:

"Around the turn of the century, the symbiotic theory was formulated. According to this, plastids (Schimper, 1883), mitochondria (Altmann, 1890), and later also the nucleus (Mereschkovskii, 1910; [cf. Lake & Rivera, 1994; Margulis, 1993a]) were proposed to be the final products of a long process of symbiotic integration.... Margulis (1981, [1993b]) has suggested that the $9(2) + 2$ microtubular apparatus arose from symbioses between spirochete-like endocytobionts and a *Thermoplasma*-like archaeobacterial urkaryote.... [An]... unsolved problem of eucyte evolution is the origin of the nucleus. According to Giesbrecht & Drews (1981), the eucytic cell's nucleus is not homologous to the mitochondria and plastids, but should probably be regarded as a composite system that developed from more or less independent karyomers, a process which can still be observed in the early oogenesis of many organisms. Each karyomer would represent a single chromosomal apparatus of a procyte" (Schwemmler, 1991).

"It is highly probable that the emerging nucleus also started as part of a permanent endosymbiosis involving the large replicon of the eukaryotes' ancestor cell and the former 'visiting' plasmids and prophages (Sonea, 1972)" (Sonea, 1991).

This curious state of affairs is rendered even more interesting by the suggestion that some so-called cytoplasmic determinants, involved in insect axis formation, which usually take up cortical positions, are actually symbionts:

"If... the endocytobionts of one generation are prevented from infecting the leafhopper eggs and thus from passing to the next generation, only head-thorax embryos (organisms lacking abdomens) result. Similarly, if the egg mitochondria are inactivated, 'headless' abdomens may sometimes develop, which grow *in vitro* like tumors" (Schwemmler, 1991).

Given the morphological complexity of some syncytial organisms, the integration of the nucleus as symbiont into the morphogenesis of organisms was thus probably a process that occurred significantly later than the symbiosis itself. Cortical components carried the morphogenesis in the meantime. Thus even the nuclear basis for differentiation in multicellular organisms may be an evolutionarily late phenomenon. Comparisons of nuclear genomes in coenocytic organisms with nuclear genomes in organisms clearly having nuclear differentiation might reveal some of this phylogenetic history.

It will, of course, be necessary to reconstruct the phylogenetic relationships between modern ciliates, their relatives, and multicellular organisms, using morphological and molecular comparisons. The latter may have to extend to molecular morphology and functioning, in cases where too many amino acid substitutions have occurred to find sequence homologies (Kafatos et al., 1987). It would not be surprising if multiple origins of multicellular organisms from primitive ciliates are found to have occurred. It may be best to consider the possibility of alternate routes to recellularization. In some cases the spirochete symbionts in primitive ciliates may have never lost their cell membranes in the first place. Golding & Gupta (1995) have obtained "...results [that] support the hypothesis of a chimeric origin for the eukaryotic cell nucleus formed from the fusion of an archaebacteria and a gram-negative bacteria" (cf. Roger et al., 1996), and the latter could well have been the gram-negative spirochetes (Margulis & Schwartz, 1987). This would explain how the bulk or all of the spirochete genome vanished from the centrosomes (i.e., into the nucleus), without much affecting cilium structure. A more specific search for possible homologies between spirochete genomes and the 'gram-negative portion' of the chimeric eukaryotic nucleus is warranted. In particular...

"...the model suggests that for some of the highly conserved proteins found in all eukaryotic species (such as actin and tubulin) a related protein may also be found in one of the parental prokaryotic species (archaebacteria or Gram-negative bacteria) [indeed, see: Erickson, 1995, 1997]. In a chimeric model for the eukaryotic nucleocytoplasm suggested by Margulis (1992), the genes encoding proteins with motility-related functions are postulated to be derived from the spirochete group of Gram-negative bacteria" (Gupta & Singh, 1994) [cf. Irwin, 1994a].

The nucleus, then, took a while to acquire the genes brought in by colonial spirochete bacteria embedded in the surface of the original ciliate-like ancestor. As it did so, it became controlled by the differentiation waves that traversed the cortex.

Proposition 125: some of the Ediacaran organisms were ciliates.

The Ediacaran organisms were sometimes huge (to 1 m), flat animals, apparently the first (Raff, 1996; but cf. Conway Morris, 1989, 1990). Adolf Seilacher's theory of the origin of the Precambrian Ediacaran fauna is suggestive of ciliate eumetazoans:

"...The bodies of Ediacaran organisms resembled giant, single cells of connected, fluid-filled compartments, much like an air mattress.... The vendobionts, as [Seilacher]... called the organisms, must have been immobile" (Monastersky, 1995b).

However, if they were ciliated, they should have been able either to move, or at least circulate currents along their surfaces. If this notion is correct, then cellularization of the Ediacaran organisms may have been the route to metazoans, contrary to...

"Adolf Seilacher [who] proposes that the quilted Ediacara suffered a mass extinction at the end of the Precambrian, leaving no descendants in the modern world. When they died, so did their unusual body architecture" (Monastersky, 1995b).

However, the Ediacara overlapped the Cambrian faunas, such as the Burgess shale (Grotzinger et al., 1995). The large size of the Ediacara is no problem: "Contemporary ciliates can be extremely large, ranging from 0.1 mm to

4.5 cm long (Lynn & Corliss, 1991)" (Herrick, 1994). Thus, at least some of the Ediacara could have been oversized ciliates, though probably not all:

"I suppose it is possible Ediacaran organisms, such as *Dickinsonia,* were giant ciliates, but my own opinion is that there is too wide a range of morphologies to include them all in a ciliate model. One has to concede that Ediacaran fossils remain exceptionally difficult to understand, but the recent work published by Fedonkin & Waggoner (1997) and unpublished work (in part carried out by a post-doc here) support the metazoan model for at least *Kimberella* and the so-called fronds. The idea that some fossils, notably the petalonamids, are non-metazoans remains popular, and here a protistan model, e.g. ciliate, could be an alternative hypothesis" (Simon Conway Morris, p.c., 1997).

The ability of microtubules to form fractal dissipative patterns at many size scales (Tabony, 1994) may make a wide range of 'cell' sizes and spacings of centrosomes attainable. Thus it is not out of the question that these unusual fossils left descendants, indeed, possibly us (amongst others).

Proposition 126: gastrulation started with the ciliates.

The evolutionary step from ciliates, with an oral apparatus that forms by "invagination... by a gastrulation-like process" (Beisson, 1994) and for whom "remodelling of the surface pattern... includes gastrulation-like movements" (Jerka-Dziadosz & Beisson, 1990), may not be too much more than the step of cellularization. The organizer experiment of Spemann & Mangold (1924a,b) may have already been done in ciliates and come out the 'same' way:

"A supernumerary implanted oral references point [in *Paramecium*] leads quickly to the development around it of a complete ventral pattern of ciliary rows, suture, and cytoproct; but this implantation has been achieved too seldom for detailed study of the sequence of events" (Sonneborn, 1975a).

If there are homologies here, then *Drosophila* may recapitulate this sequence in the cellularization of the blastoderm (Fullilove & Jacobson, 1971, 1978; Postner & Wieschaus, 1994; Rose & Wieschaus, 1992; Schejter & Wieschaus, 1993a,b; Schweisguth, Lepesant & Vincent, 1990; Simpson

& Wieschaus, 1990; Warn & Robert-Nicoud, 1990; Wieschaus & Sweeton, 1988), followed immediately by gastrulation movements (Bard, 1991a; Costa, Sweeton & Weischaus, 1993; Foe & Alberts, 1983; Kam et al., 1991; Leptin & Grunewald, 1990; Sweeton et al., 1991; Turner & Mahowald, 1977). The advantage of cellularization, and thus the advance over the ciliates, is, indeed, that it permits internalization of part of what was the external cortex. This may be the key to the evolution of diploblastic and triploblastic embryos.

Ciliate development may have *specific* significance, as the forerunner to multicellular differentiation and morphogenesis. If we take the *Paramecium* oral apparatus as more homologous to a mouth than an anus, ciliates may be more closely related to the Protostomia than the Deuterostomia, in which these openings form from the blastopore, respectively (Villee, Walker Jr. & Smith, 1963; Balinsky & Fabian, 1981). If the early ciliates evolved both ways, they may be "the primitive organism that gave rise to both [vertebrates and invertebrates]" (Travis, 1995b), postulated by Geoffroy Saint-Hilaire (1772-1844) in his "imbecility" (Gray, 1961) and penchant for the "wildest flights of fancy" (Nordenskiöld, 1928). Molecular biologists have come to his posthumous rescue in comparisons of dorsal/ventral 'patterning' genes of frogs and flies with evidence for a 'dorso-ventral inversion' early in evolution (Arendt & Nübler-Jung, 1994, 1996a,b; Nübler-Jung & Arendt, 1994, 1996; Francois & Bier, 1995; Hogan, 1995; Lacalli, 1995, 1996; De Robertis & Sasai, 1996; but cf. the 'calcichordate theory of chordate origins': Jefferies & Brown, 1995; Gee, 1996). Perhaps homologs of *sog* = *chordin* and *dpp* = *bmp-4* will be found in ciliates, if someone would care to look.

The obverse case of *Chaetopterus*, in which a multicellular organism goes far into its development under altered culture conditions, without cell or nuclear division, also shows that 'gastrulation' can occur in a unicellular organism:

"Even in animals as high as annelids... cleavage may be entirely suppressed by artificial means, yet the unsegmented egg may undergo to a certain extent differentiations similar in type to those occurring in the normal embryo, and may even give rise to actively swimming ciliated structures that have some resemblance to normal trochophore larvae. In the case of *Chaetopterus* Lillie (1902) obtained this result by treatment of the eggs, fertilized or unfertilized, by the addition to the sea-water of a certain percentage of potassium chloride. In some cases the egg becomes multinucleated, either by complete cleavage followed by fusion of the blastomeres or by division of the nucleus without cytoplasmic cleavage. In others the nucleus seems never to divide but becomes greatly enlarged, while the number of chromosomes correspondingly increases....

"Meanwhile the cytoplasmic materials of these eggs undergo a process of segregation comparable in some respects with those which precede and accompany cleavage in normal development. A part of the ectoplasm containing characteristic granules flows towards the animal pole, and there aggregates; a part remains at the vegetative pole where the polar lobe normally forms, while the entoplasm remains in the equatorial region. This is followed by a rapid overgrowth, partial or complete, of the entoplasm by the upper ectoplasm, so as to produce a 'unicellular gastrula.' The egg then develops cilia, typically over the whole ectoplasmic surface, and actively swims about. At the same time it often elongates so as to approach the form of a young trochophore larva.... All these changes recall those seen in the normal larva.... These larvae, however, never form an apical organ, mouth, anus, mesoblast-bands or coelome, and never become segmented. The main results reached in the case of *Chaetopterus* were confirmed by Treadwell (1902) in *Podarke,* and by Scott (1906) in *Amphitrite;* in both cases it was found that the cilia may be arranged in a definite band which shows a considerable resemblance to normal prototroch. Phenomena of similar type were also observed by Lefevre (1907) in *Thalassema.* [See Dan, 1960, for a contrary interpretation. Lefevre (1907) actually found the opposite: '...the parthenogenetic pseudo-trochophores, which have been described for *Chaetopterus* and other annelids, are entirely absent' in *Thalassema.*]

"These facts demonstrate that even in the multicellular animals cleavage, and even nuclear division, is not a necessary condition of either localization or differentiation [or gastrulation]" (Wilson, 1925).

This demonstration of the cortical nature of 'gastrulation' is also reviewed in Needham (1950) (who adds Lillie, 1906, and Brachet, 1937, 1938a, to follow-up work) and Karp & Berrill (1981). Dalcq (1938) states nevertheless that the "nuclear activity is permanent and important (Pasteels, 1934)". Twinning of *Chaetopterus* (Titlebaum, 1928) may be an indication of double gastrulation. See also Jeffery (1985b), Jeffery et al. (1986), Jeffery & Wilson (1983), Swalla, Moon & Jeffery (1985), and Zimmerman, Landau & Marsland (1957) for studies of the *Chaetopterus* cortex.

The apparently same phenomenon of gastrulation movements of the cortex in aging, unfertilized frog eggs was noted in Proposition 89 (Holtfreter, 1943a; Berrill, 1961). Amongst plants, *Volvox* inversion "is driven by cellular shape changes and subcellular cytoskeletal rearrangements that strongly resemble those observed in many morphogenetic events of animal development" (Kirk et al., 1982). Thus gastrulation movements seem to be a primitive cortical capability, which may indeed trace back to early ciliates.

Proposition 127: the physics of the cortex may be the key to morphogenesis.

"One may begin to look for a relationship between the universally found bilateral symmetry of organisms and the mirror symmetry between certain daughter cells at an early state of an embryo. Left or right-handedness may also be the simplest way by which nature primes one of the daughter cells, if only one of them is supposed to enter a certain pathway of differentiation" (Albrecht-Buehler, 1977a).

Consider the robustness of animal development in the face of altered cell size, number and shape (Waddington, 1938a; Fankhauser, 1945a). Is this a cortical phenomenon, demonstrating that in some fundamental sense the cells are not the center of development? The physical properties of the cortex have been but little probed (Holtfreter, 1943b), and they change with position and time (Elinson, 1980, 1990), partly due to the associated, and spatially and temporally varying cortical cytoskeleton (Elinson & Rowning, 1988; Elinson & Palecek, 1993):

"The cell cortex has the capacity to undergo local contractions, which are most clearly seen in the waves of contraction that travel over the surfaces of many eggs (Vacquier, 1981; Yonemura & Mabuchi, 1987). More generally, local contractions together with changes in the structure and composition of the cortical layer... accompany tissue morphogenesis" (Bray & White, 1988).

But this work has yet to be correlated with physical properties of the membrane (Dictus et al., 1984), or the properties and dynamics of the cytoplasmic 'determinants'. Held Jr. (1992) speaks of...

"*The Cortical Inheritance Model*.... The regulatory molecules could either reside in the cytoplasm or the cortex (Wolpert, 1971a). Many examples of 'cytoplasmic determinants' are known (Subtelny & Konigsberg, 1979; Chandebois & Faber, 1983; Malacinski, 1990a). By contrast, the eukaryotic cell membrane is usually viewed as a 'fluid mosaic' (Singer & Nicolson, 1972; Gilbert, 1991a; [cf. Fromherz, 1988a]), incapable of reliably partitioning regulatory signals because it cannot rigidly hold an array of molecules. However, epithelial cell polarities challenge this notion (Locke, 1967, 1984, 1988; Fristrom, 1988). The anisotropically pigmented cortices of many fertilized eggs (e.g. the dark animal hemisphere in *Xenopus*) are a vivid demonstration of the ability of cell surfaces to maintain large discrete domains, and there is evidence that they can stably maintain microdomains as well (Grimes & Aufderheide, 1991; [cf. Albrecht-Buehler, 1980])" (Held Jr., 1992).

This, however, is a static picture, as if the cortex were simply a map or homunculus, ready to unfold. The cortex of amoeboid and dividing cells is in motion, and some progress has been made towards understanding the physics involved, and its relationship to the cortical cytoskeleton (Oster, 1984; Bray & White, 1988; cf. Lewis & Murray, 1992). But we are very far from having a physics of a dynamic cortex in embryos, let alone the relationship of that physics to activation of underlying, localized mRNAs or other maternal gradients, to nuclear events, to cytoplasmic streaming (Flint et al., 1989), and to cleavage planes (cf. Bjerknes, 1986; Harris, 1990). Genetic studies of cell polarization, as in yeast shmoo formation (Alberts et al., 1994), while necessary, need a complementary physics approach. It is as if, absorbed in the nonspatial logic of molecular developmental biology, we have hardly begun to look at the embryo:

"Despite our inability (as yet) to provide a molecular basis for handedness in ciliates, the similarity in the dynamics underlying changes in handedness of ciliates to those encountered during pattern regulation in eggs and multicellular organisms indicates that the genic interactions and cell-to-cell signalling that are usually invoked to explain these dynamics probably do not provide a *sufficient* explanation. The cell cortex - perhaps in eggs as well as ciliates - may still hold mysteries that lie beyond current molecular biological understanding" (Frankel, 1991).

This intellectual impasse comes from the very success of the operon model (Sapp, 1991) and its nearly immediate adoption into embryology by Waddington (1962a) (cf. Gilbert, 1991b), leaving the spatial problem of differentiation essentially ignored:

"Jacob and Monod were well aware of the theoretical stakes of their work: In one of their first synthetic accounts of their work in 1961, they wrote:

'One point already seems to be quite clear: namely that biochemical differentiation (reversible or not) of cells carrying an identical genome, does not constitute a 'paradox,' as it appeared to do for many years to both embryologists and geneticists' (Jacob & Monod, 1961a).

"Ephrussi responded...:

'This statement tells us nothing about the nature of the primary causes responsible for the orderly, divergent biochemical differentiation of different cell lineages derived from a single egg (whether its mechanism be based on self-maintaining regulatory states or, for that matter, on any mechanism of differential gene activation or amplification). *The real problem is that of the origin (seat) of the asymmetrical causes which bring about these asymmetrical effects* ' (Ephrussi, 1972)" (Sapp, 1991).

The 'seat' of the matter may be differentiation waves in all organisms with a cortex, from single cells on up.

I'd like to end this discourse with a summary of how the evolution of mosaic and regulating embryos might have occurred:

1. bacteria containing microtubules colonized the surface of some kind of pre-eukaryotic cell;

2. they became integrated at the surface, providing coordinated swimming;

3. these bacteria lost their cell walls and membranes;

4. the wave of replication of these kinetochores became multiple waves of perhaps different kinds of kinetochores, leading to surface differentiations;

5. the genes that make microtubules transferred to the cell nucleus;

6. the cell became multinucleate;

7. recellularization, retaining the swimming coordination and the coordinated waves of (now cellular) replication, resulted in one nucleus and one or more of these kinetochores per cell, and produced what we call mosaic embryos;

8. cell replication became somewhat autonomous, permitting larger tissues of single cell types to form, the waves now providing coordinated differentiation, producing what we call regulating embryos.

6.00 MACROEVOLUTION

"Wallace and Darwin have propounded as the cause of modification in descent their law of natural selection. This law has been epitomized by Spencer [1864, 1898] as the 'survival of the fittest.' This neat expression no doubt covers the case, but it leaves the origin of the fittest entirely untouched" (Cope, 1887)

"In some quarters speculation concerning the origin of the adaptation of living things is frowned upon, but I have failed to observe that the critics themselves refrain entirely from theorizing. They shut one door only to open another, which also leads out into the dark. To deny the right to speculative thought would be to deny the right to use one of the best tools of research.

"Yet it must be admitted that all speculation is not equally valuable. The advance of science in the last hundred years has shown that the kind of speculation that has real worth is that which leads the way to further research and possible discovery. Speculation that leads to this end must be recognized as legitimate. It becomes useless when it deals with problems that cannot be put to the actual test of observation or experiment" (Morgan, 1903a).

6.01 Redefining Microevolution and Macroevolution

"Although it may not be possible, or even wholly desirable, to achieve a fixed meaning for scientific terms, the effort to 'control and curtail the power of language' remains a significant feature of scientific activity. The very extent to which scientists (far more than speakers of ordinary language) *aim* at a language of fixed and unambiguous meanings constitutes, in itself, one of the most distinctive features of their enterprise" (Keller & Lloyd, 1992b).

"...Most of the unexplained phenomena in macro-evolution were first minimized, then swept under the carpet and finally forgotten" (Riedl, 1978).

"The central problem is finding the mechanisms that connect genes and developmental processes to morphological evolution.... Somehow deep underlying genetic systems have remained in place in some developmental programs even though the structures constructed by genes acting downstream have undergone extensive modification" (Raff, 1996).

Proposition 128: *microevolution* consists of any change to any differentiation cascade that does not change the topology of the differentiation tree. Heterochrony is an example of microevolution.

"In the absence of other readily applied concepts, heterochrony came to provide the dominant explanation at the developmental level for evolutionary changes in morphology in fossil as well as living organisms.... Uncritical attribution of so many of the phenomena observed in the evolution of development to heterochronic 'mechanisms' may be inhibiting a more penetrating investigation of the subject" (Raff, 1996).

The simplest change that could occur to a differentiation tree in the course of evolution is to an edge, i.e., to a differentiation cascade. This could be any change to the repertoire of gene activities released by a differentiation trigger. Any such change to an edge of a differentiation tree will be call a *differentiation cascade change*. Any change occurring within a differentiation cascade I call microevolutionary. Most everyday point mutants are microevolutionary, affecting only a single differentiation cascade. Most mutations in individual proteins, cell division rate, uniparental disomy (Weiss, 1990b), etc., are changes within a differentiation cascade. In terms of previous theories, differentiation cascade changes are analogous to 'phase changes' for 'morphogenetic trees' (Arthur, 1988). The same goes for morphological changes resulting from purported mutations that bring out different components of the "discontinuous spectrum... of dissipative structures" (Katchalsky, 1971; cf. Goodwin, 1984a).

My redefinition of microevolution encompasses heterochrony, since heterochrony is presumed to involve timing changes rather than changes in which kind of tissue differentiates. Heterochrony is a change in the time until the next cell state splitting, i.e., to the duration of a differentiation cascade, which is represented in a differentiation tree as the length of an edge of the tree along the developmental time axis. This shifts the timing of all subsequent branches of the differentiation tree emanating from the edge whose length (duration) has changed (Figure 29; cf. 'distortional change' for 'morphogenetic trees' in Arthur, 1988). I thus classify all forms of heterochrony (Gould, 1977a, 1992a; McKinney, 1988; McKinney &

McNamara, 1991; McNamara, 1997) as microevolutionary changes. This is because they involve no fundamental changes in the differentiation trees, i.e., no changes to their topology.

Some readers may be upset about relegating heterochrony to microevolution:

"The most crucial issue to evolutionary biologists interested in development has been one of how morphology evolves. Both de Beer (1958) and Gould (1977a) laid out their considerations of the interface between ontogeny and phylogeny in terms of a unifying developmental mechanism that would explain the relatively easy transformation of form suggested by evolutionary histories. That mechanism was heterochrony, the concept that events in development can shift in timing relative to each other to produce new ontogenies.... In the hands of its proponents, heterochrony has continued to provide a major explanatory concept for interpreting evolutionary change in fossil as well as living organisms. The insistence that heterochrony is the dominant mechanism for evolutionary change is overdone..." (Raff, 1992a).

Raff (1996) devotes a whole chapter to the idea that: "It's not all heterochrony".

Proposition 129: *macroevolution*, the evolution of major taxonomic groups, corresponds to topological change of the differentiation tree.

"The origin of new embryonic segregates [new cell types] cannot have played a very extensive role in the evolutionary process, though a very fundamental one whenever it occurred.... The time of incidence of embryonic segregation with reference to the series of form-changes (morphogenesis) is the same in any given species, but variable when different species are compared, even to a very great extent in some cases, a fact that demonstrates it to be an independent variable in the life history" (Lillie, 1929a).

Topological change to a differentiation tree means that there is a change in the number and/or connections of the edges. (This is the same usage of 'topological change' as Nei (1987) in regard to phylogenetic trees and 'structural change' in regard to 'morphogenetic trees' in Arthur, 1988. Cf. García-Bellido, 1985a,b.) I redefine macroevolution as any topological change to a differentiation tree. Microevolutionary changes do not change the topology of the tree. The distinction between microevolution and macroevolution that I am making, in terms of kinds of changes to the differentiation tree, may be approximately what has been called for:

"...Wilson, Maxson & Sarich (1974) proposed that, 'There may be two major types of molecular evolution. One is the process of protein evolution, which goes on at about the same rate in all species. The other is a process whose rate is variable and which is responsible for evolutionary changes in anatomy and way of life. We propose that evolutionary change in regulatory systems accounts for evolution at and beyond the anatomical level.'" (Raff & Kaufman, 1983).

"...Evolutionary... toolboxes themselves are evolving in two different ways.... The fastest and most common kind of change is produced by mutations that alter the immediate function of the toolboxes, or genes on which the toolboxes act, without altering their integrity or organization.... We can call these *potential-realizing* mutations, because they realize or implement the potential of the genetic toolboxes. The second type of change occurs less often. These are mutations that change the organization of the toolboxes themselves. Let us call these *potential-altering* mutations" (Wills, 1989).

I differ from these authors in that any regulatory change occurring *within* a differentiation cascade would be classified as microevolutionary by me. The production of any protein can be regulated, and that can occur within the confines of a particular cell type. Changes to the differentiation tree (the overall regulatory system of the organism) that are topological in nature are macroevolutionary. These are changes to the hierarchy of the hierarchical genome.

For nematodes, the cell lineage tree is approximately equivalent to the differentiation tree. Sommer, Carta & Sternberg (1994) give an array of eight changes in cell lineages observed between certain nematode species, five of which would qualify as macroevolutionary since they involve changes in the topology of the differentiation tree (Figure 32). Another example is the change from one to two 'inductions' in nematode vulva development (Félix & Sternberg, 1997). As we learn the mechanisms of these evolutionary changes (cf. Fitch, 1997), we can test whether differentiation trees provide a useful means of distinguishing macroevolution from microevolution.

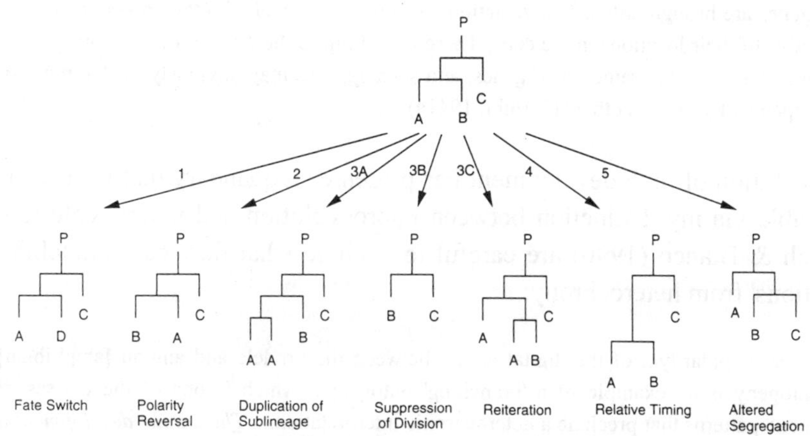

Figure 32. Nematode macroevolution. "Different types of cell lineage transformations observed between nematode species. P, a precursor cell. A, B and C, different cell types. 1. Fate Switch, cell with fate B instead has fate D. 2. Polarity Reversal, A and B are produced in different positions. 3. Changes in number of cell divisions: 3A. Duplication of A cell. 3B. Suppression of division; might change the fate of parent of A and B to A, B (as shown) or other fate. 3C. Reiteration, B has fate of its parent, i.e., an extra asymmetric cell division. 4. Change in the relative timing of divisions [heterochrony]; such a change can affect subsequent cell interactions. 5. Altered segregation.... The developmental potential to generate particular cell types might be transferred from one cell to its sister" (from Sommer, Carta & Sternberg, 1994, with permission). Transformations 3 and 4 would be macroevolutionary, while the others are microevolutionary. If we think of A, B, C as tissues rather than single cells, most of these differentiation tree changes could also occur in regulating embryos.

The number of possibilities for macroevolution via topological changes to differentiation trees is vast (Proposition 73). The combinatorics for alternative differentiation trees increases rapidly, even for trees with only a few edges. This is well known to cladists, who try to find optimal phylogenetic trees. Wright (1941a) (in Provine, 1986) might be said to have anticipated my redefinition of macroevolution and its combinatorial richness:

"...There is no theoretic difficulty with a branching hierarchic system of chain reactions in which genes are brought into effective action whenever presented with the proper substrates, irrespective of their locations in the cells. There is no limit to the number of reaction systems that can be based on the same set of genes, and such systems may obviously evolve more or less independently of each other" (Wright, 1941a).

The evolution of limb development has presented a quandary that is perhaps resolvable via my distinction between microevolution and macroevolution. Alberch & Blanco (1996) are careful to exclude what they call 'branching transitions' from heterochrony:

"The reversed polarity... of the digital arch... between the urodele and anuran [amphibian] limb ontogeny is an example of a 'branching' ontogeny... which is one of the classes of ontogenetic patterns that preclude a heterochronic interpretation.... *The unique developmental pattern of urodeles, relative to anurans and other tetrapods, cannot be interpreted from a heterochronic perspective* " (Alberch & Blanco, 1996).

Such 'branching transitions' would appear to correspond to topological changes in the limb portions of the differentiation trees of urodeles relative to other tetrapods, i.e., to a macroevolutionary difference. If the process of differentiation is viewed in terms of the topology of the differentiation tree, then we have a way of bypassing problems such as those posed by 'molecular' models of alternative ways of developing a limb:

"Nearly all molecular analyses of vertebrate limb patterning have been carried out in anurans and amniotes, [in] which... the digital arch forms in a posterior-to-anterior direction. In urodeles, however, it forms the opposite way (Shubin & Alberch, 1986). Any molecular model proposing to explain limb patterning in all vertebrates must encompass this variation; none has yet explicitly done so" (Bolker, 1995).

This is but another example of the failure of molecular approaches to adequately account for spatial components of differentiation, whether that space is the real space of the embryo or the mathematical space in which a differentiation tree is defined.

My division of mutations into microevolutionary and macroevolutionary changes may not cover all changes to the genome that can lead to speciation. For instance, changes in chromosome numbers (Duellman & Trueb, 1986) and mobile genetic elements (Temin & Engels, 1984; McDonald et al., 1987; Walbot et al., 1987) do not necessarily correspond to simple changes to an edge or to the topology of a differentiation tree. Therefore this and the previous Propositions should not be construed as excluding other mechanisms of evolution.

The whole gamut of mutations, from a single base change to 'integrated change' (Hall, 1992a), can have effects at any level within a differentiation tree, and can vary from minor to major in phenotypic effect. Thus phenotypic changes cannot, without detailed investigation, be automatically assigned as macroevolutionary or microevolutionary. Most microevolution does not produce new species, only minor varieties that are generally not reproductively isolated. This may also be true of some topological changes to the differentiation tree. What I have done, then, is move the concepts of microevolution and macroevolution from the phenotypic to the genotypic level. It is the change in the hierarchical genome that is of fundamental importance, not the appearance or functioning of the organism, though of course it is the latter, the phenotype, on which selection operates.

That the distinction between differentiation cascade changes and topological changes to the differentiation tree may have an empirical basis is seen...

"...in the elegant work by Kafatos et al. (1987), which demonstrates that similarities in regulatory mechanisms that control expression for insect chorion protein genes have persisted over considerable species divergence, even though the modern gene products controlled in the resulting expression repertoire are not [any longer] homologous" (Raff et al., 1987a).

While pigmentation changes might usually be considered as microevolutionary, the following observations are suggestive of topological changes occurring in the portions of the differentiation tree containing genes for pigmentation in *Drosophila*, i.e., for macroevolution in pigmentation between *Drosophila* species:

"*Drosophila* produces a... red-eye pigment, a pteridine called drosopterine. In some species this pigment is also present, in lesser amounts, in other parts of the body. Comparison of four major evolutionary lines of *Drosophila* shows that the primitive species of each line tend to form more pigment in the body, while the advanced species tend to restrict distribution of the pigment more effectively (Hubby & Throckmorton, 1960).... Moreover, it seems that the different phylogenetic lines have developed their control mechanisms independently, so that species in one subgeneric group restrict drosopterine distribution by one means and those of different subgeneric groups by other means. Examples such as these show that epigenetic regulation of a biochemical character does indeed have an evolutionary history" (Alston, 1967).

The availability of both macroevolution and microevolution, as redefined in terms of differentiation trees, may explain the...

"...rich and unexpected fluidity of genomes and developmental processes in evolution... Explanations of the evolution of form have to consider how form is generated, and they must account for both underlying stability and immense change in design.... [including] the freedom of early development to evolve rapidly and radically" (Raff, 1996).

The differentiation tree is a loose constraint within which much evolution of gene cascades and their alleles can occur (microevolution), and the tree itself can also change in its branching topology (macroevolution).

Proposition 130: microevolution results in movements across an adaptive landscape whose fixed dimensions correspond to the genes in the differentiation tree, while macroevolution produces a change in the number of dimensions of that adaptive landscape.

Evolution is often conceived of in terms of an 'adaptive landscape'. In Wright's original formulation (cf. Wright, 1982a,b; Williamson, 1987; Dawkins, 1996), we may take his adaptive landscape as having coordinate axes (components, dimensions) corresponding to each gene that is part of the differentiation tree:

"...Sewall Wright (1932) published a brief paper in which he pointed out a few simple genetic truths: of all the different genes, and of all the different forms each gene could take, it stood to reason that some of the combinations of those genes would produce a more vital, vigorous organism than some other combinations would create. Wright used the word 'harmonious' for those better combinations. And he drew a crude map, a topographic diagram of hills and valleys, and suggested that we could symbolize those relatively more harmonious combinations as occupying the 'peaks,' and the less harmonious combinations the 'slopes' and 'valleys,' of this 'adaptive landscape.'... Later writers were more interested in using an upscale version of the metaphor:... Each species is adapted to a 'niche,' symbolized by a peak in the adaptive landscape. The peak is a generalized representation, now, not really a bunch of harmonious genes working in concert to yield viable organisms" (Eldredge, 1985b).

A change in the number of dimensions of the adaptive landscape, which I identify with macroevolution, is barely considered in New Synthesis models of evolution:

"Small duplications have no doubt been very important in evolution in increasing the number of genes, but hardly in the abrupt origin of new species" (Wright, 1949).

This statement is reiterated in Wright (1982a), and Mayr (1991) states flatly that Wright ignored macroevolution:

"The origin of entirely new genetic factors remains unexplained.... Most of those who called themselves population geneticists worked with single, closed gene pools. Even Wright did not come to grips with the problem of the multiplication of species in his shifting-balance theory, nor with the macroevolutionary problems generated by speciation" (Mayr, 1991).

Consider, for example, the argument of Dawkins (1986) against saltational mutation. This is phrased in terms of changes in a single dimension of an organism (essentially a single gene), without considering saltational changes (large or small) in the number of genes, even though he is aware "that the total genetic capacity of a species may increase due to gene duplication". He goes on to say that "such duplications are rare enough not to invalidate my general statement that all members of a species share the same DNA 'addressing' system". But then I am saying that much speciation, macroevolution in particular (cf. Flavell, 1982; Hewett-Emmett, Venta & Tashian, 1982), is precisely correlated with changes in the DNA addressing system. That DNA addressing system is the differentiation tree. Indeed, macroevolution defined this way is a rare event, but there is plenty of geological time for the occurrence of many rare events. Perhaps now is the time to incorporate the effects of DNA duplication, sometimes substantial (Schughart, Kappen & Ruddle, 1989), and related events (King & Wilson, 1975) into an expanded theory of evolution:

"Though the occurrence of gross DNA rearrangements has been extensively described and discussed in the past fifty years, it has only recently been appreciated that deletions, duplications, inversions and translocations of DNA comprise a large proportion of 'spontaneous' mutations.... We would like to underline the evolutionary importance of mechanisms responsible for addition and deletion of sequences. The mechanisms described can account for the accumulation of DNA in higher organisms" (Dyson & Sherratt, 1985).

"Evolution is, ultimately, a complex sequence of changes in the chromosomal DNA. Some of the most eminent modern writers on evolutionary mechanisms have done no more than pay lip-service to this fact. We need to establish the chromosomes in the centre of our picture of evolution, as Darlington (1964) has rightly urged.... Such phenomena as the role of chromosomal rearrangements... the evolutionary increases and decreases of chromosomal DNA, involving on the one hand duplication of genetic material and on the other hand its loss... the various types of supernumerary chromosomes,... the heterochromatinization and dosage compensation mechanisms... the whole subject of the evolution of sex and sex-determining mechanisms... must have profoundly influenced the evolution of many groups. Yet some of the leading writers on the 'synthetic' or neo-Darwinian theory of evolution seem almost to leave them out of account, as if they were awkward facts which did not fit into their synthesis.... Every chromosomal rearrangement, every evolutionary change in a karyotype, is a genetic revolution; and once it has occurred, a whole series of point mutations will be tried out in novel combinations in the new karyotype" (White, 1973).

Indeed, every step of macroevolution, large or small, is a 'genetic revolution', within which a vast spectrum of microevolution can then occur.

Proposition 131: macroevolution can be a nonequilibrium process when there is an expanding dimensionality of the differentiation tree.

"[Henri] Bergson regarded biological evolution as the embodiment of a nonmaterial life-force striving continually to overcome matter's entropic tendency by creating ever more complex forms of order.... A perennial source of dissatisfaction with Darwinism has been its apparent failure to supply a scientifically based picture of evolution that is equally compelling. What does Darwinism have to offer in place of the life-force? And why does evolution by random genetic variation and natural selection generate ever more complex 'solutions to the problem of remaining alive'?" (Layzer, 1990).

A degree of freedom is anything which can change in the fundamental parameters of a system. In mathematics, a degree of freedom is equivalent to a coordinate in a multidimensional space, a basis function, an independent variable, a component: any point in a mathematical space is then regarded as a linear combination of elements from such an orthogonal basis set of elements. For living organisms we take these components at various hierarchical levels, the simplest being the base pair. The base pair, the codon or the gene are then degrees of freedom for an organism. For the differentiation tree, we can take a broader view, and regard each edge as one dimension. Were an organism simply a linear combination of such degrees of freedom, explanation in biology would be a simple matter. But granted that here the analogy to linear mathematics breaks down, the concept of degrees of freedom would seem to have some utility.

There is hardly any mathematics that deals with changes in the number of degrees of freedom (except, in some sense, theories of fractals, chaos, and turbulence). Basically, in a system that is expanding its number of degrees of freedom, there may generally be too little time to come to equilibrium:

"The production of new genetic variants through mutation and recombination increases the size of the genetic phase space by adding more potential locations (Collier, 1986). If the phase space expands more rapidly than [the] rate at which genetic equilibrium is attained, non-equilibrium conditions will prevail.... Given non-equilibrium conditions, both organization and complexity will increase simultaneously (Layzer, 1975; Frautschi, 1982, [1988]; Landsberg, 1984)..." (Collier, 1989).

Analogous ideas are found in Layzer (1988, 1990), Kaneko (1995), and in the relationship of ecology to evolution, which provides further expanding degrees of freedom (cf. Eldredge, 1995a):

"...The emergence of the ability to take advantage in resource competitions of an indefinite number of often infinitesimally small individual differences, creates degrees of freedom, in both the technical and the ordinary senses, well beyond what can be achieved by merely chemical and physical systems. It also creates more variables and interactions among them than can be tracked. It is impossible to reduce the components of fitness to any single language or system of variables. This situation has given rise to the notion that fitness is a supervenient property (Kim, 1978; Rosenberg, 1985; Sober, 1984a)" (Depew & Weber, 1994; cf. Weber et al., 1989).

At one level, then, the edges of the differentiation tree may be regarded as its degrees of freedom. Growth of the number of edges in the differentiation tree corresponds to an increase in the degrees of freedom for evolution. Of course, we will need actual rates of creation of new dimensions to compare with rates of equilibration in population dynamics, to see if this speculation is worth considering. This ties in with the idea of fractal evolution discussed in Proposition 196.

While I am focussing on changes in dimensionality at the level of the differentiation tree, these considerations apply at all levels. A nucleotide base addition or deletion represents a change in dimensionality. So does addition of a level of control of a gene via modifier loci and modifiers of the modifiers (C. Cristofre Martin, p.c., 1997). Gene duplication, and the evolution of gene families, need not involve duplication of the whole gene cascades in which they reside. Thus, as we see in these examples, microevolution is also a nonequilibrium process that contributes to the expanding dimensionality of the differentiation tree, by increasing the dimensionality of individual differentiation cascades.

In the next two propositions we'll consider ways in which the number of dimensions of a differentiation tree can be reduced. When this is followed by growth, the number of dimensions could be restored. In this way, macroevolution, defined as a change in topology of the differentiation tree, need not involve an increase in the number of dimensions, nor therefore, in complexity. These growth and shrinkage processes, if they occur rapidly enough, also impart a nonequilibrium character to macroevolution.

Proposition 132: macroevolution by nonterminal deletions within the differentiation tree, leaving a dangling terminal branch, would in general prune the tree.

In addition to growth of the differentiation tree, we must consider the possibility of its shrinkage. If a deletion were to occur in the middle of a terminal branch of the differentiation tree, no pointers would be available to the dangling portion. (Cf. the 'elimination of links' by deletion of intermediate edges, such as has been postulated for 'morphogenetic trees': Arthur, 1988; cf. Pritchard, 1986). If the dangling terminal branch were essential for viability, such a mutation would be lethal. Even if that branch were not essential, during the time in which any new pointer to the inaccessible terminal branch could evolve, the terminal branch itself would, due to its inaccessibility and thus disuse, randomly mutate into background DNA.

Removal of a middle portion B of a differentiation tree ABC could nevertheless be accomplished in three steps (Figure 33), each of low probability:

1. Duplication of the terminal branch to be saved, C, to give C' attached elsewhere.
2. Deletion of terminal branch C, along with the intermediate subtree B.
3. Duplication of the terminal branch C', copying C" back to A, to give AC". (C' may or may not be removed.)

Note that genes turned on in cascade B must not be essential to cascade C for this scheme to succeed. Insertions of terminal branches are also not out of the question. As always in evolution, over geological time scales, we must consider the possibility of a sequence or statistical run of rare events being the key event (Williamson, 1992; Gordon & Tyson, 1993).

We can see the possibility that pruning of the differentiation tree for segmentation has operated in the evolution of insects:

"The modification and fusion of segments in the posterior abdomen have occurred repeatedly during the evolution of modern insects (Matsuda, 1976)..." (Kelsh et al., 1994).

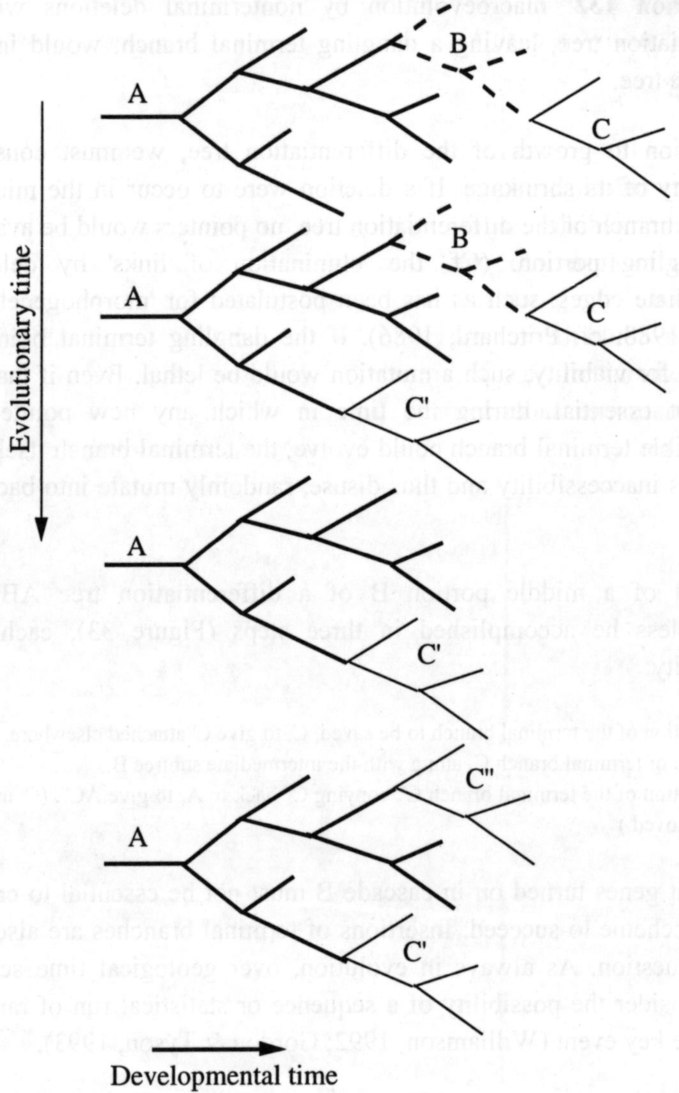

Figure 33. How to delete a middle subtree of a differentiation tree through a series of rare events.

Proposition 133: macroevolutionary simplification of a differentiation tree could occur by the fusion of two consecutive edges or differentiation cascades.

So long as the pointers are not lost between two edges of a differentiation tree, one could imagine the loss of a round of erection of cell state splitters at an intermediate stage of differentiation, causing two differentiation cascades to fuse into one. One terminal branch of the tree would be eliminated in the process. If it were a small terminal branch, viability of the organism could be retained (Figure 34). Thus we must distinguish the simplification of a differentiation tree by fusion of consecutive differentiation cascades from 'elimination of links' discussed in the previous proposition.

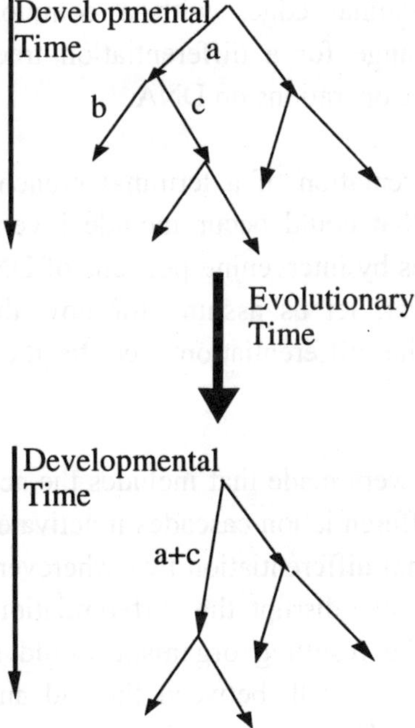

Figure 34. Simplification of a differentiation tree by fusion of two consecutive edges, **a** and **c**. The edge **b** is eliminated.

6.02 Possible DNA Mechanisms for Macroevolutionary Change of Differentiation Trees

"In general, of course, as has long been recognized (Haldane, 1932; Bridges, 1935; Stephens, 1951), duplication provides the largest single source of raw material for evolution. The duplicated sequence can evolve outside the constraints imposed by the function of the gene product of the original sequence... (Ohno, 1970). In addition, duplication provides the possibility of duplicated regulatory mechanisms..." (Mayo, 1983).

"...Developmental genes are much more than just protein-coding regions of the DNA. Their context and arrangement is essential to their function. In the process of understanding them, we have a good chance of finding the smoking gun of evolutionary facilitation" (Wills, 1989).

Proposition 134: duplication of a terminal branch of a differentiation tree, consisting of two terminal edges with a common node, is a simple macroevolutionary change for a differentiation tree, easily explained in terms of copy and paste operations on DNA.

A simple DNA representation of a terminal branch is shown in Figures 24-26. Complexities that could occur include inversions and splitting of differentiation cascades by intervening portions of DNA, or across different chromosomes. However, let us assume for now that the mapping of a terminal branch of the differentiation tree to the DNA has this neat appearance.

If a copy of the DNA were made that includes the activator of the terminal branch and the two differentiation cascades it activates, this piece would be accessible to the original differentiation tree wherever it is inserted, so long as that insertion does not disrupt the differentiation cascades in the old differentiation tree. The resulting organism would seem to generally be viable, despite some cross talk between the old and new differentiation cascades (Figure 25). The so-called 'ectopic' expression of genes in transgenic animals (Balling et al., 1989; Franks et al., 1988; Harvey & Melton, 1988) and certain other mutations (Frischer, Hagen & Garber,

1986) may be similar to such cross-talk (cf. Fisher et al., 1995). Of course, such a copy and past would be but one rare, successful event amongst many similar ones that lead to inviable individuals.

I next assume that coevolution of the new activator and the new initiators occurs, so that the old and new differentiation cascades become separated from one another, restoring the overall genome regulation to a differentiation tree structure (Figure 26):

"Is the modern vertebrate body plan, backed by 39 Hox genes, really that more complex than the body plan of the early chordate with its presumed 13 [Hox] genes? When the early vertebrate first acquired four clusters [of Hox genes] 450-500 million years ago, these would at first have generated no more than 13 compartments in the body plan, since each of the 13 kinds of Hox genes would just have been fourfold redundant. Members of paralogous subgroups would have experienced complete cross-activation and cross-repression, since all had the same enhancers. Subsequently, subgroup members diverged in two ways. With mutations in the *cis*-regulatory sequences, the location of the expression compartment of each could change individually, and as many as 39 expression compartments could evolve. On the other hand, with mutations in the coding sequences, the homeodomains could change; hence the DNA-binding specificity of the protein, and also the protein's interactions with other transcription factors could change. Thus a subgroup member would become different enough from the others to have its own set of target genes. Thus new compartments might evolve" (Gerhart & Kirschner, 1997).

This would of course take many generations and I assume that there is a fitness advantage to a clean separation of the two terminal branches of the differentiation tree (Figure 26). A gene dosage effect, from doubling the number of genes active in some of the differentiation cascades (Figure 25), is a likely source of reduction in fitness:

"...Cytogenetically visible trisomy in humans (which will usually encompass at least 40 to 50 genes) is usually associated with phenotypic abnormality, indicating that a significant minority of loci must be sensitive to 3 versus 2 dosage" (Wilkie, 1994).

Thus the simplest source of a fitness advantage to cleaning up of the crosstalk is reduction of the gene dosage effect that occurs on duplication. Use of the duplicated cascades for new tissues may also increase fitness, perhaps even beyond that of the parental stock. In some organisms, such as yeast, a simple gene duplication doubles the concentration of an enzyme, directly enhancing fitness in certain environments (Hansche, 1975), without an intermediate stage of reduced fitness:

"To the author's knowledge, duplication of the Aphtase structural gene, and its incorporation into this experimental population as an adaptive response to the short supply of an essential nutrilite, is the first direct experimental demonstration in a eukaryote that gene duplication may play its postulated role in evolution, i.e., as an intermediate adaptive step in the evolution of a new enzyme function" (Horowitz, 1965).

The sensitivity of DNA binding proteins to single amino acid changes (Treisman et al., 1989) suggests that such coevolution could be rapid:

"Li and his associates reported that gene duplication is often followed by acceleration of amino acid substitutions (Li & Gojobori, 1983; Li, 1985).... There are several other examples in which the frequency of substitutions involving amino-acid replacement change is unusually high relative to the frequency of silent substitutions (e.g., γ-crystallin genes of the rat, den Dunnen et al., 1986; histocompatibility antigen of the mouse, Lalanne et al., 1982; constant-region genes of immunoglobin of the mouse, Schreier et al., 1981, and of the rat, Sheppard & Gutman, 1981; Emorine et al., 1984; Frank et al., 1984; and variable-region genes of immunoglobin of the mouse, Bothwell et al., 1981). It is impressive to note that all the above examples are duplicated genes. Thus, gene duplication provides very favorable conditions for the acceleration of amino-acid substitutions.

There are two ways to interpret this: 1) because of gene redundancy, selective constraints are relaxed, so that a larger proportion of new mutations are neutral after duplication (Kimura, 1983; Li, 1985); and 2) positive natural selection accelerates the rate" (Ohta, 1988a).

The phenomenon of gene dosage effects gives a third interpretation to accelerated evolution after a duplication. The same accelerating effect should occur for duplicated terminal branches of the differentiation tree. Of course, the new duplicated differentiation cascades in the duplicated terminal branches could themselves also evolve new functions, so that a new pair of histologically distinguishable terminal cell types is formed (Figure 26). (An analogous model for 'morphogenetic trees' was suggested by Arthur, 1988.) This may be what creates the 'positive natural selection': the extra tissue type is an advantage. It enhances fitness. This accelerates the amino acid substitutions, providing a positive feedback loop (cf. Milsum, 1968) by yet another mechanism. It is curious that...

"...no anomaly is observed in amino-acid substitutions during the subsequent evolution of the members of established gene families (e.g., immunoglobin genes: Ohta, 1978, 1980; Kimura, 1983; Gojobori & Nei, 1984). The acceleration seems to be limited to the short period after duplication" (Ohta, 1988a).

Thus this positive feedback loop, however it operates, may cause an episode of punctuated equilibrium (cf. Section 6.03). On the other hand, a negative feedback mechanism, concerted evolution, may slow down the divergence of a duplicated gene (Li & Graur, 1991). Of course, these may operate on different time scales, perhaps explaining the transience of the acceleration of amino acid substitutions. Concerted evolution may be unimportant for duplicated terminal branches of a differentiation tree. Similarly, the distinction between the "evolution of functionally novel proteins by mutation during non-functionality (the MDN model) [attributed to Ohno, 1973]" (Hughes, 1994), would seem to be inappropriate for higher level duplications.

The duplication of terminal branches of the differentiation tree fits into...

"...the following four-stage scenario for macroevolution: (1) Liberation from the preexisting selective constraint. (2) Sudden increase or boom of neutral variations under relaxed selection. In this stage, gene duplication in addition to point mutation must play a very important role in producing genetic variations. Needless to say, their fate is largely determined by random drift. (3) Realization of the Dykhuizen - Hartl effect [Dykhuizen & Hartl, 1980; Hartl & Dykhuizen, 1981; Kimura, 1983], namely, some of the accumulated neutral mutants turn out to be useful in a new environment, which the species then exploits. (4) Intergroup competition as well as individual selection lead to extensive adaptive evolution creating a radically different taxonomic group fitted to a newly opened ecological niche" (Kimura, 1990).

Hughes (1994) claims there is no evidence for accumulation of neutral mutants, and suggests that gene duplication splits a bifunctional protein into two proteins with single functions. This does not solve the problem of the origin of truly novel functions. In duplication of terminal branches of the differentiation tree, novelty comes not so much from the opportunities for new proteins, as from the opportunities for new tissues.

Watts & Watts (1968) review mechanisms of 'primary gene duplication', distinguished from secondary duplications, which are more easily explained by homologous unequal crossing-over. These mechanisms could be applicable to duplications of differentiation tree terminal branches. See also Reanney & Ralph (1968).

Zuckerkandl (1976) anticipated what I'm interpreting as duplication of terminal branches of the differentiation tree ('subprograms' of a 'dendrogram'):

"It is true that, in the case of the Membranipora, we do not know whether the original 'gut program' has been transposed as such to another tissue, or whether a new 'gut program' has, in this other tissue, been set up so to speak from scratch. It would be surprising if a whole set of different structural genes were used for making the second gut. If many of the same structural genes are used, many of the same controller genes are undoubtedly also used, and the first alternative would apply. We would then have a case of 'reawakening,' during ontogenesis, of a given eurygenic gene complex. With respect to the ancestral state, the opening in a second tissue of an access to the same eurygenic master switch would represent an innovation in the topology of the regulatory network.... The following generalization is thus suggested: programs, subprograms, subsubprograms, etc., are under the control of master switches of different rank that can be turned either on or off and that are in relations of interdependence with respect to master switches belonging to other hierarchies of partial programs. Through such connections, a master switch in one gene complex can become indirectly the activator of a master switch in another gene complex, and thus carry the organism over to a new developmental phase. In this fashion, one may conceive of dendrograms of master switches where each commands a smaller or larger number of structural genes according to its rank.... The dendrogram would be different if we considered programs in individual tissues during ontogeny" (Zuckerkandl, 1976).

On the other hand, he attributes 'tissue duplication', at least initially, to single gene duplication. Zuckerkandl's (1976) dendrogram (his Figure 8) is more what a mathematician would call a partially ordered set (Gellert et al.,

1975), rather than a single rooted tree, or may actually go in recursive circles:

"'Earlier' master switches may be turned on again through actions mediated via 'later' master switches, thus causing a tissue to reiterate a certain succession of gene activities" (Zuckerkandl, 1976).

Evidence now exists that could be interpreted as suggesting that duplication of terminal branches of a differentiation tree does occur:

"The three paired-box and homeobox genes *paired* (*prd*), *gooseberry* (*gsb*) and *gooseberry neuro* (*gsbn*) have distinct developmental functions in *Drosophila* embryogenesis (Bopp et al., 1986; Baumgartner et al., 1987; Gutjahr, Frei & Noll, 1993; Gutjahr et al., 1993; Noll, 1993).... It was startling to discover that Prd, Gsb and Gsbn, very different in their C-terminal halves, are functionally equivalent, as this implies that the essential determinant of the function of the three genes is encoded in their specific *cis*-regulatory regions controlling their temporal and spatial expression. As the N-terminal paired- and homeodomains have been highly conserved in these genes, they must have evolved from the same ancestral gene by repeated duplication. Our results thus provide the first experimental example that, during evolution, duplicated genes may acquire new functions by changes in their regulatory regions generating an altered expression, rather than by mutations in their coding sequences.... The idea that mutations in the *cis*-regulatory rather than coding regions drive functional diversification has already been suggested (Jacob, 1977).

"We propose that during duplication, the duplicated portion of a gene, including its coding region, is juxtaposed to new *cis*-regulatory elements which change its expression and thus give it a new function - ultimately the activation of a different set of genes. Subsequently, mutations accumulating in the coding region may further contribute to an altered function of the gene. This hypothesis provides a simple explanation for what seem to be redundant functions revealed in gene knock-outs in vertebrates (Thomas & Capecchi, 1990; McMahon & Bradley, 1990; Joyner et al., 1991; Lohnes et al., 1993)" (Li & Noll, 1994).

Proposition 135: edges of differentiation trees may be distinguishable from lower levels of genetic regulation by the degree of web formation. Web formation constrains duplication to units or multiples of edges of the differentiation tree (i.e., terminal branches).

In the course of this book I have not argued for any particular regulatory arrangement of the genes within a single differentiation cascade (a single edge of a differentiation tree). At the level of the differentiation cascade, in contrast to the differentiation tree, there is no broad evolutionary constraint on the logical structure of the DNA, such as keeping induction as a secondary phenomenon for radiations. Without such constraints, regulatory webs, such as those investigated by Kauffman (1993), may form rapidly on an evolutionary time scale. Differentiation trees are thus consistent with the conclusion that "...evolution of gene networks should preferentially occur either by duplication of single genes or by duplication of all genes involved in a network" (Wagner, 1994a). A 'network' corresponds to a single differentiation cascade or equivalently, to an edge of the differentiation tree.

The idea that gene cascades are the units of macroevolution, which themselves have a kind of integrity, due to selection against web formation, may correlate with observations of chromosome band patterns:

"[J.A.] Rapaport... makes the proposal that chromosome regions in which several similar bands lie in sequence represent multiple duplications of one original band. Correlating the additive genetic effects with the chromosome changes in the [*Drosophila*] Bar series, he develops the interesting hypothesis that multiple repeats offer one possible explanation for orthogenesis - particularly cases of orthogenesis which paleontologists have regarded as involving especially rapid evolution....

"A... feature of the triple repeat case is what may be called the unit action or behavior of the repeat region. In the triple repeat, why should the three regions be the same size? In other words, why should each have remained complete and intact over such a long period of time? It looks as if the region behaves as a unit either in the sense that it can not be fractionated, which seems improbable, or that it performs a unit function for which all parts are essential. The synaptic behavior points in the same direction. Is there a primary locus or gene in the segment which acts as a controlling agent, but which requires the presence of the others for proper functioning?" (Metz, 1947).

It would be interesting to know to what extent such 'unit action' correlates with steps of differentiation, a question that should be answerable with modern data bases. It may thus be worthwhile to correlate 'conserved gene blocks' (Paterson et al., 1996) with edges of the differentiation tree.

Webs within differentiation cascades may be the norm, while weblike regulatory relationships between differentiation cascades may be the exception. The degree of webbing may thus permit us to distinguish at which level a particular regulatory gene is operating.

Proposition 136: the viability of a duplication of a terminal branch of a differentiation tree is inversely related to the size of the terminal branch.

"In order to lead to permanent changes in the caryotype, structural changes have to be capable of surviving in the heterozygous condition for a certain period of time, without lowering either the viability or the fertility of the individual to any considerable extent" (White, 1954).

The basis for this proposal is clear from Figure 25, in which there are two forms of crosstalk that must be endured until the duplicated genes can evolve separately:

1. Some crosstalk occurs at the first occurrence of duplicated activator, leading to a double gene dose for the differentiation cascades that it activates.

2. When the second copy of the duplicated activator is encountered, later in developmental time, not only does it reactivate two previously activated copies of a differentiation cascade, it also does so at the wrong time in development.

Larger terminal branches would compound these difficulties, both in terms of the number of differentiation cascades having a double gene dose, and in terms of greater discrepancies in developmental time between the times the second activations occurred (Figure 35a,b). On the other hand, the timing problem can be somewhat reduced by grafting the copied terminal branch to a terminal differentiation cascade that is activated at an appropriate slightly earlier developmental time (Figure 35c). This has three interesting consequences:

1. An unusual developmental sequence may occur in some cells, since some differentiation cascades are avoided.

2. The cells in which this happens will have a hitherto nonexistent, and thus novel set of memorons, and thus, possibly, novel histological and functional properties.

3. Double differentiation cascade gene doses are avoided.

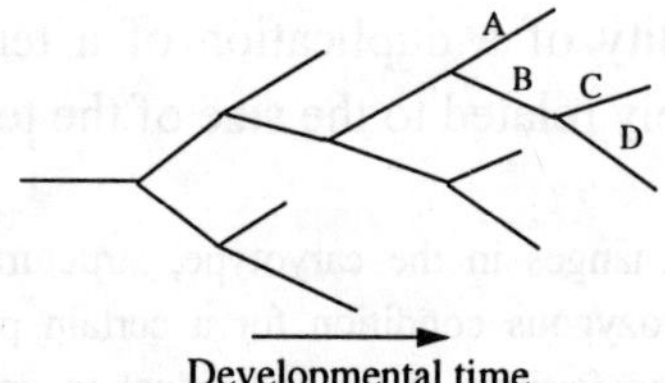

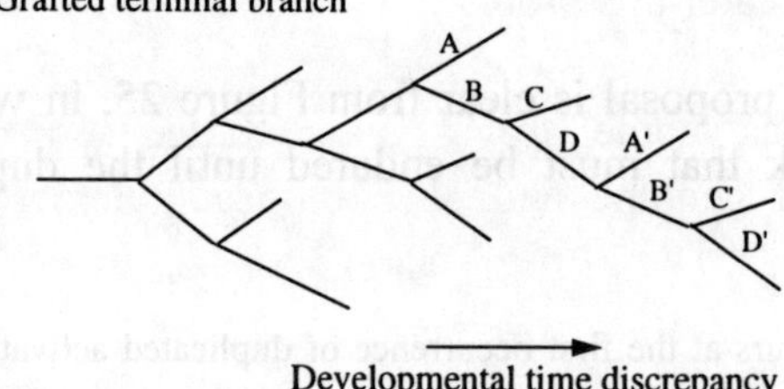

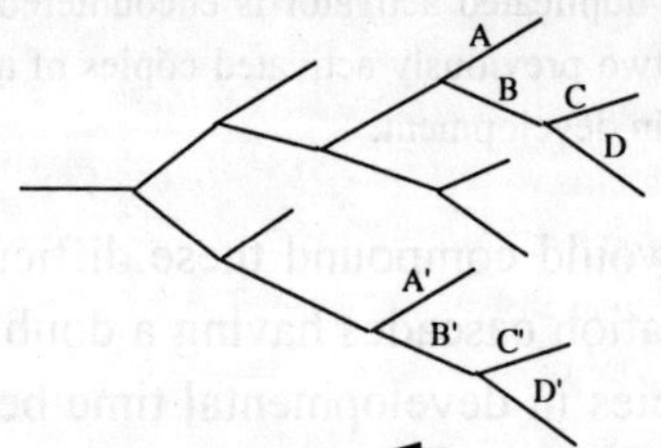

Figure 35. Reducing developmental time discrepancies. a) A differentiation tree with one terminal branch labelled. b) This larger terminal branch can be copied and inserted, but this can lead to duplicate activations at widely inappropriate times, in addition to more cascades having a double gene dose. c) The timing problem can be alleviated somewhat by making the graft to an earlier terminal cascade.

One final problem with large terminal branches is that it is rare that they could be copied as a single, contiguous strand of DNA, as suggested in Figures 24-26. This is because differentiation cascades may sometimes be somewhat disbursed over the genome, rather than being nicely localized, as I have portrayed. Clearly, the larger the terminal branch of the differentiation tree under consideration, the more likely this problem of nonlocalization will be an obstacle to its duplication. On the other hand, it might be worth checking if 'conserved gene blocks' (Paterson et al., 1996) correspond to differentiation tree branches consisting of multiple differentiation cascades.

Of course, as in any gene duplication, chromosome mismatches are more likely when more than one homologous match between parts of chromosomes is possible. The larger the stretch of a chromosome that is duplicated, the greater this problem, also consistent with this proposition.

Riedl (1978) has come to a similar conclusion based on general considerations of genome and phenotype complexity, and an attempt to quantitate them:

"One of the strangest features of evolution is the fact that heritable change occurs only in the realm of accident and the micro-world while selection happens in the realm of necessity and the macro-world. Between the two there yawns a gap in degree of complexity....

"This organizational gap between determination complexes, on the one hand, and the respective functioning phene systems, on the other hand, can now be estimated. Knowing the limits of a determination complex we can quantify the *complexity of an organized change* due to the mutation of a single gene (this **c** can be called **cg**). We can compare this with the complexity needed for the functioning of the respective system of phenes (**cp**).

"The degree of organization of genetic changes such as tetraptera, aristopedia or proboscipedia [in *Drosophila*] is 50 to 100 **cg**. But the complexity of the respective phene systems is about 200 to 500 **cg**. For the musculature of the mutated systems is grossly deficient, the innervation is lacking or confused, not to speak of those homologous paths which must be assumed for the coordinated switching of the activity of the system.... There are gaps and mistakes in the internal organization of the mutated part (as already pointed out by Waddington, 1957,...).

"Wherever there is evidence, we find that $cp > cg$. This is particularly obvious on remembering that the complexity of the organization of the larger phene systems easily reaches 10^4 to 10^5 **cp**, although genetic information complexes attain in this connection at most 10^3 **cg** to 5×10^3 **cg** and this is only for these coarse mutations, all of which are unsuccessful" (Riedl, 1978).

It is clear that there is plenty of room for a spectrum of genetic changes which need not correlate much with initial phenotypic changes. Bigger changes in either will occur with lower probabilities of success, but they will occur.

Proposition 137: the hopeful monster can be reconsidered, if we distinguish the 'hopeful genotype' from the 'hopeful phenotype'.

I have, with the possibilities for macroevolution of the differentiation tree, created a framework in which to consider potentially viable mutations of the differentiation tree of all magnitudes. The larger ones, whose rareness may be reflected in data such as the chromosomal stability of Rhinolophidae bats (Qumsiyeh, Owen & Chesser, 1988), may correspond to the 'hopeful monsters' of Goldschmidt (1940):

"The unfortunate coining of the term 'hopeful monster' to describe the product of his macromutations may have clouded the picture somewhat, as it conjures up visions of some grotesque *Missbildung* that would have almost no chance of surviving in nature, let alone finding a mate with like traits. But... Goldschmidt's systemic mutations (Goldschmidt, 1940) represented special alteration of the genome that changed the primary pattern of what he termed *reaction systems* controlling development. He was rather specific as to the kinds of mutations that were likely candidates for macromutations. These involve chromosome rearrangements which result in chromosome repatterning of the genetic material" (Bush, 1982; cf. Wallace, 1985).

We may, in retrospect, allow Goldschmidt the possibility of 'chromosome repatterning' via duplications of terminal branches of the differentiation tree.

In his own words, we can see that he might have been amenable to such an interpretation:

"As a rule, one may assume that mutants affecting developmental processes at earlier stages are likely to disturb the fabric of interwoven and closely attuned individual features of development in an adverse way, making orderly development impossible. But it is imaginable - and there are examples of this - that in rare cases such an early-acting mutation is not lethal but permits development to continue. The result is a large deviation from normal, which might affect in one step major features of development, producing a more or less divergent pattern of the entire organization. Such a macromutation might accomplish in a single step major morphological or physiological changes at the level of the higher categories from the species to the phylum.... In addition to small or large mutations of genic loci, there exists a completely different type of mutation. I called this 'systemic mutation,' meaning that a reshuffling or scrambling of the intimate chromosomal architecture, which might occur rarely by chance, will act as a macromutational agent. This means that it will produce, suddenly, a huge effect upon a series of developmental processes leading at once to new and stable form, widely diverging from the former" (Goldschmidt, 1952a).

Davidson (1982) has suggested that...

"...larger regions of the genome including a number of structural genes might be duplicated so as to produce an additional copy of a multigene regulatory unit [i.e., such as duplication of a branch of the differentiation tree], which might then be transposed to another region where it would fall under the control of its new host domain.... A major reason to pursue this line of speculation is the unexpectedly high incidence of multiple copies of closely related genes.... The cases which interest us here are those where the related genes are expressed at different times or different places in the organism, i.e., in different states of differentiation. Such genes evidentially belong to different regulatory pathways. Well-known examples abound.... Following duplication and insertion into its new regulatory domain, such a gene [or 'multigene regulatory unit'] would probably diverge independently under selective pressures other than those acting on its cousins located in different functional contexts in the genome" (Davidson, 1982).

Shepard (1987) extends Davidson's idea to actual duplication of anatomical parts:

"A complex anatomical structure that has already evolved in the service of one function might be moved or transformed as a whole, where in rare cases its already highly developed capabilities might find a new use" (Shepard, 1987).

Duplication of whole blocks of genes has indeed been observed:

"...The arrangement and function of the first four genes of the goat β globin family are very similar to those of the next four genes... [Edgell et al., 1983)]). Apparently one set of the four genes was produced by block gene duplication from the other set (Cleary, Schon & Lingrel, 1981; Li & Gojobori, 1983)" (Nei, 1987).

It is now clear, in light of the above, that Goldschmidt's conclusion that the phenotype of the hopeful monster would change suddenly is not necessary (cf. the balanced review of hopeful monsters in Grant, 1963):

"'Macroevolutionary changes in development need not be extreme' (Raff & Kaufman, 1983). This sentence implies that the hopeful monster need not be a monster at all, merely hopeful" (Wallace, 1985).

If the systemic mutation consists of a copy and paste of a terminal branch of the differentiation tree, then the immediate effects on the phenotype could, in some cases, be negligible, making it quite viable, and able to mate successfully and produce fertile offspring. Divergent evolution could nevertheless permit the genome to change markedly from the original. This is a way of having our cake and eating it too. The 'hopeful genotype' can become the 'hopeful phenotype' over the course of many generations, bringing us back to the possibility of gradualistic evolution, even in the face of large jumps in the genome. Indeed, Stern (1949) suggested how this could come about:

"Characters due to mutant genes are often different from those long established by being more variable in expression and by having a lowered adaptive value... [with] extreme variability from individual to individual or even from the left to the right... of the same specimen.... There may be an initial advantage in the variable expression of new mutants.... Low expression or even lack of penetrance will permit the character to spread more or less invisibly and to crop up phenotypically in diverse genetic backgrounds or ecological niches, some of which may happen to harmonize with the mutant toward higher adaptiveness. Variable penetrance thus may accomplish... spread under cover of unchanged phenotypes" (Stern, 1949).

On the other hand, Maynard Smith (1989a) keeps the possibility of the original phenotypic hopeful monster open:

"By a 'hopeful monster ' I shall mean here no more than an individual carrying a genetic mutation of large phenotypic effect.... A hopeful monster will need further fine tuning by selection of smaller variants before its descendants achieve detailed adaptation to the new way of life" (Maynard Smith, 1989a).

Gould (1983a) similarly concludes...

"If embryology is a hierarchical system with surprisingly few master switches at high levels, then we might draw an evolutionary message after all.... Major evolutionary transitions may be instigated (although not finished all at once as hopeful monster enthusiasts argue) by small genetic changes that translate into fundamentally altered bodies" (Gould, 1983a).

(Cf. Gould's, 1985b, discussion of Iltis', 1983, hypothesis for the 'hopeful monster' origin of domestic corn.) The distinction between the genotypic and phenotypic hopeful monster is important to clear thinking about saltational evolution.

Van Valen (1974a) has shown that extra legs ('polymely', 'accessory limbs': Szabo, 1989; Rao, 1992), albeit of questionable functionality, occur in many populations of frogs and salamanders (cf. "Why are there no six-legged vertebrates?": Slack, 1996a). The importance of this observation cannot be underestimated, for with population frequencies of up to 30% Van Valen's survey shows that...

"A normally disadvantageous phenotype is established and maintained (not necessarily fixed) in occasional populations... (for mechanisms of such establishment see Van Valen, 1960).... A deleterious phenotype is in fact sometimes common, and sometimes... remains common for ecologically long periods of time" (Van Valen, 1974a).

The fact that this phenotype does occur means that significant alterations to the differentiation tree need not be occult in the individual in which they initially appear. Of course, it remains to be seen whether extra frog legs, which sometimes might have a genetic basis (Van Valen, 1974a), are due to duplications of terminal branches of the differentiation tree. An analogous fossil reptile with four legs and two boned wings attached to the skin (Frey, Sues & Munk, 1997; Monastersky, 1997a) represents a similar challenge. It is possible that multiple legs will invariably prove to be of viral (Van Valen, 1974a) or parasitic origin (Sessions & Ruth, 1990). However, they could be phenocopies of some as yet to be found genetic mutant: if they proved advantageous in some environments, as discussed by Van Valen (1974a), coevolution with the infectious agent might propagate the phenotype, or genetic assimilation might occur. It's not exactly what Goldschmidt (1940) had in mind, but here, indeed, we have a live, phenotypic hopeful monster in our hands, which, if it found an appropriate niche (Van Valen, 1974a), might indeed evolve to functional polymely with better innervation:

"As I write this reply I am looking at seven living multilegged *Hyla regilla* in a glass jar on my desk. These have survived because they have a pair of functional limbs in addition to various numbers of extra limbs. They have just been fed *Drosophila virilis,* and they are actively leaping around to catch them and eat them. The extra limbs stick out at odd angles, and, at most, twitch pathetically. Occasionally a frog accidentally grabs one of the extra limbs in its mouth. At other times, a frog tries to jump and trips over its own legs. These are handicapped frogs, and if there is any selective advantage to anyone it is to the trematodes whose cysts infested its limb buds and caused the deformities (since the only way that the trematode's life cycle will be completed is if the infested tadpoles and frogs get eaten by a snake or other predator). We have not done histology on the spinal cords but I confidently predict that the extra limbs are very poorly innervated compared to the functional limbs" (Stanley K. Sessions, p.c., 1997).

Proposition 138: transfer of a terminal branch of a differentiation tree is less likely to be viable than its duplication.

This is because, in addition to the difficulties encountered when a duplicate of a terminal branch is inserted at a new location, loss of function of the original terminal branch occurs. In other words, some terminal tissues are lost, while others are gained. This would generally seem to be more deleterious than just a gain of terminal tissues. On the other hand, the latter correlates with a larger organism size (Proposition 179), which may not always be advantageous.

Proposition 139: the probability of gene duplication splitting essential segments of DNA, such as those used for differentiation trees, is reduced when there is a large background of functionless DNA, and may indicate an evolutionary advantage for such 'junk' DNA.

This may give a function (cf. Casimir et al., 1988; von Sternberg et al., 1992) to the so-called desert of background 'junk' or 'secondary' (Hinegardner, 1976; Cavalier-Smith, 1985b; Ohno & Yomo, 1991) DNA, described here by Ohno (1984):

"It has always been obvious that natural selection cannot afford to survey all the 3.2×10^9 base pairs of DNA contained in the haploid set of the mammalian genome.... [But cf. any image reconstruction algorithm, which does something analogous for even millions of pixels, when the equations are underdetermined: Rangayyan, Dhawan & Gordon, 1985.] Of the mammalian genomic DNA, 95% or more is thought to be ignored by natural selection.... Thus, the mammalian euchromatic region can be viewed as a barren stretch of desert in which still-functioning genes are scattered as though oases (Ohno, 1972a).... The average distance between neighboring mammalian genes was estimated as 35,000 base pairs (Ohno, 1972a).... The reason for desertification of the euchromatic region is found in the extreme inefficacy of the mechanism of gene duplication as the sole means of generating new genes with previously nonexistent functions (Ohno, 1970).... Two alternative fates await a redundant copy of the already existing gene. By accumulating all the randomly sustained mutations, it may emerge triumphant as a new gene with a hitherto nonexistent function. But its far more likely fate is degeneracy to join the rank of *junk* DNA. [A third possible fate is evolutionarily stable coexistence of the gene and its redundant copy: Nowak et al., 1997.] Mammalian ancestors at the stages of fish and amphibians had an option of utilizing the mechanism of polyploidization as a means to experiment with a redundant copy of every gene locus (Ohno, 1970 [cf. DeSimone & Hynes, 1988]).... Mammals have long forfeited the possibility of polyploid evolution. They have to rely exclusively upon tandem duplication of the already-existing genes as a means of generating new genes. Indeed, cloned mammalian DNA almost invariably reveals the presence of silent or defunct genes in the neighborhood of each still-functioning gene, thus attesting to constant experimentation with the mechanism of tandem gene duplication as well as to the extreme inefficacy of this mechanism.... The genetic distance of one cross-over unit covers one million base pairs of DNA in which there may be no more than 30 or 40 still-functioning genes. It follows then that as far as mammals are concerned, there are no closely linked genes sensu stricto..." (Ohno, 1984).

With such vast spaces between genes (on the average), recombination can proceed with relative impunity, with only occasional splitting of a gene. (The fact that the average distance between genes is large does not preclude clustering into gene blocks and thus linkages.) This fact may explain just why junk DNA 'hitchhikes' along with useful DNA, since it is then actually of use to the latter:

"Another, possibly more potent, mode of spreading for junk DNA would be by 'hitch-hiking' with useful genes (Maynard Smith & Haigh, 1974). Some support for this arises from the important studies of experimentally induced changes in genome size reviewed by Cullis (1985); the evidence suggests that the extra DNA often consists of sizable chunks of non-functional or 'neutral' DNA in which are embedded relatively small segments of DNA that are subject to strong positive natural selection. If natural selection favoured an increase in the number of copies of the embedded segment, then, if the molecular mechanism that generated extra copies did so by multiplying the whole chunk, the neutral DNA would also increase" (Cavalier-Smith, 1985b).

The strong correlation between cell volume and total genome size may serve to limit the amount of junk DNA (Cavalier-Smith, 1985c).

Lin & Riggs (1975) "suggest that 'junk' DNA provides one essential function: namely to maintain the total DNA concentration optimum for all regulatory proteins binding to DNA". For contrary opinions that junk DNA isn't junk, see A. Jeffreys in Lewin (1981) and Mantegna et al. (1994), who suggest "...that noncoding DNA sequences represent a structured language fundamentally unlike the coding of genes" (Pennisi, 1994b). "...Noncoding regions in eukaryotes display a smaller *entropy* and larger *redundancy* than coding regions, supporting the possibility that noncoding regions of DNA may carry biological information" (Mantegna et al., 1994). Kimura (1961) "notes that the amount of DNA in a cell may be ten to a hundred times the amount necessary... and he interprets this and other evidence to indicate that the DNA message is very redundant" (Williams, 1966). In any case, one problem with the concept of junk DNA is, of course, that there is less and less of it as we discover functions for some of it (cf. Eberl, Duyf & Hilliker, 1993):

"Koop & Hood (1994) sequenced and analyzed nearly 100 kb of contiguous sequence from nonvariable regions of the TCRα [T-cell receptor] complexes in the human and mouse genomes.... A comparison of sequences in this region revealed a high degree of similarity between corresponding mouse and human protein-coding and noncoding regions. These results suggest that the majority of the TCR region has been highly conserved throughout 60 million years of evolution, although only about 6% of the region contains gene-coding sequences. Until recently, many scientists believed that only 3% of the genome contained useful sequences that were embedded in vast stretches of noncoding 'junk' DNA. Recent studies are challenging that view in favor of seeing chromosomes as information organelles with complex structural and gene-control systems" (Casey, 1994).

In the computer simulation of evolution by gene duplication of Ohta (1988a): "...the site of crossing-over is assumed to be between genes, so that no 'recombinant gene' occurs." While this assumption was made for computational simplicity, the presence of the junk DNA makes it most often the reality, unless it is correct that "...the site at which crossover takes place is likely to be inside the gene, and a 'recombinant gene' may often arise" (Ohta, 1988a). Whether crossing over can occur in the junk DNA, which is likely to be somewhat nonrandom due to nucleosome structure (Trifonov, 1990b) and nucleosome arrangement (Horowitz et al., 1994; Blank & Becker, 1995), or occurs most often within active genes, should be determinable empirically. Perhaps junk DNA retains its sequence over evolutionary time (Casey, 1994), precisely because the frequency of unequal crossing over is reduced. If there are an "estimated 1×10^6 pseudogenes scattered throughout the mammalian genome (Moon & Krause, 1991)" (Krause, 1996), at, say, 1000 bases each on average, they represent roughly 1/3 of the 3.2×10^9 base pair mammalian genome (Ohno, 1984), preserving some sequence information until they decay to background.

Trusov & Dear (1996) have advanced the premise of this proposition:

"We argue... that there is a bias favouring the accumulation of non-coding DNA in close proximity to genes. This arises because deletions in such regions are liable to disrupt genes and hence are selected against. Conversely, deletions of non-coding DNA in large intergenic regions are less likely to have adverse phenotypic effects" (Trusov & Dear, 1996).

(This leads to another molecular clock: Trusov & Dear, 1996; cf. Martin & Gordon, 1995.) What I add is that splitting of genes is less likely to occur, the more junk DNA there is, so that there is also a bias towards increase of junk DNA, rather than "a gradual increase or decrease in genome size, or... a dynamic equilibrium in which the overall size remains constant" (Trusov & Dear, 1996). Yuri Trusov (p.c., 1996) now suggests that this increase is observable in many or all species, and is proportional to the time since the last speciation occurred.

Proposition 140: the requirement that the differentiation tree be a linked structure implies that a duplicated or transferred terminal branch will rarely be inserted into the middle of a differentiation tree.

I take it as unlikely that an unlinked portion of DNA can establish a link later on in evolutionary time. This is because an unlinked terminal branch is inaccessible along the differentiation tree, and is thus likely to fade into background or junk DNA, well before such a fortuitous event would occur. On the other hand, a transfer or duplication that neatly inserted itself so that both ends of the insertion form appropriate links, leaving an intact differentiation tree, while possible, is of small probability.

Of course, I could be as wrong on insisting that the linked structure of a differentiation tree is not dynamic as those who decried jumping genes (cf. Keller, 1983; Starlinger, 1993).

Proposition 141: multigene families are created in the course of duplication of branches of the differentiation tree.

Multigene families give us a glimpse of the topology of the differentiation tree and its relationship to the differentiation cascades that form its edges. They 'light up' portions of the differentiation tree, and their expression of members of the family in particular tissues shows that they can be on different terminal branches of the differentiation tree. Unfortunately multigene families don't reveal the actual branching topology, any more than lights strung on a bare tree at night show its branching pattern. The more or less random introduction of genes into the genome during gene therapy and the formation of transgenic organisms, and the great variability of gene expression that results (Morgan & Anderson, 1993), may similarly be worth interpreting as 'lights' added here or there to the edges of the differentiation tree, different subtrees of which are on or off in different differentiated cell types. (If we knew how to address particular edges, especially those that are on, we might have better targeted gene therapy.)

Studies of the multigene families for tubulin (Raff et al., 1987a; Cleveland, 1987) show that...

1. all tubulins in the same organism (and even between evolutionarily divergent organisms: Prescott et al., 1989) appear able to participate in all functions;

2. some tubulins are produced primarily in certain tissues at given stages of development;

3. the 'tissue specific' tubulins are also often expressed at lower levels in some other tissues;

4. nearly identical tubulins may have entirely different patterns of expression.

However, some tubulins do differ: in their ability to participate in dynamic equilibrium (Kreis, 1987); and in their role in microtubule nucleation (Debec, Wright & Géraud, 1994) and sperm formation (Hutchens et al., 1997). Furthermore, tubulins can even differ from one part of a cell in an embryo, to another (Houliston & Maro, 1989) and during mitosis (Oka, Arai & Hamaguchi, 1990) (perhaps suggesting that there are cell state splitter specific microtubules). Cleveland (1987) indicates that...

"...results from higher species clearly suggest that multiple tubulin gene sequences are required. But to what functional end? Two hypotheses are readily evident. First, to many the most attractive possibility was that individual tubulin genes might encode functionally divergent polypeptides which could confer some unique property to the final microtubule polymer... [as] initially proposed by Fulton & Simpson (1976).... The alternative hypothesis, most cogently advanced by Raff (1984), was that multiple polypeptides are themselves functionally equivalent but represent the products of duplicated genes which have evolved to possess different regulatory sequences for activation of transcription during alternative programs of differentiation.... Consideration of the collective findings reveals that both hypotheses for the function of tubulin multigene families in higher eukaryotes are correct in some instances" (Cleveland, 1987).

We have to take care not to attribute foresight to evolution (Dennett, 1995). This suggests that the initial duplication just happens, and the regulatory or other functions then diverge. I am merely adding that this duplication often occurs in the context of duplication of a branch of the differentiation tree. One prediction would then be the existence of unrelated, but similarly structured, multigene families. A 'phylogenetic tree' of genes in a gene family should correlate with the sequence of additions of terminal branches to the differentiation tree. This was suggested by Natalie K. Björklund in regard to the construction of cell state splitters in various tissues. Whenever a cell state splitter is regenerated rather than reused, at a node of the differentiation tree, the cell state splitter genes may themselves be duplicated. They are thus subject to genetic drift, and thus to some extent record the history of the growth of the differentiation tree.

Tropomyosin isoforms are also tissue specific (Lin et al., 1997). It is curious that, while there are 15 collagens in vertebrates, the nematode *Caenorhabditis elegans* has at least 50 (Morris, 1993), along with an approximately corresponding number of "at least 60... homeobox-containing genes..., constituting approximately 1% of the estimated total number of genes" (Bürglin et al., 1989; cf. Chalfie, 1993; Bürglin, 1995). This may be related to my suggestion that mosaic organisms have one step of differentiation per cell division, i.e., that each cell is a tissue unto itself. The corollary, suggested by this proposition, is that structural genes may be duplicated just as often as edges of the differentiation tree (and as part of the same process), and then drift apart over evolutionary time.

There are a few cases of possible duplication of differentiation tree terminal branches that can be inferred from embryological evidence. One example is the origin of iris muscles from the ectoderm, rather than from the mesoderm

(Gilbert, 1989). Another is the tail bud (cf. Gont et al., 1993; Griffith, Wiley & Sanders, 1992; Gajovic & Kostovic-Knezevic, 1995):

"Tailbud mesoderm in our model must be considered as a neural plate derivative [Figure 3 in Appendix VI]. This region remains part of the neural plate until stage 13-15 [in axolotl; Figure 4; Table 1] when it invaginates (Figure 1 in Burnside & Jacobson, 1968; Figure 23 in Jacobson & Gordon, 1976a). At this point it undergoes another contraction [wave] determining it to become tailbud mesoderm. After it is in the inside it then forms all the tissues of the tail. This suggests the mesoderm branch of the differentiation tree is duplicated from this point forward in this particular tissue to form the tail" (Appendix VI: Björklund & Gordon, 1994).

It is interesting that we would therefore come to the opposite evolutionary conclusion to that suggested by Schmitz, Papan & Campos-Ortega (1993):

"...Our observations on neurulation in the zebrafish have uncovered similarities with neurulation in higher vertebrates, although the zebrafish represents a primitive mechanism, elements of which are manifest in the process of secondary neurulation in the tail bud of higher vertebrates" (Schmitz, Papan & Campos-Ortega, 1993). [Cf. Bolker, 1995.]

In other words, we suggest that the tail bud is an 'advanced' feature, not a primitive one. Kostovic-Knezevic, Gajovic & Svajger (1991) express the puzzle presented by this anomalous tissue, whose solution may lie with our considering it as a duplication of a terminal branch of the differentiation tree:

"Instead of folding, invagination, cell migration or other 'usual' morphogenetic mechanisms, the mechanism in the tail [of rat embryos], at least formally, mimics the cell-specific aggregation of embryonic cells from mixed suspension [cf. Townes & Holtfreter, 1955; Steinberg, 1963; Gordon et al., 1975].... Some other details of morphogenesis within the tail are peculiar when compared with those which we consider as rules in embryogenesis:

a) All three axial structures of the tail develop independently of the surface ectoderm.

b) The tail bud is the source of pluripotent cells as if it retained the capacity of the primitive streak and the Hensen's node. It gives structures considered as derivatives of all three germ layers....

"The causality of the secondary body formation is poorly understood and attempts to explain it were no more than teleological speculation (Peter, 1951). Arguments of explanation in terms of evolution are also lacking" (Kostovic-Knezevic, Gajovic & Svajger, 1991).

Furthermore...

"The tail gut is a temporary structure in the embryo tail. Shortly after its formation it undergoes degeneration and disappears in an anterior-posterior direction (Butcher, 1929).... A large [unnamed] mass of surplus cells remains ventral to the tail gut as an additional structure of the tail. It has no equivalent in the trunk, does not contribute to any other structure in the tail, and disappears in massive cell death" (Gajovic, Kostovic-Knezevic & Svajger, 1993).

These authors suggest that the tail gut arises as part of the tail bud. The 'surplus' cells may be homologous with the transient vegetal yolk mass of amphibians (Appendix V: Gordon, Björklund & Nieuwkoop, 1994). Thus the tail bud does give the impression of a whole, but partially degenerating secondary, ectopic embryo attached to the primary embryo (Holmdahl, 1925), albeit forming by quite different morphogenetic mechanisms, and may represent the duplication of much, or all, of the base of the differentiation tree. (The relationship of the tail bud in amphibians to the dorsal lip of the blastopore is discussed in Section 9.06.) We thus see that it is plausible that major portions of present day embryos may be due to duplicated terminal branches of the differentiation tree. Multigene families would be created in the course of such branch duplications.

Proposition 142: the cell state splitter is often amongst those elements in a copied edge. Thus the cell state splitter's sensitivity to mechanical forces can vary, permitting, for example, tissue specific variations in the conditions needed to launch expansion versus contraction waves.

I have so far treated the cell state splitter as if it is essentially the same in each cell type. However, if the differentiation cascade for each cell type contains the coding for its own cell state splitter, we should expect some evolutionary divergence of cell state splitters. We may already be doing this in the laboratory, with mechanical consequences: one way to work out the functions of 'tissue specific' tubulins (= 'cell specific' for mosaic animals) is to mutate them:

"...<u>We</u> investigate tubulin structure and function by determining the amino acid changes in 45 *mec-7* β-tubulin mutants and correlating these alterations with mutant phenotypes. The *mec-7* β-tubulin is required for the production of 15-protofilament microtubules in the touch receptor neurons of the nematode *Caenorhabditis elegans* (Chalfie & Thomson, 1982; Savage et al., 1989).... Two mild alleles appear to increase microtubule stability and lead to the elaboration of ectopic neuronal processes in *mec-7-* expressing cells.... [These]... two mutations... result in ALM cells with unusually long posteriorly-directed processes and AVM and PVM cells with additional processes" (Savage et al., 1994).

Such stabilization might be an example of manipulation of the cell state splitter's microtubule component for that tissue (cell). Whether that is correct in this case is not the point: the method gives us a powerful tool to direct mutation at the cell state splitters of particular tissues. Many parameters of the functioning of cell state splitters may be affected. The thresholds needed to respond with launching of a contraction versus an expansion wave may thus be manipulable, for example. Assuming that such manipulation occurs naturally over evolutionary time, we may have the primary reason for the variety of tissue specific cytoskeletal components (Natalie K. Björklund, p.c.).

The *Cylla* cytoskeletal actin gene in sea urchins (Cameron, Britten & Davidson, 1989; Coffman & Davidson, 1992) may be an example of a tissue specific cell state splitter mutation, in this case affecting the aboral ectoderm.

Proposition 143: low levels of expression of genes in multigene families, outside their tissues of primary expression, may represent 'leakage' rather than 'deliberate' control.

Since the tubulins seem nearly completely interchangeable, there would be no evolutionary pressure for very specific controls that turn on only the tissue specific tubulin without turning on any other. When genes in multigene families on different edges of the differentiation tree retain the same function, crosstalk between them is at a relatively high level, while genes whose functions diverge after the initial duplication event should acquire relatively lower levels of crosstalk in the course of subsequent evolution.

This form of leakage may be similar to the chemical equilibrium leakage found in the *lac* repressor/operator complex (Zubay, Chambers & Cheong, 1970; cf. Halling, 1989; Szabo & Mann, 1994), except that it represents crosstalk between two control complexes. On the other hand, at least two sets of evolutionary possibilities must be considered:

"The duplication and change in regulation of an ancestral gene leads to two different kinds of events: the establishment of multiple gene families and the origination of novel proteins (whether by accumulated point mutations, exon shuffling, gene conversion [cf. Krieber & Rose, 1986], or by a combination of these and other mechanisms). The results of these two processes as reflected in modern gene families seem discrete, although presumably in actual events the two processes must form a continuum" (Raff et al., 1987a).

Li (1983) critically reviews "models for the emergence of new genes", pointing out that "...evolutionary opportunities may still exist because many contemporary enzymes exhibit considerable substrate ambiguity".

Proposition 144: in many cases, differentiation cascades of the differentiation tree may share the very same 'gene modules', which may be thought of as gene cascades that are accessed by, but logically separate from, the differentiation tree. Such developmental gene modules may contain most housekeeping genes.

"The existence of sets of structural genes which function together in various overlapping patterns seems to be a necessary part of any gene regulation system.... Any gene (perhaps most) could be a member of several batteries" (Davidson & Britten, 1979)

The concept of 'gene modules' may explain the dilemma of what appears to be 'intraindividual convergence' in the evolution of vertebrate limbs:

"'Genetic piracy'... is the principle that 'genes, previously unassociated with the development of a particular structure, can be deputized in evolution, that is, brought in to control a previously unrelated developmental process, so that entirely different suites of genes may be responsible for the appearance of the structure in different contexts' (Roth, 1988).... Certain anatomical observations can be counted as preliminary evidence for genetic piracy (Roth, 1988). For instance, features of a certain organ can be found in other organs later in phylogeny. The most interesting case is the convergent evolution of the tetrapod pattern in the fore- and hindlimb. The fore- and hindlimbs are derived from pectoral and pelvic fins. The origin of the tetrapod pattern common to the fore- and hindlimb is thus a striking example of intraindividual convergence.... 'Homologous structures need not be controlled by identical genes, and homology does not imply similarity of genotypes' (de Beer, 1971). Hence, the question whether the homology of morphological characters can be reduced to the continuity of gene lineages is not self-evident. Certain answers have to wait for comparative molecular data on developmental genes" (Wagner, 1989b).

The results of a gene module running in two different tissues or organs (such as forelimb and hindlimb) need not be identical, since interactions between its gene products and those of the differentiation cascades that set them off could result in different (though perhaps similar) morphologies. Since genes and differentiation cascades can function simultaneously, developmental gene modules may be thought of as being similar to modules of computer code that may be set off to run simultaneously with the main program in a parallel computer that initiates them. Examples of modules are the 'overlapping gene batteries' that are postulated to be activated by 'different homeotic gene products' (McGinnis et al., 1984a; cf. García-Bellido, 1977a; Lewis, 1978), 'gene sharing', in which one gene encodes "a protein that has two entirely different functions" (Piatigorsky & Wistow, 1989) or more than two functions (Mandrup et al., 1992). Further cases include a single tubulin isotype that is 'multifunctional' (Kemphues et al., 1982), or that "of a housekeeping gene which is regulated by the organ-specific usage of alternative promoters" (Lagrange et al., 1993; cf. Somma, Gambino & Lavia, 1991; Somma, Pisano & Lavia, 1991). In *Drosophila,* the plentiful genes that are both maternal and zygotic may fall in the housekeeping module category (Robbins, 1983, 1984, 1990; García-Bellido & Robbins, 1983), including heat shock proteins (Arrigo &

Tanguay, 1991). The *decapentaplegic* gene (*dpp*) in *Drosophila* is an example:

"The regulation of the *dpp* gene by several different groups of pattern formation genes including the dorsal/ventral group, the terminal group, the segment polarity genes, and the homeotic genes indicates that many events in embryogenesis require the cell to cell communication mediated by the secreted *dpp* protein" (Hoffmann, 1992).

In the human genome project, whole catalogs of probable housekeeping versus cell type specific genes are being compiled (Matsubara & Okubo, 1993a,b). Depew & Weber (1994) suggest "...70% of the genes... roughly corresponds to the number of constitutive enzymes (enzymes synthesized in fixed amounts regardless of the metabolic state or rate of growth of the cell) expressed in all cells in a human body". In a comparison of proteins synthesized in different imaginal discs in *Drosophila,* Santarén, Assiego & García-Bellido (1993) found they were...

"...failing to detect qualitative differences between [imaginal] discs of mature larvae.... In fact, we have found only specific quantitative differences between discs, affecting 17 polypeptides (about 1.7% of the total [gel electrophoresis] spots).... There are no obvious correlations between discs belonging to the same segment (wing and second leg, haltere and third leg) which had been expected to be under the control of the same homeotic selector genes (García-Bellido, 1977a).... Thus, we cannot ascribe disc differences to any known specific developmental property of the discs (with the possible exception of ommatidia)....

"Temporal changes have been studied in some detail in the wing disc. Whereas two stages of development of larval wing discs show no qualitative differences, large differences appear as the disc progresses through the pupal period.... All these protein changes relative to the larval discs possibly correspond to the expression and/or functional modification of histotypic differentiation genes.... Pupal disc proteins that disappear... or decrease... their rate of synthesis compared to larval discs... are likely products of genes involved in cell proliferation and morphogenesis.... The molecular identification of catalogued proteins and the study of set variations in mutant backgrounds is expected to help sort out the genes involved in morphogenesis and differentiation of imaginal discs" (Santarén, Assiego & García-Bellido, 1993).

Perhaps in rapidly developing organisms, such as *Drosophila,* gene duplication is selectively minimized in favor of gene modules. I would hope that the small differences found will someday be checked for temporal correlation with differentiation waves traversing the imaginal discs, not just with tissue cultures derived from the discs (Santarén et al., 1993). Cf. Assigeo & Santarén (1992). If 70% of all genes are in housekeeping gene modules, that leaves 30% of the genes actually on the differentiation tree. This means we have to reduce our estimate of the number of genes in a differentiation cascade from 50 to 300 (Proposition 174) down to 18 to 90. That makes each differentiation tree and the differentiation tree itself much more tractable.

Many genes are known to be under the control of multiple promoters (Schibler & Sierra, 1987). Over evolutionary time, gene modules could be duplicated and/or the copies incorporated directly into the differentiation cascades of the differentiation tree, which would lead to their eventual divergence:

"Such [shared] genes are under two or more entirely different selective constraints.... The added constraint on adaptability could also make gene sharing unfavorable in the long run. It may represent an evolutionary strategy preceding gene duplication and specialization, processes that eventually separate shared functions" (Piatigorsky & Wistow, 1989). [Cf. Bolker & Raff, 1996.]

(Cf. 'gene sharing' in Li & Graur, 1991.) As an example, we cannot presume that all homeobox genes are master genes for steps of differentiation...

"...[in] view of the inclination for some regulatory mechanisms (or 'modules') to be involved in a variety of tissue types (such as the *C. elegans Pax-6* homologue *vab-3/mab-18:* Chisholm & Horvitz, 1995; Zhang & Emmons, 1995)" (Fitch & Thomas, 1997).

In this case, homeobox genes may be functioning as housekeeping genes, or at least 'housekeeping regulators'. The balance between maximization of the use of gene modules and the tendency to duplicate and incorporate them into the differentiation tree might make an interesting line of evolution research.

Pleiotropic genes should fall into four categories, if this overall analysis is correct:

1. genes affecting most cells, because they are housekeeping genes that are always functioning;

2. genes affecting cell types at diverse locations on the differentiation tree, because they occur in gene modules that are addressed as needed;

3. genes affecting many cell types along a given subtree of the differentiation tree, since they are activated at a particular node or edge, and remain activated through more than one consecutive differentiation cascade;

4. genes that are involved in the differentiation pathway, insofar as it is the same in many or all steps of differentiation.

Of course, if particular differentiation cascades reverse the activation state of genes in any category, the pleiotropic mutant may not be phenotypically present in the corresponding cell type. This possibility may hinder classification of pleiotropies into the above categories. The same problem occurs for genes transcribed only under particular circumstances, such as heat shock proteins (though the latter can also be part of regular differentiation cascades: Arrigo & Tanguay, 1991; Tanguay & Khandjian, 1993).

Whether housekeeping genes should be regarded as a parallel operating 'second genome' (Goldman et al., 1984) or as gene modules that are accessed by most or all differentiation cascades must await future analysis. Bernardi (1989) has suggested a different way to divide the genome into two parts:

"The first compartment, the *paleogenome,* is characterized by its similarity to what it was, and still is, in cold-blooded vertebrates: the late-replicating, compositionally homogeneous, GC-poor isochores of early-condensing chromomeres contain relatively rare, GC-poor (largely tissue-specific) genes having TATA box promoters (CpG islands are scarce).... In the *neogenome,* the ancestral, early-replicating, GC-poor isochores of late-condensing interchromeres were changed into compositionally heterogeneous, GC-rich isochores that contain abundant genes (perhaps including most housekeeping genes) having G/C box promotors (genes and CpG islands are particularly abundant in the GC-richest isochores)" (Bernardi, 1989).

There may be a physical correspondence between the 'isochores' of DNA that are GC-rich and gene modules. Genes in the 'neogenome' have generally higher thermodynamic stability of their proteins (Bernardi, 1989), whereas the tissue specific proteins of the 'paleogenome' are generally of lower thermodynamic stability. This might be related to the turnover times for these proteins: protein molecules that should last longer in an organism need to have greater stability, so that selection may alter them this way. The names 'paleogenome' and 'neogenome' may be misnomers, since one would expect housekeeping genes to have evolved earlier.

DNA repair mechanisms may reflect differences between housekeeping gene modules and other genes:

"For UV-induced cyclobutane pyrimidine dimers we propose that at least three levels of repair exist: (1) slow repair of inactive (X-chromosomal) genes, (2) fast repair of active housekeeping genes, and (3) accelerated repair of the transcribed strand of active genes. These hierarchies of repair may be related to chromosomal banding patterns as obtained by Giemsa staining" (Mullenders et al., 1991).

Proposition 145: gene modules are stable over evolutionary time in proportion to the number of differentiated cell types that address them.

In order to change the 'address' of a gene module that is already in use by a number of cell types, there would have to be concerted evolution of all of the links from their differentiation cascades to that module. This could be why the basic housekeeping functions, such as the Krebs cycle or glycolysis, are not represented by redundant proteins (Przemko Tylzanowski, p.c., 1995).

Proposition 146: GC-poor isochores may correspond to edges of the differentiation tree, i.e., they may individually contain the bulk of the genes activated in the differentiation cascade for a single step of differentiation.

This speculation follows from...

1. the order of magnitude coincidence in size of isochores (">300 kb, on the average": Bernardi, 1993b; cf. Bernardi, 1995) with the average number of genes per step of differentiation (Proposition 174 , Table 3 and Figure 36);

2. "...hundreds of isochore pairs hosting... homologous genes [that]... are compositionally very close" (Bernardi, 1993b);

3. "...the increasing evidence for the conservation of gene arrangements (linkage maps) among mammals (O'Brien, 1993).... A number of genes (like the globin genes) are known to be present on different isochores in mammals and birds (Bernardi et al., 1985), but... they should represent a minority of genes" (Bernardi, 1993b).

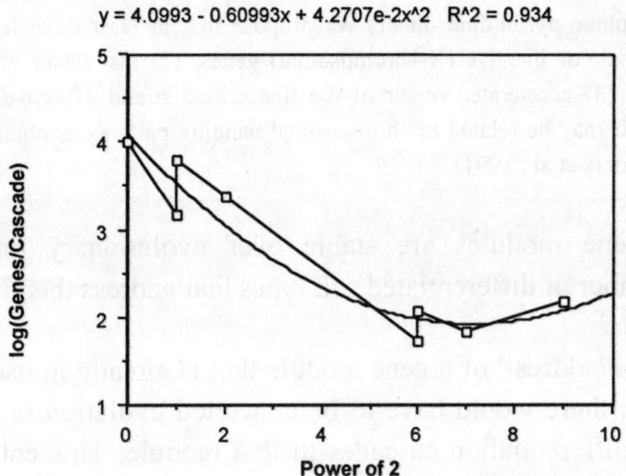

Figure 36. Genes per cascade vs number of kinds of cells. Plot of the logarithm of the number of genes per cascade versus the power of two of the number of terminal cell types (liberally interpreted from Bonner, 1988). The points are from Table 3. The curved line is the best quadratic fit.

Table 3: Estimated numbers of genes per differentiation cascade. Most of the estimates of gene numbers are from John & Miklos (1988), and some are from Li & Graur (1991). The human estimate is from Fields et al. (1994). Note that the number of genes per cascade should be roughly halved were we to take into account intermediate cell types for the animals in this list. Also, no correction is made here for genes in "housekeeping" modules. Why *Paramecium* should need more genes than *Drosophila* is a bit mysterious. It could be that enzymes in a multicellular organism can have multiple functions that can occur simultaneously due to compartmentation of the organism into cells, whereas a single celled organism must have different enzymes for each enzymatic function (N. K. Björklund, p.c.). The data above give the impression that the large number of genes in single celled organisms may have constituted an impetus to the evolution of multicellular organisms. For some reason multicellularity allowed an organism to function with fewer genes per cell type. Once this was achieved, some increase in genes per cascade became possible.

Species	Cell Types	Genes	Genes/Cascade
Paramecium	1	10,000	10,000
yeast	2	3,000	1,500
Dictyostelium (slime mold)	2	12,500	6,200
Aspergillus (filamentous fungus)	3	7,000	2,300
Drosophila	55	7,000	55
Strongylocentrotus (sea urchin)	55	15,000	120
Xenopus	120	18,000	70
Homo	500	64,000	130

The latter property may be important precisely to permit the simple kind of terminal branch duplication indicated in Figures 24-26: if the genes in a differentiation cascade were distributed all over the genome, terminal branch duplication would be nearly impossible, and evolutionary radiation would effectively cease. It is thus important to pay attention to Table 4 of correlations for GC-rich and GC-poor isochores. If the GC-poor isochores are found to closely correspond to edges of the differentiation tree, then that level of genome organization has become visible:

"It should be stressed that isochores correspond to a size range comprised between those of genes and chromosomal bands [cf. Saccone et al., 1993]. This size range is poorly known, and yet it is very important for understanding genome organization" (Bernardi, 1993b).

Table 4: Isochore correlations (cf. Bernardi, 1989, 1993b).

Isochore characteristic	GC-rich	GC-poor
fraction of genome	35%	60%
gene concentration	20-76%	4%
in chromosome	T-bands	G-bands
CpG islands	more	fewer
transcription rate	high	low
recombination rate	high	low
replication	early	late
distance from telomere	close	far
association with nuclear matrix	tight	loose
interaction with nuclear envelope	yes	no
buoyant density	higher	lower
Gly, Ala, Pro, Arg in protein	abundant	rare
thermodynamic stability of protein	higher	lower
codon variety	less	more
phylogeny	neogenome	paleogenome
function	housekeeping	tissue specific
part of differentiation tree	no	yes

I thus predict that a hierarchical organization of the GC-poor isochores, corresponding to the differentiation tree, may be found. The first eukaryotic chromosome to be completely sequenced gives hints of such a hierarchical organization (though yeast has only two cell types):

"DNA helical stability, intrinsic curvature and sequence complexity have been calculated for the complete [yeast] chromosome [III]. These features are compartmentalized at different levels of organisation. Compartmentalization of thermal stability is observed from the level delineating coding/noncoding sequences, to higher levels of organisation which correspond to regions varying in G + C content" (King, 1993a).

Hierarchical structure in *Drosophila* chromosomes has also been found:

"Using high-voltage and conventional transmission electron microscopy combined with axial tomography [cf. Bender, Bellman & Gordon, 1970; Gordon, Herman & Johnson, 1975; Guan & Gordon, 1994] and digital contrast-enhancement techniques, we have for the first time visualized significant structural detail within minimally perturbed mitotic chromosomes. Chromosomes prepared by several different preparative procedures showed a consistent size hierarchy of discrete chromatin structural domains with cross-sectional diameters of 120, 240, 400-500, and 800-1,000 Å. In fully condensed, metaphase-arrested chromosomes, there is evidence for even larger-scale structural organization in the range of 1,300-3,000 Å size" (Belmont, Sedat & Agard, 1987).

If the correlations of Table 4 hold up, the differentiation tree may actually be a physically distinguishable portion of the genome:

"The correlations constitute a genomic code (Bernardi, 1990, 1993a), namely a set of rules which are obeyed by the nucleotide sequences forming the genomes" (Bernardi, 1993b).

The distinguishability of the differentiation tree from the rest of the genome may rest on relative, rather than absolute physical properties:

"...Under conditions in which only single-copy sequence can hybridize, the GC-richest human isochores show homology not only with the GC-richest isochores of other mammals and birds, but also with the GC-richest isochores of cold-blooded vertebrates (the latter being obviously much lower in GC than the former). This indicates that the non-uniform gene distribution found in the human genome is also present in cold-blooded vertebrates.... The compositional transition which took place between reptiles and mammals or birds appears to have affected a pre-existing gene distribution pattern which was not basically modified by the compositional changes" (Bernardi, 1993b).

I would interpret this to imply that differentiation trees (roughly corresponding to the complementary GC-poor isochores) of all vertebrates are fundamentally similar.

Proposition 147: conserved autosomal segments approximately correspond to the edges of the differentiation tree.

As of 1993 "Seventy-five conserved autosomal segments between human and mouse genomes have been observed so far, some of considerable length" (O'Brien et al., 1993), up from 27 in 1984 (Buckle et al., 1984; cf. Nadeau & Taylor, 1984; Searle et al., 1987, 1989, and Weber, Schempp & Wiesner, 1986, for analogous correspondences in the sex chromosomes). This number is approaching the order of magnitude of the number of tissue types in vertebrates of 120 to 500 (Bonner, 1988). There are two considerations that would cause deviation from a one-to-one correspondence. First, if two edges happened to move together (or stay put) along the DNA in the course of evolution from the common ancestor of people and mice, they would form only one conserved autosomal segment. Second, terminal additions and prunings of the diverging differentiation trees might produce a number of tissues that simply don't correspond. Thus it would appear that the number of conserved autosomal segments will always be an underestimate (biased estimator) of the number of edges of the differentiation tree.

As with the GC-poor isochores, it seems reasonable to suggest this correspondence and that it has a relationship to the ability of a taxon to radiate. Of course, this would also suggest a correlation between GC-poor isochores and conserved autosomal segments, which remains to be explored. The reason for the existence of conserved autosomal segments has been a mystery (Phyllis J. McAlpine, p.c.). Perhaps the concept of differentiation trees and their differentiation cascades can start to unravel the higher order structure of chromosomes and the genome (cf. Filipski et al., 1990).

Proposition 148: gene modules are addressed by diffusible trans-acting molecules.

This is anticipated, since whenever a gene module is used by more than one differentiation cascade, it is likely to by physically isolated from all, or at least all but one, of them. For a review of trans-regulators, see Davidson (1986).

Proposition 149: differentiation trees and gene modules may provide a mechanistic basis for biological homology.

By distinguishing differentiation cascades from gene modules containing housekeeping genes, we can see the former as a basis for the concept of 'biological homology':

"Several attempts have been made to formulate a more inclusive concept that might be called *biological homology.* Common to all these attempts is their reference to some kind of developmental mechanism to explain the conservation of morphological patterns (Riedl, 1977, 1978; Roth, 1984, 1988; Van Valen, 1982a; Wagner, 1986, 1989b). Variability of developmental pathways forces us to be more specific in defining the relevant developmental factors. The most probable candidates are developmental constraints caused by self-regulatory mechanisms of morphogenesis acting within the organ primordium (Baltzer, 1950b; Spemann, 1915; Wagner, 1986, 1989c). Other developmental factors, like the origin of cells and inductive stimuli appear to be irrelevant.... Even though biological homology seems feasible from what is known today, its mechanistic explanation is still a challenge" (Wagner, 1989b).

(Cf. Hall, 1994a; Minelli & Schram, 1994; Minelli, 1996b.) The new field of behavioral homology (Wenzel, 1992; Greene, 1994; cf. Ekstig, 1994) may someday provide evidence that homologous behavior patterns (innate, or the capacity to learn) correspond to brain portions of the differentiation tree. The numerous hints of homologies at the molecular level (limbs of vertebrates and invertebrates: Fietz et al., 1994; diverse modes of gastrulation: De Robertis et al., 1994) might best be organized by constructing the differentiation trees of the organisms being compared. The homologous 'building blocks' of Wagner (1995) might best be seen as edges of the differentiation tree, or terminal branches. Homologies constructed at the level of differentiation trees meet the objections raised by Bolker & Raff (1996) to molecular approaches, and conform to their general definition:

"...We define homologous features as those that share a common evolutionary origin and an underlying commonality of structure, resulting from a continuity of information (Raff, 1996; Van Valen, 1982a)" (Bolker & Raff, 1996).

Proposition 150: a gene in a differentiation cascade will be activated as a subtree of the differentiation tree. It will be expressed as a subgraph of the subtree.

Let us concentrate on a gene that is activated and expressed for the first time in a given differentiation cascade, i.e., as a member of a particular edge of a differentiation tree. It has three possible fates when the cell (or its descendants) participate in the next differentiation wave that encounters them: 1) it can remain activated and expressed; 2) it can remain activated but be turned off; 3) it can be deactivated. Focussing on the activation pattern, we see that activation of the gene can be represented as a subtree of the differentiation tree (Figure 37). Here is an example of turning genes on and then off, which could be interpreted in terms of subtrees of the differentiation tree, instead of 'shifting' of expression boundaries:

"Although the boundary between *Pax-3* and *Pax-6* expression is located initially in the dorsal half of the spinal cord, it is subsequently shifted ventrally until an apparent equilibrium is reached midway between the roof plate and floor plate (Goulding, Lumsden & Gruss, 1993)" (Noll, 1993).

Another example is: "The *cyclops* gene [which]... has both positive and negative effects on the CNS of the wild-type [zebrafish] embryo" (Allende & Weinberg, 1994).

The expression graph is a subgraph of the activation subtree (Figure 37). In some cases, perhaps most, the expression subgraph will be equivalent to the activation subtree, so we may speak of an 'expression subtree'. One caution is needed: isozymes may have different subtrees, but if they are not distinguished from one another, their composite expression may not show up as a subtree.

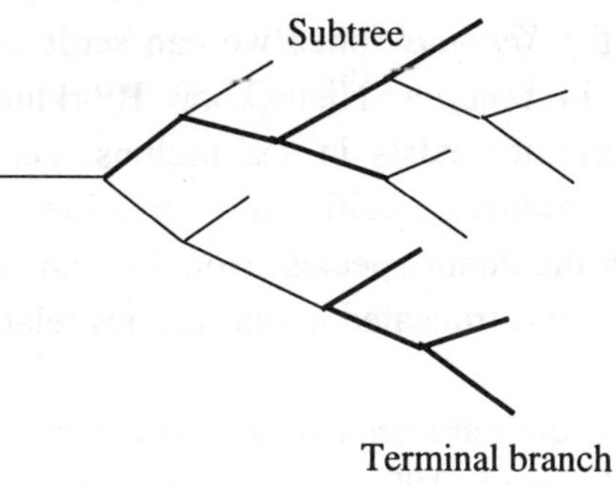

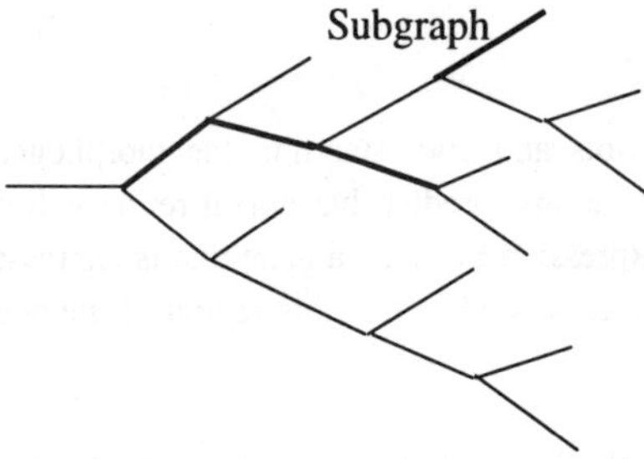

Figure 37. A differentiation tree showing a terminal branch and a subtree that is not a terminal branch. A terminal branch is a subtree ending in all terminal edges. Gene expression is likely to follow subtrees, because a gene may be activated and turned on at one edge, and turned off in various later, derived cells, at different steps of differentiation. However, to be more precise, we anticipate that it is gene activation that follows subtrees. Gene expression itself will follow a subgraph of a subtree. The gene expression subgraph need not be a connected subtree itself.

Unfortunately, most amphibian work on spatial and temporal expression of genes has been on *Xenopus,* so that we could only work by homology to check if expression always follows subtrees in the axolotl, for which we have obtained the early portion of the differentiation tree. This proposition is a strong prediction of the differentiation tree model, perhaps well worth someone's efforts to test, either by mapping gene expression in urodeles (cf. Signoret, 1989; Busse & Seguin, 1993a,b), or working out the differentiation tree for *Xenopus* (once we can settle on what constitutes a differentiation wave in *Xenopus:* Nieuwkoop, Björklund & Gordon, 1996). An analogous opportunity exists in sea urchins: gene transfer from one species to another produces additional expression (Franks et al., 1988) which is inhibited in the donor species (Hough-Evans et al., 1990), perhaps indicating a subtree that is truncated in one species relative to the other.

Sometimes new techniques are needed to ensure that all tissues expressing a gene are found (Zeller et al., 1992). Eagleson, Ferreiro & Harris (1995), in working out the relationship between placodes and gene expression patterns, express a caution that may in effect be the beginning of direct observations of subtrees:

"From following the dye spots and understanding the morphogenetic movements of the [*Xenopus*] forebrain better, we now predict that dorsal retina will have a more posterior as well as lateral origin. The expression of *Xbar,* a gene that is expressed at the tailbud stage in the dorsal-most retina is first expressed in just this region of the neural plate (Papalopulu & Kintner, 1994).

"One might wonder how well other early forebrain markers that have recently been cloned predict fate. Many of these, such as *distalless-3* and *-4, Pax6,* and *Xbf1,* are expressed in relatively simple patterns in the anterior neural plate and ridge as early as [the] stage 14 embryo [cf. Figure 4; Table 1] (Papalopulu & Kintner, 1994; N. Papalopulu, p.c.). The later expression patterns of these genes, however, appear more complex. It is our opinion that how this complexity is generated may be better understood by studying the relative movements of cells during forebrain morphogenesis. Indeed, our initial attempts to map the expression of these genes onto our fate map in detail suggests that much but not all of the later expression patterns can be explained by cell movements (N. Papalopulu & W.H. Harris, unpublished observations). Other factors, such as modification of gene expression, are likely to play important roles in the generation of the later pattern. Therefore, even though there may be a correspondence, one should be cautious in using such gene expression patterns as... fate map tools" (Eagleson, Ferreiro & Harris, 1995).

One simple place to look for whether or not gene expression is organized according to subtrees is the nematode, in which: "Many of the expression patterns involve more than one cell type" (Young & Hope, 1993).

Turning off of a gene may be just as important as turning on. For example, the homeobox containing gene XANF-1 is turned on during gastrulation in *Xenopus,* perhaps appearing everywhere on the dorsal surface. At the neural plate stage, when the anterior/posterior differentiation occurs (Sharpe et al., 1987; Sharpe & Gurdon, 1990; Zaraisky et al., 1991, 1992), perhaps due to a wave passing through part of the neural plate subsequent to the ectoderm contraction wave, the XANF-1 mRNA is found only "in the anteriormost part of the neural plate" (Zaraisky et al., 1991). Other proteins, such as the gap junction protein connexin43, while changing in their tissue pattern over the course of embryogenesis, may not change in expression during a specific step of differentiation, such as ectoderm $\Rightarrow$ neural plate + epidermis:

"By about 7.5 to 8 dpc [days post coital, in mouse], fluorescent puncta denoting the presence of Cx43-type gap junctions are seen in all three germ layers. They are increased both in size and number throughout the ectoderm. Part of the latter has thickened to form the neural plate which rolls up to form the neural tube. The labeled gap junctions, however, appear to be homogeneously distributed within the ectoderm" (Yancey, Biswal & Revel, 1992).

Thus the active genes from one differentiation cascade may continue to be active for a few of the subsequent differentiation cascades further down the differentiation tree, and then possibly be turned off many edges after they were activated. Therefore the expression subtree for a gene need not continue along any or all edges to the terminal edges.

The spatiotemporal expression patterns of genes selected by gene trapping or targeted inactivation by homologous recombination (Joyner, Auerbach & Skarnes, 1992; Meisler, 1992) and gene targeting (Joyner & Guillemot, 1994) may test the idea of subtrees.

Proposition 151: the expression patterns of assorted genes, and mutated genes, insofar as they occur as subtrees, could be used to work out the differentiation tree.

The expression pattern of a particular gene in an individual organism at a given moment in development shows in which tissues that gene is functioning. On the assumption that the tissues are related as a subtree, some idea of the structure of that subtree is discernible. If another gene has overlapping expression, then the two subtrees could be fit together to form a larger subtree of the whole (albeit unknown) differentiation tree. Repetition of this procedure for enough genes expressed as subtrees should eventually lead to the whole differentiation tree of the organism. The reconstruction of differentiation trees from overlapping subtrees would be analogous to techniques of sequencing proteins and nucleic acids from overlapping fragments (Holley, 1968; Lewin, 1970).

Duplicated genes may sometimes be or appear to be identical, because of equivalent functioning. However, a mutant in one that is deleterious can identify a subtree, and thus be useful for reconstructing the differentiation tree:

"The gene which is mutated in cystic fibrosis is believed to be a defect in a particular transmembrane conductance molecule, with sites for a protein kinase A and C, located on Chromosome 7 (reviewed in Watson et al., 1992). The mutation causes problems in a variety of functions, specifically digestion and breathing, and thus tissues. We assume that at each stage of development, an entirely new set of all the proteins that are required to regulate the cytoskeleton and send signals to the nucleus is produced from new gene sets. This would explain why there are so many isozymes of various common proteins. A mutation of any specific protein would indicate that this particular protein comes into play at a specific stage of development in all tissues when they are of the same type. Same type is defined according to their sequence of expansion and contraction waves and not necessarily by their spatial locations. We could use the cystic fibrosis defect, and others like it, to map the differentiation tree. Since this particular gene comes into play for the lining of the gut and lungs we could speculate that these linings all derive from a common earlier embryonic tissue type probably endodermal in origin. We could further speculate that in mapping the differentiation tree, we will be able to go to Chromosome 7 as a likely starting place to observe this relatively late endodermal differentiation. Other important genes may lie within the same locus. (There is a pseudogene in this region that might tell us something of the evolution of this step of differentiation for example.) Furthermore, we should also be able to explain syndromes with many symptoms attributable to a single defect as those that occur early in a differentiating cell lineage while those defects which occur in isolated tissues with less broad effects [would be explained] as occurring late in the differentiation tree. All these defects could well prove invaluable when we begin mapping the human differentiation tree as they will serve as guideposts for organisation" (Natalie K. Björklund, p.c.).

The idea of this proposition is further elaborated as the Genome Expression Map (GEM) in Section 10.12.

Proposition 152: viruses attack tissues related as subtrees of the differentiation tree.

There is, in general, a fairly narrow spectrum of tissues that a virus attacks. For example, "viral infection can activate the same cell death pathway as is used during normal development" (Sugimoto, Friesen & Rothman, 1994). Since viruses generally bind to specific receptors (Stanier et al., 1986), insofar as those receptors are not passed between cells other than by cell division, and those receptors are produced by specific differentiation cascades, this proposition follows.

Proposition 153: virus attacks on developing organisms are, in general, more devastating than attacks on adults, because more tissues of the target subtree become available in the embryo, fetus or larva.

An adult organism consists of a few stem cell types and terminal branches of the differentiation tree. An embryo, fetus or larva has or will have all of the tissues of the subtree targeted by the virus (suggested by Natalie K. Björklund).

Proposition 154: birth defects and congenital mental anomalies not involving housekeeping genes may be traced back to differentiation wave defects, i.e., defective launching, propagation, or transduction of a differentiation wave.

"It may be, then, that previous reports of phantom [limb] experiences were rejected in part because there was no conceptual framework to make sense of the data.... The idea of an innate structure for the neural basis of the phantom was, therefore, not considered. Yet this is precisely where the data point" (Melzack et al., 1997).

There are many complex, multiple birth defect associations (Kaplan, 1985; Shprintzen et al., 1985; Evans, Vitez & Czeizel, 1993; Goldberg et al., 1993), such as the...

"... VACTERL association (Quan & Smith, 1973; Khoury et al., 1983; Czeizel & Ludanyi, 1984; Weaver, Mapstone & Yu, 1986) [which] is a nonrandom clustering of vertebral, anal, cardiovascular, tracheo-esophageal, renal, and limb defects..." (Evans et al., 1989).

It has been speculated that these birth defect associations are "derivatives of nonspecific teratologic events acting on developmental fields" (Lubinsky, 1986; cf. Opitz & Gilbert, 1982; Martínez-Frías, 1994a,b; Martínez-Frías & Urioste, 1994) that sometimes do or do not have a genetic basis (Czeizel & Ludanyi, 1985). "A common denominator of the VACTERL association is suggested to be a defective mesodermal development during embryogenesis" (Khoury et al., 1983). An investigation of whether or not simple insults occurring at specific points in the differentiation tree are the basis for multiple birth defect associations may have to await animal models (Matsubara, Yamanaka & Suzuki, 1984; Mulley & Edwards, 1984; Johnson & Gentz, 1990; Levine & Muenke, 1991; de Michelena & Stachurska, 1993; cf. references in Evans, Vitez & Czeizel, 1992). The involvement of aberrant trajectories of differentiation waves is hinted in the existence of caudal duplications (Dominguez et al., 1993), sometimes unilateral penetrance (Jackson et al., 1990; Syed et al., 1990; Evans, Vitez & Czeizel, 1994), and discordant twins (Knudtzon, Jordheim & Smedsrud, 1986). This could also be the reason that: "Twinning was positively associated with NTD [neural tube defects]" (Källén et al., 1994). Some mutants, such as the deaf, circling *kreisler* mouse, lack two hindbrain segments (rhombomeres r5 and r6: McKay et al., 1994), which could be due to missing differentiation waves that failed to further divide the r4 tissue:

"An interpretation is that the cells that would normally become specified at an early stage as r5 and r6 adopt an r4 character instead, producing an excess of r4 cells that is disposed of subsequently by cell death" (McKay et al., 1994).

The earlier the differentiation wave defect, the more subscquent tissues that would be affected. On the other hand, if a wave launches and propagates correctly, but is not transduced, as in the *Drosophila shibire* mutant (Raff & Kaufman, 1983), and that failure of transduction did not occur in subsequent tissues, the development of those tissues would depend on the need for the genes expressed (or turned off) in the failed step.

We have observed a curious inducible defect in axolotl embryos compressed during a portion or all of gastrulation and then released and allowed to develop to hatching. Embryos so treated develop an upward arching of the back or kurtosis (Björklund, Gordon & Martin, 1991). Colchicine, a microtubule inhibitor, causes the same defect (John Parkinson, p.c., 1996). The same ('U shape') defect also occurs in *Xenopus laevis* embryos grown in a high osmolarity solution containing urea (del Pino, Alcocer & Grunz, 1994) (and, anecdotally, in a spawning of one of our old female axolotls). It is plausible that in these cases the trajectory of the ectoderm contraction wave has been altered, which would be easy to determine by time lapse microscopy.

We can also begin to speculate on whether specific, perhaps congenital mental anomalies, such as dyslexia (J. Hinshelwood in Benton & Pearl, 1978), autism (Courchesne, Press & Yeung-Courchesne, 1993; Courchesne et al., 1994; Courchesne, Townsend & Saitoh, 1994), and schizophrenia (Gershon & Rieder, 1992), or localized effects of teratogens (Yasuda et al., 1989), Down syndrome (Coyle et al., 1991), or domestication (Kruska, 1988), are due to incomplete unfolding of the brain's portion of the differentiation tree via differentiation wave defects, or insufficient or excessive trajectories of differentiation waves. Large brain sizes in autism (Piven et al., 1995; Wu, 1995), behavioral deficits in transgenic mice with extra growth hormone genes (Kajiura & Rollo, 1994; Lachmansingh & Rollo, 1994; cf. Shea, Hammer & Brinster, 1987; Shea et al., 1990), and growth hormone effects on the grasp reflex in human infants (Tan, 1994), may be correlated with effects on differentiation wave trajectories. On the positive side, it will be interesting to determine to what extent the evolution of the brain is a function of changing allometry (Deacon, 1988) versus elaboration of the differentiation tree:

"Evolution of the telencephalon was understood as a process which exploited the neural structures - cell groups and their connecting fiber tracts - laid down during earlier stages of evolution. Telencephalization involves selective expansion and elaboration of the front end of the neural tube" (Jacobson, 1991b).

Programmed cell death, most of it by 'apoptosis', some by necrosis (Alison & Sarraf, 1992, 1995; van der Hoeven et al., 1994), is now regarded as a terminal step of differentiation (Ellis & Horvitz, 1986; Tomei & Cope, 1994; which if reversed might "uncover ancestral cell fates": Fitch & Thomas, 1997). In schizophrenia less apoptosis occurs in prefrontal cortex (cf. Crow, 1984; Woody et al., 1987; Goldman-Rakic, 1992; Bookstein, 1996; DeQuardo et al., 1996), which further suggests that schizophrenia may be caused by incomplete playing out of the differentiation tree (Murray B. Stein, p.c.):

"A large body of research literature has demonstrated that schizophrenia is very much a physical illness.... More than 50 studies have demonstrated that patients with schizophrenia may have abnormalities in brain structure... on CT [computed tomography] scans.... The presence of these abnormalities suggests that nerve tissue either has been lost or has failed to grow normally; the bulk of the current evidence suggests that the abnormality may be neurodevelopmental..., because... the disease process in the brain antedates the onset of symptoms.... Since both [of identical] twins have schizophrenia approximately 40 percent of the time, we learn that schizophrenia has an important genetic component; since 60 percent of identical twins do not share schizophrenia, however, we also learn that genetics alone cannot explain the cause of schizophrenia.... Various factors that could serve as environmental causes include viral infections, toxins, and birth injuries" (Backlar, 1994).

"There are indications that *something* is going on in the developing brain of some people who later develop schizophrenia, but no one factor appears to be clearly associated. The pathology observed on the MRIs [magnetic resonance images] includes some dilatation of the cerebral ventricles, as well as some loss of volume of the hippocampus - amygdala complex. Both of these changes are nonspecific and, at least the latter, most likely take place during early stages of brain development.... The developmental theory postulates that any one of a number of etiological agents could initiate the causative cascade leading to schizophrenia *if* that agent affected the brain at a crucial stage of development" (A.D. D'Velupmoni in Torrey et al., 1994).

Suggestions of 'faulty wiring' in psychopathy (Hare, 1993), including reduced brain lateralization (Hare & McPherson, 1984; Hare & Jutai, 1988; Raine et al., 1990), may warrant being looked at from the point of view of uncompleted terminal branches for the brain.

The lack of a cardiovascular phase relationship to the event-related potential of the right hemisphere of the brain in schizophrenic patients (Sandman et al., 1992) may be due to truncation of the differentiation tree. Recent work has shown that the thalamus is smaller in schizophrenic brains (Andreasen et al., 1994). "The underlying cause of this and other abnormalities in schizophrenia would still be mysterious, however" (Taubes, 1994). The idea that schizophrenia is due to an incomplete differentiation tree is consistent with its genetic basis, affecting 0.8% of the human population, and perhaps with the incomplete penetrance of the gene (Gelernter & Kidd, 1991).

One curious effect in schizophrenia seems at first a contradiction of the idea that development can be analyzed in terms of differentiation waves: the direction of scalp hair whorls is more often counterclockwise than normals (Alexander et al., 1992). This suggests a more pleiotropic effect, rather than simple loss of a terminal branch of the brain portion of the differentiation tree. It may just represent a local situs inversus effect. On such small discrepancies, our whole theory could stand, fall, or (more likely) undergo revision. On the other hand, at least in the brain, schizophrenia has been attributed to "defects in [a]... single, complex brain circuit (Nancy C. Andreason in: Bower, 1996a; cf. Andreasen et al., 1996), suggesting a possible origin in a single defective placode and differentiation wave. Schizophrenia may be just one example of a broad class of birth defects:

"Developmental disorders of the brain do not always manifest themselves as morphological abnormalities at birth, but will often later bring out serious functional or behavioral defects in postnatal life" (Kameyama, 1985).

Some mental phenomena have tentatively been assigned to 'mind modules' which might correspond to placodes defined by differentiation waves:

"...Alan Leslie (1992) and others have developed evidence for... a 'theory of mind module' designed to generate second-order beliefs (beliefs about the beliefs and other mental states of others). Some autistic children seem to be well described as suffering from the disabling [or lack?] of this module, for which they can occasionally make interesting compensatory adjustments" (Dennett, 1995).

The experience of phantom limbs in congenitally limb-deficient people (Melzack et al., 1997) might be an intriguing example of mind modules.

As a final example, let's reconsider the common neural tube defects of spina bifida and anencephaly. It has recently been suggested that these defects do not occur at random along the axis, but rather present as segmented, 'multi-site' anomalies (Van Allen et al., 1993; Proposition 87). I would like to suggest that these localized failures of neural tube closure reflect failure of launching or propagation of specific neural plate differentiation waves. One prediction is then that histogenesis might stop in the affected region. Natalie K. Björklund (p.c., 1997) has interpreted the results of Moephuli et al. (1997) on the correlation of loss of apical cytoskeletal proteins in rat neural plate stage embryos with loss of differentiation in precisely this way: cell state splitters are affected when there is insufficient S-adenosyl methionine to methylate the actin, tubulin and intermediate filaments in the apical cell state splitter. This result, being obtained on late neural plate stages, is consistent with the idea that differentiation waves may indeed be propagating in the neural plate.

The segmentation of the neural plate (neuromery) is nicely illustrated in Figure 1.5 in Jacobson (1991b), based on von Kupffer (1906). One might see another contradiction in...

"Detwiler's (1936) experiments... [which] provide the best evidence that the segmentation of the spinal cord and peripheral nerves is secondary to mesodermal segmentation and that neither the neural tube nor the neural crest is intrinsically segmented. These conclusions were reached as a result of experimental grafting of parts of the spinal cord, grafting extra somites, and removal of somites. In all cases the spatial pattern of spinal nerves and ganglia corresponded with the segmentation of adjacent mesoderm" (Jacobson, 1991b).

However, the mesoderm may well be involved in determining the launching domains and trajectories of the waves of the neural plate. On the other hand, these results are contradicted by Källén (1956b):

"In Källén (1956a) it was demonstrated that the presence of neuromeres in the brain of *Ambystoma* embryos depends on the presence of surrounding mesoderm, especially the underlying notochordal structures. It was also shown that this effect is not a regionally specific one. If the mesoderm under a section of the neural tube was removed, neuromeres nevertheless developed in a normal way in this region, as long as mesodermal structures were left intact rostrally and caudally. It was then concluded, that the pattern of mitoses in the neural tube, being the basis of the neuromeres, was not the result of a 'printing' of a similar pattern in the mesoderm; the neural segmentation was not secondary to the mesodermal segmentation" (Källén, 1956b).

We see, in general, that defects in differentiation trees, i.e., 'epigenetic' errors in the propagation of differentiation waves, may be the root causes of many birth defects.

Proposition 155: childhood neuroectodermal tumors may be aberrant offshoots of the brain's subtree of the differentiation tree.

Trojanowski, Tohyama & Lee (1992) have found considerable specificity, in terms of intermediate filaments and microtubule associated proteins, in pediatric CNS (central nervous system) tumors. As the differentiation subtree for the CNS is worked out, we may start to understand when and how specific childhood tumors get started. Certainly, the new information needed is precisely what one would seek in working out the CNS subtree:

"Recent advances in understanding the molecular mechanisms of oncogenesis suggest that the genetic abnormalities leading to neoplastic transformation and tumor progression may be specific for cells of a given lineage and/or maturational state (... Cross & Dexter, 1991...).... Increments in our understanding of brain tumors have been... modest. In part, this reflects a lack of specific information on the molecular basis of neural development and cellular heterogeneity in the nervous system.... Several neuron-specific genes appeared very early in neurogenesis, and... the appearance of some of these proteins, such as NF [neurofilament] subunits, is one of the earliest molecular signals that a precursor cell has begun to 'turn on' a select group of neuron-specific genes, the products of which define and/or stabilize the mature neuronal phenotype for the life-time of the organism.... Some PNETs [primitive neuroectodermal tumors]... may be comprised of neoplastic cells with a molecular phenotype that is quite similar to neuroectodermal stem cells.... It is interesting to speculate that neoplastic transformation of the precursors of these tumors reflects the arrest of these neoplastic cells at specific developmental stages, and that this renders these cells capable of continued cell division and progression to increasing levels of malignancy" (Trojanowski, Tohyama & Lee, 1992).

Thus pediatric tumors may simply represent tissues that failed to launch or propagate a differentiation wave. Wilms' tumor may have a similar cause:

"Wilms' tumour is a childhood embryonal kidney tumour. It is believed to result from malignant transformation of abnormally persistent renal stem cells, which retain embryonic differentiation potentials" (Pritchard-Jones & Hastie, 1990).

It should be possible to prevent launching of specific waves by using cytoskeletal inhibitors topically applied to specific tissues, or heat shock, or even careful surgery. It will be interesting to note if some tissues so prevented from continuing to their terminal (adult) cell types become tumors and develop the chromosomal anomalies, etc., typical of malignancies. An epidemiological link to fetal stages has been found for breast and prostate cancer (Fackelmann, 1997). One might wonder whether competent cells that somehow miss a differentiation wave are candidates for later tumor development.

Childhood cancers may sometimes be the result of chromosome anomalies affecting specific subtrees of the differentiation tree. We may interpret the gene expression pattern of aberrant genes as subtrees. This viewpoint may shed some further light on the etiology, for example, of Wilms' tumor:

"Wilms' tumour or nephroblastoma is one of the commonest solid paediatric malignant diseases, accounting for 8% of childhood cancers. The tumour arises through aberrant differentiation of metanephric mesenchyme and thus represents a paradigm for the relationship of cancer and development.... The recently cloned Wilms' tumour gene encodes a putative transcription factor which is likely to activate or repress the expression of other genes in kidney development. We have shown by *in situ* hybridization that expression of this gene is restricted to specific cell types within the developing kidney. It is also expressed in a limited range of embryonic tissues, including the gonad, spleen and mesothelium. With the benefit of this new information, we speculate on the part played by this gene in normal kidney development, in tumorigenesis and in other aspects of Wilms' tumour; these include associated congenital abnormalities, genetic predisposition to second tumours and inheritance of Wilms' tumour" (Pritchard-Jones & Hastie, 1990).

The connection with specific chromosome abnormalities has been made for pediatric central nervous system (CNS) tumors:

"The single most commonly observed abnormality involved chromosome 6 in the form of deletions, translocations, monosomies, and trisomies" (Neumann et al., 1993).

I speculate that these large chromosomal defects disrupt a portion of the differentiation tree for those cells in which the anomaly is found. The affected subtree would extend to its terminal cell types, though development may stop well before they are reached. This may be why "CNS tumors in children differ in many ways from those in adults" (Neumann et al., 1993).

A possible contradiction of this proposal is the link between overripeness of unfertilized oocytes in frogs (leading to what seems to be a cortical or cell adhesion, rather than a chromosomal defect), and embryonic neoplasms (Witschi, 1952; cf. Briggs, 1941; Needham, 1950). On the whole, childhood tumors may have their origins in embryonic development, which implies the possibility of a significant role for differentiation waves.

Proposition 156: some genes and products on duplicated terminal branches may become vestigial, because they are not needed on the new copy of the terminal branch.

This may explain why in *Drosophila* "Several of the genes show an expansion in expression during later stages, resulting in the appearance of homeotic products in tissues where they serve no obvious developmental function" (Levine & Harding, 1989). The same thing could happen to gene modules, either when they are duplicated, or unnecessary accesses to them are duplicated.

Proposition 157: transfer of a whole terminal branch of a differentiation tree, involved in formation of a specific organ or structure, from one part of the tree to another, may explain the homology between structures in different taxa that have different embryological origins in terms of cell lineage.

Homology presents a major challenge to any developmental theory of evolution, since...

"As De Beer points out, structures as obviously homologous as the vertebrate alimentary canal are formed from quite different embryological sites in different vertebrate classes. The alimentary canal is formed from the *roof of the embryonic gut* cavity in sharks, from *the floor* in the lamprey, from *roof and floor* in frogs, and from the *lower layer of the embryonic disc,* the blastoderm, in birds and reptiles (de Beer, 1971)" (Denton, 1986).

"It does not seem to matter where in the egg or the embryo the living substance out of which homologous organs are formed comes from. Therefore, correspondence between homologous structures cannot be pressed back to similarity of position of the cells of the embryo or the parts of the egg out of which these structures are ultimately differentiated" (de Beer, 1971).

A simple solution to this problem may be available when we look at differentiation trees as the underlying structure for differentiation. Gene modules and transfer or copying of terminal branches of the differentiation tree may provide sufficient structure to unravel most problems of homology. We can thus perhaps eliminate the supposed 'failure of homology':

"Without the phenomenon of homology - the modification of similar structures to different ends - there would be little need for a theory of descent with modification.... The failure of homology to substantiate evolutionary claims has not been as widely publicised as have the problems in paleontology. Comparative embryology is a less glamorous pursuit than the biology of dinosaurs. Nonetheless, it fits into the general theme that advances in knowledge are not making it easier to reduce nature to the Darwinian Paradigm" (Denton, 1986).

A more positive view of the same problem for plants has been taken:

"I now briefly describe some results of a pilot study by Sattler & Jeune (1992) that analyzed morphological distances between a great diversity of plant structures, especially in flowering plants. In this study, principle component analysis was used.... As expected, the representative(s) of each category formed a distinct point or cluster, confirming the validity of the structural categories for the typical representatives. However, when we included atypical or controversial structures the picture changed drastically. These structures, whose homology and relationship has been discussed endlessly in the literature, occupy intermediate positions between the typical structures. As a result, we obtain a heterogeneous continuum in which the typical structures are linked to each other by intermediate atypical structures....

"In the continuum the postulated hierarchy of classical plant morphology disappears. According to this hierarchy, the plant comprises a root system consisting of root(s) and a shoot system composed of caulome(s) [stem] and phyllome(s) [leaf]. These units in turn comprise lower level units such as trichomes. If we want to recover this hierarchy, we have to restrict ourselves to the so-called typical structures, i.e., more or less arbitrary selected regions of the heterogeneous continuum. In a more global view hierarchical thinking appears inappropriate..." (Sattler, 1994).

Again, consideration of plant development from the point of view of differentiation trees might be what is needed to restore a hierarchical structure to plant homologies. While the hierarchical relationships are not to be found in morphology or even 'process morphology' (Sattler, 1994), they might be found in a comparative study of plants' differentiation trees, which may represent the underlying, more fundamental 'process morphology':

"Even when symmetry does not change during development, there is an underlying dynamic. Symmetry and any other static characteristic or character state are an abstraction from this dynamic (Hay & Mabberley, 1994). Furthermore, structure itself (not only its properties) is an abstraction from this dynamic (e.g., Chapter 7 in Woodger, 1929a). Bohm (1980) illustrated this by a vortex. Having a shape, a vortex can be seen as a structure, but first of all it is dynamic.... As long as we see the abstraction for what it is, there is no problem. But if we mistake the abstraction for the territory from which it has been abstracted, then we delude ourselves. We become victims of the 'fallacy of misplaced concreteness' (Whitehead, 1925)" (Sattler, 1994).

This problem, the failure to distinguish structure from process, has been raised by Niklas (1994a) in the context of explanations of leaf morphogenesis:

"Bird & Hoyle (1994) assert that leaf shapes are generated by 'a fractal form of logic' because they are able to simulate a variety of complex leaf shapes by their algorithms.... I am extremely uncomfortable by the sweep of this assertion.... Although no other models contend for the explanation of this global aspect of form, the biomathematician, and even more the biologist, *must* resist the temptation to reify the parameters of the algorithm on the basis of shape similarity alone.... Although the explicit modelling of shape formation is of comparatively more recent vintage than traditional experimental investigations, it is no longer so novel that it can escape tests against empirical knowledge of underlying regulatory mechanisms. Their claimed 'arithmetic' has ignored the entire botanical literature of morphogenesis. I see no reason to take their arguments seriously, no matter how closely a selection of their curves matches a selection of our leaves, until they confront more seriously the knowledge of how leaves actually grow into the forms we observe" (Niklas, 1994a).

Given the meristematic basis for leaf morphogenesis, and its obviously regulatory nature, since "...the cellular contributions made by each of the... layers to the mature leaf may vary among leaves produced by the same plant..." (Niklas, 1994a), perhaps differentiation waves are also operating here, and could provide an alternative model.

Another obvious example of the "fallacy of misplaced concreteness" is the dorsal lip of the blastopore, which we now recognize as a set of simultaneous and consecutive differentiation waves (Appendix VI: Björklund & Gordon, 1994), i.e., a process, rather than a structure (*the* 'organizer'). Homology, then, is at the level of the process of morphogenesis more than its end result. While I cannot offer a definitive solution to the problem of homology, differentiation trees offer a new perspective, and could lead to explicit investigations.

6.03 Differentiation Trees in Punctuated Equilibrium

"...If fertilization occurs, normal development may fail because of imperfect cooperation of the two heredities..." (Wright, 1982a).

"...Schindewolf (1936)... elaborates the thesis that macroevolution on a higher level takes place in an explosive way within a short geological time, followed by a slower series of orthogenetic perfections, as exemplified in the oft-quoted evolutionary series" (Goldschmidt, 1940).

Proposition 158: matings between organisms with topologically different differentiation trees (attempts at hybridization) have incompatibilities between the differentiation trees that may usually prevent development from the stage at which the incompatibility would unfold.

This proposition makes an explicit prediction about the moment in embryological development at which incompatibilities in a hybrid embryo should become apparent. It may be the basis of the "relationship between phenodeviance and developmental instability" reviewed by Williamson (1987):

"Perhaps the message of developmental biology for evolutionary studies, as already suggested by Hall (1983a), is that we should be investigating not the causation of divergence, but the mechanisms that conserve and constrain developmental pathways from the level of species stasis to the level of persistence of phyla for hundreds of millions of years. Only when the capacity to hold the status quo is understood in genetic/developmental terms will the process underlying evolutionary change over time, and the factors that govern the rates of this change, begin to be identified" (Anderson, 1987a).

If the correlation suggested by this proposition holds often enough, it may provide a genetic method for mapping out differentiation trees (cf. the subtree method in Proposition 151 and GEMs in Section 10.12).

The fact that nuclear transplant hybrids between amphibian species arrest at gastrulation or neurulation (Hennen, 1973) suggests that the underlying basis of arrest may be an inability to launch, sustain, or properly shape the ectoderm contraction wave. The last feature may be the most likely:

"When diploid blastula nuclei are transplanted between *Rana pipiens* and *Rana palustris,* the resulting nucleocytoplasmic hybrids, without exception, develop abnormally.... Thus, transfer of *R. palustris* nuclei into enucleated eggs of *R. pipiens* gives lethal embryos whose development is characterized by macrocephaly. Embryos derived from the reciprocal transfer are microcephalic. This abnormal behavior is *not* associated with cytologically detectable alterations in karyotype..." (Hennen, 1973).

The possible relationship between the size of the neural plate and the trajectory of the ectoderm contraction wave in these hybrids should be ascertainable by time lapse microscopy (at least in urodeles: Nieuwkoop, Björklund & Gordon, 1996). Since "...some die at neurula stages after forming wide neural plates and folds" (Hennen, 1973), we may have here a basis for a relationship between shaping of the ectoderm contraction wave and neural tube birth defects (cf. Nagele & Lee, 1987). Since the species source of the egg correlates with the type of defect, the cortical cytoskeleton, and thus perhaps the cell state splitters, are implicated in the incompatibility. (Since arrest does not occur at the midblastula transition, and in particular is often postneurulation, maternal genetic effects are not implicated.)

It is of interest that triploid hybrid embryos develop normally (Hennen, 1973), so that one haploid genome 'rescues' the other. Of course, we cannot be certain that the incompatibility here is due to topological incompatibilities of the differentiation trees. It may not be, since these are closely related species and arrest occurs at a relatively early embryological stage. The arrest could be due to timing (heterochronic) or other lower level genetic incompatibilities. What is important here is that these interspecific diploids and triploids may open an opportunity for genetic investigations of the basis for differentiation tree compatibilities and incompatibilities, via alterations of the 'rescuing' genome before its transplantation.

It would be curious if 'behavioral infertility' of certain hybrids has some kind of basis in differentiation tree incompatibilities, involving the brain subtree of the differentiation tree:

"Dilger (1962) carried out a comparative study of the innate releasing mechanisms of several species of parrots of the lovebird type. He then crossed lovebirds of two different species, Fischer's lovebird from Tanzania, which carries nesting material in its beak, and the peach-faced lovebird from Namibia, which carries nesting material by tucking it into its feathers. Hybrids of the two species displayed a confusing mixture of the two ways of carrying, which resulted in very little material reaching the nest site. Behavior was here contributing to the infertility that is common in hybrids" (Lea, 1984).

A measure of the validity of this proposition could come from correlating the stage of developmental arrest of spontaneous abortions with the degree of chromosomal difference between aspiring human parents with disparate chromosomes, who have high infertility rates (Makino et al., 1990). It is important to note that 10% of conceptions in these couples did result in liveborn infants, indicating that a significant fraction of chromosomal anomalies and variants are compatible with apparently normal development, despite the differences between the chromosomes of the parents. Perhaps those that come to term are chromosome hybrids of the more terminal branches of the differentiation tree.

Diploid, parthenogenetic embryos in some sense represent a null experiment, in which the differentiation trees of the two chromosome sets are presumably identical. These parthenogenones arrest at variable stages of development perhaps "due to abnormal regulation of differentiation and proliferation in both embryonic and extraembryonic lineages" (Sturm, Flannery & Pedersen, 1994), though they can be helped along to fairly advanced stages (Spindle et al., 1996). Thus, unfortunately, they do not form a normal baseline from which to measure hybrid incompatibility. Perhaps highly inbred lines would form a better baseline, since they presumably approach homozygosity for most genes.

An experimentally amenable system for testing this proposition might be found in the vast number of species and mutants of nematodes (Steve McGrew, p.c., 1997). Do greater topological differences in nematode lineage trees correspond, at least on average, to greater infertility upon attempted hybridization? When in development would they arrest?

Proposition 159: the degree of viability of hybrids is on average inversely proportional to the magnitude of the differences between the differentiation trees of the parent species.

"The chance that a mutation will be favored by selection and the chance that it will or can be integrated into a genetic system as a whole are inversely related to the effect of the mutation on the organism" (Simpson, 1949a).

There are well studied situations in which differences in the genome between mating individuals lead to their effective reproductive isolation from one another. For example, imbalances in chromosome numbers, aneuploidy and heteroploidy, etc., often lead to early termination of development (Dyban & Baranov, 1987). The study of hybrids reveals a range of levels at which reproductive isolation occurs:

"In some instances the hybrids that result are abnormal and soon die. In some cases the hybrid matures but is sterile, or incapable of reproduction. In still other instances the hybrid may mature and reproduce, but subsequent breeding with either parental species results in 'hybrid breakdown,' in which later generations of hybrids rather than the first suffer from gene incompatibility. In these instances of hybrid inviability, sterility, or breakdown there is effective reproductive isolation: little or no transfer of genes occurs between the parental types, and they are therefore distinct species even though interbreeding occurs" (Tinkle, 1982).

Analogous phenomena can occur in plant hybrids (Darlington, 1958; Grant, 1981). For an example during embryological development, here is a case for incompatibilities involving cell adhesion molecules:

"...Differential affinities, which are stable and characteristic for each species, are much impaired in hybrid embryos of amphibians (obtained by fertilizing eggs of one species with sperm of another). Gastrulation is also abnormal in these hybrids and may fail altogether. Johnson (1969) showed that there is a correlation between the extent to which the cells are able to reassort after disaggregation, and the viability of the hybrids. The more successful hybrids were able to gastrulate, and in these the disaggregated gastrula cells were also able to reaggregate and to sort out into ectoderm, mesoderm and endoderm successfully. Other hybrids which were unable to gastrulate showed an inability of their cells to reaggregate or to sort out. There were also unusually large gaps between the mesoderm cells (Johnson, 1972)" (Deuchar, 1975a).

Because N-CAM production apparently follows primary neural induction (Jacobson & Rutishauser, 1986), the reaggregation results may concern events (immediately?) after propagation of the ectoderm contraction wave, rather than its launching or propagation, a matter that could be settled by time lapse and immunofluorescence microscopy.

Species isolating mechanisms may extend down to nucleotide base composition (Forsdyke, 1995a,b,c). See Mecham (1961), Littlejohn (1969), Mayr (1970) and White (1978) for classifications of isolating mechanisms.

In the following, I interpret the "different, harmoniously balanced systems of genes" to be the differentiation trees of different species:

"Races (subspecies) of the same species differ in frequencies not of one but of several or many genes. Distinct species differ doubtless in still more genes, although in just how many cannot be determined exactly. The complexity of the species differentials is, however, demonstrable in those rather exceptional cases when fertile hybrids between species can be obtained in experiments and a second hybrid generation raised in large enough numbers and carefully studied. In such interspecific hybrid progenies an extraordinary diversity is often found; no two individuals are quite alike, and some specimens exhibit peculiar combinations of traits that render them sometimes strikingly different from both parental species and from the first generation hybrids. All this indicates that distinct species carry different, harmoniously balanced systems of genes" (Dobzhansky, 1982).

(Cf. Dobzhansky, 1937, 1951.) Mayr (1976) similarly indicates "...the gene complexes which come together are so well balanced within themselves that combinations with alien genes lead to combinations of inferior viability and are eliminated by selection". (The idea that "...only certain sets of traits could operate harmoniously together..." goes back to Georges Cuvier in 1812: Richards, 1992.) "It may well be that sexual recombination per se has mechanisms that automatically help change and stabilize different differentiation tree structures. These mechanisms could respond to mismatches between parental genome arrangements by invoking repair mechanisms that sometimes duplicate unmatched portions and sometimes trim off unmatched portions" (Steve McGrew, p.c., 1997).

This proposition refines the hypothesis of Wilson, Maxson & Sarich (1974):

"If two species have very different mechanisms for regulating gene expression during embryonic development, it is unlikely that a healthy adult hybrid organism could develop. The hypothesis that the chief molecular barriers to development of hybrid organisms are regulatory ones is consistent with observations on somatic cell hybrids as well as the phenomenon of allelic repression in organismal hybrids.... The process of embryonic development involves activation of most of the genes that were inactive in the sperm and egg (Davidson, 1968).... For successful development of an interspecific zygote, the two regulatory systems (contributed by the egg and sperm genomes) controlling the expression of such genes must be compatible" (Wilson, Maxson & Sarich, 1974).

Wilson (1992) summarizes the hybrid incompatibility problem with a hint at possible quantitation:

"When a chromosome from species A attempts to line up with its counterpart from species B, a normal procedure for the exchange of blocks of genes during the production of sex cells, the A and B chromosomes differ too much from each other to complete the maneuver" (Wilson, 1992).

A direct test of this proposition is available. Given the 0.4% human "total frequency of balanced rearrangements in the prenatal genetic studies performed with banding" (Van Dyke et al., 1983; cf. 'balanced translocations' in: Hook & Cross, 1987a, 1988; Gardner & Sutherland, 1989), it should be possible to find couples with similar chromosome rearrangements and see if this gross measure of chromosome compatibility correlates with an improvement in fertility rate over that between chromosomally disparate couples (Makino et al., 1990). The observation that specific chromosomal rearrangements appear in viable hybrid grasshoppers (Shaw, Wilkinson & Coates, 1983) may be a reflection of compatibility problems of the differentiation tree for arbitrary, random rearrangements. Hybridization rates between closely related, recently radiated species, such as wallabies (Eldridge & Close, 1993), could also provide a source of data for testing this proposition. Gamma ray (Rosenbluth, Cuddeford & Baillie, 1985), ultraviolet (Stewart, Rosenbluth & Baillie, 1991), and chemical (Rinchik, Flaherty & Russell, 1993) manipulations of chromosome rearrangement rates are available, making laboratory studies feasible. On the other hand, it will be difficult to sort out the effects of hot spots for chromosome breakage, which themselves generate nonrandom chromosome rearrangements (Eldridge & Johnston, 1993), from differentiation tree incompatibilities.

Support for this proposal comes from natural and laboratory hybrids amongst urodele species, whose differentiation trees would be relatively easy to ascertain and compare, in the parent and hybrid animals or arrested embryos:

"... Male sterility is a general finding among hybrid crosses of [newts of genus] *Triturus* (Spurwell & Callan, 1960; White, 1973; Herrero, 1991). Detailed studies of spermatogenesis indicate that the magnitude of some meiotic irregularities are proportionally correlated with taxonomic relatedness (White, 1946; Spurwell & Callan, 1960; Mancino, Matilde & Bucci-Innocenti, 1979). Other developmental anomalies are observed in hybrid crosses, including stage-specific measures of survival (Mancino, Matilde & Bucci-Innocenti, 1978; Bucci-Innocenti, Ragghianti & Mancino, 1983; Arntzen & Hedlund, 1990; Macgregor, Sessions & Arntzen, 1990), ploidy elevation (Lantz, 1947; Lantz & Callan, 1954), and sex-ratio distortion (Hamburger, 1935; Spurwell, 1953; Mancino, Matilde & Bucci-Innocenti, 1978). Some of these measures are also informative for evaluating taxonomic affinities (Macgregor, Sessions & Arntzen, 1990)" (Voss & Shaffer, 1996).

There has been a shift in evolutionary paradigm recently, away from the notion of 'reproductive isolation' and towards that of the 'specific-mate recognition system' or SMRS (specific-mate = mate of the same species: Paterson, 1993a). As for all noteworthy paradigms, it has been refuted (Coyne, Orr & Futuyma, 1988; Ryan & Wilczynski, 1991; King, 1993b) and defended (Paterson, 1993b; Goldschmidt, 1996). As an outsider to this debate, the safest position is to feign neutrality, or at least note that the matter is not yet settled. The notion of the specific-mate recognition system is attractive, so I wish to give it its due:

"True sex is found only in biparental, eukaryotic organisms.... Successful fertilization in even the simplest of unicellular eukaryotes requires the assistance of a series of adaptations which constitute what might be called the fertilization system of the organism.... This subset of adaptations, which are involved in signaling between mating partners or their cells, constitutes the specific-mate recognition system or SMRS (Paterson, 1978, 1980 [reprinted in Paterson, 1993a]). I have chosen to use 'specific-mate' instead of simply 'mate,' because it is important to distinguish between the process with which I am concerned and the quite different kind of mate recognition, 'individual mate recognition.' I also avoid the common term 'species recognition'.... [The SMRS is]... the recognition of an appropriate mating partner. ('Appropriate' here implies no more than an individual of opposite sex drawn from the same 'field for gene recombination' [Carson, 1957].)

"The response of one mating partner to a signal from the other is here regarded as an act of *recognition* (Paterson, 1982a [reprinted in Paterson, 1993a]). Recognition is thus a specific response by one partner to a specific signal from the other.... It is clear that the limits to a 'field for gene recombination' can be set by the shared fertilization system of members of a population of organisms alone.... We can, therefore, regard as a species that most inclusive population of individual biparental organisms which share a common fertilization system....

"There is no special process of cladogenesis; speciation is an incidental *effect* resulting from the adaptation of the characters of the fertilization system, among others, to a new habitat, or way of life....

"To many, the core of the evolutionary synthesis is the biological species concept. Certainly much in the synthesis is related to species theory. It follows, therefore, that a significant change in viewpoint on the genetic nature of species is likely to be an important reason to reformulate the evolutionary synthesis.... The recognition concept does constitute a new way of looking at species in genetic terms, and its adoption will thus ensure that any evolutionary synthesis built around it will be significantly new" (Paterson, 1993a).

The same could be said of the impact of differentiation trees on the evolutionary synthesis. Since changes in differentiation tree topology as a source of speciation are but specific forms of chromosomal change advocated by White (1978), "which reduces fecundity when heterozygous", the criticism of White's ideas by Paterson (1981, 1993a) should also apply to differentiation trees:

"There is no doubt that closely related species often differ from each other by fixed alternative chromosome arrangements. There certainly seems to be a correlation between speciation and the fixation event, but there is no reason to believe that the fixation of the rearrangement has caused the speciation to occur; it is more probable that a population bottleneck occurred which simultaneously favored speciation and the fixation of the rearrangement" (Paterson, 1981, 1993a).

Maybe so, but this hardly settles the matter. What is clear is that the relationship between hybrid viability and differences in differentiation trees is a fertile area for investigation. Perhaps the methods of artificial life, where hybrid inviability has been called the "'miscegenation function' (so dubbed by Richard Dawkins)" (Yaeger, 1994), could be used to explore these questions.

Proposition 160: hybrid inviability is due most often to incompatibilities between two differentiation trees at the level of intranuclear and/or intracellular interactions.

Consider the common observation of partial to normal development in interspecific chimaeric organisms (urodele amphibian + urodele amphibian: Spemann & Mangold, 1924a,b, and Raven, 1937; fish + amphibian: Oppenheimer, 1939; chick + mouse: Dhouailly, 1973; anuran amphibian + anuran amphibian: Blackler & Gecking, 1972a,b; mouse + rat: Gardner & Johnson, 1973; McLaren, 1976a; Willadsen, 1982; urodele amphibian + anuran amphibian: Slack & Forman, 1980; chick + quail: Le Douarin, 1982; assorted plant species: Ray et al., 1983 and Raven, Evert & Eichhorn, 1986), in which cells are combined without cell-cell fusion (except perhaps in syncytial tissues, such as muscle: Mintz & Baker, 1967) by various reaggregation, injection or grafting methods. The total dissociation of embryonic tissues from different species and their reaggregation as mixtures of cells often produces normal histological tissues, albeit chimaeric tissues:

"Mixtures of 7 day old chick and 15 day old mouse embryo liver cells reaggregate to form a chimaeric liver with interspersed mouse and chick cells (Moscona, 1960). Mouse and chick chondrogenic cells also reaggregate to form a single cartilage mass bounded by a common matrix. In other experiments, embryonic chick and mouse chondroblasts and chick liver cells were dispersed and mixed together. After sorting, a single chimaeric cartilage mass formed separate from the reconstituted chick liver tissue (Moscona, 1957). Thus, cells of the same histotype adhere to one another to form chimaeric tissue, whereas cells of the same genotype (chick cartilage and liver) remain separate. Clearly, both the chick and mouse cells remain stable in respect to histotype. However, one cannot generalize that cells of the same histotype but different genotype never sort out by the species of origin. Burdick & Steinberg (1969) report that chick and mouse heart ventricle cells in mixed aggregates separate after about two days. Unpublished studies of Okada (cited by Burdick & Steinberg, 1969) show that mouse and chick embryonic kidney cells also sort out by genotype" (Spratt Jr., 1971a).

These observations show a remarkable degree of 'cooperation' (or lack of 'awareness' of each other's genomes) between cells of different species when placed in juxtaposition with one another. They also indicate that incompatibility in a hybrid is probably due to nonviable interactions between the two genomes at intracellular and/or intranuclear levels, rather than at the cellular level.

Proposition 161: large topological changes in a differentiation tree may account for punctuated equilibrium.

The intervals between large successful changes to the topology of the differentiation tree are likely to be on geological time scales, given the general problems with hybrids. But they may occur. Geological time scales are enormous, and so we cannot easily dismiss rare events of small probability, because the cumulative probability of their occurrence approaches one in many cases. This struggle to come to terms intellectually with rare events can be seen in self-contradictory attitudes about whether large scale mutations are deleterious:

"Mutations of large effect are almost always deleterious due to their main effect or to pleiotropic side effects (e.g., Caspari, 1952; Grüneberg, 1963).... Direct evidence on newly arising spontaneous and induced rearrangements shows that in most higher organisms inversions, fusions, and translocations rarely produce noticeable morphological effects but often reduce the fertility of structural heterozygotes, so that cytological tests are required to confirm their existence (Lande, 1979b)" (Lande, 1980b).

The point is that sometimes large scale mutations are not so deleterious that life can't go on, and the only argument should be about exactly how often this is the case, whether a major phylogenetic event can be initiated this way, and whether the cumulative probability is sufficient to equal or surpass evolution caused by many small mutations. I don't know the answers, but it seems neither does anyone else at this time. Certainly, one could argue that, from the point of view of successful speciation, it is better for a major change at the DNA level to remain cryptic, rather than result in a 'hopeful monster', but plants and animals do not share our learned revulsions towards morphologically 'deviant' individuals. Thus whether or not a major DNA change in one individual corresponds to a major morphological change in the same individual (rather than a number of generations later, through many additional small mutations) is also an interesting but unresolved question, as much rooted in mating behavior as in chromosomal mechanics. Again, the question is whether or not we should 'allow' rare events to influence our theories of evolution:

"The popular argument that a regulatory gene affecting relative growth rates of different structures early in development can produce major discontinuous morphological change is undoubtedly correct, as many such mutations are known (Goldschmidt, 1940; Valentine & Campbell, 1975; Gould, 1977c, [1983a]; Alberch et al., 1979; Stanley, 1979). However, there are few (if any) genetically well-established cases of morphological macromutations which have been fixed in natural populations of animals" (Lande, 1980b).

Certainly, explicit examples such as the following may help clear our thinking and make time scales for change explicit, though it would be nice to have a corresponding genetic and chromosomal analysis (cf. Larson et al., 1981):

"An instructive example to consider is the evolutionary loss of limbs in reptiles. A naive observer having at hand only the fragmentary fossil record of snakes and lizards would be hard pressed to imagine that evolutionarily stable intermediate forms could exist. But the study of living teiid and scincid lizards reveals several independent genera with parallel series of many intermediate forms spanning the entire transition from typically lizardlike species with complete limbs to limbless snakelike forms.... Comparison with other genera of reptiles with fossil records suggests an average age of at least several million years. Therefore it is likely that the major morphological change to limblessness is on the whole very slow, involving many genes (Lande, 1978)" (Lande, 1980b).

Insofar as small topological changes to the differentiation tree may correspond in general (but by no means always) with small morphological differences, we see that differentiation trees are compatible with either a gradualist or a saltationist view of evolution, or anything in between. Differentiation trees may thus be the source of the compatibility that Nei (1987) sees between punctuated equilibrium and classical theory:

"...Neither Eldredge nor Gould present any convincing genetic explanation for their theory. While he is fully aware that Goldschmidt's systemic mutation is not acceptable with the current knowledge of genetics, Gould (1982b) tends to agree with Goldschmidt (1940) that macroevolution generating new species is different from microevolution that causes interracial or intersubspecific genetic differences.... In my view, however, it is not difficult to explain the observations on punctuated equilibrium in terms of classical or neoclassical theory. The only assumption necessary is that the mutational change of highly adaptive or morphologically distinctive characters (Simpson's key mutation) occurs with a much lower frequency than that of less significant ones.... If certain adaptive or morphologically distinctive mutation occurs with a very low probability, evolution cannot occur gradually because the events of mutation and fixation are stochastic" (Nei, 1987).

Cf. Futuyma (1986) and Tassy (1993), for a balanced consideration of gradualism versus punctuated equilibrium. Some microfossil lineages exhibit 'punctuated gradualism' (Malmgren, Berggren & Lohmann, 1983; Lipps, 1993). It is worthwhile, in this context, for us to take note of the now famous admonishment: "'You have,' [Thomas Huxley]... had written to Darwin on November 23, 1859, 'loaded yourself with an unnecessary difficulty in adopting *Natura non facit saltum* so unreservedly'" (Clark, 1984a). But Huxley's remark was much more casual in context: "However, I must read the book two or three times more before I presume to begin picking holes.... Looking back over my letter, it really expresses so feebly

all I think about you and your noble book, that I am half-ashamed of it..."
(in: Huxley, 1902). (See Section 10.09 for a different view of Darwin's
opinion on gradualism versus saltational evolution.)

We must also take care to distinguish the concepts of saltation as a change
in a single generation (the hopeful monster scenario) from evolutionary
changes that are rapid on a geological time scale, but nevertheless take
many generations (punctuated equilibrium: Eldredge & Gould, 1972). Either
could be considered a case of punctuated equilibrium. Thus the argument
over whether saltational change is due to a single gene or "simultaneous
changes in the frequencies of many genes with small pleiotropic effects"
(Lande, 1980b) hinges on whose definition of 'simultaneous' is being used.
When we wish to speak of mechanisms at the DNA level, we do indeed
want to consider each generation. For example, our proposed scheme for
duplicating a terminal branch of the differentiation tree (Figures 24-26)
involves an initial, one generation, large duplication of minimal phenotypic
effect, followed by numerous smaller changes that, over many generations,
restore the overall tree-like structure to the genome. However, the total
number of generations might be small on a paleontological time scale.

The historical and philosophical roots of the 'conflict' between continuity
and discontinuity are deep (Lovejoy, 1936). Another side of this attempt to
make punctuated equilibrium compatible with the New Synthesis is
expressed by Depew & Weber (1994):

"According to advocates of 'punctuated equilibrium,'... macroevolution is not gradual at all,
but is concentrated in evolutionary time around bursts of speciation (Gould & Eldredge,
1977). Between these episodes things don't change much. Perhaps, then, speciation is caused
by sudden reorganizations of largely self-organizing genetic networks, which then remain
locked in place for considerable periods of time (Gould, 1982c,d). In that case, can we be
sure that what makes it through into the phylogenetic tree gets there because of selection
(Gould, 1989b)? Cut loose from adaptive natural selection, a combination of
self-organization and sheer chance might determine phylogeny, severing macroevolution
from adaptive microevolution in the same way evolution at the protein level has been
partially decoupled from natural selection by Kimura's [1968, 1983] neutralism.

"Generally put, the threat is that the range of processes for which Darwinism, whether old or new, is approximately true might eventually be restricted to the level of organisms and populations, and that a more general theory, in which selection and adaptation play markedly diminished roles, may govern evolution in both the grand sense and at very elementary levels. Since these pressures on the modern synthesis concern what happens at the level of the very small (molecular evolution) and the very large (macroevolution), they recall similar pressures exerted on physics earlier in the twentieth century, when quantum mechanics revised received views about particle physics, and when relativity revised long established ideas about space and time" (Depew & Weber, 1994).

Wilson (1992) has written the following epitaph for punctuated equilibrium:

"In general, the continuity between microevolution and macroevolution has been upheld. Neo-Darwinian theory was not challenged in substance, only semantically - a renaming, so to speak, as opposed to a reinventing of the wheel. Punctuated equilibrium is now used mostly as a descriptive term for a pattern of alternating rapid and slow evolution, especially when the rapid phase is accompanied by species formation. Its fate illustrates the principle that in science failed ideas live on as ghosts in the glossaries of the survivors. The value of the punctuated-equilibrium challenge lay not in its claims but in the research it stimulated on evolutionary rates and in the favorable public attention it brought to evolution studies as a whole" (Wilson, 1992).

Gould & Eldredge (1993), conceding some gradualism, think punctuated equilibrium still prevails:

"The intense controversies that surrounded the youth of punctuated equilibrium have helped it mature to a useful extension of evolutionary theory. As a complement to phyletic gradualism, its most important implications remain the recognition of stasis as a meaningful and predominant pattern within the history of species, and in the recasting of macroevolution as the differential success of certain species (and their descendants) within clades.... We devised a motto: 'stasis is data.' For no bias can be more constricting than invisibility - and stasis, inevitably read as absence of evolution, had always been treated as a non-subject. How odd, though, to define the most common of all paleontological phenomena as beyond interest or notice! And, even worse, as paleontologists didn't discuss stasis, most evolutionary biologists assumed continual change as a norm, and didn't even know that stability dominates the fossil record.... Because species often maintain stability through such intense climatic change as glacial cycling (Cronin, 1985), stasis must be viewed as an active phenomenon.... Many leading evolutionary theorists, while not accepting our preference for viewing stasis in the context of habitat tracking (Eldredge, 1989a) or developmental constraint

(Maynard Smith, 1983a; Williams, 1992), have been persuaded by punctuated equilibrium that maintenance of stability within species must be considered as a major evolutionary problem" (Gould & Eldredge, 1993).

See the excellent criticism of models of stasis for crocodiles by Meyer (1984), and an overview of 'coordinate stasis' of species in a community (Monastersky, 1996). The unsolved problem of stasis remains with us (cf. Parsons, 1994; Morell, 1997; Reznick et al., 1997). Does stasis occur via environment constancy or via developmental constraints? Is there a relationship between ecosystem stasis (Plotnick & McKinney, 1993) and the morphological stasis of individual species? Whether differentiation trees will give us a clue, perhaps through irreversible web formation, remains to be seen (cf. Proposition 163).

Proposition 162: differentiation trees provide a basis for speciation.

If we interpret regulatory incompatibilities between hybrids as differences in their differentiation trees, then we essentially have the view of Nei (1987):

"The mechanism of species formation or speciation is an old but still unsolved problem.... In neo-Darwinism, it is customary to assume that reproductive isolation evolves as a by-product of adaptation of populations to different environments.... In my view, evolution of reproductive isolation can be explained more easily by neoclassical theory. Recent studies on isozyme expression in interspecific hybrids suggest that regulatory genes from one species are often incompatible with structural genes from the other and that this incompatibility of genes in developmental or gametogenetic process leads to hybrid inviability or sterility (e.g., Ohno, 1969; Whitt, Cho & Childers, 1972; Frankel, 1983a; Parker, Philipp & Whitt, 1985).... Nei (1976) and Nei, Maruyama & Wu (1983) developed a mathematical model of evolution of reproductive isolation based on this idea and showed that a variety of data on hybrid inviability or sterility, including the relationship between the extent of hybrid inviability and the time since divergence between two species (Wilson, Maxson & Sarich, 1974), can be explained by this model.... Particularly if we identify genes that control hybrid inviability or sterility and study the nature of their gene action at the molecular level, we will be able to gain deeper insight into the mechanism of speciation" (Nei, 1987).

Of course, simpler regulatory differences, at the level of the differentiation cascades and below, should also permit a range of hybrid incompatibility that overlaps the larger range of regulatory incompatibility that we anticipate between differentiation trees.

Proposition 163: evolutionary stasis may be brought about through the high frequency of reproductive isolation of an individual from its parent population when its mutation creates a topologically different differentiation tree.

Punctuated equilibrium (Eldredge & Gould, 1972, 1988; Levinton & Simon, 1980; Hoffman, 1982; Kirkpatrick, 1982; Maynard Smith, 1983a; Dawkins, 1985; Carson, 1987a) is based on the observation from the fossil record that morphologically defined 'species' (cf. Wake, Roth & Wake, 1983) can remain unchanged for even tens of millions of years ('stasis' or 'equilibrium'), and then change suddenly (on geological time scales) in perhaps as little as a few thousand years (Eldredge, 1985b):

"We might sum up, from a punctuationist point of view, with a paradox: evolution is possible because, for most of their existence, species do not evolve" (Tassy, 1993).

The idea of stasis was widely appreciated by paleontologists in the 1800's, and used to contradict Darwin's thesis (Hull, 1973; Eldredge, 1995a). On the other hand, Darwin (1872) actually considered stasis as probable (see Section 10.09).

Of course, each morphologically distinct fossil 'species' could actually represent many species, whose means of reproductive isolation from one another may have been behavioral or biochemical (as pheromones, fertilization blocks, etc.), and may have been at the differentiation cascade level: "The possibility for error has no limit, and so intrinsic isolating mechanisms are endless in variety" (Wilson, 1992). Such differences (as in cryptic species: Larson, 1989) would generally not be apparent from the fossil record.

Eldredge (1985a) considers three possibilities for the biological basis for stasis:

1. "the homogenizing influence of gene flow....,

2. "habitat recognition.... [cf. the ecological considerations of Turner, 1986],

3. "some sort of internal homeostasis (presumably of the developmental pathways of component organisms) as a source of such species stability" (Eldredge, 1985a).

Eldredge (1995a) adds, without discussing developmental constraints, that...

"Habitat tracking, combined with an appreciation of species structure, developed first by Sewall Wright [1982a], is really all that is needed to explain stasis" (Eldredge, 1995a).

I suggest that the postulated 'internal homeostasis' is due to a general incompatibility between topologically different differentiation trees, at some stage producing hybrid nonviability, as discussed in Proposition 158.

In rare circumstances, either reproductive isolation will not occur, or enough F1 individuals will be produced to establish a separate, reproductively isolated population. It is these rare events that may explain the geologically sudden speciation events of the punctuated equilibrium hypothesis.

While individuals with differentiation trees with mutationally different differentiation cascades may or may not be reproductively isolated, I thus suggest that topological differences in differentiation trees usually cause reproductive isolation, explaining extended periods of evolutionary stasis.

The basic problem in trying to explain evolutionary stasis is the long characteristic time involved, ranging up to tens of millions of years. Resolution of the cause(s) of stasis will require theories that estimate the characteristic times for each phenomenon proposed to explain stasis. These phenomena include:

developmental constraints (Williamson, 1981a; Charlesworth & Lande, 1982; Alberch, 1982a, 1985b; Holder, 1983a; Kauffman, 1983; Cheverud, 1984; Maynard Smith et al., 1985; Levinton, 1986; Wimsatt, 1986; Rasmussen, 1987; Wagner, 1988; Scharloo, 1988; Oster & Murray, 1989; Langille & Hall, 1989a; Goldsmith, 1990; Tuinstra, De Jong & Scharloo, 1990; Tabin, 1992),

and stabilizing selection (Charlesworth, Lande & Slatkin, 1982)

combined with 'developmental buffering' (Foote & Cowie, 1988),

to which may be added large population size (Maynard Smith, 1983a; Paterson, 1993a)

and nonlinear relationships between genotype and phenotype (Kerszberg, 1989).

In no case has the necessary characteristic time been demonstrated, though it may be possible to eliminate some hypotheses:

"Even very weak selection can produce rates of phenotypic evolution that are extremely fast on a geological time scale. Thus, there is no basis for claims that developmental constraints (in the accepted sense of genetic correlations; Maynard Smith et al., 1985; Oster et al., 1988) cause stasis and no support for the view that population bottlenecks are necessary for rapid morphological evolution...[though they may be sufficient: Bryant, Combs & McCommas, 1986]" (Barton & Turelli, 1989).

Rienesl & Wagner (1992) have begun a direction of research that could produce the characteristic time scale for action of a developmental constraint, by comparing the pattern of variation in populations of closely and distantly related species of newts.

The scaling of artificial life experiments with stasis that exhibit punctuated equilibrium (W. Daniel Hillis in Levy, 1992; Lindgren, 1992, discussed in Emmeche, 1994) to geological time scales is not obvious:

"Paleontologists and geneticists do not have the same concept of time. According to the punctuated equilibrium model, speciation is a sudden event, taking between 5,000 and 50,000 years. On the geological time scale, this is a rapid change that is difficult to discern in the paleontological records; but on the reproductive time scale it is a slow process" (Tassy, 1993).

Thus I feel free to add the regulatory genome hypothesis, that the general incompatibility between topologically different differentiation trees could be a source of stasis.

Proposition 164: differentiation trees explain mosaic evolution.

Differentiation trees give the embryonic lineage of an organism's tissues, and it is the presence or absence of those tissues, and their spatial shapes and arrangements, that constitute most of the characteristics used to classify organisms. If macroevolution involves mutational (initially random) changes in the topology of the differentiation tree, then multiple changes of this kind should occur with some independence from one another:

"An empirical observation of fossils reveals that it is always one characteristic or one type of characteristic that evolves at a given rhythm, rather than the entire morphology. Whether the phyletic gradualism of the ribs of Ordovician trilobites or the 'punctuationism' of eye lenses in Devonian trilobites, in each instance only a single very precise characteristic is involved. As we have seen, analyses of fossils show that in a natural group, characteristics evolve independently. This almost universal finding can easily be explained by selection processes: the adaptive response of a population to environmental changes will affect the dental apparatus, the locomotory system, or perhaps vision, depending on the case. As subsequent speciation takes place, these transformations are transmitted, become modified in turn, or even may remain unchanged. The result, within any given species, is an association of characteristics exhibiting a variety of evolutionary levels; this has been called 'mosaic evolution,' an expression proposed by de Beer almost forty years ago with reference to *Archaeopteryx.... Archaeopteryx,* the first bird, which lived 140 million years ago in Europe, represents the prototype of the intermediate fossil, since it connects two very distant large groups, the reptiles and birds. But while its 'reptilian' characteristics are extremely numerous (it has teeth, claws on its wings, etc.), its birdlike characteristics are very few. But if only because it has feathers (whose fossilized imprint is visible on several specimens discovered so far), the animal can unequivocally be considered a bird. It is an extremely primitive bird in terms of most of its characteristics; and it is by no means half reptile and half bird. In reality, its reptilian traits are still those of the dinosaurs; but in the course of the Jurassic the selection of a few skeletal features, and the conversion of scales into feathers, was enough to turn this dinosaur into a bird [a transition now in question: Burke & Feduccia, 1997; Hinchliffe, 1997; Ruben et al., 1997].... So intermediate species, 'missing links,' do indeed exist, contrary to what is occasionally said and written. The only problem is that most of the time, we are not looking for what we should be looking for, because the intermediates are not intuitive ones" (Tassy, 1993).

Mosaic evolution speaks more towards changes in the topology of the differentiation tree than in stretchings of portions of it, which I equate with heterochrony (Proposition 128). Nevertheless, one could imagine a mosaic evolution based primarily on heterochronous changes that occur in highly localized regions of the differentiation tree. The only way to clearly distinguish these cases will be to observe the differentiation trees for closely related species, and determine the actual bases for their differences.

Proposition 165: during a speciation event, the partial incompatibilities between topologically different differentiation trees may produce developmental instabilities.

Williamson (1981a) indicated...

"...[from] a morphometric analysis of mollusc lineages from a Cenozoic sequence in North Kenya... I made three key points:

(1) all lineages exhibit morphological stasis for very long periods of time (3-5 Myr),

(2) evolutionary change in each lineage is concentrated in relatively rapid speciation events (occurring over 5,000 to 50,000 years) and

(3) the speciation events are accompanied by pronounced developmental instability in the transitional populations" (Williamson, 1981a).

I am thus suggesting that developmental instability during a species transition correlates with macroevolution, as I have redefined it, i.e., a change in the topology of the differentiation tree (Proposition 129). Williamson (1981a) continues in the following vein:

"...The idea that disruption of developmental homeostasis, or constraints, is central to the speciation process is hardly new to the neo-Darwinian literature: Carson (1975) has postulated 'open' and 'closed' genetic systems, the former involved in the allele-shuffling of minor adaptive adjustments within species populations, and the latter involved in the more profound regulatory and developmental changes during speciation. Levin (1970a) has pointed out the significance of developmental disruption and instability during the speciation process. But such suggestions have been largely ignored: the implied decoupling of micro-evolutionary change and macroevolutionary phenomenon threatens the reductionist core of conventional neo-Darwinism" (Williamson, 1981a).

I would withdraw his last statement, in light of my 'reducing' disruption of developmental homeostasis to incompatibilities between differentiation trees, which are DNA structures. 'Open and closed genetic systems' obviously could correspond to my definitions of micro- and macroevolution.

Proposition 166: 'developmental homeostasis' ('canalization') does not exist, and so does not form a basis for stasis.

I am not happy with the amount of physics actually in this book. While we can now dimly see a physics of embryology, it has yet to be formulated, let alone tested. I am particularly disturbed by the lack of proof that the developmental system is robust. Yes, it is clear that evolution can fine tune the parameters to a point, by the vast throw away of unsuccessful individuals. But we have come to a view of embryogenesis where regulation is not seen as a global feedback mechanism (Gordon, 1966; Sachs, 1994), just as a means of allowing a tissue to be more than one cell (Sections 4.05-4.06). Cells do interact, and they do so via the launching and propagation of differentiation waves, corresponding to the classical notions of heterogeneous and homoiogenetic induction. But then if the system is essentially a positive feedback system, why and how does it remain robust while unfolding increasing complexity? Something is missing here: either there are negative feedbacks, such as in the concept of canalization; or there aren't negative feedbacks, but the course is straight enough that the 'momentum' of development produces a reasonably functional organism often enough to carry on to the next generation.

Is development like a bicycle, with or without a rider, or a missile, with or without a gyroscope? Which represents the embryo? I suspect the latter in each case, disagreeing with statements such as:

"If any... [homeostatic process] gets out of line, it is brought back and the original state restored. Indeed, the whole process of organization, involving not only physiology but growth and development of the unified organism, may be called a complex case of homeostasis" (Sinnott, 1966).

"What is important is that refinement can be done by a linear optimisation process which makes small random changes in any component followed by the measurement of a global value such as an energy minimum.... Thus, one begins with a messy, inaccurate 'sort of fly' which is then progressively refined into a fly" (Brenner, 1981).

There is no 'state' for an aberrant embryo to 'come back to'. Development goes on, without stopping and checking that all is well at one stage before proceeding to the next. Who checks? How does an embryo attain an overall self-image to evaluate its progress? How would corrections be instituted? I've lost the source of one quote that sticks in my mind: an abnormal mutant embryo proceeds towards its 'goal' of producing a defective organism as inexorably as a normal embryo proceeds towards its 'goal'. There is no goal, not even any control along the way, just a process of unfolding, a mapping of the differentiation tree preserved in the DNA into a process of differentiation, but no more. Strip away the purposiveness, and perhaps we shall come to see embryogenesis as a fundamentally simple process. Dennett (1995) has viewed evolution as an unfolding (albeit stochastic) algorithm. The same may be said of development:

"The algorithms we tend to talk about almost always do something interesting - that's why they attract our attention. But a procedure doesn't fail to be an algorithm just because it is of no conceivable use or value to anyone.... Some algorithms do things so boringly irregular and pointless that there is no succinct way of saying what they are *for*. They just do what they do, and they do it every time" (Dennett, 1995).

Robustness and feedback are not the same concepts: a full glass of water stays full ('robustly' retains its shape) no matter what size object is slowly dropped in it (to an obvious limit), but no energy requiring feedback (homeostatic) mechanism is involved. I suspect that the paradigm of homeostasis, so successful in physiology, has been overextended by application to embryonic development as 'developmental homeostasis' (Dobzhansky & Levene, 1955; Williamson, 1981a; Mukai, Chigusa & Kusakaba, 1982; Eldredge, 1985a):

"The simplest form of homeostasis is exemplified by reversible physiological reactions which permit the functioning of the body and the organs to continue unimpeded. Kidney function, temperature regulation, and functional hypertrophy or atrophy of organs are good examples. By insensible gradations this functional homeostasis merges with physiological reactions which result in developmental homeostasis. Developmental homeostasis, which Waddington (1942a, 1953c) prefers to call canalization, permits the development to keep to a definite path or to switch over to a closely related path. Schmalhausen's (1949) physiogenic modification is also a related concept. Lerner's (1954) *genetic homeostasis* refers to self-regulatory properties not of individual organisms but of Mendelian populations. It may seem that genetic homeostasis is a phenomenon entirely separate from functional and developmental homeostasis. But Lerner argues, and we believe convincingly, that the genetic homeostasis is brought about by essentially the same mechanisms of heterozygosity and coadaptation of gene alleles and gene complexes as those which [underlie] the other forms of homeostasis.

"Surely, the relationships between the different forms of homeostasis, between homeostasis and heterosis, and between homeostasis and adaptedness are far from completely understood. The scene is too diverse and contradictory" (Dobzhansky & Levene, 1955).

These analogies were fueled by the phenomenon of regulation, which, experimentally, always involved gross surgical insults to the embryo, from which it sometimes recovered. This is hardly what we expect a normal feedback system to have to cope with. In fact, to the contrary to such large perturbations, most embryos are isolated from fluctuations in the environment by a tough jelly coat and vitelline membrane, or by a protective maternal envelopment. Thus the whole notion that the embryo interacts with its environment may be fallacious. I would extend this to the internal environment, and question whether internal feedbacks of any sort operate during embryonic development, contrary to my earliest supposition (Gordon, 1966). This may be an extreme statement, given the possibility of regeneration at neural plate stages in urodeles (Harrison, 1947). This contrary observation should be an incitement to sort out the roles of proliferation, dedifferentiation, redifferentiation, and differentiation waves in brain regeneration, the latter which, after completing this book, I could well use.

It is perhaps worth noting that the level at which the external environment impacts on development is rather gross, and often concerned with phenomena close to hatching or birth, when learning and eating, certainly environmental activities, become important, not on the fine points of events around gastrulation and neurulation:

"For instance, the complex interactions of genetic, epigenetic, and environmental factors have been clarified in studies of the development of mammalian skeletal elements (e.g. Atchley & Hall, 1991; Atchley et al., 1991). Recent experimental studies on various vertebrate species have analyzed the interactive effects of genetic and environmental factors during development on birth weight (Jagtap et al., 1990), hatchling growth rate (Sinervo & Adolph, 1989), and size and age at maturity (Trexler & Travis, 1990a,b)..." (Blackburn & Schneider, 1994).

Consider again the concept of canalization, which adds to regulation the idea that development is 'partially self-compensating' with respect to 'developmental noise' and changes in genes and the environment (Waddington, 1956a; cf. Waddington, 1940, 1942a, 1959a; Mather, 1943; Rendel, 1959, 1967, 1979; Robertson, 1964; Scharloo, 1964; Rendel, Sheldon & Finlay, 1965; Scowcroft, Green & Latter, 1968; Scowcroft, 1973; Niklas, 1982; Kieser, Groeneveld & Preston, 1986; Clarke & McKenzie, 1987; Kieser, 1987; Kerszberg, 1989; Piñeiro, 1992a,b,c; Kovach, 1993a):

"Developmental processes are highly interactive, and interactions result in canalization [an unproven theorem]. As defined by Waddington (1942a), canalization is the buffering or homeostasis of developmental pathways such that these pathways resist distortion by either environmental or genetic perturbations" (Raff & Kaufman, 1983).

"One of the most important properties of the chreod is its stability: were a cell to deviate from its appropriate trajectory of differentiation, Waddington (1968a) suggested that there would be canalising forces to ensure that the cells were pulled back to their original route.... There are two ways in which this stability might be achieved: through a feed-back control which links some aspect of the final structure to the process of achieving it or through tight constraints on the cells while they are active" (Bard, 1990a).

We find, however, that a critical review of canalization (Scharloo, 1991) leaves us with equivocal data, alternate explanations, and a need "to analyze the developmental processes underlying the characters involved":

"The solution to perhaps the greatest problem of morphology requires a bridge to genetics, a bridge which at this time cannot yet be built. I am referring to the origin and the meaning of the great anatomical types, already known to Buffon [1749-1804, 1954, 1962] under the name 'unity of plan.' Within the mammalian *Bauplan,* for instance, such strikingly different functional types evolved as whales, bats, moles, gibbons, and horses, without any essential change in the mammalian plan. Why is the chordate type so conservative that the chorda still is formed in the embryology of the tetrapods and gill arches still in that of mammals and birds? Why are the relations of structures so persistent that they can form the basis of Geoffroy's principle of connections [Geoffroy Saint-Hilaire, 1847]? Clearly this is a problem for developmental physiology and genetics, indicated by such terms as the cohesion of the genotype or the homeostasis of the developmental system, terms which at this time merely conceal our profound ignorance" (Mayr, 1982).

Developmental homeostasis has been a convenient place to fob off the problem of stasis in evolution, but perhaps, as discussed in Propositions 161 and 163, we should look elsewhere for a solution to the stasis problem.

Let me state once more, the presumption of feedback during development may be fallacious. Perhaps we are sort of thrown together, by a rapid interaction of differentiation cascades and physics, and those who do not end up whole do not survive to breed. Sure, there are many minor feedbacks, ranging from surface tension to regulated pathways (the 'cranes' of Dennett, 1995). But perhaps there are no explicitly developmental feedbacks. Maybe this explains why we, with probably the most complex differentiation tree, have such a low post-fertilization embryonic survival rate (Roberts & Lowe, 1975; cf. Makino et al., 1992), and why, as we conquer childhood diseases, the remaining caseload has an increasing genetic component ("about a third of all children in pediatric hospitals": Mange & Mange, 1990).

Proposition 167: embryonic regulation does not exist.

"Regulation implies the presence of something to which development conforms - a norm or goal immanent in the organism " (Sinnott, 1966).

I have to state this proposition in hyperbole, to make the point that what is classically defined as 'regulation' in a regulating embryo is not at all a feedback mechanism. A feedback mechanism, such as a thermostat, requires:

1. a (usually) fixed value ('norm' or 'goal') of a parameter from which deviations are measured;

2. a means of measuring the deviation of that parameter from the goal;

3. a means of correcting the actual value to bring it back towards the goal;

4. a means of reiterating all three steps above indefinitely, which brings the value towards the goal, and corrects any further deviations that may occur.

Embryological regulation is a transient phenomenon in which an embryo appears to give normal development when we perturb it once. Shape, proportions, cell numbers, etc., may be restored, though this is rarely checked by the experimentalist in any quantitative fashion. The embryo then moves on to the next stage, with whatever residual abnormalities it may have. Sometimes it becomes a viable larva or adult, sometimes not. Not every lump of embryonic tissue that we leave after an experimental assault becomes a normal organism. Thus the success of some experiments cannot be taken as an indication of a 'norm' or 'goal' in all. At best, embryonic regulation involves some local negative feedback, such as embryonic wound healing (Stanisstreet, Smedley & Snow, 1985), or independence of trajectory shape on embryo size for differentiation waves, or local cell-cell or tissue-tissue interactions. Yes, it would be nice to know what makes the difference between viability and death, but the presumption of a 'norm' or 'goal' does not solve that problem.

Proposition 168: development consists primarily of positive feedbacks.

The feedbacks I have been referring to in Proposition 167 are all negative feedbacks. The intellectual problem with positive feedbacks is that...

"As a general rule, positive feedback tends to produce unstable systems, whereas negative feedback tends to stabilize systems, at least within certain dynamic limits.... There seem to be many biological examples... where positive-feedback inner loops are stabilized by negative-feedback outer loops (Maruyama, 1963a, for example), and functionally they belong without doubt to the class of feedback-control systems" (Milsum, 1966).

But in embryogenesis the 'negative-feedback outer loop' may not be there, which is one way of denying 'developmental homeostasis': nothing guides development from the outside. Rapoport (1968) identified part of the mental block:

"It is noteworthy... that instances of positive feedback which come most frequently and most dramatically to attention in non-technological contexts are... vicious: the poverty-apathy cycle, the unemployment-deflation cycle, the inflation explosion, the arms race, progressive drug addiction. On the other hand, examples of positively valued positive feedback are not lacking [such as the]... rise of civilizations.... Disarmament and the concomitant relaxation of tensions between rival nations or blocs could perhaps become a positive feedback cycle as easily as an arms race. However, no word has yet been coined to denote a desirable positive feedback process in the way 'vicious cycle' denotes an undesirable one.... Removing the connotation of disaster, we see that positive feedback is the embodiment of a self-enhancing process....

"In summary, positive feedback, an idea derived by engineers from problems of system control, illustrates the philosophically salubrious effect of specifying structural relations among variables in preference to categorical reasoning about 'causes and effects'. When structural relations are exhibited, questions about what causes what lose their significance, except to the extent that they refer to the accessibility of variables, to questions related to the manipulation of variables, and to the sensitivity of the system's behavior to the values of the variables. In this context, the presence of [positive] feedback loops is especially important; for it is these that make possible the attainment of very large effects by minute interventions. Social explosions (e.g. wars and revolutions), explosive chemical reactions, the present 'information explosion', all contain positive feedback loops. Progressive pathologic conditions as well as spontaneous recoveries probably also involve such loops. Therefore, the discovery of such loops is of prime importance to any one who seeks control of the course of events, either by interrupting vicious cycles or by making use of beneficent self-enhancing processes" (Rapoport, 1968).

Perhaps it is time for control engineers to get involved in embryogenesis. We may need a theory of robust, non-vicious positive feedback systems, to help us understand how we come to be (cf. Thieffry et al., 1995). Such a theory would, of course, have to include the hierarchical nature of the differentiation tree and access to the hierarchical information stored in an organism's DNA. These are not characteristics of most engineered control systems. The differentiation tree may be looked upon as a branching cascade of positive feedback loops between the physics of differentiation waves and bouts of specific gene expression via sequential and parallel differentiation cascades.

Proposition 169: the magnitude of the accumulation of developmental errors is different for mosaic and regulating organisms.

"The pseudocoelomates, which are also called aschelminths, are a bizarre collection of ten phyla, representing almost a third of all living body plans: rotifers, gnathostomulids, priapulids, gastrotrichs, kinorhynchs, nematodes, nematomorphs, entoprocts, acanthocephalans, and loriciferans.... Eutely, the extreme consistency of position and numbers of cells seen in pseudocoelomates, is a correlate of low cell numbers" (Raff, 1996).

Let's consider the pastime of setting up dominoes and letting them knock one another over (cf. the theory of avalanches: Cheng et al., 1989). Contests are held in which dominoes are arranged in elaborate patterns, some of which are robust, many of which are not. (Bryant, 1982, assigns a probability of success to each basic pattern). Now a small local negative feedback operates on every individual domino, else gravity would take it down immediately. But what distinguishes robust domino patterns from those that are not robust? The obvious answer is alignment and redundancy. If we precisely align the dominoes, we can guarantee that the pattern we produce will all fall down (chaos considerations aside). On the other hand, if a particular domino can be downed by two or more dominoes, then the falling pattern can continue even if one of the preceding dominoes does not succeed in falling. (This analysis of mine could be tested quantitatively by a physics based computer simulation of domino toppling.) Perhaps here is the fundamental difference between mosaic and regulating embryos: mosaic embryos are 'precisely aligned' via some as yet to be discovered features of their cytoskeletons and single cell differentiation waves, whereas regulating embryos use redundancy acting through multicellular differentiation waves. Mosaic embryos, like precisely aligned domino patterns, are harder to construct as they get larger, since errors in alignment may be presumed to be cumulative, even if there are local feedbacks operating in the alignment process. The accumulation of alignment errors may account for the only partial success of world record domino setups:

"***Domino toppling*** The greatest number set up single-handedly and toppled is 281,581 out of 320,236 [88%] by Klaus Friedrich, 22, at Fürth, Germany on 27 Jan 1984. The dominoes fell within 12 min 57.3 sec, having taken 31 days (10 hr daily) to set up.

"Thirty students at Delft, Eindhoven and Twente Technical Universities in the Netherlands set up 1,500,000 dominoes representing all the European Community member countries. Of these, 1,382,101 [92%] were toppled by one push on 2 Jan 1988" (Matthews et al., 1994).

Thus strictly mosaic organisms must be of relatively small size. Regulating embryos, like domino patterns with redundancy, on the other hand, can be much larger. Induction is the alignment process: in mosaic embryos it must operate at almost every cell division; in regulating embryos it need operate much less frequently. The analogy of embryogenesis to falling domino patterns may answer the call for a universal mechanism applicable to both mosaic and regulating embryos:

"It became more and more evident that other basic mechanisms must underlie the development of both types of eggs. It is the search for these basic mechanisms to which we will devote our attention" (Nieuwkoop, 1966).

We can make a further analogy with computer architectures. Dominoes arranged in a redundant fashion are analogous to parallel computers in which the individual processors do not act individually, but are, rather, used for redundancy, for example, by 'vote taking' (Museum of Science & Industry, Chicago, exhibit circa 1955), as in...

"Computation... performed by excitatory and inhibitory interactions among a network or relatively simple neuronlike units.... Eventually, thanks to statistical properties of the ensemble, the [neural] network settles into a state that reflects its particular 'tack'" (Gardner, 1985).

Let me give one more analogy, sort of a parable of mosaic versus regulative development. Suppose we had to climb a vertical ladder while building it rung by rung, and the pieces had a bit of wobble to them. We could get to some reasonable height, after which our situation would become precarious. The cumulative errors would at best do a random walk in the horizontal plane, relative to the first rung.

Now suppose there are many of us building vertical ladders parallel to one another, and we are able to lash them together as we go. We could all attain, together, a much greater height, using the very same kinds of components. Homoiogenetic induction is a form of 'lashing' of cells together into tissues, permitting larger organisms. Because of the cooperative nature of this 'lashing', we should expect the stability of the ladders or organism to increase in a nonlinear fashion with the number of ladders (or cells per tissue). Thus, for example, in a series of axolotls with increasing cell size (accomplished by polyploidy without change in adult body size: Proposition 173), we might expect some increase in fluctuating asymmetry. The increase this model predicts in error versus developmental time perhaps "...could account for the 'distal variability, proximal stability observed amongst tetrapod limbs' (Hinchliffe, 1991)" (Duboule, 1994b; cf. Alberch & Blanco, 1996).

We can make a simple model for mosaic development in terms of a few parameters that are dependent on the differentiation tree. As shown in Figure 22, these are the ratio of cell radii $c_i = a_i/b_i$, and the angle θ_i with respect to the previous cell division, probably marked by an orientation scar. The angle is written as a vector, since in three dimensional (3D) space it will have three components (the direction cosines: James & James, 1976), two of which are independent variables. (If orientation scars are chiral, an additional spin parameter would be needed describing the chirality.) I assume that surface tension, junctions, or cell adhesion forces bring the cells together and place some constraints on θ_i:

"Cells divide in an invariant pattern during early *C. elegans* development. I have found by cell isolation/reassociation experiments that the division axes of some cells (EMS and E) are controlled by contacts with specific neighbouring cells (EMS-P2 or E-P3 contact). Altering the orientation of contact between these cells alters the axis along which the mitotic spindle is established, and hence the orientation of cell division.

"I have used time-lapse videomicroscopy and anti-microtubule immunofluorescence to study centrosome movements in cell pairs. Contact-dependent mitotic spindle orientation appears to work by establishing a site of the type described by Hyman & White (1987) in the cortex of the responding cell: during centrosome rotation one centrosome moves to the cortex at the site of cell contact" (Goldstein, 1995a).

Hutter & Schnabel (1995) suggest that in nematodes a "polarity is propagated and subdivided through several divisions", an observation calling for a mechanical explanation. The basis for this may possibly be found in the centrosomes, which lead to "Loss of spatial control of the mitotic spindle apparatus in a *Chlamydomonas reinhardtii* mutant strain lacking basal bodies" (Ehler, Holmes & Dutcher, 1995). (If there is a relationship of orientation scars to centrosomes, then this would further enhance the hypothesis of the ciliate origin of multicellular organisms.) Consecutive asymmetric divisions in nematodes do not necessarily use identical mechanisms (Hird, Paulsen & Strome, 1996). Chamberlin &

Sternberg (1995) have achieved some genetic separation of the steps involved in an asymmetric division in nematodes:

"*lin-17* is necessary to establish the asymmetric division [mechanical instability conditions?], *lin-44* is necessary to properly orient the division [with respect to the last orientation scar?], and *vab-3* is necessary for the fate of one of the daughters, with its activity unequally distributed by the division [as it is attached to one cytoskeletal component segregating to one cell after the instability resolves?]" (Chamberlin & Sternberg, 1995).

More of the genes involved have been identified (Levitan et al., 1994; Cheng, Kirby & Kemphues, 1995). Similar work is in progress in yeast, in which mutant genes involved in 'polarized cell division' correlate with "an anomalous distribution of actin" (Durrens et al., 1995). The advantage of making this model explicit is that it allows us to raise succinct questions that haven't been posed before. Is there an expansion and/or contraction differentiation wave on each cell prior to or during cell division? How and where is each wave launched? Does it start from the orientation scar? How does the launching domain determine the parameters (c_i, θ_i)? What errors in the parameters will occur as a result in variations in the differentiation waves? Do steric or other interactions with neighboring cells reduce these errors, affect wave propagation, or in other ways alter the 'intrinsic' values of (c_i, θ_i)? If we think of development in terms of accumulating errors and their minimization by local feedback mechanisms (that "plethora of intercellular signals": Katz & Sternberg, 1992), we may be able to create some powerful tools for analysis, and come to understand the differences between mosaic and regulative development.

Proposition 170: differentiation trees may allow us to 'compute' at least mosaic organisms, and their developmental constraints.

This model thus guides us towards making a whole set of quantitative observations that go beyond the simple scalar questions posed by cytoplasmic 'determinants'. But we can go further. With this explicit model, we can take a given differentiation tree, such as the lineage map of nematodes (Sulston et al., 1983), and guess at values of the (c_i, θ_i). We can then create a computer simulation of the organism's morphogenesis. Such a synthesis, in modern parlance, is called 'computing a nematode' (cf. Slack, 1991a). We could imagine iterating between such a simulation and partial observations of (c_i, θ_i) to refine both (cf. Jacobson & Gordon, 1976a; Gordon & Jacobson, 1978; Gordon, 1983a). Perhaps the observed and confusing complexity of the nematode embryonic lineage (Sulston et al., 1983; Figure 38) would be more understandable when thought of in terms of the (c_i, θ_i) parameters. (Cf. the simpler analogical model of Bailly, Gaill & Mosseri, 1991.) This complexity may be similar to that of the folding of proteins, with their 20 different amino acids, specific bond rotation angles, steric interactions, and specific interactions between amino acids far apart on the peptide chain. Lineage trees and differentiation trees have the added features that they are branching structures, with the early edges 'disappearing' in the sense that a mother cell 'disappears' when it divides into two daughter cells.

A nematode simulation could also be used to test robustness of nematode morphogenesis, by adding reasonable errors to the (c_i, θ_i) and seeing how much error is tolerated before the resultant morphology diverges unacceptably. Do errors more or less get corrected at each step, or are they strictly cumulative? At this level, we are asking answerable questions about developmental constraints (cf. Raup, 1962, 1966; Raup & Michelson, 1965; Ghiselin, 1966; Raup & Chamberlain Jr., 1967; Linsley, 1978; Kemp & Bertness, 1984; Palmer, 1985). Presumably only robust species can last for geological time scales.

We can also ask design questions, such as why nematodes have little motion of cells relative to one another (but cf. Chamberlin & Sternberg, 1993, and the Enoplida, for which "A small role in development is played by the movement of blastomeres after division": Malakhov, 1994). If we introduce some motion, do the errors increase and propagate? If we extend the differentiation tree, will errors unavoidably accumulate despite corrective measures? Does the size limit correspond to the observed maximum size found in mosaic organisms? Thus we can carry out computer experiments not achievable with real organisms or even their mutants.

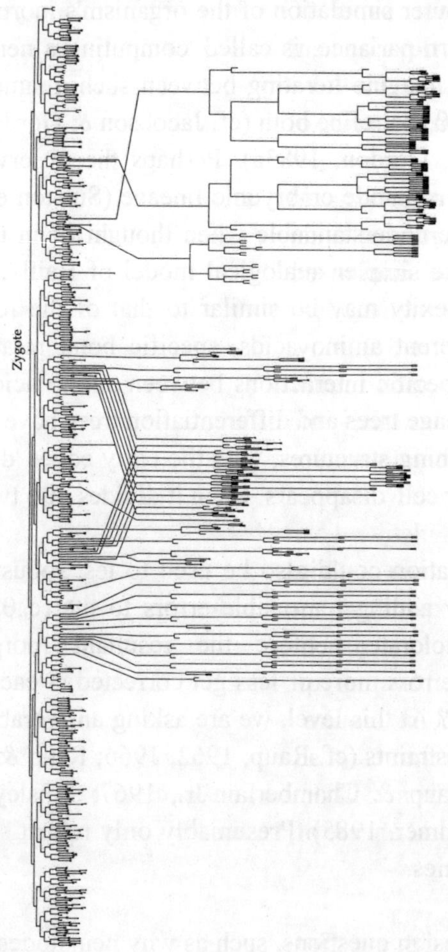

Figure 38. The lineage tree of the nematode *Caenorhabditis elegans*, which is essentially equivalent to its differentiation tree, since most cell divisions are unequal. From (Gilbert, 1991a) with permission, based on the following work:

"The embryonic cell lineage of *Caenorhabditis elegans* has been traced from zygote to newly hatched larva, with the result that the entire cell lineage of this organism is now known. During embryogenesis 671 cells are generated; in the hermaphrodite 113 of these (in the male 111) undergo programmed death and the remainder either differentiate terminally or become postembryonic blast cells. The embryonic lineage is highly invariant, as are the fates of the cells to which it gives rise" (Sulston et al., 1983).

I noted in my first paper (Gordon, 1966) that a negative feedback mechanism during development could correct for cumulative errors, and Lacalli & Harrison (1991) have looked at gradient models for morphogenesis from this point of view (cf. Lacalli, 1993).

Finally, the notion of propagating errors suggests that in the differentiation (lineage) tree for a mosaic organism, the number of steps to each terminal cell type should be minimized. Details, such as the broad, flat, bushy appearance of the nematode differentiation tree (Sulston et al., 1983; Gilbert, 1991a; Figure 38), may indeed be due to the need to minimize the number of steps to each terminal cell type. Given the existence of many species of nematodes, they may offer a group within which we could correlate developmental variation with topology of the differentiation tree.

Turning back to regulative development, we can now attempt to define the corresponding parameters (Figure 23). These would be the launching domains of the expansion and contraction waves, and the parameters affecting their trajectories, such as strains, cell motions, adhesive strengths, etc., of which we have no quantitative knowledge at this date. Waves launched from a point require three spatial coordinates, while a circular launching domain requires in addition its radius and orientation. How these values are derived from cell or cortical physics we do not yet know. Thus we have much work ahead of us to arrive at an ability to 'compute' a regulating organism:

"It has been repeatedly discovered that analysis of a physical problem in terms specific and detailed enough to permit programming of that problem for a computer gives us deeper insight into the nature of the problem itself. The same may well be true of the problem of differentiation" (Bonner, 1965b).

The concept of construction of an organism that arises from the ladders analogy is one completely lacking in goal directedness. While structures and cell types emerge from the interactions between waves through tissues and the cells they are made of, there is no global feedback as there might be if a conscious mind were sculpting something. Thus, at least for embryology, the following statement is false:

"The goal-directedness of complex biological systems seems beyond doubt" (Rosenberg, 1985).

A parallel may be made in the distinction between 'cranes' and 'skyhooks' for evolution:

"The skyhook concept is perhaps a descendent of the *deux ex machina* of ancient Greek dramaturgy: when second-rate playwrights found their plots leading their heroes into inescapable difficulties, they were often tempted to crank down a god onto the scene, like Superman, to save the situation supernaturally.... Skyhooks would be wonderful things to have, great for lifting unwieldy objects out of difficult circumstances, and speeding up all sorts of construction projects.... There are cranes, however. Cranes can do the lifting work our imaginary skyhooks might do, and they do it in an honest, non-question-begging fashion. They are expensive, however. They have to be designed and built, from everyday parts already on hand, and they have to be located on a firm base of existing ground. Skyhooks are miraculous lifters, unsupported and insupportable. Cranes are no less excellent as lifters, and they have the decided advantage of being real. Anyone who is, like me, a lifelong onlooker at construction sites will have noticed with some satisfaction that it sometimes takes a small crane to set up a big crane. And it must have occurred to many other onlookers that in principle this big crane could be used to enable or speed up the building of a still more spectacular crane. Cascading cranes is a tactic that seldom if ever gets used more than once in real-world construction projects, but in principle there is no limit to the number of cranes that could be organized in series to accomplish some mighty end" (Dennett, 1995).

A 'developmental ladder' may be seen as a series of 'cranes', and the concept of continuing differentiation may be seen as an algorithm for cascading one crane upon another, each differentiation cascade (or copied branch of the differentiation tree) corresponding to a 'crane'. However, there may indeed be a limit to how many cranes (developmental or perhaps even evolutionary - what's the difference in this case?) that can be attained by natural selection. Mosaic development may have hit this limit. Whether regulatory development, the lashing of multiple 'ladders' or 'cranes' together to attain a greater height, has hit its limit (presumably in us) is a fascinating question. Since the number of cell types in a nematode (807: Horvitz & Herskowitz,

1992) is about the same as in us (500: Bonner, 1988), perhaps we are indeed at this limit.

6.04 The Grand Sweep of Evolution

"From the very beginning of life on earth, to the larger and more complex animals and plants that live on its surface today, there has been a remarkable progression.... There has been little attempt to grapple with the great sweep of evolution..." (Bonner, 1988).

"The most prominent thing about evolution is that it is going somewhere. Organisms are becoming more complicated, their capacities are becoming more sophisticated.... Why should this be?" (Taylor, 1983).

Proposition 171: evolution has had four major stages, namely quasispecies evolution, single celled species, species with limited cell type differentiation, and species with continuing differentiation.

The mechanism of evolution of small viruses, which are presumed to be analogous to the first self-reproducing entities, is fundamentally different from that of higher organisms (but cf. Smith et al., 1997a), due to the inability of organisms with larger genomes to utilize the quasispecies mechanism:

"...The more we delve into the mathematics of adaptive landscapes [Wright, 1931, 1932; Kauffman, 1993; Macken, 1993; Stadler, 1993], the more it becomes clear that achieving global optimization via natural selection is a tenuous proposition (cf. Jacob, 1977). The problem, as reviewed, albeit in a positive light by Kauffman (1993), is that adaptive or fitness landscapes are rugged, even fractal landscapes (cf. Conrad & Ebeling, 1992; Weinberger & Stadler, 1993; [Schuster, 1995]).

"The quasispecies solution (Eigen, 1971; Demetrius, 1983), to send huge numbers of variant offspring out over the rugged landscapes, reaches a 'limit at a sequence length of a hundred to a thousand repeating units' (Eigen & Winkler-Oswatitsch, 1992). It thus generally (cf. Clarke et al., 1993) works fine for viruses (Domingo et al., 1985; Steinhauer & Holland, 1987; Epstein et al., 1991; Eigen, 1993a,b), but not for us 'higher' organisms with big genomes. Eigen & Winkler-Oswatitsch (1992) are driven to conclude 'that the optimization of individual genes must have taken place before their integration into a giant molecule, the genome', perhaps unnecessarily....

"Evolution on rugged landscapes ceases when a local maximum is attained. This poses the problem of how evolution could approach or attain a global maximum, especially for large genomes for which quasispecies [Eigen, 1971, 1993a,b; Demetrius, 1983; Domingo et al., 1985; Eigen & Winkler-Oswatitsch, 1992; Clarke et al., 1993] are ineffective. I show how increasing the dimensionality of the landscape, which occurs every time there is a gene or higher order duplication (up to polyploidy), may solve this problem. Epistasis or complementarity between the duplicated genes provides an all uphill pathway towards the global maximum. [Cf. the 'extradimensional bypass' of Conrad, 1990a,b, which 'increases the chances of finding an uphill pathway to still higher peaks': Conrad & Ebeling, 1992, and other classical ways of using epistasis to avoid the valley between adaptive peaks: Wagner, Wagner & Similion, 1994.] The evolution of hemoglobin and other dimeric and tetrameric proteins provides a testable case, since fitness is readily defined" (Gordon, 1994b).

We can characterize four stages of evolution as follows:

1. *self-reproducing entities:* quasispecies (with replication of the whole genome at a high error rate) and <1000 base pairs;

2. *single celled species:* gene and genome duplication and >1000 base pairs;

3. *limited cell type differentiation:* gene and genome duplication and ability to use different portions of the genome in different cell types;

4. *continuing differentiation:* duplication of differentiation tree terminal branches.

The ability to set up and respond to cell state splitters may have occurred at step 3 and/or 4.

Proposition 172: the number of terminal edges of the differentiation tree per gram is a monotonically increasing function of the mass of an organism.

The basic idea I wish to borrow from Bonner (1988) is that "...there is a direct correlation between complexity, as measured by the number of cell types, and size" (cf. Valentine, Collins & Meyer, 1994; Raff, 1996). From Bonner's (1988) Figure 51, I estimate that a rough linear fit would give the number of cell types proportional to the 1/6 power of the mass. Since he was referring to terminal cell types, we have the this proposition. It is interesting to note, as we should expect if the differentiation tree is the basis of genome organization, that "...gene number is positively correlated with structural complexity, while genome size is not" (Li & Graur, 1991).

This correlation may have to be modified to take into account the suggestion that the some differentiated cell types may be morphologically indistinguishable (Chater, 1993; Lewis & Wolpert, 1976).

One must carefully distinguish 'monotonically increasing' at a nonlinear power of 1/6 from 'proportional', which is linear at a power of 1. If property A is proportional to B and B is proportional to C, then A is proportional to C, and we can make straightforward arguments about the ratio of A to C. However, if A is nonlinearly monotonic to B, and B is nonlinearly monotonic to C, no simple ratio prevails of A to C. This and the following propositions are then not only crude correlations, which will have many exceptions, but also cannot be compounded one with the other in a consistent manner. Furthermore, they are meant to apply to comparisons between species or, even better, higher taxa. While the broad sweep of these generalizations invites one to make intraspecific comparisons, their validity would be even more tenuous. The broad sweep of evolution is fun to think about, but the conclusions should be taken with many grains of salt.

Proposition 173: there is a minimum size to the trajectory of a differentiation wave. In particular, this factor places a limit on the complexity of the brain, and perhaps on intelligence. It may be for this reason that 'external' food supplies (nutritive endoderm, extraembryonic yolk or placenta) for the embryo evolved.

We can assume that it takes some minimum number of cells and/or tissue mass to launch, sustain, and terminate a differentiation wave in a regulating organism:

"Various authors have established that a minimum number of cells is necessary before differentiation can take place (Andres, 1953; Lopashov, 1935a; Raper, 1941). Grobstein (1955a) even stated that 'stabilization at the tissue level seems to precede stabilization at the cell level'.... Mangold & von Woellwarth (1950) and Balinsky (1958) suggest that the formation of either one [cyclopia] or two eyes depends on the amount of eye material available" (Boterenbrood, 1962).

"One interesting feature is, that some animals are perhaps too small to get more than anterior/posterior polarity, e.g., dicyemid mesozoans. Others are too small to develop or retain segmentation, e.g. eriophyid mites (ca. 50 µm). This fact suggests a possibly interesting field of investigation, that of the grain of biological forms.... Shimizu, Sawada & Sugiyama (1993) have tried to determine the minimum tissue size required for regeneration in *Hydra;* this size turned out to be some 270-300 epithelial cells, the minimum number of cells that proved to be required for going through the critical stage of hollow sphere [cf. Chalkley, 1945], before starting re-differentiation towards the polyp form.... Weber (1992) tried to see... 'whether the necessary genetic potential exists for dense, fine grained, autonomous and localized adaptive change all over the insect wing; or whether the potential for localized remodeling is only coarse-grained and scattered here and there'. In selected cases, a very small wing region [of *Drosophila*], of less than 100 cells across, did respond to selection almost independently from the behaviour of the neighbouring cells" (Minelli, 1996b).

For example, in limb development, small dogs (Alberch, 1985b, 1986) or amphibians (Alberch & Gale, 1982, 1985) almost invariably lack an 'extra' digit compared to their larger brethren, perhaps because of one of these failures of a differentiation wave involved in homeobox related digit formation (Beauchemin & Savard, 1993; Brockes, 1989; Brown et al., 1993; Brown & Brockes, 1991; Coelho et al., 1991, 1993; Davidson et al., 1991; Dollé, Price & Duboule, 1992; Gardiner, Blumberg & Bryant, 1993; Izpisúa-Belmonte & Duboule, 1992; Morgan et al., 1992; Morgan & Tabin, 1993b; Oliver et al., 1989; Robert et al., 1991; Savard, Gates & Brockes, 1988; Spirov, 1993a; Tabin, 1991; but cf. Graham, 1994).

One evolutionary response to small size of a tissue may be to revert to a mosaic, cell by cell mechanism of differentiation. Perhaps this is what happened to create the lines of cells involved in zebrafish neurulation (Kimmel, Warga & Kane, 1994), given that the teleosts such as zebrafish are evolutionarily the most advanced fishes (Harder, 1975a,b). If so, this strengthens the case for a common mechanism for single cell and multiple cell differentiation waves.

Whatever the minimum tissue size needed for a regulating organism, it is clear that an embryo that has little yolk, upon reaching the smallest cell size in the initial cytokineses, would be at its limit for differentiation until it could eat. An embryo with a large supply of yolk, or an external food source, could produce more cells and thus more cell types, and might be more intelligent (more terminal branches in its brain differentiation tree), at least at birth or hatching:

"According to Itô's hypothesis [Itô, 1980; Itô & Iwasa, 1981] the evolution of egg size is driven, in part at least, by the problems of procurability of food by the young: when food is scarce selection will favor large, developmentally advanced young. Salamander species inhabiting stream (lotic) environments lay large eggs while those breeding in ponds (lentic) produce small eggs... But increases in egg size also increase incubation time and hence potentially increase mortality during this stage. Parental care may then evolve as a mechanism to reduce egg mortality. Nussbaum & Schultz (1989) refined the various models into a single model in which parental care and egg size coevolve" (Roff, 1992).

This view then ties in with that of the ecological advantage of rapid development as a life history evolutionary strategy (Luria, Gould & Singer, 1981; cf. Bolker, 1995), though caution in such extrapolations is advised:

"Pianka (1970) was the first to propose a set of characteristics expected to be associated with the two types of selection: r-selected organisms were supposed to have (1) rapid development, (2) high rate of increase, (3) early reproduction, (4) small body size, and (5) semelparity (single reproductive episode). K-selective organisms have the opposite characteristics, traits 1 and 2 being traded off for high population size and competitive ability. A rationale for these categories was not given. The categorization ignores the fact that in ectotherms, and to a lesser extent endotherms, development time, size at maturity, and fecundity are intercorrelated" (Roff, 1992).

Evsikov, Morozova & Solomko (1994) suggest that "...the transition to intrauterine development marks the moment when ontogeny escapes the pressure of natural selection, directed at the maximization of developmental speed".

This proposition is supported by the evidence of Roth, Blanke & Wake (1994), at least for salamanders, if not for frogs, insofar as the number of cell types depends on the number of cells available for differentiation waves to travel through:

"In general, frogs have more complex brain morphology than do salamanders.... We concentrate on the tectum mesencephali (optic tectum). Both frogs and salamanders are predators that depend on vision, and the tectum is the most important visual center for localization and identification of prey objects. In addition, the tectum exhibits the most distinctive morphology and cytoarchitecture of any part of the amphibian brain....

"In frogs, brain size is neither correlated with cell size nor with morphological complexity of the tectum.... Thus, larger brains do not necessarily have more complex morphologies. However, cell size is significantly *negatively* correlated with morphological complexity of the tectum.... Thus, frogs with smaller cells have more complex tecta (as well as other brain centers), independent of brain size....

"In salamanders, in contrast to frogs, brain size (which is positively correlated with body size...) and cell size correlate significantly... i.e., salamanders with larger brains tend to have larger cells, and those with smaller brains tend to have smaller cells. Furthermore, brain size and body size are significantly *positively* correlated with the degree of morphological complexity of the tectum.... Holding cell size constant, salamanders with larger brains have more complex tecta. In contrast, cell size is significantly *negatively* correlated with the degree of morphological complexity of the tectum. Holding brain size constant..., salamanders with smaller cells have more complex tecta....

"Brains of relatively large-celled amphibian taxa develop slowly and exhibit retarded differentiation of neuronal tissue, including degree of cell migration and formation of anatomically distinct nuclei and layers (Schmidt & Roth, 1993). As a consequence, brains of animals with large genomes and cells have an 'immature' or pedomorphic appearance when compared with related taxa having smaller genomes and cells....

"...We conclude that differences in genome size largely account for the great range in cell size and ultimately for the differences in degree of tectal morphological complexity in both frogs and salamanders..." (Roth, Blanke & Wake, 1994).

However, we must be cautious in automatically ascribing higher intelligence to animals with more complex differentiation trees, or vice versa:

"No selective advantage has been identified for the many and apparently independent increases in genome size within amphibian and dipnoan taxa [see, however, Martin & Gordon, 1995; cf. Mergen & Thielges, 1967]. Particularly telling is the case of the plethodontid salamander tribe Bolitoglossini. These salamanders are characterized by having the fastest and most precise feeding mechanisms found among amphibians (Roth, 1987; Thexton, Wake & Wake, 1977). This feeding mechanism is paralleled by highly developed stereoscopic vision (Roth, 1987; Wiggers & Roth, 1991). In addition, many bolitoglossines, particularly the arboreal species, are true acrobats, making use of limb and tail specializations (Wake, 1987). At the same time, these animals have brains that show the lowest level of morphological complexity, not only for amphibians but also among vertebrates in general. That simplification of brain morphology is advantageous under conditions of increased functional and biomechanical complexity contradicts standard functionalist views in comparative neuroanatomy. Rather, a hierarchical perspective is required that recognizes genome size increase as an independent event that has wide implications and represents a 'burden' (sensu Riedl, 1978) for brain function" (Roth, Blanke & Wake, 1994).

Nevertheless, the idea that intelligence is related to cell size and/or number has been toyed with:

"The size of the neurons is... related to their ploidy.... In experimentally produced polyploid salamanders, newts, and frog, all the cells (including neurons and glia) are greatly enlarged, but the organs (including the brain) are normal in size and shape. Although the cells are increased in volume, there is a compensatory decrease in their numbers... (Fankhauser, 1941a, 1945a,b; Gurdon, 1959; Bradom, 1960; Pollack & Koves, 1977; Jacobson, 1978d; Jacobson & Hirose, 1978; Sperry, 1988a,b). Bradom (1962) has shown that haploid salamanders also have brains of normal dimensions, since the cells, although small, are increased in number. Tetraploid mice at E14.5 and E16.5 days gestation are found to have about one-quarter as many cells as diploids of the same gestational age, but in the polyploids the nervous system contains many abnormalities (Snow, 1975).

"It is of considerable interest that the maze-learning ability of polyploid salamanders with a reduced number of very large brain cells is considerably worse than the learning ability of diploid salamanders (Fankhauser et al., 1955). In this connection, Vernon & Butsch (1957) concluded that *'polyploidy, whether triploid or tetraploid, brings about a decrease in maze-learning ability. It is not possible, however, to state whether such an effect is the result of the increase in cell size, the reduction in the number of cells, or the reduction in the number of neuronal connections that probably results from the reduced number of cells.'* Unfortunately, the learning ability of haploid salamanders, which have an increased number of neurons, has not been investigated. The relationship between brain size, number of neurons, and learning ability could also be studied very profitably in ants, in which the highest nervous centers, the corpora pedunculata, are very small in the males, larger in the females, and largest in the workers (Pandazis, 1930; Goll, 1967). *'The behavior of the three groups corresponds to their brain structure: the males may truly be called stupid, the females are far superior, and the highest faculties are those of the workers'* (Goetsch, 1957)" (Jacobson, 1991b).

One place to test these ideas would be in a series of organisms differing only in the degree of polyploidy (Muller, 1925; Sinnott, Blakeslee & Warmke, 1939; Kawamura, 1941b; Fankhauser, 1945a,b, 1952, 1955; Humphrey & Fankhauser, 1949; Sanada, 1952; Darlington, 1953; Sinnott, 1960; Astaurov, 1969; Bennett & Smith, 1972; Lewis, 1980; Evans & Davies, 1983; Licht & Bogart, 1989a,b; Vassetzky, 1991). Do differentiation trees actually quit early when the number of cells available is reduced by the larger cell size (and constant body size) usually associated with polyploidy? Neurite extension occurs at a speed comparable to differentiation waves, but has increased speed and decreased range for triploids compared to diploids (Procento & Pollack, 1989). In general, frogs develop much more rapidly than salamanders. Are there more subtle effects due to differences in differentiation wave speed, that would explain the differing results between frogs and salamanders obtained by Roth, Blanke & Wake (1994)?

Any proposition such as this is at best a broad sweep for comparing species, and is undoubtedly but one factor of many in determining intelligence, which comes in many forms, and probably cannot be summarized neatly by a scalar IQ (intelligence quotient) (Barton, Purvis & Harvey, 1995; but cf. Miller, 1992). When it comes to individual organisms, such correlations can be expected to be at best of low significance, with wide overlaps in distributions (Johnson, 1991; Anderson, 1993; Paradiso et al., 1997). Thus, I have little temptation to join the fracas over the slight statistical relationship of brain size to intelligence in humans (Cernovsky, 1990, 1991; Rushton,

1990, 1991, 1992a,b,c, 1996; Ankney, 1992; Becker, 1992a; Schluter et al., 1992; Lynn, 1993; Rushton & Ankney, 1995; Andreasen et al., 1993). Confounding factors, such as neuronal density (Witelson, Glezer & Kigar, 1995; Anderson & Harvey, 1996; Reiss et al., 1996), make such a correlation a poor basis for antisocial discrimination, for those wishing to practice it. For example, larger brain sizes are found in autistic people (Piven et al., 1995), though their corpus callosum may be smaller (Piven et al., 1997). Is there a correlation between human intelligence and number of brain branches of the differentiation tree? Maybe, but it is unlikely to be an important factor between people.

A curious experiment by Brachet (1927) may be interpreted as allowing us to alter how much the differentiation tree unfolds on one side of an animal compared to the other, providing a built in control:

"...We prick with a heated needle the gray crescent substance a little to the right or left of its middle part.... If the destruction of the gray crescent material is only partial, the formative induction can progress. But it is weaker and we can still see on the injured side a central nervous half-system, a half-chord, and a row of somites, but they are much smaller than the normal ones; they are like miniatures of normal organs. They may, nevertheless, be perfectly well made and undergo their normal histogenesis if they can acquire a sufficient bulk.... The two halves of the body are identical; they are normal in length, and have both undergone their functional histogenesis, but the reduced side remains smaller and does not win back anything it lacks. Although the conditions of recuperation if it were possible, are well provided,... the fact that organs formed with too little initial substance remain small and can never win back what they lack, shows that the materials they are made of, have a limited growing power because of their composition, that is, of the quantity of inductive substance they have received. This can only grow by nutritive assimilation within definite limits" (Brachet, 1927).

Perhaps the smaller tissues simply have too few cells to go through all terminal branches of the differentiation tree. Something like this must be going on, else we would have anticipated catch-up growth (Tanner, 1978) on the smaller side of the embryo (cf. 'compensatory growth' in mouse embryos, albeit bilateral: Power & Tam, 1993). A perhaps analogous experiment is the grafting of eyes and limbs between related species of salamanders of different sizes, in which the graft does not adapt to the size of its host (Twitty & Schwind, 1931). Understanding of later inconsistent results obtained in heteroplastic grafting of parts of the open neural plate forming the eye placodes (Rotmann, 1939; Twitty, 1940) could perhaps be worked through by observing the trajectories of the differentiation waves in normal and grafted tissues.

Once organisms made the transition from single cell differentiation waves to multicellular differentiation waves, some minimal number of cells might seem necessary to launch and propagate differentiation waves. This may be what is behind the "...well-known principle... regarding the minimal critical cellular mass required for subsequent differentiation (e.g. Deuchar, 1970)" (Desnitski, 1993). The numbers of cells in a tissue can be varied widely between individuals by ablation of early blastomeres (Huang & Moody, 1992), which gives a means for generating intact embryos with varying numbers of cells in specific, later tissues. Thus direct observation, without surgical intervention, is possible, to determine just what the critical mass is, and whether launching and/or propagation of differentiation waves occur only above that critical mass. If so, then differentiation will cease once intermediate cell types are reached, along the differentiation tree, whose masses are below the critical mass. The organism will lack certain terminal cell types, and may be expected to retain some intermediate cell types.

Proposition 174: the average number of genes per differentiation cascade has increased during eukaryotic evolution.

One might anticipate such a result merely on the basis of a general trend towards gene duplication at every level. The empirical evidence is weak, especially because the estimates for the number of terminal cell types are so crude (Bonner, 1988). Gene estimates per differentiation cascade (Figure 36) are not corrected for the (unknown) number of shared (modular) housekeeping genes (John & Miklos, 1988). The fully sequenced *"Mycoplasma genitalium,* a eubacterium thought to be the simplest known self-replicating and free-living life form" (Anon., 1995), "potentially encodes approximately 390 proteins" (Peterson et al., 1993), which can perhaps be taken as an estimate for the minimal number of housekeeping proteins for a single cell type. Correction for housekeeping genes could bring the initial part of the curve in Figure 36 down substantially. Loomis Jr. (1988) estimates...

"...that fewer than 2500 developmental genes are sufficient for the embryogenesis of complex mammals such as humans. This is only about half the number of genes required for growth of unicellular organisms. It also requires only a fairly small number of genetic events over a period of a billion years, considering that there was a storehouse full of genes [cf. Pennisi, 1994a] that could be adapted to new functions" (Loomis Jr., 1988).

Thus Figure 36 should be taken lightly. The number of differentiation cascades was estimated as twice the number of terminal cell types, after the latter was rounded to the nearest power of two. If Loomis Jr. (1988) is correct, the number of 'developmental' genes in humans would be reduced from 130 per differentiation cascade (Table 3) to 5 per differentiation cascade. The truth probably lies somewhere in between. (Cf. Proposition 144.)

For nematodes we can take the estimate of 13,100 genes (Chalfie et al., 1995) and divide it by the number of distinct cell types (at least 807 of the 949 non-gonadal cell divisions: Horvitz & Herskowitz, 1992), to get 14 to 16 genes per differentiation cascade. It is interesting to compare this with a calculation based on homeobox genes, estimated as 50-60 in number in *Caenorhabditis elegans* (Bürglin, 1995): 250 genes per homeobox gene. But the number of homeobox genes is too small for the number of distinct cell types, so we are either missing about 750 homeobox genes somehow, or there are alternative kinds of master genes, or homeobox genes can operate at more than one step of differentiation ("lack of cell or tissue specificity of homeotic proteins: Malicki, Schughart & McGinnis, 1990": Flickinger, 1994), or we are making distinctions between nematode cell types that are not 'real'. Our ability to guess these matters in advance is not great:

"An International consortium of scientists announced at the end of April that they had achieved a major goal of the Human Genome Project - the complete sequence of a eukaryote, the single-celled *Saccharomyces cerevisiae* strain S288C.... The biggest surprise of the project was that more than half the genes uncovered during sequencing were previously unknown, despite decades of intense scrutiny by yeast geneticists. Another unexpected discovery was the degree of redundancy in the genome, with several genes often appearing to have homologous sequences and functions.... [The yeast genome has] some 12 million base pairs and 6000 genes..." (Mansfield, 1996).

Given our observation of the vegetal yolk mass contraction wave (Appendix V: Gordon, Björklund & Nieuwkoop, 1994), presumably a differentiation wave that leaves the yolk mass endoderm in its wake, it would appear that master control genes other than homeobox genes must exist in vertebrates too:

"In the frog and mouse, homeobox genes are expressed in the central and peripheral nervous system and in mesodermal tissues (somites, kidneys, lungs), but not in endodermal tissues (Feinberg et al., 1987; Krumlauf et al., 1987; Dressler & Gruss, 1988; Graham et al., 1988)" (Jacobson, 1991b).

One way of estimating the number of differentiation cascades, or edges of the differentiation tree, is saturation mutation:

"A systematic effort to estimate by saturation mutation the number of pattern formation genes [in *Drosophila*] compared to the total number of loci that when mutated result in embryonic arrest has been reported by Nüsslein-Volhard, Wieschaus & Kluding (1984), Jürgens et al. (1984), and Wieschaus, Nüsslein-Volhard & Jürgens (1984). In these studies the criterion applied for the recognition of pattern formation mutations is the appearance of externally detectable morphological abnormalities at the first instar larval cuticle.... The main quantitative result deriving from this analysis is that pattern formation genes account for only about 3% of all zygotically active loci that when mutated cause developmental arrest prior to hatching (about 25% of all homozygous lethal mutations in animals developing from heterozygous parents cause embryonic, as opposed to postembryonic arrest). Identified pattern formation genes account for about 140 different loci (Jürgens et al., 1984)" (Davidson, 1986).

For this kind of rough reckoning, the figure of 140 pattern loci spread among 7000 genes in *Drosophila* yields 50 genes per differentiation cascade, which compares favorably with the 55 genes per differentiation cascade estimated using the approximate number of cell types (Table 3). (Gerhart & Kirschner, 1997, suggest over 100 homeodomain genes in *Drosophila* out of 10,000 genes, estimated by Collins, 1995, or less than 100 genes per differentiation cascade, in rough agreement, assuming such genes are all master genes.) Thus the 'pattern formation genes' may, for the most part, be precisely those active at the nodes of the differentiation tree. If the common ancestor of yeast and *Drosophila* had 6000 to 7000 genes, then the initial evolution of the differentiation tree was more a matter of reorganization of the bulk of the genes into differentiation cascades, by mutation of a relatively few 'control' genes. But this is exactly what we would expect, if we only need one pair of master genes per node of the *Drosophila* differentiation tree.

It is clear that this proposition is not applicable to single celled organisms and those of just a few differentiated cell types, which, in any case, have not achieved the capability of continuing differentiation, and thus depend on many genes for each cell type. The fit of the curve in Figure 36 indicates a slight upward trend for more complex organisms, suggesting this proposition.

Proposition 175: the average number of genes per differentiation cascade decreases inversely with branch order.

The reason that this is an inverse relationship is that 'branch order' is counted from terminal arborizations towards the main trunk for trees and streams (Leopold, 1971). If we accept the "generalization derived from comparative cytology and histology... that cell differentiation is due to a modification of a developmental stage of an evolutionarily more primitive cell type" (Flickinger, 1994), then it follows that most of the genes in embryologically later cells are luxury genes, relatively few in number. Thus the additional genome portion needed for them, i.e., the sizes of their differentiation cascades, is less than for earlier cell types. This is consistent with the following observation:

"Using genome size as a rough measure of an organism's organizational complexity, there has been a 200-fold increase from yeast [4.2 x 10^6 base pairs] to humans [3 x 10^9 base pairs], although the actual number of transcription units may have increased only 13-fold [from 4000 genes to 80,000 genes: Collins, 1995]" (Gerhart & Kirschner, 1997).

Proposition 176: the generation time of an organism with regulating embryos is a monotonic function of the maximum number of cell state splittings through which any cell could go as it traverses the differentiation tree.

Basically, on the average, each terminal tissue needs a certain minimal amount of mass in a regulating organism, if it is to have any substantial contribution to make to the physiology of that organism. I thus amplify the observation of Bonner (1988) that...

"Size is correlated with generation time in a very clear-cut way...: smaller organisms have a shorter period of development, simply because it takes them less time to build" (Bonner, 1988).

(Cf. Bonner, 1965a.) In graph theoretical terms, the generation time is dependent on the maximal pathway in a directed tree (Gellert et al., 1975). One factor that would make these relationships nonlinear is the general reduction in cell cycle time from fertilization onwards. This proposition does not apply to mosaic embryos, which have few proliferative cell divisions. It applies to frogs, but perhaps not to mammals:

"[There is a] correlation... between genome size and the duration of the larval period in anuran amphibians (Goin, Goin & Bachmann, 1968).... When we turn to development times in warm-blooded vertebrates the situation is dramatically different. The development of an elephant is spread over two years, while that of a mouse spans a mere three weeks, despite the narrow range of genome sizes found in mammals" (John & Miklos, 1988).

Proposition 177: polymacroevolution requires either an increase in the size of an organism, permitting major growth to the differentiation tree, or substantial pruning of the differentiation tree, followed by regrowth.

I have redefined macroevolution as any change in topology of the differentiation tree, no matter how small or large the change. The creation of whole new taxa may require multiple steps of such macroevolution. For this I coin the term polymacroevolution. The term may be apt, since one could think of the growth and pruning of differentiation trees as being analogous to polymerization and depolymerization of branching polymers (Allcock & Lampe, 1990). Polymacroevolution involves, then, significant additions and perhaps deletions to the differentiation tree. By the above propositions in this Section, additions generally correlate with larger organisms. This may be precisely what explains Cope's rule "that in certain vertebrate [and invertebrate] lines there is a general trend towards size increase over geological time" (Bonner, 1988; cf. Taylor, 1983):

"Cope's Law, the axiom that species within a lineage trend toward larger body size with evolutionary time, has been generally accepted in paleontology and evolutionary biology (see e.g. Newell, 1949; Stanley, 1973a; Gould, 1982a).... Stanley's (1973a) influential paper argued that Cope's Law is better viewed as evolution from small body size rather than toward large body size; most higher taxa arise from relatively small, unspecialized ancestors (see Table I in Stanley, 1973a), while their larger descendants tend to be more specialized (through scaling constraints). That adaptive breakthroughs occur primarily at small body sizes is confirmed by the observed Cambrian faunas; the first bivalves (Runnegar & Bentley, 1983...) are less than a few millimeters in maximum dimension and the Tommotian-age shelly fauna (Bengtson & Fletcher, 1983) is characterized by small size..." (LaBarbera, 1989).

"This study confirms the general pattern of Cope's Law with respect to fossil horses, although there are exceptions (the dwarfing lineages). A frequently stated corollary to Cope's Law is that rate of body size increase in various groups was gradual. The current study indicates that the first half of horse evolution (from ca. 57 to 25 ma) is characterized by relative stasis in body size. This is followed by a diversification of body size during the early - middle Miocene corresponding to the time of habitat/feeding diversification of horses. The late Miocene through Pleistocene is generally characterized by a continued increase in body size with the extinction of several smaller forms" (MacFadden, 1986; cf. MacFadden, 1992).

If pruning of the differentiation tree coincides with a decrease in organism size, we have an explanation "for the fact that extinctions often occurred in the geological past in large species, while new groups stem from small ancestors" (Bonner, 1988). Raup (1991) puts it this way, with 'a minimal load' of developmental constraint roughly correlatable with smaller differentiation trees:

"If the constraints were completely effective, how could new body plans or new physiologies ever emerge? This question has long puzzled evolutionary biologists, but there are two possible answers. First, innovations in evolution often come from the smaller, simpler, and more generalized members of the ancestral group - from species carrying a minimal load of phylogenetic [developmental] constraint. Second, bursts of speciation that often follow extinction provide many opportunities, which serves to increase the chance that at least one new body plan or physiology will succeed" (Raup, 1991).

Of course, it is not clear whether the small ancestors were derived by reduction of larger organisms, or, rather, merely represented the persistence of small species paralleling those that went through a size increase. I assume the former, at least in some cases:

"The most common effect of miniaturization on morphology is reduction and structural simplification. This is manifest in many ways, ranging from general underdevelopment to the loss of individual organs or even entire organ systems.... There is [secondarily] a consistent association between extreme phylogenetic body size decrease and morphological novelty.... A third consequence of miniaturization - increased morphological variability within species... frequently involves late-forming structures..." (Hanken & Wake, 1993).

Numerous examples of dwarfing are known (Wassersug et al., 1979; Gould, 1984a; Hanken & Wake, 1993; Quammen, 1996).

A challenge has been issued: "is Cope's rule even true at all?" (Gould, 1997a). Jablonski (1997) studied Late Cretaceous molluscs over a 16 million year period, and found that as many lineages followed Cope's law as didn't. No mention is made of any correlation between size and complexity in this series of fossils. One wouldn't anticipate much correlation in organisms of which "most... species exhibit indeterminate growth" (Jablonski, 1997), so this sampling hardly seems a fair test of Cope's law. Whether Cope insisted that *all* lineages must progress through a series of ever larger species, or only some, seems to be the argument. Clearly, the latter is assumed by Bonner (1988), and indeed Jablonski (1997) found about 30% of his lineages followed Cope's law. The argument is thus whether the cup is 30% full or 70% empty. The difficulties of reasoning such matters can be seen by simply extrapolating Gould's (1997a) conclusion:

"Most notable of... misconceptions is the false and self-serving notion that evolution displays a central and general thrust towards increasing complexity, when life, in fact, has been dominated by its persistent bacterial mode for all 3.5 billion years of its history on Earth. We should remember Little Buttercup's admonition to Captain Corcoran in *H.M.S. Pinafore,* that 'things are seldom what they seem', while we must shun the allure of bigness, for 'bulls are but inflated frogs'" (Gould, 1997a).

And frogs are but inflated bacteria? Differentiation trees provide a framework for enquiry into the relationship between size and complexity, and may indeed provide some internal drive towards larger size and complexity. Of course, other factors can operate to reduce size and complexity (dwarfism and parasitism, for example). Whether, on the whole, the lineages that do increase in size just drift that way via "diffusive or passive trends (Stanley, 1973a; McShea, 1994)" (Jablonski, 1997), or are driven that way and retarded or even reversed by other trends, will require more detailed investigations.

Proposition 178: a general correlation will be found between body size and cell size in a Cope's law series of organisms.

Since, in general, we need a larger genome to accommodate a larger differentiation tree, then (excepting allowable huge variations in the amount of junk DNA, cf. Mergen & Thielges, 1967; Martin & Gordon, 1995; Hughes & Hughes, 1995; Vinogradov, 1995, or ploidy), we expect the cell sizes to be larger because of the strong correlation between cell volume and total genome size (Olmo, 1983; Cavalier-Smith, 1985c). C.S. Minot in Thompson (1942) gives an elegant example of this in mammals:

"We get a good and even a familiar illustration of the principle of size-limitation in comparing the brain-cells or ganglion-cells, whether of the lower or of the higher animals (cf. Enriques, 1907).... Certain identical nerve-cells from various mammals, from mouse to elephant, all drawn to the same scale of magnification... are all of much the same order of magnitude. The nerve-cell of the elephant is about twice that of the mouse in linear dimensions, and therefore about eight times greater in volume or in mass" (Thompson, 1942).

For an update on nerve cell sizes, specifically in cats versus humans, see Bekkers & Stevens (1990). This correlation will be far from perfect, however:

"In mammals the largest known genome is found in the aardvark (1.7 [times] human) whereas the smallest mammalian genome is that of the deer, *Muntiacus muntjak vaginalis* (0.7 human)" (John & Miklos, 1988).

Presumably the amount of junk DNA confounds the correlation. Perhaps, in a Cope's law series, this factor will remain constant.

Proposition 179: the basis of progressive evolution may be that there is in general a greater benefit to increasing the size of the differentiation tree than pruning it as a response to a given evolutionary opportunity.

Bonner (1988) suggests...

"...that in any ecological setting, since all the intermediate sizes are represented at all times,... the only place for expansion, for pioneering, is in the upper end of the size scale. This is a place where totally new species can appear, either by selection or by drift, that will not meet stiff competition. They are pioneers that have avoided competition by finding a new niche that is above all the others" (Bonner, 1988).

Macroevolution by size increase avoids substantial pruning, and thus retains all of the benefits ('irreplaceability': Volkenstein, 1994) of the previously grown edges of the differentiation tree. This suggests that there may be a greater probability of going forward to more complex differentiation trees than backwards a bit by pruning:

"The greater tendency of large deletions compared with large duplications to be lethal would also cause a slight (selective) bias towards somewhat larger genomes" (Cavalier-Smith, 1985b).

"...Once a particular level of organizational complexity has been achieved, mutations that elaborate upon this complexity have a much greater chance of success than do mutations tending to destroy it, simplify it, or start a trend toward lower levels of complexity" (Stebbins Jr., 1969).

Ohta (1983) poses a question similar to this proposition in the context of gene families and selfish DNA:

"While the primary structure of DNA mainly evolves by random genetic drift as was first proposed by Kimura (1968), [the] organization of genes is... constantly reshuffled and tested by natural selection in any species. Such a process may increase genetic information and therefore give opportunity for progressive evolution.... The problem is whether or not self-replication is directional, since those DNA segments that have a higher chance of duplication than deletion within a genome may rapidly increase their number, even if they are not useful for the organisms.... However, the theory to analyze directional conversion at the level of populations is not yet available and is left to future study" (Ohta, 1983).

Since then, Hillis et al. (1991) produced a case that "...demonstrates that concerted evolution of rDNA (ribosomal DNA) can be driven by directional, rather than stochastic, processes, and that these directional processes do occur among (as well as within) chromosomes."

For example, 'unbreakable' inductions can lead to anastomosis of edges of the differentiation tree (cf. Figure 30) in a way that makes pruning deleterious. Thus a portion of the differentiation tree turning into a differentiation web could provide a basis for what has been called 'phylogenetic ratcheting' (Muller, 1964; Stebbins Jr., 1967; Katz, 1987; Wagner & Krall, 1993):

"If some step is added to a development it is less likely to affect the organism adversely than if a step is eliminated, and once this step is added it may be useful for other purposes.... It is easier to add than to subtract, and the additions can be the material for further developments" (Bonner, 1988).

Numerous inductions may be involved in the embryogenesis of an organism (Raff & Kaufman, 1983):

"...The epigenetic system itself may have become such a complexly interwoven nexus of cybernetic interactions that it is either extremely difficult to produce any phenotypic change at all, or the only possible changes may involve such a complete disruption of the epigenetic organization that the phenotypes produced are highly aberrant and thus probably at a strong disadvantage" (Waddington, 1953a).

Thus, when 'inexact parallelism' (Gould, 1977a) in heterochrony is prevented by selection, edges of the differentiation tree intertwined by inductions may persist. Arthur (1988) suggests that, in reference to his 'morphogenetic trees': "As terminal links are added, all earlier links become more constrained, and more 'major' in a passive way, simply due to their coming to control more and more later decisions." The same concept applies to differentiation trees, and thus offers another mechanism for ontogenetically based phylogenetic ratcheting.

Ratcheting may also be considered to occur when some terminal branches of the differentiation tree are more necessary to survival than others, and thus less likely to be pruned (or are pruned less) in the course of evolution. In this situation, these terminal branches may also provide a stable 'platform' on which further branching can occur. There are thus numerous ways to come to the conclusion that polymacroevolution will generally increase the size of the differentiation tree rather than decrease it (cf. Section 10.02).

Proposition 180: polymacroevolution proceeds by general increases of both the sizes of the differentiation trees and the physical sizes of lines of organisms, alternating with geologically sudden prunings to both smaller differentiation trees and organisms.

Bonner (1988) summarizes the grand sweep of evolution as follows, a scenario that we may interpret in terms of differentiation trees:

"If we now compare the long-term, maximum size changes that alter very slowly over billions of years and the size changes within groups that follow Cope's law, there is an important message. The maximum sizes are each attained from a different, major group of organisms: for plants they are, for instance, lycopods, gymnosperms, and angiosperms; for animals they are trilobites, eurypterids, fish, amphibians, dinosaurs, and mammals. These maximum sizes were reached by progressive size increases within each group. This means that overall size change during evolution is a result of a series of rises of different successive taxons" (Bonner, 1988).

In terms of differentiation trees, this proposition is my interpretation of what I call 'Bonner's law' (Figure 39).

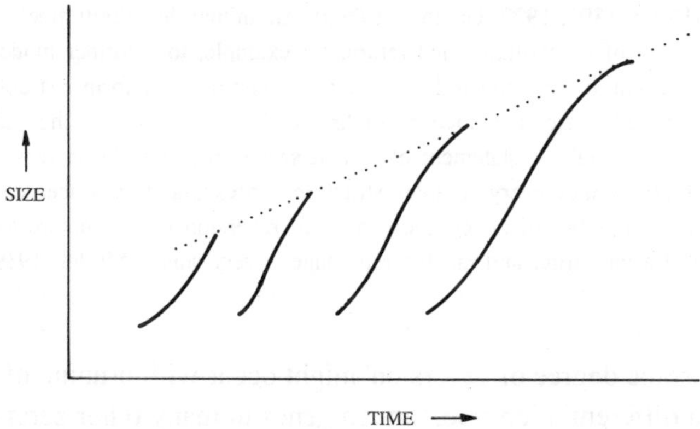

Figure 39. Bonner's Law. Schematic diagram of consecutive phylogenetic sequences following Cope's law, in which each sequence rises to a greater maximum size (which I call "Bonner's law"). From Bonner (1988) with permission.

Proposition 181: some terminal branches of the differentiation tree at a given stage of macroevolution are more expendable than others, so that the ratcheting mechanism of polymacroevolution consists in the development of more effective (higher fitness value) terminal branches and the elimination of less effective terminal branches through Cope's cycles of taxon size increase followed by pruning (Bonner's law).

The long term increase in the maximum size of organisms suggests that something more is going on than simple pruning, which I take as preservation of the best parts of differentiation trees. The pruning that is possible may be quite substantial, so that there is no need to demand an extreme form of irreversibility in evolution, i.e., 'Dollo's Law' (Gould, 1970a; Wagner, 1982; Smith & Morowitz, 1982; Brooks, Cumming & LeBlond, 1988; Morowitz, 1992b; Sanderson, 1993; Siddall, Brooks & Desser, 1993; Raff, 1996), in considering ratcheting mechanisms:

"In its literal interpretation, then, the 'doctrine' of irreversibility is obviously untenable.... The 'doctrine' is usually ascribed to the late Louis Dollo as 'Dollo's Law.' This, however, is an undeserved slur on this distinguished Belgian paleontologist. His theory was one of a more modest sort (Dollo, 1893, 1922; Gregory, 1936a). An animal, he admits freely, may change the general course of its evolution and return, for example, to a former mode of life long since abandoned; but in doing this it does not revert precisely to its former structure.... In this original and limited sense, the doctrine of 'irreversibility' seems, on the paleontological evidence, to be essentially a statement of fact. It seems further to be in reasonable accord with genetic theory. Since every complex structure is presumably produced ontogenetically by a considerable number of genes, the chance of reversion of a structure to a condition characteristic of a far earlier and far different stage is very small (Muller, 1939a)" (Romer, 1949).

Of course, some degree of 'reversion' might occur with pruning of a terminal branch of a differentiation tree, though genes in many other parts of the tree would likely have changed in the interim.

Proposition 182: the main difference between small organisms that start off consecutive Cope's cycles is that the later ones have a higher proportion of evolutionarily more effective terminal branches in their differentiation trees.

For example, a tiny, flowering plant like *Arabidopsis,* with a stripped down genome size (Marx, 1988) only slightly larger than *E. coli* (John & Miklos, 1988), and only 10% "of highly repetitive DNA that is unlikely to represent genes" (Fosket, 1994; cf. Bowman, 1993), may be poised for a Cope's law sequence of evolution. It may have a highly compact differentiation tree. The puffer fish may be in the same situation (John & Miklos, 1988; Pizon,

Cuny & Bernardi, 1984; Brenner et al., 1993; Mileham & Brown, 1994; Crnogorac-Jurcevic et al., 1997). *Drosophila* also seems to have a compact genome (Petrov, Lozovskaya & Hartl, 1996; Charlesworth, 1996a). (Organisms such as these might be best suited for ascertaining which portion of the DNA corresponds to the differentiation tree.)

A full test of this idea would require finding a group with an extant, evolutionarily older member having a compact genome, from whose ancestor it could be shown that radiation occurred.

Proposition 183: more effective edges of differentiation trees account for the phenomenon of condensation or shortening of developmental stages, which in turn allows more terminal branches to be added.

Ekstig (1994) has noted that the...

"...ontogenetic age of appearance of human developmental stages [is an inverse]... function of their evolutionary age.... This result has two intriguing aspects: first, the observation that condensation is determined by one and only one variable, viz. the age of the trait, and, second, the simple regularity of condensation.... The more condensed a developmental stage is, the harder it will be to condense it even more" (Ekstig, 1994).

It is as if the older an edge of the differentiation tree, on an evolutionary time scale, the more it has been shaken and compacted into a more efficient machine for carrying out its step of differentiation. Ekstig (1994) sees the driving force behind this process to be a continual...

"...selection pressure for early maturation.... Stearns (1992) concludes that the benefit of early maturation is demographic: organisms that mature early have a higher probability of surviving to maturity than do organisms that mature late. Organisms that mature early also have higher fitness because their offspring are born, and start reproducing, sooner. Although its magnitude varies, the benefit is always present" (Ekstig, 1994).

There is obviously a pull and shove here: earlier maturity allows a more complicated differentiation tree to evolve, provided maturity is further delayed. In a way, Cope's law may be viewed as a nonlinear oscillation between these effects, occurring on geological time scales.

Proposition 184: the fractal dimension of a differentiation tree may be related to its potential for evolutionary radiation.

As another example of a potential for predictability, note that the Hausdorff or fractal dimension of a tree is intimately related to its rate of branching (West & Goldberger, 1987). It thus may be important to determine the fractal dimension of differentiation trees, which may correlate with the fractal dimension of the corresponding portion of the phylogenetic tree, and thus the amount of species radiation.

One consequence of looking at changes in differentiation trees as fractal changes to a fractal structure is that the phylogenetic tree thereby takes on a fractal character (cf. Gould, Gilinsky & German, 1987; Gould, 1988b; Maynard Smith, 1990a), and one should not be surprised at the overall similarity between actual and simulated phylogenetic trees (Figure 14.5 in Maynard Smith, 1989b). These patterns are reflected in our simulations of diatom shell patterns, which share many qualitative features of phylogeny (Figure 40):

1. There are many initial branches (unsuccessful 'phyla', as in Gould, 1989a; cf. García-Bellido, 1985a).

2. Even major branches can exhibit extinction.

3. Only a few major 'phyla' survive to the present.

The fine, feathery branches of Figure 40 are suggestive of the notion of Schopf (1981) that all species are continually speciating, though few persist.

A fractal structure has been proposed for chromosomes (Takahashi, 1989b, 1996). Whether this has any relationship to the fractal structure of differentiation trees might be interesting to determine.

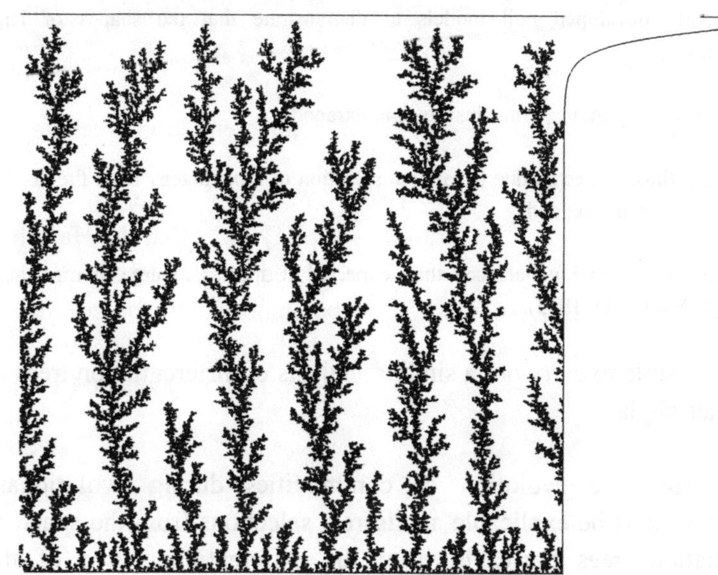

Figure 40. Computer simulation of a fractal branching pattern resembling a phylogenetic tree. What is actually simulated is totally irreversible precipitation, in which each diffusing particle (not shown diffusing) sticks where it hits (cf. Gordon & Aguda, 1988). This figure is a simulation of a portion of the shell of a pennate diatom with a moving source line of particles (colloidal silica) kept 10 units above the highest protuberance. The starting structure is a straight line at the bottom (representing the midrib). Periodic boundary conditions are used horizontally on the 960 x 960 array. The graph shows the average concentration gradient of diffusing particles as a function of distance from the moving line, which is the source of the particles. Note that most of the precipitation occurs in a relatively narrow region near the source line. Thus the structure below this zone is essentially fixed for all time. In this sense, the moving source line approximates a time line in phylogeny. The abscissa would be replaced by a multidimensional niche space. The metric for interactions, replacing the concentration gradient, would include species-species competition, etc. From Figure 9 in Gordon & Drum (1994) with permission of Academic Press.

Finally, let us note that the concept of tree topology has already been applied with some success in distinguishing evolutionarily radiating phylogenies:

"We used recently developed null models to demonstrate that the shapes of large phylogenetic trees:

1) are similar among angiosperms, insects, and tetrapods;

2) differ from those expected due to random selection of a phylogeny from the pool of all trees of similar size; and

3) are significantly more unbalanced than expected if species diverge at random..." (Guyer & Slowinski, 1993).

It should be possible to carry out a similar analysis of differentiation trees of these and other phyla.

Proposition 185: the evolution of communities during evolutionary radiation may in part be explicable as discrete selections from the space of all differentiation trees derivable from the differentiation tree of the common ancestor. This may occur in similar ways in different radiations, explaining evolutionary parallelism between unrelated groups.

Competition between species, and between species and the fluctuating environment, leads to the death of individuals and the extinction of species (cf. Bak & Sneppen, 1993; Sneppen et al., 1995). From my point of view, particular differentiation trees are thus lost. If the differentiation tree is indeed the fundamental germ-line replicator (Proposition 108), then we may view the evolution of communities as one of interactions between differentiation trees. It has been suggested that communities have discrete structures, i.e., that there is a limited set of discrete niches that can be formed, as hinted by the independent radiation of the marsupials, with its strong parallels to the placental mammals (Lima-de-Faria, 1983; Wills, 1989). In addition to this parallelism, Wilson (1992) mentions a similar case in plants:

"What selection force drives the herbs to larger dimensions and assembles the island forests? Evidence from many sources suggests that it is ecological opportunity afforded by the absence of conventional trees.... Darwin [1859] correctly deduced the process...:

'Trees would be little likely to reach distant oceanic islands; and an herbaceous plant, though it would have no chance of successfully competing in stature with a fully developed tree, when established on an island and having to compete with herbaceous plants alone, might readily gain an advantage by growing taller and taller and overtopping the other plants. If so, natural selection would often tend to add to the stature of herbaceous plants when growing on an island, to whatever order they belonged, and thus convert them first into bushes and ultimately into trees' [Darwin, 1859]" (Wilson, 1992).

Wilson (1992) also reviews the concept of "assembly rules of faunas and floras" which may be related to the marsupial/placental parallelism. On the other hand, the ability of organisms to enter and adapt to niches may not be as facile as we might imagine:

"One of the odd features about niches is how many remain empty. For instance there are plenty of sea snakes in the Pacific and Indian oceans; why are there none in the Atlantic? Or again, why are there no blood-drinking bats in Africa? There are plenty of blood-drinking bats in the tropics of the New World, and fish-eating bats too. A quite recent case is the cattle egret, which lives on the insects stirred up by grazing cattle throughout the Old World. In the 'thirties these birds reached North and South America, where they are thriving. Even before there were cattle in North America there were bison, which would have provided the required conditions. It seems that ecological vacancies may persist for millions of years without any species evolving to occupy them" (Taylor, 1983).

Similar niche vacancies occur at the level of protein evolution (Lin et al., 1997). It would be interesting to know if parallel radiations involve quite distinct differentiation trees, or whether ecologically corresponding organisms in parallel radiations have some similarities between their differentiation trees (Natalie K. Björklund, p.c., 1997).

Proposition 186: the winner in the collision of two radiations is the radiation that has previously had the most prunings of its differentiation trees.

Insofar as prunings of the differentiation tree may be detected through Cope's law series in the fossil record, a mere count of the number of such series for various radiations, prior to their collision (by formation of land bridges), might test this idea. Differentiation trees may thus form a framework for investigating this currently open question in paleontology:

"...What happens when two full-blown, closely similar dynasties meet head on? If it were possible to play God with geological spans of time to wait and watch, the ideal experiment would be this: allow two isolated parts of the world to fill up with independent adaptive radiations of plants and animals, so that the majority of species in each theater have close ecological equivalents in the other theater; then connect the two regions with a bridge and see what happens.... Two and a half million years ago the Panama isthmus rose above the sea, allowing the mammals of South America to mix with the mammals of North and Central America.... The mammalian radiation in South America was as expansive as that in Australia, and its convergence to the World Continent fauna... comprising [animals of] Africa, Europe, Asia, and North America as far south as the southern rim of the Mexican plateau... was even closer.

"The Great American Interchange [Simpson, 1980; Marshall et al., 1982; Anderson, 1984; Marshall, 1988] resulted in a sharp increase for a time in the mammalian diversity on both continents.... Why did the World Continent mammals prevail? No one knows for sure.... Evolutionary biologists keep coming back to it compulsively....

"The mammals of North American origin proved dominant as a whole over the South American mammals, and in the end they remained the more diverse. Over two million years into the interchange, their dynasty prevails. To explain this imbalance, paleontologists have forged a widely held theory, an evolutionary-biologist kind of theory, in other words a rough consensus that violates the minimal number of facts. The fauna of North America, they note, was not insular and discrete like that of South America. It was and remains part of the World Continent fauna, which extends beyond the New World to Asia, Europe, and even Africa. The World Continent is by far the larger of the two land masses. It has tested more evolutionary lines, built tougher competitors, and perfected more defenses against predators and disease. This advantage has allowed its species to win by confrontation. They have also won by insinuation: many were able to penetrate sparsely occupied niches more decisively, radiating and filling them quickly. With both confrontation and insinuation, the World Continent mammals gained the edge.

"The testing of this theory has just begun. Right or wrong, whether decisive in empirical support or not, its pursuit alone promises to link paleontology in interesting new ways to ecology and genetics" (Wilson, 1992).

Add development, and the latter is precisely the point of this book. This proposition could perhaps be tested in freshwater lakes, where, for instance, the fantastic radiation of African cichlids is now coming up against introduced species (Goldschmidt, 1996). It is tested in an irregular way on every island invaded by human vectors (Quammen, 1996).

Proposition 187: ratcheting mechanisms based on differentiation trees are consistent with physics and thermodynamics.

Ratcheting mechanisms in evolution apparently present a problem for some camps of evolutionists:

"Variation and selection work together to produce evolution. The Darwinian says that variation is random in the sense that it is not directed towards improvement, and that the tendency towards improvement comes from selection" (Dawkins, 1986).

"Paleontologists, at least since the time of the American Edward Drinker Cope [1887, 1896], were fond of espousing internal, 'vitalistic' factors as evolutionary mechanisms.... 'Orthogenesis,' a term coined by Haacke (1893; *fide* Simpson, 1944), describes a linear directional change in phylogeny, a pattern generally thought in presynthesis days [before the New Synthesis] to reflect internal evolutionary processes.... Henry Fairfield Osborn [1896, 1929], whose theory of orthogenesis (later called 'aristogenesis') saw linear evolutionary change as the result of directed mechanisms of genetic change arising from within organisms themselves, a mechanism, moreover, taking precedence over natural selection if not supplanting it altogether" (Eldredge, 1989a).

A more complete history of orthogenetic ideas (cf. Section 10.04) is given by Richardson & Kane (1988):

"As the intellectual heirs of Louis Agassiz (1849) and Georges Cuvier (1828), Hyatt (1866) and Cope [1887] [cf. Osborn, 1931] began with an orthogenetic view, locating the explanation of evolutionary patterns in the recapitulation of phylogeny by ontogeny, and denying the centrality of adaptationist explanations. By the 1870s, their orthogenesis was compromised in favor of a view emphasizing the effects of use and disuse, together with the inheritance of acquired characteristics.... By the turn of the century, American Neo-Lamarckism was in retreat.... Whatever the cause of its decline, the response of Cope's disciples, William Berryman Scott (1917) and Henry Fairfield Osborn (1907), was to turn away from the inheritance of acquired characteristics and explain the trends in the fossil record with a purely orthogenetic mechanism" (Richardson & Kane, 1988).

Volkenstein (1994) defends the directed evolution theory of 'nomogenesis' of Berg (1926) as being based on strictly physical reasoning.

Perhaps the possible physical reality of differentiation trees will make the idea of progress in evolution more palatable (cf. Section 10.05). Progress is not equivalent to detailed determination of or directionality at every step (the criticism of orthogenesis by Simpson, 1951). Mutation still provides the basic material on which natural selection acts, and as thought before, we can anticipate that most mutations are deleterious. However, the ratcheting nature of the differentiation tree may make its growth more likely than its pruning. The resultant additions to the genetic repertoire probably usually open up more directions for further evolution. The differentiation tree itself may be what Wills (1989) called "...the ability to evolve [which] adds a higher-order power to the evolutionary process and makes it much more likely that complex life forms can evolve rapidly." This idea, separable from the inheritance of acquired characteristics, goes back at least to Lamarck:

"Around the turn of the nineteenth century, the French zoologist Jean Baptiste Lamarck [1803, 1809] devised an explanation of adaptedness in which the activities of organisms in adjusting to changed circumstances, together with their ability to pass what they have learned to offspring, reflects and advances an inherent tendency of living things to become more complex, and by becoming more complex to stay adapted to environmental change" (Depew & Weber, 1994).

The resistance to the idea of such 'inherent tendencies' probably comes from a false supposition that they can only be caused by vitalistic entelechies (cf. Section 1.11), or supernatural beings, or 'directed evolution', and from a world view or paradigm inherited from a physics that considered mass to be the only (at that time) intrinsic, measurable property of an object that could influence its dynamics:

"Darwin's reliance on what he took to be Newton's methods have long been recognized (Hull, 1973). What is novel in our argument is that Darwin is also faithful to the way that objects are represented in Newtonian systems. Lavoisier [1776] and Dalton [1808] had already applied Newtonian thinking to the problem of chemical affinities. Charles Lyell [1830-1833] had applied it to the hottest science of the day, geology, the field in which Darwin had been trained. Darwin had avidly and thoroughly absorbed Lyell's Newtonian, or 'uniformitarian,' geology while he was a young man. He thus worked within a very self-conscious scientific culture in which external forces operating on bodies rather than internal drives were expected to drive change over time" (Depew & Weber, 1994).

Selection is not simply between better or worse genes, but also between fewer genes or more (cf. Shapiro, 1992c, 1993). Every time a mutant appears which yields more genes, and is also better from the point of view of selection, there is rarely a turning back. I thus suggest that progressive evolution via differentiation trees may be fully compatible with a broadened notion of the genetic basis of Darwinian selection. Ratcheting at the level of differentiation trees, therefore, does not have to invoke the concept of directed mutation, and can be fully consistent with random variation at the level of individual mutations, even though that particular concept may not be fully secure:

"The main purpose of this paper is to show how insecure is our belief in the spontaneity (randomness) of most mutations. It seems to be a doctrine that has never been properly put to the test....

"The origin of genetic variation has been the subject of bitter controversy.... Is all variation essentially random, like thermal noise? Or can the genome of an individual cell profit by experience? At its extremes, it was an argument between reductionists and romantics - between those who sought to explain the evolution and behaviour of the biosphere in terms of the laws of physics, and those who wished to make the success of evolution just another manifestation of the mysteriousness of living things" (Cairns, Overbaugh & Miller, 1988).

What seems to have been going on here is a misconception of the 'laws of physics', a confusion between the first and second laws of thermodynamics: microscopic reversibility (Keizer, 1987), or 'thermal noise' is perfectly consistent with the existence of thermodynamically irreversible processes:

"The directionality of events in nature.... A̲ll energy transactions made in nature are such that the total energy remains the same, as expressed by the law of conservation of energy.... I̲f we use the conservation law as a guide, and list the events which involve energy transformations that satisfy the conservation law, we find that we can include in this list many events that never seem to occur. To obtain such events, we need only make a motion picture film of an explosion, a ski jump, the growth of a flower, a waterfall, the decaying motion of a pendulum, etc., and then project these events backward (in time).... In fact, the conservation law has no influence whatsoever on the over-all plan according to which nature operates.... It is rather natural, then, to ask if it might not be possible to formulate another law expressing at least some feature of the plan of nature in regard to its choice of energy transactions. There is indeed such a law, and it is called the *second law of thermodynamics* " (Ingard & Kraushaar, 1960).

Proposition 188: differentiation trees may provide a basis for an irreversible thermodynamics of evolution.

Contrary to the suggestion that "...evolution has involved progressively finer tuning of a more and more recalcitrant genome" (John & Miklos, 1988), it is the pulsatile, alternate branching and pruning of the differentiation tree that may explain 'the great sweep of evolution' (Bonner, 1988; Bonner's law: Proposition 180). Perhaps differentiation trees, looked at in this dynamic way, can be thought of as analogous to dendritic structures that form spatial patterns, such as diatom shells (Gordon & Aguda, 1988; Gordon & Drum, 1994; cf. Meakin, 1986; Feder, 1988; Figure 40) or snowflakes (Bentley & Humphreys, 1931). A snowflake takes its shape in response to instabilities during its irreversible precipitation process, which gives it a fractal character. Sintering processes trim some of its growth, and historical processes, such as encounters with air of varying humidity and temperature, affect its rate of branching. Such analogies may finally provide a workable thermodynamic approach to progressive evolution (Weber, Depew & Smith, 1988; cf. Volkenstein, 1994), though even for dendritic precipitation of crystals, the theory is far from complete (Langer, 1980, 1989), and may not ultimately yield to a thermodynamic analysis, since kinetic parameters far from thermodynamic equilibrium are so important (Lotka, 1920, 1925, 1945, 1956; Kingsland, 1985; Depew & Weber, 1994):

"The idea that natural selection had a predictive thermodynamic basis was first explicitly articulated by Lotka (1922a,b), who argued that those organisms that were most successful in channeling energy flows through themselves, and at the same time increasing total flow through their ecosystems, would necessarily be selected for" (Wicken, 1984).

"Lotka's understanding of evolution as the 'history of a system undergoing irreversible changes' (Lotka, 1925) echoed a discussion of the second law of thermodynamics as a law of evolution put forth by physicist Perrin (1903)" (Kingsland, 1985).

However, if an irreversible thermodynamic explanation can be found for the growth of dendritic (fractal) crystals, then perhaps an analogous irreversible thermodynamic explanation for the growth of dendritic differentiation trees can be found (cf. Campbell, 1988). This would involve ecological considerations, in which attempts would be made to find rough correlations between the complexities of differentiation trees and their selective advantages in ecosystems (cf. Roff, 1992; Mangel & Ludwig, 1992; Wainwright & Reilly, 1994). Mental capacity may be just such an ecological parameter, which may be correlated with the size of the brain and the number of cell types in the brain (Proposition 173), i.e., with the size of the terminal branch of the differentiation tree that produces the brain:

"No IQ [intelligence quotient] measurement valid for the whole animal kingdom has as yet been proposed, nor is such a proposal likely to be imminent. On the other hand, brain volume is a measurable quantity that seems to be roughly correlated with intellectual capacities in primates.... Stebbins Jr. (1971a)... illustrates the acceleration of increase in brain volume during primate anagenetic evolution. An autocatalytic course of this evolution, at least during certain phases - an autocatalysis of intelligence - is not excluded by these data, although they are insufficient for demonstrating it" (Zuckerkandl, 1976).

Simple repeated terminal additions to a growing terminal branch of the brain's terminal branch of the differentiation tree could lead to exponential (or faster) growth in brain size, and thus to an appearance of 'autocatalysis' (Figure 41).

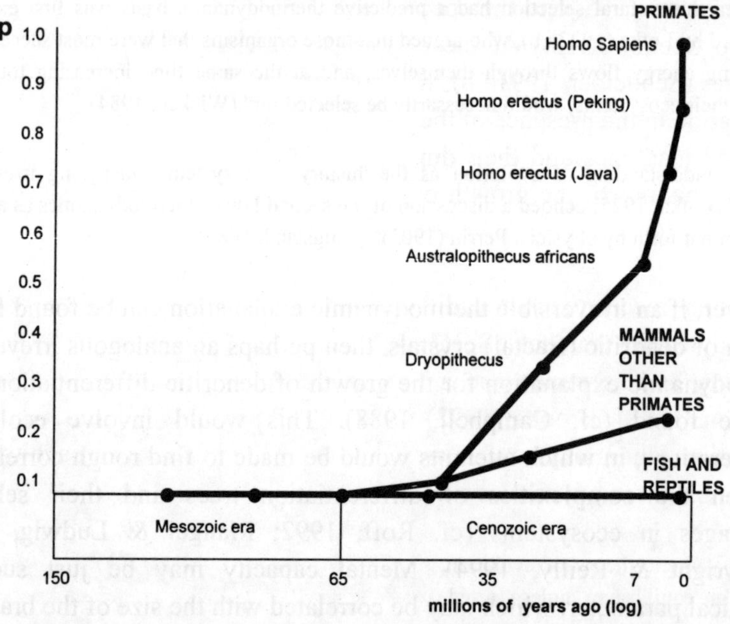

Figure 41. Evolution of brain size in mammals. While a gradual increase in brain size has occurred over evolutionary time in other mammals, primates have undergone an approximately exponential increase. p = the 'psychic factor' (brain weight/(body weight)$^{2/3}$) of Snell (1892), unfortunately not a dimensionless parameter. From Fujita (1990) with permission of the author and publisher, kindly copied and translated by Takashi Miura.

Since selective advantage is related to body size, energy flows, biomass, etc., there may indeed be an overall ecosystem thermodynamics that molds the evolution of differentiation trees (cf. Johnson, 1988, 1990a, 1992b, 1995; Schneider & Kay, 1995; Vanriel & Johnson, 1995; Gladyshev & Gladyshev, 1996; Gladyshev, 1997), expressed here in hyperbole:

"The legitimate explanation for the structure of organisms has to be sought not in physics, but in economics (Ghiselin, 1978).... When one asks how it is that what goes on in the marketplace can influence what happens on the assembly line, the answer is obvious" (Ghiselin, 1980).

(Cf. Ghiselin, 1995, and the species as 'commercial firm' in Rieppel, 1986. Cf. also Section 1.18.) At the other extreme, if the theory of molecular machines (Schneider, 1991a,b), which takes into account the preservation of information in the presence of thermal noise, could be generalized to include branched machines and their duplications, we might be able to develop a thermodynamics of the growth of differentiation trees at an actual physical level.

Ratcheting mechanisms for differentiation trees may correlate with the thermodynamic "'ratchet effect,' induced by continual interaction between symmetry and asymmetry in a fundamentally asymmetrical environment" (Johnson, 1988). Brooks (1992) hints at correlations of microevolution = equilibrium thermodynamics = New Synthesis, and macroevolution = irreversible thermodynamics = what we need:

"In the synthetic theory of evolution, the genetic structure of a population is expected to remain at equilibrium unless acted upon by an external environmental force.... New environmental forces, however, can shift the genetic structure back to its previous equilibrium. Hence, the processes of microevolution postulated by the synthetic theory are in themselves time reversible.... There are, however, many biological processes, such as reproduction, development (ontogenesis), death, speciation (phylogenesis), and extinction, that are inherently irreversible phenomena. The question of how time-asymmetric processes are to be explained and integrated into general biological theory is central to the new evolutionary biology" (Brooks, 1992).

Depew & Weber (1994) (cf. Green, 1991a) take this argument a step further:

"Our account is unique because it systematically maps... various ways in which the Darwinian tradition has used models of dynamical systems, whose home base is physics, to articulate and apply its central explanatory concept, natural selection. Darwinism's use of dynamical models has received insufficient attention from biologists, historians, and philosophers who have concentrated instead on how evolutionary biology has maintained its autonomy from physics. Yet it is only by recovering Darwin's own relationship to Newtonian models of systems dynamics, and genetical Darwinism's relationship to statistical mechanics and probability theory, that insight can be gained into how Darwinism can successfully meet the challenges it is currently facing. In our view, these challenges call for a reorientation of

Darwinism around the nonlinear dynamics of complex systems and nonequilibrium thermodynamics" (Depew & Weber, 1994).

Note that the irreversibility we seek is not the same as the irreversibility caused by losses of parts of the genome, large or small (Sanderson, 1993; Siddall, Brooks & Desser, 1993). Our irreversibility is primarily one of gain, in terms of more (functioning) genome and a presumed generally lower likelihood that the organism can survive afterwards without the gains (cf. Gordon, 1994b). It is perhaps the other side of the same coin, but the side that looks at the differentiation tree as a whole, over evolutionary time, to ask if it is generally increasing in size and complexity.

Analogies between evolution and statistical mechanics have recently been criticized as follows:

"The ergodic theorem of statistical mechanics has no analogue in [DNA base] sequence space, since selective pressures impose strong constraints on the distribution function (Eigen, 1985). There is no H theorem [Boltzmann entropy] and the volume in sequence space is not conserved - selection can lead to a marked contraction around a master sequence. The volume occupied in sequence space is therefore not a potential function.... It is the fitness of a sequence that is evaluated by selection, not its entropy.

"The problems of assuming an analogy between randomizations in phase space and genetic sequence space are best illustrated by investigating the effects of population size in each case. In statistical mechanics,... [the] larger the... system is, the better the irreversibility.... Elementary reactions are reversible.

"This situation is completely reversed in evolution. Here the strongest irreversible changes occur in small populations, e.g. founder populations. Large panmictic populations show a much higher tendency for genetic stasis. Irreversibility in evolution is not a matter of the most probable distribution of a large system, it appears already at the stage of the individual elementary reaction. Evolution is irreversible, because it is a historical process (Mayr, 1988) that conserves singular events..." (Berry, 1995).

The problem with this criticism is that it is couched mostly in terms of equilibrium statistical mechanics. Systems far from equilibrium, such as irreversible precipitation (Gordon & Drum, 1994; Figure 40), act much more like biological evolution.

Proposition 189: the growth of differentiation trees over evolutionary time corresponds to an increase in the maximal entropy and information available in living organisms.

"The open-endedness of evolution is particularly easy to understand from the perspective of differentiation trees. Open-endedness is what appeals to me about genetic programming. Once we have 'good' genetic programming structures, I think computers and software will undergo the sort of explosive evolution that life underwent in the Cambrian [cf. Soule, Foster & Dickinson, 1996; Soule & Foster, 1998]" (Steve McGrew, p.c., 1997).

An analogy may be made between the expanding universe and the expansion of the differentiation tree over evolutionary time, which may produce a thermodynamic understanding of the latter. In the expanding universe, the total amount of entropy actually increases over time:

"If the cooling occurred rapidly enough... matter and energy, initially in perfect equilibrium, never could re-achieve equilibria at successively cooler temperatures. They grew farther and farther apart, establishing exactly those disequilibria from which order could spring.... The decrease in entropy occurring when electrons 'froze out' or materialized from their substrate was more than compensated by an entropy increase woven into the fabric of space-time itself. Expansion of this fabric permitted the total quantity of entropy in the universe to increase. There was always more space requiring new information for its description. The inability to provide information as fast as expansion drove the need for more of it represented [as] increasing ignorance" (Seielstad, 1989).

An expanding differentiation tree means an expansion of the genetic landscape into realms of ever higher dimensionality (Gordon, 1994b; cf. 'open-ended evolution': Kaneko, 1995). Perhaps, in this expanding biological universe, the maximal entropy to which the system can rise also increases over time. An expanding differentiation tree could be an important component to add to, for instance, the simulation of early vascular land plant evolution by Niklas (1994c), in which he implicitly assumed that the genetic basis for the whole (albeit phenotypic) 'morphospace' was present in "The simplest and most ancient phenotype known for vascular land plants... the Silurian fossil remains of *Cooksonia.*" Evolution is not merely a walk on a landscape. That landscape must be constructed en route, and as it is constructed, its dimensionality increases. The rate of evolution, which does not appear in any natural way in static landscape models (Niklas, 1994c), might yield to understanding in terms of the evolution of differentiation trees.

The very ability to expand the genetic landscape can lead to trapping of whole classes of organisms, in which too much pruning of a highly branched differentiation tree would be needed to 'start over':

"Jacobs (1990) has proposed an ingenious model to link arrangement of developmental instructions, by what he labels as 'selector genes', and perceived constraints of morphological expression as indicated by rates of ordinal origination. He argues that arthropods and annelids with clear serial construction underwent initial exuberance of morphological experimentation in the Cambrian, but were subsequently constrained by a regulatory system that had to remain simple owing to dispersal of the 'selector' genes" (Conway Morris, 1994a).

This leads to the reasonable notion that the degree of irreversibility may depend on the 'narrowness' of the differentiation tree, not just its number of levels of branching, i.e., whether it looks more like a tall tree than a bush (Figure 38).

Proposition 190: genome doubling, a major, repetitive event in evolutionary history, does not fundamentally alter the differentiation tree.

We will see (Propositions 219 and 220) how doubling of the whole genome could account for haplontic/diplontic life cycles. However, the number of life cycles clearly has not increased with each doubling of the whole genome, demonstrated to have occurred a number of times by Sparrow & Nauman (1976) (cf. Stahl, 1967; Reanney & Ralph, 1968; Wallace & Morowitz, 1973; Shimkets, 1990; Miklos & Rubin, 1996; Wolfe, 1996), though the "most recent genome duplication [in mammals] is believed to have occurred approximately 300 million years ago, long before divergence of lineages leading to mouse and man (Ohno, 1970; Ohno, Wolf & Atkin, 1968)" (Nadeau, 1991). Thus we have to formulate an alternative relationship between genome doublings and differentiation trees, if the existence of genome doublings is not to nullify the idea of differentiation trees. The simplest hypothesis is that a genome doubling has no immediate effect (except physiologically: Sparrow & Nauman, 1976). Molecular crosstalk between two copies of a differentiation tree can probably keep them coordinated during development, so that organisms with a doubled genome have a reasonable possibility of being viable. Over evolutionary time, their differentiation cascades would genetically drift apart, making for a doubling of the richness of the repertoire available to the differentiation cascades. Thus genome doubling, while a major evolutionary event since *all* differentiation cascades are affected, is classified by me as effectively microevolutionary in its immediate consequences.

Genetic recombination may effectively remerge the two differentiation trees into a single differentiation tree. This distinguishes this case from a mere increase in ploidy.

Proposition 191: growth and pruning of differentiation trees broadens the sharp peaks of frequency of DNA per genome due to genome doublings.

If differentiation trees change over evolutionary by growth and pruning, then these processes will tend to broaden the peaks due to genome doubling. The fact that the valleys between the peaks haven't been completely filled in (Sparrow & Nauman, 1976) may reflect the rate of macroevolution by changes in the differentiation tree versus the rate of genome doubling.

Proposition 192: survival of a differentiation tree depends on its compatibility at all hierarchical levels.

At any moment in the evolution of a differentiation tree, it must meet criteria of compatibility with successful embryonic development, reproductive success, survival in the ecosystem (Vermeij, 1987; Løvtrup,

1982a), and long term balance of the laws of most and least action (Johnson, 1988, 1990a, 1995; Vanriel & Johnson, 1995). The latter may work through the constraints on body size, which has some relationship to the size of the differentiation tree, and corresponding constraints on energy flow and ability to command a share of resources. The differentiation tree must thus be compatible with at least four higher hierarchical levels: the individual, the population, the ecosystem, and the biosphere (cf. Wright, 1956; Allen & Starr, 1982). The individual must survive embryogenesis:

"Geneticists would probably agree with embryologists that the first rigorous environment to which the genome is exposed during the life of the organism, with a large number of selective pressures, is the environment of its own cells during embryonic development. One of the early 'crises' which the genome has to surmount is that of gastrulation. If it can get through this stage successfully, it has a good chance of producing a mainly normal organism, since most of the cellular interactions that occur at later stages take place within individual organs or tissues, and the rest of the organism could be relatively unaffected by them" (Deuchar, 1975a).

"Traits need to adapt to other traits in the organism, not just to the external environment" (Cheverud, 1984).

Then it must survive at all the other hierarchical levels (Eldredge, 1986). Interactions between all these levels may occur. For example, Per Bak & K. Sneppen (in Peterson, 1994; cf.: Bak & Sneppen, 1993; Bak & Paczuski, 1995; Sneppen et al., 1995) have proposed an ecological 'avalanche' model for punctuated equilibrium, which could have multiple causes (cf. runs of fatal sexually transmitted diseases in Gordon & Tyson, 1993).

Proposition 193: there may be a rough correlation between the magnitudes of topological changes to differentiation trees in the course of evolution and proposed hierarchies in the process of evolution.

Topological change to differentiation trees may occur all at once or in a succession of smaller changes. Insofar as punctuated equilibrium does suggest a hierarchy of selective events (Eldredge, 1985a; Gould, 1987a), I give this corresponding proposition. Larson (1989) has noted that...

"The importance of this [hierarchical] genetic architecture relative to nonhierarchical ones in which all loci have strictly additive effects on the phenotype, remains controversial. The question of whether emergent, epigenetic properties of organismal development impose an additional level of causality in evolutionary pattern and direction constitutes a... major controversy that separates hierarchical and nonhierarchical viewpoints (Alberch, 1980; Goodwin, 1984b; Wake & Larson, 1987)" (Larson, 1989).

Changes in differentiation trees may be the basic causal mechanism at all the purported hierarchical levels (Gould, 1982d; Eldredge, 1995a, cf. Depew & Weber, 1994). Thus differentiation trees may provide a reductionist viewpoint for such hierarchies. If we look at the differentiation tree as a self-similar branching fractal (cf. Gordon & Aguda, 1988; Gordon & Drum, 1994), then topological changes of any size are also self-similar and thus fundamentally involve the same mechanism, although in terms of the phylogenetic tree, they may seem to produce major results, such as new taxa, or minor ones, such as simple varieties. This viewpoint may allow us to reunify the current diversity of theories in evolution:

"Evolutionary theory now deploys a striking hierarchy of possible selection mechanisms.... Objects at different levels may be selected; and there may be selection for and against properties at different levels as well" (Sober, 1984a).

Proposition 194: we need an intellectual linking discipline to connect each hierarchical level to the next.

"...Each way of looking at a complex system requires its own description, its own mode of analysis, its own breaking down of the system into parts;... it is the relation of these different descriptions, which is by no means obvious, which is going to be a source of enrichment not only to biology but also to physics and to all of the sciences up to and including the human sciences" (Howard H. Pattee in: Rosen, Pattee & Somorjai, 1979).

"There is no logical reason for the existence of a snowflake any more than there is for evolution" (Eiseley, 1957).

The assumption has been made "...that events occurring at one level of a system cannot be deductively or predictively inferred from processes known to be occurring at other levels" (Depew & Weber, 1985c). Pennycuick (1992), championing Newtonian mechanics, posits the opposite point of view:

"Molecular biologists are fond of reminding those who study biology at larger scales that... all biology depends ultimately on molecular processes. However, nobody has devised a way to interpret observations of ecosystems directly in molecular terms. At each level of biology, the molecule, the cell, the organ, the organism, the population, and the ecosystem, new principles and sets of rules seem to appear, without any quantifiable connection with the level below.... I hope to show first that attention to physical concepts is very helpful in understanding biology at all these levels, and second, that Newtonian ideas provide a link allowing processes at each level to be understood in terms of the one below" (Pennycuick, 1992).

Thus the epistemological viewpoint of Depew & Weber (1985c) does not take into account the existence of scientific disciplines, such as statistical mechanics (and its subset, Newtonian mechanics), whose very purpose is to deduce the quantitative relationships between certain hierarchical levels:

"...A system such as a gas composed of many molecules is actually found to exhibit perfectly definite regularities in its behaviour, which we feel must be ultimately traceable to the laws of mechanics even though the detailed application of these laws defies our powers. For the treatment of such regularities in the behaviour of complicated systems of many degrees of freedom, the methods of statistical mechanics are adequate and especially appropriate" (Tolman, 1938).

"These two volumes will be of interest to anyone prepared to follow Hill's trail into a new era of biology, in which precise explanations of molecular, macromolecular, cellular, and even whole organism phenomena in terms of the unique blend of physics and chemistry that is statistical mechanics, is achieved. It is, after all, statistical mechanics that has shown us exactly how it is possible to intellectually bridge the gap between hierarchical levels. Terrell Hill has shown many of us not only how this can be done, but the tremendous intellectual effort and honesty it requires to achieve a lasting result. The reader will see how most of the contributors to this volume struggle with these questions of explaining the whole *from* its parts, without the crutch of the philosophical aphorism (whole > Σ parts). Statistical mechanics is the basis for this leap" (Gordon & Chen, 1987a).

We thus have a scientific paradigm for making the transition from one hierarchical level to the 'emergent' properties of another ("beneficial *serendipitous* side effects emerge": Dennett, 1995). The need for such a transition for embryology was seen in a way by Thomas H. Morgan (1901):

"The organization of the developing organism... demands more study. That study must remain the very center of embryology. For:

> 'It can be shown, I think, with some probability that the forming organism is of such a kind that we can better understand its action when we consider it as a whole and not simply as the sum of a vast number of smaller elements. To draw again a rough parallel; just as the properties of sugar are peculiar to the molecule and cannot be accounted for as the sum total of the properties of the atoms of carbon, hydrogen, and oxygen of which the molecule is made up, so the properties of the organism are connected with its whole organization and are not simply those of its individual cells, or lower units' (Morgan, 1901)" (Maienschein, 1991a).

While the parallelism of macroevolution: microevolution = thermodynamics: statistical mechanics is acknowledged (Ayala, 1985), there is in biology a general proclivity either to ignore the necessity for mastering such a *discipline* to deduce the properties of one hierarchical level from another, or to deny that such a deductive process can exist, an attitude shared with those who insist that certain phenomena are supernatural and beyond our understanding (Irvine, 1955). This may be the result of a current generation of biologists nearly devoid of formal training in the advanced methods of mathematics, physics and chemistry. The task before us is difficult:

"...A biologist, studying a clearly hierarchical system of activities like that of chromomere, chromosomal segment, chromosome, genome, would hardly expect to meet with a situation even simpler than that present in inorganic nature - namely, total action being the sum of all partial actions, as assumed in the classical theory of the gene..." (Goldschmidt, 1954).

It is not enough to appeal to the apparent diversity and complexity in biology:

"The difference between the thinking of a physicist and a biologist can be conveniently illustrated by a few recent statements of the well-known physicist and Nobel laureate, Steven Weinberg (1974)... [cf. Weinberg, 1993], 'One of man's enduring hopes has been to find a few simple laws that would explain why nature with all its seeming complexity and variety is the way it is.' Surely no biologist would ever express such a hope. It would be difficult to expect that the incredible diversity of nature, the complexity of the process of ontogenetic differentiation and of the nervous system, or the qualitative uniqueness of each kind of macromolecule, could be expressed in the form of a 'few simple general laws.'..." (Mayr, 1985).

The same could be said of snowflakes (Bentley & Humphreys, 1931), on whose complex, historical growth process plausible explanatory progress has been made (Nittmann & Stanley, 1986).

Almost the same appeal as Mayr (1985) is made by creationists:

"The simplicity that was once expected to be the foundation of life has proven to be a phantom; instead, systems of horrendous, irreducible complexity inhabit the cell. The resulting realization that life was designed by an intelligence is a shock to us in the twentieth century who have gotten used to thinking of life as the result of simple natural laws" (Behe, 1996).

It is curious that although Depew & Weber (1994) eschew Mayr's rejection of physics, they agree with him in the sense that they invoke "the complexity of the process" under the more modern and quantifiable phrase 'complex systems dynamics':

"...Complex systems are systems that have a large number of components that can interact simultaneously in a sufficiently rich number of parallel ways so that the system shows spontaneous self-organization and produces global, emergent structures.... Living things are essentially complex, and... our methods of studying them must respect that fact" (Depew & Weber, 1994).

The opposite approach and attitude is expressed by Goldschmidt (1940):

"We frequently encounter the idea that life phenomena are infinitely more complicated than those of inorganic nature and that they therefore cannot be understood on the same basis.... I cannot agree with this. If life phenomena were not based on very simple principles, no organism could exist; if embryonic development were not controlled by a few simple basic properties and laws of matter, an organism could never be developed in a series of processes unrolling with the precision of clockwork. If evolution had not been made possible by relatively simple features inherent in the material basis of organization, it would never have occurred" (Goldschmidt, 1940).

Depew & Weber (1994) claim that this is a theological argument:

"It is fortunate that a new philosophy of science is emerging just as scientists are beginning to admit that many real world systems are complex and nonlinear. These simultaneous developments reflect not only our increasing technical ability to model complex systems, but continued retreat from the crypto-theological world view that was bequeathed to scientists at the dawn of modernity. It was held then that the world *must* be composed exclusively of simple systems because the Creator's mind must be assumed to be orderly. In point of fact, Ockham's razor, which bids us to prefer, *ceteris paribus,* the simpler of two hypotheses on the ground that we should not 'multiply entities unnecessarily,' was first advanced by a theologian trying as hard as he could to make the world conform better to what Christian theology required than Aristotelians like Thomas Aquinas... Cf. Grant (1971)" (Depew & Weber, 1994).

But I do not see how a preference for complexity does anything more than make one feel comfortable in not seeking simple explanations. No objective criteria have yet been established for determining when one should give up the search for a simple solution to an enigma and settle for 'complexity' as an explanation. What smacks of religion more than the suggestion that some things are simply beyond our understanding? Complexity of a system is a property that requires demonstration. It is not a fundamental assumption. At this point, there is no proof that the development of organisms is an example of complex system dynamics. All we have is a set of analogies between the output of computer simulations of certain complex models whose relationship to the actual genome is problematic (cf. Behe, 1996), and general concepts of embryogenesis (Kauffman, 1993):

"If there is a bottom line to all this, I suppose it must be the following: Are interrelationships in molecular and developmental terms as tangled as Kauffman's model suggests? If you can answer that one, you will get my attention!... Our recent understanding of the genetics of development suggests that a hierarchical structure of gene expression carefully modulates embryogenesis.... While the system is complicated, it is not a tangled array" (Levinton, 1995).

Charlesworth (1995) is also concerned that Kauffman (1993) is "Making evolution seem complicated". The book in your hands suggests a simple view of differentiation whose testable ramifications match much of the seeming complexity of life. Have we really reached the following impasse (revelation?):

"Every so often in science, as in other human endeavors, a new way of looking at the world and the phenomena it contains emerges. Many people now argue that science is undergoing such a shift as it crosses what Heinz Pagels has called the 'complexity barrier' (Gleick, 1987; Kellert, 1993; Lewin, 1992; Pagels, 1988; Stewart, 1989; Waldrop, 1992)" (Depew & Weber, 1994).

I doubt it.

Whether the result of using statistical mechanics (cf. Totafurno, Lumsden & Trainor, 1980) or other of what we might call 'linking disciplines' between hierarchical levels should be properly labelled reductionism (Depew & Weber, 1985c) or not, is an issue in the philosophy of science that I shall not attempt to resolve here. The theory of statistical mechanics itself not only involves a "contraction of the description" as "all but a few variables are ignored", but the choice of which variables can be eliminated to produce "self-contained contracted descriptions of matter" has no theory of its own. Furthermore...

"...one is left in the awkward position in mechanical ensemble theory of having to introduce postulates about the nature of initial distribution functions which cannot themselves be measured" (Keizer, 1987).

Nevertheless, modern statistical mechanics does provide a hierarchy of descriptions, which may be linked each to the next (Keizer, 1987), and thus may point the way to linking the hierarchies in biology:

"Despite the accelerating move away from wholism and toward reductionism in embryology,... [we] still cannot fully account for the remarkable ability of the organism to regulate and regenerate during development. We may eventually find an explanation in terms of the chemical makeup and physical processes in the organism.... Whether we can in fact achieve this reduction remains an open question, and whether we can do so in principle remains a matter of conviction rather than rigorous logical demonstration.... There was something about the ability of the organism to regenerate that led most American embryologists around 1900 to believe in an organization of the whole that amounted to more than the sum of the parts.... The current bias toward reductionism has not settled the issue, any more than has the morass of writing on reductionism by philosophers of science..." (Maienschein, 1991a).

To be concrete, consider the dynamics of tensegrity structures such as the cell state splitter or the mitotic spindle. For the latter, Pickett-Heaps, Forer & Spurck (1997) suggest that there are "limitations of the molecular approach to explaining global issues of biological organization..." and label such an approach 'reductionist', on the basis of observations of spindle 'regulation' such as this:

"...when a few or all k-fibres [kinetochore fibers] in one half spindle in newt [lung epithelial] cells are severed by UV-microbeam irradiation, not only is the lesion repaired (including movement of the pole toward the equator) but also the *unirradiated* half spindle soon shrinks, adjusting itself to match the length of the repaired half spindle (Spurck et al., 1990; Forer, Spurck & Pickett-Heaps, 1997; Spurck, Forer & Pickett-Heaps, 1997). It is very difficult to explain plausibly how localised biochemical reactions alone could detect and then control these global events. Rather, to paraphrase Ingber et al. (1994), the structural stability if the spindle depends upon the dynamic balance of mechanical forces acting upon its components; the latter constantly rearrange globally and the responses cannot be explained in terms of characteristics of its individual parts" (Pickett-Heaps, Forer & Spurck, 1997).

Yet, if we wish to understand the operation of a supramolecular tensegrity structure such as the spindle, surely we will examine it in terms of the interactions between its components. Is this reductionist? I think yes. It is merely that we are including the spatial component of macromolecular interactions, so that the structure is there from the beginning, and is accounted for in our math or computing. Reductionism is not reduction to the scalar chemistry of early biochemistry, but to physical/chemical interactions in space and time, explained in terms of molecules that occupy space and exert and react to mechanical forces in addition to undergoing chemical reactions. Thus while this is surely 'synthesis', I cannot agree that:

" 'Morphogenesis defies 'reduction...' ' (Harold, 1995). Mitosis is in essence a special type of morphogenesis: the issues outlined in Harold's paper, while directed at morphogenetic principles at and above the level of the whole cell, apply directly to the spindle, and thus mitotic mechanisms" (Pickett-Heaps, Forer & Spurck, 1997).

On the contrary, synthesis is precisely the way to demonstrate that reduction to components and their interactions works. Statistical mechanics provides the linking discipline, in this case from macromolecules to a tensegrity structure in the cell, so long as *all* the physical forces involved are identified and included. When we go about demystifying the mechanical component of living phenomena (His, 1888), we need not turn around and raise the mechanics to a new form of mysticism. Now, of course, there is a fundamental problem to be solved here, namely how it is that...

"*...the spindle is essentially the mechanical system that generates two cytoplasts from one* " (Pickett-Heaps, Forer & Spurck, 1997).

"Cell division can be viewed in terms of a single tensegral entity becoming separated into two entities, with kinetochores playing the key role of attaching the replicated genetic elements [chromosomes] accurately to each" (Pickett-Heaps, Forer & Spurck, 1996).

Until we solve the mechanics of cell division, there is a bit of mystery involved. Perhaps the answer will be found in instability properties of the spindle viewed as a tensegrity structure, analogous or homologous with the instability of the cell state splitter. If so, it will be a reductionist explanation, though one fully in the spirit of a linking discipline, taking us from the macromolecular to the cellular size scale. The 'abstract chemistry' approach of Fontana & Buss (1996) is an explicit attempt to provide a new linking discipline from molecules to biologically relevant entities:

"*...We believe that biology, particularly molecular biology, has a pressing need for support from a new kind of theoretical chemistry....* We have basically no clue as to even the nature of the abstract space of possibilities for novelty [above the level of point mutations]" (Fontana & Buss, 1996).

The ongoing dispute between the 'naturalists' and the 'ultra-Darwinians' expounded by Eldredge (1995a) (cf. Gordon, 1996b) can be expressed in terms of the concept of linking disciplines. According to Eldredge (1995a), the ultra-Darwinians, "guilty of a form of 'physics envy'", would like to derive all scales of evolutionary phenomena from population dynamics, or, failing that, wish them away. The naturalists observe hierarchies of species, ecosystems, etc., but, as I see it, lack explicit linking disciplines to explain them, one from the other. There is a perceived conflict between the 'hierarchical perspective' and reductionism:

"Ours is a hierarchical perspective. As such it dovetails nicely with parallel developments over the past few decades in other sciences. There are even connections with chaos theory. Ultra-Darwinians, in striking contrast, seem hell-bent on establishing a highly reductionist, old-fashioned physics model of simple lower-level determinism" (Eldredge, 1995a).

"We conceive... our true goal... as the full development of a comprehensive, nonreductionist hierarchical theory (with independent, though interacting, levels and complex effects in upward and downward causation between them),... helping us to understand the full grandeur of this view of life" (Gould & Eldredge, 1986).

This conflict between hierarchy and reductionism could possibly be resolved by explicit attention to linking disciplines. These are the disciplines that explain how the 'cranes' of Dennett (1995) (Proposition 170) are constructed. Such disciplines are needed to allay the...

"...realistic fear that the greedy abuse of Darwinian reasoning might lead us to deny the existence of real levels, real complexities, real phenomena. By our own misguided efforts, we might indeed come to discard or destroy something valuable" (Dennett, 1995).

I would also push the starting point for the hierarchy down a bit further, at least to differentiation trees. But even more levels than we are used to may be needed, since there has been some attempt to look at individual protein structures as developmental hierarchies (Crippen, 1978; Rose, 1979).

6.05 Neutralist Theory

Proposition 195: the neutral theory of molecular evolution should be extended to encompass DNA changes of all magnitudes. Neutral changes, including morphologically neutral changes to the topology of the differentiation tree, can form the basis for speciation.

In the words of the originator of this theory...

"...'selectively neutral' means selectively equivalent, that is, neither advantageous nor disadvantageous. In other words, the mutant forms involved can do the job equally well in terms of the survival and reproduction of individuals. The theory does not deny the role of natural selection in determining the course of adaptive evolution, but it assumes that only a minute fraction of molecular (i.e., DNA, RNA, and protein) changes are adaptive.... The proposal of the neutral theory (Kimura, 1968) led to a great deal of controversy, with strong opposition expressed by the neo-Darwinian establishment. This is often referred to as the 'neutralist - selectionist controversy,' which stimulated much research not only in molecular evolution but also in population genetics. The neutral theory also triggered reexaminations of the orthodox synthetic theory of evolution that had earlier appeared to be firmly established" (Kimura, 1992).

(Cf. Kimura, 1983; Berry, 1996.) There is nothing in this definition that restricts neutralist theory to molecularly small changes in the DNA. If whole terminal branches of the differentiation tree can be cut, copied and pasted with impunity (on occasion), then such changes constitute an expanded chapter of neutralist theory.

The case for 'chromosome speciation' has been heavily documented by King (1993b), whom I interpret to distinguish three major cases of chromosome change:

1. none:

"The traditional views on allopatric speciation by isolation remain widely accepted and in some cases justifiably so. There are numerous examples of species which have speciated without any significant change in chromosome morphology" (King, 1993b).

2. polymorphisms, in which chromosome rearrangements do not interfere with reproduction, so that a population contains a mix of different chromosome arrangements.

3. chromosome changes which cause speciation, because of a barrier to reproduction via reduced fertility of crosses, which can occur in the F1 or F2 crosses or might also involve sex ratio distortion.

Those particular types of chromosome rearrangements that lead to changes in the topology of the differentiation tree should be considered as special cases of either polymorphisms or chromosome speciation. One of the remarkable conclusions of King (1993b) is that there is little evidence for morphological change concurrent with the moment of speciation: "Morphological change itself is not directly responsible for the primary speciation event" (King, 1993b).

Thus, as I've argued in Proposition 137, we may anticipate that a change in differentiation tree topology will often be morphologically neutral, though, especially if it is a duplication of a branch of the tree, fraught with potential for divergence of the duplicated or original branch, or both (i.e., the hopeful genotype rather than the hopeful phenotype). Then, as King (1993b) argues, not only 'genic differentiation' (change in frequency of alleles) will arise *after* an act of chromosome speciation, but also 'morphological differentiation'.

Most of the evidence reviewed by King (1993b) deals with either chromosome rearrangements visible by light microscopy or alternate alleles for enzymes. The vast range of chromosome structure between the karyotype and the single gene, i.e., the domain of the differentiation tree, has yet to be explored as to its role in speciation. Perhaps most chromosome speciation will be found to occur via topological changes to the differentiation tree, i.e., what I have redefined macroevolution to be.

Proposition 196: an expanded neutral theory of evolution leads to the concept of fractal evolution, in which each individual is an incipient species.

From the point of view of the individual, its problem becomes to find a reproductively compatible mate.

We come in a circle at this point, and return to the problem of stasis. As pointed out by Schopf (1981), given the large number of newly discovered mechanisms by which the genome can change, rapidly on a geological time scale, and the biases in interpretation of the fossil record, "...prolonged periods of stasis... are likely to be an illusion...." He suggests average species durations of 10^3 to 10^4 years, rather than 10^7 years. In this case, speciation would resemble fine branching of fractal growth patterns, such as we have produced in first order models for the morphogenesis of diatom shell patterns (Figure 40; Gordon & Aguda, 1988; Gordon & Drum, 1994; cf. Daoud, 1987; Vicsek, 1992; Bassingthwaighte, Liebovitch & West, 1994; Kaandorp, 1994). We could call such a model 'fractal evolution', to distinguish it from phyletic gradualism and punctuated equilibrium (Vrba, 1980):

"Metaphorically we run a leg of an ever-branching relay race, passing on a baton of order that will both shape the next runner in the relay and present to him an information pool to which he can add [or subtract] before handing it to his successor. The race is peculiar because the baton, not the runners, drives it onward. Life's spread of order resembles a cancer: once started, its growth cannot be stopped" (Seielstad, 1989).

In the extreme case of fractal evolution, every individual may have a distinct, nonallelic difference from the DNA of others and could be regarded as an incipient species. (Spatial separation may still be necessary: Carson, 1989a.) This concept reverses the notion of 'species as individual' (Rieppel, 1986) to 'individual as species'. Each individual organism is then faced with the problem of finding a mate that is sufficiently genetically compatible to produce viable offspring. I express this as hyperbole to make the point. Nevertheless, it is a mere (pervasive) assumption that differences between individuals are differences in alleles of the same genes. It could be that a closer examination will reveal variability between individuals in other aspects of the genome, and the prevalence of such nonallelic variations (cf. "the uniqueness of the individual": Medawar, 1957):

"The Russian geneticist Sergei Chetverikov [1926, 1959; cf. Adams, 1968, 1980]... [proposed] a view of genetical natural selection that insisted not only on variation - creation, but on banking it for a rainy day. That nature stores variation in populations that are chock full of genetic diversity was for Dobzhansky a good argument for resisting the enthusiasm of many earlier Darwinians, including Fisher, for 'eugenics,' an array of suspiciously anti-democratic ideas about enhancing the fecundity of those who were assumed to be superior and discouraging, sometimes physically preventing, people who were assumed to be less fit from having offspring. Dobzhanksy's hypothesis about the width of genetic variance in natural populations, and the role of natural selection in maintaining it, has been spectacularly vindicated by several generations of his students, who have dominated recent American evolutionary biology [cf. Adams, 1994]. Indeed, so great is this variation that it has become something of an embarrassment to the Darwinians who discovered it, since it is increasingly difficult to maintain that natural selection is primarily responsible for maintaining that much diversity" (Depew & Weber, 1994).

Differentiation trees do not solve the evolutionary conundrum of why genetic diversity is maintained, but they do add a potentially major source of individual variation. As seen in Proposition 173, our hypotheses about evolution have had major impacts in fueling concepts of 'racial purity', or dampening them, and on questions of maintaining genetic diversity in crop plants, etc.

Certainly, given the amount of work presented to DNA repair mechanisms, presumably including to some extent the sequestered germ cells, it is not unreasonable to suppose that every individual is unique for more than reasons of recombination:

"Each cell throughout your body has about 10,000 damaging events occur within it every day, damaging DNA by depurinating, de-aminating, and so forth..." (Smith, 1995c).

In her computer simulations of the evolution of gene families by gene duplication, Ohta (1988a) found "...the resultant gene arrays were very different under the same set of parameter values" from one Monte Carlo run (cf. Gordon, 1980c) to another, exhibiting 'inflation of variance'. "Starting from a single gene, my simulation results showed... the process of gene accumulation is highly chancy and many different gene families may evolve under similar conditions" (Ohta, 1990a). This phenomenon, which would also occur with duplications of terminal branches of differentiation trees, thus exhibits fractal evolution. If the self similar or fractal structure of the thermal stability of chromosomes (King, 1973; cf. Takashi, 1996) corresponds, in multicellular organisms, to the self similar structure of the differentiation tree, then there may be a physical basis for fractal evolution.

Lest the reader think that my speculations on the individual as species have gone over the edge, consider the curious (or exemplar?) case of 'ecological release', a form of evolutionary radiation occurring at the level of the individual organism:

"Ecological release is a common phenomenon on remote islands with small faunas and floras - I have observed it many times in ants, for example - but it occurred to spectacular degree in the Cocos Islands finch. The birds, all members of a single freely breeding species, occupy niches ordinarily divided among species, genera, and even entire families of birds. They range from shore to hilltop, forage within the forest from ground to canopy, and feed variously on insects, spiders, and other arthropods, mollusks, small lizards, seeds, fruit, and nectar. In this respect the Cocos finch far exceeds any one species of Darwin's finch in the Galápagos [Lack, 1947, 1949; Grant, 1986; Quammen, 1996]. Most striking of all, each individual bird specializes on a particular kind of food and maintains the habit for at least several weeks, perhaps for its entire life. This microcosmic adaptive radiation appears to be based on observational learning. During a ten months' visit to Cocos,... Werner & Sherry (1987), the biologists who discovered the release phenomenon, watched juvenile finches approach and imitate the distinctive feeding behavior of yellow warblers and sandpipers. No doubt the young birds also copy older birds of their own species. Like medieval apprentices selecting masters within specialized guilds, the young birds appear to prosper with personalized instruction.

"Cocos finches have bills roughly intermediate in size and shape to those of warblers and finches. With the tendency of the birds to diversify in feeding habits individually, the stage is set for rapid species formation and adaptive radiation in the Galápagos mode, if circumstances permitted it. But circumstances do not permit it. The place is too small and too remote from other islands to allow the formation of new species. The Cocos finch radiation therefore remains stalled in an embryonic state, one species to the island" (Wilson, 1992).

Cf.:

"In a word, it seems that it was the *behaviour* of the finches which brought about, via selection, the structural change" (Taylor, 1983).

Griffin (1992) might even add that this could be conscious behavior:

"There is a strong tendency among contemporary behavioral scientists to assume that conscious mental states, on the one hand, and learning or evolutionary selection, on the other, are mutually exclusive alternatives. But this is by no means self-evident. An animal may or may not be conscious, and its behavior may be influenced to varying degrees by genetic programming... The customary assumption that if some behavior has been genetically programmed, it cannot be guided by conscious thinking, is not supported by any solid evidence.... To paraphrase Karl Popper [cf. Popper & Eccles, 1978], animals that think consciously can try out possible actions in their heads without the risk of actually performing them solely on a trial-and-error basis" (Griffin, 1992).

While *Homo sapiens* is presently touted as a single species (especially in the battle against racism), our 3 to 15% rate of couples' involuntary infertility (i.e., couples for whom both would be fertile with different partners: Webster, Cook & Garner, 1990; cf. Gordon & Tyson, 1993) might suggest that we too consist of incipient, sympatric species. Such incipient speciation is mostly 'cryptic': idiopathic, couples involuntary infertility runs at up to 1% of couples, with no morphological or race correlation except, perhaps, when large average size differences occur, affecting success of childbirth (prior to the use of Cesarean deliveries: John A. McCoshen, p.c.).

I will not pursue fractal evolution further here, but will rather continue as if the idea of a species has validity. Green (1991a) has also concluded that phylogenies are fractal, but by analogy of evolution with dynamic systems (cf. Depew & Weber, 1994; Fussy, Grössing & Schwabl, 1997), rather than by analyzing the relationships between development and evolution:

"The products of the evolutionary process are organisms and taxa (diversity) and their history (phylogeny). But the products are also part of the process. Replication is the principal evolutionary characteristic of living things. Cells, species and DNA molecules arise from previously existing cells, species and DNA molecules. Thus, change in a genetic constituent of character can be written in the general form of a finite-difference equation.... Dynamic systems represent the physics behind 'just history'.... Evolution, as a dynamic system, has inherent behaviour predicated by the properties of the fractal set of all living things, and produces a fractal history, phylogeny, as the consequence of generating that set. This can account for the hierarchy we perceive in nature (Vrba & Eldredge, 1983) without requiring the erection of hierarchical theories of process. Hierarchy is an emergent characteristic of the history of the system, not a property of the system, and that hierarchy is infinitely nested. There is no fundamental need to separate micro- from macroevolution when the 'butterfly effect' (Lorenz, 1963) of a dynamic system - its sensitivity to precise conditions and its ability to amplify the effects of small perturbations exponentially - offers a bridge between these perceived levels" (Green, 1991a).

See also Minelli, Fusco & Sartori (1991). The two views of why evolution is fractal are probably complementary.

Proposition 197: the concept of fractal evolution, in which numerous differences between differentiation trees are arising frequently, may help solve some outstanding problems in evolution, such as the definition of species, cryptic species, and the advantages of sex.

1. Quantitative Traits

"The mechanisms responsible for genetic variation in quantitative traits in natural populations remain a mystery" (Gillespie & Turelli, 1989) that could find its resolution here.

2. Definition of Species

Species may be nothing more than transient correlations in random walks (cf. Bookstein, 1988), more or less distinguishable because of transient 'cohesive' phenomena (Collier, Wiley & Brooks, 1991), such as the 'specific-mate recognition system' (Paterson, 1993a,b). The length of the 'transient' of course defines whether or not we consider that we are dealing with stasis.

"...<u>M</u>any tropical species, particularly insects, are found only in one small area, perhaps a single tree.... <u>W</u>e are faced with the problem of explaining how a substantial portion of the tremendous diversity of the present rain forests evolved in less than fifty thousand years. This seems an incredible speciation rate..." (Raup, 1991), which is perhaps explicable by fractal evolution. The continuing discovery of cryptic species in lungless salamanders of the Plethodontidae family, essentially indistinguishable on morphological criteria, may be another facet of fractal evolution: "...genetic isolation... [or] speciation proceeds at a higher rate than does the origin of morphological novelties" (Larson, 1989). Eldredge (1995a) suggests that:

"Fledgling species have a much greater chance of survival if they are ecologically differentiated from the parental species.... And their ecological distinctiveness is mirrored in their anatomies, which bear the traces of evolutionary, adaptive change.

"Speciation... does not necessarily beget evolutionary adaptive change. But *successful* speciation - ...survival long enough to show up in the fossil record - does. That's why adaptive change is so tightly correlated with speciation....

"Reproductive systems aren't about keeping members of other species away so much as they are about finding mates and reproducing successfully. But consider the simple consequence of Paterson's [1985] SMRS [Specific-Mate Recognition System] concept [cf. Paterson, 1993a,b]. Speciation cannot possibly be an evolutionary process designed to partition genetic information and keep gene pools apart. Speciation must really be an accident.

"Paterson says that only reproductive attributes - the anatomies, behaviors, and physiologies pertaining to successful reproduction - actually count in speciation. The more traditional view says speciation is a function of change in all manner of characteristics" (Eldredge, 1995a).

The process of cryptic species formation may be one of duplication of a terminal branch of the differentiation tree that leads not just to reproductive isolation, but also anatomical/ecological differences that convert the cryptic species to a fledgling species. Of course, this could be a two step 'accident', involving two different terminal branches of the differentiation tree, one reproductive and another somatic. The presence of cryptic species amongst the parent population increases the effort involved in finding a compatible mate. On the other hand, this may be an anthropocentric viewpoint:

"Cryptic or sibling species, in fact, are not species that have yet to diverge more fully. They are normal genetical species in every way and cause difficulty only because human taxonomists use their optical sense in identifying species. If humans were as well provided for with other senses as insects are, species would not be cryptic.... Cryptic species are likely to be found in any group of organisms that use mainly auditory, chemical, tactile, or transient optical signals in the critical, discriminating stages of their SMRS [Specific-Mate Recognition System]" (Paterson, 1993a).

The intraspecific variation of nuclear volume of plants with latitude (Stebbins Jr., 1966; Mergen & Thielges, 1967) makes one wonder whether here too we are dealing with cryptic species. This is likely to be a metabolically related variation in amount of junk DNA (Martin & Gordon, 1995), which may not include a variation in the differentiation trees themselves. This is a reminder that while changes in differentiation trees may be a significant source of speciation, there are undoubtedly many other possible mechanisms of speciation.

3. Differences between Specialized and Generalized Species

We can take the question of fledgling species a bit further, by noting the differences between generalized and specialized species:

"As a systematist, [Elizabeth] Vrba [1984a] has worked intensively to clarify the evolutionary relationships among antelope species.... Species of the wildebeest group (Aepycerotini) typically rely on a narrow range of plant species for forage, and are likewise strongly dependent on reliable water supplies.... There are some six or seven species in the wildebeest group, but only one impala species. Impalas range over vast segments of the African landscape, occupying a wide range of habitats. Not dependent on any single plant species, nor restricted by constant reliance on ready availability of water, impalas are ecological generalists. Physically, impalas are very generalized antelopes, retaining many primitive features.... Ecologically generalized species tend to retain many of the primitive features that were present in remote evolutionary ancestors.... Such species typically last a very long time....

"Lineages consisting of species who remain generalized have fewer species, tend to resist extinction, and accumulate little or no major adaptive change. Vrba counts just two species of impala over the past six million years, while there have been a minimum of thirty species in the wildebeest group over that same span of time.... Vrba's wildebeests are among the most specialized (even downright anatomically bizarre) of all antelopes - fossil or modern....

"...The rate of speciation itself was higher among the wildebeests than the impalas. Ecological specialization, it seems, confers higher rates of speciation, and thus of development and accumulation of adaptive change..." (Eldredge, 1995a).

The 'anatomically bizarre' species suggest the possibility of a more developed differentiation tree. It could be that if we were to work out the differentiation trees for these antelopes, the generalized impalas would have topologically distinct trees from the wildebeests. The question that could then be answered is whether the rate of speciation is a function of the topology of the differentiation tree (for instance, presence of webbing). Changes in embryonic development obviously underlie and correlate with adaptive change, and we need to know to what extent the manner of that development, exemplified by the differentiation tree, is causative in higher order phenomena like speciation. While a convincing ecological argument can be made, there is an embryological reality underlying it:

"...With broadly adapted species, a newly isolated fledgling species must quickly develop a relatively great amount of adaptive change - taking it outside the range of adaptive, ecological behaviors and functions already present in the ancestral species. With narrowly adapted species, on the other hand, relatively less adaptive divergence is required for the fledgling descendant to become ecologically different enough from its ancestor that it stands a chance of survival" (Eldredge, 1995a).

As Eldredge (1995a) indicates, these ideas are important to understanding biodiversity, especially in the tropics.

4. Advantages of Sex

One could imagine that the 'cost of sex' may be counterbalanced by "individual-level advantages to sex" (Charlesworth, 1989; cf. Lyons, 1997;

Proposition 233) that are involved in solving the problem of an individual leaving enough viable offspring, when it is part of a fractally evolving population that is 'speciating' every generation, perhaps with every individual. The 'repair function of recombination' (Bernstein, 1977; Bernstein et al., 1985a; Bernstein & Bernstein, 1991) may have as its purpose keeping genomes and differentiation trees sufficiently similar to permit viable matings. In this sense, it may actually act to reduce diversity, rather than enhance it:

"...We develop the idea that the two basic aspects of sex, physical recombination and outcrossing, are fundamental adaptive responses to DNA damage and mutation. We argue that recombination repairs DNA damage and outcrossing masks mutations in reproductive systems which have recombination.... In dealing with damage, recombination produces a form of informational noise, allelic recombination, as a by-product.... As some lines of descent became more complex, their genome information increased, leading to increased vulnerability to mutation. The diploid stage of the sexual cycle, which at first was transient, became the predominant stage in some lines of descent because it allowed complementation, the masking of deleterious recessive mutations" (Bernstein, Hopf & Michod, 1987).

6.06 A Universe Aware of Itself: Differentiation Waves and the Brain

"The means whereby a nervous system succeeds in developing from a few precursor cells in the embryo to a fully differentiated, specialized, wired-up neuronal machine is beginning to be understood. This knowledge is likely to play a crucial role in theories of how brains learn, how they represent the world, and what sort of business learning really is - what constraints there are, and what degrees of freedom. Understanding how neuronal order arises out of such modest beginnings is bound to help us understand the fundamental principles of brain function" (Churchland, 1986).

"Some have... asked what is the neuroscientific equivalent of the discovery of DNA structure.... The equivalent, at the level of mind-producing brain, has to be a *large-scale outline of circuit and system designs,* involving descriptions *at both microstructural and macrostructural levels* " (Damasio, 1994).

Proposition 198: evolution of the central nervous system is primarily a matter of growth via bifurcation of its portion of the differentiation tree, which may be greater than 50% of the whole.

Complex mental abilities presumably have a genetic basis related to the construction of the nervous system (John & Miklos, 1988; Price, 1989). This complexity is reflected at a molecular level:

"It has been estimated that over half of the genome [i.e., > 50%] codes for proteins specific to the nervous system and that the young adult mouse brain contains about 140,000 mRNAs (Hahn, Van Ness & Chaudhari, 1982) [cf. Chaudhari & Hahn, 1983; Hahn et al., 1983; Hahn & Chaudhari, 1984; Takahashi, 1992]; other RNAs are undoubtedly found in the rest of the nervous system and the developing nervous system" (Mullen, 1984).

"Over a million neurons are present in the [*Drosophila*] adult fly and they express several thousand [kinds of] mRNA molecules, some at very low levels (Palazzolo et al., 1989; Jacobson, 1991b; Truman, Taylor & Awad, 1993)" (Rodrigues et al., 1995).

"The diversity of mRNA molecules in the brain is biologically consistent with the extensive cellular heterogeneity and microdifferentiation of the various cell types that occur in this organ" (Chaudhari & Hahn, 1983).

For instance, many forms of tubulin appear in the brain:

"We have seen that each major event during neurogenesis is related to a specific series of modifications of the MT [microtubule] components. It remains to be determined if there is a causal or just a correlative relationship between the appearance of specific isotypes and the occurrence of specific events and/or functions. We have also to determine the exact spatial and temporal relations among the different isotypes of MT proteins, tubulin, and MAP [microtubule attachment proteins]" (Meininger & Binet, 1989).

At the fine level of the synapse, "SIF [still life gene] proteins may regulate synapse differentiation through the organization of the actin cytoskeleton by activating Rho-like GTPases" (Sone et al., 1997). Perhaps it is the other way around, as we postulated for the cell state splitter (Appendix IV: Björklund & Gordon, 1993b), and 'colocalization' implies instead 'control' of SIF proteins by actin.

Lumsden (1993) describes CNS development as follows:

"The vertebrate central nervous system [CNS] develops from an apparently uniform sheet of cells in the early embryo, the neural epithelium, that eventually produces many hundreds of cell phenotypes each with characteristic morphology, neurotransmitter content, axonal trajectory, and synaptic specificity. Distinct neuronal types appear at predictable times and at precise positions within the forming CNS.... Little is known about the developmental processes involved in the definition of spatial domains in the CNS and the generation of their specific neuronal cell types" (Lumsden, 1993).

It is easy to imagine subtree(s) of the brain differentiation terminal branch in the following descriptions:

"It appears that the mammalian cerebral cortex uses an intermediate strategy: some precursors are multipotent (Price & Thurlow, 1988; Walsh & Cepko, 1988), but the precursors appear to become restricted in potential near the end of the lineage (...Luskin, Pearlman & Sanes, 1988; McConnell & Kaznowski, 1991; Parnavelas, Barfield & Luskin, 1990; [Parnavelas et al., 1991])" (Wetts & Fraser, 1993).

"The [chick] brain does not differentiate in one smooth coherent wave, but instead five separate primary AChE-activation zones are detected: the first originating at stage 11 [40-45 hrs, staging according to Hamburger & Hamilton, 1951: Paul G. Layer, p.c., 1996] ('rhombencephalic wave'), the second at the same time ('midbrain wave'), the third at stage 15 [50-56 hrs] ('tectal wave'). A fourth zone develops later, at stage 18 [65-69 hrs], from the bottom part of the telencephalon to the top. Retinal development also starts at stage 18. In a given area, it appears that AChE-development shortly precedes that of the formation of major fiber tracts. AChE [acetylcholinesterase] might therefore represent a prerequisite for fiber growth and pathfinding" (Layer et al., 1988).

With a width of about 100 μm for each 'zone', these waves travel roughly 0.1 μm/min, consistent with the speed of ultraslow waves (Jaffe, 1995). The overall impression the above studies give, then, is that the brain forms multiple cell types by a hierarchical mechanism, perhaps using a number of hierarchically nested differentiation waves, though no correlation with mechanical waves was sought out. In mosaic organisms, the brain, formed mostly by asymmetric cell divisions, might be considered as consisting of multiple tissues, each one cell in size, formed in a nested hierarchy of single cell differentiation waves:

"The *Drosophila* embryonic central nervous system (CNS) is derived from a stereotypic array of progenitor stem cells called neuroblasts (NBs)... ~25 NBs per hemisegment.... The NBs undergo a series of asymmetric divisions to produce, on average, 5-10 ganglion mother cells (GMCs), each of which divides to generate two neurons; studies on the grasshopper have led to the conclusion that each NB gives rise to a unique cell lineage, forming a characteristic family of neurons specific to that NB. The same is assumed for *Drosophila* because its nervous system is highly homologous to that of the grasshopper (Thomas et al., 1984)" (Yang et al., 1993).

There are experimental difficulties in delineating areas of the brain ('brain nuclei') in juvenile and adult organisms, since...

"...the cytoarchitectural, connectional and cytochemical features... mature independently.... [However,] developmental studies suggest that a brain nucleus is defined early in development before it can be detected by cytoarchitectural criteria, that is, the parcelation of even small brain areas... occurs early in development" (Gahr, 1997).

'Parcelation' may occur via differentiation waves, and if so, the earliest definition of brain nuclei could be done by observing these waves.

Proposition 199: pathfinding by developing neurons is secondary to the hierarchically nested placodal structure of the brain based on differentiation trees and differentiation waves.

The general notion of the outgrowth of axons or migration of neurons as the brain develops has been that each one is guided in its fate and to its destination:

"The development of the nervous system is based on a sequential interplay of neuronal genes with the embryonic environment" (Patterson, 1979).

"Borrowing from earlier work on morphogenesis in sponges [starting with Wilson, 1907, reviewed by Wilson, 1925; Trinkaus, 1984a], Roger Sperry (1963) proposed that the patterning of neuronal networks reflected processes of cellular individuation and recognition enormously more refined than but in principle the same as those underlying specific aggregation of sponge cells. What followed from this was an extended period in which investigators reported a variety of experimental observations both consistent and inconsistent with the notion that neurons would form synapses only at locations corresponding to their targets" (Grobstein, 1988).

For a number of reviews of Sperry's 'chemoaffinity theory' see Trevarthen (1990). To the contrary, there is now new evidence of "an unusually large, pure patch of neuron progenitor cells... [that travel] without the aid of tracks" (Pennisi, 1993b; cf. Parnavelas et al., 1991; Luskin, 1993), suggesting that cells within a "patch" may all be equivalent, from the point of view of differentiation, before migration:

"When neurons must undergo an extended (long distance) migration from their site of genesis to their final destination, most progenitor cells do not appear to be multipotential by the onset of neurogenesis (Luskin, Parnavelas & Barfield, 1993; Luskin, Pearlman & Sanes, 1988; Grove et al., 1993; Parnavelas et al., 1991; and reviewed in Luskin, 1994a)" (Luskin, 1994b).

This provides a possible link between differentiation waves and later brain development by axon outgrowth or neuron migration, as such patches or placodes may correspond to the trajectories of waves through a confluent sheet of cells:

"Several lines of evidence have suggested that the fate of cortical neurons is determined prior to migration and may be independent of cues present in the cell's environment either along its migratory path or in its final position in the cortical plate (Caviness, 1982; Kriegstein & Dichter, 1983; McConnell, 1988, 1991; Schwartz, Rakic & Goldman-Rakic, 1991).... Thus it appears that during the time prior to migration, a developing neuron makes a number of decisions, and in doing so its developmental potential becomes progressively restricted.... It will be intriguing to study the lineage relationships between the various subtypes of cortical neurons" (Parnavelas et al., 1991).

This speaks for a model of brain development based on differentiation trees rather than inductive interactions (as in Patterson, 1979), and thus, I would venture, on differentiation waves. I do not mean that cell-cell interactions during migration are unimportant, as they may at least help produce the nerve tracts and axonal pathways:

"A discrete region of the anterior part of the subventricular zone (SVZa) generated an immense number of neurons that differentiated into granule cells and periglomerular cells of the olfactory bulb - the two major types of interneurons. Thus, the SVZa appears to constitute a specialized source of neuronal progenitor cells. To reach the olfactory bulb, neurons arising in the SVZa migrate several millimeters along a highly restricted route. Guidance cues must be involved to prohibit widespread dispersion of these migrating neurons" (Luskin, 1993).

Smith & Frank (1988) argue for a balanced view:

"One hypothesis proposed to explain the development of... highly specific connections is that sensory neurons receive instructive signals from their peripheral targets, which they innervate first (Ramón y Cajal, 1911; Windle & Baxter, 1936), that specify their central [nervous system] targets (Weiss, 1936, 1942; Sperry & Miner, 1949; Miner, 1956). According to this hypothesis, developing sensory neurons initially may be pluripotent. An alternative possibility is that sensory neurons acquire affinities for particular peripheral and central targets before they form axons; their fates would be predetermined (Miner, 1956).... Perhaps a more prudent approach is to be open to both types of explanations" (Smith & Frank, 1988).

Some of these groups of neurons may differentiate as early as neural plate stages (cf. Rowe, Messenger & Warner, 1993) or even gastrulation stages, though we have not correlated them with differentiation waves yet:

"In lower vertebrates, for example, some neuronal populations complete their final DNA synthesis by gastrulation and prior to neurulation (Vargas-Lizardi & Lyser, 1974; Lamborghini, 1980; Mendelson, 1986; Hartenstein, 1989). These 'primary' neurons, such as Mauthner's neurons, Rohon-Beard cells, and reticulospinal neurons, are necessary for the functional circuitry of free swimming larvae.... We find that beginning as early as the definitive streak and early head process stages [of chick], there is a population of cells that completes its final cell division and subsequently expresses neurofilament proteins.... These results suggest that the instructive cues governing reticular neuron birth and differentiation are imparted during gastrulation and early neurulation" (Sechrist & Bronner-Fraser, 1991).

However, if homeobox genes are the master genes (Appendix IV: Björklund & Gordon, 1993b), then the division of the neural tube into neuromeres may precisely represent a division that occurs via successive differentiation waves:

"The possibility that segmentation of the neural tube could underlie the regulatory mechanisms specifying the central nervous system pattern was suggested by embryologists at the turn of the century, who observed that repeated bulges characterize and subdivide the early neural tube of vertebrates. These bulges, named neuromeres by Orr (1887), were rapidly considered as morphogenetic units in which the cells are crowded together and may not mix with cells from adjacent neuromeres (Orr, 1887, cited from Vaage, 1969). The neuromeric theory has recently come to light again since the discovery of numerous homeobox-containing genes, sharing homologies with known morphogenetic loci of *Drosophila,* expressed in the vertebrate neural tube from the beginning of neurogenesis, whose limits of expression are usually coincident with interneuromeric boundaries (Lumsden & Keynes, 1989; Wilkinson & Krumlauf, 1990) [cf. Nieuwkoop, 1947a].... The importance of neuromeres during the specification of the neural tube is emphasized also by experiments of single-cell labelling which have demonstrated that the rhombomeres (the neuromeres of rhombencephalon) are true segments separated by clonal restriction boundaries (Fraser, Keynes & Lumsden, 1990).... [I find]... that a territory, whose alar plate gives rise to the primordia of cerebellum (metencephalon), isthmic region, and mesencephalic - metencephalic (mes-met) territory, corresponds to the domain where the homeobox-containing gene *Chick-En,* (the chick *En2* gene), homologous to the *Engrailed* gene of *Drosophila,* is expressed (Gardner et al., 1988; Patel et al., 1989).... The cerebellar primordium... extends on both sides of the so-called 'mesencephalic - rhombencephalic' constriction.... The expression of *En2* in other vertebrate species such as mouse (Davis & Joyner, 1988; Davis et al., 1988), zebrafish (Fjöse et al., 1988; Hatta et al., 1991a), *Xenopus* (Hemmati Brivanlou, Stewart & Harland, 1990) and humans (Logan et al., 1992) also extends on both sides of the mesencephalic - rhombencephalic constriction" (Alvarado-Mallart, 1993).

The bulges and constrictions of neuromeres may correspond to the differentiation waves and their launching domains, respectively, that set off the neuromeres one from the other.

Another elegant example of differentiation starting in groups of cells is the recent discovery of the central nervous system (CNS) stem cell (Reynolds & Weiss, 1992; Morshead et al., 1994; Reynolds, Leonard & Weiss, 1994; cf. Fults et al., 1992; Baetge, 1993). The unexpected existence of these cells, showing that at least part of the adult mammalian is not terminally differentiated, implies incredible new opportunities (Reynolds & Weiss, 1993a; Weiss et al., 1996). "...Several studies have reported that new neurons continue to be added into the brain of adult fish, frogs, reptiles, birds and mammals" (Alvarez-Buylla & Lois, 1995); cf. Alvarez-Buylla & Kirn, 1997). The CNS stem cells are located apparently as a contiguous tissue in the subependymal zone and that appears to derive from a single placode of the neural plate (cf. Alvarez-Buylla & Lois, 1995). In tissue culture, balls of these cells can give rise to neurons, astrocytes, and oligodendrocytes (Reynolds & Weiss, 1996; cf. differentiation *in situ:* Lim, Fishell & Alvarez-Buylla, 1997). The size of the ball at which this happens varies considerably (Samuel Weiss, p.c.). I would suggest that the event which leads to differentiation within a ball of cells is accompanied by a differentiation wave covering part of the ball or affecting its surface cells only. If so, this *in vitro* system could provide an excellent opportunity to elucidate the relationship between growth factors, which bias which cell type is produced (Reynolds, Tetzlaff & Weiss, 1992; Reynolds & Weiss, 1993b; Vescovi et al., 1993; Hammang et al., 1994), and differentiation waves.

The brain, therefore, may be modularly and hierarchically constructed by consecutive binary processes corresponding to the brain branch of the differentiation tree. First the placodes are set aside by differentiation waves. Then the axons grow out from them or neurons migrate in specific directions. The neuron tracking that occurs along the way may, like embryonic induction, be a secondary phenomenon in its construction. Again, 'secondary' here does not mean unimportant, only that first, differentiation occurs, second, axon outgrowth or cell migration occurs. Thus differentiation waves do not directly specify the directions nor the destinations of axon growth or neuron migration, though the new differentiated state of the cell undoubtedly has something to do with what happens during those processes.

Proposition 200: explanations for components of behavior, and mental phenomena such as mental rotation, could be approached in terms of differentiation waves.

The parts of the differentiation tree devoted to the nervous system may be the basis for the link between development and perception that Sinnott (1963) was seeking (cf. Jacobson & Sater, 1988), or perhaps equivalently, the source of "the beginnings of reflective thought, paralleling the ascent of man" (Morowitz, 1985). The suggestion that the microtubule/intermediate filament/microfilament cytoskeleton is at the basis of brain plasticity (Aoki & Siekevitz, 1988), and possibly memory, hints at a link between differentiation via cell state splitters and experience mediated 'differentiation' of neonatal brain cells. In rats that have been given an experimental stroke (cerebral ischemia), if it lasts more than 10 min, their memory is impaired, even though by MRS (magnetic resonance spectroscopy) criteria, there is no metabolic change in the affected tissue (Garnette R. Sutherland, p.c.). 10 min is enough for microtubules to depolymerize, and microtubules are associated with synapses (Larrivee, 1991). Insofar as there may or may not be "...laws of thought established through the millennia of evolution" (Miller, 1984), these may be transmitted from generation to generation via the 'mental portion' of the differentiation tree.

The phenomenon of mental rotation, suggesting an internal representation and manipulation of three dimensional objects (Shepard & Cooper, 1982), extends at least down to pigeons and "...is exciting because it provides us with a better understanding of the evolutionary precursors of human cognition" (Hauser, 1993). Thus we may be able to look forward to a day when we can extend our analysis from a unification of development, genetics and evolution to behavior itself. We have made a minor foray into this area with a new theory of background extinction (Gordon & Tyson, 1993), which is supported by hereditability of monogamous and polygamous behavior (Shapiro et al., 1991; Insel & Shapiro, 1992).

Eccles (1989) looks forward to anatomical correlations of higher mental functions. His 'patches' may correspond to the trajectories of differentiation waves, and their 'specialization' to growth of the brain portion of the differentiation tree:

"So in hominid evolution the neural tube of the hominoid, with its neocortical generation, would come to have patches for production of neo-neocortex. It can be assumed that these patches were small in Australopithecines, but the greatly enlarged brain of *Homo habilis* would be produced by large neo-neocortical generating zones. With further stages of hominid evolution these zones would grow to be larger than the neocortical generating zones.

Eventually with *Homo sapiens sapiens* the neural tube would be largely taken over by neo-neocortical generating areas. However, relative size is not the unique feature of the neo-neocortical generating areas. Rather it is the wide range of specialization for gnostic functions of the most diverse kinds" (Eccles, 1989).

Salamanders may be an excellent group in which to relate brain differentiation waves to brain structure (cf. Roth, Blanke & Wake, 1994) and behavior, since they may have gone through a simplification of the brain, perhaps followed by "compensating mechanisms [that] have evolved that may restore or even enhance brain function" (Roth et al., 1993). Their ability to regenerate a severed spinal cord (Piatt, 1955; Polezhaev, 1972; Clarke, Alexander & Holder, 1988; Muneoka, Bryant & Gardiner, 1989; O'Hara, Egar & Chernoff, 1992; Zammit et al., 1993; Chernoff, 1996) or portions of the brain (Burr, 1916; Harrison, 1947) and to take brain grafts (Pietsch & Schneider, 1969; Pietsch, 1981) suggests a wealth of experimental possibilities. The caecilian or apoda *Ichthyophis* may be even better for direct observation of higher order neural plate differentiations, since "regions of the brain begin to differentiate before the neural tube is closed posteriorly" (Duellman & Trueb, 1986; cf. Sarasin & Sarasin, 1887-1890). If this observation is valid, this third, strictly tropical order of amphibians presents an opportunity to extend the exploration of differentiation waves, using simple external surface time lapse observations, into stages of brain formation that are well beyond those reached by urodeles, anurans, birds and mammals at neural tube closure.

Proposition 201: cerebral hemispheric specialization involves an asymmetric unfolding of the differentiation trees between the two hemispheres. The hierarchical nesting of the differentiation waves amplifies small left-right differences.

In cerebral hemispheric specialization or cerebral 'dominance', different functions, regarding language, emotions, music, etc., appear to be taken on by the two sides of the brain (Bryden, 1982; Dunaif-Hattis, 1984; cf. critique of Churchland, 1986). We can expect an explosion of data in this field, as functional imaging takes off (Churchland, 1986; Mullani & Volkow, 1992; Turner, 1992; Barnes, O'Tuama & Tzika, 1993; Neil, 1993; Liotti, Gay & Fox, 1994). We will no longer have to depend on accident victims and split brain patients (Sperry, 1977; Gazzaniga & LeDoux, 1978; Harth, 1982, 1995; Geschwind, 1984; Beaton, 1985; Gardner, 1985; Eccles, 1989), and can also investigate stimulus/'response' inside the brains of nonverbal animals (cf. Griffin, 1981, 1984, 1992; Dawkins, 1993). In humans some functional brain asymmetry has been found at birth (Witelson & Pallie, 1973) to 10 months old (Ornstein & Thompson, 1984). Morphological asymmetries (reviewed in: LeMay, 1984; Best, 1988; Bianki, 1988; Iaccino, 1993), even in the speech area, are observed in newborns and human fetuses, and in the homologous region in chimpanzees (Bower, 1998; Gannon et al., 1998). Cerebral hemispheric specialization is essentially independent of *situs inversus* and whether a person is left or right handed (Penfield & Roberts, 1959; Annett, 1985; cf. Lewis, 1981b). The inheritance of left-right asymmetries is problematic:

"For example, Collins (1985) reports that although individual mice show a very consistent paw preference, the full range of left to right preference is found in a pure genetic strain. Even selective breeding within paw-preference subgroups yields offspring demonstrating the entire range. What does seem to be transferred through reproduction is the strength of asymmetry. Bryden (1987) finds similar results with humans: that strength of asymmetry but not direction is related between generations" (Molfese & Segalowitz, 1988).

Clearly there must be some early asymmetry, which may occur as early as gastrulation (as seen in asymmetric gene expression in chick, which subsequently turns with one lateral side against the allantoic membrane: Levin et al., 1995; cf. *Xenopus:* Danos & Yost, 1995, mammals: Brown, McCarthy & Wolpert, 1991). Asymmetry may occur as early as the inner cell mass stage in mammals, as suggested by discordant X-chromosome inactivation in human female twins (Jørgensen et al., 1992). What I want to suggest here is only that, once some asymmetry between the hemispheres occurs, when a tissue is divided into two tissues by a pair of differentiation waves, this bias could make the trajectory of one particular differentiation wave larger on one side of the brain than the other.

Once such an event occurs (Section 8.02), its effects could be cumulative as the tissues become hierarchically subdivided into the hundreds of tissues that may constitute the brain (by my definition of 'tissue'), so that by the end of brain development (perhaps at puberty or early adulthood), the differences between the two hemispheres could be substantial. Growth of each tissue, by just about any allometric formula (Huxley, 1932), would also accentuate the left-right differences. Nieuwkoop (1947a) (his Figure 5) gives a model of mechanical force balance between two embryos in a *duplicitas* experiment, which is unstable, and gives the essence of what I am suggesting could occur between hemispheres:

"With a symmetrical position of the two rudiments, the forces from both sides are equal. With small lateral deviations from the symmetrical position, however, the forces acting on one side predominate, since the distance between the two centres has been decreased on the side towards the deviation and increased on the other side. Equilibrium has been disturbed, and the secondary rudiment is displaced towards the primary. This results in a still more marked disturbance of equilibrium. In this way the secondary embryo shifts laterally and, in some cases, even to the dorsal side" (Nieuwkoop, 1947a).

The amplification or accumulation of differences between the two sides of the brain may be similar to differences arising between individuals in a monozygous litter. Armadillos typically produce 4 or 8 and up to 12 monozygous embryos from a single fertilization. While these become mechanically separated, their biochemistry varies considerably from one individual to the next, which is attributed to...

"...*the extent to which each of the numerous types of differentiated cells proliferates* [due to] small variations in initial size and constitution (apart from chromosomal content)... How many of each kind of cell and how much of each tissue will be produced during development are of paramount importance, because they may govern the most fundamental and important characters of the organism" (Karp & Berrill, 1981).

Such a model of accumulating differences in the trajectories of differentiation waves might also account for the plasticity of the brain in response to damage in childhood: some of each needed tissue is perhaps always present on each side, and might compensate by hypertrophy or hyperplasia (Goss, 1969). This model would also account for the ability of some right handed patients to partially recover language after surgical removal of the left hemisphere (hemispherectomy):

"...It might be safest to postpone concluding that language is strictly a left hemisphere operation....

"Smith (1966, 1972) has offered the intriguing suggestion that when the hemispheres are intact, there is a complex cooperative relation between the RH [right hemisphere] and the LH [left hemisphere], with the LH exerting dominance and control and inhibiting RH language. These inhibitory effects continue so long as there is LH tissue but are diminished or extinguished with left hemispherectomy and with commissurotomy, whereupon the RH is permitted some participation in linguistic business. Initially Smith's hypothesis received derisory rejection from orthodox opinion, but is now being taken seriously as a testable hypothesis (Sperry, 1982).... Textbook accounts of split-brain research naturally strive to present a clean package, uncluttered with details of exceptions and variations... the real value of the split-brain research may lie less in what it reveals concerning anatomical organization - which hemisphere is dominant for which capacities - than in the light it sheds on the *cognitive organization* of the brain" (Churchland, 1986).

Smith's (1966, 1972) inhibitions may fit the general physiological compensatory hypertrophy that occurs on reduction or removal of one of a pair of bilateral organs, as in renal hypertrophy, for which the search is on for a 'renotropic hormone' (Meyer, Scholey & Brenner, 1991). My notion that the differentiation trees develop completely though unevenly in the two hemispheres can accommodate both this suggestion and the point of view of Sperry (1977):

"Space in the intracranial regions is tight, and one wonders if this premium item could not have been utilized for better things than the kind of right-left duplication that now prevails. Evolution, of course, has made notable errors in the past, and one suspects that in the elaboration of the higher brain centers evolutionary progress is more encumbered than aided by the bilateralization scheme which, of course, is very deeply entrenched in the mechanisms of development and also in the basic wiring plan of the lower nerve centers" (Sperry, 1977).

It also fits the notion that brain plasticity is not a matter of altering the gross morphology of the brain:

"...No one has yet succeeded in demonstrating anatomically a single fiber or fiber connection that could be said with assurance to have been implanted by learning. In this same connection, it is entirely conceivable (though not particularly indicated) that the remodeling effects left in the brain by learning and experience do not involve the addition or subtraction of any actual fibers or fiber connections but involve only physiological, perhaps membrane, changes that effect conductance or resistance to impulse transmission, or both, all within the existing ontogenetically determined networks" (Sperry, 1977).

Galaburda (1991) states flatly "it is unlikely that area A on the left does one thing while homologous area A on the right does another".

Best (1988) has suggested that the differences between left and right sides of the brain accumulate in a particular order, explaining the morphological twist to the whole brain (reviewed in Hellige, 1993):

"...We should be able to 'read' adult asymmetries in brain structure as a record of the forces in embryological growth... a dynamic, developmental gradient in the lateral right-to-left axis of the embryo.... We need to conceptualize a *growth vector* cutting through at least three dimensions over time; this vector represents a *wave of leading growth activity*... a growth gradient that moves in the following direction: from primary motor and sensory zones to secondary association areas, and finally to tertiary association zones.... The effect of this growth vector is counterclockwise torque evident in the shape of the developing as well as the adult brain.... The growth vector would be expected to have the following influence on the development of cortical gray and white matter: For earlier-developing cortical regions relatively greater growth emphasis would be seen in the earlier-emerging deeper cell layers, which should result in smaller ratios of gray matter to white matter and greater regional width/volume.... The possibility of hormonal influences on growth gradients may aid our understanding of the development of sex differences in brain organization and cognitive functions, which may be mediated by sex differences in maturation rates.... These sex differences are most apparent in the extreme cases of sexual anomalies..." (Best, 1988).

This empirically observed 'growth vector' may represent left-right differences between the left-right pairs of corresponding differentiation waves. Its change over time would then represent the same for left-right pairs of corresponding consecutive differentiation waves. The hierarchical gene expression and time delays observed by Levin et al. (1995) may be consistent with these left-right asymmetrically expressed genes being components of the differentiation cascades triggered by consecutive differentiation waves. Whether the total asymmetry of these early, gastrulation events is consistent with a lesser asymmetry of morphology, as in cerebral hemisphere asymmetry, remains to be seen.

A casual perusal of drawings of individual brains in dorsal or ventral aspect reveals that 'lower' animals, such as fish (Harder, 1975b), geese, frogs and alligators (Romer & Parsons, 1977), cats (Weichert, 1965), and early human fetuses (Romanes, 1964) have brains whose gross anatomy is essentially left-right symmetric, whereas 'higher' animals, such as horse (Romer & Parsons, 1977), 6 to 7 month human fetus (Bailey & Miller, 1911), and human adult (Posey & Spiller, 1906; Ranson & Clark, 1947; Romanes, 1964) have grossly asymmetric brains. It would be nice to know if these left-right variations, and variations from one individual to another, or between females and males (see Best, 1988, for some references), correspond in any more or less direct way to the individual playing out of the differentiation trees in the two hemispheres. This might, of course, revive a new form of the now ridiculed 'science' of phrenology (Ornstein & Thompson, 1984; Damasio, 1994), and put our computed tomography and MRI scanners to hitherto unimagined nonmedical uses. Intrinsic marker techniques, such as X-chromosome methylation ratios (Azofeifa, Waldherr & Cremer, 1996), might also be useful in exploring the embryological origin of brain asymmetries.

Proposition 202: learning is primarily a matter of extension of the differentiation tree beyond its inherited components.

The molecular biology of learning is uncannily similar to that of primary neural induction:

"...There is already evidence linking PKC [protein kinase C] translocation with changes in gene expression. PKC and *calmodulin dependent protein kinase* (CaM kinase) activation (simultaneously triggered by a Ca^{2+} ion flux) are both implicated in changing gene expression in *Hermissenda* [sea slug] nerve cells (Connor & Alkon, 1984). PKC and CaM kinase regulate both short term associative memory acquisition and longer term maintenance of the associative memory (Matzel, Lederhendler & Alkon, 1990):

> '....Associative learning in *Hermissenda* results in a specific induction of a distinct set of at least 21 mRNAs, rather than in a generalized increase in synthesis of all mRNA, thus resembling in some respects a differentiation-like response' (Nelson & Alkon, 1990).

"Matzel, Lederhendler & Alkon (1990) clearly link the activation of PKC (and CaM kinase) with storage of associative memory and subsequent changes in the external morphology of neurons. The type of prolonged PKC translocation found in *Hermissenda* is also consistent with '...slow but relatively persistent...' translocation of PKC that occurs at primary neural induction (Otte et al., 1988). (Otte et al., 1988, in fact compare it to such a translocation.) Induction of cellular changes that normally appear with associative learning resulted from the use of an artificial PKC activator (Matzel, Lederhendler & Alkon, 1990). Artificial activation of PKC also induced neuralisation in competent *Xenopus* ectoderm (Otte et al., 1988). In both cases, changes in gene expression occur because of a prolonged PKC translocation" (Appendix IV: Björklund & Gordon, 1993b).

(Cf. Kandel & O'Dell, 1992.) Furthermore...

"...the pattern of homeogene expression... is not restricted to the early periods of development but is also observed around birth and in some cases throughout adulthood (Awgulewitsch & Jacobs, 1990; Porteus et al., 1991; Thor et al., 1991; Price et al., 1992; [Price, 1993]). The physiological significance of this late expression of homeogenes is not established. However, the finding that unc-4, a nematode homeogene mutation, alters the synaptic input to ventral cord motor neurons suggests that homeogenes play a role in the establishment of neuronal networks (Miller III et al., 1992; White, Southgate & Thomson, 1992)" (Bloch-Gallego et al., 1993).

We could speculate that the key to extension of the differentiation tree beyond that built into the DNA is based on extending the differentiation code. As we have seen, the differentiation code is probably stored in each cell in a form involving presence or absence of various non-DNA molecules. While the inherited differentiation tree may operate by binding of these molecules to the DNA, it is conceivable that an additional set of molecules may be formed that do not require such DNA binding. It is not essential that these molecules be transmitted faithfully during cell division, for the simple reason that little cell division occurs, at least of neurons in the mature mammalian brain (Jacobson, 1991b). Thus learning may require differentiation without cell division (as in manipulated *Chaetopterus:* Lillie, 1902), perhaps at the synapse level (Atwood & Cooper, 1996), and our dependence on it may actually explain why the brain has so little ability to regenerate and why brain nerve cells are, in general, terminal cells (not just terminal tissues).

Here I use an extended definition of differentiation: cells are 'learning-differentiated' from one another if they have different subsets of genes turned on. Thus cells within a single, embryologically terminal tissue, as I have defined it, can come to differ from one another. Whether this operates at the level of genes being activate/inactive, or sequestered/exposed, is not obvious, though the former is easier to imagine. Now, of course, there are many genes in a tissue that respond to environmental stimuli. For most tissues in the body, the environment would be the same, so in this sense the whole tissue responds, and we think of the process as homeostatic, merely physiological. However, in the brain, the environment of each cell includes a possibly unique set of inputs from the unique subset of cells it is 'wired' to, plus whatever in it that (partially) records its history or 'memory'. Thus the combinatorics of learning-differentiation in the brain may far exceed that of embryological differentiation.

It would be interesting to find out if organisms that can regenerate brain tissue, such as the axolotl (Harris, 1989), can retain memories in the proliferating cells. The "...emerging view that the brain contains a plethora of cell groupings devoted to different tasks, rather than a general mechanism that orchestrates all sorts of learning" (Bower, 1993a) may be consistent with a correlation between learning and cell differentiation. On the other hand, this consensus is being challenged:

"How do minds emerge from developing brains? According to 'neural constructivism,' the representational features of cortex are built from the dynamic interaction between neural growth mechanisms and environmentally derived neural activity. Contrary to popular selectionist models that emphasize regressive mechanisms, the neurobiological evidence suggests that this growth is a progressive increase in the representational properties of cortex. The interaction between the environment and neural growth results in a flexible type of learning: 'constructive learning' minimizes the need for prespecification in accordance with recent neurobiological evidence that the developing cerebral cortex is largely free of domain-specific structure. Instead, the representational properties of cortex are built by the nature of the problem domain confronting it. This uniquely powerful and general learning strategy undermines the central assumption of classical learnability theory, that the learning properties of a system can be deduced from a fixed computational architecture. Neural constructivism suggests that the evolutionary emergence of neocortex in mammals is a progression toward more flexible representational structures, in contrast to the popular view of cortical evolution as an increase in innate, specialized circuits. Human cortical postnatal development is also more extensive and protracted than generally supposed, suggesting that cortex has evolved so as to maximize the capacity of environmental structure to shape its structure and function through constructive learning" (Quartz & Sejnowski, 1997).

It may be possible to have it both ways, using the DNA dependent differentiation tree for 'selectionist' aspects of learning, and the DNA independent extension of it for 'neural constructivism'.

Perhaps it is not far fetched to envision a forthcoming clear distinction between instinctual and learned behavior, which is presently a matter of controversy (Wenzel, 1992), based on differentiation trees. If the first major step towards metazoans was the invention of continuing differentiation, perhaps the next major step towards mental phenomena was the invention of the ability of the individual to extend its own differentiation tree as part of the functioning of the nervous system. Whether or not differentiation trees present a sufficiently rich structure for 'programmability' (Conrad, 1989a) of a learning brain could perhaps be the subject of a formal analysis. The molecular complexity needed for such programmability does apparently exist:

"The structural and neurochemical complexity of the nervous system is somehow encoded for by the genome. The amount of genetic information required to generate this complexity has been estimated to be greater than that required for any other tissue. About 65% of brain sequence complexity of poly $(A)^+$ mRNA is brain specific... (Caudhari & Hahn, 1983;

Sutcliffe et al., 1984).... 10,000 [genes] may be brain specific. However, we know very little about the distribution of these unique transcripts, most of which are present at low abundance....

"...Gene amplification following subtractive hybridisation (Liang & Pardee, 1992, [1995]), and genomic subtraction for cloning DNA corresponding to deletion mutations (Lisitsyn, Lisitsyn & Wigler, 1993) are methods currently being employed to detect brain genes. Differential cDNA cloning has more recently been used in the context of brain function to detect plasticity related genes in the hippocampus (Nedivi et al., 1993). Plasticity related gene expression was induced by injection of kainate into the hippocampus, producing LTP [long-term potentiation] and seizures. A subtracted cDNA library from the hippocampus of kainate injected animals was differentially screened for LTP/seizure-induced cDNAs. This type of functional approach in the context of subtractive differential screening could be most readily applied to regions of the brain involved in behaviour. Unilateral activation of limbic brain regions in behavioural contexts would refine the molecular work by allowing each animal to serve as its own control. However, molecular genetics is never that simple, and as a sobering illustration some 500-1000 candidate plasticity related genes were estimated to have been activated in the hippocampus by kainate-induced LTP/seizures" (Keverne, 1994).

Whether in terms of total brain transcripts or just hippocampus plasticity related genes, we see that there are probably enough for a few hundred genes per differentiation cascade, or its learning equivalent.

It is interesting to note that the highly conserved structure of calmodulin may trace back to the behavior of ciliates, from which both continuing cell differentiation and learning may have later been derived:

"...We were greatly surprised when we traced the defects of some of the *Paramecium* behavioral mutants to the Ca^{2+}-binding loops, hydrophobic patches, and charges in loop-flanking helices of CaM [calmodulin] and found the mutants to be viable.... They grow, multiply, feed, secrete, excrete, mate, and swim about. This suggests that general metabolism, cell-cycle progression, exocytosis, cortical development, and axonemal motility, do not absolutely require the conserved features in the calmodulin.... This molecule has essentially been held constant since the emergence of modern eukaryotes some two billion years ago. What are these conserved features for, if not for survival? The *Paramecium* work reviewed here indicates that these features are required for normal membrane excitation....

"Perhaps it is no accident that Ca^{2+}-CaMs are known to activate pairs of targets, which enigmatically catalyze opposite reactions: e.g. cyclase and phosphodiesterase, kinase and phosphatase. Here we have another pair: the Na^+- and the K^+-channel in *Paramecium*. Perhaps the bipartite arrangement of a single pivot, calmodulin, is important in how it orchestrates these targets" (Kung et al., 1992; cf. Ling et al., 1994).

We proposed that Ca^{2+}-CaMs and their catalysis of seemingly opposite reactions serves as a mechanism for contraction wave launching and propagation (Appendix IV: Björklund & Gordon, 1993b). While we proposed that Ca^{2+} levels are regulated by a Ca^{2+}/Na^+ exchanger, this work on *Paramecium* suggests a need to look at the role of K^+-channels in the same context (Natalie K. Björklund, p.c.), especially since, as discussed in Section 5.05, the ultraslow waves on *Paramecium* (Iftode et al., 1989) are differentiation waves.

To go beyond the vague suggestion of this proposition, we would need to know how memory actually works, a question that is far from settled (Morris, Davis & Butcher, 1990; Kano, 1995). The relationship between learning extensions to the differentiation tree and the mental storage of so-called 'memes', units of cultural memory (Dawkins, 1976, 1982, 1986), might be great fun to explore:

"...A human mind is itself an artifact created when memes restructure a human brain in order to make it a better habitat for memes" (Dennett, 1995).

Proposition 203: instinct is genetic assimilation of an extension of the differentiation tree corresponding to a learned behavior.

As we have seen in the Cocos Islands finches (Wilson, 1992; Proposition 196), the behavior of individual animals smacks very much of the start of new species. What makes these behaviors possible? Let us suppose that the previous proposition is correct. Then each incipient finch species may be setting off specific differentiation cascades involved in its particular behavior. We assume that these gene activations are extensions of the differentiation tree. But we must now qualify the phrase 'beyond its inherited components' in the previous proposition. Obviously, the capability to go beyond the differentiation tree must itself be inherited. This means that certain differentiation cascades, which already exist in the genome, come into play only when environmentally triggered.

We can now see instinct in the light of differentiation trees. All we need is that genetic assimilation (Waddington, 1952b, 1953b, 1956b, 1959a, 1961b; Taylor, 1983; Sara, 1989; Gerard, Laffort & Vancassel, 1992; Gerard, Vancassel & Laffort, 1993) occurs, i.e., mutations arise that cause these extensions of the differentiation tree to be triggered automatically during development, rather than requiring an environmental trigger. For instance, for nematodes...

"...cell fate determination by environment or lineage may involve similar underlying mechanisms that cause and maintain cell division asymmetry (Jan & Jan, 1995)" (Fitch & Thomas, 1997).

(Cf. Wolpert, 1994b.) Under appropriate circumstances, such mutations might be favored over the 'mere' ability to learn the behavior, since, so to speak, the organism is ready to go at birth or hatching, without having to learn. This is an old hypothesis, used to rescue Lamarck's concept that acquired characters are heritable (Lamarck, 1803, 1809; Cannon, 1958; Mayr, 1972; Burkhardt Jr., 1977; Barthélemy-Madaule, 1982; Corsi, 1988) in the context of Darwinian mechanisms:

"Learned behaviors, abilities, and skills are often essential for survival or reproduction. If such skills are so important, why have they not become genetically programmed? Undoubtedly, over evolutionary time, this can indeed occur; learned behaviors may become instincts (Papaj, 1993), and learned sensory abilities may become hard-wired (Lettvin et al., 1959; Hinton & Nowlan, 1987). J.M. Baldwin (1896) proposed how learning can be advantageous to individuals whose behavior is only partially genetically determined. First, learning allows an individual to complete a partially programmed behavior, increasing survival probability. For example, the rat's brain is partially genetically 'hard-wired' such that learning particular associations, such as between a taste and gastric illness, is facilitated (Garcia & Koelling, 1966; Garcia, Ervin & Koelling, 1966). The retina of a frog preprocesses the visual field such that small, rapidly moving ('fly-like') objects are greatly enhanced

(Lettvin et al., 1959). Individual learning supplements these genetic predispositions to increase fitness. Baldwin further proposed how, over evolutionary time, this individual advantage of learning guides the process of evolution: 'the variations...' in genetic components of behavior 'which were utilized for ontogenetic adaptation in the earlier generation, being thus kept in existence, are utilized more widely in the subsequent generation' (Baldwin, 1896). The process by which learning assists the integration of genetic components of behaviors into the gene pool is variously referred to as 'genetic assimilation', 'canalization', or the 'Baldwin effect' (Baldwin, 1896; Morgan, 1896; Waddington, 1942a; [Simpson, 1953]; Maynard Smith, 1987a)" (Anderson, 1995a). [Dennett (1995) additionally attributes the idea to H.F. Osborn.]

"It has been conventional to argue, from the fact that 'acquired characters' are not inherited, that the development of adaptive modifications during the lifetime of an organism is irrelevant to the evolution of similar genetically determined phenotypes. The theory of genetic assimilation, and the practical demonstration that the process can occur, shows that this argument is misguided, and provides a new way of accounting for all those evolutionary facts for which, in the past, some authors were tempted to advance a 'Lamarckian' explanation" (Waddington, 1961b).

"Recent models of the Baldwin effect have focussed on how learning increases the rate of convergence of evolutionary processes in fixed environments (static fitness landscapes) (Hinton & Nowlan, 1987; Fontanari & Meir, 1990; Ackley & Littman, 1991, 1994).... But learning has no selective advantage in fixed environments, since, presumably, once the optimal genotype is found, exploration away from this optimum only reduces fitness (Stephens, 1993; Via, 1993) [presuming other optima are not accessible, such as via gene duplication: Gordon, 1994b].... In this study, I develop a quantitative genetics approach to modeling the interactions between learning and evolution in variable environments" (Anderson, 1995a).

(Cf. Belew & Mitchell, 1995; Dennett, 1995; Littman, 1996; Menczer & Belew, 1996; Turney, Whitley & Anderson, 1996.) Waddington (1953c, 1957, 1961b) gives some history and argues for separating the Baldwin effect from genetic assimilation. Wolpert (1994b) invokes the Baldwin effect repeatedly in a model of the origin of embryos. Obviously, under other evolutionary circumstances, such as, perhaps, variable or heterogeneous environments, learning may be favored over instinct. As the old nature/nurture controversy shows us, the lines between learning and instinct are sometimes hard to find. But that doesn't mean they aren't there:

"...It is doubtful... that beavers learn to build dams.... The fact cannot be emphasized too often that the central problem of instinctive behavior is one of development (Maier & Schneirla, 1935)" (Schneirla, 1964).

Baldwin (1902), in a pre-Mendelian genetics summary of his work along these lines, was a bit more forthright than many of our contemporaries about...

"...the larger problem which may be called *psychophysical evolution* - the evolution of mind and body together.... We do not have one series of genetic forms, the mental, evolving under shorthand formulas of its own; and another series, the organic, doing the same thing under different formulas. On the contrary, the two sets of facts really go together in the one set of formulas.... It is an equally embarrassing thing for the scientific enquirer in one case to be a monist to such a degree as to deny one of these lines altogether - the mental in favour of the physical 'shorthand,' or the physical in favour of the mental - and in another case insist upon the separateness of what nature shows us always joined together, to the extent of refusing to use facts from one series to illumine and even to explain facts in the other.... To refuse... is suicidal to genetic science....

"Accommodation, when all is said, is a positive thing, a vital and mental functional process supplementary to the hereditary impulse. It must be considered a positive factor in evolution, a real force emphasizing that which renders an organism fit; whereas natural selection, while a necessary condition, is yet a negative factor, a statement that the most fit are those which survive. If it be true that those variations which can accommodate, either very much or very little, to critical conditions of life are the ones to survive, and that such variations will be accumulated and will in turn progressively support better accommodations, then it is the accommodations which set the pace, lay out the direction, and prophesy the actual course of evolution. This meets the view of the Lamarckians that evolution does somehow reflect individual progress; but it meets it without adopting the principle of Lamarckian inheritance....

"...The endowments of Moses would have been quite ineffective, despite his salvation by the women, had not opportunities arisen for him to use his gifts. His actual performance is what counted in history; and so it is with the humblest organism which accommodates itself to the environment, in so far as it makes effective contribution to the characters of the generations which follow after it" (Baldwin, 1902).

Rensch (1967) finds that...

"Generally, the learning achievements show a successive increase parallel with the phylogenetic complication and differentiation of the central nervous systems.... We can state that the absolute size of the central nervous systems and the number and differentiation of the neurones is sometimes more important than the phylogenetical level.... The fact that... among related animals the larger species show a greater learning capacity and longer [memory] retaining period is an important contribution to the explanation of *Cope's rule* of successive increase of body size in many lines of descent, especially in nonflying vertebrates" (Rensch, 1967).

If learning is reducible to a form a gene expression, we will need to investigate whether all forms of learning are accessible to genetic assimilation, or just certain kinds. After all, natural selection is a sledge hammer approach to biological change, requiring many individuals and much culling. The subtle differences in learning, between individuals, may evade selection and thus not be available for genetic assimilation.

Proposition 204: instinctual behavior is evolutionarily more advanced than learned behavior.

To pursue questions of instinctual behavior at a cellular/developmental level, we may need to compare two closely related species with the same behavior, one learned and one instinctual. The learning species should then be found to have the smaller differentiation tree, with 'differentiation' waves and differentiation cascades that respond to appropriate environmental cues that trigger that learning, causing some of the cells in the brain to 'learning-differentiate' further. The instinctual species should have the larger differentiation tree, and be born or hatched with those same cells already differentiated.

The same point may be seen in Williams' (1966) argument against the idea...

"...that natural selection must be supplemented by another process, which [Waddington (1956b, 1958)]... calls *genetic assimilation*.... The process starts with a germ plasm that says: 'Thicken the sole if it is mechanically stimulated; do not thicken it if this stimulus is absent,' and ends with one that says: 'Thicken the soles.' I fail to see how anyone could regard this as the origin of an evolutionary adaptation. It represents merely a degeneration of a part of an original adaptation.... It must, as a general rule, take more information to specify a facultative adaptation than a fixed one.... <u>As</u> a general class, facultative adaptations represent more difficult evolutionary attainments than obligate adaptations... [See] Warburton (1955)... Underwood (1954)..." (Williams, 1966).

This objection notwithstanding, we do need a constructive theory for facultative adaptations, which may be provided by the growth of differentiation trees. Williams (1966) himself does not look towards embryology since he insists that "...morphogenesis... is adequately understandable without recourse to creative evolutionary forces that would not be predictable outcomes of selective gene substitution".

Of course, if this scenario proves correct, then the species with the more complex differentiation tree will actually prove to be less 'advanced', if it has traded off some of its ability to learn in gaining some automated behavior. In other words, instinct, rather than being a 'primitive' feature, may actually represent an 'advanced' specialization. (The same can be said about the evolution of 'natural' antibodies from 'adaptive' antibodies: Anderson, 1996b.) On the other hand, species that both genetically assimilate a particularly useful behavior and fully retain the ability to learn other behaviors may be the best adapted. This could be accomplished via duplication of a particular 'learning' terminal branch of its differentiation tree followed by genetic assimilation of just one copy.

It is important to note that Baldwin (1909) also postulated the reverse process from this proposition, namely "...the decay of congenital characters (e.g., of instincts)...; in cases where the intelligent or other adjustive factor is on the whole of greater utility, variations toward the disintegration of the instinctive congenital part, would be selected." However, it is hard to conceive of a hard-wired behavior becoming 'genetically disassimilated'. From the point of view of differentiation trees, and their terminal branch duplication, instincts and learned behaviors need not be mutually exclusive, and could build on one another.

Differentiation trees may provide a new, physical approach to resolving the nature of instinctual behavior, whose existence has gone in and out of favor over the centuries:

"**Instinct** was regarded by early writers as the natural origin of biologically important motives. Thus Thomas Aquinas wrote that animal judgment is not free but implanted by nature.... The associationists appeared to reject all notions of instinct, although Locke did write of '...an uneasiness of the mind for want of some absent good.... God has put into man the uneasiness of hunger and thirst, and other natural desires... to move and determine their wills for the preservation of themselves and the continuation of the species.' (Locke, 1690).... The concept of instinct was seized upon by the new rationalists, such as Reid [in] 1785 [cf. Reid, 1815; Lehrer & Beanblossom, 1975], Hamilton [posthumously in] 1858 [cf. Murray, 1870], and James [1890], as a convenient vehicle for the nonrational elements of behaviour. Thus man's nature was seen as a combination of blind instinct and rational thought.... Darwin (1859) was the first to propose an objective definition of instinct in terms of animal behaviour. He treated instincts as complex reflexes, which were made up of units that were compatible with the mechanisms of inheritance. Instincts were thus the product of natural selection, and evolved together with other aspects of the life of the animal.. This is very much the view propounded by the early students of ethology. Lorenz (1937a,b) [cf. Evans, 1975] claimed that there were many fixed action patterns that were determined by an animal's genetic make-up and were characteristic of species.... Tinbergen (1951) proposed that the different sources of action-specific energy were arranged in a hierarchy.... The concept of instinct has undergone many changes in its history. In its early usage, instinctive behaviour was seen to be inborn, reflex-like, and driven from within. Modern research has separated out the innate, the reflex, and the motivational aspects of apparently instinctive behaviour, and regards these as separate issues. This is probably the main reason for the diminution in the importance of the concept of instinct" (McFarland, 1987).

Obviously, this vacillation is tied up with our desire to believe in free will, and thus with existential questions. Nevertheless, this proposition suggests a route for further enquiry. Maynard Smith (1987a) succinctly summarizes the issues from an evolutionary and developmental point of view:

"In a recent paper, Hinton & Nowlan (1987) suggest how a combination of evolution and learning may make possible the evolution of a structure that is of value only when perfect.... If individuals vary genetically in their capacity to learn, or to adapt developmentally, then those most able to adapt will leave most descendants, and the genes responsible will increase in frequency. In a fixed environment, when the best thing to learn remains constant, this can lead to the genetic determination of a character that, in earlier generations, had to be acquired afresh each generation. The idea goes back to Baldwin (1896) [cf. Baldwin, 1917], Morgan (1896) and Waddington (1942a). It has not always been well received by biologists, partly because they have suspected it of being lamarckist (a suspicion that Waddington was curiously reluctant to allay), and partly because it was not obvious that it would work. What Hinton & Nowlan (1987) have done is to answer these objections. They consider the following simple model, of whose biological unreality they are well aware. Imagine an organism with a neural net in which 20 connections must be correctly set ('on' and 'off') if its fitness is to be increased: if even one connection is wrong, it is no better than if all were wrong. This is the worst type of fitness surface, as there is no slope leading to the summit [cf. Gordon, 1994b]. There are 20 gene loci, each with three alleles, 0, 1 and ?. The first two specify on and off, and the third that the connection can be varied. During learning, an individual tries out a succession of settings for these variable connections. If, by chance, it tries out a correct set, and its genetically specified connections are also correct, it is rewarded, and does not alter its settings again. An individual makes 1,000 trials in its life.... <u>An</u> individual that is born with the correct settings genetically fixed has a fitness 20 times that of one that never learns" (Maynard Smith, 1987a).

What this model shares with my hypothesis is a one to one correlation between inherited features and features needed for learning.

When I was a graduate student, we had much fun with a casual 'journal' called the *Worm Runner's Digest,* which mixed spoofs and apparently serious work on planaria and their supposed ability to cannibalistically transfer memories, speculated then to be carried by RNA. I have not kept up with that work, nor reviewed it for this book, but must note the following transfer of an inherited 'behavior' by brain grafting of specific primordia, which just might correspond to tissues set off by differentiation waves:

"Species-typical crowing behavior could be transferred from the quail donor to the chicken recipient by means of brain transplants (Balaban, Teillet & Le Douarin, 1988; [cf. Balaban, 1997]).... In the Fayoumi chicken, a spontaneous recessive autosomal mutation (F.Epi) is responsible for high susceptibility to [epileptic] seizures that are especially inducible by intermittent light stimulation.... Microsurgery was performed *in ovo* at the 12- to 15-somite stage, when the encephalic vesicles are well defined by constrictions [launching domains?], which were used as limits for the operations.... We have found so far that only Ch.Epi birds into which at least both [proencephalon] and mesencephalon have been implanted presented the full spectrum of epileptic manifestations" (Teillet et al., 1991; cf. Teillet et al., 1995).

How much of our behavior is hard wired and transferable in this manner will be fascinating to learn. The following is consistent with the possibility that, during such transplantation of discrete brain parts, the 'planar inductions' deduced in the neural plate are actually consecutive differentiation waves, and rather than a threshold for a fading morphogen, the extent of each wave is determined by its trajectory:

"In this study we have analysed the expression of *Hoxb-4, Hoxb-1, Hoxa-3, Hoxb-3, Hoxa-4* and *Hoxd-4* in the neural tube of chick and quail embryos after rhombomere (r) heterotopic transplantations within the rhombencephalic area. Grafting experiments were carried out at the 5-somite stage, i.e. before rhombomere boundaries are visible.... When a rhombomere is transplanted from a caudal to a more rostral position it expresses the same set of Hox genes as in situ. By contrast in many cases, if rhombomeres are transplanted from rostral to caudal their Hox gene expression pattern is modified. They express genes normally activated at the new location of the explant, as evidenced by unilateral grafting. This induction occurs whether transplantation is carried out before or after rhombomere boundary formation....

"We noticed that one requirement is that the grafted fragment of tissue be perfectly incorporated within the plane of the host's neuroepithelium.... The inductive signal is transmitted from caudal to rostral via the neuroepithelium itself.... One can imagine that the cells induced to express Hox genes also acquire the capacity to produce the inductive signal and that a progressive decrease in the amount of substance generated at each step of its progression from caudal to rostral is responsible for fixing the anterior limit of Hox gene expression. Our experiments do not provide any insight concerning the nature of the signal" (Grapin-Botton et al., 1995).

The 'nature of the signal' might be found to be differentiation waves, were time-lapse observations to be made.

To summarize, we see that instinctual behavior may be rooted in distinct, sometimes transplantable parts of the brain, which are defined by differentiation waves, and that these may have derived from earlier learned behaviors fixed by genetic assimilation.

Proposition 205: sleep, an evolutionarily primitive feature of animals, lasts for hours because that is how long differentiation waves involved in learning take to complete their trajectories.

It appears that learning and memory are correlated with changes in gene expression (Connor & Alkon, 1984; Goelet et al., 1986; Lisman & Goldring, 1988; Tsukada, 1988; Byrne et al., 1991; Davis & Dauwalder, 1991; Brennan, Hancock & Keverne, 1992; Pfaff et al., 1992; Robertson, 1992; Qian et al.,1993; Skoulakis, Kalderon & Davis, 1993; Alberini et al., 1994), and indeed, we have already suggested a possible relationship between learning and differentiation waves involving changes in cAMP concentration (Appendix IV: Björklund & Gordon, 1993b; Proposition 202). Short term and long term memory are now distinguished by lack and presence, respectively, of protein synthesis, with the latter triggered by...

"...elevated levels of cAMP.... Recent studies in systems ranging from two mollusk neurons in culture to complex behaviors in mammals have revealed a molecular mechanism by which memories are generated. The landmark papers by Bourtchulade et al. (1994) and Yin et al. (1994) demonstrate that a particular transcription factor, CREB, is critical for [long term] memory formation. That this protein and its role in memory are highly conserved through evolution suggests that the mechanism of protein synthesis-dependent memory is also well conserved" (Frank & Greenberg, 1994).

In considering how short term memory becomes long term memory, Anderson (1998) has suggested the existence of a...

"...reinforcement signal [that] is likely to carry only general, non-specific, information. Thus, it could be neural or chemical (hormonal) in origin" (Anderson, 1998).

This reinforcement signal was conceived as a response to an impasse that had developed in the literature on neural networks:

"Hinton (1989), one of the pioneers of neural networks, acknowledges that 'as a biological model, back-propagation is implausible'.... In his article 'The Recent Excitement about Neural Networks,' Francis Crick (1989a) writes:

> 'It is hardly surprising that such achievements have produced a heady sense of euphoria. But is this what the brain actually does? Alas, the back-prop nets are unrealistic in almost every respect... Obviously what is really required is a brain-like algorithm which produces results of the same general character as back propagation.'

"In this paper we describe such an algorithm:... a random walk through the space of synaptic weights. It relics on Gaussian trial fluctuations, which can be generated locally at the synapses, and which are then fixed by a global signal 'To all synapses: Freeze the last trial fluctuation and make it the starting point for new trials.'" (Bremermann & Anderson, 1991).

A similar analog model of a learning circuit was based on a dendritic crystal version of the iron wire model for nerve signal transmission:

"An iron wire immersed in concentrated nitric acid assumes properties strikingly resembling those of nerve fibers. Iron does not dissolve in this system owing to the formation of a passive oxide film on its surface. If the oxide is destroyed at any point, activating the iron, the metal dissolves rapidly.... The shiny surface... becomes matte and darkens while gas bubbles evolve. The process begins to spread over the wire, and after brief activation the wire is again passivated....

"This phenomenon was already known in the last century.... Ostwald (1900) pointed out the analogy between this phenomenon and the conduction of excitation along a nerve cell axon. The first studies of this problem (Heathcote,..., 1907) were carried out under his direction. However, full and consequent study of this phenomenon was carried out by Lillie (1918, 1920a,b, 1925, 1928, 1929b, 1931, 1935, 1936).... Lillie showed that... there is refractoriness - insensitivity to repeated activation for a period of time after preceding activation.... Lillie observed that when impulses 'collide', they destroy one another.... Bonhoeffer ([1948]) elucidated the electrochemical mechanism.... Yamagiwa (1948a,b, 1949a,b,c, 1950a) [studied]... conduction of activation impulses in geometrically more complex structures.... Yamagiwa also discovered an interesting effect of a slight acceleration of an impulse on a wire parallel to a refractory wire (i.e., a wire on which an activation potential had passed just before).... Stewart (1965a,b) studied the propagation of activation waves in three-dimensional porous structures (mixtures of iron and glass balls covered with nitric acid)" (Markin, Pastushenko & Chizmadzhev, 1987).

W. Ross Ashby (p.c., 1965; cf. Ashby, 1956, 1960) made a 'stochastic switch' out of mixtures of copper and glass balls between two copper plates (without the nitric acid). This is an example of a percolation system (Hammersley & Handscomb, 1964; Forgacs et al., 1989) or 'random electrical network' (Grimmett, 1989; cf. Forgacs, 1995a). Robert M. Stewart (p.c., 1966; cf. Stewart, 1965a,b) further grew iron dendrites under hot, pressurized nitric acid, with electrical and optical inputs and outputs at the dendrite tips:

"On the technological side we are hopeful of a breakthrough of the *complexity* barrier.... We believe that it is possible to construct a pliable electrochemical machine to be used for such things as pattern recognition, in which the fine structure and packing density of cells compares favorably with that of the brain. That this should be possible in an electrochemical system rests primarily with three unique characteristics:

> (a) Distributed energy supply for signal impulse propagation (a la *Lillie iron-wire nerve model*),

> (b) Fine structure or *dendrite* growth by electrodeposition and corresponding *plasticity* of function, and

> (c) Intrinsic active metal surface behavior which results in hypersensitivity to gross electrical control fields or massive shocks in recently active regions as compared to recently passive regions.

On the biological side, interest centers on the possibility of developing a meaningful model technique for the study of the behavior of large cell assemblies, especially with respect to the parallel questions of:

> (a) Intercellular wave propagation and interactions, 'synaptic transmission,' and ephemeral memory,

> (b) Long-term memory substrate, and

> (c) Learning mechanisms" (Stewart, 1965b).

'Learning' was accomplished by a direct current passed through the dendrites immediately after propagation over subtrees of the dendritic tree had occurred. If the 'output', the subset of terminal branches over which propagations occurred, was deemed desirable by the experimenter, then the direction of the direct current was chosen such as to cause some electroplating, which occurred preferentially along the pathways that the activation current had just followed. If the 'output' pattern was not desirable, then the direct current was applied in the reverse polarity, causing some dissolution of the iron along the activated pathways, thereby weakening them. In the course of this 'training', the iron dendrite 'learned' the pattern desired by the investigator.

Differentiation waves could fill the role of such a 'global' signal.

Given their slow speed, ultraslow waves could correlate with and perhaps explain the process of consolidation of memory during sleep. The major item missing in the puzzle of why we sleep would then be why reduced behavioral activity is necessary for differentiation waves to either propagate or propagate and be effective in memory consolidation (cf. Jouvet, 1991.) The empirical correlation between sleep and memory consolidation is sometimes negative (Carskadon & Dement, 1994) or equivocal (Bonnet, 1994). Nevertheless there is widespread support for a presumed correlation between sleep and memory consolidation (Roth et al., 1988; Leconte, 1989; Rotenberg, 1992a,b; Guerrien, 1994; Crick & Mitchison, 1995; Gardner-Medwin & Kaul, 1995; Giuditta et al., 1995; Sejnowski, 1995; Smith, 1995a, 1996; Dotto, 1996; Kavanau, 1997a), and since I am postulating that memory is a form of cell differentiation, we might well entertain the hypothesis that learning-differentiation waves propagate best in the still of night.

If we stick to our presumption that differentiation waves all have physical components, involving cell contractions and/or expansions, then in appropriate preparations, these waves might be directly observable (cf. Glanzman, Kandel & Schacher, 1990; Bacskai et al., 1993): "In the case of astrocytes, it has been postulated that the cell-to-cell propagation of Ca^{2+} waves may form the basis of long-range signalling in the brain (Cornell-Bell et al., 1990; Cornell-Bell, 1991)" (Thomas, Renard & Rooney, 1992). Alternatively, such waves might be imagined to operate through the postsynaptic cytoskeletal organelle known as the 'postsynaptic density' (Kennedy, 1993), which might actually be a form of cell state splitter. Even axolotls sleep, so our favorite animal might be useful here:

"**Evolution of sleep**. The phenomenon of sleep, as we know it in man and other mammals, can be identified with a high degree of certainty in the majority of familiar vertebrates. Highly analogous phenomena can also be identified in some invertebrates, such as molluscs and insects. This suggests that sleep goes back a long way in the EVOLUTION of animals. It is possible that many of the complex aspects of sleep in vertebrates constitute specializations of a more primitive mechanism which divided an animal's TIME into periods of activity and inactivity....

"By international convention... two states are distinguished by the names quiet sleep (QS) and active sleep (AS).... The functions of AS and QS are poorly understood.... It is now well established that sleep appears in the developing animal before it is born. Experiments using foetal animals in utero appear to show that AS matures before QS. Moreover, young animals spend a very high percentage of their sleep in AS (up to 100%), whereas in adulthood this percentage is usually low (not exceeding 25%). The suggestion has therefore been made that AS plays a crucial role during development....

"Although studies of amphibia are few, sleep is known to occur in tiger salamanders (*Ambystoma tigrinum*).... Axolotl (*Ambystoma mexicanum*) have been observed to sleep in groups suspended in the water supported by plants. At these times they are relatively unresponsive, and have a very slow gill-stroke rate" (Meddis, 1987).

Whether differentiation waves can be pinned down to particular stages of sleep would be worth investigating, especially if the distinction made in embryology between determination and specific gene expression carries over to our purported waves for learning. Rial et al. (1993) have suggested "...that the reptilian waking state and the mammalian slow wave sleep are homologous states.... We also propose that instead of looking at the polygraphic sleep as a new evolutive acquisition of mammals and birds, it seems more convenient to look at the full waking state; the 'advanced wakefulness' as the true new evolutionary acquisition of these animals." It would be interesting to know if 'advanced wakefulness' has a basis in the central nervous system terminal branch of the differentiation tree.

7.00 THE BIOGENETIC LAW

"Does evolution proceed mostly by changes in the terminal stages of ontogeny? How much information about the history of an organism is encoded in its sequence of developmental transformations? More than a century since Haeckel proposed that ontogeny recapitulates phylogeny, most experts cannot provide definitive answers to these questions. Haeckel has been dismissed, his theories rejected, but, as any comparative study of embryology would illustrate, some subtle parallel between ontogeny and phylogeny exists.... There are many valid reasons for the downfall of Haeckel's views, but some relate to a generic prejudice of the neodarwinian synthesis against internalist approaches to evolution" (Alberch & Blanco, 1996).

7.01 'Ontogeny Recapitulates Phylogeny' Revisited via Differentiation Trees

"For many years - let us say between 1850 and 1900 - embryologists were engrossed with the idea that development of higher forms recapitulated the entire historical path over which their evolution had passed.... The historical appeal was irresistible, especially if one believed that what he was seeing and describing was the history of 'creation'..." (Morgan, 1934).

"Unreliable rules for evolutionary inferences include the assumptions that primitive traits are correlated in extant animals, that evolution proceeds from simple to complex structures, that earlier stages in development are more conservative, and that structures that develop earlier evolved earlier" (Strathmann, 1993a).

Proposition 206: terminal topological additions to differentiation trees consisting of duplicates of terminal branches are more often compatible with reproductive success than other topological changes to the differentiation tree, because they are relatively frequently evolutionarily neutral or of selective advantage.

The idea that ontogeny recapitulates phylogeny has a long and rugged history. In its simplest form it conveys the idea that evolution occurred in a linear sequence, along the ladder of life (Sections 10.02, 10.03, 10.05, 10.09), and each embryo, during its development, makes the same ascent, up to its ordained rung. Some authors abandon the idea, while others retain the nagging sensation that there must be something to it. In this chapter I will try to reformulate the concept in terms of differentiation trees. This does not leave us with the simple, naive conception of the so-called 'biogenetic law', but rather a richer, and perhaps exploitable way of thinking about comparative embryology.

This proposition is one expression of the general assertion that much of evolution seems to occur via gene duplications (Ohno, 1970; Jacob, 1977; Savageau, 1986; Ohta, 1987a,b, 1988a,b; John & Miklos, 1988) that could be initially selectively neutral:

"The existence of duplicate genes was first established upon genetic evidence when Sturtevant (1925) described the behaviour of the Bar eye locus in *Drosophila*..... Huxley (1942) pointed out the possibilities for divergent specialization of such genes.... Lewis (1951) was among the first to state the role of duplicate genes in the evolution of new proteins: that one duplicate might evolve while the other maintained function" (Watts & Watts, 1968).

"It may... turn out to be of the nature of nucleic acids and the chromosomal apparatus that they tend spontaneously to proffer genetical variants - genetical solutions of the problem of remaining alive - which are more complex and more elaborate than the immediate occasion calls for..." (Medawar, 1967).

Perhaps whole differentiation cascades (edges) or even terminal branches of the differentiation tree may be added in a benign fashion:

"Terminal addition of gene programs should be particularly 'easy,' in that they can hardly interfere with the programs for preceding developmental stages" (E. Zuckerkandl in Gould, 1977a).

Since these additional 'gene programs' would thus initially be identical copies of existing terminal branches (Figure 25), this proposition could be regarded as an extension of the neutral theory of evolution (Kimura, 1983; Provine, 1990; Takahata, 1994; Section 6.05).

It is important to note that I postulate that the easiest changes to the differentiation tree not only come from terminal branches, but also acquire terminal locations on the changed differentiation tree.

The differentiation tree thus provides a 'scaffold' on which to hang the idea that adding to the genome is easier than taking away, which Mayr (1982) suggests goes back at least to Darwin:

"How does the modern biologist explain the presence of gill arches in the ontogeny of mammals? To be frank, until the physiology and biochemistry of developmental systems is better understood, only a tentative answer is possible.... Darwin's [1872] thesis that evolutionary new acquisitions are superimposed on the existing genetic structure, even though frequently attacked, has a correct nucleus" (Mayr, 1982).

Modern evidence on the relative frequencies of terminal additions versus deletions is equivocal, but perhaps the question could be resolved if comparisons were made at the level of the differentiation tree:

"The hypothesis of 'terminal modification' is non-committal regarding the relative manifestation of terminal addition (peramorphosis) vs. terminal deletion (paedomorphosis).... Studies empirically support evolution by terminal modification in an extraordinarily high frequency (up to 70-80%) of cases (Krauss, 1988; Mabee, 1993).... It is not possible to state that peramorphosis is more common than paedomorphosis, although Krauss (1988) implies the former to be more common" (Alberch & Blanco, 1996).

The following is consistent with the idea that one can't delete an intermediate edge of a differentiation tree without disconnecting the terminal branches that are attached to it (Proposition 132):

"Once the genetic basis of a structure is thoroughly incorporated into the genotype and forms part of its total cohesion, it can be removed only at the risk of destroying the entire developmental system. It is less expensive to keep the complex regulatory system of mammalian embryogenesis intact, even though (as a by-product) it produces unneeded gill arches, than to break it up and produce unbalanced genotypes" (Mayr, 1982).

Finally, the following notion of Mayr (1982) is consistent with the idea of parts of differentiation trees becoming, in some evolutionary lines, differentiation webs:

"Our understanding of developmental regulation is far too incomplete to rule out the possibility that late evolutionary acquisitions are indeed 'added' to the genotype more loosely than characters inherited from remote ancestors" (Mayr, 1982).

We can thus see that there are some subtle, high order questions about the structure and evolutionary dynamics of the genome, for which differentiation trees give at least a working model. Note that I have not precluded microevolutionary changes of earlier edges of the differentiation tree by this proposition. It is only macroevolutionary changes of early edges that cannot take place, unless they include all attached terminal branches. It would be interesting to know if some of the many presumed simple (microevolutionary) heterochronies identified as objections to the existence of a 'phylotypic period' in development (Richardson, 1995; Richardson et al., 1997) are actually macroevolutionary changes in differentiation tree topology. If not, Ernst Haeckel would be partially exonerated of so-called 'fraud' (Anon., 1997c; Pennisi, 1997a), at least in the abstract sense of an underlying commonality in differentiation tree topology for early development. He is accused of omitting features and exaggerating similarities between early embryos. I suspect that Haeckel was simplifying for teaching purposes, as his artistry and attention to detail (Haeckel, 1862, 1904) far exceeds that in the offending drawings:

"Haeckel (1891) published a series of comparative drawings showing different animals arising from virtually identical somite-stage embryos. It is surprising to find that these famous old drawings are still the main evidence for a conserved [phylotypic] stage in vertebrate development. the plates are very influential and are still reproduced in textbooks (Alberts et al., 1994; [also Gerhart & Kirschner, 1997]).... As Gould (1977a) has pointed out, Haeckel (1891)... was essentially a popular work..." (Richardson, 1995).

Proposition 207: except for genetic drift, the differentiation tree will preserve the history of its production by duplications of previous terminal branches.

This is just a generalization of the concept that:

"Some genes of hormones and their receptors form supergene families which contain one or more domains of common origin.... Indeed several important domains must have been repeatedly utilized in evolution, and it has been suggested that the number of such basic domain structures in the whole biological world is not very large, probably 10^3 or even less (Doolittle, 1981; Neurath, 1986)" (Ohta, 1988b).

Proposition 208: macroevolution proceeds primarily by addition and deletion of terminal edges of the differentiation tree.

This is because reproductive isolation should be least for organisms whose differentiation trees differ by terminal edges. This idea is quite similar to that of Cope:

"How, then, do new genera evolve? Cope argues that generic characters originate as additions to the end of ancestral ontogeny (although genera can also evolve retrogressively, by the loss of stages).... 'Every change by complication of structure is by addition; every simplification is by subtraction' (Cope, 1872 [in Cope, 1887]).... The stages of ancestral ontogenies are repeated in successively shorter intervals, leaving time for the addition of newly acquired characters. This is the law of 'acceleration'; it is responsible for all progressive evolution.... These principles of terminal addition and acceleration are the preconditions of recapitulation" (Gould, 1977a).

(Cf. Gould, 1992b.) Essentially the same idea had been put forward by Darwin:

"In order to account for the preservation of embryonic similarity, Darwin advanced two specific principles of adaptation.... We might call them 'principles of terminal addition.' He specified them to be: (1) that the successive modifications by which species have attained their structure 'supervened at a not very early period of life' and (2) that variations occurring in the adult recurred at a 'corresponding age in the offspring.' (Darwin, 1859).... (Exceptions to the two principles would occur, as Darwin noted, if the young led an independent life.).... As a result of species descent and the operation of the two principles of adaptational inheritance, the embryo must go through the stages of its progenitors. This is a conclusion that Darwin expressly drew.... The principle of embryological recapitulation was woven into the original fabric of Darwin's conception of evolution. By 1859 the principle had become a deductive consequence" (Richards, 1992).

The recent investigations of alternative developmental strategies in sea urchin embryos may be related to Darwin's caveat (Raff, 1987).

Unfortunately, the idea of terminal additions has been derided of late:

"This is precisely the principle that Gould (1977a) attacks when Haeckel used it for the same theoretical purpose. Gould thinks that the principle is bound to the Lamarckian idea of the inheritance of acquired characteristics, which Haeckel certainly endorsed. This connection, then, condemns the principle of terminal addition and ultimately recapitulation theory in neo-Darwinian eyes" (Richards, 1992).

It is important to recognize that I am proposing that additions and subtractions occur to the differentiation tree, rather than to anatomical structures. But if there is some rough correlation between anatomy and differentiation trees, then the "concordance of ontogeny with phylogeny" (Swan, 1990) may occur in the following way. Imagine a real, growing tree, with so full a complement of leaves that we cannot see its branches. Neighboring boughs will nevertheless, in general, be more closely related, because they are more likely to have recent branches in common. Thus, to some extent, the external appearance of a tree reflects the underlying branching structure that generated it. The differentiation tree is similarly invisible, only its latest products being apparent. It is even less discernible than the hidden branches of a real tree, because its earlier edges generally have no cells representing them, since they went on to become more differentiated tissues. Nevertheless, some of the underlying topology of the differentiation tree is preserved in the anatomy of the organism, especially because (with a few major exceptions, such as germ cells, the neural crest: Hörstadius, 1950; Le Douarin, 1982; and "towing of... organs [such as]... the tongue and the diaphragm [and testicles]": Jacobson, 1991b; cf. flatfish eye migration: Morgan, 1991), most tissues retain their relative positions throughout development (Wetts & Fraser, 1989). Even neural crest cells, however, retain some spatial relationship, and 'memory' of, the rhombomere from which they originate (Hunt et al., 1991b).

Of course, one must be skeptical in evaluating each case of a potential correlation between the genome level differentiation tree and the tissue level morphology. Even in single celled organisms, the correlation between genome and morphology is questionable (Simola, 1991, 1993; Sandgren & Smol, 1991; cf. Gordon & Drum, 1994):

"It is perhaps intriguing, and possibly heretic, to state the problem: How much transfer of genetic information really occurs through the long path from the nuclear DNA code, via silicalemma [of diatoms], to the vesicular deposition of silica that produces the taxonomically valued structure" (Simola, 1991). [Cf. Gordon, 1996a.]

We have yet to answer the analogous question for multicellular organisms. If the differentiation tree is the key to the genotype to phenotype algorithm, then we should be able to test, by comparative 'anatomy' of differentiation trees, whether macroevolution is primarily confined to terminal branches.

Proposition 209: the differentiation tree is in effect the Bauplan of an organism.

The somewhat vague concept of a Bauplan, the 'construction plan' of a phylum of organisms, has been thoroughly reviewed by Hall (1992a) (cf. Arthur, 1997; Erwin, Valentine & Jablonski, 1997). The multiple definitions of the Bauplan, and its hierarchical nature within the life cycle of a single organism, fit the structure of the differentiation tree. It may be possible to combine artificial life simulations of entire ecologies, such as PolyWorld (Yaeger, 1994), with differentiation tree representations of the ALife organisms' genomes, to simulate evolution of a variety of bauplans. This might then show us just how the massive radiation of phyla found in the Burgess Shale came about (Conway Morris & Whittington, 1979; Conway Morris, 1989; Gould, 1989a; Briggs, Erwin & Collier, 1994; Plotnick, 1997).

Proposition 210: recapitulation is fundamentally a matter of degree of correspondence of the topologies of the differentiation trees of the organisms being compared. Given two differentiation trees, when we compare them by matching roots, either: one tree is a subtree of the other, or there is a common subtree of both.

In the first case, we could, except for the microevolutionary differences, speak of, say, organism A as effectively being like the ancestor of B (Figure 42).

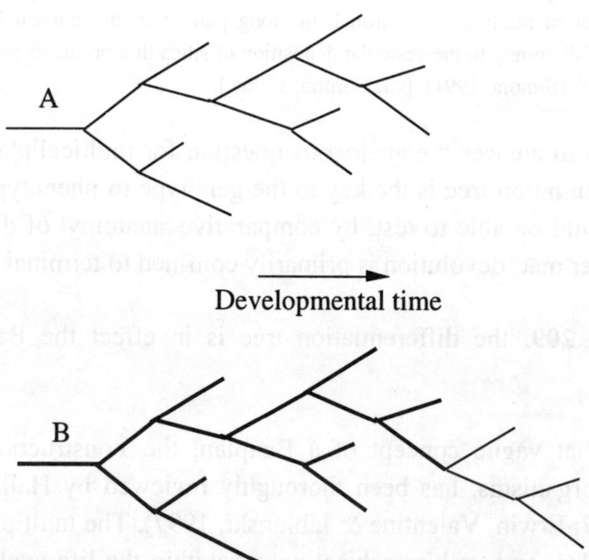

Figure 42. Differentiation tree of a common ancestor. The differentiation tree of organism A is similar to a subtree of the differentiation tree of organism B. Thus A may be from the common ancestor of A and B, and be similar to that common ancestor.

In the second case, if C is the rooted subtree common to A and B (Figure 43), then C in some sense represents the 'archetype' (reviewed in Hall, 1992a) of A and B. It is unlikely that any organism actually existed with differentiation tree C, due to tree prunings that may have occurred. Moreover, since the topology of the differentiation tree is not all there is to it (given the 300 or so alleles that can vary for each edge of the tree, and heterochronic variations), we can hardly say that C tells us what the archetype looked like, except in the broadest sense.

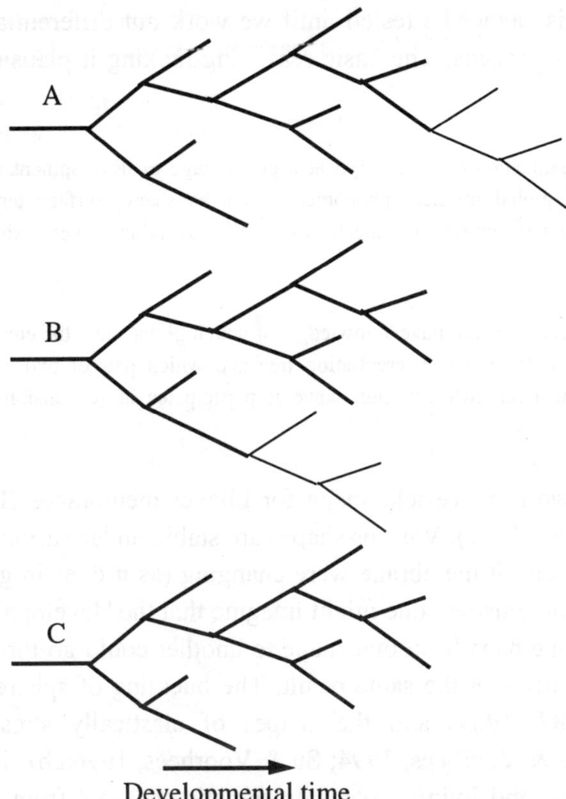

Figure 43. Differentiation tree of an archetype. Differentiation trees A and B have a rooted subtree C in common. While C may not correspond to any organism that ever existed, it can be taken as the 'archetype' or Bauplan of organisms A and B.

The nested nature of bauplans (Hall, 1992a) is thus accounted for. Furthermore, the old disputes on recapitulation, such as whether present or past or archetypical organisms appear in the sequence of morphologies of an embryo are resolved as partial victories for all the embattled.

Proposition 211: closely related organisms that have morphologically distinct pathways of morphogenesis nevertheless can have similar differentiation trees.

This hypothesis cannot be tested until we work out differentiation trees in such pairs of organisms. The basic reasoning making it plausible goes like this:

1. The global organization of an embryo at a given stage of development may sometimes have to do with global physical phenomena, such as shape, surface tension, pressure, orientation of polar structures, or trajectories of differentiation waves, exhibiting distinct, alternate states.

2. An individual cell does not have 'knowledge' of which global state the embryo is in. All it 'knows' is where it is in the differentiation tree and which pair of two 'choices' will be determined by the next differentiation wave it participates in (expansion or contraction wave).

Consider the work on vesicle shape for bilayer membranes (Berndl et al., 1990; Miao et al., 1991). Various shapes are stable under various conditions, and, if the amount of membrane were changing (as it does in growing cells or developing organisms), one might imagine that the 'development' of these closed membrane bags from one shape to another could go through various pathways and arrive at the same result. The buckling of spheres (Brodland & Cohen, 1987, 1989) and the shapes of elastically stressed crystals (Thompson, Su & Voorhees, 1994; Su & Voorhees, 1996a,b) similarly have alternate modes, and living organisms sometimes 'snap' from one form to another (Kobayashi & Akai, 1974a,b; Glen Klassen, p.c.; cf. Proposition 56), suggesting as much a mechanical component of the relationship between the forms as one based on cellular differentiation:

"The metamorphosis of the actinotrocha larva takes place rapidly, requiring, according to Ikeda, only 15 to 20 minutes. As MacBride (1914) also records, 'on one occasion we left an advanced Actinotrocha in a watch-glass, left the room for a short time, and on coming back found a young *Phoronis* '" (Kumé, 1968).

"At metamorphosis... of the polychaete *Owenia fusiformis*..., the juvenile trunk evaginates in less than 30 seconds..." (Williamson, 1992, with permission for this and all other quotations from same).

Although once their silica shell is formed diatoms are stuck with their shape, three species of polymorphic *Stephanodiscus* "form valves [shells] with two strikingly different structures and few, if any, intermediates" (Stoermer & Sicko-Goad, 1985). The single celled ciliated organism *Tetrahymena* goes through different pathways in arriving at a stable shape. Each intermediate state can be assigned an 'energy' and the alternate pathways can be modelled in terms of alternate relationships between the relative energies, depending on total size and other parameters (Brandts, 1993a; cf. Brandts & Trainor, 1990).

The systematic variation of mode of gastrulation of jellyfish with change in a simple parameter, egg size, is suggestive of an underlying, common mechanism for the two modes of gastrulation:

"The eggs of certain scyphomedusans develop directly into medusae; those of others never do so, and in some the developmental course is dependent on the size of the egg, which is variable. Eggs of the sessile medusae *Haliclystus* and *Lucernaria* are only 0.03 mm in diameter; of the permanently pelagic *Pelagia,* 0.30 [mm]; of *Cyanea,* 0.12-0.15 [mm]; and of *Aurelia,* 0.15-0.23 [mm] (Berrill, 1949a). To a great extent, the size of the egg determines the manner of gastrulation, which in turn sets the later course of development. In general the smallest eggs gastrulate by unipolar ingression of cells, filling the blastocoele; the largest eggs, by invagination alone; and the intermediate eggs, by a combination of invagination and ingression in varying degree according to species. In light of the fact that there has been considerable disagreement about the mode of gastrulation in *Aurelia,* it is significant that its egg is the most variable in size.... In the species with the largest eggs, the egg develops directly into a young medusa (ephyra) and in *Haliclystus,* the smallest-egg species, into a larva too small to form even a polyp directly" (Berrill, 1961).

A similar phenomenon occurs in reaggregated balls of hydra cells, which regenerate a number of polyps roughly proportional to the surface area (Chalkley, 1945), and in sea urchins:

"Indirect-developing sea urchins produce eggs of 65-320 μm in diameter, with most about 100 μm. Direct developers produce eggs ranging from 300 to 2,000 μm in diameter.... Direct developers have larger genome sizes than sister taxa with indirect development.... An adaptive role for a larger genome size is not apparent [but may be related to cell size: Martin & Gordon, 1995]" (Raff, 1996). [Cf. Sinervo & McEdward, 1988.]

What all of these mechanisms may have in common is the existence of alternate states or modes that depend on nonlinear phenomena on a sphere. (Cf.: "...Alternative stable states, among which evolutionary transitions have occurred several times during the phylogenetic history" of volvox: Larson, Kirk & Kirk, 1992, and the size dependency of asymmetric cell divisions in volvox: Kirk et al., 1993. See also the "robust flow into one or a few morphologies" of: Goodwin, Kauffman & Murray, 1993.)

Taking all of this together, whole embryos can have alternative physical pathways that have nothing to do, at least directly, with the underlying logic of development. Of course, that logic must be consistent with alternate physical pathways, and perhaps natural selection has acted to ensure this happens. On the other hand, these requirements may not be too stringent. I therefore suggest that the differentiation tree is often sufficiently robust to accommodate at least two alternative physical pathways to the same configuration of adult tissues. In fact, if we assume that during evolution individuals are generated that use one physical pathway or another, there may be an evolutionary constraint on differentiation trees that keeps them robust in this regard. It is as if differentiation trees have alternate ways of 'folding' into a three dimensional organism.

In trying to swallow this, the reader should keep in mind that according to our concept, a cell never knows where it is in an embryo, only the sequence of expansion and contraction waves it has experienced. Given the highly nonlinear nature of the mechanics involved, it would be surprising if only one sequence of physical states of the whole embryo were compatible with a specific differentiation tree, under all physical constraints (such as temperature and size of egg). Perhaps the alternative modes of gastrulation in vertebrates (Waddington, 1952a) will find an explanation here. It will require some careful comparative work to sort out, for instance, whether phenomena such as shear are essential to alternate modes of gastrulation:

"Spemann (1938) in describing the movements in amphibian gastrulation says: 'The cells, while they form in line and move together, pile up, and then, having passed through the narrow opening [blastopore] into the interior, spread out again.' In contrast to this, in the gastrulating embryo of [the sand dollar] *Dendraster,* the cells retain their places in the single-layered mosaic" (Moore, 1941).

Niklas (1994c), in the context of a model for the evolution of vascular land plants, has expressed the idea of multiple states in terms of engineering design:

"...Engineering theory shows that the number of equally efficient designs for an artifact generally is proportional to both the number and the complexity of the tasks that an artifact must perform (Meredith et al., 1973) because the efficiency with which each of many tasks is performed must be relaxed due to unavoidable conflicting design specifications for individual tasks (Gill, Murray & Wright, 1981), and, as the number of tasks increases, the number of configurations that achieve equivalent or nearly equivalent performance levels increases (Brent, 1973). If such relationships hold true for organisms, these relations may account for the morphological and anatomical diversity seen among even closely related species" (Niklas, 1994c).

Embryos have the additional 'conflicting design specification' that they must function while they are being constructed, not just when they are finished. If this general theory applies, we should not be surprised if the "functional tasks that an organism must simultaneously perform to grow [and] survive" during embryogenesis are compatible with a few alternative paths of construction, even when the logic of the 'tasks' (the differentiation tree) is invariant. The numerous examples of "variable development of homologous characters where the variant developmental pathway does not have any substantial consequences for the adult pattern" (Wagner, 1994b; cf. Wagner & Misof, 1993) need to be reexamined in the light of comparative study of the differentiation trees involved.

Raff (1996) decides, without proof, that the...

"...idea ...[of] an upwardly expanding and branching tree whose initial starting point is the fertilized egg (Arthur, 1988)... suggest[s] strongly that early development must be refractory to substantial evolutionary change.... The hypothesis of a hierarchical ontogeny of expanding complexity requires that it be difficult, if not impossible, to make substantial modifications in early development" (Raff, 1996).

In assuming that variable early development is primarily genetic, he is left in a quandary:

"How it is that genes that are so highly regulated in sites of expression can substitute so freely for one another among related species remains unexplained" (Raff, 1996).

What Raff (1996) ignores is the nonlinear physics of embryogenesis, with the possibility of retaining the same hierarchical development of cell types, while producing alternate spatial modes of development. Of course, beyond this, there is no reason that substantial microevolutionary changes to the individual differentiation cascades couldn't occur, producing evolutionary divergence. But when a single species has two modes of morphogenesis, it is unlikely that the underlying hierarchy of differentiation is altered. In any case, this conflict of viewpoints is open to experimental test.

Curiously, this proposition allows us to reconsider Haeckel's 'gastraea theory' as possibly having a basis in differentiation trees:

"Together with the biogenetic law, the gastraea theory formed the core of Haeckel's evolutionary edifice.... The existence of so many different modes of gastrula formation... cast doubt on a central claim of the theory, that all gastrulae were derived from a common ancestor and therefore were homologous. If they came from a common ancestor, then, following Haeckel's logic, they should show the same mode of ontogenetic development.... Haeckel's theory and the criticisms of it by his colleagues provided a rich set of problems for investigators to pursue" (Nyhart, 1995).

Indeed, Haeckel is still belittled to this day, perhaps a measure of his lasting impact.

Proposition 212: direct and indirect development use the same differentiation tree.

In light of the previous proposition, it is hard to accept the conclusion in the following, since it implies that every change to direct development involves a major reorganization of whatever it is that drives embryogenesis:

"Direct development has evolved in a large number of lineages; the phenomenon is seen at every taxonomic level from species to phylum.... Sometimes two or three developmental patterns exist in one species (Bonar, 1978).... Another consequence of lecithotrophy is that in some cases the size of the egg becomes so large that there is a reorganization of the cleavage stages of development so that the embryo shows some form of partial or superficial cleavage. This has occurred in several phyla including the cnidarians, the molluscs, the arthropods, the echinoderms, and the chordates. In the forms in which the cleavage pattern has changed dramatically, it is frequently apparent that it would be difficult to make developmental mechanisms, which have been shown to exist in related groups in which cleavage has not been modified, function after the cleavage pattern has been altered" (Freeman, 1982).

It seems much easier to think that something fundamental, such as perhaps the differentiation tree, is the invariant in the presence of such changes, especially because, in some cases, these changes can go back and forth from one generation to the next (Berrill, 1961; Bonar, 1978). Raff (1992a) gives a detailed comparison of two species of sea urchins that diverged 10 million years ago, one with direct and the other with indirect development:

"...It is important to note that evolution shows that parts of the gastrulation process are dissociable from each other. Some features of [the direct developing] *Heliocidaris erythrogramma* gastrulation have been conserved, and others have changed in substantial ways (Wray & Raff, 1991). Thus, the position of gastrulation initiation, invagination by involution, and timing of primary mesenchyme cell ingression have been conserved. The number of primary mesenchyme cells, the origins of cells contributing to the archenteron, the mechanism of archenteron elongation, the symmetry of cellular movements, and the origin of coelemic cells have all substantially changed. In total the evolution of the novel ontogeny of *H. erythrogramma* [compared to the indirect development of the pluteus larva producing species *H. tuberculata*] has resulted from a suite of changes encompassing changes in timing of developmental events, cell fates, cell-cell interactions, and gastrulation movements" (Raff, 1992a).

All of these changes would seem to fall in the category of morphogenesis, i.e., perhaps they are consequences of differentiation cascades of essentially the same differentiation tree being played out differently in the two species. Despite what appears as a resulting gross change in morphology between the two species, Raff (1992a) sees an underlying common theme:

"The dissociability revealed by comparative studies is a manifestation of what developmental biologists have long recognized as the regulative abilities of embryos subjected to experimental perturbations. The phenomenon of regulation is still not understood mechanistically, but evolution gives us an insight into its potentially great importance.... Cell types and structures that appear to be homologous arise from quite different cell lineage precursors and different developmental processes. The temptation to seek a developmental basis for homology is always strong. However, early as well as late development proves to be an unreliable guide to homology, a point well realized by earlier embryologists (Wilson, 1896; de Beer, 1971). If a developmental basis for homology is to be defined, it must lie in the genetic regulatory systems that [underlie]... particular developmental features" (Raff, 1992a).

Indeed, Emlet (1995) finds "remnants of indirect development in the direct developing sea urchin *Heliocidaris erythrogramma* " and suggests...

"The term 'direct development' can be misleading, because it implies complete loss of larval features that are part of indirect development. While there is no doubt that development of *H. erythrogramma* is highly modified relative to a developmental program that includes a functional pluteus larva, certain key features, including larval spicules and bilateral symmetry of the primitive pattern, have been retained" (Emlet, 1995).

I would suggest that this underlying homology, the 'genetic regulatory system' held in common between direct and indirect developing species, is the differentiation tree.

An alternate hypothesis (Natalie Björklund, p.c.) is that different differentiation subtrees, such as may occur in metamorphosis (Section 7.02; Alberch, 1987; Martin & Gordon, 1995), are the basis for the distinction between direct and indirect development. In this case, the adult tissues would be the terminal branches of different portions of the whole differentiation tree, in the two cases. As such, genetic drift should set them apart from one another at the molecular level. The place to test this idea is perhaps in the scyphomedusans, discussed in the previous proposition. Since some species, such as *Aurelia* (Berrill, 1961), have alternate modes of development depending on egg size, the adult tissues could be compared for (preferably cloned) individuals that developed by different routes. Urodele amphibians may also prove fruitful for studies of direct development versus egg and genome size (Wake & Hanken, 1996), especially since we can anticipate easy visualization of their differentiation waves.

Proposition 213: in regulating embryos, so long as a cell sheet is reconstituted which can support the next pair of differentiation waves, it does not matter what kind of morphogenetic movements occur in between, including dissociation and reassociation of its cells.

One of the mysteries of development has been the wide range of morphogenetic movements that can be used to accomplish gastrulation:

"...The essential message is that there is a great deal of diversity, and although the movements can be classified into broad types, such as invagination, involution, individual ingression etc., there seem to be no constraints on what is possible. It is sometimes said that although gastrulation movements look very different, they are 'topologically' equivalent in different taxa, but this is quite incorrect. Topological equivalence must involve a conservation of connectivity in cell sheets, but cells do not have to respect any topological constraints since they are able to migrate as individuals and two populations can interpenetrate or move through one another. The truth is that although some sort of gastrulation is a universal feature of animal development, we are faced with massive diversity of the gastrulation movements themselves" (Slack, 1994a).

The underlying universal mechanism for gastrulation may be the differentiation tree. If, as we assume, all cells are equivalent in a tissue, it does not matter if they temporarily disperse, so long as they come back together into a confluent sheet for the next step of differentiation. This supports my suggestion that morphogenetic movements are secondary features of development, just like all other tissue specific events that are part of the differentiation cascade within one step of differentiation. Of course, if the cells disperse, they must 'find' one another later to reconstitute an epithelium. Cell sorting mechanisms may come into play here.

Proposition 214: the differentiation tree and morphogenetic movements can evolve independently of one another.

This follows as a corollary of the previous proposition. The consequence is that the morphogenetic movements in early stages of embryogenesis can diverge widely, while preserving the differentiation tree. Thus, conservation of the differentiation tree may represent the underlying unity, while the cell movements, which catch our eye, can come to differ considerably between even closely related species. This may be, then, what permits variations in developmental mechanism with egg size (Berrill, 1961) and amount of yolk (Slack, 1994a).

In the latest definition of 'animal', I would substitute 'differentiation waves' for 'pattern of gene expression':

"What is an animal?... Since the time of Geoffrey St. Hilaire [1772-1844, see Nordenskiöld, 1928], there has been no morphological concept of what an animal really is. We propose that an animal is an organism that displays a particular spatial pattern of gene expression, and we define this pattern as the zootype" (Slack, Holland & Graham, 1993).

If we work out the differentiation trees of a number of animals, and compare them to those of the "non-animal multicellular eukaryotes such as plants, fungi and slime moulds" (Slack, Holland & Graham, 1993), it may become apparent what 'zootypes' exist, and their phylogenetic relationships. Whether the portion of the differentiation tree representing the "early, pre-phylotypic, stage of development" (Slack, 1994a) evolves more slowly, in some sense, than morphogenetic movements, would also be learned by such a study, though, if the pattern of heterochronies found for the 'phylotypic' somite stage (Richardson, 1995; Richardson et al., 1997) holds up for earlier stages, this is now doubtful.

Proposition 215: juvenile adaptation and heterotopy can be explained in terms of macroevolutionary changes to the differentiation tree.

Gould (1992a) notes two features arising from comparative embryology that at first sight would seem incompatible with differentiation trees:

"Haeckel recognized two main categories of cenogenesis. The first, and most important, was *juvenile adaptation* - the interpolation of new features into early ontogenetic stages for their own immediate evolutionary utility. Adaptations of marine larvae for eating and staying afloat ranked highest in this category of regrettable disturbances to recapitulation [the 'definition' of cenogenesis].... _Heterotopy_ [is]... change in place of differentiation (a term that has disappeared from use). As a primary example, Haeckel cited the current development of reproductive organs from mesoderm in most animals. Because mesoderm is, itself, a belated development in animals (for primitive creatures had but two layers of endoderm and ectoderm), and because ancestral animals with only two layers must have possessed reproductive tissues, these organs must have originated in endo- or ectoderm and later changed position to differentiate in mesoderm" (Gould, 1992a).

Juvenile adaptation may be a result of duplications of any portion of the differentiation tree being grafted to early terminal branches (Figure 35c). In particular, it is not out of the question that adult terminal branches could sometimes be what is duplicated and transferred. Of course, we will have to learn to distinguish juvenile adaptations that are due to changes in the topology of the differentiation tree from microevolutionary changes, involving alterations in early differentiation cascades themselves.

I am thus suggesting that juvenile adaptation is merely ordinary macroevolution that happens to involve terminal branches of the differentiation tree expressed during embryonic and larval stages. It could, for example, occur by a multistep deletion (Figure 34) or insertion of edges between existing edges, as discussed in Proposition 140, but this is a less likely scenario.

Heterotopy could be explained by invoking a terminal branch that was duplicated and grafted, and the original terminal branch subsequently lost by pruning (Figure 44).

Proposition 216: phylogeny reconstruction or taxonomy should be based on the differentiation trees of organisms.

a) Initial differentiation tree

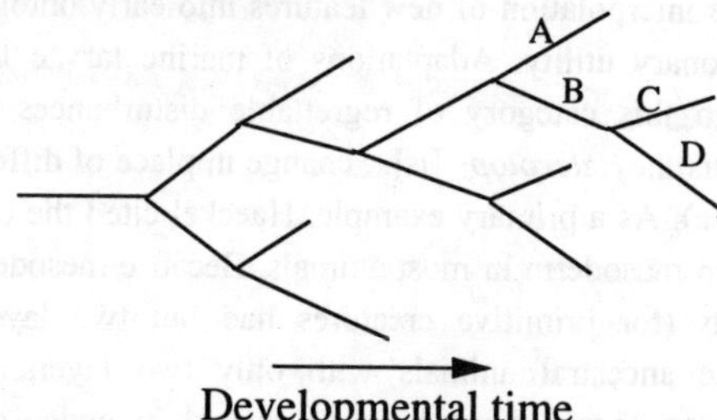

b) Branch grafted to an earlier terminal cascade

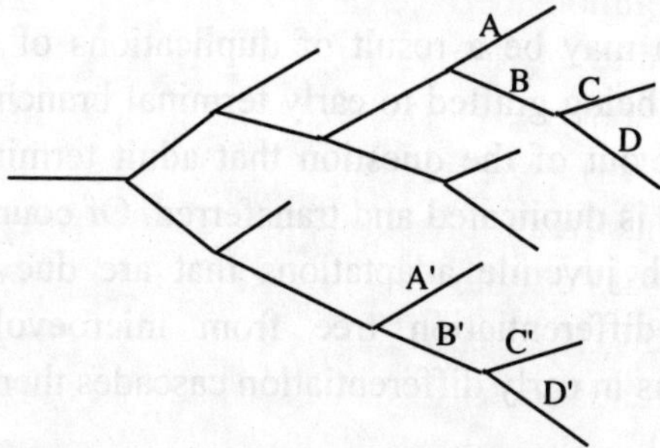

c) Pruning of the original branch

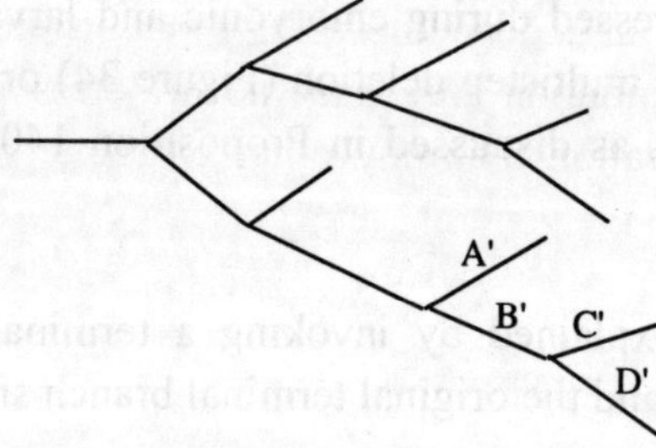

Figure 44. Heterotropy, or the change in place of a set of differentiation events, could occur via terminal branch duplication in a differentiation tree, followed by pruning of the original terminal branch.

"A taxonomist attempts to assess the relative importance of a large number of properties that cannot easily be expressed in terms of numbers. In this way he determines the species to which an organism belongs. Unfortunately, many taxonomists denounce this intuitive aspect of their work. Roaming around the margins of science and often looked down upon by experimental researchers, they long for recognition from the 'hard-core' natural scientists. Experimental biologists often forget that without the expertise of experienced taxonomists their hands would be tied. Despite this, everything is being done to render the taxonomist extinct - the taxonomist, the only one who knows the names of endangered species. Before long we will be left only with the statisticians and gene freaks and there will be no biologists around to identify species or observe organisms or plants" (Goldschmidt, 1996).

We now have a solution to the age old question of the extent to which 'ontogeny recapitulates phylogeny' (Gould, 1977a). Taxonomically closer organisms will have relatively larger subtrees of their differentiation trees in common (cf. Figure 43). (This is similar to the approach to taxonomy of Ho, 1990.) Subtrees thus may provide the 'essential characters' on which cladistics (Hull, 1988a) or other methods of classification (Mettler, Gregg & Schaffer, 1988; Hillis, Huelsenbeck & Cunningham, 1994) could be based. Insofar as the differentiation cascade differences between the earlier parts of their trees are small, early embryos of different organisms will appear to be similar. On the other hand, major differentiation cascade differences in the early stages of embryonic development, or differentiation cascade fusions (Figure 34), could mask an underlying similarity of differentiation trees. The topology of the differentiation tree may nevertheless prove to be more fundamental to classification than any other character of organisms. This conclusion is contrary to the general presumption:

"As the database for molecular comparisons expands, we predict that phylogenies [of nematodes] will become increasingly resolvable, not because molecules are necessarily better than morphology for phylogenetic inference, but because the potential database for independent characters is so enormous" (Fitch & Thomas, 1997).

Despite such optimism, there has been a failure of molecular taxonomy to equal or surpass morphological taxonomy:

"...Molecular systematics has not yet produced phylogenetic trees of broad phylum relationships more robust than those based on morphology" (Raff, Marshall & Turbeville, 1994).

"As morphologists with high hopes of molecular systematics, we end this survey with our hopes dampened. Congruence between molecular phylogenies is as elusive as it is in morphology and as it is between molecules and morphology.... The *Homo - Pan* pairing, though suggested by many molecular studies since the 1960s, has achieved decisive molecular support only within the last year (Bailey et al., 1992; Horai et al., 1992), from ca. 30 kb of aligned sequences.... We end with a vision of the future... that happy day when molecular systematists achieve the goal of adequate sampling in terms of both taxa and sequence length..." (Patterson, Williams & Humphries, 1993).

The need for large, 30 kilobase pieces of genome may hint at a resolution of this dilemma (cf. the use of "57 different enzymes [that] were used to determine the divergence times of the major biological groupings": Doolittle et al., 1996). The problem may be analogous to that of trying to compare two noisy, digital images, one of which is a distortion of the other (Mazur, Mazur & Gordon, 1993). If we were to align and compare all of the pixels, we would have a reasonable chance of finding the transformation relating the two pictures. However, if we were to select only a few pixels, and ignore the rest of each picture, we could hardly expect to recover the whole transformation, and we might even mismatch some of the individual pixels. This would seem to be the current state of affairs in most molecular systematics. An alternative approach is to extract and match the broad features of each picture (Sivaramakrishna & Gordon, 1997b). The differentiation tree corresponds to such broad features. Details, such as individual genes, can be compared afterwards.

David H.A. Fitch (p.c., 1995) is concerned that no single characteristic records the phylogenetic history of an organism, so that "differentiation trees will... [at least] provide another important set of characters that could be very informational about not only relationships among organisms, but also about the constraints that may limit the variation available to selection during their evolutionary history!" Actually, the differentiation tree is not one characteristic, but many rolled into one. The earlier edges of differentiation trees may be taken to correspond to cladistic 'primitive' characters, and the terminal branches to 'derived' characters. Of course, for species found to have the same differentiation tree, one would next turn to the molecular level.

For nematodes, in which the cell lineage nearly corresponds to the differentiation tree, this new basis for taxonomy in differences between differentiation trees has already begun:

"Comparison of the cell lineages of *Caenorhabditis elegans* with lineages that have been partially elucidated in other nematode species, such as *Turbatrix aceti, Panagrellus redivivus,* and *Aphelencoides blastophorus* (Sulston et al., 1983; Sternberg & Horvitz, 1982), reveals striking similarities in developmental pattern. Moreover, morphological differences by which one species is distinguished taxonomically from another have been traced in several instances to focal differences in otherwise homologous cell lineages, such as the death or terminal differentiation in one species of an identified cell, whose homolog undergoes further divisions in another species" (Stent & Weisblat, 1985).

"Comparative biology seeks general principles by studying similar features or processes in different species. For example, the largely invariant cell lineages of nematodes related to *C. elegans* have allowed detailed interspecific comparisons of cell lineages (von Ehrenstein & Schierenberg, 1980; Sternberg & Horvitz, 1981, 1982; Ambros & Fixsen, 1987; Skiba & Schierenberg, 1992; Sommer, Carta & Sternberg, 1994)" (Fitch & Thomas, 1997).

At another level, the problem of whether multicellular organisms have a single or multiple evolutionary origin from single celled organisms might be elucidated by comparing not only their symbiotic cell components (Margulis, 1993a), but also their differentiation trees. This adds a criterion to those of Morris (1993) for checking on the monophyletic origin of multicellular phyla:

"Development in all animals is characterized by the folding of epithelial sheets and transitions between mesenchymal and epithelial cell types.... I propose that the complex (collagen, proteoglycan, adhesive glycoprotein, and integrin) system that mediates cell motility and transitions between epithelial and motile cell types is central to multicellularity in animals. I further propose that the extracellular matrix is a deep rooted homology that unites the kingdom Animalia into a monophyletic group of multicellular organisms" (Morris, 1993).

Of course, the observation of the differentiation tree is a major undertaking for each organism, so if this proposition is correct, I've guaranteed employment for taxonomists so inclined for another generation or two.

Proposition 217: because of the possibility of heterochrony, recapitulation need not occur in a strict chronological order.

Differentiation trees provide a way around the current impasse, in which the insights of Haeckel have been belittled by his (and our present day) hyperbole:

"...The biogenetic law is basically false (see Gould, 1977a). By the closing years of the nineteenth century Haeckel's program had become a source of much ridicule [cf. Morgan, 1919a, and see Morgan, 1903a, for an excellent history], and the negative platform for launching other approaches to embryology - particularly the *Entwicklungsmechanik* of Wilhelm Roux and Hans Driesch.... So pervasive were embryonic and juvenile adaptations, and shifts in timing and placement of features relative to each other, that any potential signal of complete palingenetic repetition was completely swamped into imperceptibility" (Gould, 1992a).

In fact, differentiation trees allow us to see how to combine *Entwicklungsmechanik* with (an albeit watered down) biogenetic law.

What is important in comparing organisms is the topologies of their differentiation trees. This is more fundamental than the set of events that might have occurred up to a given moment in embryological time, which depends on their heterochronic relationships. Two topologically identical trees differing by a heterochronic shift can look somewhat different until development is complete (Figure 45).

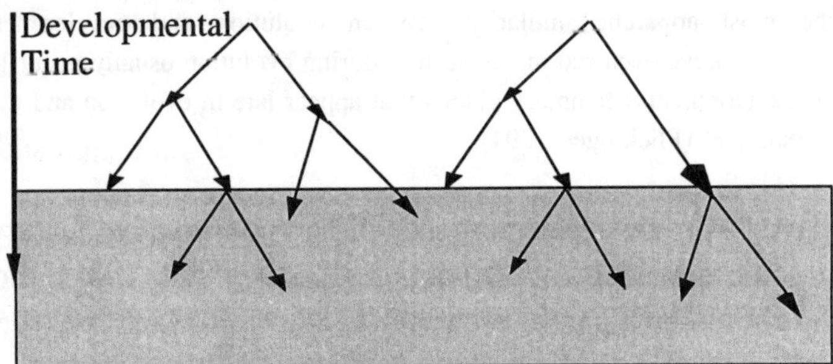

Figure 45. Heterochrony and differentiation trees. Two differentiation trees with identical topologies, related to one another by a heterochrony, can appear to be of different topologies if one does not wait for completion of development.

Proposition 218: more conserved terminal cell types differentiate via fewer differentiation waves.

This proposition is a variation on a conclusion drawn from a wide body of evidence by Flickinger (1994):

"Examples of earlier differentiation of more conserved cells are seen in the developing embryo. Cartilage, which first appears in squid, differentiates earlier than bone, which is found only in vertebrates.... Germ cells arose very early in evolution. *Volvox* was one of the first organisms to produce sperm and eggs. If cell types appearing early during evolution are then determined after fewer cell divisions, germ cells should be the best example. This seems to be true.... It appears that the temporal sequence of differentiation of various cell types corresponds to their order of appearance during evolution with... qualifications.

"...Hybridization experiments... reveal that heterogeneous nuclear RNA transcribed at the late gastrula - early neurula stage is more conserved, i.e., shows greater reaction with fish DNA, than at the tailbud stage (Flickinger & Shepherd, 1981).... Lower vertebrates sometimes have individual cells that synthesize more than one tissue-specific gene product. For example, cells in cyclostomes [lampreys] synthesize both gastrin and glucagon (Van Noorden & Pearse, 1974). These products are produced by separate cell types in higher vertebrates.... Evolutionarily more primitive organisms have less division of labor. Cnidarians, brachiopods, and bryozoans have epithelio-muscular cells that both secrete and contract (Welsch & Storch, 1976). By the flatworm stage of evolution these functions usually are performed by separate cell types...."

"Perhaps the most apparent similarity between evolution and development is that differentiations that have occurred more recently during evolution usually occur later during the course of development. Mammary glands that appear late in evolution and development are one such example" (Flickinger, 1994).

Thus we have "...progressive emergence of less conservative features during early embryonic development" (Flickinger & Shepherd, 1981), though this relationship should be a partial ordering (Birkhoff & MacLane, 1953) along the differentiation tree (not counting duplicated terminal branches), rather than necessarily a linear ordering versus developmental time. This implies that the length of the differentiation code for more conserved cell types should be shorter.

7.02 Organisms with Two Differentiation Trees

"...A water-soluble factor released into the environment by the predacious phantom midge larva *Chaoborus americanus* (Diptera: Chaoboridae) causes embryos of the waterflea *Daphnia pulex*... (Crustacea: Cladocera) to develop into a form called *Daphnia minnehaha....* *Chaoborus* larvae are unable to eat the *D. minnehaha* form as readily as the *D. pulex* form" (Krueger & Dodson, 1981).

Proposition 219: the haploid and diploid phases probably use exactly the same differentiation tree in plants with isomorphic generations.

If macroevolution sometimes proceeds by whole genome duplication (Sparrow & Nauman, 1976), then that means that on occasion the whole differentiation tree is duplicated. If these genome duplications are not simply cases of polyploidy, then the duplicated differentiation trees may play themselves out in different parts of the life cycle. In this section I will examine this question, along with the similar question of what happens when two whole genomes or differentiation trees are concatenated.

Consider the hypothesis that higher plants evolved from one of the algae with alternating, isomorphic generations (or, better, 'phases': Stewart & Rothwell, 1993), for which...

"...the gametophyte and sporophyte generations are identical in appearance. They differ only in the fact that the gametophyte is haploid and produces gametes and the diploid sporophyte produces meiospores" (Stewart, 1983).

Note that identity of the haploid and diploid phases suggests that the very same DNA is used. Otherwise genetic drift would make them different.

Proposition 220: the first step in an evolutionary shift from an isomorphic to a diplontic life cycle was a DNA duplication involving the whole differentiation tree.

At this point, the two differentiation trees, for the haploid and diploid phases of the life cycle, could begin to diverge:

"Subsequent evolution in a desiccating land environment would produce an elaborated, free-living sporophyte and a reduced gametophyte. The reason for the emphasis on the progressive evolution of the sporophyte is clear. Sexual reproduction by the gametophyte generation of algae is dependent on the presence of free water. In transmigration to a land environment, access to water for sexual reproduction would be occasional at best, so the emphasis on the reproductive process shifted to that generation not requiring the aquatic medium" (Stewart, 1983).

The apparently abrupt 'phase change' in plants is akin to metamorphosis in animals:

"At some point in time, the plant makes the transition from juvenile to adult development [Poethig, 1990]. Thereafter the apical meristem forms adult structures, which then occupy the upper regions of the stem.... In many cases the juvenile - adult phase change will affect nearly every aspect of the morphology and anatomy of the plant" (Fosket, 1994).

Duplication of the whole differentiation tree may have occurred at other major evolutionary steps at which whole genome duplication is inferred (Sparrow & Nauman, 1976).

A possible contradiction to this proposition is that the haploid and diploid phases of some plants, whose close relatives are diplontic, are capable of producing exactly the same morphology (as the common field mustard, *Brassica campestris,* in: Niklas, 1994b), and, therefore, they have not diverged from one another at all, which would imply that they use exactly the same differentiation tree (cf. Proposition 219). If plant evolution occurred this way (see Stewart & Rothwell, 1993 for this scenario and one alternative), then it should be possible to obtain molecular evidence for or against the massive DNA duplication and the subsequent divergence (C. Cristofre Martin, p.c.).

Proposition 221: diploidy and polyploidy may often allow mutant differentiation trees to become established much faster than differentiation trees of haploid organisms or generations, thus permitting faster evolutionary radiation.

Diploidy and polyploidy may protect or buffer evolutionary 'experiments' with the differentiation tree, permitting an altered differentiation tree to be carried along for a while in a somewhat flawed state, whereas, if the morphogenesis of an organism were dependent on the flawed tree, as in the haploid state, it might fail immediately. We thus see that the rate of evolution, which has been an unsolved theoretical problem, may be understood in part via differentiation trees:

"...The lower limit of time required for speciation has indisputably been reduced to a few thousand generations... in direct conflict with Haldane's calculations (Haldane, 1957) that an average gene substitution requires 300 generations.... If two species differ by 1000 genes, as he estimates, 300,000 generations would be required for a new species to evolve. This discrepancy between observed rates of speciation and calculated rates of gene substitution came to be known as 'Haldane's dilemma." Escape from the dilemma has proven elusive, but gene interactions and regulatory hierarchies [as differentiation trees] may provide part of the explanation" (Mettler, Gregg & Schaffer, 1988).

In fact, high rates of evolution in plants are correlated with polyploidy:

"Perhaps it is no mere coincidence that the two vascular groups showing most rapid evolution today, flowering plants and ferns, also show the highest incidence of polyploidy. Exactly how or why polyploidy increases tolerance of environmental extremes [extending range] is not known, but the high incidence of polyploids in arctic and desert habitats has been well-documented. It has been estimated that 25 to 33% of the angiosperms are polyploid" (Weier et al., 1982).

The same may be true of fish (Schultz, 1980) and amphibians (Bogart, 1980). Mammals are a blatant exception.

Proposition 222: organisms that undergo metamorphosis may use separate differentiation trees for each part of the life cycle.

Bonner (1965a) points out that organisms with life cycles that pass through two 'points of minimum size' have two sets of 'chains of steps', i.e., what we may refer to as two differentiation trees (cf. Arthur, 1984, 1988). This may be what permits independent evolutionary divergence of larval and adult stages of insects in which the egg leads to a larva, and the imaginal discs, which "really begin an embryonic process all over again" (Bonner, 1965a), lead to an adult. This theme will be expanded in the following Propositions in a full consideration of metamorphosis. Precisely the same separation of adult and larval tissues, based on differentiation trees, was postulated by Martin & Gordon (1995) for urodele amphibians. Social insects (Hölldobler & Wilson, 1990) might also represent organisms with distinct subtrees for each form (worker, soldier, drone, queen) in some cases "...each caste being determined by the activity of a special set of genes" (Noirot (1990) (cf. Sakagami, 1982), while in others simple allometric processes may be at work (Huxley, 1932, who considers both possibilities). Two different differentiation trees may be the underlying cause of the "canalization [that] has made the paedogenic [anuran] tadpole genetically unlikely" (Wassersug, 1974). Cf. Cohen & Massey (1983).

At least one case of cyclomorphosis or seasonal polymorphism (reviewed by Hall, 1992a) involves a predator/prey (carnivorous rotifer/rotifer) interaction that "...acts prior to cleavage, probably during oogenesis..." (Gilbert, 1966) and may thus involve alternative differentiation trees. Other cases are not so clear, and may involve even single alleles (Hall, 1992a).

Moore (1957) successfully crossed sea urchins with sand dollars, so, from our point of view, two different differentiation trees ('scleroblastic patterns') were present and perhaps operating simultaneously, instead of consecutively, as in metamorphosis:

"There is... considerable irregularity in skeletal structure. It may be suggested that this is the result of conflict between the two types of genetically determined scleroblastic patterns" (Moore, 1957).

Proposition 223: cataclysmic metamorphosis occurs from either a single tissue within a larva or from tissues that dedifferentiate back to a single tissue.

Williamson (1992) has presented an interesting case, based primarily on comparative anatomy, for his idea that many organisms with larval stages are concatenations of two or even three (when there are two larval stages) organisms' genomes that can be from different phyla (cf. 'nuclear merger' in Cnidaria: Shostak, 1993, the 'synology' of Gogarten, 1994; Gogarten, Hilario & Olendzenski, 1996, and species formation via the combinatorics of 'genome acquisition' in symbiosis: Margulis, 1993a, "by definition a form of macroevolutionary saltation": Bermudes & Back, 1991). This puts flesh on the "...generally accepted current belief... that echinoderm larvae (Fell, 1948), ascidian tadpoles (Berrill, [1947], 1955a), and nemertean larvae (Smith, 1935) are evolutionary novelties 'interpolated' into an originally more-direct developmental cycle" (Berrill, 1961) and on the concept of 'action systems':

"The ascidian larva is a dual organism in which there is a sharp demarcation between temporary organs of the larva and the rudiments of the permanent organs of the adult (Berrill, 1935, 1947; Grave, 1944). It may be thought of as containing two quite distinct action systems, one of which has a full but brief period of functioning during the larval phase only, whereas the other is maintained in a suppressed condition until the beginning of metamorphosis. Both of these action systems begin their embryonic development together [Scott, 1952], so they share a common phase of development up to and including gastrulation. From that point onward, however, two independent developmental mechanisms are operating side by side, the development of the larval structures proceeding virtually independently of that of the permanent ascidian organization" (Barrington, 1968).

I take the two 'action systems' to be two differentiation trees. I only have a problem with Williamson's argument that:

"If all stages in the development of the Enteropneusta and the Echinodermata have evolved by descent with modification from a common ancestor, then the two utterly different processes of metamorphosis that occur in the two groups must have evolved from the process practiced by that ancestor" (Williamson, 1992).

One must consider that in scyphomedusans entirely different modes of gastrulation correlate simply with egg size (Berrill, 1961). While gastrulation isn't metamorphosis, it is important to be aware that development is a highly nonlinear process, and nonlinear equations often have two or more solutions, as discussed in Proposition 211. More to the point, in "echinoderms... increased egg size is correlated either with absence of the larval type characteristic of a class or with modification or partial suppression of that type" (Berrill, 1961). Thus what could be happening is that one genome produces two different morphological histories, without consisting of two concatenated genomes. The two cases are distinguishable experimentally: loss of the second morphology entails loss or suppression of a substantial chunk of DNA, if the organism contains concatenated genomes, while the first morphology can persist. If the two morphologies are alternate 'states' or phenotypes of one genome, then loss of one means loss of the other, i.e., all the DNA is essential to existence. Of course, if these organisms actually contain one, rather than two genomes, we would still be left with the problem of explaining the similarities between larvae in widely different phyla, which Williamson's hypothesis handles so well.

This general reservation aside, the evidence for concatenation of species presented by Williamson is impressive. (See also Williamson & Rice, 1996, but cf. Strathmann, 1993a,b.) Similar taxa will exist with and without a larval stage. Similar larval stages will precede the development of very different adults. The manner of development of the larva and the adult are often incongruous., etc. For example, a switch from mosaic to regulative development occurs...

"...in the ascidian *Clavellina lepadiformis*.... In larvae the capacity for reparative regeneration is very limited and low, while in the adult forms it is high; a part of their body can regenerate into an entire animal [Schulz, 1907]" (Polezhaev, 1972).

Nevertheless, for other friendly caveats, see Blackstone & Dick (1993) and Hart (1996). For a review of 150 years of theories of phylogenies of invertebrates with larvae, see Willmer (1990).

Proposition 224: different forms of metamorphosis can be distinguished on the basis of the organisms' differentiation trees.

In reviewing Williamson's data and the general literature on metamorphosis, I find that five types of metamorphosis can be distinguished:

1. *Continuing differentiation metamorphosis:* almost all the larval tissues are used in forming the adult by a process of further differentiation at metamorphosis (Figure 46).

2. *Pulsatile metamorphosis:* the adult develops from a 'larva' through many discrete steps (Figure 47).

3. *Single tissue metamorphosis:* the adult develops from a single tissue in the larva (Figure 48).

4. *Dedifferentiation metamorphosis:* the adult develops from a few larval tissues that undergo dedifferentiation (Figure 49).

5. *Deferred metamorphosis:* the adult develops from a few larval tissues that are set aside, and from some larval tissues that persist into the adult stage (Figure 50).

These metamorphosis categories overlap, but are not identical to, the four categories of Weisz (1966) (cf. Spratt Jr., 1971a). I would like to show how Williamson's results fit naturally into the rubric of differentiation trees.

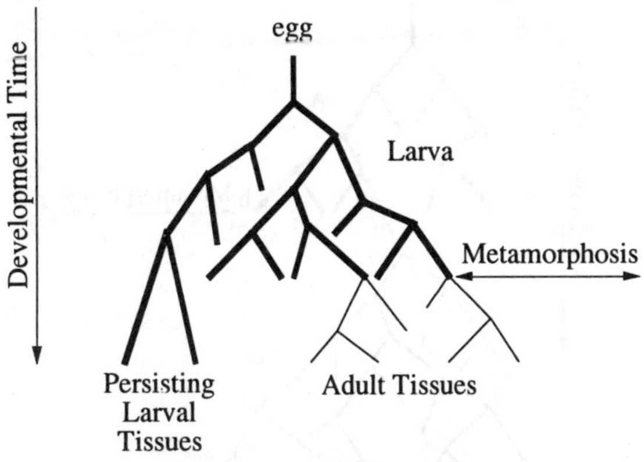

Continuing Differentiation
Metamorphosis

Figure 46. In *continuing differentiation metamorphosis,* the larval stage represents a pause in the differentiation tree. Metamorphosing amphibians (Etkin & Gilbert, 1968; Martin & Gordon, 1995) and the invertebrate Enteropneusta (Williamson, 1992) fit this category.

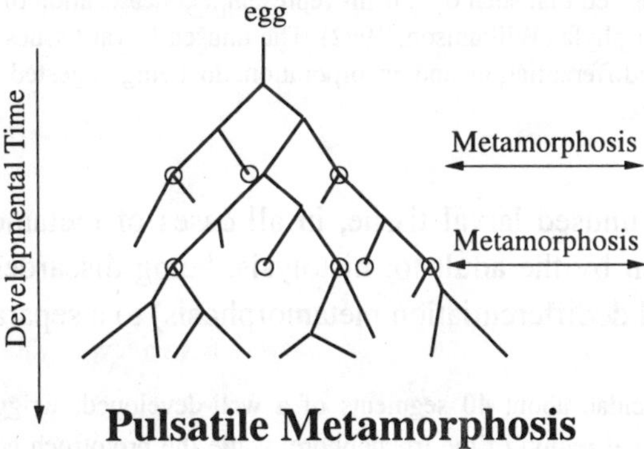

Pulsatile Metamorphosis

Figure 47. In *pulsatile metamorphosis* many steps of differentiation (or termination of differentiation) occur simultaneously at intervals. The nodes (or termini) involved in two metamorphosis steps shown here are circled. It is possible that other events of differentiation occur in between these steps. Other names for this form of metamorphosis are "gradual" (a clear misnomer) and "incomplete", referring to the lack of a larval stage. The hemimetabolous insects (Etkin, 1955) undergo pulsatile metamorphosis.

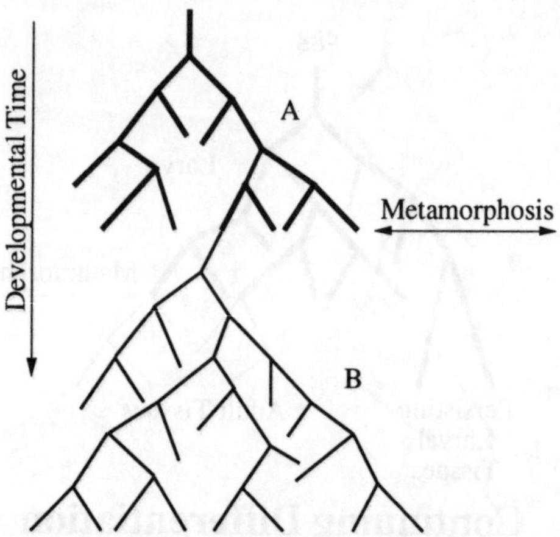

Single Tissue Metamorphosis

Figure 48. In *single tissue metamorphosis* a larva develops, as show here by differentiation tree A, and one of its "terminal" tissues becomes the adult, indicated by differentiation tree B growing out of one terminus of tree A. A number of marine invertebrates follow this pattern, including echinoderms, and it has been hypothesized that such organisms represent a concatenation of two genomes from different phyla (Williamson, 1992). The unused larval tissues have various fates, from dedifferentiation and incorporation, to being ingested or an independent existence.

The fate of the unused larval tissue, in all cases of metamorphosis, varies from being eaten by the adult to: histolysis, being discarded, or, rarely (in single tissue and dedifferentiation metamorphosis) to a separate existence:

"In some Phyllodocidae about 40 segments of a well-developed, wriggling juvenile may protrude from the anal region of the trochophore, while the prototroch continues to provide pelagic transport....

"As metamorphosis approaches, the juvenile [nemertine] shows independent movements within the larva and eventually emerges from it. The pilidium, now without its quasiparasite, goes on swimming for some time, demonstrating an independence of larva and juvenile comparable with that found in the echinoderms" (Williamson, 1992).

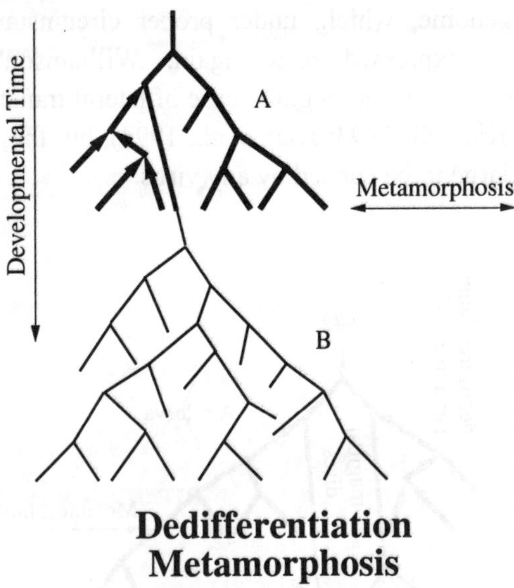

Figure 49. In *dedifferentiation metamorphosis* a few or most of the larval tissues (differentiation tree A) dedifferentiate (reverse arrows) and then redifferentiate into the adult (differentiation tree B). The remaining tissues may contribute in a nutritive fashion. Ascidians, which have larvae with a notochord and are presumed to be close to our chordate ancestors, represent a case of dedifferentiation metamorphosis, and another possible case of the concatenation of two very different genomes (Williamson, 1992). Thus the ascidian larva may be close to our common ancestor, but the ascidian adult would not be.

These startling events emphasize the concatenated nature of the larval and adult life cycles in single tissue and dedifferentiation metamorphosis, as if they really are, at least initially, two organisms coexisting in one concatenated genome, taking turns at expression. While chimaeric animals (Gardner & Johnson, 1973; Lewis & Rossant, 1982; Rossant et al., 1982) have given us a basis for the myth of the Minotaur, Williamson's concatenations ('sequential chimeras': Williamson, 1991) are like the snakes forming Medusa's hair or "Minerva (Pallas Athene), the goddess of wisdom,... the offspring of Jupiter... [who] sprang forth from his head completely armed" (Bullfinch & Klapp, 1942). In such a case, if Williamson's hypothesis is correct, the larva freed of its adult continues to carry the adult genome, which, under proper circumstances, should be capable of being expressed once again. Williamson's hypothesized concatenations would represent a grand case of lateral transfer of genes (i.e., between species: Kidwell, 1993; Diaz et al., 1994), but the amount of DNA transferred is too large to be carried by any virus.

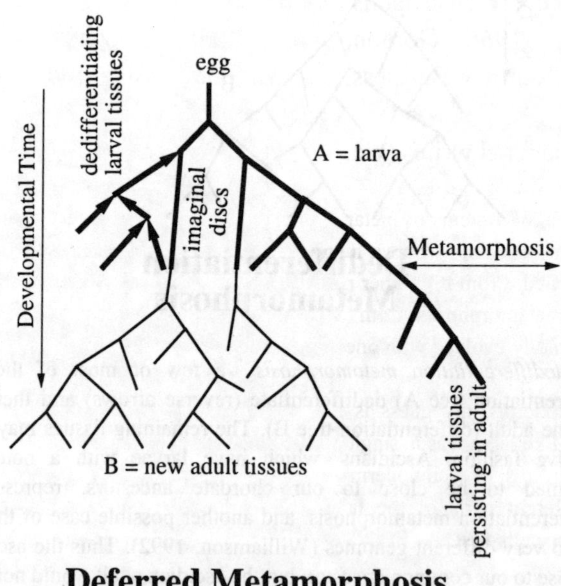

Deferred Metamorphosis

Figure 50. *Deferred metamorphosis* is the most complex case of metamorphosis. Some larval tissues, called imaginal discs, continue to differentiate after metamorphosis into adult parts, such as wings and legs. Other larval tissues, such as the brain, persist in the adult. This is the pattern of holometabolous insects, whose larvae (caterpillars) may be concatenated genomes (Williamson, 1992). In this case, we would postulate that an original adult differentiation tree broke into a set of terminal branches that were transferred to a number of larval terminal edges, perhaps starting from a case of single tissue metamorphosis (Figure 48) or dedifferentiation metamorphosis (Figure 49). In the meantime, perhaps by heterochronic changes (Figure 29), some of the larval tissues came to persist in the adult. Some dedifferentiation, as indicated by reverse arrows, may also be involved.

Now that we have seen the five cases of metamorphosis sketched in terms of plausible topologies of their differentiation trees (Figures 46-50), it is clear that many nuances and qualitatively different types of metamorphosis could, and indeed may, exist. A search for evolutionarily and ecologically reasonable and unreasonable varieties of metamorphosis can now begin, analogous to the understanding brought to the used and unused portions of parameter space for gastropods (Raup, 1962, 1966; Raup & Michelson, 1965; Ghiselin, 1966; Gordon, 1966; Raup & Chamberlain Jr., 1967; Linsley, 1978; Kemp & Bertness, 1984; Palmer, 1985; Stone, 1996).

Donald I. Williamson writes (p.c., 1993):

"Your classification of systems of metamorphosis is very helpful in drawing attention to the enormous range of processes that we lump together as metamorphosis and in showing that they can be classified. From my rather restricted point of view, I wish a distinction could be drawn between metamorphoses from acquired larvae (acquired by larval transfer) and evolved larvae (wholly evolved with one adult), but I cannot draw one....

"Regarding echinoderms, my guess is that the original form of metamorphosis from the acquired tornaria-like larva was a 'single tissue' metamorphosis, from larval mesenchyme. In time, different groups of echinoderms evolved (in the Darwinian sense) schemes for incorporating different parts of the larval endoderm into the juvenile, but they had to 'dedifferentiate' the cells first.

"I am glad that, in general, you seem well disposed to my heretical ideas. I have my critics too, of course, but criticism stimulates discussion and is also to be welcomed. The reviews of the book [Williamson, 1992] have varied from near total condemnation to enthusiastic praise, so I am well satisfied."

Proposition 225: there are many examples of the five categories of metamorphosis.

1. *Continuing differentiation metamorphosis* (Figure 46).

Amphibian metamorphosis fits into this category (Dent, 1968; Duellman & Trueb, 1986; but cf. Alberch, 1987), and it applies to the Enteropneusta:

"An adult *Balanoglossus* is very different in form and way of life from its tornaria larva, but the transformation is practically all accomplished by cell division and differentiation.... The orientation of the juvenile, with respect to anterior and posterior, dorsal and ventral, left and right, is the same as that of the larva, and the mouth, alimentary canal, and anus of the tornaria develop into the same organs of the settled juvenile. Virtually the only parts of the larva to be discarded during metamorphosis are its natatory cilia. A tornaria larva can certainly be said to 'develop into' a juvenile enteropneust..." (Williamson, 1992).

2. *Pulsatile metamorphosis* (Figure 47).

The best examples are...

"...Grasshoppers, cockroaches, bugs, and lice, [which] undergo *gradual metamorphosis.* The young of such insects resemble the adults [and presumably have essentially the same tissues], except that their body proportions are different.... They go through a series of molts during which their form gradually changes and becomes more and more like that of the adult, largely as a result of a differential growth of the various body parts. They have no pupal stage and experience no wholesale destruction of the immature tissues" (Keeton, Gould & Gould, 1986).

It is clear that the usual name for this form of metamorphosis, 'gradual', is a misnomer, meant to distinguish it from more dramatic, one step metamorphoses. Given the more correct use of 'gradual' in distinguishing gradual from punctuated phylogenetic trees, and my desire to speak about metamorphosis in terms of differentiation trees, it seems that 'pulsatile metamorphosis' is a better name. 'Pulsatile' refers to the simultaneous entry of a number of tissues into their next step of differentiation. ('Punctuated' does not imply simultaneity, and so would not be appropriate here.) Pulsatile metamorphosis has also been called 'incomplete' metamorphosis (Whitten, 1968), again, as a negative name, to contrast it with other forms.

The very existence of pulsatile metamorphosis, without a distinct larval form, in hemimetabolous insects (Etkin, 1955), and "within the insects, of the absence of metamorphosis in the groups Collembola, Diplura, and Thysanura" (Whitten, 1968), supports Williamson's suggestion that in the holometabolous insects (Etkin, 1955) with 'complete' metamorphosis (Whitten, 1968), caterpillars might be a grafted larval stage.

Pulsatile metamorphosis may represent more a case of allometry (Huxley, 1972; Gould, 1977a; Reiss, 1989) which is stepwise because of the exigencies of growth in the confines of an exoskeleton that gets shed once in a while. Even the use of 'metamorphosis' to describe pulsatile metamorphosis may be an exaggeration, but removing this phenomenon from metamorphosis would go too far beyond previous usage. Whether it is a generalization of continuing differentiation metamorphosis, or an evolutionarily independent development, remains to be seen.

3. *Single tissue metamorphosis* (Figure 48).

Williamson lumps together what seem to be two separable cases of 'cataclysmic' metamorphosis. The first case, development of the adult from a single larval tissue, is suggested by the following descriptions:

"Some features of the metamorphosis of echinoderms are much the same whether the larva is a bipinnaria, brachiolaria, pluteus, auricularia, or doliolaria.... The left mesocoel sac is usually bigger than its right counterpart, and as it grows still larger it develops five lobes, which soon become arranged radially around the sac. This is the first indication of the pentaradial symmetry of the adult.... The five-lobed rudiment around the left mesocoel sac will grow to form most of the radial structures of the developing juvenile [small adult], but it will receive some juvenile radial components from one or occasionally both metacoel sacs.... [I presume that all three sacs are parts of the same tissue, per my definition of a tissue, Proposition 30.]

"In the Echinodermata growth of the mesenchyme that surrounds the coelomic pouches determines the shape and symmetry of the juvenile and hence the adult.... The larval cells of this mesenchyme are also probably the only ones to produce adult cells by direct descent....

"Much more rarely the left and right mesocoel sacs are of similar size, and then twin juveniles may develop. Such a case in the development of the common British sea-urchin, *Echinus esculentus,* was described by MacBride (1911).... [When]... the juveniles had... reached a fairly advanced stage of development, they apparently had not incorporated any parts of the larval epidermis or stomach.... This case again emphasises the independence of the larva and the juvenile (or juveniles), and it is fully consistent with the suggestion that the only larval feature necessary for the development of a juvenile is one or more coelomic sacs of suitable size and shape....

"The segmented polychaete may develop by the gradual proliferation of segments at the posterior end of the trochophore to produce a second larva, the polytrochula or nectochaete....

"In some genera,... such as *Owenia*, the extrusion of the segmented body of the next phase may be quite explosive.... The trunk segments of the developing juvenile first form between the mouth and anus of the larva....

"Some members of the class Aplacophora... hatch as a lecithotrophic test cell larva that has a single ciliary band and apical tuft, as in a trochophore, but resembles a protobranch larva in having only one exterior opening, the posteriorly situated 'pseudoblastopore'. The juvenile develops from cells around this aperture, at first internally, then externally, and produces its own ring of cilia, the telotroch" (Williamson, 1992).

4. *Dedifferentiation metamorphosis* (Figure 49).

This second case of 'cataclysmic metamorphosis is more tenuous, because we only have descriptive accounts of dedifferentiation:

"In the Holothuromorpha... classes with larvae, any parts of the larval gut that are incorporated into the juvenile are first broken down, and the cells lose their orientations and special functions before being redifferentiated to the juvenile condition....

"It has been established in an asteromorph and in an echinomorph that no larval ectodermal or endodermal cells are incorporated into the juvenile until they have undergone histolysis... and redifferentiation, and this may well be the case throughout the phylum (Chia & Burke, 1978)....

"After settling, all bryozoan larvae embark on a cataclysmic metamorphcsis, contracting under a layer of ectoderm into a rounded or oval mass in which all the tissues undergo histolysis. Part of the ectoderm invaginates, then closes off to form a vesicle, and the inner part of this chamber then constricts off a second vesicle, which remains in contact with the first. The cells surrounding the outer vesicle give rise to the adult lophophore and pharynx, and those surrounding the inner vesicle produce the remainder of the gut....

"Some ascidians develop without a tadpole stage, and, where a tadpole does occur in the life history, its metamorphosis has all the criteria of a larva acquired from another group. Although the larva and the juvenile ascidian are both bilaterally symmetrical, the juvenile does not adopt the orientation of the larva, and most of the larval tissues and organs are discarded at metamorphosis (Brien, 1948)" (Williamson, 1992).

In one organism we seem to have a combination of single tissue metamorphosis and dedifferentiation metamorphosis:

"...A juvenile nemertine develops round the stomach of a pilidium like a parasite within the larva, with a totally new orientation (Gontcharoff, 1961). Seven or eight discs, representing only a small proportion of the larval ectoderm, invaginate and cut off from the rest of the ectoderm and migrate inward. They then grow, spread, flatten out, and finally fuse together to enclose the larval gut and form the epidermis of the developing juvenile. Although the larval gut is incorporated into the juvenile, the cells undergo considerable redifferentiation, and these, together with the derivatives of the ectodermal discs, are the only larval cells that contribute to the juvenile" (Williamson, 1992).

What would be nice to have is a demonstration that...

1) all dedifferentiated cells constitute a single tissue;

2) all dedifferentiated cells are from a single edge of the larval differentiation tree.

The latter is necessary if a single tissue, per our definition of tissue (all cells that traversed the same path along the differentiation tree), is to be derived from multiple tissues via our speculation on dedifferentiation in general (Section 5.03). One consequence is that...

3) redifferentiation (formation of the adult) will occur off an edge of the larval differentiation tree that was not used during larval differentiation.

5. *Deferred metamorphosis* (Figure 50).

The case of deferred metamorphosis ('complete' metamorphosis: Whitten, 1968), is that of the holometabolous insects, in which the adult develops from imaginal discs and "the dorsal blood vessel and the central nervous system [which] are little affected and pursue an uninterrupted course of differentiation (Imms, 1946)". Williamson (1992) keeps this option open:

"Perhaps, however, the caterpillar may be regarded as a phase in development transferred from an onychophoran or a myriapod. It could help to explain the occurrence of this type of larva not only in all lepidopterans but also in some hymenopterans" (Williamson, 1992).

Alberch (1987) might place amphibians in this category:

"Our results support the hypothesis that larval chondrocytes undergo cell death and do not participate in the formation of the adult cartilage epibranchial during the metamorphosis of the hyobranchial skeleton in *Eurycea* [a salamander/urodele amphibian]. The adult chondrocytes appear to come from the connective tissue in a localized region of the perichondrium in the larval first epibranchial. Therefore, of the two possible developmental pathways..., the one with separate larval and adult developmental programs appears to be the correct one. I refer to this duplicate system as being *compartmentalized*. Compartmentalization, thus, refers to the fact that at some time during early development prospective larval and adult cells acquire different fates. Larval cells undergo morphogenesis, while prospective adult cells remain relatively undifferentiated (as connective tissue) during the complete larval period. At the time of metamorphosis larval and adult cells respond differently to global cues, most likely hormonal. Then, larval cells, already differentiated, have only two possible fates: They can either die or remain unaffected....

"...There are limits to how much dedifferentiation and remodelling of existing structures can occur during metamorphosis.... Organisms seem to have been able to circumvent this constraint by developing redundant systems in which prospective larval and adult cells are apportioned early in development and the two developmental programs unfold sequentially at different times..." (Alberch, 1987).

Proposition 226: concatenation of distinct phyla occurs in two steps: hybridization followed by a recombination event that grafts one differentiation tree to the other.

Williamson shows that living hybrids between far flung phyla can be obtained on occasion merely by bringing ascidian eggs and sea urchin sperm together at high concentrations (cf. Deuchar, 1975a, in Section 9.02):

"There is no evidence that sea squirts and sea-urchins have ever hybridized in nature, and crossing them experimentally does not prove that an ancestor of any animal group ever acquired a larval form by cross-fertilization. The experiments show, however, that the first step in the process of transferring a larval form from one species to a very distantly related one by cross-fertilization is entirely feasible. It shows that a heterozygote can give rise to a paternal-type larva rather than to a chimera. The experiments also show that a hybrid larva is capable of metamorphosis, although, in this case, all juveniles were of the paternal form.

Perhaps the echinoderm juvenile and adult body always take precedence over both the tadpole larva and the adult of a urochordate" (Williamson, 1992).

This concept of 'precedence' or 'priority', while interesting, may not be the important point, since the presence of both genomes has yet to be demonstrated in these experimental cross fertilizations (Blackstone & Dick, 1993) and rare events more complex than simple hybridization could have occurred over the available geological time:

"It may be rather daunting for the experimenter to consider the billions of times eggs and sperm of different species must have come into contact in the sea over hundreds of millions of years, yet all the cases of anomalous embryos and larvae outlined in this book can be explained as the result of no more than 25 [individual] successful cross-fertilizations" (Williamson, 1992).

The rarity of these events is not addressed by the 'critical' experiment of Hart (1996) supposed to contradict Williamson (1992). Rather than trying to synthesize hybrids (Williamson, 1992; Hart, 1996), a better approach might be analysis of the active parts of the genome, in larvae and adults, to see if they have distinctive properties (such as 5S rRNA spacer sequences: Klassen, Balcerzak & de Cock, 1996). I would like to add, then, one rare event besides cross fertilization to produce the initial concatenated individual in each line: grafting of one differentiation tree to another via an appropriate crossover event. The rare graft that is needed to produce a smoothly functioning, concatenated organism is like the one in Figure 48, where one differentiation cascade from organism A runs into the root of the tree of organism B. If the differentiation cascade in A is terminal (at least for the larva of A) then single tissue metamorphosis is generated. "Conserved molecules used as gene regulatory signals (with same or different function) in different phyla are the obvious grafting points" (Steve McGrew, p.c., 1997).

If the graft is to an earlier terminal branch (Figure 49), or to the root of organism A, then dedifferentiation of some or all of the tissues, triggered by metamorphosis, would be required as a third rare step to provide cells at the stage needed to kick off the grafted B tree, to produce dedifferentiation metamorphosis.

Like an initial gene duplication, the initial acquisition of a new larval stage may be close to neutral in its effect on fitness, since "once they had completed their metamorphoses they would be at no disadvantage" (Williamson, 1992). However, the concatenated organism would now have a planktonic larva with better dispersal: "...A planktonic larval phase is particularly valuable in times of rapid climatic change because it greatly increases the chance that some individuals could reach conditions suitable for growth and reproduction" (Williamson, 1992). Furthermore...

"Once an individual has acquired a new larval form and has survived through metamorphosis to reproduction, all its descendants will have the same larval form (or one subsequently modified by Darwinian evolution)" (Williamson, 1992).

Proposition 227: after the grafting of one tree onto another, subsequent transpositions of terminal branches of the adult differentiation tree could occur over evolutionary time, dispersing the adult tree over some of the terminal branches of the concatenated larva. This could be the origin of the pattern of deferred metamorphosis from a few tissues without their dedifferentiation, as in holometabolous insects.

We need only postulate movement of whole terminal branches, as shown in the course of differentiation tree evolution depicted in Figure 49 to produce the differentiation tree in Figure 50. Of course, we'll need a suggestion for the phyletic origin of caterpillars, if this is what happened. Presumably, the adult in such a graft came from a nonmetamorphosing insect, rather than one with continuing differentiation metamorphosis. It should be possible to test this by molecular taxonomy.

Proposition 228: grafting of one whole differentiation tree onto another is more likely between distantly related phyla, providing a mechanism for one step macroevolution.

The concept of "some minimum degree of genetic difference" (Williamson, 1992) to avoid mixed features in a hybrid (such as the chimeras found in many interspecific crosses of sea urchins, reviewed by Williamson, 1992) may be required because of the multiple crossing over that could easily occur at all hierarchical levels when two similar genomes find themselves in the same meiosis. Genomes that are very different will stay reasonably separate, being unlikely to cross over anywhere, at any hierarchical level, so that genome grafting becomes the most probable means of their combination. Thus Williamson's mechanism actually selects for major evolutionary saltations, and is a reasonable way of producing macroevolutionary change in one step, partially exonerating Goldschmidt (1940) and his 'hopeful monsters' (cf. Gould, 1982b; Bateman & DiMichele, 1994), at least of the Minerva variety (Bullfinch & Klapp, 1942).

Williamson (1992) considers...

"...the possibility that the implied genetic transfers may have also occasionally affected adult form.... The partial bilateral symmetry of adult holothurians may be the result of genetic factors that produce bilateral symmetry in the larvae continuing to exert some influence in the adults" (Williamson, 1992).

Such transfers could happen via any mutational mechanism available to ordinary differentiation trees. The oppositely directed influence, from adult to larva, is also possible. Of course, any unused edges of either differentiation tree in the graft would eventually decay to background DNA (cf. Martin & Gordon, 1995) or otherwise be eliminated. Nevertheless, Williamson (1992) ends with the intriguing suggestion:

"The adult morphologies of some groups are probably so different that any attempt to mix them is doomed to failure, but can we deny the possibility that any group had its origin in a hybrid between two different groups?"

In principle, new organisms could be generated by grafting any subtree of one organism onto any subtree of another that includes the root. Undoubtedly many restrictions apply, but we have much work ahead of us to understand what they are:

"Introduced genetic material, I believe, produces new forms of development in the target lineage, starting from the fertilized egg and extending a variable distance into subsequent phases of development" (Williamson, 1992).

This concept can be quantitated via differentiation trees.

Proposition 229: the phylogenetic tree is, in part, a web.

This is a corollary of Williamson's rather graphic statement:

"This phylogenetic tree has been a familiar part of the zoological landscape throughout my lifetime, and old trees, whether phylogenetic or botanical, undeniably have their attractions. They should, however, be inspected periodically to see if they are still healthy, and in some cases the only responsible course of action is to cut them down. Perhaps the reader should be warned that, if my views are accepted, little but the stump of this tree will remain intact" (Williamson, 1992).

New trees grow out of many an old stump or root, and it may well be that the base of the phylogenetic tree, and perhaps some of its higher branches, have undergone anastomosis via 'horizontal' gene (Lewin, 1982; Syvanen, 1985; Kidwell, 1993) to genome (Williamson, 1992) transfers. The outstanding problem of who were the original metazoans (Willmer, 1990), those organisms at or evolving directly from Williamson's 'stump', may require a combination of much molecular taxonomy, comparative anatomy, construction and comparison of numerous differentiation trees (both larval and juvenile), and heretical hypotheses such as Williamson's.

The distinctions between phylogenetic trees and webs have been elucidated by Maynard Smith (1990a):

"How far is a tree, and the associated hierarchical classification, the appropriate image for classifying natural objects? Clearly, it is not always so - Mendeleyev's name would not be remembered if he had insisted on classifying the elements cladistically. A tree is the appropriate image only if the objects to be classified have arisen by a branching process. The oddest thing about the pattern cladists is that they wish to combine the idea, silly but not illogical, that the pattern of the living world must be fully described before questions about its evolution can be addressed, with the assumption (illogical in the absence of an evolutionary hypothesis) that the appropriate method of description is hierarchical. For the members of an asexually reproducing population, a tree is the appropriate image: If reproduction is by binary fission, the tree is dichotomously branching. For the members of a sexual population, the appropriate image (which, unhappily, cannot be drawn on a flat piece of paper) is a multidimensional net. At higher taxonomic levels, however, a tree is again the appropriate image (although dichotomous branching cannot be justified), provided that reproductively isolated species exist. This suggests a distinction between a 'fractal tree,' which is tree-like at all scales of magnification, down through individual reproduction and cell division to DNA replication, and a 'large-scale tree,' which is tree-like only at low magnification.... It is important to mention a third possible type of tree, which I will call a 'local continuum tree.' Imagine a 'genus' of sexually reproducing plants (the oaks and the potentillas of North America may approximate to this image) with the following properties:

(i) Any individual can cross successfully with others that are genetically not too distant (if dioecious, the partners must obviously be of opposite sex), but cannot cross with genetically more distant members of the group.

(ii) There are few discontinuities, so that, even if A cannot mate with E, A can mate with B, B with C, C with D, and D with E. Hence, gene flow can occur throughout the whole taxon. However, such local continuity could be combined with occasional discontinuities, arising for accidental historical reasons, so that the living world would be divided into a large number of large taxa between which gene flow could not occur but within which gene flow is possible via a series of closely related intermediates" (Maynard Smith, 1990a).

As a final note on concatenated genomes, perhaps they could be involved in the phenomenon of phenotypic plasticity, in which the phenotype is a function of the environment the organism finds itself in (Schlichting, 1986; Scheiner, 1993), as aerial leaves of aquatic plants.

7.03 Winding Up Evolution

"We can understand evolution without really knowing what life itself is! This view, held by the great biologist Haldane (1932), was characteristic of a half century of evolutionism which produced models and calculations in the absence of any intimate knowledge of cellular logic.... Our aim now is to tackle a problem which not so long ago seemed insoluble: how to evaluate, having taken into account the state of organization of life at any given moment, the various possibilities which are apt to manifest themselves" (Ninio, 1982).

"Perhaps the main respect in which the biological picture is more complex than the physical one, is the way in which time is involved in it.... One has to consider [life]... as affected by at least three different types of temporal change, all going on simultaneously and continuously.... On the largest scale is evolution; any living thing must be thought of as the product of a long line of ancestors and itself the potential ancestor of a line of descendants. On the medium scale [is]... life history. It is not enough to see that horse pulling a cart past the window as the good working horse it is today; the picture must also include the minute fertilised egg, the embryo in its mother's womb, and the broken-down old nag it will eventually become. Finally, on the shortest time-scale, a living thing keeps itself going only by a rapid turnover of energy or chemical change.... In the biological picture towards which we are finding our way, the three time systems will have to be kept in mind together.... Most of our theories are still only partly formed because they leave one or other of the time scales out of account" (Waddington, 1957).

"Fundamentally, the evolutionists resolve all their problems by pushing them over to the geneticists and assuming that they are capable of dealing with them" (Taylor, 1983).

Proposition 230: since there is a finite set of topological changes of a given magnitude that can be made to a given differentiation tree, we may be able to predict the species that are immediately derivable from an existing species, i.e., to compute the developmental constraints on a given species.

If we accept that the starting point for speciation is an existing differentiation tree, and that descendent species will have similar differentiation trees, then to some extent we can make evolution a predictive science. This may answer the objection of Lima-de-Faria (1988) that evolution is not a predictive science:

"In evolution there are no laws that formulate in precise mathematical terms the interrelationships between organisms, or define the mechanisms involved in their transformation. As a consequence no predictions of any kind can be made in evolution. No one can predict what species will emerge from *Homo sapiens* or from any other animal or plant" (Lima-de-Faria, 1988).

Of course, insofar as vast numbers of potential differentiation cascade changes may also lead to speciation, the practical force of this proposition may be moderated. If there are too many possibilities of evolution from a single differentiation tree, then the range of possibilities available, while mathematically restricted, is hardly what one would consider predictive. We may have here a problem of point of view: if we agree that the differentiation tree provides significant developmental constraints ('You can't get there from here') on the course of evolution, then 'the cup is half full'. On the other hand, if there are still 'too many' possibilities despite these constraints, then 'the cup is half empty' (cf. McKitrick, 1993), and the berated assumption of randomness of mutations ('Anything is possible') in neo-Darwinism (Lima-de-Faria, 1988) may be justified. The possibility that the increasing degrees of freedom of a genetic system (Collier, 1989), an expanding universe of genetic possibilities (by gene and genome duplication), creates macroevolutionary irreversibility beyond the capacity of equilibration processes to keep up (Section 10.08), supports the latter view. The alternate extreme position in favor of developmental constraints has been formulated by Goldschmidt (1940):

"The selection of the direction in which genetic changes may push the organism is therefore not left to the action of the environment upon the organism, but is controlled by the surroundings of the primordium in ontogeny, by the possibility of changing one ontogenetic process without destroying the whole fabric of development.... Thus what is called in a general way the mechanics of development will decide the direction of possible evolutionary changes. In many cases there will be only one direction" (Goldschmidt, 1940).

Charlesworth (1995) gives the perspective of an evolutionary biologist:

"...A fully explanatory theory of phenotypic evolution requires two kinds of knowledge that we currently lack: knowledge of the spectrum of possible phenotypes that can be produced by the genetic variability available to a given evolving lineage, and an understanding of how fitness is related to phenotype.... It is... tempting to suppose that the first kind of knowledge can be provided by developmental biology. A predictive theory of the phenotypes which can be realized by mutation would supplant our current reliance on the empirical study of natural genetic variability for determining the nature of the raw material for evolution.... But it is hard to make non-trivial predictions about the nature of the limitations on such phenotypes, and about the resulting constraints on the ability of selection to control the outcome of evolution.... The literature on molecular evolution, which Kauffman [1993] regards as a paradigm for the *NK* model, also provides strong evidence against Kauffman's assumptions.... Indeed, there seems to be no limit to the ability of gene duplication, mutation and selection to develop ever-increasing varieties of new gene functions, with a corresponding increase in organismal complexity.

"There has been a long history of attempts to integrate developmental and evolutionary biology, but the net results of these efforts have, for good reason, had little impact on the thinking of most evolutionary biologists. A major reason for this has been the more or less overt hostility of the proponents of these theories to Darwinian concepts of stepwise adaptive evolution, from Richard Goldschmidt and D'Arcy Thompson in the 1940s, to Pere Alberch, Stephen Gould and Brian Goodwin in the 1980s" (Charlesworth, 1995).

There will be no resolution of the half full/half empty viewpoints until we can compare differentiation trees between wild type and mutant individuals, and between species. We need something in the cup to talk about. ("Nowhere are Mother Nature's hidden constraints *written down* in a way that can be read without the help of the interpretative rules of artifact hermeneutics": Dennett, 1995.) In the meantime, the question of developmental constraints will remain empirical or mostly an abstraction.

Perhaps a good place to start, if we could somehow guess (Jablonski, Gould & Raup, 1986) at their differentiation trees, would be with the early metazoans (cf.: Erwin, Valentine & Sepkoski Jr., 1987):

"...If powerful processes of genomic and morphologic change exist, why have most Baupläne [bauplans] originated so early?... During the late Precambrium, metazoan adaptive space was relatively unoccupied - there were many vacant peaks in the adaptive landscape - and novel adaptive types found little interference.... Genomes may have evolved so as to become less forgiving of major restructuring (Wright, 1982a).... If those genomes were significantly more compartmentalized, with fewer epistatic complications, than those of today, then major alterations in a given part of the developmental system would be less likely to invoke deleterious side effects in other parts. Such compartmentalization would tend to permit extensive changes to occur with less cost to fitness and adaptation and, therefore, more rapidly. If these explanations are on the right track, then there is a possibility of understanding the factors which have determined the number of Baupläne produced. Basically, Bauplan numbers may be determined by the rate at which new Baupläne are developed relative to the rate at which 'old' Baupläne are modified and diversified, as adaptive space fills (Valentine, 1980)" (Valentine, 1986).

I take such 'compartmentalization' to correspond to different branches of the early metazoan differentiation trees, with the caveat, recognized by Valentine (1986), that they were not locked together in webs of unbreakable inductive relationships. Because of the relatively small sizes of these early differentiation trees (which are more explicit descriptions than bauplans), the combinatorics may be within our grasp. If this combinatorics is sufficiently restricted, we may be able to explain (i.e., predict) why we had only a certain number of major taxa. Of course, many of these major early evolutionary experiments became extinct (Gould, 1989a), leaving us with a relative poverty of major taxa.

Clearly, what is needed is quantification of differences between differentiation trees, and measurement of these differences for related existing species. Differentiation trees are rooted trees with labelled edges, in terms of graph theory (Biggs, Lloyd & Wilson, 1976; Berge, 1985). Unfortunately, there does not appear to be any mathematics developed that would provide the metric that we need (however, cf. Day, 1983; McShea, 1996). Hints of the directions to go include the relationship between rooted trees and partial orderings (Birkhoff & MacLane, 1953; Nash-Williams, 1967; Rival, 1984) or the theory of stream orders (Leopold, 1971) or branching processes (Jagers, 1975). The phylogenetic tree is a tree whose nodes are trees (differentiation trees), apparently a mathematically unexplored structure. If we are to quantify the relationships between taxa based on differentiation trees as nodes of the phylogenetic tree, we will need appropriate metrics. Then we could possibly answer the objection of Vermeij (1987):

"The principal difficulty with the concept of biological progress is that no operational criteria for recognizing and studying it has ever been outlined" (Vermeij, 1987).

If the acceleration of evolution after a gene duplication (Li, Luo & Wu, 1985; Proposition 134) includes further duplication of the very same genes, then a positive feedback mechanism may exist that could cause one terminal branch of a differentiation tree to grow faster than others.

Proposition 231: some of the major events in evolution have been those that permitted larger size without decreasing the interface surface to volume ratio of the organisms' living cells.

In animals, this has been done by internalization of surface area via the circulatory and respiratory systems (cf. Mayr, 1976), or by externalization, in the form of colonial organisms with great contact surface to the environment (cf. Propositions 123 and 126). In plants, externalization, via branching and flat structures such as leaves, seems to be the major mechanism:

"Where any considerable size is reached in a plant or animal, it is always attended with structural devices for increasing surface in relation to volume. A good deal of the complicated anatomy of higher animals is a result of extending surfaces which serve the fundamental vital processes of absorption, excretion, digestion, and respiration.... In organs that absorb food [sic] the same principle is abundantly illustrated. Consider the surface of a large tree... with its numerous leaves..." (Holmes, 1948).

(Cf. West & Goldberger, 1987.) There seems to be a rough correlation between these two solutions to the interfacial surface/volume problem and the number of tissue types: externalization requires few terminal tissue types, while internalization requires many. Thus we make some weak contact with the complexity of the differentiation tree. A fundamental constraint on higher organisms may be retention of the interfacial surface to volume ratio for each living cell. This hypothesis has been critically reviewed in Reiss (1989). Niklas (1994b) attributes the idea originally to Ryder (1893).

Proposition 232: the evolution of perception can be approached via the biogenetic law.

Hoffman (1998) dissects human visual perception into a number of simple rules, many of which are attributable to specific regions of the brain. Each of these regions may be constructed in the course of embryonic development via differentiation waves in the neural plate and the hierarchically organized genome. If so, the corresponding differentiation tree provides a framework for discussing the embryology and evolution of perception. We can take a more general lesson from Hoffman (1998): perception is not a single entity, but rather a perhaps hierarchically arranged, modular set of functions, some of which have been added at later stages during the course of evolution. In a way, then, if we can reconstruct this hierarchy, we also thereby roughly reconstruct its evolution. This is but Haeckel's "ontogeny recapitulates phylogeny" approach.

Root-Bernstein & Dillon (1997) have proposed that molecular complementarity, the lock and key approach to specificity in molecular interactions (Proposition 270), applies at higher than molecular levels, and offers a hierarchy for understanding "the origin and evolution of life". On the embryological side, they only discuss the phenomenon of embryonic cell sorting (Steinberg, 1963) as one of these higher order complementary interactions. It probably has a direct molecular basis (Section 1.10). But lock and key is in a way the beginning of perception. If we can bridge the gap between human visual perception and molecular complementarity, through embryology, we may be well on our way to understanding the evolution of perception and its feedback to evolution itself. After all, you can't eat what you can't perceive, and thus the evolution of perception may be tied to the evolution of predator/prey relations, from the molecular level of chemotaxis on up. In a way, perception (not just of 'specific mate': Proposition 159) may be the key to emergence and its evolution. Hierarchical levels in ecology and evolution may be defined by the perceptual abilities of organisms.

Proposition 233: genetic recombination in sexually reproducing organisms causing topological changes to the differentiation tree provides a basis for more rapid speciation than in species with asexual reproduction.

Eldredge (1995a) succinctly summarizes the current 'conundrum' of sex in evolution (cf. Proposition 197):

"Geneticists... tended to see sex as a good thing, affording the variation that enabled rapid spurts of evolution.... All of a sudden sex became a big problem. That central claim of ultra-Darwinism - that nature is organized around a competitive compulsion to leave as many copies of one's own genes to the next generation - is incompatible with the mere existence of sex. If the name of the game is to leave as many copies of your genes behind as possible, it is pure folly to mix them with someone else's on a 50-50 basis" (Eldredge, 1995a).

Differentiation trees may help resolve the problem of why most species maintain sexual reproduction (Margulis & Sagan, 1986a; Maynard Smith, 1989a,b; Charlesworth, 1989; Hamilton, Axelrod & Tanese, 1990), without invoking group selection (Farley, 1982). To see how this might be done, let us accept the view of punctuated equilibrium that...

"...speciation is where the action is, and this is where we must look for the central role of sex.... It seems safe to say that divergent speciation would be difficult without genetic recombination. This means that divergent speciation is difficult for sexless creatures.... Without sex there is no real speciation, only the very gradual divergence of clonal lines of descent, as these accumulate occasional mutations.... Asexual 'experiments' [parthenogenetic species] do not in general fare well in the great game of species selection [i.e., which ones radiate].... Unfortunately for them, they do, nonetheless, participate in the negative aspect of species selection: like sexual species, they are subject to extinction" (Stanley, 1981).

Recombination may be the basis for sexuality as an 'evolutionary driver' (Campbell, 1987a). Since gene duplication is likely to be involved in many topological changes to the differentiation tree, and chromosome rearrangements commonly occur between repeated portions of DNA (Maynard Smith, 1989b), an actual mechanistic basis for the difference between speciation rates of sexual and asexual species may exist. This proposition may lie at the bottom of the phenomenon of 'genetic homeostasis' via heterozygosity (Lerner, 1954; Soulé, 1979; Gillespie & Turelli, 1989). The basis for the ability to achieve divergent evolution may lie in gene duplication, as seen in...

"...the acceleration of the evolution of new genes by gene duplication with subsequent differentiation by sexual recombination. Although this advantage of sexual recombination has been thought to be important by molecular biologists (see a standard text book of molecular biology, Alberts et al., 1989), population biologists appear not to recognize it (e.g., see Michod & Levine, 1988 [cf. Kondrashov, 1993, and other articles in the same issue on 'Evolution of Sex'; Maynard Smith, 1978b; Williams, 1975a])…. Results of simulation studies carried out so far show that the time for acquiring a new gene in a diploid population with sexual recombination is roughly 1/5 - 1/2 of that for the haploid population under realistic values of parameters (Ohta, 1988c). This is because, for diploids, when two or more beneficial mutant alleles exist at the same locus in the population, their frequencies may increase by selection, and the two alleles may be combined onto one chromosome by unequal crossing-over at sexual recombination (Spofford, 1969)" (Ohta, 1990a).

(Cf. Spofford, 1972; Ohta, 1990b). Thus we see that topological changes in differentiation trees, such as duplication of terminal branches, would seem to allow an order of magnitude greater change than simple genetic recombination, and thus provide an even better basis for divergent speciation via sex than single gene duplication. The mechanism can involve either so-called…

"…'legitimate' recombination between long homologous sequences (Anderson & Roth, 1977, 1981)… [or] 'illegitimate' recombination, between sequences of little or no homology (Franklin, 1971; Anderson, 1987b; Ehrlich, 1989; Meuth, 1989). Illegitimate recombination is important, since it… can… affect any region of the genome. It is probably ubiquitous, since it has been observed in all organisms where it was sought" (Ehrlich et al., 1993).

Proposition 234: if the genetic recombination rate increases, each differentiation cascade of the differentiation tree becomes more likely to be disbursed across the genome, and radiation slows down or ceases.

It is clear from Figures 24-26 that cut, copy and paste operations on DNA that lead to successful growth of the differentiation tree are more likely if the genes in each differentiation cascade remain together along the DNA as a physically contiguous unit. Recombination tends to disburse them across the genome. Thus a higher recombination rate acts to reduce the speciation rate by macroevolution, i.e., it limits the rate of change of topology of the differentiation tree. Very high rates of recombination are achieved in some single celled organisms, as high as 1% per cell division in the protozoan *Giardia lamblia,* for instance (Le Blancq, Korman & Van der Ploeg, 1992), and thus perhaps occur in multicellular organisms. What controls the rate of recombination? Is there an inverse correlation between rate of recombination and rate or production of new species? Are differentiation trees more stable in organisms with higher recombination rates? These are empirically approachable questions.

Proposition 235: there is an inverse relationship between the number of chromosomes and the rate of evolution of differentiation trees.

Chromosomes have seemed to have an evolution independent of that of the organism. For example, their numbers can vary widely within a single taxon, with little or no consequence to morphology or development. Leigh Van Valen (p.c.) has pointed out that "macroevolutionary switching of [terminal] branches of the [differentiation] tree seems to require... a physical contiguity of developmental determinants, on a chromosome". Indeed, homeobox genes occur in a physical order corresponding to the temporal order of activation (Krumlauf, 1994). However, Van Valen's point is well taken. If a terminal branch of a differentiation tree becomes split amongst two or more chromosomes, and/or along two or more separate segments of the same chromosome, duplication of that terminal branch, as a whole entity, becomes problematic.

The problem is lessened, however, if there are fewer chromosomes. Thus I suggest that those taxa that have fewer chromosomes, for a given total genome size, are able to evolve more rapidly because they can duplicate more and larger terminal branches of their differentiation tree. Species that experience catastrophic dispersion of their genome amongst numerous chromosomes, such as the Chinese muntjac, *Muntiacus reevesi*, a deer with many more chromosomes than its morphologically close relative, the Indian muntjac, *M. muntjak* (Wurster & Atkin, 1972; John & Miklos, 1988), while perfectly viable as individuals, are probably evolutionary dead ends.

On the other hand, if the notion that all the chromosomes are linked as one continuous circular strand of DNA is correct (reviewed by Takahashi, 1996), then these problems are alleviated.

Proposition 236: the ability of a population of organisms to undergo evolutionary radiation may depend mostly on the topology of its differentiation tree.

"The extent of the difference between speciating and nonspeciating lineages will not be appreciated until a true science of comparative genetic systems has emerged" (Carson & Templeton, 1984).

"Van Valen [1973] has pointed to a number of examples where significant evolutionary adaptive innovation seems definitely not to be linked to either an immediately preceding extinction event or to the overt and demonstrable invasion of new environments.... George Gaylord Simpson [1944] pointed out years ago that many such instances hinge on the development of what he called 'key innovations' - adaptative modifications that seem to trigger environmental explosions" (Eldredge, 1995a).

Vrba (1983) suggested that "...there are characters of genomes and organisms that confer characteristic probabilities of speciation...." In the absence of detailed knowledge of the structure of differentiation trees, I would like to suggest that evolutionary radiation may depend somehow on their overall topologies, some particular kind of topology allowing easy change and thus evolutionary radiation (cf. Proposition 185). Arthur (1984) has similarly suggested that...

"...in as much as details of the [morphogenetic] tree turn out to be different for different higher taxa, this should lead to predictions of which groups should exhibit saltational change most readily..." (Arthur, 1984).

I have already mentioned one kind of topological feature, namely webbing between branches of the differentiation tree via inductions that have become obligatory (Figure 30). Future investigations may make it possible to ascertain whether there are other topological or logical properties of differentiation trees that allow for the most radiation. This is one sense in which evolution could become a predictive science:

"Contrary to many statements attacking the ability of microevolutionary theory to explain such macroevolutionary phenomena as the origin of morphological novelty and directionality in evolution (Gould, 1980a; Alberch, 1980; Wiley & Brooks, 1982),... standard microevolutionary theories of phenotypic evolution... (Lande, 1979a, 1980a) take account of developmental constraints through the genetic variance/covariance matrix. These quantitative genetic constraints can significantly alter the rate and divert the path of evolution from the optimal rate and path defined by directional selection" (Cheverud, 1984).

The topological structure of the differentiation tree may determine that "genetic variance/covariance matrix". Just why is it that evolutionary radiations of frogs and mammals were quite different from one another?, Do mammals have some special features of their differentiation trees, the highest level 'regulatory' system, that permits especially rapid speciation?:

"We think the most likely explanation for the marked molecular restriction on hybridization among mammals is that the ratio of regulatory evolution to protein evolution is higher for mammals than for frogs. Mammals may have experienced unusually rapid regulatory evolution; indeed, this could be the factor responsible for their unusually rapid anatomical evolution" (Wilson, Maxson & Sarich, 1974).

In considering the predictability of evolution, it is interesting that evolution sometimes accelerates after a gene duplication (Li, Luo & Wu, 1985; Proposition 134). Duplication of terminal branches of the differentiation tree should thus have the same effect, speeding radiation.

Proposition 237: if we could relink detached terminal branches of differentiation trees, we could recover organisms similar to some extinct forms.

This would have to be done relatively soon after the terminal branch became detached. For example, the difference between facultative and obligate neotenic salamanders may lie in the decay of the adult terminal branch of the differentiation tree, either out of disuse, or due to actual detachment from the larval portion of the differentiation tree (Martin & Gordon, 1995). The random accumulated mutation of unused terminal branches of the differentiation tree should cause them to have many single base substitutions, deletions, etc., which perhaps could be repaired first to produce viable organisms. Given the 10^5 to 10^6 species of nematodes (Malakhov, 1994), material for testing this and related ideas is at hand:

"If the primary differences between species reside in the topology of the differentiation tree, and if those differences are due to cutting and pasting or copying and pasting of the genome, then it should be possible to discern the differences between the genomes of related species of, say, nematodes, and from those differences possibly determine the points at which segments have been moved or copied to produce the observed phenotypic changes" (Steve McGrew, p.c., 1997).

Proposition 238: genetic manipulations at the pointers may be one of the quickest ways for us to design new organisms.

"With care and with growing knowledge, we should be able to reach back into the past and restore much of the diversity of animal and plant life - or at least a perfectly serviceable simulacrum of that diversity - that we are now needlessly squandering" (Wills, 1989).

If we could create opportunities for genetic manipulation of the differentiation tree, they may lead to a new, artificial diversity of organisms, a potential opportunity to partly reverse our current devastation of the biosphere. This kind of manipulation is the basis for genetic programming (Koza, 1992a), which may lead to one form of artificial life (Singleton, 1994).

Proposition 239: defragmentation of the genome would permit a new bout of radiation.

Chromosome rearrangements tend to disburse each differentiation cascade, causing it to consist of numerous fragments. If we learn how to chromosome paint all genes corresponding to a single differentiation cascade, this disbursement may become visible. Gene expression maps (GEMs, described in Section 10.12) are an alternative way to see this fragmentation of a differentiation cascade.

I have made an analogy between chromosomes and a computer disk: the hierarchical structure of the differentiation tree is mapped across the linear

DNA in a fashion analogous to the mapping of a hierarchical document across a linearly organized computer disk (Proposition 111). Let us take this analogy further, by noting what happens to a computer disk as it is used:

"When a computer writes to a [disk] platter, it begins writing to the first available location. If it runs out of space before it runs out of data, it moves to the next available location, which may not be on a contiguous block. Spreading a file across non-contiguous blocks is called file fragmentation....

"Fragmentation is not very apparent when a drive is new or freshly formatted. It gets worse over time as the [disk read/write] head grabs whatever sector space is available" (Fong, 1996).

There are many implications of this point of view. For instance, if we had a measure of the rate of fragmentation, we would have yet another 'molecular clock', i.e., a 'genome fragmentation clock'. What I would like to concentrate on here is the prospect of defragmenting the genome:

"There are two ways to defragment a [disk] drive. The first is with a dedicated optimizer. These software utilities can reorganize files on the drive and put them in contiguous sectors. The second method involves copying the contents of your hard drive onto a second, essentially empty (although formatted) hard drive, and then recopying the files back onto your hard drive" (Fong, 1996).

Once a whole genome is sequenced, and we then learn which parts correspond to each differentiation cascade, we could, indeed, consider synthesizing a new whole defragmented genome. The main problem with the direct approach is the enormous magnitude of such a DNA synthesis. The first approach is more rapid. It could be emulated by using molecular biology techniques such as reverse transcriptase (RNA-directed DNA polymerase) to create double stranded DNAs from isolated RNAs active in given cell types, perhaps at an acceptable error rate (Williams & Loeb, 1992; Johnson, 1993). The DNAs could then be strung together in the desired order (cf. DNA reordering in ciliates: Mitcham, Lynn & Prescott, 1992) using a ligase, and, with appropriate regulators added, inserted into YACs (yeast artificial chromosomes: Lewin, 1994). In fact, with the yeast genome now fully sequenced (Mansfield, 1996), reconstruction of its genome in any desired order could be tried out. DNA defragmented organisms could only mate with one another, since recombination with DNA fragmented chromosomes would snarl everything up.

Why would we want creatures capable of further radiation? One reason might be to accelerate the recovery of the biosphere from our current mass extinction (Koopowitz & Kaye, 1983; Myers, 1989; Leakey & Lewin, 1995; Quammen, 1996). Desirable organisms that are stuck in stasis could be an excellent source of new evolutionary material, if freed from stasis by defragmentation. Perhaps something like this occurs in "the sudden flowering ('explosive evolution' [or radiation]) of certain previously long-stagnant evolutionary lines [i.e., in stasis]" (Mayr, 1975, 1984). Another reason to create such organisms would be to terraform (Okuda et al., 1994) and rapidly populate a planet somewhat different from ours. If we put the differentiation cascades back together to emulate a differentiation tree unlike that of the original organism, we could immediately create new life forms. Of course, all these DNA reorderings would rapidly test how sufficient the differentiation tree is as a model for the genetic program. If this all works, we will have made a transition from genetic engineering to genome engineering.

Proposition 240: artificial life should be based on differentiation trees.

The differentiation tree and its growth over evolutionary time (compressed in simulated and robotic automata) may be the key to the fabrication of 'artificial life' (Levy, 1992), both in its construction and in its behavioral aspects. Apparently no one has yet attempted to represent the genome in a-life as a growing differentiation tree, though some have come close:

"Koza (1992b) began to work on a system to breed computer programs genetically.... His breakthrough was deciding to identify the units of crossover not as single characters, or even as lines in a computer program, but as symbolic expressions (S-expressions) written in the LISP syntax.... These S-expressions were essentially subroutines, which were commonly viewed as tree structures. These subroutines could be successfully crossed over so that in a reasonable percentage of matings, the offspring computer program would conform to syntax at least as well as its parents did. Another way of viewing it was that the S-expressions formed tree-shaped 'chromosomes.' Crossover was the equivalent of swapping branches.

"Evolutionary biologist William Hamilton [p.c.] thought this was the sort of advance that could make artificial life work significant. 'Of course, biology has never produced a tree-shaped chromosome, they're all linear...' he says. 'But Koza is able to get remarkable results by using natural selection on tree-shaped chromosomes, by transferring little twigs of trees from one to another. And surely that is a very interesting thing. I think tree-shaped chromosomes are the coming wave. It may also be good for biologists to understand the limitations which may be forced on life by the fact that it always has to work with linear chromosomes" (Levy, 1992).

(Cf. Anon., 1994 in which "...Gene Levinson... devised a computer simulation of an evolutionary process that takes into account the effects of recombination... [which] 'is more potent than mutation as a source of evolutionary change'". Cf. also genetic recombination in genetic algorithms: Holland, 1975) Of course, my point is that chromosome physical linearity notwithstanding, the logical organization of real chromosomes is a hierarchical tree. Moreover, the terminal branches ('twigs') of the differentiation tree cannot only be transferred by crossing over, but can also grow by unequal crossing over (Ohta, 1988b) that not only leads to gene duplication, but to duplication of whole terminal branches (Finkle, 1993; Finkle, Gordon & McAlpine, 1994). Without duplication, branch swapping is analogous to optimizing phylogenetic trees in cladistics (Duncan & Stuessy, 1985).

Perhaps differentiation trees will take us a step further towards reality in our simulations:

"...We are painfully ignorant about the details of molecular evolution, and computer simulations, by necessity, incorporate gross simplifications of evolutionary processes. The most complex of computer simulations still make sweeping generalizations and simplifying assumptions about how organisms evolve. This will likely always be the case, because a computer simulation that did not simplify evolutionary processes would have to be as complex as a real functioning organism" (Hillis & Huelsenbeck, 1994a).

Certainly, it is embryological development that is needed to fill in the gap between genotype and phenotype in artificial life simulations incorporating neural networks (cf. Proposition 16):

"The central problem in trying to apply crossover is the difficult mapping between the genetic code string (genotype) and the [neural] network's weight matrix (phenotype).... If we want the schemata theorem to hold for our problem, the genotype structure should be a binary string. But how do we map a network onto a string?" (Menczer & Parisi, 1992).

Spirov (1996a,b) has made a step towards this by simulating a rough approximation to evolution of the gradient model for *Drosophila* segmentation, including gene duplication (cf. commentary by Gordon, 1996d). If we can show how neural networks arise from the differentiated tissues of the brain, and those arise from the differentiation tree, which does map onto the genotype, we may be able to solve this 'central problem'.

Once a differentiation tree is represented in some form of artificial life hardware, with its built in digital computer(s), nearly "as complex as a real functioning organism" (Hillis & Huelsenbeck, 1994a), the distinction between the inherited and learned portions of the tree is a mere formality:

"Now, the theory goes, we are ready for a second genetic takeover [cf. Cairns-Smith, 1982]: the silicon-based organisms of a-life will replace carbon-based life, including human beings. The new life forms would have certain advantages. Physically, they would be more protean: their bodies could be made of any materials and in any shape. They could be more durable.... These new organisms would also be able to evolve by two forms of evolution: Darwinian natural selection and Lamarckian inheritance of acquired characteristics. Because their essence would be information held in malleable form of silicon bits and not in the hard-wired molecules of DNA, one could tinker with one's own genetic code and integrate what one learned during the course of one's lifetime - or even what others learned during the course of their lifetimes" (Levy, 1992).

This can lead to problems, however, since: "...the [learning] genes atrophy, as evolution removes the necessity to learn something that is now an inherent trait" (Levy, 1992; cf. Ackley & Littman, 1991). In other words, too much genetic internalization of learned material might actually select against the ability to learn, at least under conditions where natural selection continues (cf. Propositions 202-204).

8.00 THE HOMEOBOX

8.01 Why Insects and Vertebrates Share Homeobox Domains

"Ten years ago it was difficult to say anything about biochemistry and morphogenesis. Today we have to record remarkable advances in knowledge centering round the biochemical nature of the 'morphogenetic hormones' which act in normal embryonic development, forming a hierarchy of inductors" (Needham, 1939).

"The Spemann organizer in amphibian embryos is a tissue with potent head-inducing activity, the molecular nature of which is unresolved" (Glinka et al., 1998).

"The gold rush of the past few years that has been the cloning and decoding of genes involved in developmental processes has created the mirage that we understand.... In the larval epidermis of *Drosophila,* a closer and critical look at our harvest shows that although we have glimpsed at regulatory genes and their interactions, we still know little about how these regulatory networks of gene activity are transformed into the pattern of the cuticle.... Dealing with the processing of information in cell assemblies might require new conceptual frameworks..." (Martinez Arias, 1993).

"Sure, the homeobox performs an important function, and (from the genes chosen as bedfellows) it probably has something to do with the embryology of spatial organization. But why should that something be the actual control of segmentation itself...?" (Gould, 1985a). [Cf. Raff & Raff, 1985; Brown & Greenfield, 1985; Raff, 1996.]

Proposition 241: insects and vertebrates have a common ancestor whose tissues differentiated by the use of differentiation waves.

"The dorsal-ventral pattern of the *Drosophila* embryo is established by three sequential signaling pathways.... Although genetic analysis has provided the outlines of each signaling pathway, we do not yet understand what positional information is generated at each stage of the process" (Morisato & Anderson, 1995).

Homeobox genes have been shown to be universally involved in morphogenesis (Bachiller et al., 1994; Bürglin, 1994a). They have been best studied in the fruit fly *Drosophila,* which is a difficult organism to accept as a prototype (model organism) for all others: it is an advanced insect; it has a syncytial blastoderm; it is highly mosaic. Yet there must be an underlying unity, which I would like to try to get at.

There is a widespread tendency to look upon the genetic analysis of segmentation in *Drosophila melanogaster,* with its 'morphogen' gradients, hierarchically interacting genes and their DNA binding proteins, as a success and a *fait accompli,* and a full exoneration of the concept of positional information (Wolpert, 1969). I would like to raise some doubts. The recent survey of the 'genetics of design' in *Drosophila* by Lawrence (1992) lacks any discussion of the relationship of segmentation to morphogenetic movements and cytoskeletal events. The latter have indeed been studied to some extent in *Drosophila* (Zalokar & Erk, 1976; Turner & Mahowald, 1976; Fullilove & Jacobson, 1971, 1978; Schejter & Wieschaus, 1993a,b; Sullivan & Theurkauf, 1995; Guichard, Bergeret & Griffin-Shea, 1997). However, even in recent textbooks (Gilbert, 1991a; Browder, Erickson & Jeffery, 1991) these are discussed as if they were separate topics. An exception is Bearer (1991a), who suggests that...

"Recent genetic manipulations have revealed that the cytoplasm of the early *Drosophila* embryo contains localized information that specifies the future embryonic axes. It is the restricted distribution or activity of particular gene products, either messenger RNA or protein, that is crucial for this specification. While some of the genes responsible for this information have been sequenced and the nature and distribution of their products examined, it is not known how this localization is established or maintained. The actin-based cytoskeleton is a likely candidate for the formation of a cytomatrix that would allow such distributions and yet no direct evidence has yet been found that implicates actin in positional cue localization. In this review I summarize what is known about actin filament behavior in *Drosophila* embryos and compare it to the distribution of positional cues. My purpose is to juxtapose these two bodies of information such that the relationship between them may be revealed" (Bearer, 1991a).

Lawrence (1992) admits quite freely that...

"Pattern formation in the early embryo takes advantage of the free diffusion of proteins between nuclei, but after gastrulation this cannot happen because cell membranes have been completed" (Lawrence, 1992).

If we accept this conclusion, then what do we make of the presence of segmentation homeobox genes in vertebrates (reviewed in: Browder, Erickson & Jeffery, 1991; Gehring, 1993; Krumlauf, 1994, and shown to correspond to late, postneurulation tissues (waves?) in Chisaka & Capecchi, 1991; Chisaka, Musci & Capecchi, 1992; Carpenter et al., 1993; McGinnis & Kuziora, 1994) and more primitive insects, such as the African desert locust, in which "segment patterning appears after cellularization" (Kelsh, Dawson & Akam, 1993; cf. Akam & Dawes, 1992)? Lawrence (1992) answers that...

"Transcription factors cannot be directly used as morphogens in multicellular systems because they cannot pass from cell to cell; some intermediary mechanism that might involve intercellular signals and receptors may be needed" (Lawrence, 1992).

This is the notion of a chemical inducer in modern dress. There is actually evidence that the membranes in *Drosophila* are not an absolute impediment to protein transport, so a mechanism of patterning based on diffusion may be operating (cf. Gurdon et al., 1994):

"The segment polarity gene wingless encodes a cysteine rich protein which is essential for pattern formation in *Drosophila*. Using polyclonal antibodies against the product of the wingless gene, we demonstrate that this protein is secreted in the embryo and that it is taken up by neighbouring cells [cf. Mathies, Kerridge & Scott, 1994]. The protein can be found two or three cell diameters away from the cells in which it is synthesized" (Gonzalez et al., 1991).

This approach may thus be worth following, including suggestions of extracellular diffusion (Bryant, 1988; Ferguson & Anderson, 1992a). In an upbeat but inconclusive review of alternative gradient models compatible with syncytial and cellular segmentation in insects, Tautz & Sommer (1995) include the possibility that...

"...transcription factors might cross cell walls to form diffusion-controlled gradients, even in short-germ insects. While it seems unlikely that this could occur via the normal pathway of protein secretion, as this would require the presence of both signal peptides in the proteins and fairly specific receptors in the neighbouring cell, it is interesting to note that *in vitro* studies of at least one transcription factor, the product of the homeotic gene *Antennapedia,* have shown that it can enter cells by a novel pathway (Le Roux et al., 1993)" (Tautz & Sommer, 1995).

However, if we are right that induction is, in the first instance, a mechanical phenomenon, then we should look afresh at *Drosophila* to see if its early development could involve differentiation waves instead of being driven by gradients. Davidson (1993) warns that...

"We are at present inundated with descriptions of the transient regional expression of various transcriptional regulatory factors during development.... However, it is apparent that for organisms that are cellular rather than syncytial at the time a transcription factor pattern appears, the relevant *cis-/trans*-interactions must in some crucial measure be subject to control through intercellular signaling processes. Unfortunately, on present evidence, the relations between signaling and regional transcription factor expression are rarely obvious or explicit.... Among the key control functions required are those that at the transcriptional level establish the signaling boundaries. In its subsequent development the territory of the progenitor field is typically subdivided, and new internal boundaries are formed" (Davidson, 1993).

We already know that the first differentiation wave to be discovered, at least thrice (*shibire* bars: Suzuki, 1974; 'transverse groove': Melamed & Trujillo-Cenóz, 1975; 'morphogenetic furrow': Ready, Hanson & Benzer, 1976; maybe even attributable as the 'eye sulcus' to: Sánchez y Sánchez, 1919, as described in Melamed & Trujillo-Cenóz, 1975) and subsequently ignored as part of the causal process for two decades, is the furrow preceding ommatidium development in *Drosophila* :

"The eye does not develop all at once; the posterior ommatidia are formed first and the anterior last: as they mature in succession a furrow, like the front of a wave, sweeps slowly across the eye imaginal disc" (Lawrence, 1992).

It is unfortunate that the eye imaginal disc wave's role, mechanics, and ultrastructure have not received much attention while a superb analysis of the gene products and their interactions has proceeded apace (reviewed in Lawrence, 1992; Yamamoto, 1996).

Some ultrastructural work has been done on eye imaginal disc (Waddington & Perry, 1960; Waddington, 1962a; Perry, 1968a,b; Wehman, 1969). Fristrom & Fristrom (1975, 1993), and Fristrom & Rickoll (1982), as discussed in Proposition 1, did uncover what I interpret as a cell state splitter, though no connection with the morphogenetic furrow was made (cf. wing discs: Reinhardt, Hodgkin & Bryant, 1977; Reinhardt & Bryant, 1981). Woods et al. (1996) illustrate paraxial microtubules in the *Drosophila* wing imaginal disc (their Figure 8c; cf. analogous paraxial microtubules in newt neural plate cells: Burnside, 1971, 1973a,b). Immunofluorescence staining shows actin at its highest concentration in the morphogenetic furrow and in forming ommatidia (Figure 2C in Tio & Moses, 1997).

We thus see that what should be a gnawing lack of a relationship between the morphogenetic furrow, which I take to be a differentiation wave, and eye differentiation, is at best expressed parenthetically:

"An indentation forms at the posterior margin of the [eye] imaginal disc, and this furrow begins to travel towards the anterior of the epithelium. As the furrow passes through a region of cells, those cells begin to differentiate in a specific order. The first cell to develop is the central (R8) photoreceptor. (It is not yet known how the furrow instructs certain cells to become R8 photoreceptors.)" (Gilbert, 1991a)

Here is the closest it seems that anyone has come to noting that cell wave mechanics could trigger this differentiation:

"Events in the [*Drosophila* eye imaginal disc] morphogenetic furrow lay the foundation for all subsequent eye development.... No causal link has been established between these cellular events and the onset of pattern formation in the furrow, but the coincidence is provocative. If apical membrane receptors are potentiated when concentrated in massed arrays, the approximately fifteen-fold reduction in apical surface area could facilitate cell communication by 'corralling' these receptors into high density [if the apical membrane actually decreases its surface area rather than just folds: Löfberg, 1974a; Bolker, 1993a]. A heightened level of communication might be critical for parceling the epithelium into periodic multicellular domains [ommatidia]" (Wolff & Ready, 1991a).

Even the latest models for *Drosophila* eye imaginal disc development do not invoke the morphogenetic furrow in their biochemical reaction schemes, in either a causative or consequential manner (Brown et al., 1995), as if it were an epiphenomenon, a 'Ghost in the Machine' (Koestler, 1967). It somehow tracks along with the travelling gradients of gene activation and gene products like an unexplained shadow. Phylogenetically, the eye wave probably dates over 500×10^6 years ago when insects and crustaceans had a common ancestor with ommatidia (Osorio & Bacon, 1994). It thus casts a rather long shadow.

Proposition 242: gradients of morphogens are irrelevant to differentiation.

"It could even be, that the gradient-concept is a dead end, distracting us from a search in other directions" (Albers, 1987).

The refusal to let go of gradients as causative agents in embryogenesis can be seen, for example, when, for the *Drosophila* eye imaginal disc, Ma et al. (1993) suggest that...

"Two mutually exclusive classes of explanation have been proposed for the forward progress of the [morphogenetic] furrow: furrow 'pushing' and furrow 'pulling' models. Their initial observations of the furrow led Ready, Hanson & Benzer (1976) to propose a pushing model:... the newest column of ommatidia acts as a template to position the next column.... This type of model was supported by grafting experiments done in other insects, such as the mosquito *Aedes aegypti* (White, 1961, 1963), the cockroach *Periplaneta americana* (Hyde, 1972), and the hemipteran *Oncopeltus fasciatus* (Shelton & Lawrence, 1974; Green & Lawrence, 1975)....

"The simplest form of pulling model proposes an anterior source of some inhibitory influence (or factor) that might decay as a monotonic gradient toward the posterior margin of the eye imaginal disc, without any posterior influence. The eye field contains cells that are competent to respond to the factor. A low threshold value for this anterior factor permits such competent cells to enter the furrow. The lowest value for the factor is at the posterior margin, so this is where the furrow begins. With time, the factor decays so that the furrow moves forward.... Some signal from the posterior margin must be responsible for furrow initiation. As the edges are contiguous extensions of the posterior margin, we propose that the furrow is initiated at the posterior margin and then continuously reinitiated at the edges, while the center progresses by a different mechanism (which requires *hh* [*hedgehog*])....

"...In the center of the eye, *hh* protein is induced behind the furrow, and it diffuses forward, perhaps for a considerable distance (30% of the eye field or about 20 μm [probably corresponding to initiation of contraction of cell state splitters]), where it binds a receptor (possibly the *ptc* [*patched*] protein) and induces the receiving cells to begin a process that will later result in their entry into the furrow" (Ma et al., 1993).

Here again, the furrow is incidental, an epiphenomenon, like a shadow; its cytoskeletal and mechanical properties are ignored and irrelevant (as again in Gerhart & Kirschner, 1997), and rather than a self-propagating wave in an active medium, another gradient (hypothetical) is invoked whose thresholds choreograph the morphogenetic furrow and everything else. But even this gradient requires continuous, coordinated assistance from an 'induction' phenomenon at the margins, which continuously launches the edge component of the furrow. Furthermore, the equal concentration contours of such a gradient would not seem to match the shape of the morphogenetic furrow. (Nor, I suspect, could one construct a gradient that could lead the axolotl ectoderm contraction wave through its focus to focus trajectory: Figure 17 in Appendix V: Gordon, Björklund & Nieuwkoop, 1994.) Again, as with epicycles for planetary motions (Sarton, 1959). we build gradients upon gradients, genes upon genes, and explanation never achieves closure.

The gradient 'pushing' model reviewed by Heberlein & Moses (1995) similarly does not explain why there is a morphological feature, namely the furrow, that supposedly gets pushed along. The induction of ectopic eyes (cf. Boterenbrood, 1958), in tissue that is clearly embryologically competent for such, shows how we need to distinguish the problem of launching of a differentiation wave from the mechanism of its propagation:

"The *Drosophila decapentaplegic* (*dpp*) gene, encoding a secreted protein of the transforming growth factor-β (TGF-β) superfamily, controls proliferation and patterning in diverse tissues, including the eye imaginal disc. Pattern formation in this tissue is initiated at the posterior edge and moves anteriorly as a wave; the front of this wave is called the morphogenetic furrow (MF). *Dpp* is required for proliferation and initiation of pattern formation at the posterior edge of the eye disc. It has also been suggested that *Dpp* is the principal mediator of *Hedgehog* function in driving progression of the MF across the disc. In this paper, ectopic *Dpp* expression is shown to be sufficient to induce a duplicated eye disc with normal shape, MF progression, neuronal cluster formation and direction of axon outgrowth. Induction of ectopic eye development occurs preferentially along the anterior margin of the eye disc" (Pignoni & Zipursky, 1997).

Consider another *Drosophila* gradient. Lawrence (1992) claims that...

"...the cells of the mesoderm invaginate because they have already differentiated from the surrounding cells, as indicated by expression of the *twist* gene" (Lawrence, 1992).

(Cf. Bate, 1993.) However, he gives no proof that this is the cause and effect sequence. His Figure 3.7 shows the *twist* gene and cell columnarization at the same time and place. Further on in the text he makes a less definitive statement:

"The *twist* gene product is expressed at late stage 5 ["in which the cephalic furrow is formed" in *Drosophila melanogaster*: Imaizumi, 1958a] in the ventral band of cells that will form the mesoderm. In *twist* ⁻ embryos the process of gastrulation begins but soon goes wrong. The *twist* protein sequence indicates a helix-loop-helix motif and the protein is localised to nuclei; it is presumably another transcription factor. Initially, the boundaries of *twist* expression are not sharp but, by gastrulation, the borders become more definite and it is the *twist*-expressing cells that roll in. The *twist* gene remains active in the mesodermal cells until they begin to differentiate and so may be required to specify mesoderm" (Lawrence, 1992).

But this suggests that *twist* is not needed to start gastrulation, merely to continue it, i.e., the mechanics of gastrulation precedes *twist* expression. Perhaps the same doubts could be raised about the order of cause and effect between genes and mechanics for other genes:

"In *fork head⁻*, the invaginating gut is missing.... The expression pattern is very precisely confined to just those cells that invaginate, which suggests that the expression of this gene (like other selector genes such as *engrailed* and *Ubx*) may **determine** the cell state, the capacity to invaginate and make gut" (Lawrence, 1992).

Or could it be vice versa? Again, the same potential problem arises with the parasegments of *Drosophila:*

"...The first anatomical signs of metamerisation in the insect embryo are little grooves that appear after the germ band has extended. These grooves are at the boundaries between parasegments" (Lawrence, 1992).

We must ask whether these grooves are consequences of, or causes of, the bands of purported morphogens? Perhaps in this case the truth lies in between, and the grooves are involved in the process of band sharpening? Perhaps they are the launching domains of differentiation waves.

It should be noted that the following result was later used by Spemann (1938) to argue against gradients as determiners of the axis of the embryo:

"On some of the first series of 'organizer' experiments it had been noted that the induced embryos did not necessarily show the same axis orientation as the host embryos. Some were oriented almost perpendicular to the host axis, and intermediate [angular] positions were observed also (Spemann, 1927a). Cases of this sort indicate that the orientation of the secondary complex cannot be merely the work of activated intrinsic patterns of the host, because such patterns most likely would correspond in their orientation to the basic patterns of the host. It is reasonable, therefore, to conclude that the orientation of the secondary embryos has been the work of orienting influences of the graft, at least in certain cases" (Weiss, 1935).

(But cf. Jargiello & Caplan, 1983, who revived the concept of physiological gradients for morphogenesis: Child, 1914, 1941; Needham, 1931b.)

Proposition 243: the earliest 'morphogen' gradients in the syncytial blastoderm stage of *Drosophila* are secondary phenomena.

"Model systems may share adaptations for rapid development that are not present in more slowly developing species" (Bolker, 1995).

Certainly, the very first gradient across the whole *Drosophila* early embryo has an air of reality:

"The identification of *bicoid* is important because it is the first morphogen to be discovered and it plays a crucial part in the patterning of the egg. Antibodies against the *bicoid* protein show it is absent from eggs when they are laid but soon appears, close to the [anteriorly localized] RNA which encodes it. By the time pole cells are being formed it is present in a concentration gradient that extends from 100% to about 30% Egg Length, far beyond the RNA source from which it presumably spreads by diffusion. The protein has a short half life, probably less than half an hour - otherwise the egg would just fill up with protein and the gradient would be obliterated" (Lawrence, 1992).

(Cf. Driever, 1993. See Bohrmann & Biber, 1994, for a step towards seeing how this gradient is formed.) On the other hand, invertebrate comparative embryology trivializes the role of this initial, maternal gradient:

"In *Schistocerca* [the African desert locust], a fully-formed embryo can be generated in an isolated posterior half egg. Maternal factors appear to define where the embryonic primordium will form, and may define its polarity, but appear to play little or no further role in patterning the embryo (Sander, 1976). Indeed, in these lower insects the germ-line derived nurse cells that synthesize the *bicoid* and *nanos* determinants of *Drosophila* are simply not present (Mahowald, 1972). In the ovaries of these insects, only the oocyte nucleus is available to provision the egg with maternal RNA.

"Even among the endopterygote insects, the requirement for maternal information can be eliminated.... For example, in [the parasitic wasp] *Ageniaspis* the egg is small and contains little yolk. It undergoes complete cleavage, to yield a loosely-packed ball of cells that resembles the mammalian morula. As cleavage continues, the morula falls apart into as many as 400 separate clusters of cells, each of which then organizes itself into a separate embryo (Martin, 1914, reviewed in Ivanova-Kasas, 1972). It is hard to believe that localized maternal information has much of a role in the generation of pattern in this embryo" (Akam & Dawes, 1992).

In fact, the polyembryony of one parasitic wasp, *Litomastix truncatellus*, can exceed 1000 embryos (Gillott, 1995; cf. Patterson, 1927; 1700 embryos are reported by Ivanova-Kasas, 1972), belying the fundamental importance of any maternal gradients that might have existed (cf. Cabrero, Pascual & Camacho, 1996). While Grbic, Nagy & Strand (1996) "favor the... hypothesis that [poly]embryonic primordia must have some intrinsic polarity" left over, in at least some cells, from the multiple early cleavages of the egg, Grbic et al. (1996) have thrown up their hands:

"The polyembryonic wasp *Copidosoma floridanum* produces up to 2000 individuals from a single egg. During the production of individual embryos the original anteroposterior axis of the egg is lost and axial patterning must subsequently be reestablished within each embryo. The mechanism by which this occurs is unknown.... In spite of the absence of a syncytium, the elements of the *Drosophila melanogaster* segmentation hierarchy are conserved.... It is believed that these regulatory interactions depend on diffusion of DNA-binding factors in the syncytial blastoderm. However, such a patterning mechanism is also thought to be incompatible with the cellularized environment in which the majority of other metazoans develop (Wolpert, 1994a).

"It has also been suggested... that patterning in primitive insects does not rely on diffusion within a syncytium (reviewed by Patel, 1994a; Tautz & Sommer, 1995).... We have shown that an injected dextran dye of 3×10^3 M_r does not diffuse between cells subsequent to the early cleavages of *C. floridanum*.... [Furthermore]... our results indicate that *C. floridanum* undergoes long germband development [like *Drosophila*]" (Grbic et al., 1996).

Even in *Drosophila* a case occurs which leads one to wonder about the importance of the *bicoid* gradient. The maternal effect *Drosophila funebris* mutant *spheroidal* (Crew & Auerbach, 1936), in which homozygous "males and females hatched from spheroidal eggs are wild type in appearance" (given as an example of a mutant affecting early development without affecting the adult: Kacser, 1957), might be worth investigating for distortion or presence of a *bicoid* gradient, if any. Why does it develop normally?

Schröder & Sander (1993) came to the conclusion that "The differences between *Calliphora* and the other blowflies [*Lucilia* and *Phormia*] in the *bcd* [*bicoid*] rescue assay are likely to indicate that the molecular patterning system, including some key gene(s), must have undergone considerable divergence even within the morphologically rather uniform Calliphoridae". But one might equally conclude from their data that the purported 'patterning system' is not the actual one.

In *Drosophila,* the work of Hülskamp et al. (1989)...

"...demonstrates that the main function of the posterior system is to control maternal *hb* [hunchback] activity, rather than to provide morphogen for the spatial organization of the abdomen.... It was reasonable to assume that the *nos* [nanos] gene product would act as a morphogen rather like *bcd* [bicoid] in the anterior region of the embryo (Nüsslein-Volhard, Frohnhöfer & Lehmann, 1987). In the absence of maternal *hb* activity, however, *nos* is dispensable. This observation rules out the possibility that the *nos* gene product acts as a morphogenetic molecule.... The establishment of the posterior segment pattern would thus be mechanistically very different from the spatial organization of the anterior region by the morphogenetic *bcd* protein gradient (Frohnhöfer & Nüsslein-Volhard, 1986; Driever & Nüsslein-Volhard, 1988a,b)" (Hülskamp et al., 1989).

All told, it has become obvious that *Drosophila* blastoderm development does not provide a model for much of anything else: the paradigm it presented is falling apart. *Drosophila* is likely to be a special, extreme case of a more general mechanism, that works for both cellular and syncytial organisms, perhaps differentiation waves. Perhaps a set of differentiation waves is what cellular and syncytial blastoderm blowflies (Schröder & Sander, 1993) have in common, and in common with *Drosophila.* Certainly, at least when cellularization sets in, the components of cell state splitters are present in or near the blastoderm cortex, namely microfilament rings (Warn, Bullard & Magrath, 1980; Warn & Magrath, 1983; Warn, Smith & Warn, 1985) or arrays of microfilament caps (Warn, Magrath & Webb, 1984; Warn & Robert-Nicoud, 1990) and microtubules (Warn & Warn, 1986; Warn, Flegg & Warn, 1987). Perhaps the following observations are actually of differentiation waves:

"The cortex of the early [*Drosophila*] embryo is subject to a periodic contraction during each round of nuclear division (Fullilove & Jacobson, 1971; Foe & Alberts, 1983). This process begins during the initial stages of mitosis and consists of a strong inward contraction of the yolk mass at both poles of the embryo, coupled to a poleward flow of periplasm. As mitosis nears completion, both yolk and periplasm motions reverse direction and soon reestablish the original positional conformation. The periodic nature of the contraction suggests a role for the mitotic spindles in their initiation. Consistent with this notion is the finding that colchicine treatment abruptly halts both nuclear divisions and the flow of yolk and periplasm. A more direct role for microfilaments in bringing about the yolk contractions is suggested by the effects of cytochalasin, which arrests the contractions without disturbing the division cycle (Foe & Alberts, 1983). The position and mechanistic properties of such an actin-based system are yet to be identified" (Schejter & Wieschaus, 1993).

These contraction waves would seem to coincide with the *Drosophila* mitotic waves described in Section 9.04, which travel at 10-20 µm/min. Thus we have the possibility that there are, indeed, differentiation waves in the *Drosophila* blastoderm, which may contribute to (control?) the segmentation at this stage. Furthermore, since they may be mostly cortical in nature, such waves would work in cellularized insect embryos too. The only caveat is that:

"Microfilament concentrations are also observed in the interior of the [*Drosophila*] egg (Warn, 1986; Hatanaka & Okada, 1991a), but the exact nature of these structures requires more detailed description" (Schejter & Wieschaus, 1993b).

Proposition 244: differentiation waves can proceed along single file rows of cells.

If we throw our taxonomic net a bit wider than in the last propositions, we find that, while...

"In both flies and vertebrates, *Hox* gene expression is strongly influenced by the position of a cell in the embryo (for reviews see Lawrence, 1992; Riddle et al., 1993). In contrast, in the leech, an interesting experiment suggests that positional signals do not initiate *Hox* gene expression (Nardelli-Haefliger, Bruce & Shankland, 1994). Here *Hox* genes are switched on one by one as new segments are added to the posterior. Segments are composed of parallel 'bandlets' of cells that bud at different rates from precursor cells at the posterior pole.... If the growth of one bandlet of cells is delayed temporarily by photoablating a cell, then that bandlet never comes into register with the other bandlets, and neither does its *Hox* gene expression pattern.... This rules out a requirement for a positional signal located within the region of *Hox* gene expression. Instead, timing is tightly regulated so that cells destined to end up in the same segment express the same *Hox* gene" (Kenyon, 1994).

In mosaic nematodes...

"...when early cell interactions are prevented, precursor cells generate perfect copies of lineages normally generated by other blastomeres. These lineage copies are generated in abnormal positions in the embryo, suggesting they do not require positional signals to develop (Mello, Draper & Priess, 1994; Hutter & Schnabel, 1994).... In summary, it appears that embryos can use either positional, temporal, or lineal strategies (or some combination of these) to set up the conserved pattern of *Hox* gene expression" (Kenyon, 1994).

A simpler explanation, not requiring diffusion of morphogens to set up gradients of positional information, nor precision in timing, nor the assumption of different mechanisms for different organisms with homologous genes and gene expression, may be that 'waves' of differentiation proceed along each bandlet or sublineage, which, in these mosaic animals, may consist of tissues of one cell each, strung out in lines.

The parallel between these bandlets and the 'clonal strings' of cells in the zebrafish central nervous system (CNS: Kimmel, Warga & Kane, 1994) may be worth pursuing in this light (cf. also ciliates in which "lateral connections between the surface stripes and structures [of basal bodies] are weaker - ...permit[ting]... slippage - than axial, tandem joinings of stripe components": Tartar, 1960, and note that these develop axially in waves: Iftode et al., 1989). Jaffe (1998) notes, from the work of Tio, Ma & Moses (1994), that the morphogenetic furrow in the *Drosophila* eye imaginal disc consists of "progress straight down the middle of the disc 'forward', i.e. from starting mid-posterior point to ending mid-anterior one. From successive central positions it then repeatedly spreads 'laterally' to the edges of the disc" (Lionel F. Jaffe, p.c., 1997), perhaps a double example of single file wave propagation. That differentiation waves can propagate along a single clonal string is also suggested by the observation in zebrafish...

"...that, in single clones making both CNS and mesoderm, the anterior part of the cell string makes CNS and the posterior part makes mesoderm. This posterior part is closer to the zone of involution [invagination] in the gastrula (Warga & Kimmel, 1990) and is captured by involution [invagination]. The string folds back upon itself: the anterior part stays in an outer ectodermal primordium and the original posterior part folds into an underlying mesodermal primordium" (Kimmel, Warga & Kane, 1994).

The 'clonal line wave' here should be homologous with one or more of the blastopore waves in urodele amphibians (Appendix V: Gordon, Björklund & Nieuwkoop, 1994; Appendix VI: Björklund & Gordon, 1994).

Stomata in monocotyledonous plants also form in strings (Boetsch, Chin & Croxdale, 1995), perhaps also via line differentiation waves that 'travel down files':

"Distal to (above) the proliferative zone and proximal to (below) the stomatal development zone is the region of stomatal patterning where no divisions occur. According to Charlton's [1988, 1990] theory, cells in this region receive the patterning signal and each cell in a stomatal string rather than a single founder cell responds to the signal.... Stomata develop synchronously.... A single group never spans two regions of stomatal development.... Stomata within each string in *Tradescantia* developed in concert with one another, consistent with these cells maintaining a degree of synchrony in the division cycle.... Since the length of the stomatal differentiation zone showed independence by cell file, the patterning signal, whatever that agent might be, may travel down files rather than across them. The appearance of stomata in different developmental stages opposite one another in adjacent files also indicates an independence of patterning by file that reflects the asynchrony of the cell cycle in cells of adjacent files" (Chin et al., 1995).

As with the dorsal lip of the blastopore, the wave of 'stomatal patterning' is stationary in the laboratory coordinate system, and thus is not perceived as a wave travelling through the tissue. The stomatal wave is much like the furrow in the eye imaginal disc, as hinted by Boetsch, Chin & Croxdale (1995), so it may be worthwhile overcoming the technical obstacles to time lapse observations that they had with the use of peels of dried cyanoacrylate glue, which killed the immature leaf cells. Direct, sped up time lapse observations would indicate whether or not there is a single file wave present with a mechanical component.

Proposition 245: gene expression boundaries are determined by the trajectories of differentiation waves.

Chasan & Anderson (1993) point out that an "...injected ligand or RNA would seem unlikely to diffuse into a gradient that mimics precisely the normal gradient". Even in normal development, Driever (1993) raises the question:

"What mechanism allows Bicoid to define sharp posterior borders of expression domains...? Further biochemical studies will be necessary to determine whether cooperative binding [cf. Wilson et al., 1995] of Bicoid to multiple binding sites within target gene promoters is responsible for the generation of sharp on/off decisions" (Driever, 1993).

Pankratz & Jäckle (1993) raise numerous quantitative questions about...

"...precisely where the borders of a given gene's expression domain lie along the longitudinal axis of the blastoderm embryo... [due to] the uncertainties concerning the detection methods used to localize the gene products in the embryo" (Pankratz & Jäckle, 1993).

The same problem occurs in vertebrates: what causes the 'coincidence' of homeobox gene expression domains with segmentation?:

"*Hox* genes have limits of expression that coincide with rhombomeric boundaries..." (Mark, Rijli & Chambon, 1995).

"If we consider... the morphogenesis of the vertebral column, it is... fair to speculate that the coupling between the anterior-posterior (AP) progression of somite formation and the sequential activation [temporal colinearity] of the Hox genes (in presomitic mesoderm) determines the combination of Hox genes expressed at a given level of the sclerotome and, consequently, the shape of the future vertebra. In this view, it is not because more posterior metameres are sequentially produced that posterior Hox genes become successively active, but instead, because subsequent Hox genes are turned on that the newly appearing structures can acquire distinct identities. Hence, any variation in the relative speeds of the two processes would lead to mis-identifications of structures" (Duboule, 1994b).

"In human, four clusters are designated HOX1, HOX2, HOX3, and HOX4 and are located in chromosomes 7, 17, 12, 2, respectively. Boncinelli et al. (1988) have speculated that having four sets might provide 'a finer-grained spatial control' than would be needed in organisms such as *Drosophila,* which have only one set and perhaps fewer requirements for setting up the body segments than have the higher vertebrates" (Lewis, 1992).

Moreover, the kinds of interactions between homeobox genes that occur in the syncytial stages of *Drosophila* are apparently not important in vertebrates (Dollé et al., 1993c), suggesting a cleaner, hierarchical structure (i.e., a differentiation tree) is operating in cellularized embryos.

This dilemma is recognized by Lawrence (1992):

"...How is it that the boundaries of *Ubx* expression at its limits coincide exactly with parasegment boundaries...? ....It is likely that the *Ubx* gene responds to the parasegment borders as laid down by the anterior margins of *ftz* and *eve*. This might be the only way to get the cells that express a selector gene to be exactly the same set that belong to a particular parasegment by lineage" (Lawrence, 1992).

Perhaps DNA binding strengths of the assorted DNA binding proteins, and gradient overlaps, can indeed generate focussed gradients that then lead to groove formation (Lawrence, 1992):

"The bands of gap gene expression have graded edges which may overlap with adjacent bands. The graded parts of the bands and the overlaps are used to activate other genes even more precisely in narrower stripes, which define the position and polarity of the parasegments.... The purpose of the molecular interaction is to read the gradient like a **vernier** scale - to transform a gradual change in concentration over a long distance into a sharper response of a gene that occurs at a particular concentration threshold.

"...The parasegment border has to become established as a long-lived feature with specific cellular properties. The activation of the *engrailed* gene has to mature from a provisional pattern into a definitive one: eventually, after several steps, those cells expressing *engrailed* become locked into an autoregulatory feedback loop (Morata & Lawrence, 1975; Heemskerk et al., 1991)" (Lawrence, 1992).

But it seems that we should leave open the possibility that the mechanics of groove formation could have a role in, or even be the cause of, this gradient sharpening process. Certainly grooves and furrows of all descriptions, some 'very transient' (Martinez Arias, 1993), correlate with many steps of *Drosophila* differentiation and morphogenesis. Consider the analogy made here with vertebrate organizers, i.e., with the process of induction, which we attribute to differentiation waves:

"...It is likely that the Wg/En group of cells represents an interactive and transient source of developmental instructions (Ingham & Martinez Arias, 1992...), which in the epidermis generates cell diversity (Lawrence et al., 1987; Ingham, 1988; Martinez Arias, 1989; Ingham & Martinez Arias, 1992; Vincent & O'Farrell, 1992), specifies the adult precursors (Cohen, 1990), and demarcates the segment boundary. Because the singling out of these cells or the future position of the groups of cells takes place always at defined positions relative to the parasegment boundary, it is possible that information arises as a function of distance from the boundary measured, for example, as the local concentration of a diffusible molecule. Therefore, the parasegment boundary... reflects the location of a group of cells... acting much as the organizers of vertebrate embryos, i.e., as transient sources of molecular information that instruct cells toward different developmental pathways; cells located at different distances, perhaps conditioned in different ways, respond differently to this information (Martinez Arias & Bate, 1993)" (Martinez Arias, 1993)

In fact, this analogy goes back at least another decade:

"A refinement of the cell state model (Meinhardt, 1983; Lawrence et al., 1987) postulates that the boundaries between states are the real organizing centers from which patterns are specified. Pattern elements are arrayed around segment and parasegment borders. Experimental evidence that these borders are organizing centers for the pattern, rather than elements of the pattern, is lacking. However, molecular analysis of segment polarity genes suggests that borders are established early in development, and their integrity is essential for pattern formation within the segment" (Hooper & Scott, 1992).

Indeed, the ventral furrow in *Drosophila* gastrulation seems to involve what may be both expansion and contraction waves:

"The ventral furrow region comprises two overlapping domains of cell shape changes: All of the cells of the ventral plate flatten on their apical sides (facing the vitelline membrane) and the central-most, or midventral, cells also subsequently constrict their apices (Leptin & Grunewald, 1990; Parks & Wieschaus, 1991; Sweeton et al., 1991).... It is not clear whether apical flattening has a function" (Costa, Sweeton & Wieschaus, 1993).

That functioning might be initiating a step of differentiation. These authors also review the posterior and anterior midgut invaginations, comparing their apically constricting cells (Anderson, Bokla & Nüsslein-Volhard, 1985; Sweeton et al., 1991) to bottle cells in the dorsal lip of the blastopore in amphibians (Holtfreter, 1943a,b; Perry & Waddington, 1966a; Landström,

Løvtrup-Rein & Løvtrup, 1976; Keller, 1978; Hardin & Keller, 1988; Keller, 1996), which we have identified as cells in contraction differentiation waves (Appendix V: Gordon, Björklund & Nieuwkoop, 1994; Appendix VI: Björklund & Gordon, 1994). Cephalic furrow formation in *Drosophila* also appears to involve a wave of 'change in cell shape' that is a contraction wave:

"Unlike the ventral furrow and posterior midgut invaginations, the infolding of the cephalic furrow is not preceded by a surface flattening (D. Sweeton, unpublished). Although the cells that invaginate eventually change from columnar to bottle-shaped, the change in cell shape is not associated with apical constriction or membrane ruffling on the apical surface.... The infolding cephalic (CF) cells... narrow their apices only slightly and do not elongate.... Analyses of time-lapse cinematography and sectioned material indicate that formation of the furrow proceeds as a sequence of cell shape changes and nuclear migrations (Turner & Mahowald, 1977; E. Wieschaus, unpublished).... For any individual cell, the process appears to be very rapid" (Costa, Sweeton & Wieschaus, 1993).

On the assumption that genes control gastrulation, and not vice versa, Costa, Sweeton & Wieschaus (1993) face a dilemma:

"Mutagenesis experiments have nearly saturated the *Drosophila* genome for maternal and zygotic effect lethal mutations in noncomplementing loci (Gans, Audit & Masson, 1975; Mohler, 1977; Nüsslein-Volhard & Wieschaus, 1980; Jürgens et al., 1984; Nüsslein-Volhard, Wieschaus & Kluding, 1984; Wieschaus, Nüsslein-Volhard & Jürgens, 1984; Schüpbach & Wieschaus, 1989). In addition, screening large genetic deficiencies that collectively cover almost the entire genome has failed to uncover any new zygotic genes whose null phenotype includes defective gastrulation events (Wieschaus, Nüsslein-Volhard & Jürgens, 1984; Merrill, Sweeton & Wieschaus, 1988; Wieschaus & Sweeton, 1988). Yet, there obviously must be many more genes that control aspects of the cell shape changes driving gastrulation events.... There must... be regulatory components that are synthesized zygotically in spatially and temporally localized patterns in the embryo. These gene products would activate changes in the structural proteins ubiquitously present in the embryo to control the cell movements with respect to position and developmental time.... Why has genetic analysis failed to identify many genes directly involved in the mechanics of cell shape changes and cell movements? ...Once more of these genes become identified, biochemical approaches as well as mutagenesis screens for suppressors and enhancers of mutant phenotypes may lead to the isolation of additional gene products that physically interact with... [cytoskeletal] components to mediate or regulate their activities" (Costa, Sweeton & Wieschaus, 1993).

Perhaps these genes don't exist because they are not needed by differentiation waves, which may use pleiotropic cytoskeletal components to access whatever master genes are ready and waiting.

In summary, there are many observations suggestive of multiple differentiation waves in *Drosophila,* which could perhaps explain the sharp boundaries to gene expression. Some of the observed grooves and furrows may represent the launching domains of these waves, or the waves themselves.

Proposition 246: segmentation involves a sequence of differentiation waves.

Now let us return to *Drosophila* segmentation. When segments themselves are being set up, they are set up sequentially:

"...At some unknown point in development the gradients of positional information are established and must eventually take over the patterning of the segments. Some support for this hypothesis comes from locust embryos, where *engrailed* is activated **progressively**, first in cells at the parasegmental borders and then backwards from them. Such an observation seems more in key with a gradient model than a digital and combinatorial one - but the argument is weak. It is important to find out exactly how *engrailed* is switched on in the right cells" (Lawrence, 1992).

Thus there may be a differentiation wave that coincides with, or perhaps even causes, segmentation. Sequential waves, one setting up the segmentation, and the next set occurring within each segment, forming the grooves, may solve the same dilemma of the coincidence of parasegment boundaries with boundaries of gene expression (Lawrence, 1992).

If both the physical segmentation of somites and the expression of Hox genes are caused by differentiation waves, then their speeds will be precisely and always identical, and 'mis-identification' (Duboule, 1994b) could not occur (nor has such been documented).

The clues may be in the morphology of the *Drosophila* embryo itself:

"While the ventral furrow is closing, several shallow transverse furrows (the cephalic, anterior dorsal, and posterior dorsal furrows) are formed by invagination. These transverse furrows provide convenient markers for fate-mapping experiments but have no apparent embryonic function and eventually disappear" (Browder, Erickson & Jeffery, 1991).

Indeed, these furrows appear to parallel the segmentation pattern that appears later, just like the ectoderm contraction wave in axolotls precedes the appearance of the neural plate, and their 'apparent embryonic function' may well be that they represent launching or propagation of differentiation waves. Gastrulation in *Drosophila* also involves a furrow (Sweeton et al., 1991), though apparently without a 'wave of contractions' (Bard, 1991a).

In general, it appears that no cognizance has been taken of the simultaneity of major changes in cytoskeletal proteins, such as actin, myosin and spectrin (Kiehart, 1991) and the patterning of the purportive morphogens, as if they were independent phenomena. In fact, the general presumption of such studies is that "...pattern formation... genes regulate the structure and function of the membrane skeleton" (Kiehart, 1990), without any mention of the reverse possibility. The situation may be changing, with ideas that "Interplay between... cytoskeletal complexes and signaling pathways may regulate morphogenesis" (Barth, Nathke & Nelson, 1997) and "proteins of the actin cytoskeleton, including unconventional myosins, play active roles in the segregation of differentiation factors and mRNAs" (Fan & Sokol, 1997).

'Waves' of gene expression have been recognized in the *Drosophila* development literature (Baumgartner et al., 1987). We will need to see if they occur in the wake of differentiation waves. One particular locus, *string,* may allow a separation of the cell division aspects of morphogenetic furrows from propagation of cell differentiation:

"Mitoses before interphase 14 run on maternal products, and occur in metasynchronous waves. Mitoses after interphase 14 require zygotic transcription, and occur asyncronously in an intricate, highly ordered spatio-temporal pattern. Mutations at the *string* (*stg*) locus cause cell-cycle arrest during this transition, in G2 of interphase 14, yet do not arrest other aspects of development. This phenotype suggests that *stg* is required specifically for initiating mitosis. We describe the cloning of *stg,* and show that its predicted amino acid sequence is homologous to that of *cdc25,* a regulator of mitotic initiation in the yeast *S. pombe.* In addition, we show that zygotic expression of *stg* mRNA occurs in a dynamic series of spatial patterns which anticipate the patterns of the zygotically driven cell divisions. Therefore we suggest that regulated expression of *stg* mRNA controls the timing and location of these embryonic cell divisions" (Edgar & O'Farrell, 1989).

Of course, I would reinterpret these observations by saying that each wave is a differentiation wave that 'controls the timing and location' of differentiation, and *string* is probably a pleiotropic gene involved in the cell state splitter or the nuclear state splitter.

If differentiation waves have a relationship to segmentation, we must bear in mind that segmentation itself appears to involve two consecutive steps:

"The body of arthropods is divided into regions (tagmata), such as the head, thorax and abdomen of insects and the prosoma and opisthosoma of spiders. The number of tagmata is always very small and quite independent of the number of body segments.... Articulating the body into regions means to fix a small number of strong markings along the main axis of the body. These markings act as 'hot spots' (Minelli & Schram, 1994) for the subsequent expression of whole alternative networks of structural genes" (Minelli, 1996a).

This would suggest that tagmata are formed by differentiation waves that divide tissues into pairs of tissues, within which further nested waves complete segmentation.

Proposition 247: the genes involved in the mechanism of continuing differentiation, particularly those of the cell state splitter and the differentiation pathway, are highly conserved and pleiotropic.

There are many cases of pleiotropic genes (Davidson & Hill, 1991) that fit the idea that they could be components of the differentiation pathway(s):

"The *msh*-like genes are an ancient family, represented in animals ranging from coelenterates to mammals.... The mouse genes, *Msx1* and *Msx2* (previously *Hox7* and *Hox8*), and their homologues in other species... are expressed during embryogenesis and organogenesis.... During early development, both genes are transcribed in the mesoderm and ectoderm of the primitive streak. Later, expression becomes restricted to the lateral surface ectoderm, lateral mesoderm and the neuroepithelium that will form the dorsal part of the neural tube, including the region that gives rise to migratory neural crest cells. From these early domains develop more complex, focussed patterns of expression in diverse organs. These temporal and spatial patterns of expression are highly conserved in mammals (Hill et al., 1989; Robert et al., 1989) and birds (Takahashi & Le Douarin, 1990) and, at least, [are] broadly similar in amphibians (Su et al., 1991).... *Msx* expression can be activated by tissue interactions that are otherwise known to induce differentiation, for example, in the limb bud (Davidson et al., 1991; Robert et al., 1991), facial processes (Takahashi, Bontoux & Le Douarin, 1991; Brown et al., 1993), tooth buds (Jowett et al., 1993) and epithelium of the perinatal Mullerian duct (Pavlova et al., 1994).... *Msx1* and *Msx2* expression precedes apoptosis at several locations, including the limb bud and the hindbrain. A tight correlation has been established in the chick hindbrain through an elegant series of experimental manipulations, which permit, or prevent, the apoptosis that occurs specifically in the neural crest region of rhombomeres 3 and 5 during normal development (Graham et al., 1994; Graham, Heyman & Lumsden, 1993).

"...It is paradoxical that neither of the *Msx* mutations reported, so far, displays phenotypic effects other than on the development of bones and teeth. If *Msx1* and *Msx2* do have wider roles, why is this not revealed in the phenotypes of existing mutants? Clearly, this could be a result of redundancy, either 'functional' (the function of the mutant gene is substituted by another process or by a related gene), or 'non-functional' (the gene products have no function in some regions where the gene is transcribed: Erickson, 1993). Alternatively, the phenotypes reported thus far might not represent the null condition - more subtle abnormalities might remain to be discovered or the mutant proteins might retain some functional capabilities....

"The next generation of problems concerns the extent to which entire molecular pathways that underlie supracellular organization are subject to change or are static modules, 'developmental units', in evolution" (Davidson, 1995).

Refinement of the observations of *Msx* gene expression, correlating it with differentiation waves, might go a long way towards solving its role in differentiation. The gene *p53* is yet another candidate for a pleiotropic gene (Rotter et al., 1994; cf. Fukasawa et al., 1996).

How will we recognize pleiotropic genes? Well, conservation has two consequences: between species, and from one stage of differentiation to another within a species. The latter can take two forms, which are not mutually exclusive: reuse of the same genes from one step of differentiation to another, and duplication of genes so copies are available for other steps of differentiation. Reuse implies pleiotropy, as in the *lit-1* gene involved in most asymmetric divisions (cell state splittings) in mosaic nematodes (Kaletta, Schnabel & Schnabel, 1997). We can see all of these possibilities operating in the...

"...fibroblast growth factor (FGF) family of ligands.... On the basis of the wide spectrum of tyrosine kinases representing the receptor [R] and nonreceptor class that were shown to be conserved in the genome of *Drosophila melanogaster* (for review, see Shilo, 1987; Hoffmann, 1990), we decided to search for an FGF-R homolog as well. The structure of over 10 *Drosophila* tyrosine kinases reveals that the major gene duplications in this large and diverse family have occurred prior to the arthropod - chordate split. Each *Drosophila* homolog is clearly most similar in sequence and overall organization to its vertebrate counterpart. In vertebrates, additional duplications have occurred, generating small subfamilies. In contrast, each of the members in *Drosophila* appears to be represented by a single gene, stemming perhaps from the lower complexity of this organism, requiring a lesser degree of microdiversity" (Glazer & Shilo, 1991).

Indeed, I would interpret the 'lower complexity' and 'microdiversity' as meaning that *Drosophila* has a smaller differentiation tree than vertebrates. While the roots of the trees are similar, the additional terminal branches of vertebrate differentiation trees contain 'additional duplications'. The functions of these molecules suggest they might be parts of the differentiation pathway:

"The identification of tyrosine kinases in *Drosophila* opened the way to the elucidation of their biological roles. Isolation of mutations in the genes encoding several homologs provides the first insights into their function (Hafen et al., 1987; Gertler et al., 1989; Price, Clifford & Schüpbach, 1989; Schejter & Shilo, 1989; Sprenger, Stevens & Nüsslein-Volhard, 1989; for review, see Hoffmann, 1990; Pawson & Bernstein, 1990). These tyrosine kinases appear to control a diverse set of processes involving communication between cells, such as determination of photoreceptor number and identity in the compound eye, determination of dorsoventral and terminal polarity of the embryo, and formation of the central nervous system. Some tyrosine kinases are expressed at a wide range of developmental phases and were shown to have pleiotropic roles..." (Glazer & Shilo, 1991).

We shall see other examples of pleiotropic genes as I develop this line of thought.

Proposition 248: some mutations can be classified as to whether they interfere with the propagation of a differentiation wave or with the differentiation pathway it sets off.

A mutant of *Drosophila* isolated earlier has pleiotropic properties, and a distinct link to differentiation waves:

"...*shibire* (meaning paralyzed in Japanese)... was originally isolated as a sex-linked, temperature sensitive paralytic lesion. Adult male and female flies carrying this mutation have normal mobility at 22°C. If these adults are shifted to 29°C, they immediately fall to the bottom of the culture vessel totally paralyzed. If the flies are subsequently returned to 22°C they begin to move in a few minutes and soon appear normal in all respects....

"However, this is not the only defect observed in *shibire* mutants. In order to determine if there was any paralysis of larvae and possible developmental defects, Poodry, Hall & Suzuki (1973) performed temperature shifts on developing embryos and larvae. They found that a shift up to 29°C during any stage of development resulted in paralysis and death. By performing pulse experiments, it was shown that a shift of > 18 hours to 29°C was sufficient to kill developing larvae and embryos. Heat pulses for shorter periods (two, four, and six hours) revealed unexpected developmental defects. Six-hour heat pulses revealed six critical periods in development when the *shibire* [+] gene or its product is needed, or death occurs. One period of high sensitivity occurs at gastrulation, when a two-hour heat pulse is sufficient to kill the animal....

"The heat pulse experiments also revealed temperature-sensitive periods for several visible defects in the adult fly. The most striking of these was the production of a vertical 'scar' on the eye facets caused by a disruption of these structures. The scar is produced by a heat pulse of three to six hours administered during a period from about 48 hours before the pupal stage until just after the pupa is formed. By a judicious spacing of these pulses, it is possible to create a fly with a double scar on the eye and demonstrate that the position of the scar moves across the eye in a posterior-to-anterior direction.... Interestingly, in this period direction of movement and position in the eye corresponds to a wave of cell divisions that moves across the developing eye in a manner analogous to the scar.

"The formation of bristles and hairs on the thorax and head of the fly is also affected by heat pulses, but the sensitive period is slightly later than that for the eye. Early heat pulses produce duplication of bristles, whereas later pulses result in the deletion of the same bristle types. As with the eye scar, these effects show a posterior-to-anterior movement with progressively later pulses.

"The tentative conclusion from these studies is that the *shibire* [+] gene product is a membrane component that is necessary for some types of cell-to-cell interaction or communication.... We have a case of direct pleiotropy and the demonstration that a single gene can affect several seemingly disparate developmental events" (Raff & Kaufman, 1983).

It would appear that Poodry, Hall & Suzuki (1973) (cf. Suzuki, 1974) made the first time lapse images of a differentiation wave, using the eye itself as the recording medium (cf. eye 'scar' formation in Bishop 3d & Corces, 1988). (The bristle and other pleiotropic effects may represent additional differentiation waves, in which the *shibire* gene is involved.) The history of this discovery, which I take as the first discovery of a differentiation wave, and its links with earlier work, has been recorded by Suzuki (1974):

"Some of the heat pulse experiments yielded surviving flies at certain [developmental] times and Clif Poodry decided to inspect them phenotypically. He immediately noted a wide range of phenotypic defects in the eyes, wings, legs and bristles. The defects often showed an anterior-posterior polarity reflecting the time of the heat pulse.

"The most striking defect noted by Poodry and [Linda] Hall was a third instar heat pulse-induced vertical strip in the eye in which facets are replaced by a 'scar'.... A heat pulse administered early in the TSP [temperature sensitive period] for this phenomenon causes a scar in the posterior rim of the eye, whereas successively later pulses result in a scar moving towards the front rim....

"Anne Junker asked whether the rate of movement of the scar can be measured by applying two 4-hour pulses separated by varying intervals at 22°C. Among the survivors, many exhibited two eye scars separated by the number of facet rows reflecting the 22°C interval between the pulses. She showed that the scarring moved at a constant rate of 1 facet row every 4 hours. This movement was reminiscent of the movement of a polarized vertical facet disruption reported for the *ts* [temperature sensitive] mutant, N^{60g11} (Foster & Suzuki, 1970). In fact, a *ts* allele of the *Notch* locus, which is known to cause embryonic death by nerve hypertrophy (Poulson, 1940), exhibits many of the defects found with shi^{tsl} (Schellenbarger, 1971). Interestingly, it has been reported that a vertical wave of cell division in the normal organism proceeds across the eye disc at the same time as the TSP for the eye scar.

"In addition to the above eye defects, it was found that heat-pulsed pupae yield adults which exhibit leg defects (such as improper elongation) and clipped wing edges. However, the most striking defect is an anterior-posterior wave of disruption in bristle development. Initially, a wave of duplication of large bristles (macrochaetae) passes to the posterior; this is followed by duplication of the microchaetae, proceeding in the same manner. The TSP for duplication is followed by a wave of bristle loss. The bristles of the eye exhibit this same multiplication, then loss" (Suzuki, 1974).

The *Drosophila* eye wave was apparently independently rediscovered a few years later, this time recorded in cross section as a moving 'morphogenetic furrow', by Ready, Hanson & Benzer (1976) (cf. Wolff & Ready, 1991a, 1993). It was also found by Melamed & Trujillo-Cenóz (1975) in Muscoid flies. As the *shibire* locus has been linked to membrane recycling (Masur, Kim & Wu, 1990; Poodry, 1990; Sapp, Christianson & Stark, 1991), this process may have a role in the differentiation pathway. Perhaps most important, *shibire* has also been directly linked to the cytoskeleton, and thus, perhaps, with the cell state splitter:

"Dynamin is a GTP-driven mechanochemical enzyme related to mammalian mx-proteins and to the yeast *vps* 1 gene product. Because the *shibire* gene product and dynamin have extensive similarity, we propose that they are cognate homologues. Dynamin causes microtubules to slide along each other in vitro and in extracts it is associated with a distinct, but so far uncharacterized, membrane fraction. In light of the *shibire* phenotype, we suggest that these proteins provide the motor for vesicular transport during endocytosis" (van der Bliek & Meyerowitz, 1991).

The link might be that *shibire* has a role in maintaining the apical end of the cell, the part containing the cell state splitter (Mays, Beck & Nelson, 1994). Further work on the *shibire* locus includes: Swanson & Poodry (1981); Kim & Wu (1987); Hummon & Costello (1987); Kessell, Holst & Roth (1989); Masur, Kim & Wu (1990); Poodry (1990); Chen et al. (1991, 1992); Sapp,

Christianson & Stark (1991); van der Bliek & Meyerowitz (1991); Shpetner & Vallee (1992); Damke et al. (1994); Hinshaw & Schmid (1995); Clark et al. (1997); Urrutia et al. (1997). The pleiotropic effects of *shibire* extend to the cellularization of the blastoderm stage in *Drosophila* (Dornan, Jackson & Gay, 1997), which adds to the hints that that stage is fundamentally similar to later cellular stages.

Earlier work on the assorted temperature dependencies of ommatidia (facet) number in the *bar, ultrabar* and *infrabar* mutants of *Drosophila,* and their crosses (Driver, 1931a,b; Luce, 1931; Schmalhausen, 1949), may reflect changes in speed and/or trajectory of this differentiation wave versus temperature. Alternatively, the effect may be primarily on size of the eye imaginal disc (Steinberg, 1941). This effect could be due either to limited cell proliferation or perhaps a limited trajectory of an earlier differentiation wave setting off "that portion of the cephalic complex that will eventually give rise to the eye disk" (Steinberg, 1943).

Another gene involved in the differentiation wave traversing the eye imaginal disc is *hedgehog:*

"In *Drosophila,...* the *wingless* (*wg*) and *hedgehog* (*hh*) genes are expressed in cells anterior and posterior to the parasegment boundary; proximity of the two cell types expressing these genes is essential for maintenance of the parasegment boundary and, ultimately, for an appropriate pattern of cellular differentiation (Hooper & Scott, 1992). The proteins encoded by *hh* and *wg* are secreted and appear to act as messengers carrying information across the boundary from one cell type to the other (Gonzalez et al., 1991; Lee et al., 1992; Tabata & Kornberg, 1994; Taylor et al., 1993; van den Heuvel et al., 1989).... The observed change in intracellular localization suggests that a primary effect of *wg* signaling is to alter adhesion properties (Peifer et al., 1993)" (Chiang & Beachy, 1994). [Cf. Barth & Wilson, 1995; Basler & Struhl, 1994; Roelink et al., 1995.]

"We identify here a gene specifically required for furrow progression, *hedgehog* (*hh*). We show that *hh* expression posterior to the morphogenetic furrow is continuously required for its progression. We propose that forward diffusion of *hh* protein induces anterior cells to enter the furrow" (Ma et al., 1993).

"We have investigated the mechanism by which this front advances, and our results suggest that developing retinal cells drive the progression of morphogenesis utilizing the products of the *hedgehog* (*hh*) and *decapentaplegic* (*dpp*) genes.... We show that *hh*, synthesized by differentiating cells, induces the expression of *dpp*, which appears to be a primary mediator of furrow movement" (Heberlein, Wolff & Rubin, 1993).

In this case we are dealing with genes involved in the propagation of the wave itself, rather than the signalling from the cell state splitter to the nucleus. Another ancient (Martinez, Givel & Wahli, 1991) pleiotropic gene that, for example, occurs in all imaginal discs with "expression... restricted to the beginning of their differentiation" (Pauli et al., 1990) is the heat shock protein hsp27 (a chaperone: Jakob et al., 1993) that is involved with regulation of microfilament contraction (Bitar et al., 1991; Yamada et al., 1995) and its 'stabilization' (Lavoie et al., 1993a,b, 1995). With enough such genes (cf. Riedl, 1978), and their interrelationships, we might be able to flesh out the differentiation pathway. It is curious that many genes involved in the eye imaginal disc are located on the same chromosome (Baker et al., 1992), perhaps because they are all part of the same differentiation cascade (edge of differentiation tree). Of course, not all genes associated with the eye imaginal disc morphogenetic furrow are pleiotropic (Hinton, 1988; Awasaki et al., 1994; Yoshida et al., 1994).

As might be expected, if the mechanism of propagation is reasonably universal, both between tissues, and between organisms, these genes should be found elsewhere than in the *Drosophila* eye imaginal disc:

"We have identified three members of a mouse gene family related to the *Drosophila* segment polarity gene, *hedgehog* (*hh*). Like *hh*, they encode putative secreted proteins and are thus implicated in cell-cell interactions. One of these, *Sonic hh* (*Shh*), is expressed in the notochord, the floor plate, and the zone of polarizing activity, signaling centers that are thought to mediate central nervous system (CNS) and limb polarity" (Echelard et al., 1993).

"The results of Hiromi et al. (1993) reveal that ectopically expressed *seven-up* is capable of inducing neural development in addition to directing photoreceptor identity" (Warren, 1993).

Cf. Lee et al. (1992). Differentiation wave and differentiation pathway genes may be pleiotropic (in gene modules) or in some cases may be duplicated so there are distinct copies in some or all differentiation cascades. These possibilities will be sorted out as we come to understand the molecular bases for differentiation waves.

Proposition 249: in the formation of spacing patterns, a differentiation wave can consist of both contracting and expanding cells.

Natalie K. Björklund (p.c., 1994) has interpreted the *shibire* observations in terms of the nuclear state splitter (Figures 51-52):

"1) Dynamin has ATPase activity (Shpetner & Vallee, 1992). Conversion of ATP to cAMP by such an ATPase (assuming that is the ATPase activity of dynamin) would result in high levels of cAMP in the cell. High cAMP levels would inhibit the contraction of microfilament rings in a CSS (cell state splitter) thereby preventing a contraction signal from the CSS to the nucleus. This would also serve as a chemical reinforcement to any mechanical bias towards microtubule polymerisation. In Otte et al. (1989), it was shown that a calcium dependent protein kinase (PKC) translocation was followed by a rise in cAMP levels during neural induction in *Xenopus*. But if cAMP was artificially raised without PKC translocation, no induction occurred. In Björklund & Gordon (1993a,b) [Appendix IV] we suggested that a slowly rising cAMP level following a prolonged PKC translocation [cf. Blobe et al., 1996] was the signal that ended microfilament ring contraction after the nucleus had received the contraction signal. In primary neural induction, where the force due to the microfilament ring dominates over that due to the apical microtubule, annular mat, we suggested that a calcium flux would induce a microfilament ring contraction while simultaneously inhibiting microtubule polymerisation.

"2) In Björklund & Gordon (1993a,b) [Appendix IV] we suggested that the signaling to the nucleus from microtubule polymerisation is likely done through the GTP pathway but couldn't speculate further due to lack of general information on this pathway during development. The fact that dynamin binds GTP (Chen et al., 1991), suggests that this original speculation may well be correct and gives a possible route for further research into the nature of the signal from an expansion wave to the nucleus.

"3) If dynamin in the *shibire* mutant is defective to the degree that it allows microtubule elongation (and therefore the expansion wave), but does not correctly signal the expansion wave activity to the nucleus, this may explain why *shibire* appears to be a mutant in which the wave travels through the tissue, but the nucleus fails to respond with an appropriate round of new gene expression. The result would be no differentiation in regions traversed by expansion waves while the temperature is raised (Poodry, Hall & Suzuki, 1973).

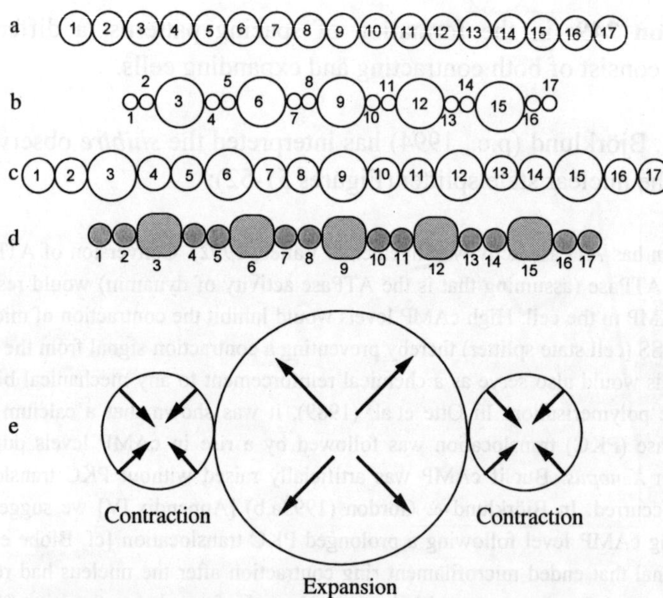

Figure 51. The *Drosophila* morphogenetic furrow modelled as a compound wave in which expansions and contractions alternate along the furrow. a) Prewave: all cells are the same. b) During wave: the net mechanical forces on a single cell result in expansion of some cells and contraction of others. c) Immediately post wave: contracted cells return to normal while expanded cells remain expanded. d) After differentiation: regular spacing pattern with two cell types has formed as a result of a single wave. e) The net effect of the forces on a single cell can result in either a contraction or an expansion of that cell while it and its neighbors are participating in a single differentiation wave. Cell divisions are not taken into account here.

a

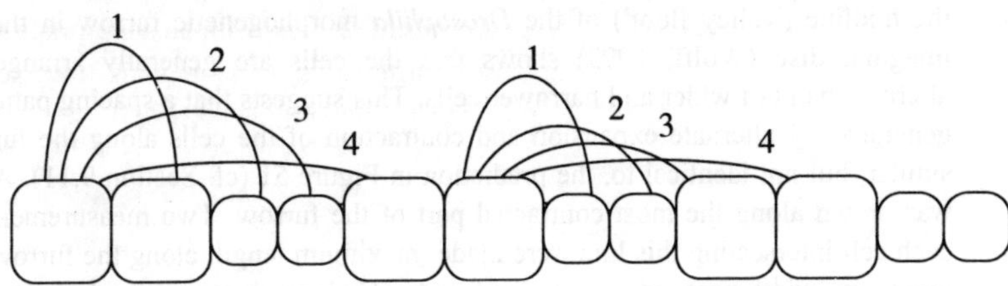

b

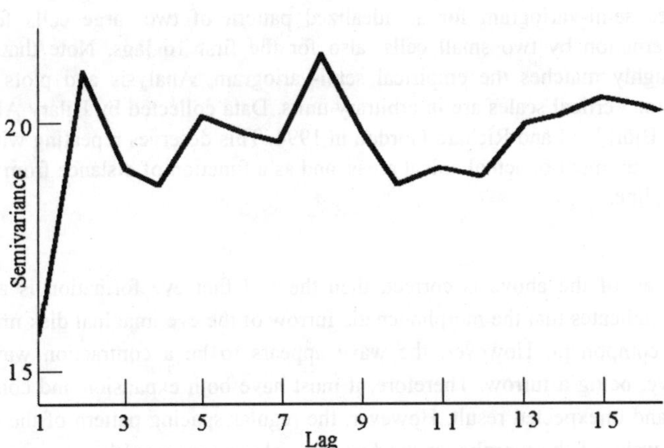

c

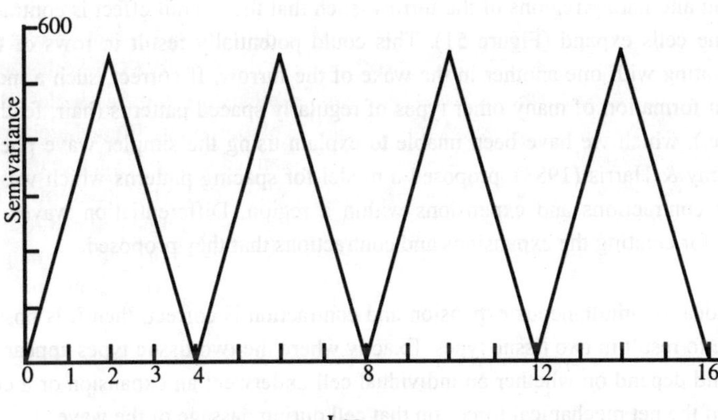

Figure 52. a) Variogram analysis (Woodcock, Strahler & Jupp, 1988a,b) of the cells along the midline ('valley floor') of the *Drosophila* morphogenetic furrow in the eye imaginal disc (Wolff, 1993) shows that the cells are generally arranged in alternate pairs of wider and narrower cells. This suggests that a spacing pattern is generated by alternate expansion and contraction of the cells along the furrow, similar, but not identical to, the prediction in Figure 51 (cf. Section 9.11). A line was drawn along the most contracted part of the furrow. Two measurements of each cell intersecting this line were made: maximum length along the furrow and maximum width perpendicular to the furrow. Their product is a rough measure of the apical 'area' of the cell. The 'lags' are taken in number of cells, rather than physical distance. The first four lags are shown here, counted off from two different cells. b) The empirical area semi-variogram for the first 16 lags. c) The area semi-variogram for an idealized pattern of two large cells followed in alternation by two small cells, also for the first 16 lags. Note that the shape roughly matches the empirical semi-variogram. Analysis and plots by Hilary Alto. Vertical scales are in arbitrary units. Data collected by Hilary Alto, Natalie K. Björklund and Richard Gordon in 1994. This deserves repeating with accurate measurement of actual apical areas, and as a function of distance from the furrow midline.

"4) Assuming all of the above is correct, then the fact that eye formation is affected by *shibire* clearly indicates that the morphogenetic furrow of the eye imaginal disk must include an expansion component. However, the wave appears to be a contraction wave, not an expansion wave, being a furrow. Therefore, it must have both expansion and contraction, a contradictory and unexpected result. However, the regular spacing pattern of the ommatidia formed in the wake of the morphogenetic furrow is also unexpected by our one wave, one tissue model. Perhaps this contradiction is explained by a combination of contraction and expansion in alternating regions of the furrow, such that the overall effect is contraction even though some cells expand (Figure 51). This could potentially result in rows of two tissue types alternating with one another in the wake of the furrow. If correct, such a model could also explain formation of many other types of regularly spaced patterns (hair, feathers, plant stomata, etc.), which we have been unable to explain using the simpler wave phenomenon. Oster, Murray & Harris (1983), proposed a model for spacing patterns which was based on the idea of contractions and expansions within a region. Differentiation waves provide a mechanism for creating the expansions and contractions that they proposed.

"5) If this idea of simultaneous expansion and contraction is correct, then it is possible for a single wave to result in two tissue types. Exactly where the two tissue types appear in a wave furrow would depend on whether an individual cell underwent an expansion or a contraction as a result of the net mechanical forces on that cell during passage of the wave.

"6) Why would some waves result in one cell type (homoiogenetic waves) and others (heterogenetic waves) result in two cell types? This would have to be due to differences in the mechanical and/or the chemical signalling properties of the entire cell sheet through which a wave passes. This may be the mechanical result of differing numbers or types of cell 'rivets' (such as desmosomes). A chemically based explanation could be the presence or lack of gap junctions for the passage of chemical signalling molecules (such as IP_3 or calcium ions) which would act to propagate a homoiogenetic wave signal. Cells that lack a purely chemical means of passing a wave signal must respond solely to the mechanical action of the cells around them. These types of cells may be more likely to produce spacing patterns. Some combination of differences in both chemical wave propagation systems and mechanical properties within a cell sheet is also likely to have evolved over evolutionary time. Such combinations could explain the many variations in spacing patterns that we know exist. Comparisons of the differences in cell sheets that are known to produce either a single cell type or a spacing pattern could yield important clues about the nature of wave signal transfer between individual cells within a single sheet."

As we shall see, our own evidence (Figure 52) and that of other does suggest that the morphogenetic furrow is a compound wave, with both contraction and expansion components.

Proposition 250: many spacing patterns could be due to contraction/expansion differentiation waves.

Compound differentiation waves, containing both contraction and expansion components, provide an alternative model to traction and reaction-diffusion models for spacing patterns (Nagorcka, Manoranjan & Murray, 1987; Nagorcka, 1986a,b, 1989; Nagorcka & Mooney, 1992). Spacing patterns are also seen between mammalian hairs (look at your arm; cf. whiskers: Scarisbrick & Jones, 1993), feathers (Brush, 1996), scales (Nardi & Magee-Adams, 1986) and insect bristles (Wigglesworth, 1940, 1966). (In statistics, spacing patterns are called 'over-dispersed' compared to Poisson random distributions: Jones, 1984). In the case of stomata in plants (well spaced pairs of guard cells), whose development involves microtubules (Galatis & Mitrakos, 1979), and thus, perhaps, mechanical effects, it has been possible to discriminate against diffusion based models:

"...There are two types of cells: guard cells forming stomata and other epidermal cells. The distances between the stomata vary greatly, but they appear to be more uniform than would be expected if the distribution were random. The leaf is therefore a relatively simple example of orderly, or patterned differentiation. The type of pattern found in the epidermis was called a 'spacing pattern' by R. Gordon (Wolpert, 1971a) [a phrase probably original with Bonner & Hoffman, 1963]....

"The meristematic cells which form the epidermis go through a stage in which they can take one of two alternative developmental pathways: one leading to the formation of epidermal cells and the other leading to the formation of both epidermal cells and stomata.... The relative position of a stoma and the epidermal cells... produces stomata which are separated by one or more epidermal cells, the exact number depending on the species. This orderly development is the main basis for the formation of spacing patterns of stomata....

"...We may consider the possibility that a stoma has a direct effect on the differentiation events in its vicinity. This is the idea of 'inhibition', suggested by Bünning & Sagromsky (1948) and Bünning (1948, 1965), which is commonly accepted as obviously correct.... A phenomenon that would appear to contradict the idea of inhibitory interactions between stomata is the formation of groups or clusters of stomata in many plants (Weber, 1949), such as the genus *Begonia*. Within these groups, however, the stomata are spaced and do not touch one another" (Sachs, 1978a; cf. Sachs, 1991a).

"Stomata are spaced differently in different species, but always such that no two stomata are touching.... One class of model proposes that lateral inhibitory fields [cf. Wigglesworth, 1966; Baker, Mlodzik & Rubin, 1990] are set up between developing stomata, and thus nearby cells are inhibited from differentiating as stomata. Another class of model proposes that the formation of stomata is lineage dependent, and several observations support the idea that lineage does play a role in establishing the pattern of the stomata in the epidermis. Observations of cell divisions in young leaf primordia indicate that cells divide in a stereotypic pattern, beginning with an unequal cell division [a single cell contraction/expansion 'wave'?] that establishes a smaller parental cell, which in turn divides to give the paired cells that form the stoma (Sachs, 1984b). Evidence supporting the critical role of cell lineage in this terminal differentiation pattern has come from mosaic plants, which have epidermal sectors with different stomatal spacings. Lateral inhibition models [cf. Usui & Kimura, 1993] are based on diffusion mechanisms to establish pattern, and thus would predict a distortion of stomatal spacing at the borders of sectors. In contrast, even at the sector boundaries in these mosaic plants, stomatal spacing remains characteristic of its epidermal type (Szymkowiak, 1990)" (Irish, 1991a,b).

Stomata are sometimes oriented with respect to one another (Bünning, 1965, his Figure 11; Sachs, 1984b, his Figure 16), suggesting the possibility that this alignment reflects traversing of a differentiation wave, though, as I discussed for monocots (Proposition 244), such waves appear to move along single file lines of cells (Chin et al., 1995). Some aspects of the pattern not accounted for by the cell lineage theory, such as "the collective stomatal pattern along the cell files, at the one-dimensional level of patterning" (Croxdale et al., 1992) might be attributable to a differentiation wave. On the other hand, the suppression of some immature stomata (Kagan & Sachs, 1991; Sachs, Novoplansky & Kagan, 1993) has yet to be accounted for. This suppression, leading to 'arrested initials' that differentiate much like normal epidermal cells rather than as stomata, occurs along straight line files of stomata, but not across files, at least in monocotyledenous plants, and is a function of distance (Boetsch, Chin & Croxdale, 1995). This implies an anisotropic mechanism, supporting, for example, the idea of an alignment of microtubules from cell to cell along files as possibly being causal, or diffusion channels that only occur between cells in files, or both. The "maturation of stomata in patches" (Sachs, 1978a) suggests that stomatal differentiation waves, if present, do not progress in an orderly fashion across whole leaves. However, time lapse studies of stomata development have only been carried out for small regions of leaves (Kagan, Novoplansky & Sachs, 1992; Sachs & Novoplansky, 1993), leaving the open question of long range, propagating waves of stomatal development.

The spacing pattern of trichomes on *Arabidopsis* leaves also occurs as a propagating wave, whose launching requires a minimum leaf size of 100 μm (Larkin et al., 1996): "Trichome development starts near the tip of the leaf and proceeds basipetally.... Stomatal differentiation also proceeds basipetally, starting at the tip of the leaf." Cf. Marks (1997).

The spacing pattern of heterocysts differentiating (Gallon & Chaplin, 1988; Golden et al., 1988) along chains of cells in the prokaryotic blue green algae (cyanobacteria) is probably due to a diffusion mechanism (Wolk, 1967; Mitchison & Wilcox, 1972; Wilcox, Mitchison & Smith, 1973a,b; Wolk &

Quine, 1975; cf. spore or akinete patterns: Hirosawa & Wolk, 1979; but note the possibility of microfilaments and microtubules in these organisms: Jensen & Ayala, 1976; Bermudes, Hinkle & Margulis, 1994; Jensen, 1994). On the other hand, apparent cell-cell coordination involved in the unknown mechanism(s) of gliding motility (Halften & Castenholz, 1971; LeLeng & Margulis, 1978; Burchard, 1981; Hernandez-Muniz & Stevens Jr., 1987) or even swimming (Brahamsha, 1996; Pitta et al., 1997) in cyanobacteria, suggest the possibility of more elaborate developmental coordinating mechanisms too, than simple diffusion.

While the *Drosophila* eye imaginal disc morphogenetic furrow may appear to represent a contraction wave passing through all cells of a given tissue, the furrow actually represents a wave that nearly simultaneously acts as a contraction wave for some cells, while allowing other cells right within the furrow to actually be undergoing expansion (Figure 51). A set of apparently evenly spaced contractions along the length of the eye imaginal disc morphogenetic furrow has been observed:

"The following description summarizes pattern formation as viewed on the apical surface. The regular spacing of the ommatidial array originates in the furrow. The earliest recognizable periodic form, a rosette, evolves in the furrow and consists of a ring of about 10-15 uniformly shaped cells surrounding a central quartet or quintet of cells. Although this is the earliest organized form, aggregates of disorganized cells (not shown), recognized by their more darkly stained membranes [i.e., apically contracted], frequently occupy sites where rosettes subsequently form. It is not known how or if these aggregates are transformed into rosettes, but the observation that their placement approximates the spacing of rosettes suggests that some of these cells are likely to be incorporated into rosettes" (Wolff & Ready, 1993).

The cells not undergoing these initial contractions may be the ones Björklund (in Proposition 249) postulates are undergoing expansion. Indeed, a correlation analysis of a line of cells along the middle of the furrow pictured by Wolff (1993) suggests an alternation of groups of different apical areas (Figure 52). This empirical observation of ours gives at least qualitative confirmation of the model of Banerjee & Zipursky (1990):

"...We propose that cellular determination takes place when cells are in a linear array at the posterior edge of the morphogenetic furrow.... An evenly spaced array of determined R8 cells is resolved, perhaps through lateral inhibitory interactions.... While in the linear array, the R8 cell could then induce its only neighbors, the R2 and R5 precursors, which in turn induce their only neighbors, the R3 and R4 precursor cells.... In summary, the organization of the fly visual system appears to result from a cascade of inductive interactions initiated at the morphogenetic furrow" (Banerjee & Zipursky, 1990).

Without a time lapse sequence, this analysis could be confused by the simultaneously occurring wave of cell division. However, "second-wave mitoses are typically seen between two and eight rows behind the furrow" (Wolff & Ready, 1993). These probably correspond to the 'founder' cells:

"In its wake, the furrow leaves immature five-cell preclusters that have formed around founder R8 cells by inductive interactions.... An early aspect of ommatidial development is establishment of the precise spacing between ommatidial units. The prevalent model for establishment of regular arrays during development assumes that founder cells arise among equivalent cells by a stochastic mechanism and these then inhibit their neighbors from assuming a similar fate by a process of 'lateral inhibition.' Three known genes, *scabrous* (*sca*), *Notch* (*N*), and *Ellipse* (*Elp*) are thought to be involved in these events (Baker & Rubin, 1989; Cagan & Ready, 1989b; Baker, Mlodzik & Rubin, 1990;... Mlodzik, Baker & Rubin, 1990)" (Zak & Shilo, 1992).

The subsequent partitioning of the *Drosophila* eye imaginal disc cells into other types has been presumed to be driven by local inductions, but the evidence is equivocal:

"Ectopic expression of boss protein is sufficient to transform the cone cells to an R7-like identity via activation of the *sevenless* pathway (Van Vactor et al., 1991), indicating both the pluripotent nature of the undifferentiated cone cells and the inductive nature of the *boss-sev* interaction.

"Curiously, similar photoreceptor-unique genes have not been found for the other photoreceptor cells. How other photoreceptor subtypes are specified is a remaining question" (Warren, 1993).

Note that the possibility of both apical cell expansion and contraction in the eye imaginal disc was postulated by Fristrom & Rickoll (1982) via a cytoskeletal apparatus that we interpret as a cell state splitter (see Proposition 1).

The segment polarity gene *wingless* in *Drosophila* is another candidate for a pleiotropic differentiation pathway component, as are *Delta* (Parody & Muskavitch, 1993) and *Notch* (Shellenbarger & Mohler, 1978; cf. van den Heuvel et al., 1989, and Noordermeer et al., 1992). So is the "*Drosophila* epidermal growth factor (EGF) receptor homolog (DER)" (Shilo & Raz, 1991) and the "*Drosophila miti-mere* gene, a member of the POU family" (Bhat & Schedl, 1994). One challenge ahead will be to distinguish pleiotropic genes that are involved in differentiation wave launching from those that are involved in differentiation wave propagation. Another will be to integrate them into the differentiation pathway, and bring it beyond our working model (Appendix IV: Björklund & Gordon, 1993b).

A spacing pattern need not be perfect when it is initially formed. Nardi & Magee-Adams (1986) observe a somewhat irregular spacing pattern of single, larger moth wing scale primordial cells amongst contiguous, smaller epithelial cells, which somehow then form parallel rows, perhaps using long cell extensions that interact, and 'oriented cytoskeletal elements' causing anisotropic tension. Morphogenetic movements may thus improve a pattern initiated as a compound contraction/expansion differentiation wave.

In the following example, the 'preparatory phase' may correspond to a differentiation wave. This then becomes an example of the temporal separation of the differentiation wave from morphogenetic movement, as expected if development is describable using a differentiation tree:

"The first event of *Drosophila* gastrulation is the formation of the ventral furrow. This process, which leads to the invagination of the mesoderm, is a classical example of epithelial folding.... We find that the ventral furrow invaginates in two phases. During the first 'preparatory' phase, many prospective furrow cells in apparently random positions gradually begin to change shape, but the curvature of the epithelium hardly changes. In the second phase, when a critical number of cells have begun to change shape, the furrow suddenly invaginates. Our results suggest that furrow formation does not result from an ordered wave of cell shape changes, contrary to a model for epithelial invagination in which a wave of apical contractions causes invagination. Instead, it appears that cells change their shape independently, in a stochastic manner, and the sum of these individual changes alters the curvature of the whole epithelium" (Kam et al., 1991).

If we were to look along a *Drosophila* eye morphogenetic furrow, the exact order in which the cells expand or contract may have a random component. Yet they end up reasonably well spaced. The "maturation of stomata in patches" (Sachs, 1978a) is another example.

8.02 The Development of Bilateral Asymmetry

"We do not know how the initial left-right orientation is established or what the initial asymmetric cell-to-cell signal is" (Yost, 1995).

Proposition 251: the breaking of bilateral symmetry, as in the formation of the heart, is due to cumulative deviations, from perpendicularity of their wave fronts, of spacing pattern differentiation waves.

"The identification of a cascade of asymmetrically expressed genes which play a role in LR [left-right] patterning of a vertebrate embryo makes possible significant progress in this field. The genetic pathway characterized in these experiments, however, represents roughly the 'middle' third of LR patterning.... A currently unknown mechanism is able to pick out a L (or R) side relative to the other two axes" (Levin, 1997).

Ultimately, when we try to explain the asymmetric positioning of the heart and its *situs inversus* abnormalities (Yost, 1990; Brueckner et al., 1991; Hoyle, Brown & Wolpert, 1992; Seo et al., 1992), we are looking for a mechanism that relates the chirality of some molecular structure to the breakdown of bilateral symmetry (Brown & Wolpert, 1990; Levin et al., 1995; Wolpert & Brown, 1995; Levin, 1997; Levin & Nascone, 1997). In other words, the problem is how to intellectually bridge the gap from microscopic to macroscopic asymmetry (Wolpert, 1991c), and I go along with the presumption that this is the correct line of reasoning. In this section, I will present a few proposals for the mechanism of bilateral asymmetrization, which are not mutually exclusive. The important point is not the models, but the kind of physical reasoning behind them, though they do each make testable predictions. Let us just call them 'working models' that may help get us to the actual mechanism.

I have suggested that some differentiation waves leave in their wake spacing patterns. In some cases these spacing patterns consist of or contain oriented microtubules in the form of bristles (*Drosophila*) or cilia (axolotl epidermis; initial heart tubes in vertebrates). For heart development, the mechanism of bilateral asymmetrization appears to involve the extracellular matrix (Yost, 1991, 1992), in particular the fibronectin, perhaps in a transmembrane interaction with the cortical cytoskeleton (Shiraishi, Takamatsu & Fujita, 1995). The analogies or homologies with ciliates (Frankel, 1989) may prove important in solving the origin of left-right asymmetry (whose developmental effects extend, for instance, to left-right asymmetries in drug-induced limb defects: Sanders & Stephens, 1991). Certainly, the bilateral symmetry of ciliates is 'imperfect', since there is actually a left-right asymmetry that is functionally important, which exhibits phenomena analogous to *situs inversus* (Shi & Frankel, 1990; Shi, Qiu & Frankel, 1990), and which is the primitive condition in both ciliates (Baroin-Tourancheau et al., 1992) and the Bilateria (Jefferies, 1991). For amphibians, Yost (1991) places the primary event right during the supercooperative phase transition (Gordon & Brodland, 1988) of cortical rotation (Gerhart et al., 1989), and puts the cause(s) in the cortex:

"In approximately 25% of the embryos that are UV-treated and manually tilted to rescue the subcortical rotation, the left-right axis is fully reversed. Thus, the presence of the microtubule array during cortical rotation [Elinson & Rowning, 1988] is not necessary to generate left-right asymmetry, but is required to align left-right asymmetry consistently with the other body axes. What mechanism might bring the left-right axis into register with the dorsoventral axis? Polymerized microtubules are helical in structure. In theory, a parallel array of helical molecules could provide a handedness to the subcortical rotation. This would cause an unequal distribution of cytoplasm from left to right across the concomitantly established dorsoventral midline. Interestingly, dynein defects in humans [cf. dynein in ciliates: Schroeder, Fok & Allen, 1990] may lead to situs inversus (for review, see Brown & Wolpert, 1990). Since situs inversus occurs in less than 50% of the embryos in these experiments, other cellular structures or events during the first cell cycle that may also contribute to a left-right asymmetrical rotation of the subcortical cytoplasm are currently being sought" (Yost, 1991).

There has actually been evidence for 'an unequal distribution of cytoplasm from left to right' (Yost, 1991) and its possible impact on left-right asymmetry for quite a while, if we assume that the following normal and experimental observations are related:

"Stage 2 [*Xenopus*]... Advanced two cell stage.... blastomeres not equal in size. First plane of cleavage more or less coinciding with plane of bilateral symmetry" (Nieuwkoop & Faber, 1956, as reprinted in Nieuwkoop & Faber, 1994).

"In 60-70% of the eggs of *Ambystoma* and *Xenopus* the first cleavage plane coincides with the plane of bilateral symmetry (Ancel & Vintemberger, 1948; Ubbels et al., 1979)....

"Spemann & Falkenberg (1919) and Fankhauser (1930) observed normal *situs viscerum* in larvae developing from left halves, and *situs inversus* in 50-75% of those developing from right halves of medially constricted amphibian eggs" (Nieuwkoop, Johnen & Albers, 1985).

One might question whether the 90% common orientation of microtubules found in natural cortical rotation (Houliston & Elinson, 1991b) is also found in UV treated and manually tilted embryos, i.e., perhaps it is orientation and not mere presence of the microtubules that is important. How oriented, cortical bundles of microtubules lead to left-right asymmetry in amount of cytoplasm is a good challenge for theory. Perhaps the microtubules rotate about their axes (Jarosch, 1992), or perhaps they are not responsible for this phenomenon but it is the first cleavage spindle that is actually asymmetric. Why do 90% have a common orientation? A hint comes from the work of Baas, White & Heidemann (1987), who "speculate that microtubule organization accompanying neurite outgrowth is not the result of a microtubule organizing structure but of a process, the interaction of microtubule assembly with the motility of the cell cortex". If one end of a microtubule 'binds' to the cortex better than the other end, during motion of the cortex relative to the underlying cytoplasm, then a common orientation would be achieved. Such a mechanism, combining a chemical difference between the ends of a chiral supramolecular structure, and the mechanics of the whole cell, may then be the fundamental means by which the left-right asymmetry of an embryo is generated. If the mechanism operates in neurites, then it may be available at any stage of development, and in particular able to explain left-right asymmetry in embryos that acquire it in a multiple cell state. Perhaps we need some of the single particle tracking methods for membrane bound proteins (Saxton & Jacobson, 1997) applied to the microtubules beneath.

Note that I am making an assumption, namely that the vegetal cortical microtubules that orient during cortical rotation (Elinson & Rowning, 1988) extend, at lower concentration, throughout the cortex of the animal hemisphere. This would solve the problem that, later on in development, "it is hard... to imagine how each cell knows which way is 'anterior'" (Levin, 1997). Cortical rotation certainly occurs here, with no indication that it is a passive response to active rotation only in the vegetal hemisphere:

"Selective fluorescent labelling of yolk granules revealed considerable movement in the deep cytoplasm of the animal hemisphere concomitant with the cortical rotation... of fertilized *Xenopus* eggs (Ubbels et al., 1983; Ubbels & Vermeulen, 1986; Danilchik & Denegre, 1991; Denegre & Danilchik, 1993) (also seen in the anuran *Discoglossus pictus:* Klag & Ubbels, 1975)" (Ubbels, 1997a).

Parallel arrays of microtubules, probably extending the full length of the cell, are one of the components of the ciliate cortex (Lynn, 1981; Adoutte et al., 1991), and: "In the model [of Brown, McCarthy & Wolpert, 1991], a tethered asymmetrical molecule has a long-range asymmetrical effect. Gary Grimes (1982a) has evidence for something he calls an intracortical communication system [in hypotrich ciliates]" (J. Frankel in: Brown, McCarthy & Wolpert, 1991). Thus I suggest that the fundamental study of left-right asymmetry should perhaps be carried out on what may be differentiation waves in ciliates (Iftode et al., 1989; Section 5.05). Here may lie the answer to Wolpert's (1991c) query:

"If life had used D-amino acids instead of L-amino acids, would we all have had our hearts on the other side? And would the laterality of the brain be the other way around?" (Wolpert, 1991c).

Note that Brown, McCarthy & Wolpert (1991) and Levin & Nascone (1997) hypothesize 'extracellular fibres' and 'chiral centrioles' (cf. Holmes & Dutcher, 1989), respectively, without explaining how they got there or became oriented. In Levin & Nascone (1997), an "asymmetric placement of ion pumps on the cell surface, established by a *dynein*-like mechanism", is postulated to establish global orientation. I propose that the global orientation comes first, by a mechanical effect, at least in amphibians, via the cortical rotation. The resulting structure is inherently left-right asymmetric, since the cortical microtubules, which have their own chirality, are mostly oriented the same way (cf. Yost, 1991). Our problem is then simply to postulate mechanisms for translating this left-right asymmetry of supramolecular structure to the left-right asymmetry of gene expression.

Suppose that there is some slight interaction at the centrosome level, so that the forming bristle or cilium can slightly deviate the local portion of the ectoderm contraction wave, to the left or right of its local direction, but with a small, consistent bias towards one of the directions (which should be observable by time lapse microscopy). The deviation would accumulate as the wave moved from cell to cell, and the trajectory of the wave would lose its initial bilateral symmetry. Such an effect has been modelled for action potential wave propagation on the heart, in the presence of oriented muscle fibers (Rogers & McCulloch, 1994a). In effect, the refractive index of the tissue becomes anisotropic, so that ordinary rectilinear propagation of the differentiation wave does not occur. This is analogous to double refraction in anisotropic crystals (Morgan, 1953), except for the fact that in an excitable medium only the faster of the two waves, the 'ordinary wave' and the 'extraordinary wave', will propagate. For the extraordinary wave, the Huygens' wavelets are ellipsoids (Morgan, 1953), or ellipses in a two dimensional medium. Mutants could be sought that alter the relative speeds of the ordinary and the extraordinary differentiation waves, and thus alter whether or not bilateral symmetry is broken (or how much: cf. Molfese & Segalowitz, 1988). A centrosome mechanism is implicated in the left-right asymmetry of *Paramecium* (Iftode & Adoutte, 1991; Le Guyader & Hyver, 1994), whose differentiation waves may be worth looking at from the point of view of double refraction.

Of course, other explanations of the breaking of bilateral symmetry may be possible, such as an asymmetric buckling phenomenon during heart looping (cf. Manasek, Burnside & Waterman, 1972; Yost, 1990; Manner, Seidl & Steding, 1993), biased by a slight mechanical anisotropy of the tissue, perhaps due to myofibrillar patterns (Price et al., 1996). While "asymmetry exists long before morphogenesis of organs" (Michael Levin, p.c., 1998), similar phenomena could occur earlier. Any of these proposed mechanisms will have to explain away the apparent 'independence' of pattern formation and polarity (= alignment: Proposition 268):

"In the original formulations of gradients in the insect segment (Stumpf, 1966; Lawrence, 1966), the pattern of differentiation and the orientation of structures such as bristles and hairs were believed to be separate manifestations of the same underlying gradient.... However, patterns of differentiation and of polarity can be affected separately in mutants of *Drosophila* (Gubb & García-Bellido, 1982; Held Jr., Duarte & Derakhshanian, 1986) and can be dissociated in rotated cuticular grafts in the bug *Dysdercus* (Nübler-Jung & Grau, 1987; Nübler-Jung, 1987a).... The polarity pattern is strictly dependent on local cell-to-cell interactions, possibly related to concordance in orientation of microtubule organizing centers (Whitten, 1973; Nübler-Jung, Bonitz & Sonnenschein, 1987).... Nübler-Jung, Bonitz & Sonnenschein (1987) have compared the hierarchy of control in *Dysdercus* to the situation encountered in *janus Tetrahymena* [a ciliate], in which dissociation of local and global asymmetry was first reported (Jerka-Dziadosz & Frankel, 1979)" (Nübler-Jung & Grau, 1987).

The link between the two levels of organization could be weak, in the form of a bias, so that instabilities at the global level that are bistabilities are not resolved 50:50 (cf. Le Guyader & Hyver, 1994). The interactions between differentiation waves and chiral molecular structures might be of this nature. Such a view may help account for the spectrum of outcomes or "gradation from normal situs through isomerism to situs inversus" (Burn, 1991).

Michael Levin (p.c., 1995) raises the difficulty of a differentiation wave model for left-right asymmetry accounting for the sharp left-right expression of genes in chick embryos (Levin et al., 1995; cf. Tsuda et al., 1996), which is not likely to be the product of the accumulation of small deviations (nor, perhaps, of "gradients of left-right positional information": Morgan, 1991). Further models in the following Propositions may meet this objection.

One problem is posed (Steve McGrew, p.c., 1997) by assuming a residual orientation of microtubules (Section 3.10) due to the cortical rotation: what is the relationship between these microtubules and those in the cell state splitter? Are they one and the same, so that the cell state splitter is not axially symmetric about the axis of the columnar ectoderm cell? Or are they different? The paraxial microtubules (Burnside, 1971, 1973a) are alternate candidates for retention of the orientation at cortical rotation, and, as we shall see, may have their MTOCs at the apical ends of the ectoderm cells, directed basally, though the existence of their MTOCs, let alone orientation, has yet to be investigated.

Proposition 252: asymmetric penetrance is due to failure of launching of one of a pair of differentiation waves on one side of a bilateral organism.

Incomplete penetrance is, in general, an unsolved problem (Sang, 1963). Sometimes a tissue appears on one side, but not another, of an organism, a result called asymmetric penetrance. We could attribute this to somatic mutation, or to bizarre circumstances analogous to the asymmetric positioning of the heart (*situs inversus:* Yost, 1990; Brueckner et al., 1991; Hoyle, Brown & Wolpert, 1992), brain asymmetry (Levy, 1988a; Hellige, 1993) or to stochastic processes, as postulated by Chambon (1993) in knockout mouse mutants of retinoic acid receptors (RAR):

"Congenital malformations in RARγ null offspring included agenesis of the ocular Harderian glands and malformations of cartilage-derived structures. The variation in expressivity of the Harderian gland agenesis was particularly remarkable, since the alteration corresponded to either a bilateral absence of the glandular epithelium, or to a partial or total absence of the epithelium in one gland whereas the epithelium of the contralateral gland was normal.... *The variability in the expressivity* most probably results from stochastic variations (Ko, 1992) in the amount of a given redundant receptor (e.g., RARα1) within identical symmetrical cells.... It is tempting to speculate that such stochastic variations could be used during embryogenesis to initiate a development program within a population of apparently identical cells" (Chambon, 1993).

There may, indeed, be a stochastically approached threshold of whatever phenomenon it is that launches differentiation waves, in which case any of the four cases could occur:

1. (no wave on left, no wave on right);
2. (wave on left, no wave on right);
3. (no wave on left, wave on right);
4. (wave on left, wave on right).

Thresholds are anticipated for initiating wave propagation in any excitable medium (Murray, 1989), so a subthreshold stimulus should at most launch a wave that damps out (but yet may be visible by time lapse microscopy). Mechanical imbalances resulting from slight differences in differentiation wave speeds between the left and right halves of an embryo might cause the 'random generation of asymmetry' required in the theory of Brown & Wolpert (1990). The asymmetries that result from a left-right temperature gradient across an embryo (Spemann, 1938), discussed in Proposition 264, may be relevant here.

Asymmetric penetrance may be an extreme intraorganism version of the interorganism variation in development rates found, for example, in insects (Lamb & Loschiavo, 1981; Lamb, MacKay & Gerber, 1987). In caecilian amphibians asymmetric lung development seems to be the norm for some species:

"*Ichthyophis orthoplicatus* (Sri Lanka) has paired lungs, but the left is a quarter the size of the right.... *Hypogeophis rostratus* (Seychelles Is) is reported to have a well-developed right and a rudimentary left lung, whereas in *Boulengerula taitanus* (Kenya) only a single lung has been found (Maina & Maloiy, 1988) [cf. unilateral agenesis of kidneys in humans: Potter, 1972]. Terrestrial species, as a group, usually have an elongate right lung which may be

50%-75% of total body length (Wake, 1974) and a left lung that is no more than 10% of body length" (Stebbins & Cohen, 1995).

While human lungs are asymmetric, this observation shows that a structure such as the lung can have extreme left-right asymmetry, even in an amphibian. This may be related to the compound nature of fractal morphogenesis of lungs (Proposition 41), permitting an accumulation of tissue more on one side than the other. Unilateral agenesis of kidneys leaves a right kidney more often than a left kidney (Emanuel et al., 1974).

Proposition 253: incomplete penetrance of morphological features is due to a differentiation wave whose launching sometimes fails.

The concept of incomplete penetrance of a feature assumes that that feature corresponds to a single gene, but that the gene may or may not get to be expressed in a given individual:

"...Timoféeff-Ressovsky [1931] introduced the terms 'penetrance' and 'expressivity'. Penetrance means the extent to which every animal carrying the gene is affected by it (percentage phenotypic effect). Expressivity means the extent of the effect itself" (Needham, 1950).

If we presume that this would be the case for genetically identical organisms (a clone), then something other than genetics determines whether or not the gene 'penetrates' in a given individual. My best guess is that it is the launching of a wave, i.e., whether it does or does not get started, that is the explanation of penetrance. How big its trajectory is, compared to normal, might account for the separable phenomenon of expressivity.

All of the waves that we have investigated so far, except the ending of the wave forming the axolotl eye placodes, and some of the blastopore lip waves, are single waves. Morphological cases of incomplete penetrance in *Drosophila* include the genes *eyeless, bicoid,* and homeotic genes (Sang, 1963, 1984), plus an assortment of wing mutations (O.L Mohr in: Needham, 1950). For vertebrates we have incomplete penetrance in:

mice: visceral inversion or *situs inversus* (Tihen, Charles & Sipple, 1948; Kalter, 1968), *disorganization* (Hummel, 1958; Kalter, 1968), *curly-tail* causing spina bifida (Kalter, 1968; Brook et al., 1991; van Straaten et al., 1992), *fused* vertebrae and ribs (reviewed in Kalter, 1968), perhaps homologous to pair rule genes in *Drosophila* (Lawrence, 1992), plus a few others in mice, rabbits and cattle (Kalter, 1968).

humans: genes for "bent and stiff little finger" (Lerner, 1968).

Teratogens (Kalter, 1968; Seller, Perkins & Adinolfi, 1983), nutrition (Cockroft, Brook & Copp, 1992), and other genes (Mayr, 1970; Estibeiro, Brook & Copp, 1993) can alter the degree of penetrance, and thus may allow direct manipulation of the expression of genes involved in the launching of differentiation waves. In any case, my notion that launching of a wave correlates with penetrance is directly testable.

Proposition 254: left-right asymmetry is encoded in the lateral membrane cortices of ectoderm cells as left or right pitched macromolecular helices during propagation of the ectoderm contraction wave.

The exact timing of the earliest event of left-right asymmetry may vary between species, as we saw in considering the origin of twinning in armadillos versus humans (Proposition 119). Here I will consider the possibility that one such left-right determining event happens early in gastrulation. (More than one event could occur, as in skin versus heart in armadillos: Newman, 1917.)

The development of left-right asymmetry sometimes appears to coincide precisely with propagation of the ectoderm contraction wave. A recent review (Levin, 1997) confirms earlier literature (also reviewed in: Nieuwkoop, Johnen & Albers, 1985) that determination of this axis occurs during gastrulation (see also Smith et al., 1997b):

"...I have expanded some of the classical transplant experiments by fate mapping [*Xenopus*] transplanted tissues. First, wounding the blastocoel roof of late blastulae or early gastrulae results in some cardiac situs inversus and some visceral situs inversus. To fate map the wounding sites precisely, small pieces of ectoderm in pigmented gastrulae were explanted and replaced with similar pieces of ectoderm from albino gastrulae. To eliminate anteroposterior and dorsoventral biases, these axes had been eliminated in the albino (ectoderm-donor) embryos by UV-irradiation of the vegetal hemisphere during the first cell cycle (for review, see Gerhart et al., 1989)....

"...Fate mapping of the transplanted ectoderm reveals a strong correlation between the site of ectoderm transplantation and the orientation of the underlying heart and viscera. In cases in which the transplanted ectoderm does not appose an organ, the organ always develops in the normal left-right orientation. When the transplanted ectoderm directly apposes an organ, orientation of the left-right axis occurs randomly, that is, in half the cases the organ develops in the normal situs and in half the cases the organ is situs inversus....

"...One possibility is that a signal arising from the vegetal part of the embryo crosses the plane of the ectoderm toward the animal pole [i.e., the ectoderm contraction wave?]. This signal might direct the modelling of left-right axial information in the ectoderm and its ECM [extracellular matrix].... Researchers studying neural induction have suggested that in blastulae and early gastrulae signals pass from the dorsal lip region through the plane of the animal cap ectoderm to participate in neural tissue formation (Spemann, 1938; Slack, 1983; Jacobson & Sater, 1988; Dixon & Kintner, 1989). Signalling through the ectoderm up to the animal pole is a possible mechanism for assuring that the dorsoventral and left-right axes are in register throughout the embryo.

"Transplantation of ectoderm appears to cause situs inversus most effectively when performed on early gastrulae, coincident with the period of active ECM deposition.... The results suggest that left-right axial information is sequestered on the basal surface of animal pole ectoderm and/or in its ECM during the period in which the matrix is initially deposited, and that this information can be transmitted to underlying presumptive cardiac mesoderm and visceral endoderm" (Yost, 1991).

ECM molecules are huge (Trelstad, 1990), approaching the dimensions of cells, and if oriented, as they certainly can be (Scott, 1989a), could well carry left-right orientation at the basal ends of the ectoderm cells, and perhaps further. The orientation may be determined by "binding sites on cell surface receptors (Ruoslahti & Pierschbacher, 1987)" (Yost, 1991), which I shall presume are, somehow, oriented in the basal membrane, as in the common orientation of intracellular, cortical actin and fibronectin (Hynes &

Destree, 1978). The problem we must solve, then, is how an orientation at the apical surface of the cell can be transmitted to the basal end.

I would like to suggest that what conveys this orientation consists of one or more helical wrappings in the cortex of the lateral membrane of the ectoderm cells:

"[*W.B.*] *Wood:* Some biological structure is going to depend on interactions between sheets of helices [cf. Lepescheux, 1988]. In nematodes, the cuticle has at least two layers of helical fibres arranged with a particular pitch. A mutation in a particular collagen gene locus can produce animals where these sheets are shifted relative to each other (Kramer et al., 1988). The animal itself may have a left-handed helical twist, for example, and move in an abnormal way; these are called 'left-rollers'. Another allele at the same locus can produce a right-roller phenotype, where the helices are shifted a little differently (Kusch & Edgar, 1986; [cf. Félix, Sternberg & De Ley, 1996]). These are single amino acid substitutions that presumably result in different interactions of the sheets with each other to affect the entire morphology of the animal.

"*Galloway:* An area which I didn't touch on at all is this idea of helicoidal structures, where flat sheets interact face-to-face, and you may get stacks of many hundreds, if not thousands, with a gradual turn [cf. chick eye: Coulombre, 1965]. Insect cuticle is a good example of that. Plant cell walls also often have a mechanism of that kind to change from one structural layer to another. These are helicoidal; they are flat sheets arranged with a definite screw sense" (Galloway, 1991).

In plants there are parallel, helical cortical bundles of microtubules affecting cell wall deposition (Lloyd, 1983; Green, 1984a; Emons, Derksen & Sasen, 1992), so the cortical matrix can be involved in such helicoidal arrays. The lateral molecules could be microfilaments, given their demonstrated presence in newt neural plate cells (Burnside, 1973b) and the known helical organization of smooth muscle cells (Bagby, 1983, 1986). The "different interactions of the sheets" (W.B. Wood in: Galloway, 1991) might account for left-right asymmetries in gene expression (Levin et al., 1995), by exposing different chemical surfaces to the cytoplasm (cf. Scott, 1989a). Whether the paraxial microtubules found in neural plate cells (Burnside, 1971, 1973a) (and presumably earlier in ectoderm, and later in nerve cells: Heidemann, 1996) play any role is not clear. Bilaterally asymmetric sheets of collagen, i.e., sheets with the same local symmetry on both left and right, are found in chick eye:

"In submammalian forms, the [collagen] fibrils become aggregated into sheets or lamellae, which lie in the plane of the cornea and extend to the limbus at all points (Coulombre & Coulombre, 1961; Polack, 1961). The fibrils in each lamella are roughly at right angles to those in adjacent lamellae. There is, however, a systematic, clockwise, angular shift in this orthogonal ply between the outer and inner surface of the stroma in the chick embryo. The shift becomes less pronounced as the deeper strata are approached. This angular shift is in the same direction in both right and left eyes, and is, consequently, asymmetric around the body midplane" (Coulombre, 1965).

I would like, then, to suggest that when an ectoderm cell participates in the ectoderm contraction wave, the following sequence of events occurs:

1. The wave interacts with apical MTOCs and/or apical microtubules emanating from them, that have at least a residual orientation from the cortical rotation that occurred before first cleavage (90% of the microtubules have a common orientation at that time: Houliston & Elinson, 1991b).

2. The interaction initiates growth of a helical structure down the lateral membrane.

3. The direction of turn of the helix (clockwise or counterclockwise, looking down on the apical surface of the cell) is biased by the apical microtubules, perhaps by a double refraction mechanism of wave propagation, so that, in general, those cells on the left spiral one way, and those on the right the other way.

4. Some chemical difference in the helical surfaces exposed to the cytoplasm versus membrane leads to specific left-right differences in gene expression. (This might be due to the supercoiling of the helices, exposing different active sites.)

5. Cooperative interactions between the lateral cortices of neighboring cells further bias the net helical pitch and tend to cause it to be the same for all neighboring cells on one side of the midline.

6. At the basal end, the lateral helix imparts an orientation to the ECM receptors.

This should, of course, only be regarded as a crude and incomplete working model for the origin of left-right asymmetry. Steps 1, 3 and 5 would seem to correspond to those of Brown, McCarthy & Wolpert (1991):

"A model of left-right specification. The fundamental assumption is that handedness is signalled by a molecule (or larger structure) which is itself handed and can be fixed in a particular orientation in relation to the anteroposterior and dorsoventral axes. We make this assumption in the absence of plausible alternatives. Magnetic fields induced by an electric current down the midline [cf. Jaffe & Stern, 1979] have been proposed (Huxley & de Beer, 1934), but this hypothesis does not fit well with known cellular processes [however, cf. Ho et al., 1992b]....

"The model has three components: *conversion,* in which the molecular asymmetry is transmitted to the cellular level; *random generation of asymmetry,* which can be biased by *conversion* to produce handed asymmetry; and *interpretation* in which individual organs use left-right information" (Brown, McCarthy & Wolpert, 1991).

What happens at the midline is not clear, though this is the region of the notoplate (Gordon, 1983a; McKanna, 1993) and other midline effects (Opitz & Gilbert, 1982; Lubinsky, 1987). (This region will be incorporated in the model in Proposition 256.) If we presumed a lack of influence across the midline, then it would be reasonable to expect some left-left and right-right organisms could develop. The advantage of this model is that it invokes arrays of macromolecules constructed on the scale of the whole cell, and all its predicted components and interactions should be readily observable. The model accounts for the timing and the local nature of the effects observed by Yost (1991): local random orientations, leading to 50% situs inversus, would be due to some failure in the sequence of events, resulting in local random orientation of the ECM receptors.

A curious aspect of Yost's (1991) observations is that they provide a case in which induction proceeds from the ectoderm to the mesoderm, i.e., in a direction opposite to that presumed for neural induction. Of course, both are possible, if the initial event of neural induction is launching of the ectoderm contraction wave by contact with the underlying mesoderm (perhaps pharyngeal endoderm: Hama et al., 1985), and the wave subsequently affects the underlying mesoderm. As I pointed out in Proposition 89, the wave is twice as deep as the cells it travels through, so significant ectoderm to mesoderm inductive effects would not be surprising (although mesoderm is not always present everywhere under the ectoderm: Section 9.24).

If the change in gene expression is actually within the ectoderm cell, then the lateral, helically supercoiled fibrils or microtubules represent a form of bistable cell state splitter. However, there is no dynamic equilibrium that is resolved to one of the two states (i.e., helical pitches), since the pitch is determined by whether the cell is on the left or right.

Proposition 255: the angle between the advancing wave front of the ectoderm contraction wave and the microtubule orientation remaining from cortical rotation prior to first cleavage divides the bilaterally symmetric neural plate into 8 regions related by colored symmetry.

Left-right asymmetry is usually thought of as simply involving a difference between the two lateral halves of an embryo. Empirically, this is clearly not the case. As we have seen in Proposition 254, in the experiments of Yost (1991), there is a certain amount of local autonomy in the determination of sidedness. Let us take the model for orientation of the direction of differentiation wave propagation, relative to cortical rotation, seriously, and see what it predicts. In Figure 53 Natalie K. Björklund (p.c.) and I have worked out predicted consequences if we assume a simple, discrete angle relationship between the cortical rotation and the direction of wave propagation. The letter F is used to indicate a handed molecule (as in Figure 2 in Brown, McCarthy & Wolpert, 1991), such as the microtubules or the MTOCs (microtubule organizing centers), presumed to be globally oriented to each other by the cortical rotation (Houliston & Elinson, 1991b).

The crossing of the latitudes of oriented microtubules with the wave fronts of the ectoderm contraction wave leads to 8 zones, the posteriormost corresponding to presumptive notochord, the next to presumptive spinal cord, and the top two corresponding to the brain, with the sense plates at the margin (Figure 53). Zones opposite each other relative to the anteroposterior axis are different, so this left-right asymmetry is an example of 'antisymmetry', part of the more general concept of 'colored symmetry':

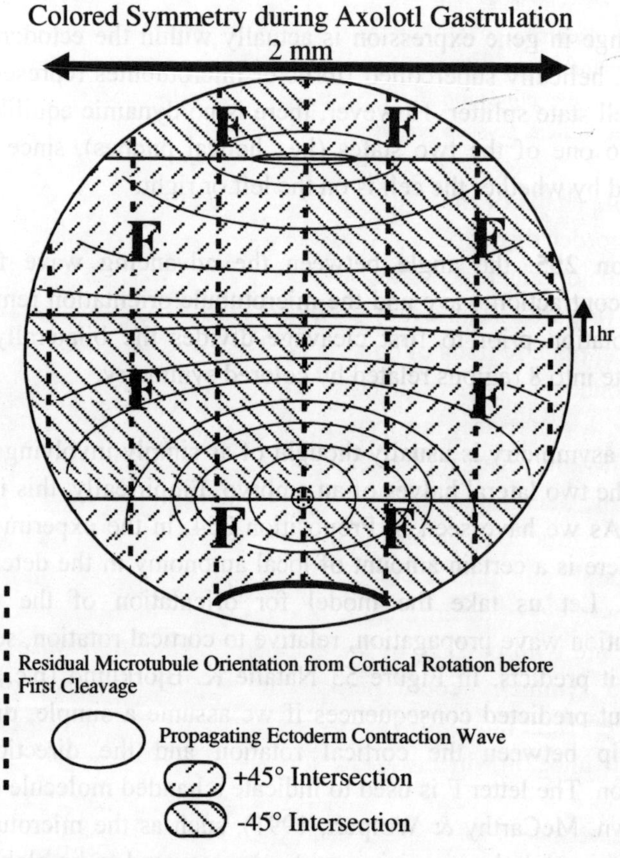

Figure 53. Microtubule/wave colored symmetry. The ectoderm contraction wave (Figure 17 in Appendix V: Gordon, Björklund & Nieuwkoop, 1994) is shown, along with parallel, aligned rows of microtubules, whose orientation is presumed to be preserved from the initial cortical rotation (Figure 6 in Appendix V). The Fs indicate the handedness of the microtubules. The diagonal shading blocks off regions in which the angle between the microtubules and the advancing wave front is closest to +45° or -45° from the midsagittal plane. Since bilaterally placed regions have opposite shading, this is a case of colored symmetry.

"Many of [M.C.] Escher's periodic drawings are based on the principles of colour symmetry. The simplest aspect of this, so-called black-white or anti-symmetry, was introduced in the crystallographic literature about 1930, on the occasion of a symposium on liquid crystals. Practically immediately afterwards, the notion of colour symmetry fell into a twenty years 'sleeping beauty' slumber, from which it was restored to life very vigorously by Shubnikov's (1951) book on symmetry and antisymmetry [cf. Shubnikov & Belov, 1964; Loeb, 1971; Shubnikov & Koptsik, 1974]. Polychromatic symmetry made its first appearance in the scientific literature about 1956.... [Escher's] notebook dates from 1942. Thus, in particular the possibilities of the polychromatic groups were explored, and their symmetry elements marked, before official crystallography even thought about them" (MacGillavry, 1965).

The keyhole shape of the neural plate (Stage 15 in Figure 4; Table 1), with its second order discontinuity separating anterior from posterior, correlates with the major anteroposterior boundary between the regions of colored symmetry (Figure 53), and may be a secondary morphogenetic consequence of it. Neural plate regions of corresponding 'color' might be exactly the ones that are connected later by nerve tracts (Figure 9.10 in Jacobson, 1978d), explaining, for instance, the ipsilateral and contralateral wiring of the eyes and the left-right reversal of muscle control. (These neural mappings of one region of the neural plate to another may be the basis for the hypothesized visual 'reconstruction screen' of Gordon & Hirsch, 1977 or the recursive "LGN [lateral geniculate nucleus]... screen, or stage, or sketchpad" of Harth, 1995.) The stages of determination of the visuotectal map (Jacobson, 1978d) may correlate with passage of the ectoderm contraction wave. The concept of colored symmetry could perhaps be applied to some of the (albeit probably left-right randomly) generated pigment patterns of some allophenic or chimeric mice, in which:

"Bands on either side of the midline need not be continuous in color, suggesting that right and left patterns develop autonomously" (Grant, 1978). [Cf. Mintz, 1967.]

Clearly, more complex schemes than that of Figure 53 could be envisaged. For example, in this figure, an intersection of a wave at +45° from the anteroposterior axis is regarded as identical to an intersection at -135° from that (vertically drawn) axis, and -45° is taken as equivalent to +135°. Perhaps, instead, angles 180° apart are not equivalent. Also, all gradations in the intersection angle are ignored. Thus it is premature to define the exact symmetry group of embryonic left-right color symmetry.

It is curious to note that, while Brown, McCarthy & Wolpert (1991) depend on gradients and diffusion-reaction schemes to explain left-right asymmetry, they invoke a microtubule based system as an example:

"Cells become polarized with respect to the midline; for example, a cell component becomes more concentrated at the side of the cell opposite to the midline, perhaps in response to a diffusible substance produced by midline cells.... We depict the handed molecule as an 'F', which is held in a specific orientation with respect to both anteroposterior and dorsoventral axes. This breaks the mirror symmetry across the midline. An appealing example of such an orientation of a handed structure is seen in the cilia of tadpole tails. These are fixed in the same orientation on left and right sides, so they all beat caudally.... For the *generation of random asymmetry*... [we] favour a simple mechanism provided by a gradient in a morphogen produced by reaction-diffusion (Kauffman, Shymko & Trabert, 1978). This could occur across the midline and left-right asymmetry would be established, the concentration rising on one side and falling on the other" (Brown, McCarthy & Wolpert, 1991).

Of course, a problem for both our model and the model of Brown, McCarthy & Wolpert (1991) is how any of these cell differences are turned into left-right specific gene expression. Levin & Nascone (1997) postulate that dynein, riding microtubules (not oriented parallel to the prior cortical rotation), shoves some determinant to one side of a cell. This sidedness is expanded to organism sidedness by a mechanism involving gap junctions (Levin & Mercola, 1998b). A possible 'switch' could be either the exposure of different chemical groups in the lateral cortex of ectoderm cells in the at least bimodal 'interactions between sheets of helices' of W.B. Wood (in Galloway, 1991; Proposition 254). Alternatively, analogous effects could occur in the extracellular matrix, via the surface exposure of either hydrophobic or hydrophilic groups (Scott, 1989a). (Cf. the similar hydrophobic/hydrophilic exposure of the inside of a chaperone upon a 90° twist: Strauss, 1997.) In this case, either the lateral or basal membranes might be said to contain their own form of cell state splitter.

There is a certain danger in using a two dimensional figure such as the letter F to represent the left-right asymmetry of the 'extracellular fibres' (Brown, McCarthy & Wolpert, 1991) at the beginning of gastrulation. If they are microtubules, for instance, then the F's would spiral around the axis of the microtubule, and from a given point of view, be equally represented on the left and right of any single microtubule. Either the 'extracellular fibres' are indeed some other sort of molecule that lies flat against the apical and lateral membranes, unable to flip over, or something more subtle is the root cause of this asymmetry.

One possibility is that the spiraling of microtubules down the lateral cell membrane is actually a coil of a coil (an open, rather than tight supercoil), and that coiling in one direction is concordant with the microtubule's direction of coiling, while coiling in the other direction is opposite to that of the microtubule. This could possibly lead to a gentle twist of the microtubule that exposes different active groups in each case (John Parkinson, p.c., 1996). These in turn could initiate the expression of different master genes. To see how this model behaves mechanically, consider two cylindrical cells with a microtubule spiraling down along the inner wall of each. But let them be left-right mirror images, as in bilaterally symmetrically placed cells in an epithelium (like vertebrate ectoderm), so that the spirals are in opposite directions. Now, grab the microtubule at the top and turn it say clockwise on its axis. (If it's going to turn by some cellular mechanism, the chirality of the microtubule and attached motor proteins should dictate which way, not the direction of supercoiling.) Let's presume that the microtubule can be thought of as a spring with a given chirality. The chirality is the same for both microtubules (formally breaking the left-right symmetry). But now it has a mechanical consequence on the scale of the whole cell. On one side the microtubule will just tighten against the cell membrane. On the other it will buckle towards the middle of the cell. You can try this with a long, not too stiff spring inside a cylinder, such as a tin can with the ends cut out, or a graduated cylinder.

An alternative geometry is possible. The microtubules in a left-right matched pair of cells could supercoil in the same direction. However, the shear, considered in Proposition 256, could turn them in opposite directions, again yielding tightening and buckling, respectively.

Proposition 256: the coupling between the shear gradient along the midline and the apical microtubules oriented at cortical rotation is the major mechanism for left-right asymmetrization.

The remarkable cases of left-right asymmetry in gene expression found so far occur just adjacent to the midline of the primitive streak in chick embryos (Levin et al., 1995). This is probably a region of high gradient of mechanical shear, such as we found near the uniformly elongating midline in newt neural plates, both by computer simulation and empirically (Jacobson & Gordon, 1976a; Gordon & Jacobson, 1978; Jacobson, 1978a,b). Although "...it seems reasonable to regard the pre-primitive streak thickened area of the chick blastoderm as the symbolic homologue of a blastopore..." (Patten, 1951), the shape changes of the tissue forming the primitive streak parallel the 'convergence and extension' of the notochord and notoplate in amphibians (Jacobson & Gordon, 1976a; Gordon & Jacobson, 1978; Jacobson, 1978a,b; Keller & Danilchik, 1988):

"The primitive streak first becomes visible in the hindmost part of the area pellucida and is designated the **short primitive streak**.

"The primitive streak now elongates by the concentration of more and more material from the sides towards the midline in front of the original short primitive streak.... In later stages it contracts in a transverse direction, becoming narrower and quite sharply delimited, and is called the **definitive primitive streak**. In the process of this transformation, the primitive streak elongates and its anterior end is pushed even farther forward, though the amount of this forward thrust has been a matter of controversy among embryologists. (See Waddington, 1952e.)" (Balinsky & Fabian, 1981).

A shear gradient is likely to generate some torque on cells just lateral to the midline. While these torques are bilaterally symmetrical, they could twist a left-right oriented molecule (represented by an aligned string of F's) differently on either side of the midline (Figure 54). This is elaborated in Figures 55-56, which show that the F's will all face towards the cytoplasm if the fibrils spiral down the lateral membrane in cells on one side of the midline, while the F's will all face away from the cytoplasm in cells on the other side of the midline.

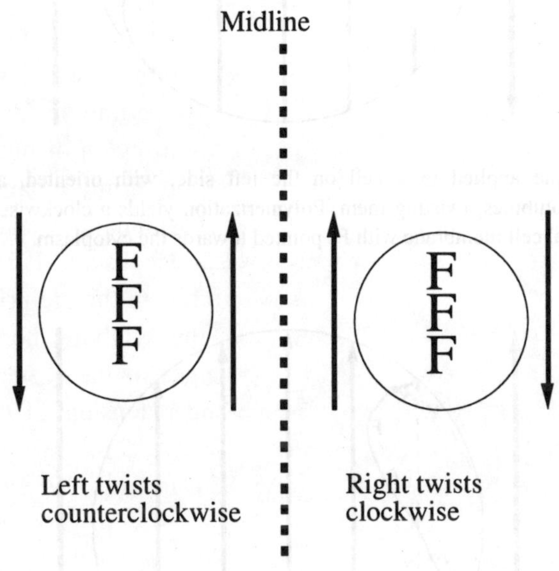

Figure 54. Bilaterally symmetric shear couples twist around corresponding, left-right cells containing oriented, asymmetric fibrils, such as microtubules, in opposite directions.

Lack of a notochord in amphibian embryos (Lehmann, 1934; Hoadley, 1938; Clarke et al., 1991) might be expected to reduce the shear gradient and therefore lead to more situs inversus.

Proposition 257: paraxial microtubules and their interactions with the nucleus may be involved in bilateral asymmetry.

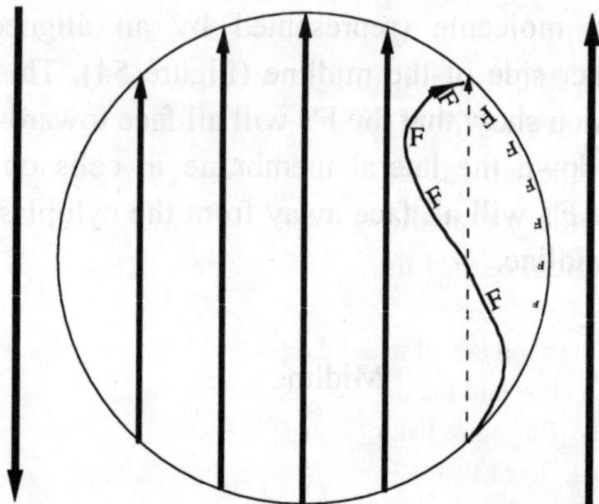

Figure 55. Torque applied to a cell on the left side, with oriented, anchored, apical microtubules, twisting them. Polymerization yields a clockwise helix down the lateral cell membrane with Fs pointed towards the cytoplasm.

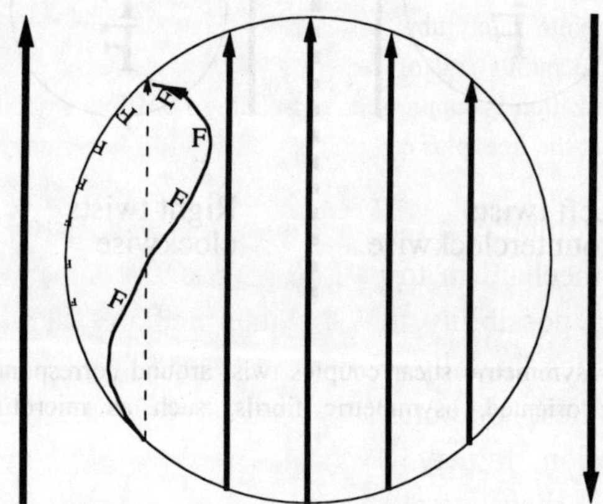

Figure 56. Torque applied to a cell on the right side, with oriented, anchored, apical microtubules, twisting them. Polymerization yields a counterclockwise helix down the lateral cell membrane with Fs pointed away from the cytoplasm, with supercoiling opposite to that of Figure 55. The supercoiling will thus enhance the intrinsic coiling of the microtubules on one side, while partially canceling it on the other side. This can lead to a left-right difference in gene expression via differential binding of, say, dynein, to the positively and negatively supercoiled microtubules.

The paraxial microtubules in newt neural plate cells have been assumed to have a role in cell shaping and possibly force generation (Burnside, 1971, 1973a,b). Since neurons grow from the neural tube lumen outward, and the lumen surface is the apical surface of the cells, we can assume (until someone tests the idea) that the orientation of the paraxial microtubules is minus end at the apical end of the cell; plus end at the basal end of the cell:

"The microtubules in the axon are all oriented the same way, with the plus end pointed in the direction of the growth cone and the minus end anchored in a microtubule-organizing center located in the cell body (Burton & Paige, 1981; Heidemann, Landers & Hamborg, 1981) [cf. Heidemann et al., 1984; Joshi et al., 1986; Baas, White & Heidemann, 1987; Heidemann, 1991]" (Jacobson, 1991b).

This orientation is precisely what is needed for polymerization of the microtubules from the apical end towards the basal end of the ectoderm cell, in which I presume paraxial microtubules are present:

"Assembly of tubulin into microtubules in cells occurs in head-to-tail order (Bergen & Borisy, 1980), with the minus end of the filament anchored in a microtubule organizing center, where polymerization is suppressed, and addition of tubulin subunits to the filament occurring exclusively at the free, plus end (Kirschner, 1980)" (Jacobson, 1991b).

If the paraxial microtubules do not impinge upon the lateral membrane, then we need another mechanism to couple gene expression to supramolecular torque. There is a possibility that the intermediate filaments binding the paraxial microtubules laterally are involved, and perhaps also the nucleus, which may ride these microtubules in the apical/basal direction during interkinetic migration. It may be that the polymerization of the paraxial microtubules and/or their lateral contact occurs while the cell is participating in the ectoderm contraction wave. It may be that nuclear rotation is involved, especially if the nuclei are oriented with respect to the apical surface, as in *Drosophila* (Hiraoka, Agard & Sedat, 1990).

Proposition 258: left-right asymmetry develops in bottle cells at the blastopore.

There is one problem with the idea that left-right asymmetry is caused by the ectoderm contraction wave: our preliminary evidence suggests that much of the propagation of that wave is through ectoderm that has no underlying tissue (Williot et al., 1998). Natalie K. Björklund (p.c., 1997) has proposed that the site of left-right asymmetrization is the blastopore itself (cf. "whatever the molecular mechanism, left-right patterning is unlikely to be due to a [chick] blastodermal prepattern but rather is initiated in a streak-autonomous manner" Levin et al., 1997; and Nascone & Mercola, 1997, place the event in the formation of the organizer in *Xenopus*). The margin of the blastopore probably consists of a number of expansion and contraction waves:

"Some of the mesoderm travels over the lateral lips of the blastopore and participates in bottle cell formation. Some of the mesoderm slips in below the dorsal lip without participating in bottle cell formation" (Appendix VI: Björklund & Gordon, 1994).

The pharyngeal endoderm cells similarly arise after going through a bottle cell shape (Figure 10 in Appendix V: Gordon, Björklund & Nieuwkoop, 1994). In particular, we can then conclude that at least some of the cells forming the heart itself go through a bottle cell phase:

"The heart in vertebrates develops from the mesoderm forming the ventral edges of the lateral plates in the pharyngeal region of the body.... During gastrulation in the amphibians, the sheet of mesoderm advances forward from the blastopore...[after invaginating]" (Balinsky & Fabian, 1981).

Now, when a mesoderm or endoderm cell goes through a bottle shape, its apical end becomes narrow, sinuous, and extended. The narrowing would tend to bring the paraxial microtubules into contact with the inner cell membrane, or at least with the cell cortex. The sharp curvature of the blastopore, coupled with probable shearing motions (cf. shear to the point of separation of some cells, which may become germ cells: Appendix V: Gordon, Björklund & Nieuwkoop, 1994), and the left-right lateral turning of cells as they invaginate (Figure 146 in Balinsky & Fabian, 1981, taken from Vogt, 1929a), could be exactly the situation in which a bottle cell itself becomes committed to being left or right handed, by the mechanism outlined here. This might even be reflected in left-right twisted shapes of bottle cells. Although they are not bottle cells, Antone G. Jacobson and I may have observed this kind of twisting in notoplate cells, alluded to indirectly:

"As the shape of the notochord region [notoplate] becomes more elongated, most internal units [simulated cells] become part of the perimeter. Extensive rearrangement is necessary. That cell neighbors do in fact change in this region in real embryos is illustrated by tracing the movements of some cells in the region.... How this rearrangement is accomplished with the basal ends of the neural plate cells attached to notochord cells remains an enigma" (Jacobson & Gordon, 1976a).

It would be curious if the very geometry of these cells proves to be left-right asymmetric, i.e., deviating on average from mirror symmetry. Quantitative measures of the curvature and torsion (James & James, 1976) of the symmetric axis (Blum, 1973, 1979; Blum & Nagel, 1978; Pizer, Oliver & Bloomberg, 1987) of cells (a space curve: James & James, 1976) at corresponding left-right positions may be needed to test this idea, which could be checked by 3D imaging or reconstruction.

At the molecular level, new evidence of a mutant dynein corresponding to randomization of left-right asymmetry is morphologically and temporally in the right place, occurring in mice at the homolog for the amphibian blastopore, during gastrulation:

"...We report the positional cloning of an axonemal dynein heavy-chain gene, left/right-dynein (lrd), that is mutated in both lgl [legless] and iv [inversus viscerum]. lrd is expressed in the node of the embryo at embryonic day 7.5, consistent with its having a role in LR [left-right] development. Our findings indicate that dynein, a microtubule-based motor, is involved in the determination of LR-handed asymmetry and provide insight into the early molecular mechanisms of this process" (Supp et al., 1997; cf. Travis, 1997e).

Such a motor molecule could be precisely what tightens or loosens supercoiled microtubules against the lateral membrane of a bottle cell. The proposed signal transduction pathway from microtubules through a dynein/ZIC3 complex (Gebbia et al., 1997) to the nucleus (Figure 2b in Srivastava, 1997) lacks any distinction between left and right:

"The intricacies of how asymmetry of gene expression is established and maintained is a mystery, but it is suspected that signals emanating from the midline notochord may play a role (Danos & Yost, 1995; Lohr, Danos & Yost, 1997)" (Srivastava, 1997).

The left/right difference in supercoiling of microtubules in bottle cells, near the midline, which may be the very cells that are becoming notochord as they participate in one of the blastopore contraction waves (Appendix V: Gordon, Björklund & Nieuwkoop, 1994; Appendix VI: Björklund & Gordon, 1994), may be the solution to this mystery. One need only assume that (or test whether) dynein binds to and/or transports differentially along microtubules that are tightened (one side of the embryo) or loosened (on the bilaterally opposite side) by the torque due to invagination. With modern methods of exerting forces on single microtubules while they are binding and transporting single motor molecules (Ashkin et al., 1990; Kuo, Ramanathan & Sorg, 1995; Gittes et al., 1996; Higuchi et al., 1997; Kojima et al., 1997), even a direct test of binding (actin: Nishizaka et al., 1995; kinesin to microtubules: Romberg & Vale, 1993; Vale et al., 1996) versus degree of supercoiling of a microtubule may be possible.

There may be feedback effects of this binding on the structure of the microtubule itself (Huyett et al., 1998), so we could imagine that a small effect on a left twisted microtubule versus a right twisted microtubule would be amplified, especially if multiple dynein molecules were involved. Analogies with sperm and cilia movement, in which bound 'motor' proteins alter microtubule structure (Lindemann, 1996), might be worth exploring. The buckling of microtubules (Kurachi, Hoshi & Tashiro, 1995) can vary by four orders of magnitude (10,000 times) under compression, depending on the presence of laterally bound proteins (Brodland & Gordon, 1990). "Interestingly, the critical load of an elastically supported compressive member does not depend on its length when it is more than approximately its buckling wavelength, λ, long" (Brodland & Gordon, 1990). This implies that when lateral support falls below this wavelength, which we estimate as $\lambda = 0.9$ μm, buckling will occur. Therefore the linear density of motor proteins along a supercoiled microtubule could act like a mechanical switch as a function of its length, and the switching threshold might be quite sensitive to whether the supercoiling was tightened or loosened, which we have postulated as the primary left-right difference. It is also possible that lateral binding between neighboring attached motor proteins could accentuate this effect via one (Hill, 1960) or two dimensional (Gordon, 1980c) Ising lattice phase transitions.

The advantage of placing the development of left-right asymmetry in mesodermal bottle cells is that no induction from another tissue is needed. It is their very participation in a blastopore differentiation wave that confers this property. Randomization of left-right heart orientation in notochordless embryos may be consistent with this location and timing of left-right asymmetrization, though the authors conclude that the event occurs at later neurulation stages (Danos & Yost, 1996). Alternatively, given that microtubule motor molecules are at work during cortical rotation in amphibians (Larabell et al., 1996), we must consider the possibility that left-right asymmetry starts before first cleavage:

"Cytoplasmic transfer experiments suggest that dorsal determinants are transported 90 degrees from the vegetal pole to the dorsal equator, even though the cortex rotates only 30 degrees. Here we show that, during rotation, small endogenous organelles are rapidly propelled along the subcortical microtubules toward the future dorsal side and that fluorescent carboxylated beads injected into the vegetal pole are transported at least 60 degrees toward the equator" (Rowning et al., 1997).

Wood (1997) gives three criteria for a satisfactory theory of left-right asymmetry:

"A workable model for establishment of L/R differences in development must (*a*) identify this component or process, (*b*) explain how it can function to generate observed handed asymmetries, and (*c*) rationalize in terms of this component or process the known mutations that result in randomization or reversal of laterality" (Wood, 1997).

At least for amphibians, a theory based on the initial cortical rotation, followed by differences in supercoiling of left-right microtubules setting off different gene cascades, would appear to meet all three criteria. Wood's (1997) own theory, based on centrioles or centrosomes in organisms in which left-right asymmetry develops at first cleavage, could be related to the microtubule supercoiling theory in some way. For instance, cortical motions during first cleavage could lead to torques on the surface of each new cell that twist supercoiled microtubules (in the cortex?) in opposite directions. At this level, we may be dealing with single cell differentiation waves, and so the mechanisms of left-right asymmetry and asymmetric cleavage may ultimately be one and the same.

8.03 Facets of Embryogenesis

Proposition 259: the differentiation pathway may be similar or identical for all steps of differentiation.

The broad pleiotropy of the *Drosophila shibire* mutant, and the fact that it uncouples the differentiation wave from differentiation of the ommatidia, suggests that this gene is part of the differentiation pathway, rather than of the cell state splitter. Perhaps then the differentiation pathway is identical for all edges of the differentiation tree. The *shibire* mutant may affect microtubules in the cell state splitter, perhaps their binding to the cell membrane (somehow involved in membrane recycling: Hummon & Costello, 1987), and have some cross-phyla universality:

"Dynamin was discovered in bovine brain tissue as a nucleotide-sensitive microtubule-binding protein of relative molecular mass 100,000. It was found to cross-link microtubules into highly ordered bundles, and appeared to have a role in intermicrotubule sliding in vitro.... *Drosophila melanogaster* contains multiple tissue-specific and developmentally-regulated forms of dynamin, which are products of the *shibire* locus previously implicated in endocytic protein sorting" (Chen et al., 1991).

This potential universality may extend down to a common metazoan ancestor:

"...The presence of a single *Hox/HOM* cluster in the red flour beetle *Tribolium castaneum* (Beeman et al., 1989), the nematode *Caenorhabditis elegans* (Kenyon & Wang, 1991; Salser & Kenyon, 1994), the crustacean *Artemia* (Averof & Akam, 1993), and some primitive chordates (Pendleton et al., 1993; Holland et al., 1994) is most consistent with the concept of a single ancestral cluster..." (Krumlauf, 1994).

Another *Drosophila* mutation that may also uncouple the differentiation wave from differentiation has been discovered:

"It should be noted that in *ato* mutants, the MF [morphogenetic furrow] still exists (Jarman et al., 1994; Nadean L. Brown, unpublished data), suggesting that proneural determination and cell shape changes are separable processes" (Brown et al., 1995).

We can anticipate that the differentiation pathway might be duplicated in part or whole at each edge of the differentiation tree. Thus this proposition represents the extreme statement of no such duplication. Where the truth lies will have to await further enquiry. Miklos & Rubin (1996) argue for a set of 'core' pathways, which could include the differentiation pathway.

Identity of cell state splitters between states of differentiation could only obtain if all the specificity is in the nucleus, i.e., if the nucleus can indeed expose just two master genes, one to respond to a contraction wave, and another to respond to an expansion wave, fully sequestering all other pairs that could respond equally well. Since the pathway itself can be present in duplicate copies, which would diverge, I would not anticipate complete identity. Thus I expect that 'the' differentiation pathway would vary a bit between cell types, and, of course, between organisms.

Proposition 260: critical periods in embryonic development correspond to periods when differentiation waves are moving.

This follows by generalization of the results on the *shibire Drosophila* mutant described in Proposition 248. In human embryos: "Each organ has a critical period during which its development may be deranged" (Moore, 1977). (Cf. Erzurumlu & Killackey, 1982.)

Proposition 261: differentiation waves, being slow moving, generate long lasting gradients (i.e., gradients are epiphenomena).

Part of the conceptual problem with leaning on gradients of morphogens for *Drosophila* patterning may be a predilection to believe that a gradient can only be the product of diffusion. However, in a cellular system, and perhaps even in a syncytium, a slow wave, by activating nuclei in succession, could leave behind it a gradient of gene products (Appendix V: Gordon, Björklund & Nieuwkoop, 1994; Propositions 12, 50), as it obviously does in *Drosophila* eye development (Kimmel, Heberlein & Rubin, 1990; Moses & Rubin, 1991; Lawrence, 1992). Thus the mere presence of a gradient does not solve the problem of cause and effect:

"As I write, gradients are respectable again, people are now happy to resort to them on the least provocation. However, we still do not know how gradients work in true multicellular systems, although there are some promising leads" (Lawrence, 1992).

Differentiation waves, by generating gradients, thus permit us to tie up both ends with the middle in the ancient dispute between the preformationists and the epigeneticists:

"From the beginning of our subject there has been a tension between the 'left wing' and the 'right wing'. On the right I see those who thought that pattern formation, the organization of spatial differentiation, was something internal to the cell with little influence of cell interactions. Those people believed in cytoplasmic localization and autonomous cell lineages generating diversity between cells, following in the tradition of A. Weismann [1889] and E.B. Wilson [1925]. What I regard as the left wing were the people who thought that things were much more global, more interactive, in the tradition of Hans Driesch [1914]. I think that distinction is still present today - on the one hand are those who pursue the lineage cytoplasmic line and on the other hand are those who prefer gradients and more global interactions. In the middle are those who think in terms of local interactions, such as induction.

"It is curious how few ideas we have in our field. We have a lot of data, but I feel I have already mentioned... the ideas.... It would be nice if there were some coming together between them, for few... are necessarily mutually exclusive.

"...I would hope that the 'left' and the 'right' and the 'middle' will ultimately come together. I would be very sad if it turned out that there were lots of different mechanisms of morphogenesis that bear no relation to each other whatsoever. I hope that general mechanisms for going from genes to flesh and blood are going to emerge. But I may just have to face the fact that I am going to be disappointed " (Wolpert, 1989b).

Spemann (1927b) puts the argument against gradients for determination of the neural plate rather succinctly, in the context of the Siamese twin experiment of Spemann & Mangold (1924a,b):

"It is a significant fact that the secondary embryo need not have the same orientation as the primary. If it had, this would have pointed to some intimate structure of the host, in accord, perhaps, with Child's [1914] idea of gradients. Now such gradients may exist in the host and may be effective; there are reasons to believe it. Yet their influence is surely not decisive. The induced secondary embryo may even lie at right angles to the primary one" (Spemann, 1927b).

In summary, a differentiation wave initiates a differentiation cascade consisting of 50 to 300 or so gene products on average (Proposition 174, Table 3 and Figure 36). Each of these will be expressed in a gradient in the wake of the wave. The shape of each gradient will depend on the time course of synthesis and degradation of each particular gene product.

Proposition 262: the homeobox 'code' and regional differentiation can be explained in terms of consecutive differentiation waves with nested, overlapping trajectories.

Coletta, Shimeld & Sharpe (1994) have expressed the need for a level of control above that of the homeobox genes:

"In order to provide positional information in the embryo, the Hox genes themselves must be regulated to produce spatially restricted patterns of gene expression. The expression domains of the Hox genes are central to their function in specifying positional identities, it being equally important to express Hox genes in the correct cells as it is to prevent their expression in the wrong cells. How are the spatially and temporally restricted patterns of Hox gene expression created?" (Coletta, Shimeld & Sharpe, 1994).

The answer given is that "...the combined effects of an unknown number of cis-elements and trans-acting factors must result in expression patterns that are unique to each gene and define its domains of spatial and temporal expression" (Coletta, Shimeld & Sharpe, 1994) without any mechanism being proposed for how this works, though one might suppose that a Turing mechanism is implied. I suggest that the required 'regulatory' mechanism is the passage of differentiation waves, whose trajectories' shapes and timing define the "spatially and temporally restricted patterns of Hox gene expression".

In addition to somites and segmental organization of the main body axis (Akam, Dawson & Tear, 1988; Gaunt, Sharpe & Duboule, 1988; Harvey & Melton, 1988; Murphy, Davidson & Hill, 1989; Wilkinson et al., 1989b), homeobox containing genes have been implicated in limb (Brockes, 1989; Gardiner & Bryant, 1996), tooth (MacKenzie, Ferguson & Sharpe, 1992), head (Holland, 1988; Mahaffey, Diederich & Kaufman, 1989; Balling et al., 1989; Lonai & Orr Urtreger, 1990), vertebrae and ribs (Kessel & Gruss, 1991; Le Mouellic, Lallemand & Brûlet, 1992; Gehring, 1993), hindbrain (Nieto et al., 1992) and neural crest (Hunt et al., 1991a) development (reviewed in Hunt & Krumlauf, 1992). These genes are expressed in overlapping sequences:

"Antp-like homeogenes are expressed in overlapping longitudinal areas of the neuroectoderm and mesoderm. The genomic organization of these genes is most uniquely connected with their expression. They are organized into unlinked clusters of 8-10 loci. The linear order of genes within the clusters is highly conserved both among clusters within a species and among clusters of distant species. Expression of these genes along the longitudinal body axis uniformly follows a 5'-posterior-3'-anterior rule. Their anterior border of expression stops abruptly at the hindbrain. A hypothesis is proposed, which suggests that much of the mid- and hindbrain, as well as the craniofacial structures, are newly acquired in the evolution of vertebrates, and do not utilize the more ancient Antp-like homeogenes" (Lonai & Orr Urtreger, 1990).

The spatial and temporal overlap has, of course, yet to be related to differentiation waves. There are four waves of nuclear division ('mitotic waves') that occur after the nuclei reach the surface of the embryo and before cellularization sets in, as described here for *Drosophila melanogaster:*

"After fertilization, embryogenesis proceeds rapidly, with the first instar larva hatching from the eggshell after 24 h. Nuclear division cycles occur extremely rapidly, with nine nuclear doublings in the first 90 min of development. These nuclear divisions take place in a syncytium, and are not accompanied by cell division. After the ninth nuclear division, the nuclei migrate from the central yolk region of the egg to the periphery, where they go through four more syncytial divisions in the next hour. After the 13th nuclear division, cell membranes form between the nuclei. As soon as cellularization is complete, at about 3.5 h after fertilization, gastrulation begins. Transcription in zygotic nuclei cannot be detected until the nuclei reach the periphery, and the rate of transcription per nucleus remains low until the end of the syncytial divisions (Edgar & Schubiger, 1986)" (Anderson, 1989b).

If each mitotic wave is a differentiation wave, turning on and off sets of genes, then the combinatorics of the 'homeobox code' (Baumgartner & Noll, 1990; Kessel & Gruss, 1990, 1991; Hunt & Krumlauf, 1991; Hunt et al., 1991a,b,c; Holmgren & Engel, 1992; Hunt & Krumlauf, 1992; Shimeld & Sharpe, 1992; Pellerin et al., 1994; Burke et al., 1995; Lumsden, 1995; Johnson & O'Higgins, 1996; Lemaire & Kessel, 1997) might be accounted for. (Note that some are beginning to doubt such a code exists, which would suggest that homeobox genes may not always be master genes or, more radically, that master genes do not exist: Carrasco & Lopez, 1994; Godsave et al., 1994; Graham, 1994; Zákány et al., 1997). Some systematic time lapse observation, and correlation of the observed mitotic waves with the expression of homeobox genes, is clearly warranted. Do these waves correspond to the waves of gene expression seen in *Drosophila* (Skeath & Carroll, 1994)? Do they correlate with the ionic currents that enter the two ends of the *Drosophila* preblastoderm embryo (Overall & Jaffe, 1985)?

It is of possible significance that primary neural induction in amphibians is also accompanied by a wave of cell division:

"Our results with gastrulae, and results of other workers, all indicate that FUdR causes abnormalities of development through inhibition of DNA synthesis. It is perhaps significant that in *Xenopus* a wave of DNA synthesis passes along the dorsal ectoderm during gastrulation (Maleyvar & Lowery, 1973). This wave appears to be one of the first responses of the presumptive neural tissue to the stimulus of primary induction" (Maleyvar & Lowery, 1981).

This correlation will be elaborated in Section 9.04. But the key issue here, yet to be experimentally investigated, is whether any set of observable waves correlates with the hierarchy of homeobox genes. If so, and the waves are nested, then regional differentiation would also be accounted for.

Proposition 263: the boundaries (grooves) for each level of segmentation in a segmenting or compartmentalizing organism are launching domains for differentiation waves.

"It has become increasingly clear that the cues ['control' proteins] in *Drosophila* are not the actual fundamental system for general segment formation" (Hunding & Engelhardt, 1995).

This is a simple prediction, which the reader is welcome to confirm or deny by direct observation of segment formation, especially in *Drosophila* (cf. Proposition 242). It replaces modelling in terms of gradients:

"In my view, the lineage boundaries established by *ftz, eve* and *engrailed* in the *Drosophila* embryo are likely to register gradients of positional information, but the form of the gradient landscapes and their relation to segments and parasegments is unknown. There are a number of possibilities, but the one I prefer is a sawtooth gradient landscape with alternating direction of slopes in the anterior and posterior compartments; changes of slope occurring at the boundaries" (Lawrence, 1992).

Instead of assuming that at each compartmental boundary a set of cells "...act as a source of morphogen that has a longer range [than 1 cell] and provides detailed positional information to both anterior and posterior compartments" (Lawrence & Morata, 1994; cf. Basler & Struhl, 1994), I would suggest that a differentiation wave will be found to depart from such a compartmental boundary, leaving in its wake both cells going through the next step of differentiation, and the purported gradients of so-called morphogens (Propositions 12, 50, 261). Differing trajectories of differentiation waves might account for whether or not overlap occurs in the domains of homeotic genes (Mahaffey, Diederich & Kaufman, 1989). Waves could be launched in either or both directions from a segment boundary. Asymmetries at the launching domain, due to cell polarity and/or underlying tissues, could account for asymmetric launching.

We have seen in axolotls that waves can be launched along circles or circular arcs (vegetal yolk mass contraction wave and blastopore waves: Appendix V: Gordon, Björklund & Nieuwkoop, 1994). However this is accomplished, it makes launching of a wave from a groove on a *Drosophila* embryo plausible. "How can a wave be launched from an arc or line, as opposed to a point? Somehow the whole set of cells on the arc needs to 'fire' synchronously" (Steve McGrew, p.c., 1997). Possibilities include electric currents and mechanical buckling, to achieve such synchrony.

Differentiation waves, which we could envisage as being transmissible either prior to or after cellularization, because they are essentially cortical phenomena, solve the following paradox, caused by gradient thinking, and preserve the obvious homologies between organisms with cellular and syncytial segmentation:

"Molecular probes reveal similar patterns of segmentation and homeotic gene expression in the mature germ bands of *Schistocerca* [African desert locust] and *Drosophila* embryos - representatives of insects with short and long germ type embryos respectively. This raises a paradox, for the analysis of segmentation in *Drosophila* has revealed a mechanism that depends primarily on the intracellular diffusion of transcription factors within the syncytial environment of the blastoderm. Such a mechanism cannot operate to generate segments within the cellular environment of the growing germ band in short term insects. Different, and probably non-homologous, mechanisms must have evolved to generate homologous structures in these different insect groups" (Akam, Tear & Kelsh, 1991).

Cf.:

"...The segment patterning mechanism that acts in an open [syncytial] blastoderm in *Drosophila* works in a similar way in the cellularized *Tribolium* embryo" (Sommer & Tautz, 1993).

Are we to believe, from the following, that *Tribolium* switches halfway along its length to a 'non-homologous' mechanism?:

"In *Drosophila,* all of the pair rule stripes are formed during the syncytial stage of development. However, it has been recently demonstrated that in the beetle *Tribolium* only the first two *h* stripes are formed during [the] syncytial blastoderm [stage], and the more posterior stripes are generated in a cellular field of segment primordia (Sommer & Tautz, 1993)" (Hartmann et al., 1994).

Anderson et al. (1992), to solve this quandary, suggest that the morphogens form extracellular gradients, though the sharp distinctions between adjacent male and female cells in gynandromorphs (Gilbert, 1991a) and the 'cell-autonomous' action of *Notch,* for instance (de Celis, Marí-Beffa & García-Bellido, 1991), might be taken as evidence against this idea. Hunding & Engelhardt (1995) accept the possibility that...

"...the homeobox genes are not 'pattern' genes as such, but rather convenient control genes linked to another set of fundamental pattern forming processes [such as differentiation waves?].... Stripe control by combinations of diffusing large molecules as proteins (cues) seems unlikely. This recent discovery [of similar genes in cellularized insects] has reopened the debate regarding the evolutionary significance of the cue control system found in *Drosophila*.... There may... be several chemical control systems around that have generated repetitive patterns in early evolution, and it may be that various segmentation processes are genetically related only superficially (Riddihough, 1992a,b)" (Hunding & Engelhardt, 1995).

Despite major efforts at supercomputing of gradient/Turing models (Hunding, Kauffman & Goodwin, 1990; Hunding, 1991, 1993; cf. Nagorcka, 1988), Hunding & Engelhardt (1995) pin their hopes for an explanation on the generic nature of nonlinear reaction-diffusion mechanisms with...

"...high cooperativity in the gene regulatory system.... Such systems may easily yield chemical time-oscillations, which may be used to generate segments.... Turing's mechanism may be generalized... with cell-to-cell communication... replacing simple diffusion" (Hunding & Engelhardt, 1995).

Given the evidence cited here, it seems more likely that there is, indeed, one universal mechanism, perhaps differentiation waves, rather than two unrelated mechanisms that uncannily give the same phenotype. Besides, reaction-diffusion models are in trouble:

"Since it was first reported by Hafen, Kuroiwa & Gehring (1984), the beautifully simple periodic pattern of pair-rule stripes in *Drosophila* embryos has intrigued theoretical and experimental biologists alike. A primary concern among theoreticians has been the question of whether pair-rule stripes are, to any extent, Turing structures, sustained by far-from-equilibrium kinetic processes. Current opinion among experimentalists tends to the view that they are not" (Lacalli, 1993).

The pair rule genes in cellular chicken embryos, for example, are expressed in a way perhaps consistent with their being involved in the launching of differentiation waves:

"Our study is focused on the paired-box containing genes, which have been implicated in delineating boundaries early in development.... Expression of paired-box genes occurs early in development as shown by Northern analysis, and is localized by in situ hybridization to the edge of each somite, a patch at the central core of each somite, and the periphery of the neural tube. This specific spatial pattern of expression is consistent with the hypothesis that the pair-rule genes function as effectors of border formation in the early embryo" (Love & Tuan, 1993; cf. Smith & Tuan, 1995).

In fact, the kinds of "cross-regulatory interactions between homeotic genes, which have been shown to play an important role in the maintenance of their expression domains during *Drosophila* development" do not occur in vertebrates: rather, whole tissues fail to develop (Dollé et al., 1993c). (Even in *Drosophila,* the importance of such cross-regulatory interactions has been 'challenged': González-Reyes et al., 1990.) The 'temporal colinearity' of genes in homeotic gene complexes occurs in vertebrates but not *Drosophila* (Duboule, 1994b; cf. Dollé et al., 1991b; Izpisúa-Belmonte et al., 1991a). But even *Drosophila* exhibits temporal colinearity at later, cellular stages (Jagla et al., 1994), suggesting that it might also occur in cellularized pattern formation in insects.

We thus see that there are many empirical challenges to the primacy of homeobox genes as controllers of pattern formation. The physical act of segmentation is unexplained, but it seems to exactly correlate with boundaries of homeobox gene expression. It is simpler to suppose that the boundaries are the launching domains of differentiation waves, but we'll have to see if these predicted waves actually exist.

Proposition 264: differentiation waves provide the common morphogenetic mechanism between cellular and syncytial organisms.

A situation of syncytial versus cellular mechanisms (Proposition 124), analogous to what we have seen in insects, occurs in plants:

"Multicellularity is not a prerequisite for the expression of complex morphology,.... nor does it intrinsically limit the absolute size of a plant (Kaplan, 1987a,b). Indeed, some aquatic unicellular plants rival many bona fide terrestrial plants in size and shape.... The propositions and corollaries of the cell theory are not legitimized by studies of plant development or by our current understanding of plant systematics and evolution.... 'Plants make cells; cells do not make plants' [Heinrich Anton de Bary: cf. Nordenskiöld, 1928].... The organismal theory essentially views the organism as a continuous mass of protoplasm (the symplast) that may or may not be incompletely partitioned into cells during the plant's ontogeny. Plant and animal cells are viewed as the result of ontogeny, not its cause, and unicellular and multicellular organisms are placed in parity with one another - the latter is the septated equivalent of the former.... Multicellularity must be viewed as a highly specialized expression of development in which the division of the protoplasm and the division of the nucleus are highly correlated" (Niklas, 1992; cf. Niklas, 1994b).

While I do not agree that cellularization is an entirely incidental phenomenon, the notion that differentiation waves travel through the cortex, in insects, in plants, and in ciliates, independently to some extent of the cellularization of the protoplasm within, may provide the unifying concept for the morphogenesis of multicellular and unicellular organisms or developmental stages. This idea was well expressed by Whitman (1893):

"Whitman's notable essay on 'The Inadequacy of the Cell Theory of Development' is often cited as opposed to the cell concept of the organism. However, this paper is in essence a plea for the recognition of a microstructure as the basis of organization, and is in no sense the negation of the cell as a living unit.... 'If the formative processes,' says Whitman (1893), 'cannot be referred to cell-division, to what can they be referred? To cellular interaction? That would only be offering a misleading name for what we cannot explain; and such an answer is not simply worthless, but positively mischievous, if it put us on the wrong track.... The answer to our question may be difficult to find, but we may be quite certain that when found will recognize the regenerative and formative power as one and the same thing throughout the organic world. It will find, as Wiesner [cf. Wiesner, 1892] has so well insisted, a common basis for every grade of organization, and it will abolish those fictitious distinctions we are accustomed to make between the formative processes of the unicellular and multicellular organisms'" (Harrison, 1969).

Here is an amphibian example in which the extent of cellularization seems to have little effect on development:

"J.S. Huxley (1927) exposed frog's eggs in early stages of development to a temperature gradient, by bringing them, in their jelly, between two tubes, through one of which flowed hot water and through the other, cold. The heat was applied to the eggs either on their animal or their vegetative pole, and thus the primary gradient of activity either made steeper... or flatter.... The immediate effect was the expected one and it was very striking indeed: in the first case the difference in size between blastomeres was increased, whereas in the latter it was diminished. The final effect, however, was surprising to Huxley himself on account of the insignificance of the abnormality of development" (Spemann, 1938). [Cf. Drury, 1941.]

Hans Spemann appears to have thought about induction mostly in terms of supracellular regions or tissue than in terms of cells (with the emphasis gradually reversed by Johannes Holtfreter: Oppenheimer, 1970a). It may be important to note that the views of an organismal theory versus a cellular theory may have their roots in our ability to intertwine our political views with our theories of life:

"...The multicellular organisms of the ordinary macroscopic world came to be regarded as collectives. For Schwann (in Jacob, 1973) the multicellular organism was a cellular state in which 'each cell is a citizen.' Indeed, in such a cellular state there is a distribution of tasks, or division of labor. The whole depends upon the cooperation of the parts" (Ravin, 1977).

Not only our words (Section 1.18), but our metaphors, may obscure clear thinking in biology.

The gradient paradigm is in trouble, when its promulgators are willing to throw out the evolutionary baby (phylogeny) with the bath water, when comparing syncytial with cellular organisms:

"If we are to believe that *Drosophila* has invented a whole new mechanism to make segments, then what was the primitive mechanism that insects used to build their segments? The only hint of an answer to this question comes from *Drosophila* itself. After gastrulation, the segment pattern that was defined by the pair-rule genes becomes dependent on a completely new set of regulatory interactions" (Akam & Dawes, 1992).

"It is currently difficult to imagine the engineering of signal transduction machinery that could underlie reliable, but different, responses by early embryonic cells to several successive concentration ranges of an extracellular ligand. How well can the effective numbers of cell surface receptors or concentrations of presumed phosphorylated intermediary messengers, per cell nucleus, be controlled among the rapidly dividing, variable-sized blastomeres of the early frog embryo, for instance?... Is the vertebrate situation formally different from [*Drosophila*]..., as well as necessarily using *trans*-cellular transduction machinery?... There are cogent reasons for doubt that vertebrate primary body patterning will turn out to rely upon direct reading of initial morphogen gradients to give five or more alternative cell states..." (Cooke, 1995).

This in not to say that such gradients don't exist, but rather, that they are simple consequences of differentiation waves ('epiphenomena': cf. Proposition 261, Section 9.24). The problem of the difference in sequence of formation of the segments between long-germ and short-germ insects may simply be one of timing (Akam, 1994), once the cellular/syncytial intellectual bottleneck is overcome. The suggestion that "vertebrate myogenesis and *Drosophila* neurogenesis [represent] a common theme" (Skeath & Carroll, 1994) works both ways: if differentiation waves prove to be the basis of pattern formation in vertebrates, then we should be looking for them in fruit flies.

If one considers the history of modelling of stripes of gene expression in the *Drosophila* blastoderm, one is struck by the difficulties that beset those who try to test these models quantitatively:

"It was... widely believed for years that the maternal genes *bicoid* and *nanos* provided gradients from the anterior and posterior ends of the egg, respectively, which in turn were used for activation of three gap genes. These gap genes were believed to inhibit each other. This model is now abandoned and replace with a model in which at least one more gap gene is added, and their mutual interactions are more complex....

"...A mechanism similar to the one that activates the gap gene level could also operate on the primary pair-rule gene level. This requires a sufficient number of gradients provided by the gap level to define at least seven distinct stripes and it is not clear that the gap and maternal genes could provide sufficient information for this. How a number of independent particular stripe generators could cooperate to form the observed equally spaced stripes is a much more serious problem.... Moreover, the formation of stripes begins before the gap genes reach their final positions. One would expect to see the initial stripes move around until they found their final positions, but this is not observed. Also, stripes would be expected to narrow or broaden in gap mutants....

"...*hb* protein may be crucial even in regions where current techniques barely are able to detect it, a feature that further complicates the use of available expression patterns for model building....

"It is difficult to see what common striping mechanism could be present if it does not have its origin in cues set up by the gap (and maternal) genes.... The experimental studies that have given rise to the present cue model are quite impressive. Nevertheless, the cue model as it stands is more a program for further research than an actual model....

"...Stabilization of zebra stripes is by no means trivial even with a Turing model, when the stripes, as here, are short wavelength with respect to the characteristic length of the embryo. Thus, unwanted more or less patchy patterns obtain as a rule if the stabilization of stripes from the maternal and gap level is ineffective....

"Because the present state of knowledge of the proposed cue model does not specify control of more than a few of the stripes, and only in quite preliminary terms, the number of possible interactions... (5^6) [is]... impossible to exhaust.... Computed patterns for the most successful model so far are given....

"What emerges so far is that the control of any pair-rule gene is highly intricate.... This may be taken to mean that the cues in *Drosophila* are not the fundamental system for segment formation. One may speculate that the cues are evolutionary late additions to a fundamental segmentation mechanism (Webster & Mansour, 1992), which has been obscured by the addition of gap and pair-rule genes....

"This may thus indicate that the fast life cycle of *Drosophila,* which is an asset for laboratory work, actually may be a disadvantage from the point of view of those now trying to understand the basic mechanisms behind pattern-forming processes in biologic systems" (Hunding, 1993).

There is a sense in which *Drosophila* experimentalists have produced qualitative models based on apparent gene and gene product interactions and gene expression patterns, and presumed that the spatial component of embryogenesis would just take care of itself. The theoreticians trying to make sense of all this, in terms of gradients, have been stymied. The fundamental problem here may be that concentrations of molecules are scalars, and Turing reaction/diffusion modelling is inadequate to parlay them up to the real three dimensional, tensor structure of real organisms, syncytial or cellular. Differentiation waves, if present, may fill the gap.

Proposition 265: the hierarchy of segmentation genes is reflected in and triggered by the hierarchy of segmentation differentiation waves.

This is a consequence of our general approach that gene expression should follow, rather than precede, differentiation waves. This effect should be directly observable by time lapse with *in vivo* immunofluorescence of any of the numerous mRNAs and their gene products found in *Drosophila* segmentation. This may solve the problem of gradient sharpening, as gradients would be produced by waves. It may also explain why there is "not a functional hierarchy... [in the] cross-regulatory interactions among homeotic genes" (Lamka, Boulet & Sakonju, 1992).

Proposition 266: launching domains may represent the fundamental homologies in animals.

The fact that most, though not all, launching domains are circles or circular arcs, suggests that a new comparative anatomy, transcending phyla, may be keyed to these launching domains. We have essentially no information on what determines the shapes of launching domains. Thus the excitement in working out phylogenetic relationships may shift to an exploration of the physics of launching domains of differentiation waves, since they are the primary determinants of how an embryo is carved up into tissues.

In the *Drosophila* eye imaginal disc, some genetic information on 'initiation', i.e., launching, of its 'morphogenetic furrow' has recently come available, though it would be nice if it were combined with mechanics, ultrastructure, temperature effects, and a search for ionic currents:

"Initiation of furrow movement must occur by a different mechanism than its propagation.... Since *dpp* [*decapentaplegic*] is expressed around the edges of the eye disc before initiation (Masucci, Miltenberger & Hoffmann, 1990), it may be involved in this process. The *eyes absent* (*eya;* Bonini, Leiserson & Benzer, 1993), *sine oculis* (*so;* Cheyette et al., 1994) and *dachsund* (*dac;* Mardon, Solomon & Rubin, 1994) genes are also likely to act in initiation (Thomas & Zipursky, 1994). All of these genes can mutate to eyeless phenotypes, [and] are expressed in a similar early pattern at the margins of the disc and encode nuclear proteins.

"The *wingless* (*wg*) gene encodes [a]... secreted protein that is required for segmentation of the embryo and development of the imaginal discs (Rijsewijk et al., 1987; Baker, 1988a). It is a member of the *Wnt* gene family (Nusse & Varmus, 1992); vertebrate *Wnt* genes are involved in axis specification (McMahon & Moon, 1989a), determination of the midbrain (Thomas & Capecchi, 1990; McMahon & Bradley, 1990) and limb development (Parr & McMahon, 1995)... The [*Wnt*]... proteins appear to bind to the extracellular matrix and to act over a short range (Bradley & Brown, 1990; Papkoff & Schryver, 1990; van den Heuvel et al., 1993).

"In the eye disc, *wg* is expressed at the dorsal and ventral margins in the regions that will form head cuticle (Baker, 1988b). We show here that it acts to prevent *dpp* present in these regions from initiating a wave of photoreceptor development.... Thus *wg* and *dpp* interact to define the region in which the morphogenetic furrow can initiate.... The effect of *wg* on furrow initiation could be described as a decision between head cuticle fate and eye fate..." (Treisman & Rubin, 1995). [Cf. Heberlein & Moses, 1995.]

The launching domain from the edge of the eye imaginal disc may be homologous with the launching domain of the eye contraction wave (axolotl eye 'morphogenetic furrow') that we have observed in axolotl:

"**Natalie**: The eye anlage contraction wave begins as a circle on the edges of the rising neural ridges of the anterior neural plate at stage 14 [Figure 4; Table 1]. Upon reaching the central area of the anterior neural plate the circle breaks into an arc and the two ends close into two separate circles corresponding with the two eye rudiments. The circle vanishes and the region disappears laterally under the rising neural ridges" (Appendix V: Gordon, Björklund & Nieuwkoop, 1994).

wingless is a pleiotropic gene, perhaps also involved in launching a differentiation wave for the *Drosophila* leg, this time from a point-like domain rather than an arc:

"In the leg imaginal disc, *wg* is expressed as a stripe along the ventral half of the anterior-posterior compartment boundary (Baker, 1988b). At the point where it abuts a stripe of *dpp* along the dorsal half of the A-P boundary, the homeobox genes *distalless* and *aristaless* are induced and direct distal outgrowth of the leg (Campbell, Weaver & Tomlinson, 1993; Diaz-Benjumea, Cohen & Cohen, 1994)" (Treisman & Rubin, 1995).

In fact, this description smacks of a possible wave-wave interaction, such as we've speculated may coordinate the launching of consecutive waves, providing independence of development from temperature over a wide temperature range (Proposition 82). Furthermore:

"These results have cast doubt on the theory that [*wingless*]... acts as a graded morphogen and suggest that additional components may contribute to determining ventral cell fates" (Treisman & Rubin, 1995).

We thus have hints of launching domains, and a few genes that may be required for them. A broader investigation might reveal elements of universality and a basis for predicting homologies.

Proposition 267: the only 'morphogenetic' gradient not set up by differentiation waves is the original gradient of the oocyte, represented by *bicoid* mRNA in *Drosophila* and the animal/vegetal hemisphere difference in amphibians.

The *bicoid* protein gradient (Driever & Nüsslein-Volhard, 1988a,b; Micklem, 1995), discussed in Proposition 243, has been interpreted in terms of its concentration gradient leading to threshold determination of segmentation (cf. Dubnau & Struhl, 1996; Gibson, 1996). However, it could just as well directly or indirectly affect physical properties of the cortex of the fertilized egg, leading to the launching of differentiation waves from specific regions. The involvement of microtubules in *bicoid* distribution (Pokrywka & Stephenson, 1991; Stephenson & Pokrywka, 1992) indicates that the cytoskeleton is, indeed, implicated: "*bicoid* mRNA localisation appears to occur via active transport towards the minus-ends of microtubules" (Micklem, 1995). Mechanical effects should thus be anticipated. Kaimanovich, Krupitski & Spirov (1994) imply that these effects might be electromechanical, with microtubule assembly mediated by intracellular ionic currents.

Proposition 268: cell alignment in epithelia is correlated with the direction of propagation of differentiation waves.

Unlike water waves, differentiation waves are intrinsically asymmetric and irreversible, since they leave further differentiated cells in their wake (kink waves: Section 1.15). Many structures exhibit patterns of alignment:

bristles on insects (Lees & Waddington, 1942; Maynard Smith & Sondhi, 1961; Lawrence, 1966; Perry, 1968b; Held Jr. & Bryant, 1984; Vinson & Adler, 1987; Adler et al., 1990; Mitchell, Edens & Petersen, 1990; Held Jr., 1991; Orenic et al., 1993; Wong & Adler, 1993; Krasnow & Adler, 1994; Sharp, Martin & Adler, 1994);

scales on butterfly and moth wings (Nardi & Magee-Adams, 1986);

scales on fish (Zylberberg, Bereiter-Hahn & Sire, 1988) and reptiles;

scales and feathers on birds (Rawles, 1963; Sengel, 1975, 1990);

hairs on mammals (Nagorcka & Mooney, 1982; Nagorcka, 1983; Mooney & Nagorcka, 1985; Nagorcka & Mooney, 1985, 1988; Alexander et al., 1992);

cilia on amphibian embryos (Assheton, 1896; Woerdeman, 1925; Twitty, 1928; Holtfreter, 1933d; Tung & Yeh-Tung, 1940; Tung, Yeh-Tung & Chang, 1949; Steinman, 1968;

Waterman, 1972; Billett & Courtenay, 1973; Kessel, Beams & Shih, 1974; Löfberg, 1974b; Landström, 1977);

cilia on mammalian embryos (Sulik et al., 1994);

'hairs' in inner ears (Lenzi & Roberts, 1994);

pairs of daughter cells (Jacobson & Tam, 1982) and their microtubule filled 'connecting cords' ("elongated telophase bridges persisting between separating daughter cells": Everaert et al., 1988; cf. Bellairs & Bancroft, 1975; Nagele & Lee, 1979).

Although I cannot propose a single, universal mechanism for the generation of such alignments, one could at least, in each case, look for the passage of a differentiation wave and check if the subsequent aligned structures are, in every case, perpendicular to the wave fronts (cf. Propositions 254-255). Twitty & Bodenstein (1941) found that cilia orientation in late grafts of amphibian epidermis can change:

"This 'redetermination' of the original pattern of ciliary polarity [= alignment: epithelia are apical/basal polar]..., when it occurs, is accomplished by gradual and orderly shifts in the direction of the original currents" (Twitty & Bodenstein, 1941).

Twitty (1928) noted that mechanical effects may be involved in redetermination, which raises the suspicion that the initial determination also has a mechanical component:

"Contact [of the amphibian embryo] with a surface is effective, then, not only in inducing altered action, but also in coordinating the beat of the cilia. However, the pattern of action is sometimes atypical.... The specific conditions under which [a]... vortical pattern of action becomes induced have not been investigated, but one may sometimes observe such behavior, on the side of the embryo next the dish, after a bridge has been removed from the middle of the trunk region. The bridge taken away, the embryo spins around on its side on a stationary axis, as one might expect. Although of odd pattern, the action cannot be said to lack unified plan....

"...Determined grafts, transplanted in reverse orientation on younger embryos, may have their determination reversed by the influence of the host" (Twitty, 1928).

Similar redetermination of the orientation of aligned scales and epidermal ripples occurs in grafts on insects (reviewed by Hooper & Scott, 1992), but, perhaps due to the need to wait for a molt to see the results, externally applied mechanical forces have not been considered. Thus there is still much to learn about this process, especially interesting if we believe it might be the primitive patterning process dating back to a common ancestor with ciliates, whose surfaces are covered with aligned structures.

The following observations on the lack of a direct relationship between the genetics of a cell and its alignment suggest that alignment may have a long range mechanical component, possibly attributable to the mechanics of one of the differentiation waves that the cell has participated in:

"Bristles and hairs [in *Drosophila*] normally point posteriorly on the body and distally on appendages. These consistent orientations demonstrate that each cell has a distinctive polarity reflecting the overall polarization of the epidermis in a given region. Each hair originates from an actin-rich vertex in each epidermal cell (Adler et al., 1990; Mitchell, Edens & Petersen, 1990). Thus, hair and bristle polarity may reflect an underlying polarity of the cytoskeleton. Several mutants have been studied that disrupt the nonrandom orientations of bristles and hairs, including *prickle* (*pk*) and *frizzled* (*fz*) (Gubb & García-Bellido, 1982).... Recessive mutants of polarity genes have the common property that in small homozygous mosaic patches, they take on the phenotype of the surrounding heterozygous normal cells (are nonautonomous), but in large mosaic patches, they express the recessive mutant phenotype (are autonomous) (Gubb & García-Bellido, 1982). Moreover, large patches of *pk/pk* or *fz/fz* cells have the added property that they disrupt the polarity of neighboring wild-type cells. The disruptive effect of *fz* is particularly interesting because it extends on wings only into neighboring regions distal to the mosaic mutant patch but not into those proximal to the patch (Vinson & Adler, 1987). One speculative interpretation of these observations, in terms of cytoskeletal structures, is that the precise orientation of the cytoskeleton in one cell, whether normal or abnormal, can influence that of a neighboring cell" (Fristrom & Fristrom, 1993).

The interpretation offered is incomplete, in that it would not explain the results of Vinson & Adler (1987), whereas a differentiation wave propagating in the proximal to distal direction would do so:

"[Our] data suggest that fz has two mutably separate functions in establishing hair polarity on the wing. One function involves the transmittance and/or generation of a polarity signal along the proximal-distal axis of the wing. The second function involves the cellular interpretation of a polarity signal" (Vinson & Adler, 1987).

We need only assume that the alignment is determined by the mechanical properties of the wave, and that these are determined as a nonlinear combination of the wave propagation properties of cells within a certain (mechanically effective) distance from one another.

An apparently alternative hypothesis is that of Nagorcka & Mooney (1988), in which the asymmetry is presumed to be generated...

"...as a cephalo-caudal progression... [in which] the RD [reaction-diffusion] system is switched on first of all in the crown and then progressively over the rest of the epidermis (Nagorcka, 1986a).... The gradient [of one of the two morphogens needed in a Turing, 1952, mechanism] causes the mitotic activity to be higher on one side of the primordial [hair] follicle than on the other, which accounts for the fact that the follicle develops obliquely to the skin surface" (Nagorcka & Mooney, 1988).

However, the hypothesized 'cephalo-caudal progression' smacks very much of a 'non-specific' differentiation wave:

"A substance S, of dermal origin, which can diffuse across the basement membrane is also required to 'switch on' the RD system, e.g. by modifying the interaction [between the Turing morphogens] X and Y so that spatial inhomogeneous distributions in X and Y may arise spontaneously. S provides positional information on a regional scale. Although S was not explicitly introduced in the mechanism proposed for hair follicle initiation (Nagorcka & Mooney, 1985) the need for it was suggested to account for the spatial spread of the first wave of initiation across the skin. Both S and the RD system are assumed to be common to reptiles, birds and mammals. S is, in fact, the first morphogenetic message of dermal origin suggested by Dhouailly (1975) and described as a 'triggering factor that is apparently non-specific and can therefore be understood... by a foreign epidermis from another zoological class'.

"Since in the theoretical mechanism X and Y are confined to the epidermis, it is necessary to introduce another substance of epidermal origin, denoted E_i, which can diffuse across the basement membrane. Unlike S, E_i is characteristic of zoological class so that the subscript i may take the values r (reptile), b (bird) and m (mammal)" (Nagorcka, 1986a).

The assumption (and possibly direct observation) of a universal dermal or epidermal differentiation wave in the first place would make it unnecessary to build morphogen upon morphogen, in a manner analogous to epicycles (Sarton, 1959).

Lawrence & Morata (1994) suggest that...

"The model of positional information, which derives from grafting experiments on... insects, makes the scalar and the vector at any point in the gradient landscape define both the type of differentiation and the polarity of the cells at that point (see Lawrence, 1992; Sampedro & Lawrence, 1993). The main virtue of a gradient hypothesis over others is that it offers a single mechanism that engenders both detailed pattern and cell polarity" (Lawrence & Morata, 1994).

Differentiation waves may have the same advantage, while simultaneously explaining gradients. It will be curious to see how gradients of auxin relate to differentiation waves in plants and the development of alignment, perhaps through the mechanics of the cytoskeleton:

"A clear correlation emerges: auxin acts as a correlative signal of very young developing leaves [cf. Sachs, Novoplansky & Cohen, 1993].... [Other signals]... are especially likely in earlier stages of meristem development (Young, 1954; Sachs, 1972) as well as in developing embryos.... [The] commonly held view of orderly development (Meinhardt, 1982a; Wolpert, 1989a) might not suffice.... The specification of orientation raises problems not apparent from differentiation alone: orientation could not be specified by a non-vectorial property, such as signal concentration. This means that cells must be sensitive to the gradients or the flux (Sachs, 1981, 1991a) of correlative signals, possibilities that are often ignored in research on the cellular basis of development. Furthermore, gene expression is not known to have a directional component [but cf. Hiraoka, Agard & Sedat, 1990], so the determination of various cell axes - certainly a form of differentiation - could be expected to be relatively distant from gene expression. There are inconclusive indications that the cellular basis of this orientation depends on cytoskeletal elements (Gunning & Hardham, 1982; Sylvester, Williams & Green, 1989; Hush, Hawes & Overall, 1990; Emons, Derksen & Sasen, 1992) and these could be a bridge between the genetic basis of orientation and its control by signals from neighbouring tissues" (Sachs, 1993).

If differentiation waves directly cause alignment, then their launching domains determine "...the correct handedness as well as the correct orientation in relation to the main body axes" (Meinhardt, 1991). In other words, a launching domain is similar to an 'organizing border':

"An organizing border would be a boundary region in terms of Wolpert's concept of positional information, i.e. the source region of a morphogenetic gradient" (Meinhardt, 1991).

The main mathematical difference between a differentiation wave and diffusion of gradients will be in the dynamics. Differentiation waves will follow some kind of nonlinear wave equation, while morphogens should follow some kind of (perhaps nonlinear) diffusion equation. It might be easier to distinguish the two on the basis of mechanics: differentiation waves use the mechanics of the cytoskeleton; no postulated morphogens use anything but diffusion.

If this proposition is correct, then hairs and bristles on *Drosophila*, for example, record the direction of propagation of a differentiation wave. One might think this corollary is contradicted in mosaic animals in which "hairs near [a]... clone boundary have an abnormal polarity and contain multiple hair cells" (Krasnow & Adler, 1994). However, differentiation waves, such as the ectoderm contraction wave, do not progress completely evenly (Appendix II: Brodland et al., 1994), so that local variations and inconsistencies in alignment could occur. We should also anticipate discontinuities in mechanical properties at clonal boundaries, which might affect both the local direction of wave propagation and the mechanics of hair erection.

Proposition 269: alignment and perhaps single cell polarity can be better related to the propagation of a wave than to a gradient.

For the mathematically sophisticated reader, the embryological definitions of the concepts of field, gradient, alignment and polarity may cause some confusion. An 'embryological gradient' is a spatially nonuniform concentration of some (unknown) substance. This is consistent with the term 'concentration gradient' in chemistry. In mathematics, the correct term is 'scalar field' (Gellert et al., 1975). An 'embryological field' is a contiguous (connected), bounded, usually convex region of an embryo all of whose cells have the potential to form some particular tissue. Those that do so become 'determined', while those that don't are usually ignored by the investigator. In my terminology, a field is a tissue, which (perhaps by differentiation waves) is divided into two tissues, neither of which should be ignored. A mathematician's scalar field f has a gradient:

$$grad \ f = \frac{\partial f}{\partial x}i + \frac{\partial f}{\partial y}j + \frac{\partial f}{\partial z}k$$

where (i, j, k) are Cartesian unit vectors. The entity grad f is a particular kind of 'vector field'. As such, it ascribes a direction or vector at each point (x, y, z). This direction is effectively the alignment' of the cell at (x, y, z), or at least, this is presumed to be the case, in gradient models of alignment, which apparently originated with Thomas H. Morgan:

"In... studying the regeneration of *Tubularia* Morgan (1905a)... tries to explain....

> 'I have assumed that the stem of *Tubularia* is not homogeneous, but that from the hydranth to the base there is a graded difference and this gives the order or stratification of the material.... How does such a view differ from the old assumption of a 'polarity' in the material? On my view there is no such directive force residing in the material as the term polarity suggests, but the polarity is only a name for the gradation of the material and on this as a basis the formative changes are carried out.'" (Wolpert, 1991b).

The 'old' concept of polarity that Morgan (1905a) refers to is thus equivalent to the idea that the embryo has a vector field intrinsic to its cells, rather than a vector field derived from a scalar gradient (grad f). Whether the underlying field is scalar, vector, or nonexistent has yet to be resolved, but clear mathematical formulation of the problem may aid in its solution.

One popular model for polarity of single cells is based on the membrane capping of ion pumping channels in the brown alga *Fucus* egg with subsequent intracellular electrophoresis of cell components and their stratification (Jaffe, 1968, 1969; Nuccitelli, 1984, 1988a). This has been expanded into a possible basis for differentiation:

"An early determining event in the polarization in zygotes of brown algae is the appearance of an oriented electrical current (Jaffe & Nuccitelli, 1977; Jaffe, 1981a). It has been suggested that this current is both localized by and further localizes specific channel proteins at one pole of the cell membranes. Measurements have since shown that electrical currents are characteristic of polarized growth in plant systems (Weisenseel, Dorn & Jaffe, 1979). Currents are also characteristic of the early stages during the organization of new embryos in [plant] tissue cultures (Brawley, Wetherell & Robinson, 1984; [Overall & Wernicke, 1986]; Gorst, Overall & Wernicke, 1987; Rathore, Hodges & Robinson, 1988). The same principles of an oriented 'flow' which reinforces itself and leads to overt differentiation appear in relation to both electrical currents and auxin transport. It is an intriguing possibility that auxin transport, which occurs even in embryos (Fry & Wangermann, 1976) and appears to play a role in their organization (Schiavone, 1988), is but a late manifestation of an earlier, less-specific polarity expressed by electrical currents.... Electrical phenomena may transmit developmental information (Pickard, 1973)" (Sachs, 1991a).

However, the fundamental premise that electrical phenomena are primary causal factors in morphogenesis is still unproven (cf. Harold, 1991):

"A transcellular ionic current begins to flow into the future rhizoid site and out the presumptive thallus as the axis is forming (Nuccitelli, 1978), but whether it is a cause or consequence of polarity is uncertain" (Kropf, 1992).

"While this evidence that ionic currents play a primary role in the establishment of polarity in fucoid eggs is quite strong, we are reminded by Waaland & Lucas (1984) that we cannot generalize.... They carefully examined the current pattern around elongating rhizoids and repair shoot cells of the marine red alga, *Griffithsia pacifica,* and concluded that the ionic currents were neither sufficient nor necessary for the maintenance or reinitiation of sites of localized growth and organelle accumulation.... Therefore, each case must be examined independently to determine if the causal link between the current and morphogenesis is present" (Nuccitelli, 1988a).

"...There is strong and growing evidence that the cytoskeleton and its molecular motors, rather than an electrical gradient, is responsible for most (if not all) of the cytoplasmic transport processes that occur in the nurse-cell/oocyte syncytium of *Drosophila* " (Bohrmann & Schill, 1997).

Perhaps we should look for waves on *Fucus* eggs? Certainly, the cytoskeleton is involved in their initial asymmetrization (Kropf, Berge & Quatrano, 1989; Quatrano et al., 1991; Bouget, Gerttula & Quatrano, 1995), and there could well be a wave that propagates through the *Fucus* cortex (Kenneth R. Robinson, p.c., 1998).

Proposition 270: a set of higher order, key-like 'homeokey' memoron molecules may exist, each of which stabilizes the state of determination of a particular cell type, and may explain the colinear organization of homeobox genes in a given homeobox gene complex, and the mechanism of dedifferentiation.

"As Abercrombie (1965) has said: '...there is a kind of cell or tissue heredity about differentiation, which ensures a lasting change of properties'. This would suggest that during metazoan evolution there has been developed a series of repressors which are able to bind particularly strongly with their relevant genes or operators" (Bullough, 1967).

One of the unsolved mysteries to emerge from molecular developmental genetics has been the colinearity within groups ('complexes') of homeobox genes:

"A distinguishing hallmark of the *Hox/HOM* complexes is the correlation between the physical order of genes along the chromosome and their expression/function along the anteroposterior axis of the embryo. This property was recognized by Lewis (1978) and his colleagues in *Drosophila* BX-C and referred to as colinearity.... However, despite the widespread interest and importance of colinearity in formulating ideas about function, there is still no coherent explanation of how it is achieved. It remains one of the major unsolved problems, and a good deal of effort is being put into the dissection of regulatory mechanisms for the complexes in attempts to discover the molecular basis of colinearity" (Krumlauf, 1994).

"A hallmark of the primary homeotic genes is that in all animals they are clustered and their order is conserved, a conservation that would therefore appear to be important for function. Nevertheless, in *Drosophila* the clustering is not essential; for example, the complex can function if split (Struhl, 1984a). *Ubx* can be put on another chromosome, where it works well, and there are other examples from both the BX-C and the ANT-C.

"Why is it that the order of the genes on the chromosome is colinear with the patterns of expression in the body, with the most proximal gene on the chromosome having the most anterior role and the next gene along being expressed in the next-most anterior region? This rule is impressive, for it holds true from *lab* on the left side of the ANT-C to *Abd-B* on the right side of the BX-C; it also applies to vertebrates and nematodes and not only orders the protein coding parts of the genes, but also the disposition of the 3' regulatory elements. This question we cannot answer; we merely do our duty by posing it yet again!" (Lawrence & Morata, 1994).

"None of the known mechanisms for regulating *Hox* gene expression has been shown to depend on the position of a *Hox* gene in the genome. For example, *Drosophila* gap and pair-rule proteins can still act on *Hox* control regions that have been excised from the complex and inserted elsewhere in the genome (Busturia & Bienz, 1993)... Lewis (1978) originally proposed that the genes in the cluster were regulated by a substance distributed in a gradient along the body axis. The cluster, in turn, would contain response elements arranged in order of binding affinity, either because of their intrinsic properties or because of the local chromatin structure (Peifer, Karch & Bender, 1987) [cf. Wolffe, 1992]. So far, there is no good candidate for this type of graded regulatory molecule in *Drosophila;* candidates for such factors in vertebrates include components of the *hedgehog* pathway (Riddle et al., 1993) or possibly retinoic acid (Kessel & Gruss, 1991; Kessel, 1992; Marshall et al., 1992), as both can reposition *Hox* expression patterns" (Kenyon, 1994).

"The conservation of a set of clustered genes over half a billion years is difficult enough to accept, but colinearity with body axis defies credibility. Yet it's true" (Raff, 1996).

 (Cf. Manak & Scott, 1994.) Indeed, the colinearity correlation has been extended to a corresponding sequence of affinity constants for the homeoboxes of one murine homeobox complex (Pellerin et al., 1994), which may be related to the temporal response of homeobox genes to retinoic acid (Simeone et al., 1991). Since retinoic acid receptors (RARs) have little or no effect on early embryogenesis, as shown by double knockout experiments in mice (Lohnes et al., 1994; Mendelsohn et al., 1994), they may be linked to terminal differentiations, and thus RARs and colinearity may be related to the maintenance of the terminally differentiated state. Contradictions also occur: colinearity of *HoxA* genes in axolotl limbs occurs during normal limb development, but not during limb regeneration (Susan V. Bryant and David M. Gardiner in Travis, 1997d).

It should now be possible, especially given the same colinearity phenomenon in amphibians (Dekker et al., 1992, 1993) and nematodes

(Wang et al., 1993a), to work out a relationship of homeobox gene expression to differentiation waves, and thus to the spatial and temporal aspects of those waves, and to any gradients they leave in their wakes. But this will not solve the problem of colinearity.

To posit a function for colinearity, I would like to go back to the concept that a cell retains a 'memory' of its state of differentiation (Proposition 36). I suggest in this Proposition that the homeobox proteins may represent the 'memorons' of a cell. Now there are two components of stability of the differentiated state: 1) it survives cell division; 2) it persists, in terminally differentiated cells, to the end of the life of the organism. If the stability of the differentiated state of a cell depended on a set of bindings between single sites on the DNA and a few DNA binding proteins, one could imagine that after a few decades, especially with occasional cell divisions (as often as once a day in some of our tissues), a cell might transform to another cell type. The immune system wouldn't pick this up, and we would be sprouting ectopic structures in untoward places. Why doesn't this happen? (Perhaps it does occasionally, yielding teratomas, tumors with an assortment of differentiated tissues: Mintz & Illmensee, 1975; Balinsky & Fabian, 1981).

The answer may be that a much more robust control system has been superimposed to keep the genes that were involved in development from becoming active under normal circumstances (i.e., when regeneration is not active). Consider, for instance, three consecutive homeobox genes. Call them H, I, J, in the on state and h, i, j in the off state. Eight possible tissues can be produced involving these presumed master genes: HIJ, HIj, HiJ, Hij, hIJ, hIj, hiJ, hij. Of course, by our model for development, these correspond to eight different differentiation wave sequences, say CCE, CCE, CEC, CEE, ECC, ECE, EEC, EEE, where C = a contraction wave and E = expansion wave.

Let us suppose, by some trick of DNA looping (Irvine, Helfand & Hogness, 1991) and/or folding, that the three consecutive homeobox genes or their

DNA bound proteins can wrap around what I will call a 'homeokey' memoron molecule, and that there are, in this example, eight such molecules, one for each possible sequence of the homeobox genes in their on/off states. (Homeokey memorons are somewhat analogous to the hypothesized 'locking molecules' of Zuckerkandl, 1974.) The specificity, and stability, of binding of this molecule to the three sites may far exceed that of any of the individual sites. (Cf. "The weak and rather nonspecific DNA binding of the homeodomain proteins suggests that they must acquire specificity by interacting with other proteins": Gerhart & Kirschner, 1997). The more homeobox genes are involved, the greater is the stability of this arrangement. I make an analogy with a key, because the more complex the teeth on a key, the more specific it is for a given lock. The individual homeobox genes, in this model, are analogous to the complements of single teeth on a key.

The homeokey is an example of the concept of an 'adaptor' molecule of Conrad (1990c, 1993a) in the 'lock-key paradigm' (Conrad, 1992b):

"Adapter enzymes pick out shape features associated with different subsets of input patterns and link these to the output of the cell (or device) by activating specific effector proteins.... The advantage of the self-assembly model of computing is that it converts a symbolic pattern recognition problem into a free-energy minimization process. This is due to the representation of the input signals as specific molecular shapes" (Conrad, 1992a).

I could further speculate that the terminally differentiated cell depends on this homeokey molecule for its gene regulation, such as maintenance of its specific gene products, rather than on the combination of homeobox genes that initiated it into its differentiated state. Errors, such as transdetermination (Propositions 97-99), which can be tolerated in an organism that has many offspring, such as *Drosophila,* may be caused when an aberrant combination of the homeobox genes, perhaps due to the stochastics of binding of their homeobox proteins, leads to a fluctuation allowing another homeokey molecule to dominate.

Colinearity, then, is maintained during evolution because the homeobox genes or DNA bound proteins wrap around the homeokey molecule in a particular order, and the function of this molecule is diminished or lost when this cannot happen. Normal development is not interfered with, but long term stability of the differentiated state may be lost. A prediction of this model is that transdetermination rates should increase when the colinearity is genetically disrupted. The model of Coletta, Shimeld & Sharpe (1994), in which "essential regulatory sequences are shared by more than one HOM gene... puts a constraint on the preservation of the physical linkage of the genes" should, in contrast, lead to abnormal development if this proposed common regulation were disrupted.

Abercrombie (1965) has gone further in suggesting that ongoing cell interactions (differentiation waves?) may be needed to maintain a tissue in its differentiated state:

"Interactions concerned with differentiation by no means cease at this point; the maintenance and modulation of existing differentiations may be the work of a persistent traffic of signals between cells" (Abercrombie, 1965).

Indeed, this may be the function of (fast) waves found in adult liver (Robb-Gaspers & Thomas, 1995; cf. Section 9.19). Finally, we must take account of the following observations and their implications:

"A particularly striking finding was that fusions between more or less differentiated cells typically extinguished the luxury functions of the more differentiated cells (Ephrussi, 1972). Sometimes, e.g., in fusions with melanocytes, loss of chromosomes derived from the less differentiated strain in the resultant hybrid cells restored the lost functions (Davidson & Yamamoto, 1968; Klebe, Chen & Ruddle, 1970). Weiss & Chaplain (1971) obtained similar results....

"...Such cases provided evidence of 'an extraordinary stability of the *differentiated state itself*' since the capacity to resume the luxury function remains present, though deactivated, in the hybrid cells (Ephrussi, 1972). Using terminology borrowed from Abercrombie (1967), Ephrussi (1972) maintained that the 'epigenotype' of differentiated cells - 'that part of the total genome which, under appropriate conditions, *can be* expressed in a given cell type to the exclusion of those limited to other cell *types*' - was retained in the face of the phenotypic dedifferentiation that often resulted from hybridization. Thus, different cell types express different epigenotypes, and the question of regulation can be put in terms of the control of epigenotype expression" (Burian, Gayon & Zallen, 1991).

In other words, strictly speaking, the homeokey molecules stabilize the state of determination, not the state of differentiation. Ephrussi's observation is reflected in the wording of this proposition. Note that an epigenotype corresponds to the differentiation cascade represented by an edge of a differentiation tree. Steve McGrew (p.c., 1997) raises an interesting question:

"If two cells were taken from different locations in the differentiation tree and were hybridized, would the resulting cell type revert to the 'lowest common denominator': the most differentiated common ancestor [along the differentiation tree] of the two cells?" (Steve McGrew, p.c., 1997).

The separation of functions between homeobox and homeokey DNA binding molecules is consistent with differentiation waves and differentiation trees, but apparently not with reaction-diffusion gradient models:

"...Strong binding to DNA may be incompatible with morphogen function, and this may preclude the primary pattern-forming gene products being homeobox proteins" (Lacalli, 1990).

Proposition 271: there are a few candidates for homeokey molecules.

"To assert that the genome of a cell may by itself be incapable of *maintaining* a fixed state of differentiation once it is achieved may sound unorthodox..." (Bennett, Boyse & Old, 1972).

One could object that the size of the homeokey molecule required would be unwieldy (Natalie K. Björklund, p.c.). However, the size needed depends on the number of homeobox genes that are stabilized by the homeokey molecule, and the ability of the DNA they occupy to physically wrap around it. Addressing the former point, I might suggest that only the last few homeobox genes need be tied up by a homeokey molecule to ensure sufficient stability of the cell type. In fact, this may be what determines the number of steps back a cell can (and does) take in dedifferentiation for subsequent redifferentiation and regeneration (cf. Section 5.03). Furthermore, this would permit the evolution of separate homeobox gene complexes: only the terminal branches would be stabilized, but they could (on rare evolutionary occasions, because of their size) be duplicated and move around in the differentiation tree independently of one another (perhaps explaining the four HOX gene clusters in humans: Boncinelli et al., 1988; Lewis, 1992). A rough analogy can be made with braided hair: while there is some degree of adhesion along its length, only the end need be tied to keep it from unraveling. Here is the same analogy:

"As early as the 1930s, biologists knew that the tips of chromosomes had specific roles. Like aglets, the short pieces of plastic that protect the ends of shoelaces, the tips seemed to preserve the integrity of chromosomes. Without these structures, which became known as telomeres, chromosomes would fuse to each other or go astray when a cell replicated itself" (Travis, 1995c).

The *Polycomb* group and *trithorax* genes found in *Drosophila* (cf. Chinwalla, Jane & Harte, 1995; Kennison, 1995), in many aspects, fit what would produce my hypothesized homeokey molecules, including their being huge molecules (Stassen et al., 1995):

"Although many segmentation genes are only transiently expressed, their products fading away by the completion of gastrulation, the homeotic genes maintain their initially established patterns of expression through later development, both among and within segments. Therefore, additional mechanisms and factors are required to preserve the homeotic gene expression patterns initiated at blastoderm.

"This complicated function can be subdivided into two parts: prevention of ectopic homeotic gene expression and the maintenance of expression where it is appropriate. Posterior repression of each homeotic gene outside its normal expression domain is mediated by cross-regulatory interactions among the homeotic genes themselves (Struhl, 1982; Hafen, Levine & Gehring, 1984; Harding et al., 1985), while anterior boundaries are maintained by the products of the *Polycomb* group (*Pc*-G) genes. *Pc*-G genes are not required to establish the initial patterns of expression, but in the absence (or haplo-insufficiency) of one or more members of the group, strong anterior toward posterior segmental transformations occur, resulting from anterior expansion of homeotic gene expression (reviewed by Paro, 1990).

Protein products of several members of the *Pc*-G bind to multiple sites on salivary gland chromosomes, and two of them appear to be constituents of a multimeric protein complex (Zink et al., 1991; DeCamillis et al., 1992; Rastelli, Chan & Pirrotta, 1993; Martin & Adler, 1993, [cf. *Polycomb* and polyhomeotic colocalization on *Drosophila* salivary gland polytene chromosomes, reviewed in: Swedlow, Agard & Sedat, 1993]). Genetic and molecular evidence suggests that maintenance of proper levels of homeotic gene expression depends on the activity of the *trithorax* group genes (reviewed by Kennison, 1993, [1995]) and, in some tissues, on positive autoregulation of transcription by their own protein products (Kuziora & McGinnis, 1988a; Bienz & Tremmi, 1988)" (Sedkov et al., 1994).

"The ten *Polycomb* group genes... begin expression at the segmented germ band stage, and their encoded proteins bind at specific DNA sites in the Hox cluster, but do so only if the gene is being specifically repressed at the time. Those genes are then permanently repressed by a Polycomb-mediated change in chromatin structure... (Pirrotta, 1995).... Acting in the opposite sense, proteins encoded by the *trithorax* group of genes (Tamkun et al., 1992) stabilize activated and nonrepressed genes in a state of permanent activation or activatability" (Gerhart & Kirschner, 1997).

Paro (1993) is a bit more explicit:

"Once established,... determined states need to be maintained over developmental time.... It has been proposed that early transcription factors encoded by segmentation genes tether Pc-G proteins to the corresponding initiation elements of homeotic genes, and thus trigger the transition between establishment and maintenance (Zhang & Bienz, 1992). ... Ectopic activity... would result in an irreparable change of the determined state of a cell, and thus, in an aberrant body pattern.... In the case of misregulation, the resulting loss of cellular memory might lead to onset of tumorigenesis.... Multimeric Pc-G proteins exert their function at the higher-order chromatin level, as the disruption of one member of the group results in a cytologically visible alteration of chromosome organization.... With increasing size of the [Pc-G/chromatin]... structure, the repressor capabilities seem to become more effective (Irvine, Helfand & Hogness, 1991; Zink et al., 1991; Simon et al., 1993). Indeed, this might be one of the reasons why homeotic genes have been retained in large genetic clusters during evolution" (Paro, 1993).

Maintenance of the differentiated state is a fascinating problem (Moehrle & Paro, 1994), especially given the need to both repress and activate genes, apparently done by *Polycomb* and *trithorax,* respectively (Orlando & Paro, 1995), and the link I am trying to establish to dedifferentiation may be important in its understanding. The link to "alteration of chromosome organization" (Paro, 1993) is suggested in Section 10.13.

Here is a description of another complex including homeotic proteins that may also be a homeokey:

"The DNA binding specificity and the functional activity of homeodomain proteins also depend upon combinatorial interactions with other transcriptional factors. These interactions may occur between the free proteins in solution or alternatively when the proteins are bound to the DNA. The operator in this case may not be simply a DNA segment, but rather a DNA-protein complex (see Johnson, 1992a). The DNA binding specificity can also be increased in proteins containing two or more separate DNA-binding domains. So far, the structural analysis has been focused on the homeodomain as a module. Future studies will have to consider this module in the context of the entire protein and its interactions with other regulatory proteins" (Gehring et al., 1994).

Homeokey molecules, and, indeed, protein/DNA complexes in general, may fit the requirements for a kind of lock and key 'molecular computation' (Lahoz-Beltrá & Hameroff, 1995). If, for example, the last three differentiation states a cell has been through are X, Y, and Z, then an appropriate homeokey molecule accomplishes the logical operation of 'X and Y and Z'.

Proposition 272: the on state for a homeobox gene corresponds to a contraction wave, while the off state corresponds to an expansion wave (or *vice versa*).

It would be nice if we could label each edge of the differentiation tree of an organism with a tag or color showing whether it is triggered by a contraction wave or an expansion wave. If we could, some patterns might appear. For now we can make use of a partial correlation of anatomy with homeobox genes, which I speculate might be complete if we made the correlation with type of differentiation wave instead of anatomy:

"There is... a reasonable correlation in the nature of the phenotypic changes... in *Drosophila*. In *Drosophila*, for the *Antp, Ubx, abd-A* and *Abd-B* genes, loss-of-function mutations result in anterior homeotic transformations, while gain of function induces posterior transformation (Lewis, 1978; Akam, 1987a; McGinnis & Krumlauf, 1992). This is not an invariant rule because loss-of-function mutations in *Deformed* and *Scr* result in posterior transformations" (Krumlauf, 1994).

Krumlauf (1994) reviews similar incomplete anatomical correlations in the mouse, "...but in general the phenotypes are much less severe than would be predicted by the expression patterns". Since the expression patterns should correspond to the wave trajectories, observation of these trajectories may reveal something other than a genetic explanation. The monotonic sequence of binding constants of the homeobox proteins (Pellerin et al., 1994) establishes a uniformity that hints at the same differentiation pathway being used repeatedly, and thus suggests that the cell has participated in a sequence of waves that are all of the same type.

Gehring (1994) is careful to indicate that homeobox genes are "involved in the genetic control of development" rather than blatantly claiming that they control development. He recognizes that...

"...much remains to be learnt about the mode of action of homeotic genes. For example, the timing of heat induction [Schneuwly, Klemenz & Gehring, 1987; Gibson & Gehring, 1988] is of crucial importance. The target cells seem to be sensitive only during critical developmental stages when the determination of the respective structures normally takes place. Once the developmental decision is made, the state of determination is quite stable and the ectopic overexpression of a single master control gene like *Antennapedia* is no longer effective in changing cell fate" (Gehring, 1994).

Chow, Hall & Emmons (1995) note that in nematodes a "possibility for *mab-21* action is as a component of the pattern-formation mechanisms that set the expression levels of the HOM-C/Hox genes in the seam cells". It can still be said: "Largely unknown, however, are the signaling molecules that provide the positional cues for the precise establishment and maintenance of the Hox code" (Oh & Li, 1997). So something else controls the homeobox genes, something that transiently exists during critical stages, that makes the developmental 'decisions', i.e. differentiation waves.